Statics and Strength of Materials

George P. Kraut

Reston Publishing Company, Inc.
A Prentice-Hall Company
Reston, Virginia

This book is dedicated to my mother

MARTHA FREESE KRAUT

who, in her love of knowledge, industriousness, organization, and precision, serves as my exemplar.

Library of Congress Cataloging in Publication Data

Kraut, George P.
Statics and strength of materials.

Includes index.
1. Statics. 2. Strength of materials. I. Title.
TA351.K7 1984 620.1′03 83-17739
ISBN 0-8359-7112-0

10 9 8 7 6 5 4 3 2 1

Printed in the United States of America.

Contents

Preface

This text is designed for use in noncalculus based courses at the associate degree level of engineering technology. The emphasis throughout is on applications with sufficient foundation in principles to accomplish those applications.

Two major variations from many similar texts are the lack of separate chapters on equilibrium and friction. The concepts of equilibrium, free body diagram, and reaction are introduced in the first chapter and then carried throughout the text as the complexity of the work increases. Friction has been dealt with as an application since students have been at least introduced to the concept in physics courses. Friction is therefore dispersed throughout the earlier chapters of the text.

The text stresses graphical solution more than most texts do for three reasons. First, the method has a definite validity and will appeal to some individuals for regular use. Second, graphical depictions can frequently clarify a concept. Finally, in the process of problem solving, a graphical sketch is extremely useful for both conceptualizing the problem and making an initial estimate.

The text may appear lengthy at first glance. However, it is designed to be suitable for any engineering technology with little or no supplementation. In any given program, certain chapters and topics within chapters might well be deleted.

Appendix A provides a series of prerequisite review units. Nationwide testing has indicated that students arrive at their first mechanics course with a wide variation in preparation. The instructor's manual designed to accompany this text includes a pretest keyed to the review units. The manual also includes some notes on individually paced instruction and augmentation with audiovisual aids.

Throughout the text and in Appendix B, a comprehensive set of tables is provided to facilitate the applications orientation of the course.

Every attempt has been made to give the S.I. system relatively equal footing with the U.S. customary. Explanations, examples, and problems are all done in both systems.

Objectives are provided at the beginning of each chapter, and a readiness quiz appears at the end of each chapter. Students should therefore be able to ascertain where they are going before they start studying and whether or not they have arrived before they take an exam. Sample exams are provided in the instructor's manual. Answers to readiness quizzes, review unit problems, and selected text problems follow Appendix B.

George P. Kraut

Chapter 1

Introduction

1.1 OBJECTIVES

Upon completion of the work relating to this chapter, you should be able to perform the following.

1. For the words listed below:
   (a) Select from several definitions the correct one for any given word.
   (b) Given a definition, select the correct word from a choice of words.

action
collinear
concurrent
coplanar
direction
dynamics
equilibrium
force
force system
free-body diagram
kinematics
kinetics
kip
line of action
magnitude
mechanics
moment
newton
parallel
pound
reaction
scalar
sense
statics
strength of materials
ton
transmissibility (force)
vector

2. When shown several force systems, identify the one that is
   (a) coplanar and concurrent
   (b) coplanar and nonconcurrent-parallel
   (c) coplanar and nonconcurrent-nonparallel
   (d) noncoplanar and concurrent
   (e) noncoplanar and nonconcurrent-parallel
   (f) noncoplanar and nonconcurrent-nonparallel
3. Given a system of forces acting on a body, make a force diagram for the system.
4. Given a simple collinear force system, identify any unknown reaction, make a free-body diagram, make a force diagram, and solve for the unknown reaction.

1.2 THE STUDY OF STATICS

Statics is part of a larger body of knowledge called mechanics. This study of mechanics is usually divided into two broad areas called solid mechanics and fluid mechanics. The broad field of solid mechanics consists of the study of rigid bodies and deformable bodies. No body is really rigid but for purposes of analysis it can be assumed to be and simplifies the analysis. Deformation does indeed occur in all bodies but frequently can be considered separately. A simple example is the design of the legs of a chair. The general shape of the chair and the largest expected load on it will permit determination of the forces acting on every piece of the chair. Such an analysis involves statics, a part of the study of rigid bodies. Selection of a material for and the sizing of the parts of the chair is a separate problem. This involves strength of materials, part of the study of deformable bodies. Statics provides a foundation not only for strength of materials, but also for some aspects of fluid mechanics, and for dynamics, the study of rigid bodies in motion. Figure 1-1 illustrates the relationships among these various aspects of mechanics. You will probably be taking one more course in the general field of mechanics, and quite possibly several more.

The study of statics involves a rather small number of laws, all of which you have already encountered in physics. Statics simply requires the use of these laws as well as algebra, geometry, and trigonometry to solve more complex problems than those you have previously encountered. The laws involved are:

1. *Newton's first law.* A force is required to change the motion of a body. A body at rest will remain at rest and a body in motion will move uniformly in a straight line unless acted upon by a force.

2. *Newton's third law.* If one body exerts a force on a second body, the

second body must exert an equal but opposite force on the first. (For every action there is an equal and opposite reaction.)

3. *Newton's law of gravitational attraction.* Two particles are attracted to each other, along a straight line connecting them, with a force whose magnitude is directly proportional to the product of the masses, and inversely proportional to the square of the distance between them. The most common application of this law is the attraction between the earth and the "particles" on its surface.

4. *Friction.* When one solid body slides over another, a resistance to motion occurs called the frictional force, which is proportional to the pressure force between the bodies, independent of the area of contact and of the sliding velocity.

The quantities encountered in static analysis are quite simple: mass, force, length, and time. All of these should be thoroughly familiar to you from physics. The most common systems of units in use today are the Système International (SI) and the English (often called customary). In the SI system mass is in kilograms, force in newtons, length in meters, and time in seconds. In the English system, mass is in slugs, force is in pounds, length is in feet, and time is in seconds. Conversions between SI and English units are found in Appendix B-1.

An area of occasional confusion that bears clarification is the relationship between force and mass. The relationship is based upon Newton's

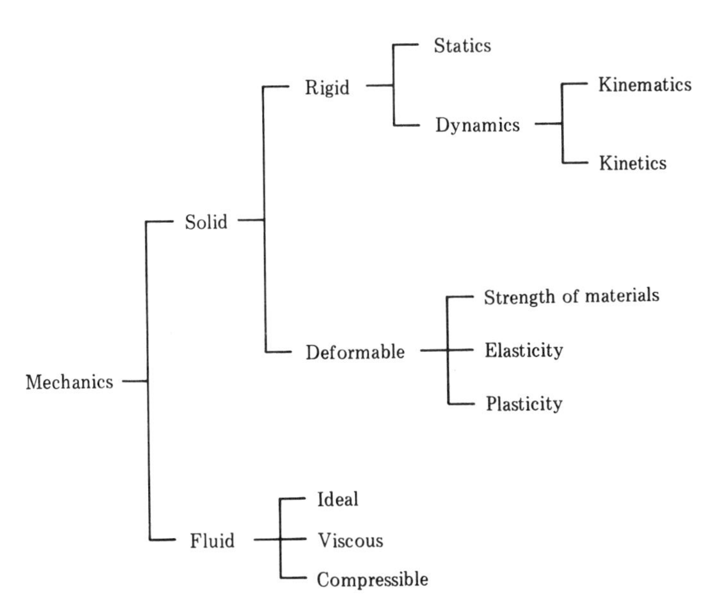

Figure 1-1. The field of mechanics.

second law, which states that an unbalanced force acting on a mass will cause that mass to accelerate. In equation form, this is shown at left below. When the force is due to gravitational attraction, it is generally called weight and the resultant acceleration is denoted by the letter g. This is shown at right below.

$$F = ma \qquad W = mg$$

If one recalls that the acceleration due to gravity at the earth's surface can be taken to be 32.2 ft/s^2 in the English system and 9.81 m/s^2 in the SI system, correlating mass and force becomes quite simple. On occasion, the unit slug or kilogram is encountered in an equation where it will not cancel with any other units. Analysis of the above relationship will show that the units lb s^2/ft can be substituted for slug and N s^2/m for kilogram to solve the dilemma.

1.3 EQUILIBRIUM

The concept of equilibrium is based upon Newton's first law. Essentially, a rephrasing of that law tells us that if all the forces acting on a body add up to zero, then the body is either standing still or moving at a constant velocity. In the study of statics, we concern ourselves only with static equilibrium, the case where the body is standing still. Any one force may attempt to cause the motion of a body; but as long as one or more forces counteract the initial force, no motion will occur.

A person standing on a floor provides an excellent example of static equilibrium. Let us consider a person weighing 150 lb. If the person is to stay

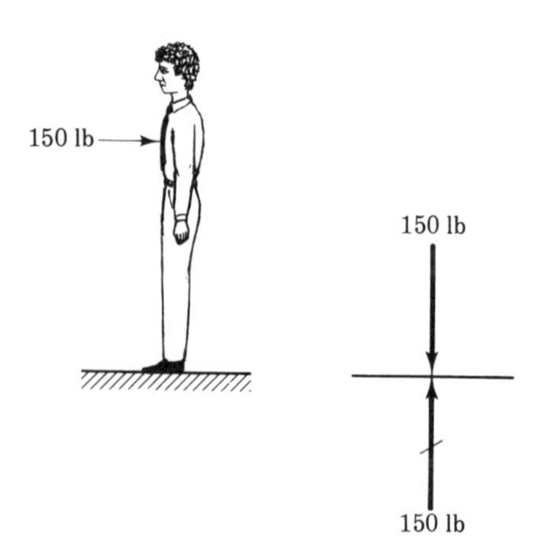

Figure 1-2. Equilibrium in one dimension.

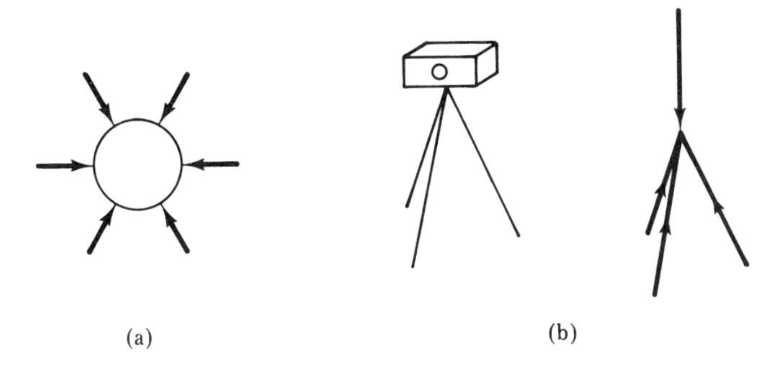

Figure 1-3. Equilibrium in two and three dimensions.

in place vertically, then the floor must push up with a force of 150 lb. Thus the 150-lb upward force results in a net force of 0 lb. This is illustrated in Figure 1-2. It should be noted that throughout this text the basic cartesian coordinate system is used (up is +, down is −, right is +, left is −, towards one is +, away is −).

The above example was in one dimension, but the same principles prevail in a two- or three-dimensional situation. In a school gymnasium where several students are pushing on a large ball, usually called an Indian ball, the forces applied by the students are parallel to the floor and all in one plane. The problem becomes two-dimensional (if we neglect the mass of the ball) but if the ball is not moving then all of the applied forces must add up to zero. A simple three-dimensional problem is the camera on a tripod. If we examine the forces acting on the tripod head, we find that because of the angles of the tripod legs, the weight of the camera and the forces in the tripod legs are not all in one plane. The problem thus becomes three-dimensional; but for the camera to remain stationary, the forces must still add up to zero.

Another condition must be met for static equilibrium to occur. Not only must all the forces acting on a body add up to zero, but all the moments acting on the body must also add up to zero. This means that when a body is in static equilibrium, it neither moves in a linear manner nor does it rotate in place about any axis. The concept of moments will be dealt with thoroughly in Chapter 3 and applied thereafter.

1.4 FORCE SYSTEMS

Systems of forces are classified by their geometries. Basically, the simpler the geometry, the simpler the problem is to analyze. One method of classification groups the force systems into three types depending on whether they are one-, two-, or three-dimensional in nature. These are

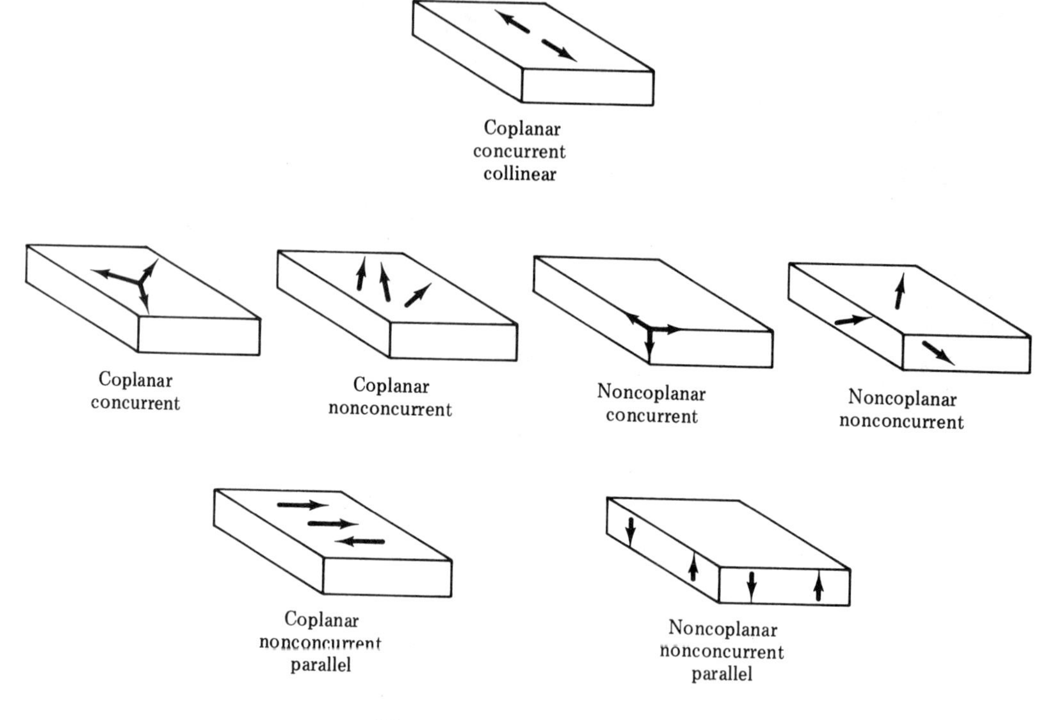

Figure 1-4. Classification of force systems.

called, respectively, collinear (sharing one line), coplanar (sharing one plane), and noncoplanar and are illustrated in Figure 1-4 using a boxlike figure to indicate that three dimensions may be involved. It can be seen that a collinear force system is really a special case of the coplanar force system since a single line must lie in a single plane.

Another means for classifying force systems groups them into two categories depending on whether or not all the forces in a system pass through a single common point. If they do, it is called a concurrent force system. If they do not, it is called a nonconcurrent force system. Once again, the collinear system is a special case, this time of the concurrent-type system. Care must be taken in examining complex systems before assuming that they are nonconcurrent. This is because it is legitimate for purposes of analysis to move a force from its point of application as desired *along its line of action.* This concept will be examined in more detail in Section 1.8 of this chapter.

Two special cases besides the collinear warrant attention. These are cases where the lines of action of all the forces in a system are parallel to one another. This may occur in either a coplanar system or a noncoplanar system. An example of a coplanar system is a tug of war, where all the forces involved are essentially parallel to the ground and for practical purposes in a single vertical plane. An example of a noncoplanar parallel force system is the chair previously mentioned. The forces acting upon the legs and the force of gravity acting down on the chair are all parallel to one another but certainly do not fall in a single plane.

This chapter will only consider collinear force systems in detail. However, subsequent chapters will deal with all of the other systems mentioned.

1.5 ACTION AND REACTION

The nature of a force is frequently categorized in two different ways. Establishing and understanding these categories requires that we first establish what the force is acting on. The "body" which we wish to deal with may be as large as a single beam or a single link in a machine. It may even be as small as a single point on a building or a machine. The choice of whether to deal with a large body or a small portion of it depends on two things. First, what is known, and second, what do we wish to know. Our entire analysis, including our choice of how much of a body we wish to deal with, is a matter of selecting the shortest path between the known and the unknown.

Once we have chosen what is to be the body in our analysis, all forces can be categorized as being internal or external forces. In the case shown in Figure 1-2, if the entire person is considered to be the body for our problem, then the force of gravity acting on the body and the force of the floor acting on the body are considered to be external forces. The forces occurring in each leg are considered to be internal forces. Frequently, even though we are actually interested in the internal forces acting on individual parts of a structure or machine, we must first solve to find all the external forces before proceeding to the internal.

Examining the external forces acting on a body reveals that these can be grouped into two categories, acting forces and reacting forces. In the case of the person standing on the floor, the force of gravity is an acting force while the force of the floor on the body is a reacting force. Thus an acting force can be a gravitational force, a magnetic force, or a mechanical force being applied to a body. A reacting force is not independent; it is, as its name implies, a reaction to one or more acting forces. In our simple case above, the floor would not be pushing up if gravity were not pulling down on the person. The floor is reacting to the force of gravity acting on the person's mass.

The nature of the reaction that occurs at the points of contact of the body being analyzed with the rest of the world depends in part on the nature of the connection at that point. Our person standing on the floor is probably the simplest case possible. There is only one acting force which is straight down; therefore the reacting force must be equal and opposite or straight up.

Since we have assumed a level or horizontal floor and no horizontal forces, friction does not enter into the problem.

A second simple case was just alluded to and that is friction. In simple sliding friction, the friction force is equal and opposite to the force being applied to overcome friction up to the point of motion. Right at that point, the maximum static friction force is reached and is equal to the coefficient of friction times the normal force. Friction force, thus, is a type of reaction force. There is no resistance to an attempt to overcome friction unless that attempt is made.

The above cases are shown in Figure 1-5 along with three other cases, one of which has a variety of subtle variations that is helpful to recognize. This is particularly true since the nature of the action and reaction in this case is very simple and identical despite the many minor variations in the support.

If we consider first a simple roller, a wheel or cylinder, and assume that there is no friction between the roller and the surface it rests on, the analysis of action and reaction becomes very simple. If we were to attempt to exert an acting load that is not perpendicular to the surface, then one of two things would have to occur. Either the roller would roll in the direction of the applied force (thus eliminating static equilibrium) or there would have to be a frictional force large enough to resist motion (but we just assumed a friction-free setting). Thus, for static equilibrium to occur with a friction-free roller on a flat surface, both the acting force and the reacting force must be perpendicular to the flat surface. The acting force of the roller is always pressing toward the flat surface and the reacting force of the flat surface is always acting toward the center of the roller. In the cases of the slot and the collar additional analysis must determine which flat surface to deal with. Studying the variations on the roller shown in Figure 1-5 shows that the above analysis applies equally well to a series of rollers mounted on a bracket with a pin point, to a rocker (part of a roller), to a member or part with a rounded end (part of a wheel), and to a collar with a pin point.

Another case is where the connection at the support consists of a pin or hinge as on a door. In many, but not necessarily all, cases the acting force may be collinear with the link or member. In later chapters we will encounter both types of situations and will learn to distinguish between the two. The one case where the acting force is always collinear with the member (and therefore the reacting force is also) is when the member is a cable.

The last type of case we shall consider is the fixed support. You have seen this type of connection in a fence post in the ground, a flagpole standing on the ground or attached almost horizontally on a building, and a theatre marquee. It differs from the other supports in that it not only reacts to an acting force in any direction but can also react to a twist or moment. Looking back at the other types of supports shows that any application of twist, torque, or moment would result in a turning or rolling motion.

Simple reaction — reaction opposite to force of gravity.

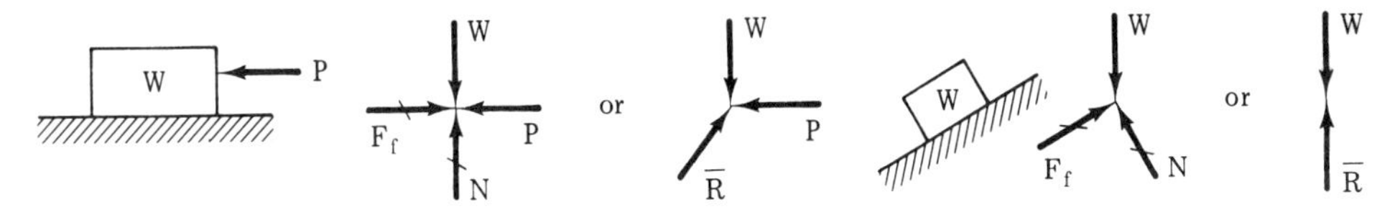

Friction — total reaction is sum of normal force and friction force.

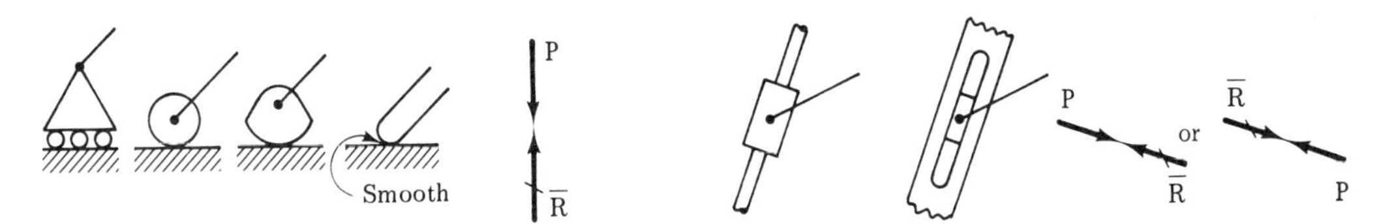

Rollers, rockers, sliders — reaction is normal (perpendicular) to the surface acted on (assuming zero friction).

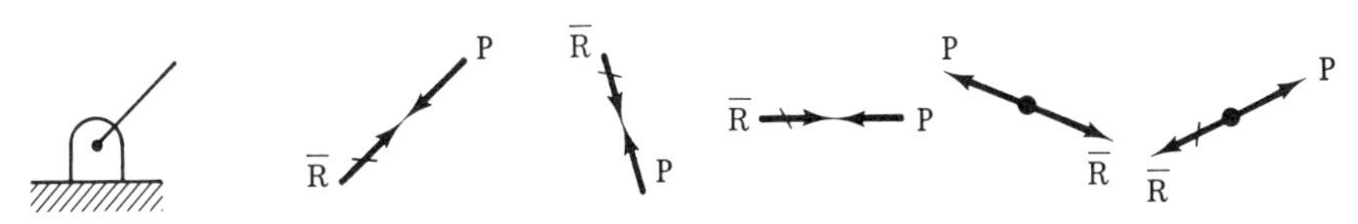

Pin (hinge) — reaction can be in any direction in the plane normal to the pin axis.

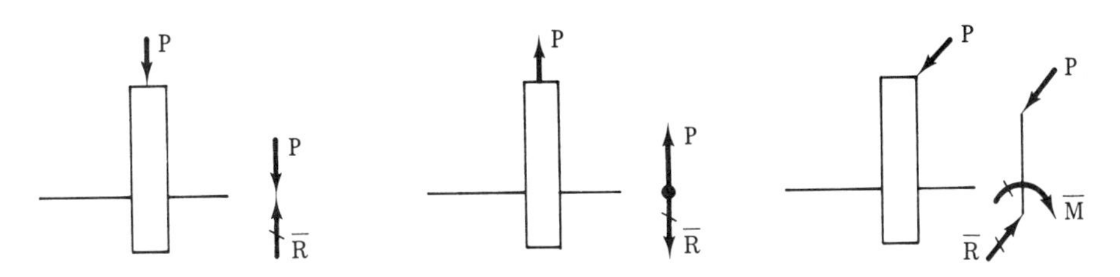

Fixed support — reaction can be in any direction and may include a moment reaction.

Figure 1-5. Coplanar support reactions.

The above discussion and Figure 1-5 deal entirely with coplanar forces. The concepts are the same but the reactions somewhat more involved for noncoplanar forces and will be considered in Chapter 8.

1.6 FREE-BODY DIAGRAMS

An important tool in static analysis is the "freeing" of the body to be analyzed from its supports. In place of the supports we put the appropriate reaction forces. This leaves us with a sketch of the body with all external forces shown. Once this step is completed, analysis of the system of external forces acting on the body can begin.

Example 1-1: Draw the free-body diagram for the railroad flat car shown in Figure 1-6a showing the reactions for each truck.

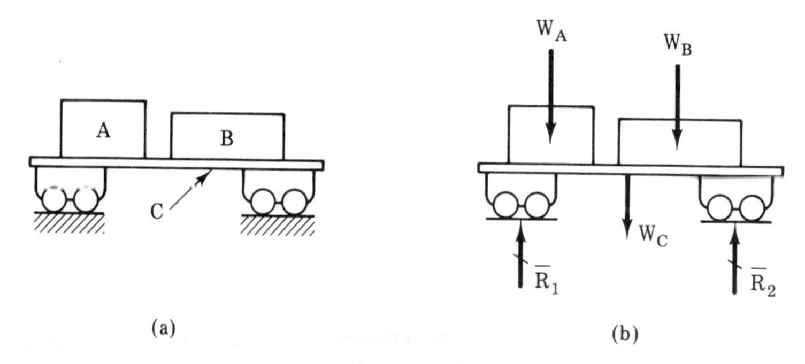

Figure 1-6

Solution:

Step 1 Redraw the problem as shown in Figure 1-6b with the supports (the ground in this case) removed.

Step 2 Recognize that the reaction force of the ground on the wheels is straight up.

Step 3 Since the problem asked us to deal with each truck and not each axle, we will assume that the reaction forces are located exactly between the axles on each truck.

Step 4 Draw in the reaction forces at the correct location and in the correct direction. (Note the slashes through the vectors for the reaction forces. This is a fairly common practice to differentiate reaction forces from acting forces.)

Problem 1-1: Draw the free-body diagram for the tractor trailer shown in Figure 1-7.

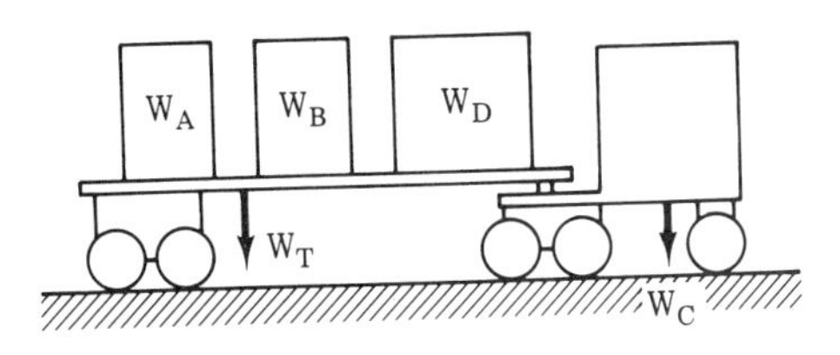

Figure 1-7

Example 1-2: Draw the free-body diagram for the structure shown in Figure 1-8a.

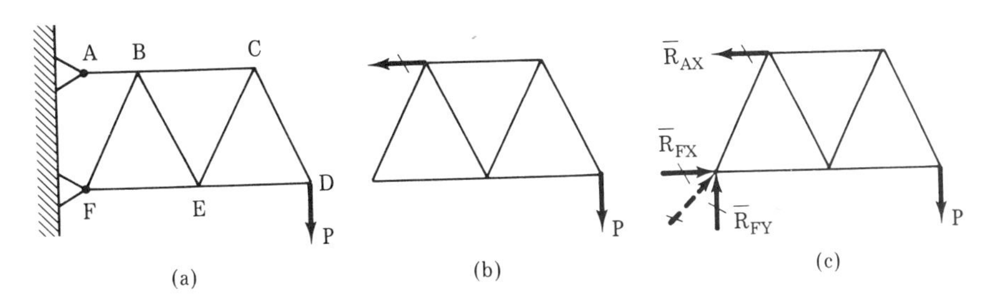

Figure 1-8

Solution:

Step 1 Redraw the problem as shown in Figure 1-8b with the supports (two pin joints) removed.

Step 2 Recognize that the cable at *AB* must be in tension both because a cable cannot be in compression and from a general review of the nature, loading, and support of the structure.

Step 3 Draw in the reaction force at *A* as shown in Figure 1-8b. (You would not normally make two free-body diagrams as shown here in Figure 1-8b and 1-8c but it has been done to show how Step 4 follows on Steps 2 and 3.

Step 4 Recognize that static equilibrium for this structure must be achieved by a reaction force at *F* since there are no other external forces to be drawn in. This reaction force must have a horizontal component equal and opposite to the reaction at *A* so that the structure will not move horizontally. This reaction force at *F* must also have a vertical component equal and opposite to the load *P* so that the structure will not move vertically.

Step 5 Draw in either the components of the reaction at *F* or the sum of the components (as shown by the dashed line) as indicated in Figure 1-8c.

Problem 1-2: Draw the free-body diagram for the structure shown in Figure 1-9.

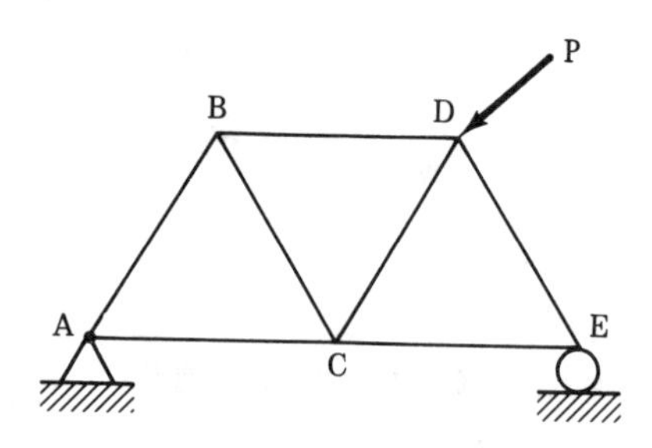

Figure 1-9

With the completion of the free-body diagram, all the external forces acting on a body have been identified in terms of location and usually direction. We are now ready to determine any magnitudes of forces that are unknown.

1.7 FORCE DIAGRAMS

A force diagram is simply a vector representation of a system of forces. If static equilibrium does not exist, then this vector representation will add up to a resultant force as shown in Figure 1-10. Note that the order in which the vectors are summed does not change the magnitude and direction of the resultant. If static equilibrium exists, then the vector representation must close adding up to zero as shown in Figure 1-11.

Such force diagrams are useful even if only sketched for they can permit the estimation of direction and magnitude of unknown forces. Actual solution can then be accomplished by a combination of algebraic and trigonometric techniques. However, if desired, the force diagram can be drawn to scale providing a graphical solution for unknowns. All of these techniques will be used throughout this text.

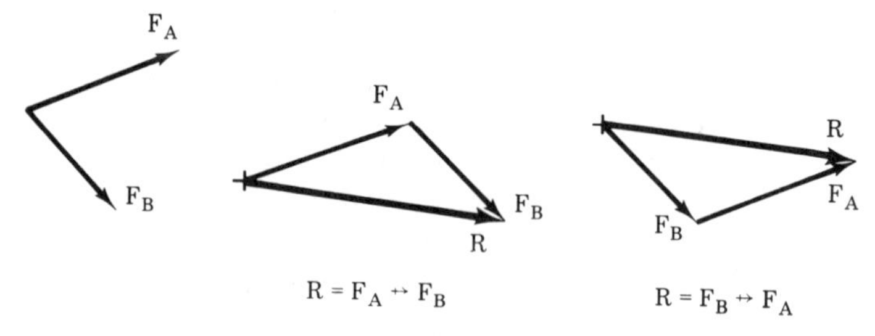

Figure 1-10. Force diagrams for nonequilibrium.

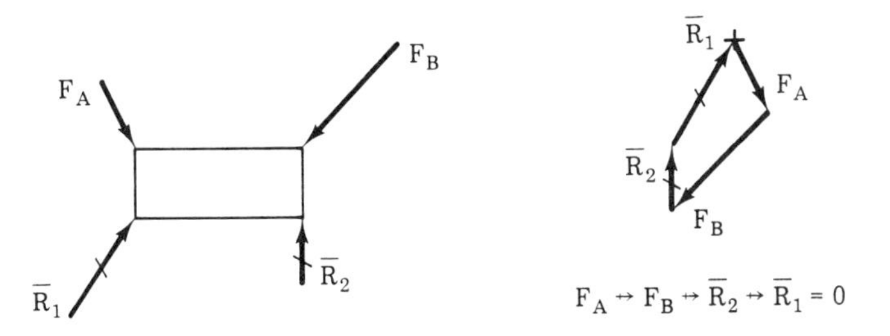

Figure 1-11. Force diagram for equilibrium.

1.8 TRANSMISSIBILITY OF A FORCE

A force may be moved along its line of action without changing the effect of the force. This statement describes the transmissibility of a force. Recall that the line of action of a force is a line drawn through (collinear with) the direction vector of the force and extended as far in either direction as desired. An example of this is the moving of a stalled car. Although there may be other factors that must be considered, technically it makes no difference whether the car is pushed or pulled. The direction and magnitude of the force required are the same in either case and lie on the same line of action. Another example is shown in Figure 1-12.

Gravity is pulling on the mass hanging from the cable. Not matter what the length of the cable is, as shown by the three locations for the mass, nothing else in the situation has changed.

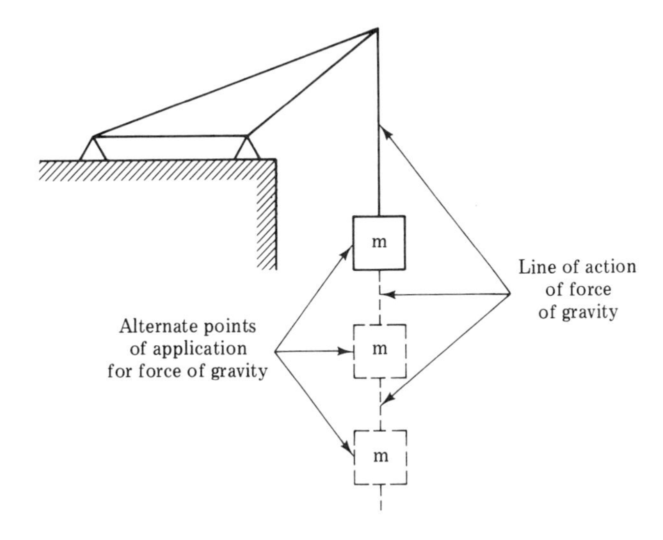

Figure 1-12. Transmissibility of a force along a cable.

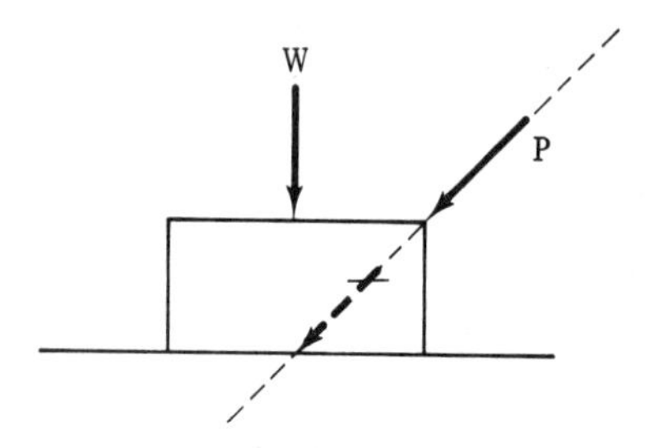

Figure 1-13. Transmissibility of a force through a body.

Another example is shown in Figure 1-13. Here a force P is applied to a box at an angle to attempt to slide it along. The vertical component of this force P will add to the normal force pressing the box to the floor. The horizontal component of this force P will attempt to overcome friction to move the box. However, none of this is changed by moving the force P from its initial point of application to the location shown by the dashed line. The shift may well make visualizing and implementing a solution easier.

Frequently in the problems encountered in statics, analysis can be simplified by shifting a force along its line of action. Applications of this will be seen throughout the text.

1.9 COLLINEAR FORCE SYSTEMS

The simplest type of force system is the collinear system in which all the forces lie on one common line of action. The person standing on the floor discussed earlier is an example of a collinear force system. At this point, we will bring together much of the material discussed in this chapter and apply it to a variety of collinear force systems.

Example 1-3: The winch shown in Figure 1-14a pulls an ore bucket up from the mine below. The ore bucket weighs 1000 lb empty. It can carry up to 40 tons of ore. If the cable weighs 15 lb/ft, what is the tension in the horizontal portion of the cable?

Solution:

Step 1 Recognize that this is indeed a collinear force system. Since no friction at the pulley was mentioned, it can be assumed to be negligible. The pulley, therefore, changes only the direction and not the magnitude of the tension in the cable.

Step 2 Recognize that only the vertical portion of the cable contributes appreciably to the tension in the cable. Gravity pulls vertically across the horizontal portion, and the horizontal portion is quite short; therefore, its weight contributes negligibly to the stress in the cable.

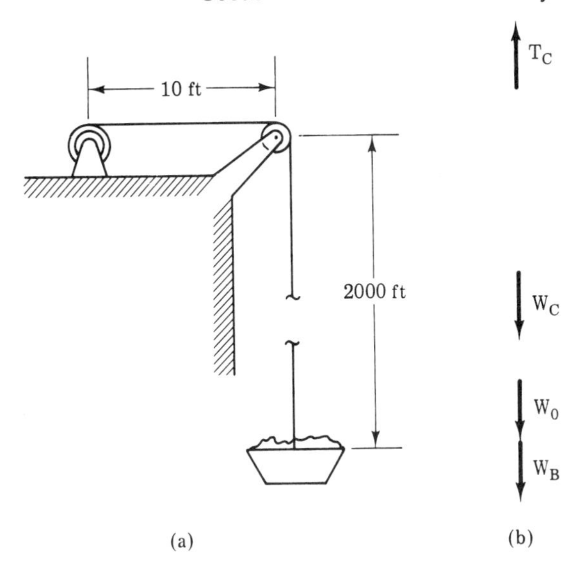

Figure 1-14

Step 3 Draw the free-body diagram as shown in Figure 1-14b.

Step 4 Sum the forces to zero as required for equilibrium (note that all collinear force problems can be solved algebraically since the vectors of all the forces lie on a common line of action).

$$\Sigma F = 0$$

$$T_C + W_C + W_B + W_O = 0$$

$$T_C = -(W_C + W_B + W_O)$$

$$W_C = 40 \text{ tons}(2000 \text{ lb/ton}) = 80{,}000 \text{ lb}$$

$$T_C = -(-30{,}000 - 1000 - 80{,}000) = +111{,}000 \text{ lb}$$

Note that correct signs were used throughout and that the tension in the cable came out opposite in direction to the three loads as shown in the free-body diagram.

Problem 1-3: A large area is roofed over as shown in Figure 1-15 with only a single cylindrical support at the center. The roof measures 30 m by 70 m by 0.5 m thick and has an average density of 700 kg/m^3. During downpours, the roof must also support 5 cm of rainwater (density = 1000 kg/m^3). The support is 5 m tall and 4 m in diameter and has a mass density of 2500 kg/m^3. Determine the force the ground exerts on the base in kN.

Example 1-4: Two teams are engaged in a tug of war. Team *A* is pulling to the left and consists of Frank, who is pulling with a force of 100 N; Pete,

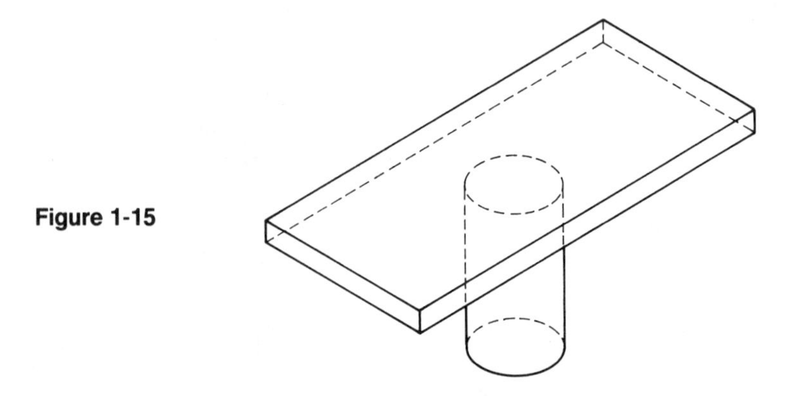

Figure 1-15

120 N; Carmella, 80 N; and John, 110 N. Team B is pulling to the right and consists of Joe, who is pulling with a force of 90 N; Dave, 95 N; Eric, 105 N; and Ernie, 130 N. Which team is winning, and what is the tension in the portion of the rope between the two teams?

Solution:

Step 1 Recognize this as a collinear force system that is oriented horizontally.

Step 2 Recognize that the sum of the forces may not equal zero, that a lack of static equilibrium may be causing motion to the left or right.

Step 3 Draw the free-body diagram as shown in Figure 1-16.

Step 4 Solve for the sum of the forces.

$$\Sigma F = F_{A1} + F_{A2} + F_{A3} + F_{A4} + F_{B2} + F_{B3} + F_{B4}$$

$$\Sigma F = (-100 \text{ N}) + (-120 \text{ N}) + (-80 \text{ N}) + (-110 \text{ N}) + 90 \text{ N} + 95 \text{ N} + 105 \text{ N} + 130 \text{ N}$$

$$\Sigma F = +10 \text{ N}$$

Therefore, Team B is winning.

Step 5 Recognize that the tension in the section of the rope between the two teams will be equal to the force exerted by the losing team (the net force of 10 N found above is going into moving the masses of the two teams).

Figure 1-16

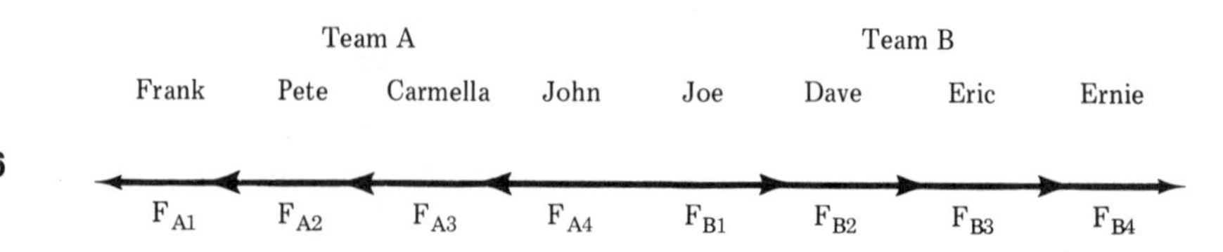

Step 6 Solve for the force applied by Team A.

$$\Sigma F_A = F_{A1} + F_{A2} + F_{A3} + F_{A4}$$

$$\Sigma F_A = (-100 \text{ N}) + (-120 \text{ N}) + (-80 \text{ N}) + (-110 \text{ N})$$

$$\Sigma F_A = -410 \text{ N}$$

Tension in central portion of rope is 410 N.

Problem 1-4: A force of 50 lb is being exerted on the lever shown in Figure 1-17. The tubing is 1 in. in diameter and is filled with water (weight density is 62.4 lb/ft³). What is the force acting at A if the weight of the piston and all friction can be neglected?

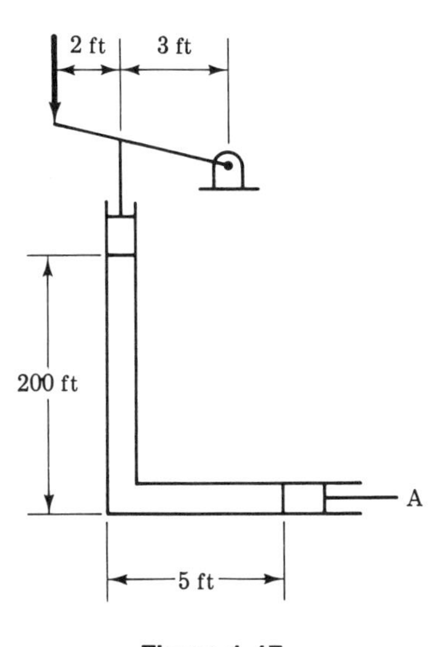

Figure 1-17

Example 1-5: Will the block shown in Figure 1-18a move?

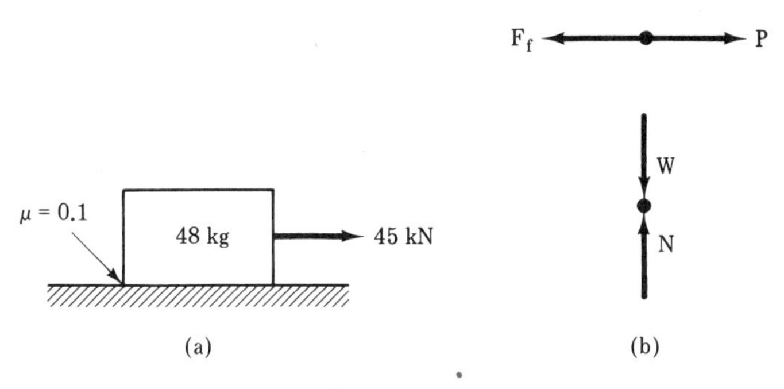

Figure 1-18

Solution:

Step 1 Recognize this as a basic sliding friction problem.

Step 2 Recognize that a basic sliding friction problem can be viewed as two collinear force sytems at right angles to each other.

Step 3 Draw the free-body diagrams for the force systems as shown in Figure 1-18b.

Step 4 Solve as follows:

$$W = 48\text{ kg}(9.81\text{ N/kg}) = 471\text{ N}$$

$$\Sigma F_Y = 0 \qquad W + N = 0 \qquad N = -W = -(-471\text{ N}) \qquad N = +471\text{ N}$$

$$\max F_f = \mu \text{N} = 0.1(471\text{ N}) \qquad \max F_f = 47.1\text{ N}$$

$$\text{act } F_f + P = 0 \qquad \text{act } F_f - P = -(+45\text{ N}) \qquad \text{act } F_f = -45\text{ N}$$

Answer: No. Applied force must be greater than maximum friction force for motion to occur. Actual friction force in this case is 45 N, equal and opposite to applied force.

Problem 1-5: Will the block shown in Figure 1-19 move?

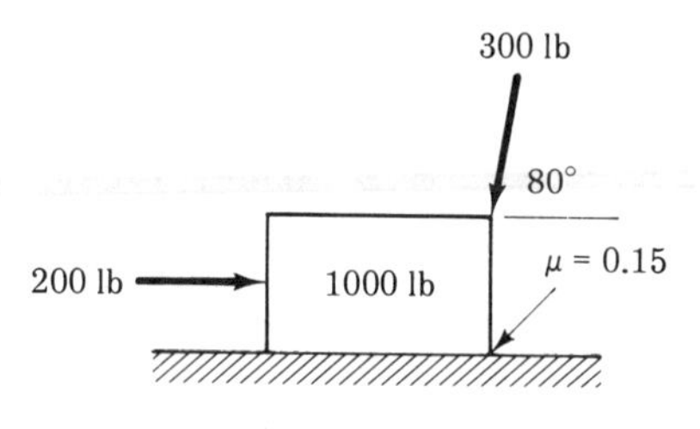

Figure 1-19

If you have studied the above material including use of the glossary as needed and successfully completed the five problems provided, you should be prepared to demonstrate that you have accomplished the objectives stated in Section 1.1 of the chapter. You can easily check yourself by taking the readiness quiz that follows. If you have questions or feel unsure, see your instructor as soon as possible. You may wish to do some of the supplementary problems at the end of the chapter.

1.10 READINESS QUIZ

1. When two or more forces share a common line of action, they are said to be

(a) noncoplanar
(b) collinear
(c) parallel
(d) all of the above
(e) none of the above

2. One thousand pounds is equivalent to a

(a) kip
(b) ton
(c) newton
(d) kilogram
(e) none of the above

3. Statics is part of a larger body of knowledge known as

(a) kinetics
(b) fluids
(c) strength of materials
(d) all of the above
(e) none of the above

4. A quantity of any measure that has magnitude but no direction is said to be

(a) scalar
(b) vector
(c) arithmetic
(d) geometric
(e) none of the above

5. Several forces acting on a single body are known collectively as a

(a) coordinate system
(b) force system
(c) reaction
(d) all of the above
(e) none of the above

6. A ton of force is equivalent to

(a) 2000 lb
(b) 2 Kips
(c) 8900 N
(d) all of the above
(e) none of the above

7. The force system shown in Figure 1-20 is

(a) coplanar and concurrent

(b) coplanar and nonconcurrent, parallel
(c) coplanar and nonconcurrent, nonparallel
(d) all of the above
(e) none of the above

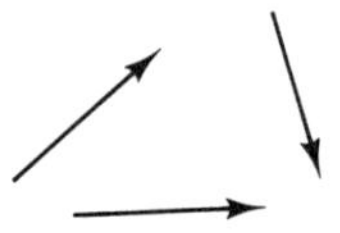

Figure 1-20

8. The study of mechanics includes

(a) statics
(b) dynamics
(c) strength of materials
(d) all of the above
(e) none of the above

9. The transmissibility of a force provides that a force can be moved __________ without changing its effect.

(a) so that its direction is reversed
(b) along its line of action
(c) to any location on the same body
(d) any of the above
(e) none of the above

10. In order for two or more forces to be coplanar they *must*

(a) pass through a common point
(b) lie on the same line of action
(c) never meet
(d) all of the above
(e) none of the above

11. The force of gravity acting on 20 kg of mass is slightly less than

(a) 10 N
(b) 200 N
(c) 1000 N
(d) 2000 N
(e) none of the above

12. A free-body diagram shows a body with its __________ removed

(a) forces
(b) reactions
(c) supports
(d) all of the above
(e) none of the above

13. If a body is in equilibrium, then

(a) it is standing still
(b) it is moving at a constant velocity
(c) all the forces acting on it add up to zero

(d) any of the above
(e) none of the above

14. The force system shown at right below is

(a) collinear
(b) coplanar
(c) noncoplanar
(d) any of the above
(e) none of the above

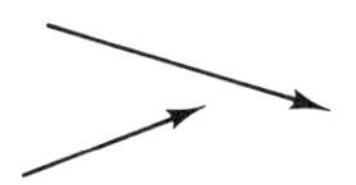

Figure 1-21

15. On a free-body diagram, the supports have been replaced with

(a) resultants
(b) parallels
(c) reactions
(d) any of the above
(e) none of the above

1.11 SUPPLEMENTARY PROBLEMS

1-6 Draw the free-body diagram for the structure shown in Figure 1-22.

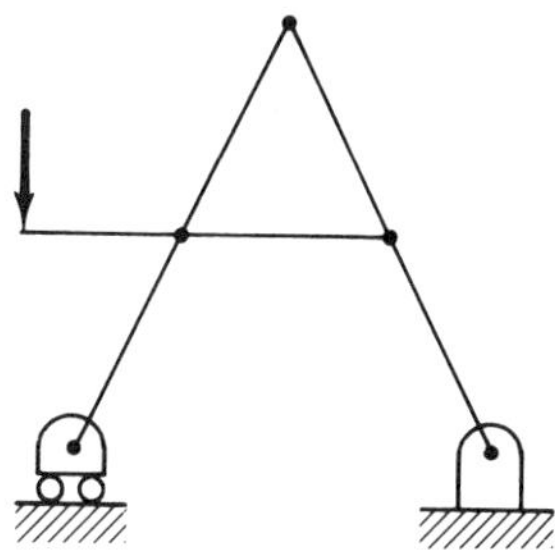

Figure 1-22

1-7 Draw the free-body diagram for the structure shown in Figure 1-23.

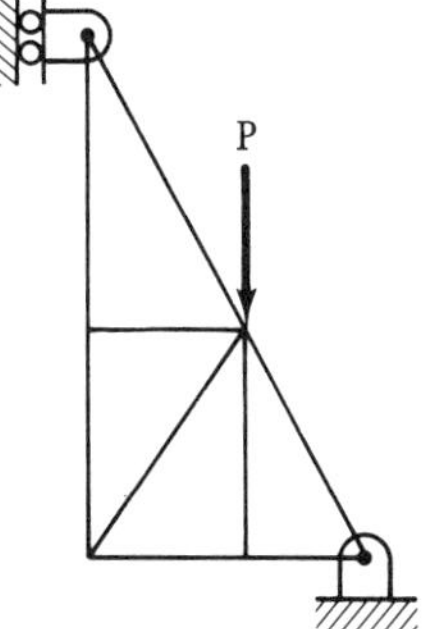

Figure 1-23

1-8 What minimum force must be applied at A for the block in Figure 1-24 to start to move?

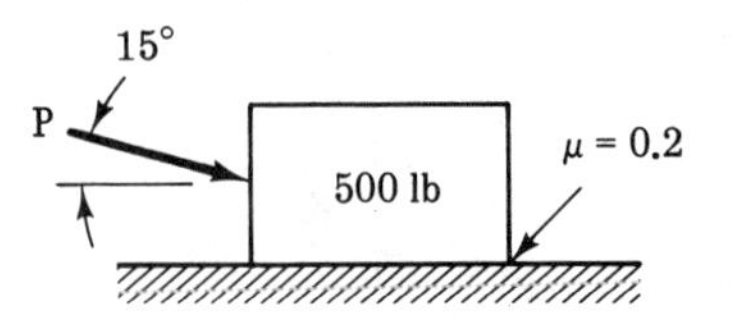

Figure 1-24

1-9 If 20 lb of force is required to accelerate the block in Figure 1-25, what is the maximum tension in the cable?

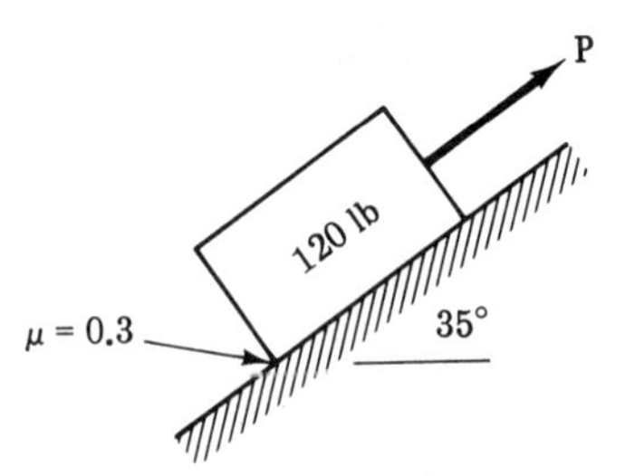

Figure 1-25

1-10 A utility pole has two cross arms weighing 200 lb each. Four cables are attached to the cross arms through insulators. Each cable pushes down with a total force of 400 lb. The pole is 40 ft long with 5 ft of the length embedded in the ground. If the pole weighs 50 lb/ft and the weight of the insulators and friction between the ground and the pole are neglected, what is the force of the ground on the bottom of the pole?

Chapter

2

Coplanar, Concurrent Force Systems

2.1 OBJECTIVES

Upon completion of the work relating to this chapter, you should be able to perform the following.

1. For the words listed below:
   (a) Select from several definitions the correct one for any given word.
   (b) Given a definition, select the correct word from a choice of words.

component	normal	resultant
concurrence	perpendicular	string polygon

2. When given a problem involving a concurrent force system of two or more forces, determine the resultant and reaction by the
   (a) method of components
   (b) parallelogram method
   (c) graphical method
3. Resolve a force into rectangular components referenced to any direction.

4. Resolve a force into two nonrectangular components whose directions are known.

2.2 PRINCIPLE OF CONCURRENCE

When two forces whose lines of action intersect are added together, their sum or resultant also has its line of action passing through that intersection or common point. This is illustrated in Figure 2-1a and 2-1b. If equilibrium is desired, then a reaction force equal and opposite to the resultant must be applied as shown in Figure 2-1c. Since the resultant and reaction are equal and opposite and pass through a common point, these two forces are both collinear and concurrent. If we now substitute the original two forces for the resultant as shown in Figure 2-1d, we have the basis for the principle of concurrence. *If three nonparallel forces acting on a body are in equilibrium, their lines of action must meet in a common point.* We will make use of this principle throughout the text where the direction of some unknown force is not immediately apparent.

2.3 TWO CONCURRENT FORCES: GRAPHICAL METHOD

A simple and quick means for determining the resultant and reaction for two concurrent forces is the graphical method. With only moderate care, the force diagram for the problem can be constructed to scale and the values for the resultant and reaction read using a scale and protractor. The accuracy required by most problems is well within the accuracy of this method.

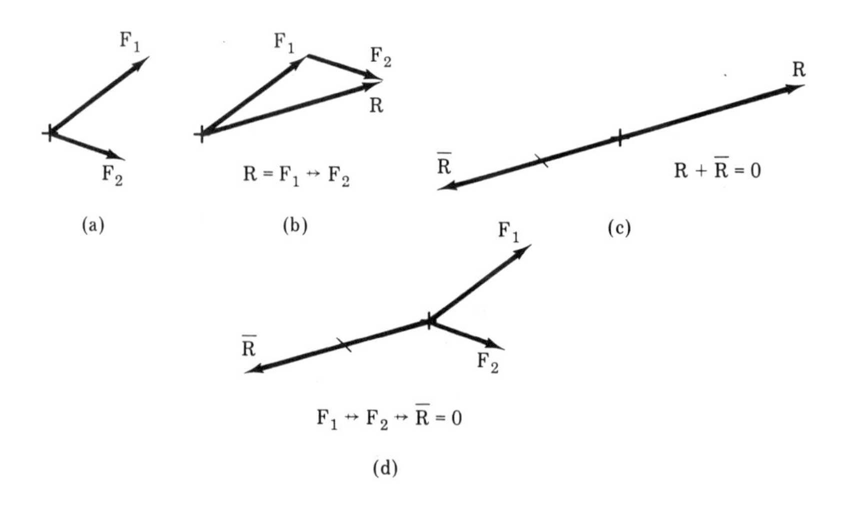

Figure 2-1. Principle of concurrence.

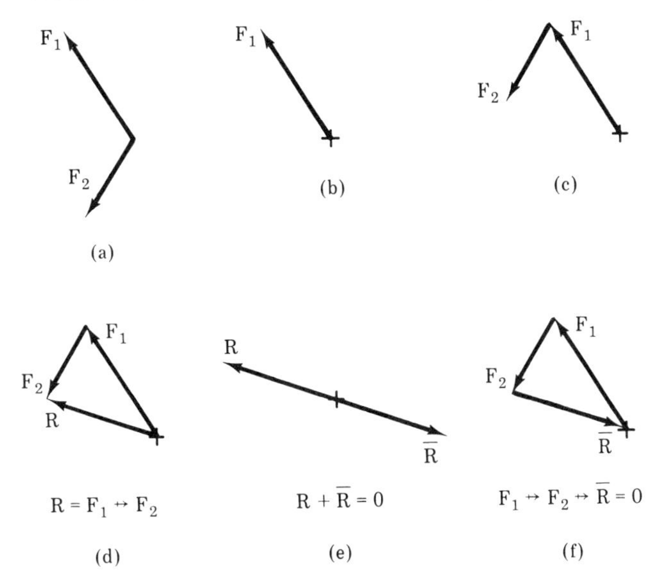

Figure 2-2. Graphical solution.

However, the method also has further usefulness. Sketching the problem by this method to only a rough scale provides a "feel" for the problem and an approximation of the answer even though other techniques are used for actual solution.

In constructing the force diagram several rules must be observed. First, mark a point on the paper as the origin. Second, lay out the first known force with its tail on the origin. Third, lay out the second known force with its tail on the head of the vector for the first force. The sum or resultant of these two forces is then a straight line drawn with its tail at the origin and its head at the head of the last force added. The reaction will simply be equal in length and opposite in direction to the resultant. If we view the force diagram in terms of the original forces and the reaction, we see that the last force added, the reaction, comes back to the origin. This indicates that the sum of the forces adds up to zero, a necessary condition for equilibrium. The above discussion is illustrated step by step in Figure 2-2.

Example 2-1: Determine the resultant of and reaction to the two forces acting on the eye bolt shown in Figure 2-3a, using the graphical method.

Solution:

Step 1 Recognize that these are two coplanar forces that are nonparallel and therefore concurrent.

Step 2 Recognize that by applying the principle of transmissibility of a force the intersection or point of concurrence of the forces appears to be at the center of the ring.

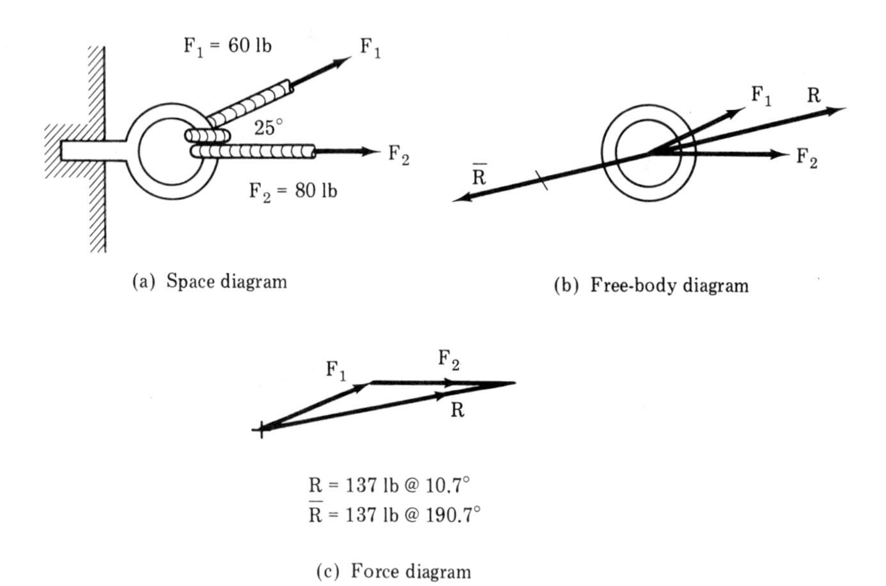

Figure 2-3

Step 3 Draw the free-body diagram for the problem as shown in Figure 2-3b.

Step 4 Draw the force diagram to scale for the problem as shown in Figure 2-3c including the resultant.

Step 5 Measure the magnitude and direction of the resultant.

Step 6 Recognize that the reaction (denoted by $\overline{R}$ as opposed to a plain R for the resultant) will be equal in magnitude but opposite in direction to the resultant.

Problem 2-1: Determine the resultant and reaction to the two forces acting on the sled as shown in Figure 2-4, using the graphical method. If the sled and its passenger have a combined mass of 50 kg and the coefficient of friction is 0.05, is the sled moving?

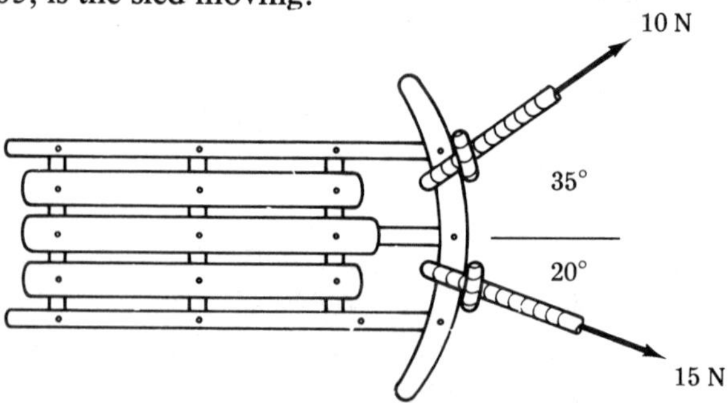

Figure 2-4

2.4 TWO CONCURRENT FORCES: METHOD OF COMPONENTS

In Chapter 1, we considered a basic sliding friction problem and analyzed it as two separate collinear force systems that were perpendicular to each other. The method of components simply extends this type of thinking to forces in any direction by resolving each force into its horizontal and vertical components. We can then consider the problem as two separate force systems and determine resultants and reactions independently for each. If desired, these individual resultants and reactions for the vertical and horizontal force systems can be combined to determine a total resultant and reaction for the overall problem.

The initial steps in the solution of a problem by this method are similar to those in the graphical method. However, when it comes to the force diagram, it is only sketched to estimate the magnitude and direction of the resultant. Each force is then resolved into its vertical and horizontal components. This is done by multiplying the force by the sine of its direction angle to obtain the vertical component and multiplying the force by the cosine of its directional angle to obtain the horizontal component. It is extremely important to state the direction angle in its correct quadrant measured from 0° being to the right of the origin. Proper quadrant identification is what will correctly determine the sign of the components.

Example 2-2: Determine the resultant of and reaction to the forces acting on the bearing support in Figure 2-5a, using the method of components.

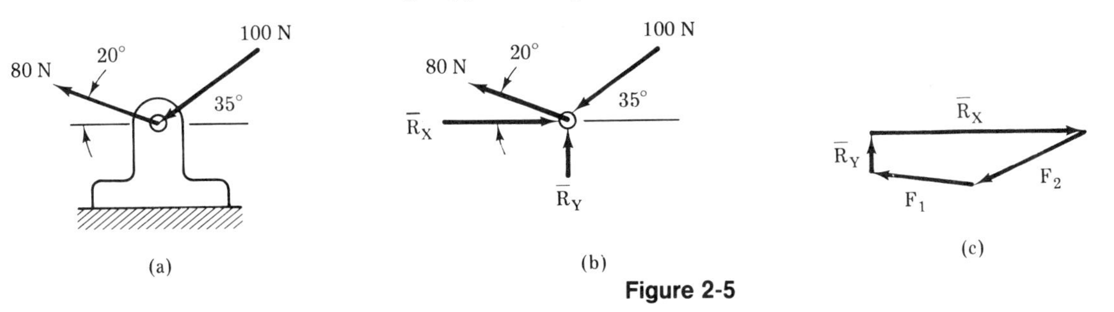

Figure 2-5

Solution:

Step 1 Recognize that these are two coplanar forces that are nonparallel and therefore concurrent.

Step 2 Recognize that the 100-N force while appearing to be located in the first quadrant is actually pointed or directed in the third quadrant.

Step 3 Draw the free-body diagram for the bearing as shown in Figure 2-5b.

Step 4 Sketch the force diagram for the bearing as shown in Figure 2-5c. You may wish, as shown, to ghost in with dashed lines the components of each force including the resultant.

Step 5 Proceed with the solution as follows:

$$R = F_1 \leftrightarrow F_2 \qquad R_Y = F_{1Y} + F_{2Y} \qquad R_X = F_{1X} + F_{2X}$$

$$F_{1Y} = F_1 \sin \theta_1 = 80 \text{ N} \sin 160° \qquad F_{1Y} = +27.4 \text{ N}$$

$$F_{2Y} = F_2 \sin \theta_2 = 100 \text{ N} \sin 215° \qquad F_{2Y} = -57.4 \text{ N}$$

$$R_Y = F_{1Y} + F_{2Y} = +27.4 \text{ N} + (-57.4 \text{ N}) \qquad R_Y = -30 \text{ N}$$

$$F_{1X} = F_1 \cos \theta_1 = 80 \text{ N} \cos 160° \qquad F_{1X} = -75.2 \text{ N}$$

$$F_{2X} = F_2 \cos \theta_2 = 100 \text{ N} \cos 215° \qquad F_{2X} = -81.9 \text{ N}$$

$$R_X = F_{1X} + F_{2X} = -75.2 \text{ N} + (-81.9 \text{ N}) \qquad R_X = -157.1 \text{ N}$$

Since R is the hypotenuse of a right triangle whose sides are R_Y and R_X, Pythagoras' theorem can be used to find R.

$$R^2 = R_Y^2 + R_X^2 \qquad R = \sqrt{R_Y^2 + R_X^2}$$

$$R = \sqrt{(-30 \text{ N})^2 + (-157.1 \text{ N})^2} \qquad R = 160 \text{ N}$$

The magnitude of R is now determined, but its direction is not. Review of the force diagram shows that the direction of R can be determined from the right triangle R, R_Y, R_X.

$$\theta_R = \arctan \frac{R_Y}{R_X}$$

(You may not be familiar with the notation "arctan" meaning "the angle whose tangent is" but it has the same meaning as invtan and $\tan^{-1}$.)

$$\theta_R = \arctan \frac{-30 \text{ N}}{-157.1 \text{ N}} \qquad \theta_R = 191°$$

If you followed this on your calculator, you probably had an answer of 11°. This is because most calculators cannot differentiate between − divided by − and + divided by +. Therefore, it is up to you to determine the correct quadrant. The same problem occurs in dealing with the first and third quadrant.

$$\text{If } R = 160 \text{ N} \quad \text{at} \quad 191°$$

$$\text{Then } \overline{R} = 160 \text{ N} \quad \text{at} \quad 11°$$

Problem 2-2: A barge is being pulled by two tugs as shown in Figure 2-6.

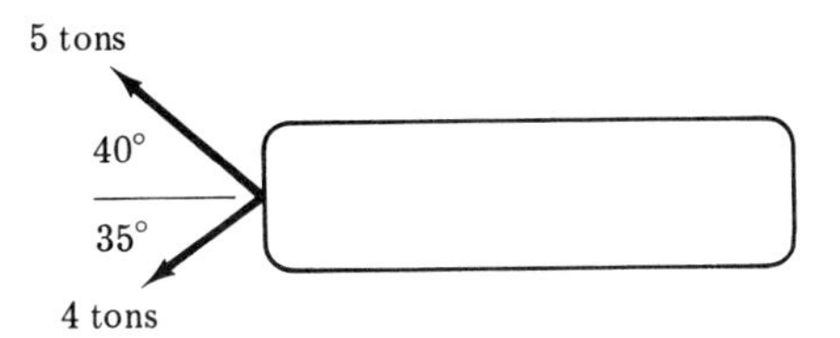

Figure 2-6

Determine the resultant of and the reaction to the two applied forces. Keep in mind that at any given moment the line of action of the total force acting on the barge need not be collinear with the axis of the boat.

2.5 TWO CONCURRENT FORCES: NON-RIGHT-TRIANGLE TRIGONOMETRY METHOD

A third method sometimes used for two force problems utilizes the law of cosines to determine the magnitude of resultant and the law of sines to determine the direction of the resultant. The method is slightly cumbersome which is why we will not apply it to problems dealing with more than two applied forces. However, many people use it and you should be familiar with it.

Example 2-3: In Figure 2-7a, cables *AB* and *BC* are connected to boom *BD* but not to each other. Determine the forces in cable *AB* and boom *BD* and the reactions at supports *A* and *D*.

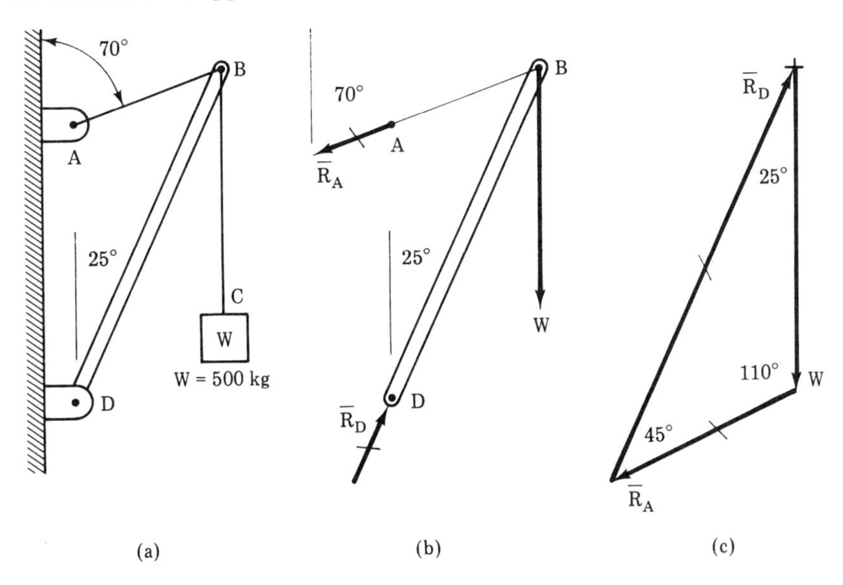

Figure 2-7

Solution:

Step 1 At first glance, this problem may appear to be too complex to handle with what you have learned to date. However, let us recognize that cable *AB* and boom *BD* are both simple two-force members. This means that the forces acting on each end of each of those members must be collinear with the number. Since the joints or connections are all essentially pin or hinge joints, any net force not collinear with the member would tend to rotate the member.

Step 2 Based on the thinking of Step 1, draw the free-body diagram as shown in Figure 2-7b.

Step 3 Sketch the force diagram for the problem as shown in Figure 2-7c, recalling that the principle of concurrence states that three concurrent forces must be in equilibrium. A slightly different force diagram could have been drawn since the order in which the vectors are drawn does not matter as long as they are all drawn head to tail. Note that the word resultant has not been used here. This is because there is only one applied force, and the reaction to it is the sum of the two individual reactions being determined.

Step 4 Applying some basic geometry, determine the angles indicated in Figure 2-7c.

Step 5 Since all the angles and one side of the triangle are known, the law of sines will suffice for solution.

Step 6 Proceed with solution as follows:

$$W = 500 \text{ kg } (9.81 \text{ N/kg}) \qquad W = 4905 \text{ N}$$

$$\frac{a}{\sin A} = \frac{b}{\cos B} = \frac{c}{\cos C}$$

$$\frac{\overline{R}_D}{\sin 110°} = \frac{\overline{R}_A}{\sin 25°} = \frac{W}{\sin 45°}$$

$$\overline{R}_A = W\frac{\sin 25°}{\sin 45°} = 4905 \text{ N}\frac{\sin 25°}{\sin 45°} \qquad \overline{R}_A = 2930 \text{ N at } 225°$$

$$\overline{R}_D = W\frac{\sin 110°}{\sin 45°} = 4905 \text{ N}\frac{\sin 110°}{\sin 45°} \qquad \overline{R}_D = 6520 \text{ N at } 65°$$

Note that each reaction force was totally defined by including its direction angle. Since $\overline{R}_D$ is *pushing* on *BD* and *BD* is a simple two-force member, *BD* must be undergoing a *compressive* force of 6520 N. Since $\overline{R}_A$ is *pulling* on *AB* and *AB* is a simple two-force member, *AB* must be undergoing a *tensile* force of 2930 N.

Example 2-4: Determine the resultant of and reaction to the forces acting on the welded joint shown in Figure 2-8a using non-right-triangle trigonometry.

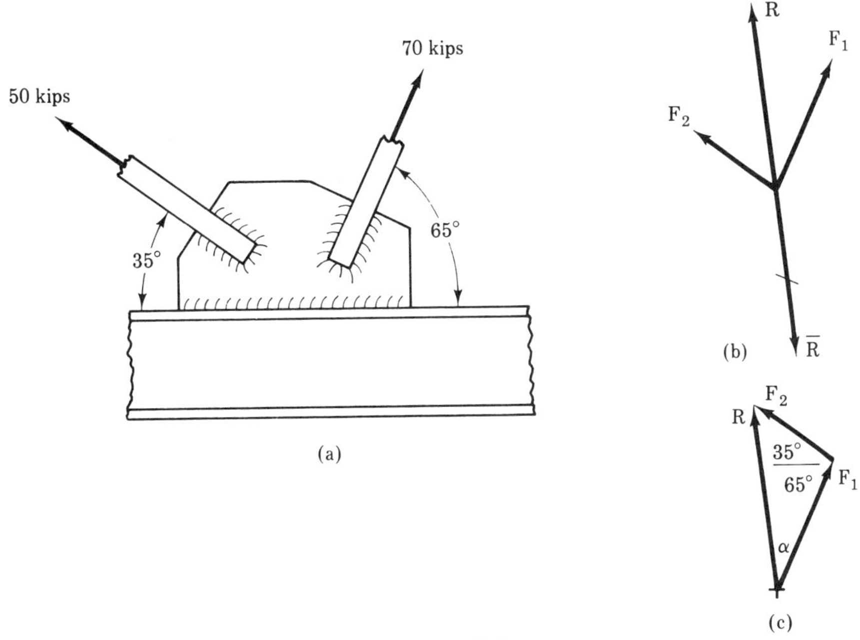

Figure 2-8

Solution:

Step 1 Recognize that these are two coplanar forces that are nonparallel and therefore concurrent.

Step 2 Draw the free-body diagram for the welded joint as shown in Figure 2-8b.

Step 3 Sketch the force diagram for the welded joint as shown in Figure 2-8c.

Step 4 By simple geometry, determine the interior angle between F_1 and F_2 to be 100°.

Step 5 Proceed with the solution as follows:

$$R = F_1 \leftrightarrow F_2 \qquad \overline{R} = -R$$

Since two sides and their included angle are known, the law of cosines will be used to find the third side R.

$$R = \sqrt{F_1^2 + F_2^2 - 2F_1F_2 \cos\theta}$$

$$R = \sqrt{(70)^2 + (50)^2 - 2(70)(50)\cos 100°}$$

$$R = 92.8 \text{ K} \quad (\text{K} = \text{kip})$$

If you were following this on your calculator and did not get the same answer, check to be sure that you observed the negative sign under the radical *and* used the correct sign for the cosine, in this case negative.

Determining the direction of R will require two steps. First, use of the law of sines to determine the angle between F_1 and R on the force diagram, and then adding that to 65° to find the direction of R.

$$\frac{F_2}{\sin\alpha} = \frac{R}{\sin 100°}$$

$$\sin\alpha = \frac{F_2}{R}\sin 100° = \frac{50}{92.8}\sin 100° = 0.531$$

$$\alpha = 32°$$

$$\theta_R = 32° + 65° = 97°$$

$$R = 92.8 \text{ K} \quad \text{at} \quad 97°$$

$$\overline{R} = 92.8 \text{ K} \quad \text{at} \quad 277°$$

Problem 2-3: Determine the resultant of the forces acting on the structure shown in Figure 2-9 using non-right-triangle trigonometry.

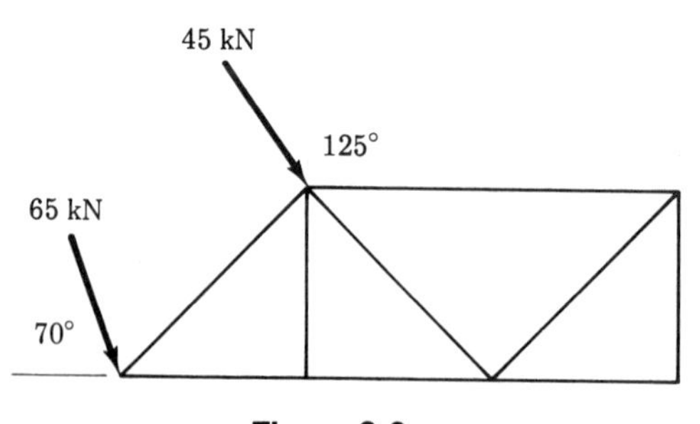

Figure 2-9

2.6 THREE OR MORE CONCURRENT FORCES: GRAPHICAL METHOD

Application of the graphical method to more than two concurrent forces is a simple extension of the approach applied to two forces. When several forces are involved, careful scrutiny may be required to assure that the system is indeed concurrent. Also, during solution care must be taken to assure that no forces are either neglected or duplicated. Solutions may be for one or two unknowns if in the latter case they do not fall on the same line of action.

Example 2-5: Determine forces in members *AD* and *AE* in the welded connection shown in Figure 2-10a, using the graphical method.

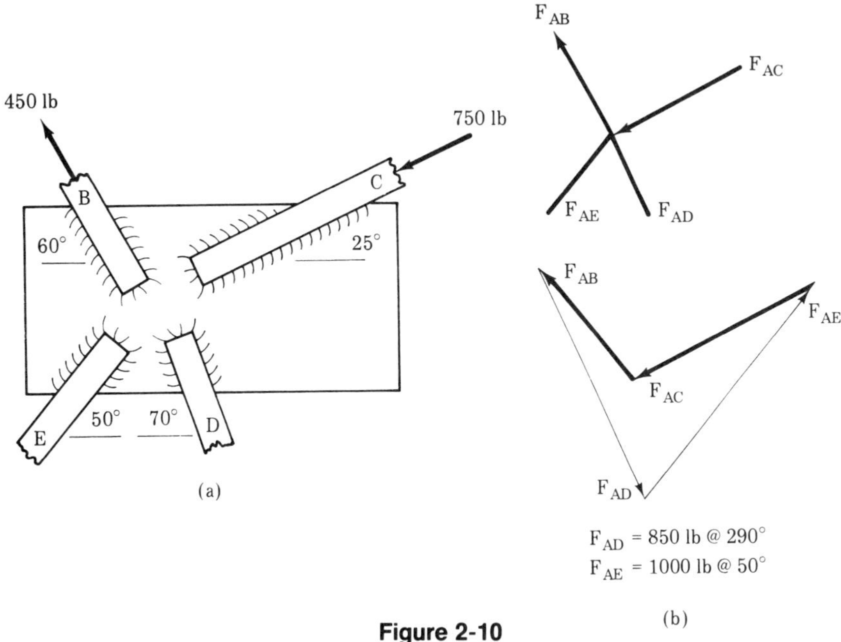

Figure 2-10

Solution:

Step 1 Recognize this as a coplanar, concurrent force system.

Step 2 Draw the free-body diagram for the welded joint as shown in Figure 2-10b. Note that the directions of the forces in *AD* and *AE* are not readily apparent.

Step 3 Draw the force diagram (sometimes called a force polygon or string polygon) to scale for the problem. There is no need in this problem to speak of a resultant and reaction, although it is quite possible that the given loads are acting forces and the unknown reacting forces. The known forces are drawn in tail to head as before; and since the unknown forces must result in equilibrium, one of them must have its tail at the head of the last known vector and the other must have its head at the origin. Which one is drawn where does not affect the solution.

Step 4 Measure the magnitude of the two unknown forces and determine whether the members are in tension or compression.

Example 2-6: Determine the resultant of and the reaction to the forces acting on the plate in Figure 2-11a, using the graphical method.

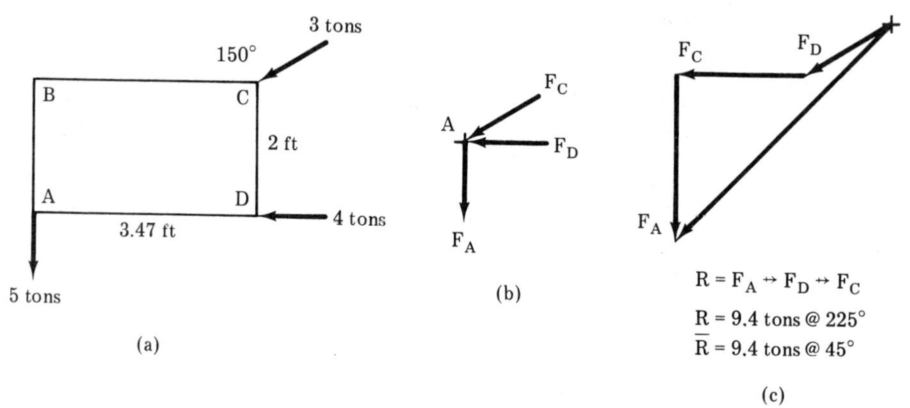

Figure 2-11

Solution:

Step 1 Recognize that this may be a concurrent system. Check by determining if the angle of the plate diagonal is the same as the direction angle of the inclined force.

$$\text{angle of diagonal} = \arctan \frac{2}{3.47} \text{ ft} = 30°$$

Step 2 Draw the free-body diagram for the plate as shown in Figure 2-11b.

Step 3 Draw the force polygon to scale as shown in Figure 2-11c.

Step 4 Measure the magnitude and direction of the resultant.

Step 5 Recognize that the reaction will be equal in magnitude but opposite in direction to the resultant.

Problem 2-4: Three forces act on a fixed support as shown in Figure 2-12. If the resultant must be vertical, determine the magnitude and direction of F_3 and the magnitude of the resultant using the graphical method.

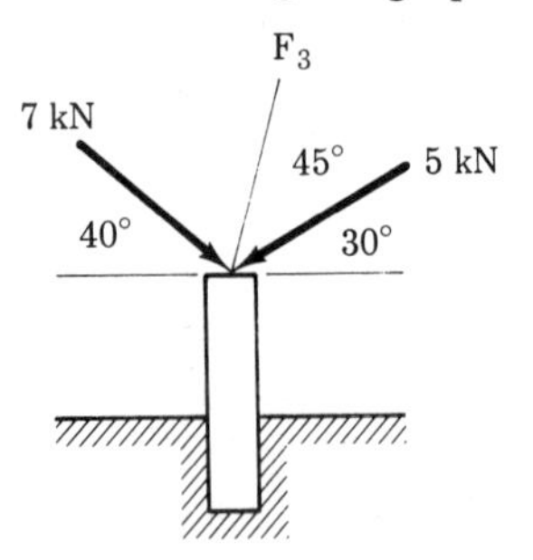

Figure 2-12

Problem 2-5: Find the resultant of and the reaction to the force system acting on the bearing block shown in Figure 2-13 using the graphical method.

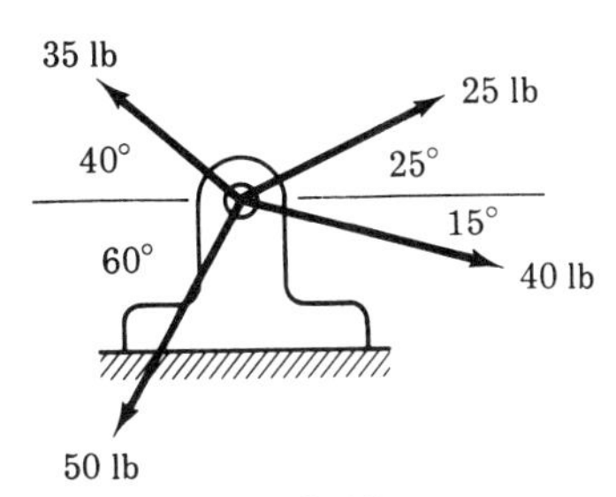

Figure 2-13

2.7 THREE OR MORE CONCURRENT FORCES: METHOD OF COMPONENTS

The basic approach for three or more forces is the same as for two forces. However, it may be useful to tabulate the information rather than write it in linear form when many forces are involved. It is also possible to deal with two unknowns in any direction. However, if neither of the two unknowns is either vertical or horizontal, the use of substitution or simultaneous equations is required.

Example 2-7: Determine the resultant of and the reaction to the forces acting on the machine link in Figure 2-14a, using the method of components.

Solution:

Step 1 Recognize that this may be a concurrent force system. A quick check of the dimensions shows that the slope of a line drawn through A and C is the same as the slope of the force applied at C making the system concurrent.

Step 2 Draw the free-body diagram as shown in Figure 2-14b.

Step 3 Sketch the force polygon to provide an estimate of R as shown in Figure 2-14c. If desired, components may be sketched in also.

Step 4 Proceed with the solution as follows:

$$R = F_1 \leftrightarrow F_2 \leftrightarrow F_3$$

$$R_Y = F_{1Y} + F_{2Y} + F_{3Y} \qquad R_X = R_{1X} + R_{2X} + R_{3X}$$

$$F_{1Y} = F_1 \sin \theta_1 = (35 \text{ N}) \sin 180° \qquad F_{1Y} = 0$$

$$F_{2Y} = F_2 \sin \theta_2 = (45 \text{ N}) \sin \arctan \frac{-4}{-3} \qquad F_{2Y} = -36 \text{ N}$$

$$F_{3Y} = F_3 \sin\theta_3 = (20\text{ N})\sin 120° \qquad F_{3Y} = +17.3\text{ N}$$

$$R_Y = F_{1Y} + F_{2Y} + F_{3Y} = 0 + (-36\text{ N}) + 17.3\text{ N} \qquad R_Y = -18.7\text{ N}$$

$$F_{1X} = F_1 \cos\theta = (35\text{ N})\cos 180° \qquad F_{1X} = -35\text{ N}$$

$$F_{2X} = F_2 \cos\theta_2 = (45\text{ N})\cos \arctan\frac{-4}{-3} \qquad F_{2X} = -27\text{ N}$$

$$F_{3X} = F_3 \cos\theta_3 = (20\text{ N})\cos 120° \qquad F_{3X} = -10\text{ N}$$

$$R_X = F_{1X} + F_{2X} + F_{3X} = -35\text{ N} + (-27\text{ N}) + (-10\text{ N}) \qquad R_X = -72\text{ N}$$

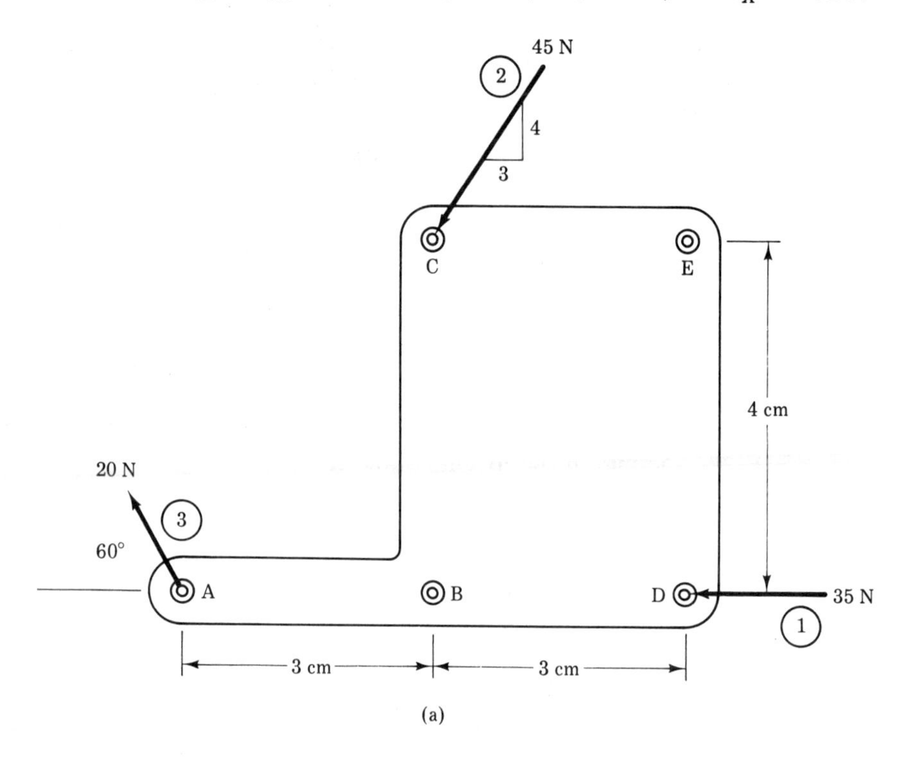

(a)

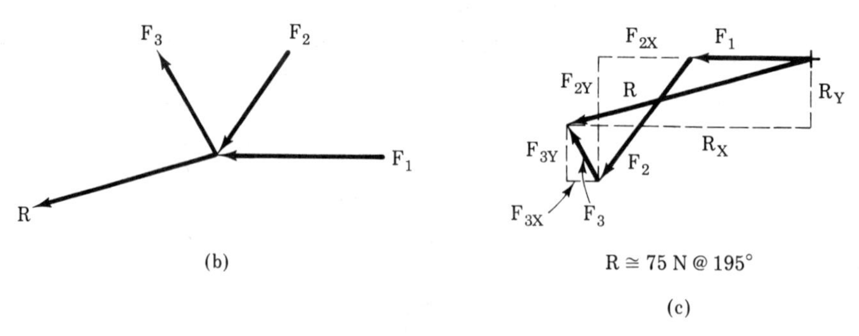

(b)

R ≅ 75 N @ 195°

(c)

Figure 2-14

In the above calculations involving force 2, care must be taken to notice that the angle whose tangent is -4 divided by -3 is in the third quadrant, making both its sine and cosine negative.

$$R = \sqrt{R_Y^2 + R_X^2}$$

$$R = \sqrt{(-18.7\ \text{N})^2 + (-72\ \text{N})^2} \qquad R = 74.4\ \text{N}$$

$$\theta_R = \arctan \frac{R_Y}{R_X} = \arctan \frac{-18.7}{-72.0} \qquad \theta_R = 194.6°$$

$$\text{If } R = 74.4\ \text{N} \quad \text{at} \quad 194.6°$$

$$\text{Then } \overline{\text{R}} = 74.4\ \text{N} \quad \text{at} \quad 14.6°$$

Example 2-8: Determine the resultant of and reaction to the system of forces shown in Figure 2-15a using the method of components.

F_1 = 200 lb, F_2 = 150 lb, F_3 = 250 lb,
F_4 = 100 lb, F_5 = 300 lb

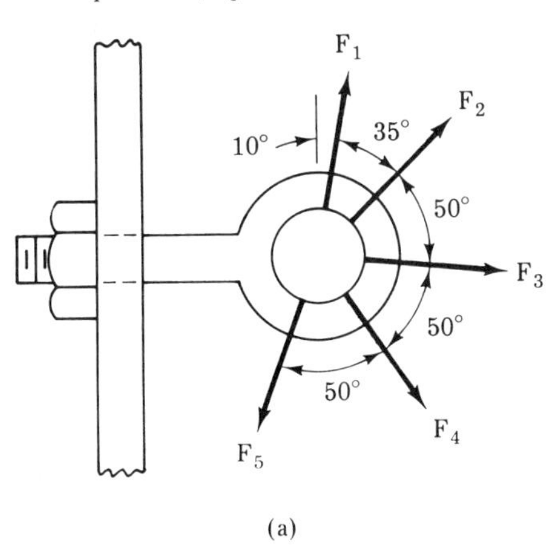

(a)

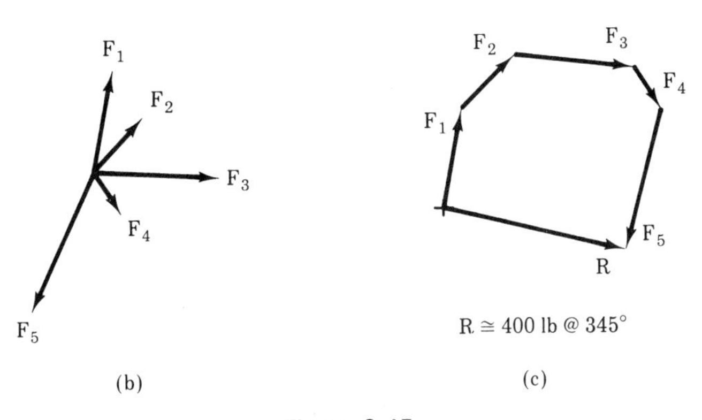

(b) (c)

Figure 2-15

Solution:

Step 1 Recognize this as a coplanar-concurrent force system.

Step 2 Draw the free-body diagram as shown in Figure 2-15b.

Step 3 Sketch the force polygon to provide an estimate of R, as shown in Figure 2-15c.

Step 4 Proceed with the solution as follows:

$$R = \Sigma F \qquad R_Y = \Sigma F_Y \qquad R_X = \Sigma F_X$$

$$F_Y = F \sin \theta \qquad F_X = F \cos \theta$$

No.	F	θ	F_Y	F_X
1	200	80	197	35
2	150	45	106	106
3	250	355	−22	249
4	100	305	−82	57
5	300	255	−290	−78
			−91	+369

$$R = \sqrt{R_Y^2 + R_X^2}$$

$$R = \sqrt{(-91)^2 + (369)^2} \qquad R = 380 \text{ lb}$$

$$\theta = \arctan \frac{R_Y}{R_X} = \arctan \frac{-91}{369} \qquad \theta_R = 346°$$

$$\text{If } R = 380 \text{ lb} \quad \text{at} \quad 346°$$

$$\text{Then } \overline{R} = 380 \text{ lb} \quad \text{at} \quad 166°$$

Example 2-9: Three tugs are pulling a tanker as shown in Figure 2-16a. If tug *B* pulls with a force of 250 kN and the resultant of the effort of all three tugs must be 500 kN at 5°, how hard must tugs *A* and *C* be pulling?

Solution:

Step 1 Recognize this as a coplanar-concurrent force system.

Step 2 Draw the free-body diagram as shown in Figure 2-16b.

Step 3 Sketch the force polygon as shown in Figure 2-16c to provide an estimate of the forces applied by tugs *A* and *C*. Note that, as in the case of Example 2-5, the directions of these two vectors are known but their magnitudes are not, so the line of action of one was drawn through the head of the vector for tug *A* and the line of action for the other was drawn through the head of the resultant.

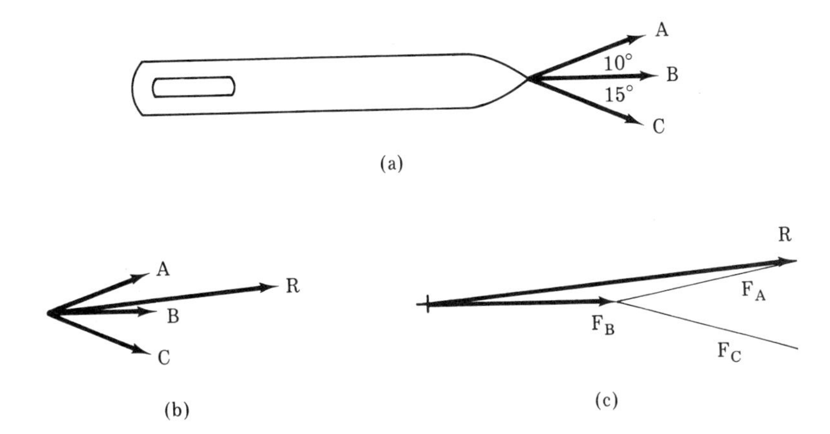

Figure 2-16

Step 4 Proceed with the solution as follows:

$$R = F_A \leftrightarrow F_B \leftrightarrow F_C$$

$$R_Y = F_{AY} + F_{BY} + F_{CY} \qquad R_X = F_{AX} + F_{BX} + F_{CX}$$

$$F_{AY} = F_A \sin\theta_A = F_A \sin 10^\circ \qquad F_{AY} = 0.174F_A$$

$$F_{BY} = F_B \sin\theta_B = (250\text{ kN})\sin 0^\circ \qquad F_{BY} = 0$$

$$F_{CY} = F_C \sin\theta_C = F_C \sin 345^\circ \qquad F_{CY} = -0.259F_C$$

$$F_{AX} = F_A \cos\theta_A = F_A \cos 10^\circ \qquad F_{AX} = 0.985F_A$$

$$F_{BX} = F_B \cos\theta_B = (250\text{ kN})\cos 0^\circ \qquad F_{BX} = 250\text{ kN}$$

$$F_{CX} = F_C \cos\theta_C = F_C \cos 345^\circ \qquad F_{CX} = 0.966F_C$$

We end up with the same two unknowns, F_A and F_C, when we sum both vertically and horizontally. However, recall that in this case the magnitude and direction of the resultant are known, so the total vertical and horizontal forces can both be determined.

$$R_Y = R\sin\theta_R = (500\text{ kN})\sin 5^\circ \qquad R_Y = 43.6\text{ kN}$$

$$R_X = R\cos\theta_R = (500\text{ kN})\cos 5^\circ \qquad R_X = 498\text{ kN}$$

Now the summations can be written.

$$R_Y = F_{AY} + F_{BY} + F_{CY} \qquad 43.6\text{ kN} = 0.174F_A + 0 + (-0.259F_C)$$

$$R_X = F_{AX} + F_{BX} + F_{CX} \qquad 498\text{ kN} = 0.985F_A + 250\text{ kN} + 0.966F_C$$

Solving the first of these equations for F_A yields

$$F_A = 1.49F_C + 251 \text{ kN}$$

Substituting this for F_A in the second equation yields

$$498 \text{ kN} = 0.985(1.49F_C + 251 \text{ kN}) + 250 \text{ kN} + 0.966F_C$$

Solving this for F_C yields

$$F_C = 0.3 \text{ kN}$$

Substituting this in the last equation for F_A above yields

$$F_A = 251.5 \text{ kN}$$

Essentially, tug C must almost slack off while tug A pulls about as hard as tug B. Looking back at the original problem, this seems quite reasonable since the resultant bisects the angle between A and B and equals twice B.

Problem 2-6: The tension in the cable in the derrick in Figure 2-17 is equal to neither the load nor the 800-lb force. If the force in the boom is 4000 lb, find the load P and the tension in the cable using the method of components.

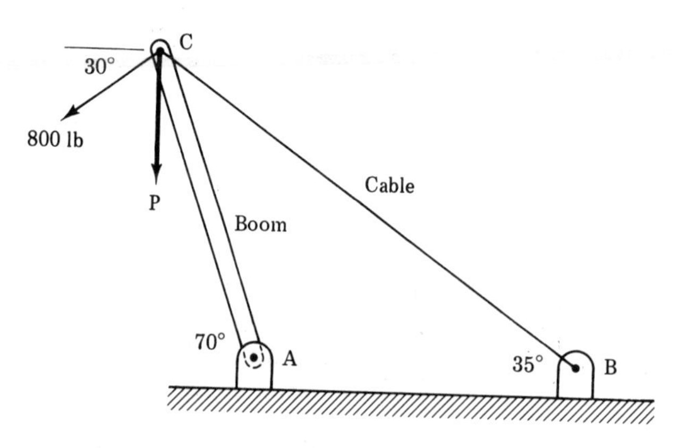

Figure 2-17

Problem 2-7: The engine of the 30,000-kg aircraft shown in Figure 2-18 provides a thrust of 500 kN resulting in a lift (perpendicular to direction of flight) of 300 kN. If the drag force is 30 kN, determine the resultant of all forces and its components parallel to and perpendicular to the line of flight.

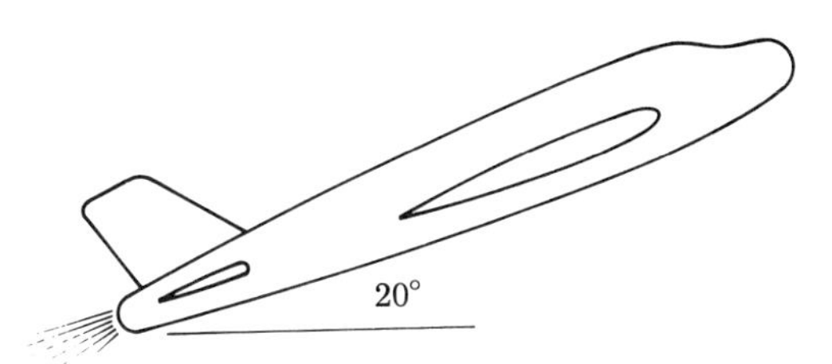

Figure 2-18

2.8 APPLICATIONS TO FRICTION

Previous problems involving friction in Chapter 1 were handled by considering the forces involved as two different collinear systems. Since these systems were perpendicular to each other, we were really applying the method of components to a concurrent force system. Figure 2-19a shows the free-body diagram for Example 1-5 while Figure 2-19a shows the free-body diagram for Problem 1-5. Note that in both cases the horizontal forces are not really collinear and that in Problem 1-5 the vertical forces are not collinear either. Therefore, such problems are not truly concurrent, however, in most cases the lack of concurrence can be neglected. Figure 2-19c shows the type of case where the lack of concurrence may be too great and the object involved may tip rather than slide. Such cases will be dealt with in Chapter 3.

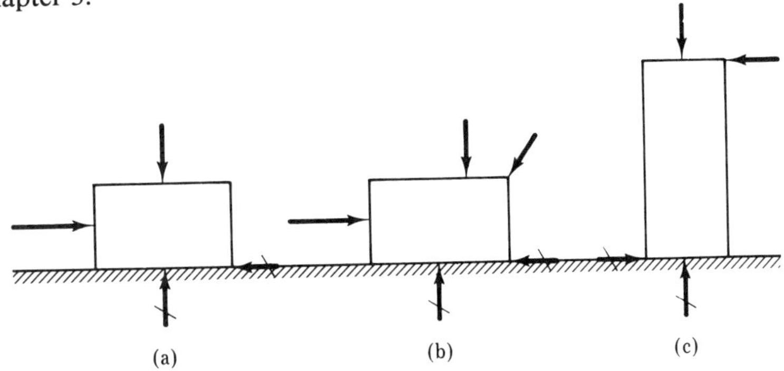

Figure 2-19. Friction free bodies.

Before examining some illustrations of static friction, let us review some basic facts about it.

1. The frictional force that develops between two surfaces is proportional to the force pushing them together.

2. The reaction force to the force acting to push the two surfaces together is often called the normal force and both are perpendicular to the frictional force.

3. The frictional force that develops between two surfaces is independent of the contact area between the two surfaces.

4. There is a maximum static friction force that is proportional to the normal force with the proportion being a function of the roughness of the two surfaces. The proportion is known as the coefficient of friction. If the maximum friction force is exceeded, relative motion between the two surfaces occurs.

5. If we examine Figure 2-20 and the accompanying equations, we can see that the resultant of the normal force and the friction force forms an angle with the normal force whose tangent is equal to the friction force divided by the normal force.

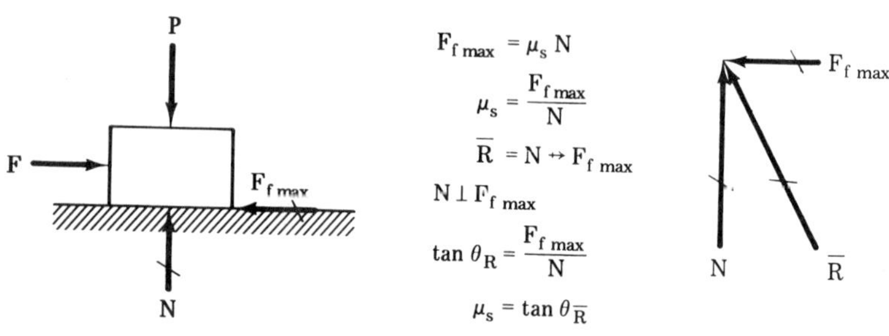

Figure 2-20. Sliding friction.

Table B-2 in Appendix B provides the coefficients of friction for some common combinations of materials.

The examples and problems that follow will illustrate the treatment of friction as a concurrent force problem.

Example 2-10: Determine the force required to start the 500-kg block shown in Figure 2-21a moving up the incline if the coefficient of friction is 0.3. Also, determine the reaction force at the surface.

Solution:

Step 1 Recognize that this may be approximated as a concurrent force problem with the forces in equilibrium.

Step 2 Recognize that there are really only three forces involved: the force of gravity acting straight down, the force that is to start moving the block acting parallel to the incline up and to the right, and the reaction force acting up and to the left against the block at an angle to the incline whose tangent equals the coefficient of friction. (While the force of gravity could be resolved into components parallel and perpendicular to the incline and the

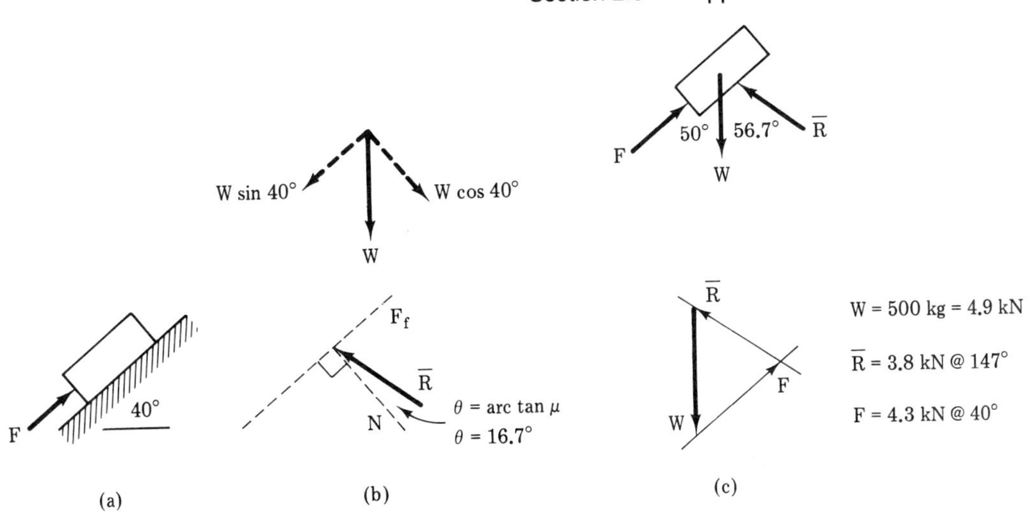

Figure 2-21

reaction force could be resolved into the normal force and the friction force, as shown in Figure 2-21b, there is no need to do so.)

Step 3 Draw the free-body diagram as shown in Figure 2-21c.

Step 4 Write the equilibrium equation as follows:

$$\Sigma F = 0 \qquad F \leftrightarrow W \qquad \overline{R} = 0$$

Step 5 Draw the force diagram to scale thus achieving a graphical solution. The force of gravity is drawn first; then the line of action for either of the unknowns can be drawn through either the head or tail of W, and then the line of action for the other unknown through the other end of W. The appropriate arrowheads are put in for F and $\overline{R}$, the magnitude and direction measured and the solution is complete. This force triangle could have been simply sketched and the solutions found using the law of sines. Also, as might be suggested by Figure 2-21b, a rectangular coordinate system could have been set up parallel to and perpendicular to the incline and the solutions obtained by the method of components.

Problem 2-8: A force of 200 lb is acting on a 350-lb block as shown in Figure 2-22. If the coefficient of friction is 0.2, which way is the block moving?

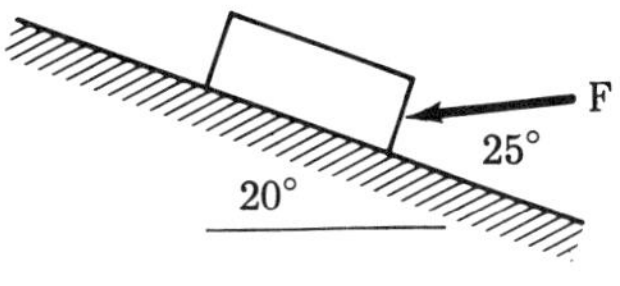

Figure 2-22

Example 2-11: Determine the force F required to act on the wedge to start the load in Figure 2-23a moving upward.

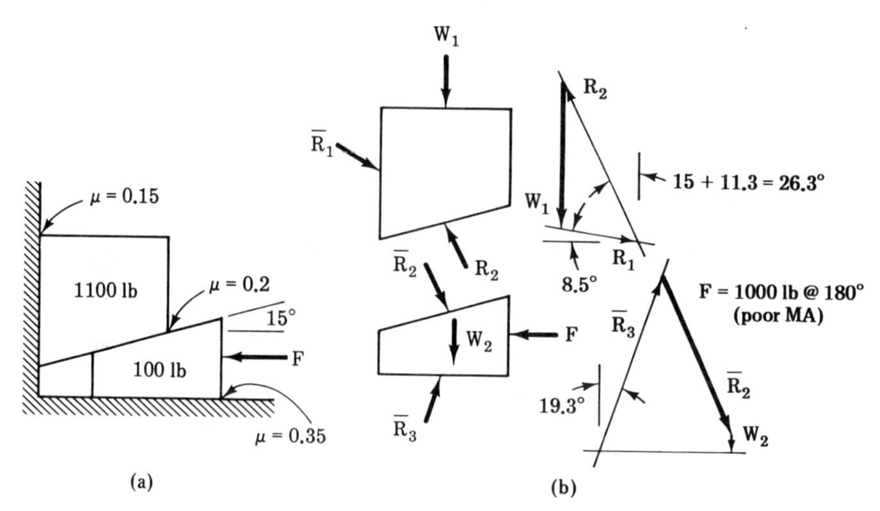

Figure 2-23

Solution:

Step 1 Recognize that this may be approximated as two related concurrent force problems with the forces in equilibrium.

Step 2 In determining reaction force directions, recall that the normal force component is always perpendicular to the surface in question and that the friction force component is always opposite in direction to the impending or actual motion.

Step 3 Arbitrarily label the three pairs of surfaces 1, 2, and 3 for the vertical, inclined, and horizontal.

Step 4 Determine the friction angle for each pair of surfaces.

$$\theta_1 = \arctan 0.15 \qquad \theta_1 = 8.5^\circ$$

$$\theta_2 = \arctan 0.2 \qquad \theta_2 = 11.3^\circ$$

$$\theta_3 = \arctan 0.35 \qquad \theta_3 = 19.3^\circ$$

Step 5 Draw the free-body diagram for each block as shown in Figure 2-23b keeping in mind that the action of the wedge on the load at surface 2 is equal and opposite to the reaction of the load on the wedge at that surface.

Step 6 Examination of the free-body diagrams show that the wedge diagram is initially insoluble for there are three unknowns. However, the load diagram shows only two unknowns and can be solved. This has been done graphically in Figure 2-23c. Since $R_2 = -\overline{R}_2$, there are now only two unknowns in the wedge diagram and a solution can be achieved graphically as shown in Figure 2-23c.

Problem 2-9: Determine the force required to start the load moving in Figure 2-24. Consider the rollers to be frictionless.

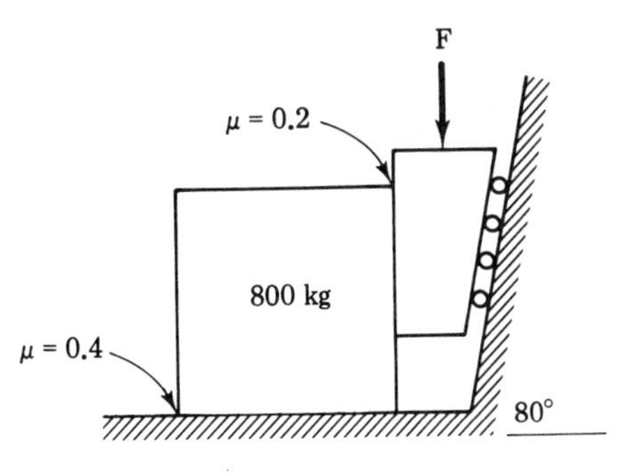

Figure 2-24

2.9 APPLICATIONS TO PULLEYS

Pulleys are frequently used to change the direction of the force in a cable, rope, belt, or chain. Three possibilities present themselves and are shown in Figure 2-25: a single pulley, a system of pulleys for obtaining mechanical advantage, and a system of pulleys for power transmission. The latter two cases will be considered in subsequent chapters. Here, we will devote ourselves strictly to the first case.

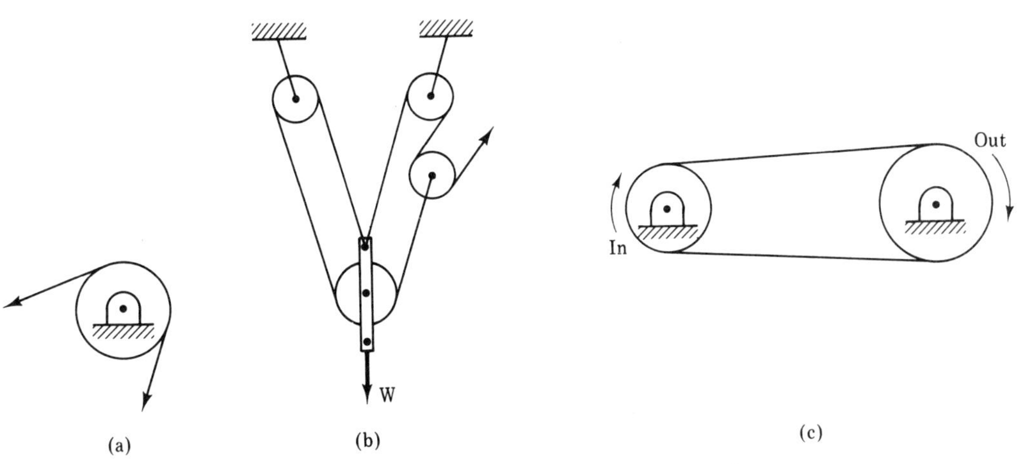

Figure 2-25. Pulley applications.

In many practical applications, pulley operation is highly efficient and quite simple. Three assumptions must be made to achieve these ideal conditions, and it will be seen that these are neither unreasonable nor unusual. First, no slippage, that is, perfect friction between the cable and the pulley is assumed. Second, zero friction at the pulley or pulley shaft bearings is assumed. Finally, it is assumed that no torque is being applied to or by the pulley shaft. Given these assumptions, the tension in the cable will be the same on both sides of the cable as indicated in Figure 2-26a. If the lines of action of these two forces are extended until they intersect and the two are added together, it can be shown geometrically that the line of action of the resultant will pass through the center of the pulley support as shown in Figure 2-26b. The reaction at the support would be equal and opposite to the resultant as shown in Figure 2-26b, and a free-body diagram of the pulley with the reaction resolved into components would appear as shown in Figure 2-26c. A simple pulley problem is therefore a simple concurrent force problem as will be illustrated in Example 2-12.

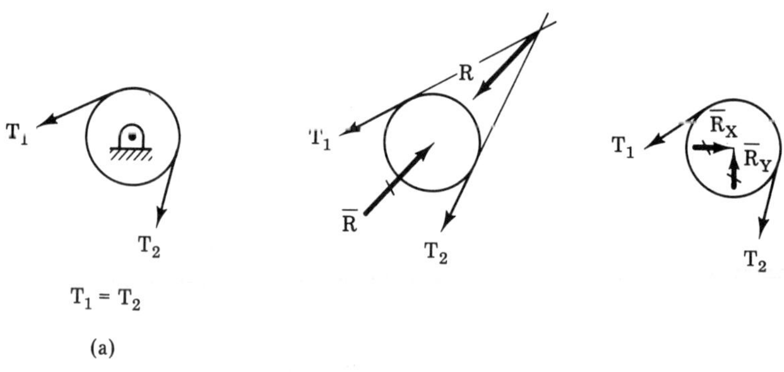

Figure 2-26. Idealized pulley.

Example 2-12: Determine the support reaction for the pulley shown in Figure 2-27a if the tension in the cable is 2500 N.

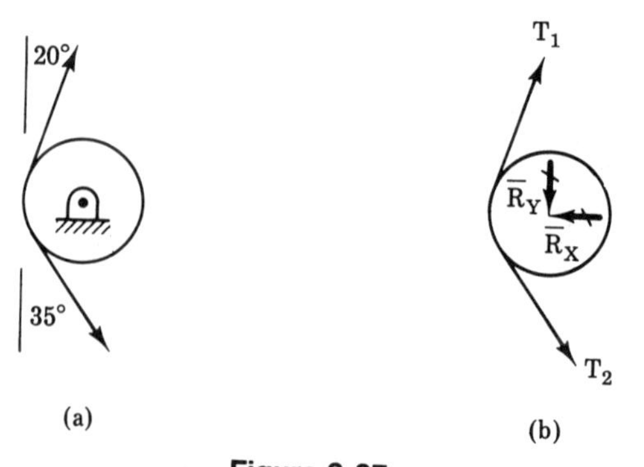

Figure 2-27

Solution:

Step 1 Recognize this as a concurrent force problem.

Step 2 Draw the free-body diagram as shown in Figure 2-27b.

Step 3 Determine the support reaction by the method of components as follows:

$$\Sigma F_X = 0 \qquad T_{1X} + T_{2X} + \overline{R}_X = 0$$

$$2500 \cos 70^\circ + 2500 \cos 305^\circ + \overline{R}_X = 0 \qquad \overline{R}_X = -2289 \text{ N}$$

$$\Sigma F_Y = 0 \qquad T_{1Y} + T_{2Y} + \overline{R}_Y = 0$$

$$2500 \sin 70^\circ + 2500 \sin 305^\circ + \overline{R}_Y = 0 \qquad \overline{R}_Y = -301 \text{ N}$$

$$\overline{R} = \sqrt{\overline{R}_X^2 + \overline{R}_Y^2} = \sqrt{(-2289)^2 + (-301)^2} \qquad \overline{R} = 2328 \text{ N}$$

$$\theta_{\overline{R}} = \arctan \frac{\overline{R}_Y}{\overline{R}_X} = \arctan \frac{-301}{-2289} \qquad \theta_{\overline{R}} = 187.5^\circ$$

Problem 2-10: Determine the support reaction for the pulley shown in Figure 2-28 if the tension in the cable is 15 lb.

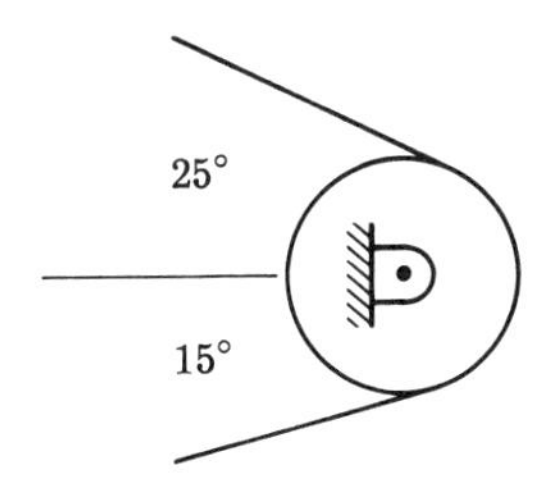

Figure 2-28

2.10 READINESS QUIZ

1. Vectors can be added
   - (a) graphically
   - (b) by method of components
   - (c) by parallelogram method
   - (d) any of the above
   - (e) none of the above

2. Forces acting along different lines of action but passing through a single point are said to be
   - (a) collinear
   - (b) concurrent

(c) parallel
(d) coexistent
(e) none of the above

3. The summation of two or more forces is called their

(a) resultant
(b) vector
(c) system
(d) all of the above
(e) none of the above

4. A quantity of any measure that has magnitude and direction is called a

(a) coordinate
(b) vector
(c) scalar
(d) all of the above
(e) none of the above

5. The graphical method of adding vectors is sometimes also called the

(a) method of components
(b) parallelogram method
(c) string polygon method
(d) any of the above
(e) none of the above

6. Of the force systems shown, the concurrent one is

(a) ↑↓ (b) ↙↑ (c) ←↑↑ (d) all (e) none

7. Replacing a force with its rectangular components is called ________ the force.

(a) resolving
(b) solving
(c) coordinating
(d) all of the above
(e) none of the above

8. The word normal is sometimes used instead of

(a) coordinate
(b) perpendicular
(c) scalar
(d) direction
(e) none of the above

9. Resolving all forces in a system into their components is the first step of the
   (a) string polygon method
   (b) parallelogram method
   (c) graphical method
   (d) all of the above
   (e) none of the above

10. The law of cosines must be used in the
    (a) method of components
    (b) graphical method
    (c) parallelogram method
    (d) all of the above
    (e) none of the above

11. The resultant of the forces shown in Figure 2-29 is
    (a) 5 N, to the right and up
    (b) 5 N, to the right and down
    (c) 5 N, to the left and up
    (d) 5 N, to the left and down
    (e) none of the above

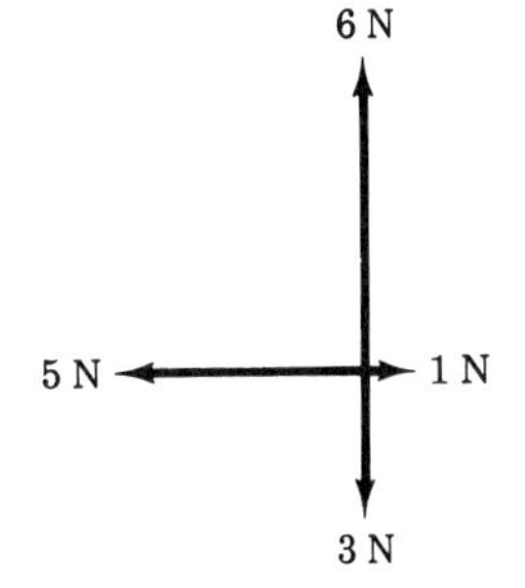

Figure 2-29

12. The components of the force shown in Figure 2-30 are
    (a) 30 lb left and 40 up
    (b) 40 lb right and 30 down
    (c) 40 lb left and 30 down
    (d) 30 lb right and 40 up
    (e) none of the above

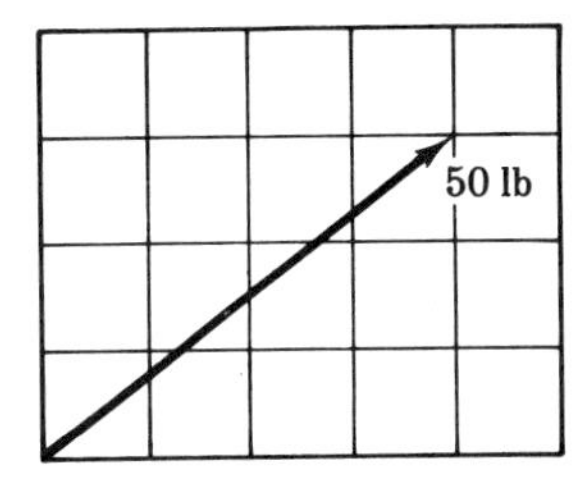

Figure 2-30

13. The resultant of the forces in Figure 2-31 lies in the ________ quadrant.

(a) first
(b) second
(c) third
(d) fourth
(e) indeterminate

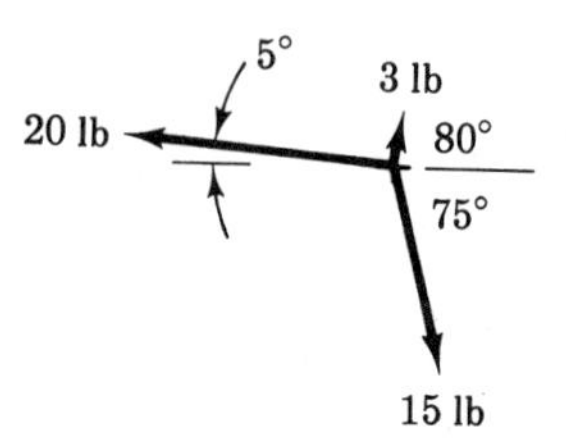

Figure 2-31

14. Of the force systems shown, the concurrent one is

(a) (b) (c) (d) all (e) none

15. The reaction at the axle of the pulley shown in Figure 2-32 is

(a) strictly horizontal
(b) strictly vertical
(c) vertical and horizontal
(d) any of the above
(e) insufficient information given

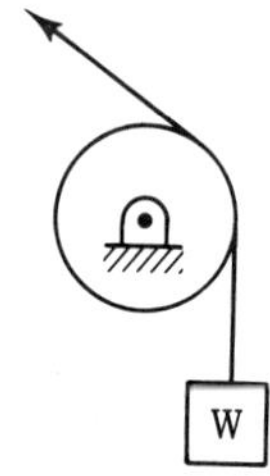

Figure 2-32

16. For a simple pulley to be considered as a concurrent force problem, it must be assumed that there is no

(a) slippage between the pulley and the cable
(b) friction at the pulley bearing
(c) torque applied to or by the pulley
(d) all of the above
(e) none of the above

17. Pulleys are used

(a) to change the direction of a tensile force in a cable
(b) in groups to provide mechanical advantage
(c) in pairs to transfer power
(d) all of the above
(e) none of the above

18. A sliding block friction problem can be considered to be a concurrent force problem if

(a) the forces involved are small
(b) the normal force is large
(c) no tipping is occurring

(d) any of the above
(e) none of the above

19. If the components of a force are both negative, the direction of that force must lie in the

(a) first quadrant
(b) second quadrant
(c) third quadrant
(d) fourth quadrant
(e) insufficient information given

20. The problem shown in Figure 2-33 can be solved by treating it as __________ concurrent force problem(s).

(a) one
(b) two
(c) three
(d) four
(e) five

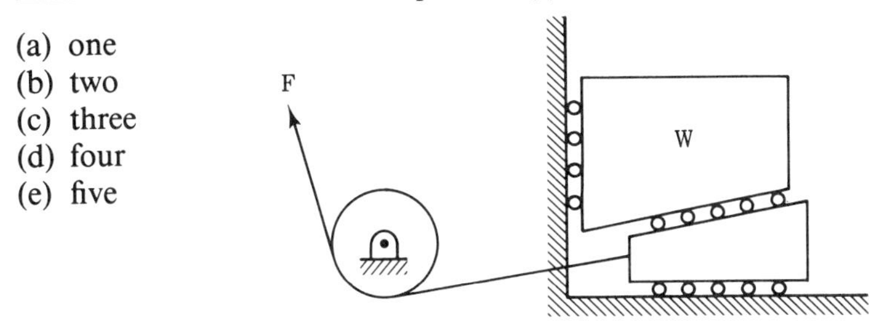

Figure 2-33

2.11 SUPPLEMENTARY PROBLEMS

2-11 Determine the resultant of the forces shown acting in Figure 2-34 by the graphical method. Determine the support reaction.

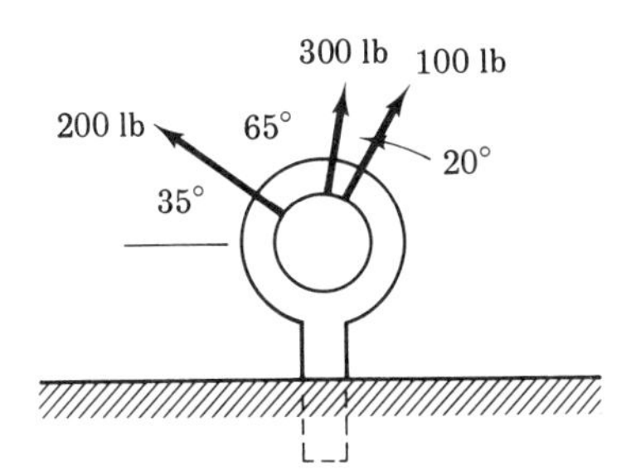

Figure 2-34

2-12 Determine the resultant of the forces shown acting in Figure 2-35 by the method of components. Determine the support reaction and its components.

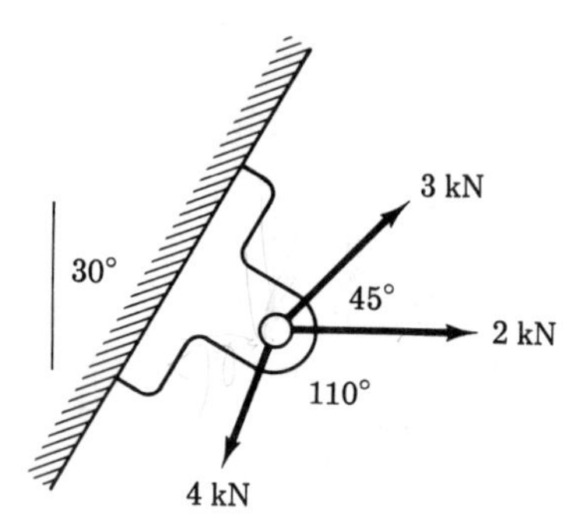

Figure 2-35

2-13 Determine the resultant of the forces shown acting in Figure 2-36 by the parallelogram (non-right-triangle trigonometry) method.

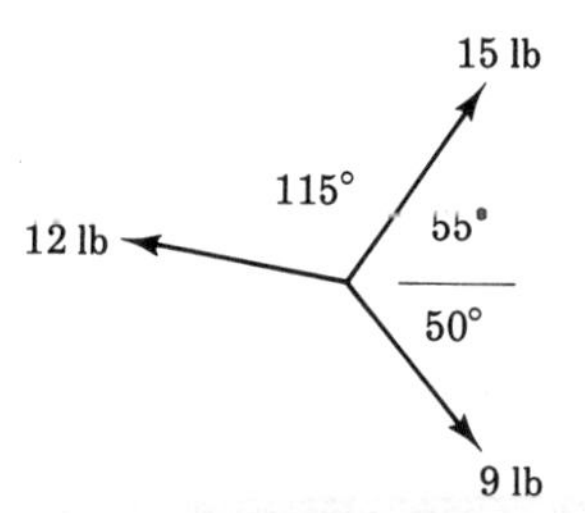

Figure 2-36

2-14 Determine graphically the reaction to the forces shown acting in Figure 2-37.

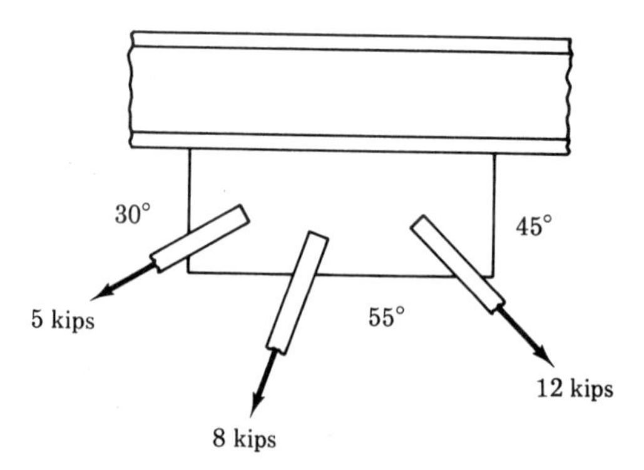

Figure 2-37

2-15 Determine by the method of components the reaction to the forces shown acting in Figure 2-38.

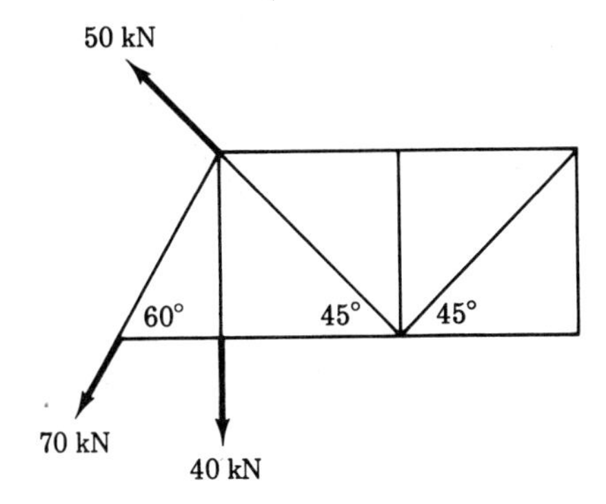

Figure 2-38

2-16 Determine the force F required to just start motion in the problem shown in Figure 2-39.

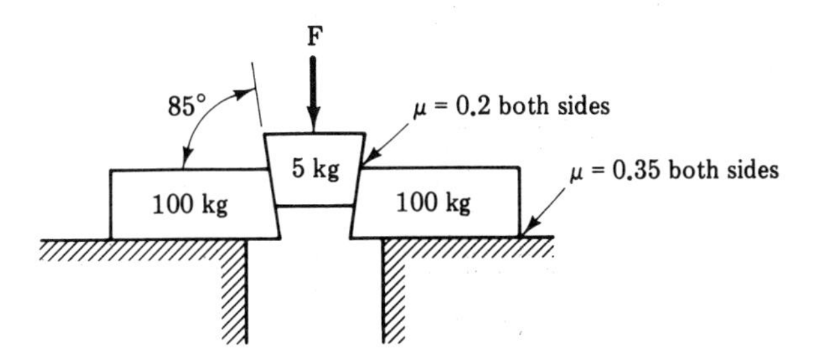

Figure 2-39

2-17 Determine the support reaction for the pulley shown in Figure 2-40.

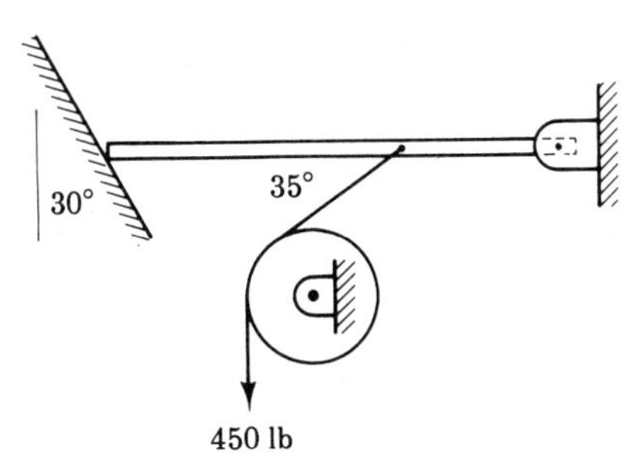

Figure 2-40

2-18 Determine the support reaction for each pulley in the system shown in Figure 2-41.

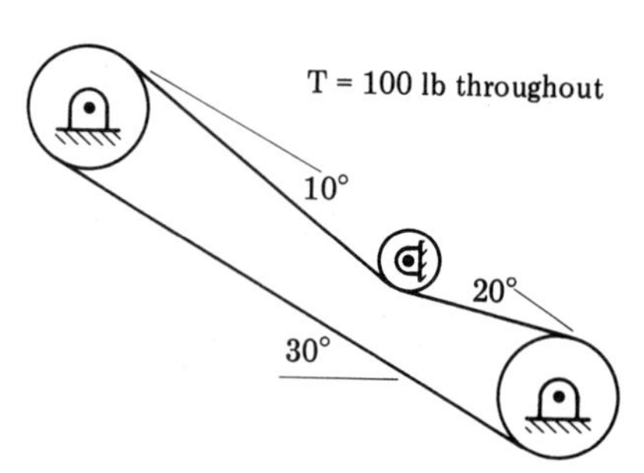

Figure 2-41

Chapter 3

Moments, Parallel Forces, and Couples

3.1 OBJECTIVES

Upon completion of the work relating to this chapter, you should be able to perform the following.

1. For the words listed below:

   (a) Select from several definitions the correct one for any given word.
   (b) Given a definition, select the correct word from a choice of words.

concentrated load	lever arm	rotation
couple	moment	torque
distributed load	moment arm	

2. Given a fully defined force located with respect to a known point

   (a) Find the moment arm of the force with respect to the point
   (b) Determine the moment of the force about the point

3. Given a system of parallel forces located with respect to a known point

   (a) Determine the total moment of the system about the point

(b) Determine the resultant force for the system and its location with respect to the point by each of the following methods:

(1) moments
(2) graphical-dummy forces
(3) graphical-proportion

4. Given a series of loads, differentiate between concentrated loads and distributed loads.
5. Given a uniformly distributed load, determine

(a) the magnitude of the total load.
(b) the location of the line of action for the load.
(c) the moment of the load about any given point.

6. Given a parallel or concurrent system of concentrated and distributed loads acting on a body, determine the magnitude, direction, and location of the resultant of the system.
7. Given a series of force systems, identify those that contain a couple.
8. Determine the moment acting on a body as a result of one or more couples.

3.2 PRINCIPLE OF MOMENT

In physics and in the first two chapters of this book, you studied simple forces and learned to add concurrent forces so that they could be replaced with a single resultant force. Whenever the resultant equaled zero, or whenever we added a reaction force equal and opposite to the resultant, equilibrium occurred. If the forces did not add up to zero and no balancing reaction force was applied, then, motion would occur in the direction of the force. These two possibilities are shown in Figure 3-1.

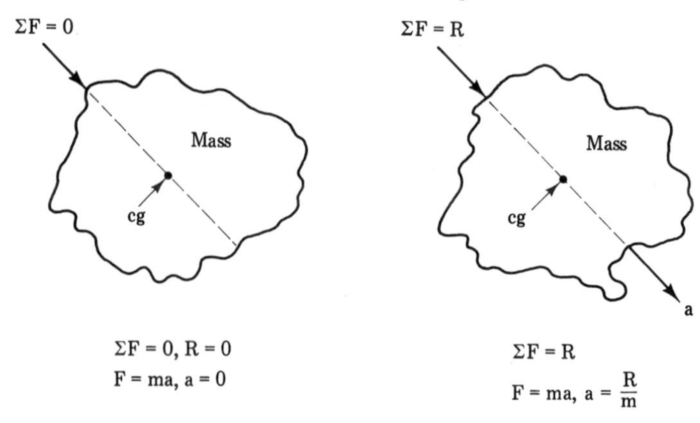

Figure 3-1. Equilibrium versus nonequilibrium.

This is exactly what happens as a child pulls straight forward on a sled or a wagon.

If, however, the mass is fixed at some point, such a force tends to cause the mass to rotate about that point rather than travel in a straight line. This phenomena is illustrated with a playground merry-go-round in Figure 3-2.

In the cartesian coordinate system, six directions of force are possible: $+X, -X, +Y, -Y, +Z$, and $-Z$. These are illustrated in Figure 3-2 in that order by (a) through (f), with the line of action of the X forces passing through the center of rotation of the merry-go-round. Assuming that the merry-go-round is well mounted on a sturdy vertical shaft, the X and Y forces will cause no motion. The X forces will *attempt* to bend and shear the shaft while the Y forces will *attempt* to bend the shaft and raise or lower the merry-go-round. Only the Z forces (or the Z component of any force at a compound angle) will cause the merry-go-round to turn. This may appear to be a very complex analysis of something most children discover by simple experience. However, it lies at the heart of all actual or attempted angular motion and is known as the moment of a force.

A force can have a moment only with respect to some point. Unless such a point is identified, there can be no discussion of a moment. The moment discussed also applies only to that point. Further, for the force to have a moment about a point, the line of action of the force cannot pass through the point. This was evident with the X forces acting on the merry-go-round. The Y forces caused no rotation for another reason—the merry-go-round was not free to rotate in a vertical plane.

The moment of a force can be determined by multiplying the magnitude of the force by a distance known as the moment arm (occasionally called the

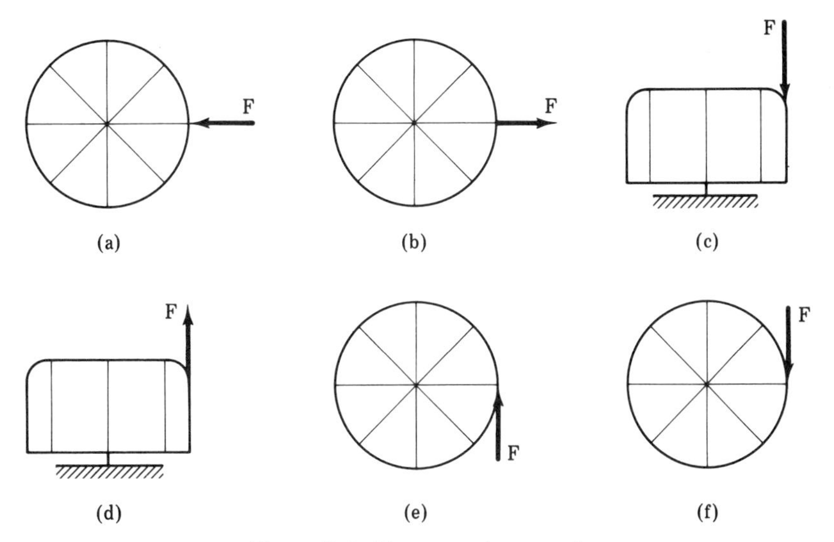

Figure 3-2. Forces and moments.

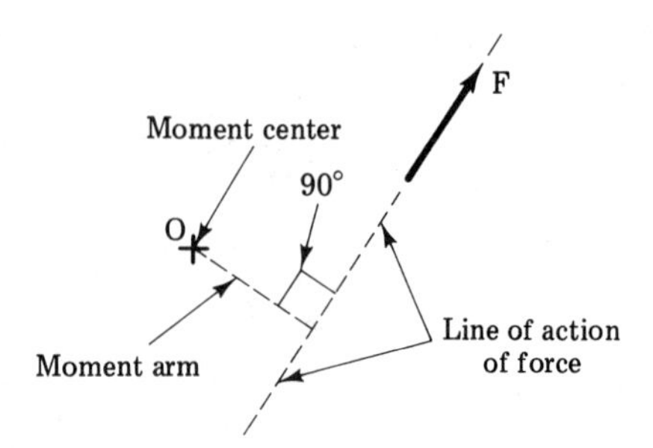

Figure 3-3. Determining moment from force.

lever arm). Its units, thus, are always units of force times units of length, for example, lb ft or N m. The length of the moment arm is the shortest distance from the moment center to the line of action of the force. It can be shown by geometry that this shortest distance always lies on a line drawn perpendicular to the line of action of the force. Thus, to find the length of the moment arm of a force, it is simply necessary to draw such a perpendicular as shown in Figure 3-3 and either measure it if drawn to scale or calculate it by use of trigonometry.

In practical everyday application, the above relationship between force and distance can be verified by attempting to open a heavy door as shown in Figure 3-4a and 3-4b. A hand exerts a force upon the door knob trying various directions shown from 1 to 4 in 3-4a. The amount of force required to open the door will vary depending upon the direction of the force. A force along direction 4, parallel to the door will produce negligible rotational effect, no matter how large the force may be. Additionally, what little effect there is will tend to close the door, not open it. The most favorable attempt is that along direction 1, perpendicular to the door. The conclusion is that the greater the moment arm, the less force is required.

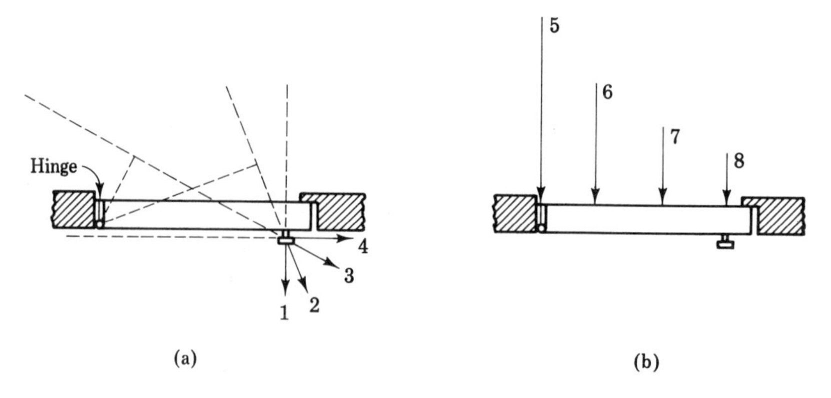

Figure 3-4. Effect of point of application and direction of force on moment arm.

Similarly, if the door is pushed at various places along the door and perpendicular to the door as shown in 3-4b, the least force is needed at position 8. As one approaches the hinge more and more force is needed as the moment arm becomes shorter. A force applied at the hinges, position 5, will not produce rotation, no matter how great the force may be.

In the previous units, we treated forces as vector quantities having both direction and magnitude. The same is true of the moments of forces. However, the matter of direction is simpler in the case of moments in that there are only two possible directions, clockwise and counterclockwise. In the illustration above, we were trying to turn the door in a clockwise direction about its hinges. Generally, clockwise is given a negative designation and counterclockwise a positive designation. *Do not try to equate force direction and the direction of the moment for the same force.* For example, the force shown in Figure 3-5 would generally be given a negative value since it is directed toward 270°. The moment of that force *taken about point A* is clockwise and would conventionally be considered negative. However, the moment of that same force *taken about point B* is counterclockwise and would conventionally be considered positive.

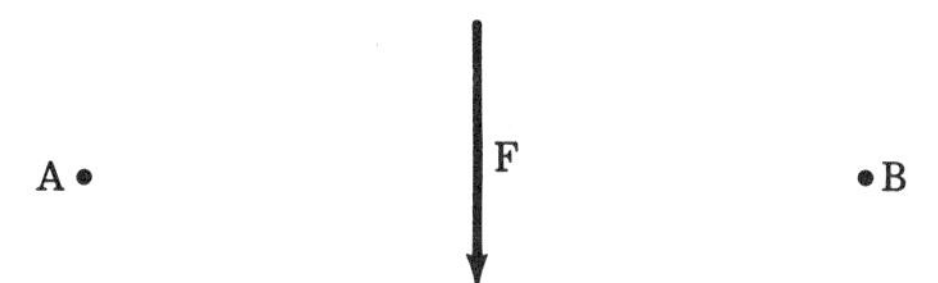

Figure 3-5. Moment direction versus force direction.

Frequently, the extension of the line of action of a force and the construction of its moment arm as shown in Figure 3-3 is not the fastest route to a solution. This occurs whenever the force does not lie parallel to one of the naturally horizontal or vertical axes of the problem. The difficulty can often be minimized by resolving the force into vertical and horizontal components at the point of application. Another alternative is to move the point of application along the line of action of the force until it lies on either a horizontal or vertical line running through the moment center. Thus, there are three different approaches to solving the same problem. The first is based upon the basic principle of moment. The second adds to this the theorem that *the moment of a force is equal to the sum of the moments of its components.* The third further adds the principle of the transmissibility of a force. Example 3-1 illustrates all three approaches.

Example 3-1: Determine the moment of the force shown in Figure 3-6a about point *A*.

Solution 1:

Step 1 Extend the line of action of the force back well past *A* as shown in Figure 3-6b.

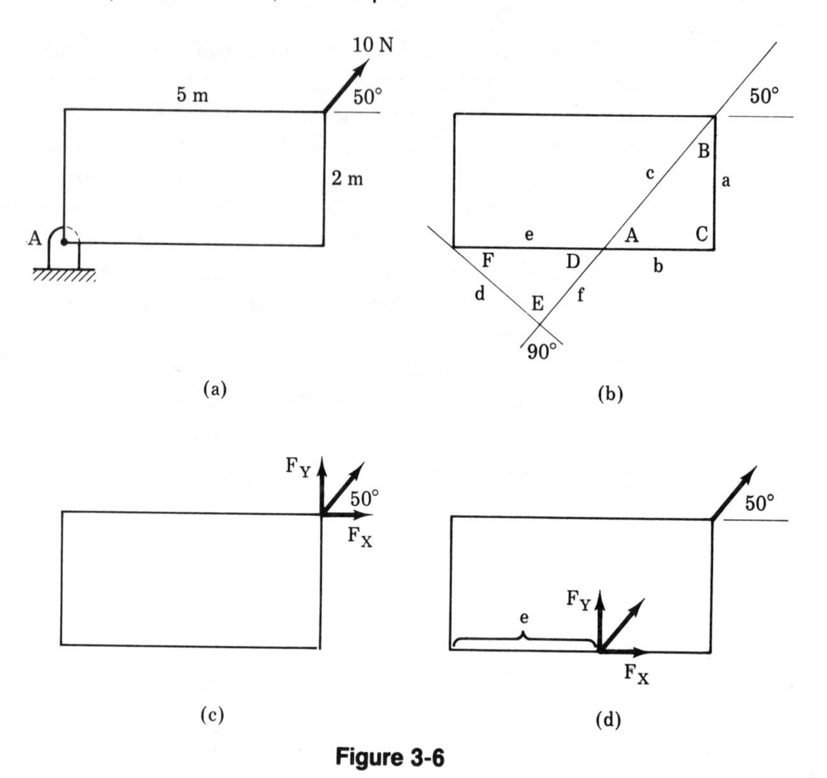

Figure 3-6

Step 2 Construct the moment arm as shown in Figure 3-6b.

Step 3 Label the angles and sides of the triangles formed by steps 1 and 2.

Step 4 Determine the length of the moment arm by use of right-triangle trigonometry as follows:

$$\tan A = \frac{a}{b} \qquad b = \frac{a}{\tan A} = \frac{2\text{ m}}{\tan 50°} \qquad b = 1.68\text{ m}$$

$$e = (e + b) - b \qquad e = 5 - 1.68\text{ m} \qquad e = 3.32\text{ m}$$

$$\sin D = \frac{d}{e} \qquad d = e \sin D = (3.32\text{ m})\sin 50° \qquad d = 2.54\text{ m}$$

Step 5 Determine the moment of the force as follows.

$$M_A = F \cdot d = +\,(10\text{ N})(2.54\text{ m}) \qquad M_A = 25.4\text{ N m} ↺$$

Note that the subscript A was applied to the symbol M to indicate what point the moment was being taken about. Further, the moment was recognized to be counterclockwise and, therefore, given a positive notation both in the calculation and in the answer. This was further emphasized by applying a curved arrow, directed counterclockwise, to the answer. Note, also, that no attempt was made to place any negative or positive direction on either the force or the moment arm.

Solution 2:

Step 1 Draw the problem as shown in Figure 3-6c with the force resolved into its components.

Step 2 Determine the components of the force as follows:

$$F_X = F\cos\theta = (10\text{ N})\cos 50° \qquad F_X = +6.4\text{ N}$$

$$F_Y = F\sin\theta = (10\text{ N})\sin 50° \qquad F_Y = +7.7\text{ N}$$

Step 3 Determine the total moment of the two-force system established in Step 2 as follows:

$$\Sigma M_A = F_Y d_Y \leftrightarrow F_X d_X = +(7.7\text{ N})(5\text{ m}) - (6.4\text{ N})(2\text{ m})$$

$$\Sigma M_A = +38.5\text{ N m} - 12.8\text{ N m} \qquad M_A = +25.7\text{ N m}$$

Solution 3:

Step 1 Move the force back along its line of action until the point of application is on a horizontal line running through point A as shown in Figure 3-6d.

Step 2 Resolve the force into components and determine those components exactly as done in Steps 1 and 2 of Solution 2.

Step 3 Determine the moment arm for F_Y exactly as done as part of Step 4 in Solution 1 (length e).

Step 4 Determine the total moment of the two-force system established in Step 2 as follows:

$$\Sigma M_A = F_Y d_Y \leftrightarrow F_X d_X = +(7.7\text{ N})(3.32\text{ m}) - (6.4\text{ N})(0\text{ m})$$

$$\Sigma M_A = +25.6\text{ N m} - 0\text{ N m} \qquad M_A = +25.6\text{ N m}$$

Comparing the answers obtained by the three methods, we find that they vary from 25.4 to 25.7 N m, a range of 0.3 N m with an average of 25.55 N m. This variation of approximately $\pm\frac{1}{2}$% can be considered negligible indicating that the three methods are wholly interchangeable.

Problem 3-1: Determine the moment about B of the force shown in Figure 3-7 using each of the methods demonstrated above.

As you might already suspect, what we have just done with one force

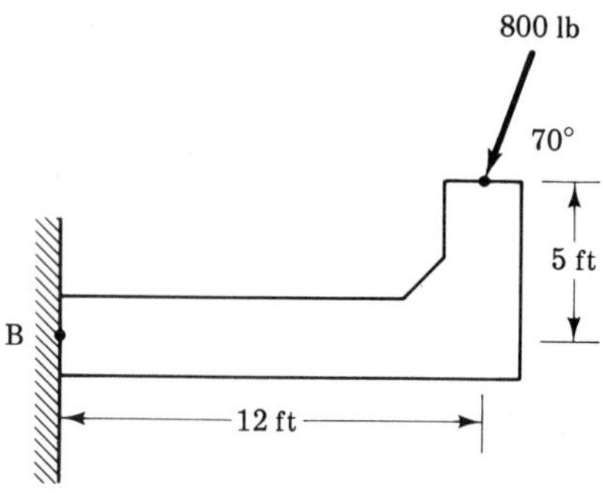

Figure 3-7

can also be done with two or more. Restating the theorem, *the total moment of a force system is equal to the sum of the moments of the individual forces* (and, therefore, also equal to the sum of the moments of all the components of all the forces). This is illustrated in the following example:

Example 3-2: Determine the total moment about D of the force system in Figure 3-8a.

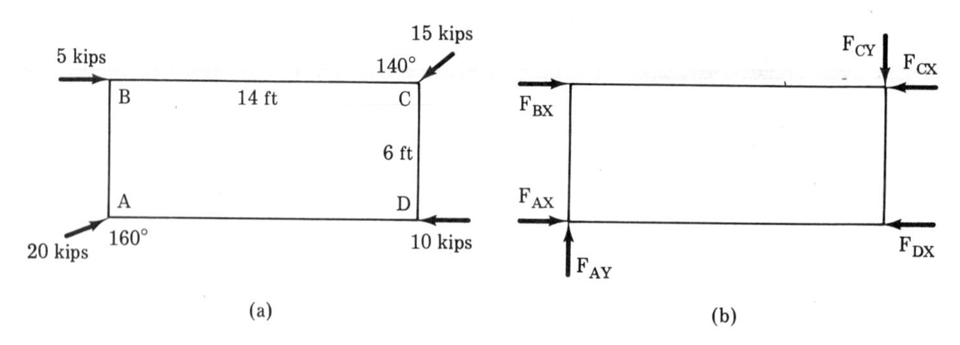

Figure 3-8

Solution:

Step 1 Choose a method, for example, the components approach.

Step 2 Resolve all the forces into components and label them as shown in Figure 3-8b.

Step 3 Determine the components of all the forces as follows (K = kip):

$$F_{AX} = F_A\cos\theta_A = 20\text{ K}(\cos 20°) \qquad F_{AX} = +18.8\text{ K}$$
$$F_{AY} = F_A\sin\theta_A = 20\text{ K}(\sin 20°) \qquad F_{AY} = +6.8\text{ K}$$
$$F_{BX} = F_B\cos\theta_B = 5\text{ K}(\cos 0°) \qquad F_{BX} = +5.0\text{ K}$$
$$F_{BY} = F_B\sin\theta_B = 5\text{ K}(\sin 0°) \qquad F_{BY} = 0\text{ K}$$
$$F_{CX} = F_C\cos\theta_C = 15\text{ K}(\cos 220°) \qquad F_{CX} = -11.5\text{ K}$$
$$F_{CY} = F_C\sin\theta_C = 15\text{ K}(\sin 220°) \qquad F_{CY} = -9.6\text{ K}$$
$$F_{DX} = F_D\cos\theta_D = 10\text{ K}(\cos 180°) \qquad F_{DX} = -10\text{ K}$$
$$F_{DY} = F_D\sin\theta_D = 10\text{ K}(\sin 180°) \qquad F_{DY} = 0\text{ K}$$

(Actually, the calculations for force B and force D could have been skipped and the components simply determined by inspection.)

Step 4 Determine the total moment of the force system by summing the moments of the components about D as follows:

$$\Sigma M_D = F_{AX}d_{AX} + F_{AY}d_{AY} + F_{BX}d_{BX} + F_{BY}d_{BY} + F_{CX}d_{CX} + F_{CY}d_{CY} + F_{DX}d_{DX} + F_{DY}d_{DY}$$

$$\Sigma M_D = +(18.8\text{ K})(0) - (6.8\text{ K})(14\text{ ft}) - (5\text{ K})(6\text{ ft}) + (0)(14\text{ ft}) + (11.5\text{ K})(6\text{ ft}) + (9.6\text{ K})(0) + (10\text{ K})(0) + (0)(0)$$

$$\Sigma M_D = -56.2\text{ K ft}\circlearrowright$$

Note that, once again, the directions of the forces were not used in the moment equation, only the directions for each moment (moments that turned out to be 0 were arbitrarily labeled +, but, of course, it would have had no effect on the problem either way). As you gain experience, you might not make a second sketch of the problem, and recognizing that some components and some moment arms are zero, you would probably write the moment equation as follows:

$$\Sigma M_D = F_{AY}d_{AY} + F_{BX}d_{BX} + F_{CX}d_{CX}$$

However, until you gain experience, it might be well to write complete equations and then let terms drop out when you discover that their value is zero.

Problem 3-2: Determine the total moment about B of the force system shown in Figure 3-9 using the first method learned (sum of the moments of the unresolved forces).

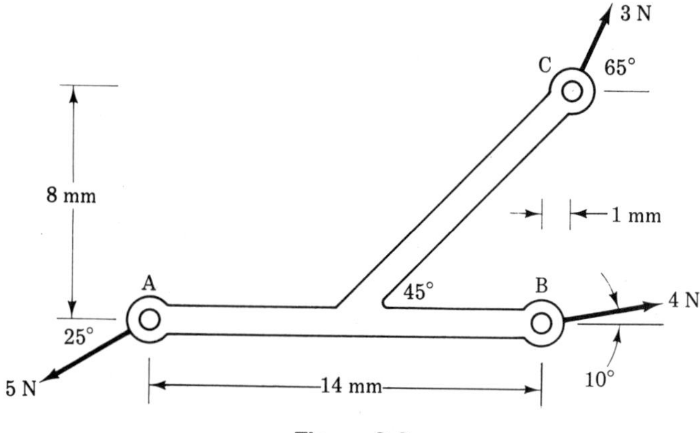

Figure 3-9

Problem 3-3: Determine the total moment about A of the force system shown in Figure 3-10 using the third method learned (sum of the moments of the components *after* moving one or more points of application to more convenient locations.

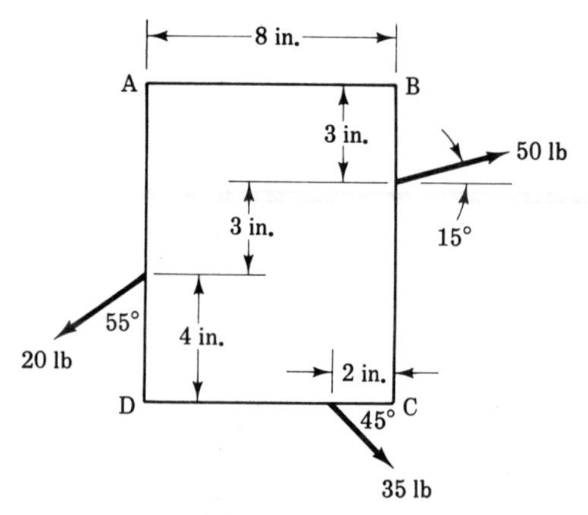

Figure 3-10

3.3 PARALLEL FORCES—METHOD OF MOMENTS

Parallel-force systems constitute a rather special type of system. It is a nonconcurrent system, which means that the lines of action of the forces do not intersect at one common point. However, they are even a special type of nonconcurrent force system for, since the lines of action are parallel to each other, they never meet.

The sum of the forces in a parallel system can be determined very readily for the vector equation for the sum of the forces becomes an algebraic equation since all forces have either one direction or a direction 180° opposite to the first. However, the location of the resultant is not so simply obtained. With concurrent force systems, the resultant passed through the point of concurrence. With parallel systems, there is no such point. We must, instead, rely on the fact that the resultant of any force system must not only produce the same force effect as the system of forces it replaced, it must also produce the same moment effect. Mathematically, the following equation must be satisfied (in addition to the force equation):

$$\Sigma M_0 = M_{01} + M_{02} + M_{03} + \cdots$$

where 0 here simply represents some unknown moment center (in an actual problem you will select a convenient moment center).

Example 3-3: Determine the magnitude, direction, and location (point where it crosses AB) of the resultant of the force system shown acting on AB in Figure 3-11.

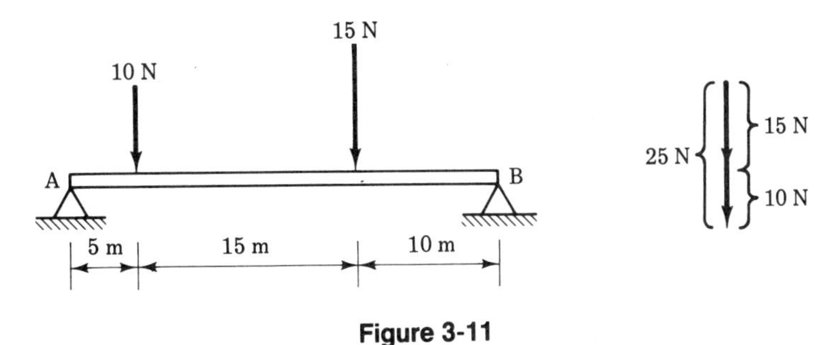

Figure 3-11

Solution:

Step 1 Recognize this as a nonconcurrent, parallel force system.

Step 2 Determine the direction and magnitude of the resultant by summing the forces as follows:

$$R = \Sigma F = F_1 \leftrightarrow F_2$$

Since both are vertical, we can simply assign conventional signs to up and down and proceed algebraically.

$$R = F_1 + F_2 = (-10\text{ N}) + (-15\text{ N}) \qquad R = -25\text{ N}$$
$$\text{or} \quad R = 25\text{ N} \quad \text{at} \quad 270°$$

Step 3 Determine the location of the resultant by the method of moments, taking moments about point A (B could also be used).

$$\Sigma M_A = M_{A1} + M_{A2} = F_1 d_{1A} + F_2 d_{2A}$$

$$\Sigma M_A = -(10\text{ N})(5\text{ m}) - (15\text{ N})(20\text{ m}) \qquad \Sigma M_A = -350\text{ N m} \circlearrowright$$

Thus, we know the magnitude and direction of the moment about A for the two-force system. The resultant force which replaces the two-force system must also provide this exact same-moment effect. Since we already know the magnitude and direction of the resultant force, we need only to set that force at the correct distance from A to provide the required moment.

$$\Sigma M_A = R d_R \qquad d_R = \frac{\Sigma M_A}{R} = \frac{350\text{ N m}}{25\text{ N}} \qquad d_R = 14\text{ m}$$

Note that this calculation was done with absolute numbers only. This is necessary for the reasons discussed in Section 3.2. Whether the resultant crosses AB at a point 14 m to the right of A or 14 m to the left of A must be determined by jointly evaluating the directions of both the force effect and the moment effect of the resultant. The resultant is acting downward and must provide a clockwise moment about A. Examination shows that a downward force can provide clockwise moment about A only by being located to the right of A. All of this may have been quite obvious to you in a simple problem such as this. However, you will see in the next example that solutions are not always so obvious. Our complete answer in this case should read

$$R = 25\text{ N} \quad \text{at} \quad 270°, \quad 14\text{ m to the right of } A$$

Example 3-4: Determine the magnitude, direction, and location (point where it crosses CD) of the resultant of the force system acting on CD in Figure 3-12.

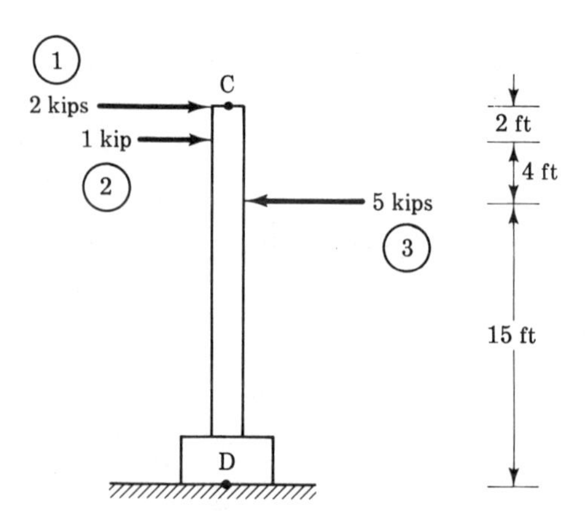

Figure 3-12

Solution:

Step 1 Recognize this as a nonconcurrent, parallel force system.

Step 2 Determine the direction and magnitude of the resultant by summing the forces as follows:

$$R = \Sigma F = F_1 \leftrightarrow F_2 \leftrightarrow F_3$$

$$R = F_1 + F_2 + F_3 = 2 + 1\text{ K} + (-5\text{ K}) \quad R = -2\text{ K or } R = 2\text{ K at } 180°$$

Step 3 Determine the location of the resultant by the method of moments, taking moments about D.

$$\Sigma M_D = M_{D1} + M_{D2} + M_{D3} = F_1 d_{1D} + F_2 d_{2D} + F_3 d_{3D}$$

$$\Sigma M_D = -(2\text{ K})(21\text{ ft}) - (1\text{ K})(19\text{ ft}) + (5\text{ K})(15\text{ ft})$$

$$\Sigma M_D = +14\text{ K ft} \circlearrowleft$$

$$\Sigma M_D = R d_{RD} \qquad d_{RD} = \frac{\Sigma M_d}{R} = \frac{14\text{ K ft}}{2\text{ K}} \qquad d_{RD} = 7\text{ ft}$$

Since the resultant is to the left and its moment is counterclockwise, the resultant must be located above D.

$$R = 2\text{ K} \quad \text{at} \quad 180° \quad \text{located 7 ft above } D$$

Problem 3-4: Determine the magnitude, direction, and location (point where it crosses BC) of the resultant of the force system acting on BC in Figure 3-13.

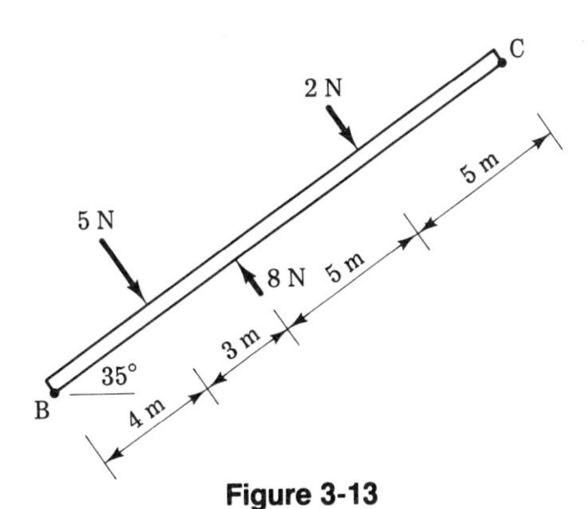

Figure 3-13

3.4 PARALLEL FORCES—GRAPHICAL DUMMY FORCE METHOD

As indicated earlier, graphical methods are quite valid on their own merit but, perhaps even more important, are useful in helping to visualize a problem and even estimate a solution. Parallel force systems present a problem, for, while addition of the forces graphically by the head-to-tail approach is simple enough, it does not yield a location for the resultant. There are, however, several special techniques that will yield a complete solution.

One of these techniques is the dummy force method. It is based upon a concept that you are quite familiar with, namely that the addition of equal quantities to both sides of an equation does not change the equality of that equation. If we modify this to add equal but opposite quantities to one side of the equation and then apply this to a vector equation, we have the basis for the dummy force approach.

$$R = F_1 \leftrightarrow F_2 \qquad D_1 \leftrightarrow D_2 = 0 \qquad D_1 = -D_2$$

$$F_1 \leftrightarrow D_1 = R_1 \qquad F_2 \leftrightarrow D_2 = R_2$$

$$R_1 \leftrightarrow R_2 = F_1 \leftrightarrow D_1 \leftrightarrow F_2 \leftrightarrow D_2$$

but since $D_1 \leftrightarrow D_2 = 0$

$$R_1 \leftrightarrow R_2 = F_1 \leftrightarrow F_2 = R$$

Thus, at the end, the dummy forces that were introduced are eliminated again. However, as you will see in Example 3-5, in the graphical process of going through the above steps, the location of the resultant is found.

Example 3-5: Determine the magnitude, direction, and location of the resultant of the force system shown acting in Figure 3-11.

Solution: Trace the following steps in the graphical solution shown in Figure 3-14.

Step 1 Draw the body involved to some scale.

Step 2 Draw the parallel forces in to some scale with the heads of the vectors at the points of application.

Step 3 Select a dummy force size and draw in the equal and opposite dummy forces with the heads of the vectors at the points of application of the initial parallel forces and their direction perpendicular to the initial parallel forces.

Step 4 Graphically sum up each initial force and its accompanying dummy force. The value of making the latter perpendicular to the former becomes apparent here.

Step 5 Extend the lines of action of the two resultants obtained in Step 4 until they intersect.

Step 6 Move both resultants along their lines of action until their heads touch the intersection found in Step 5.

(a) A
B Step 1, 1 cm = 3 m

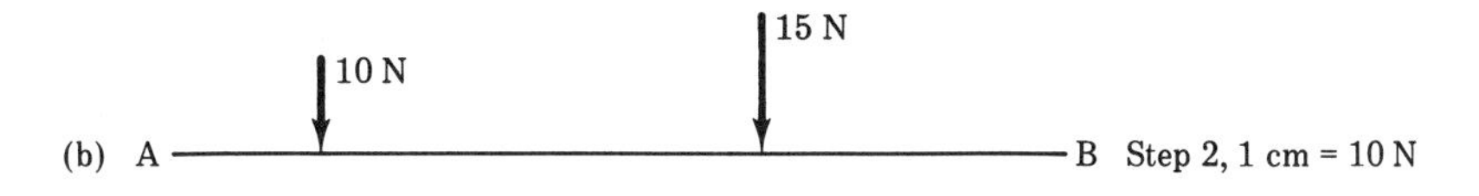
15 N
10 N
(b) A
B Step 2, 1 cm = 10 N

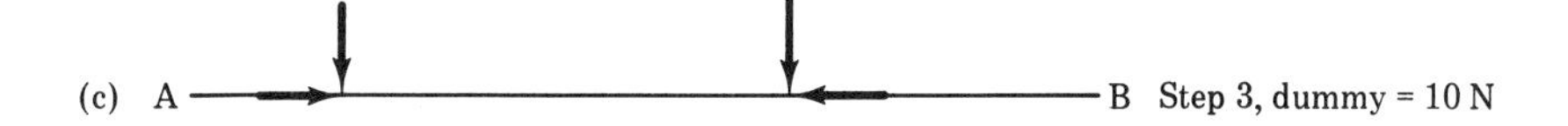
(c) A
B Step 3, dummy = 10 N

(d) A
B Step 4, sum

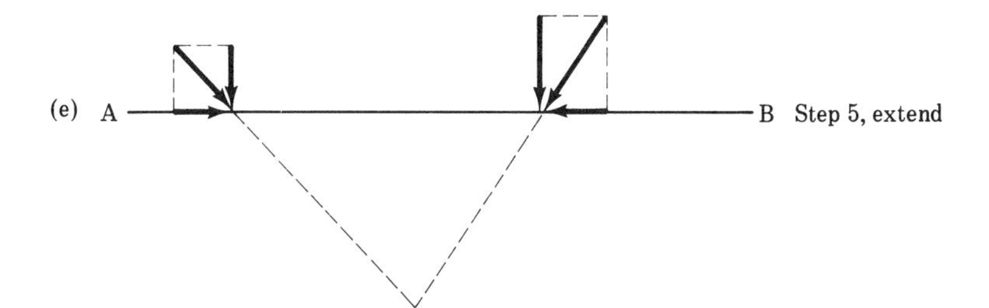
(e) A
B Step 5, extend

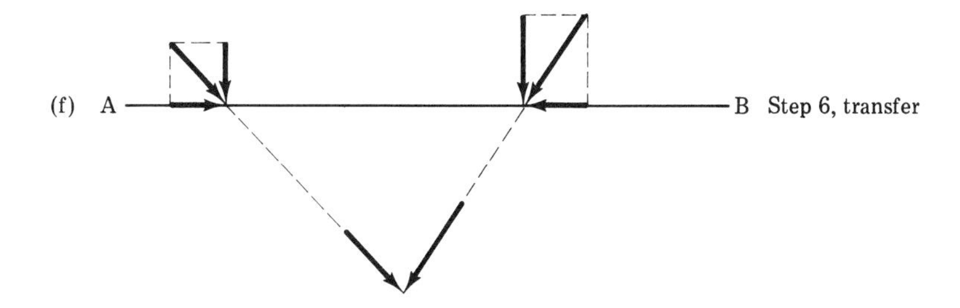
(f) A
B Step 6, transfer

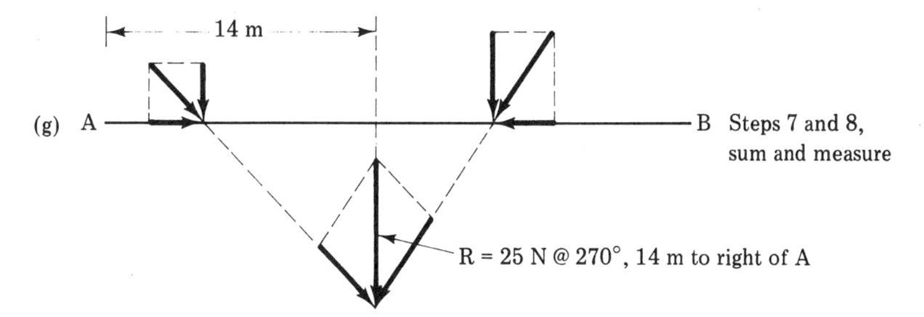
14 m
(g) A
B Steps 7 and 8, sum and measure
R = 25 N @ 270°, 14 m to right of A

Figure 3-14

Step 7 Graphically sum the two resultants to form a total resultant and extend its line of action back across AB.

Step 8 Measure the magnitude of the resultant, note its direction, and measure the distance from A or B at which it crosses AB.

$$R = 25 \text{ N} \quad \text{at} \quad 270° \quad \text{located 14 m to the right of } A$$

Fortunately, as seen in Figure 3-14, the method is much simpler in application than it is in description. Selection of scales and dummy force size are a compromise between accuracy and paper size. However, careful work on $8\frac{1}{2}$ by 11 paper will usually provide more than enough accuracy. The method can become cumbersome if several forces are involved. It does work equally well if the initial forces are in opposite directions, as you will find when you work Problem 3-5.

Problem 3-5: Determine the magnitude, direction, and location of the resultant of the force system shown acting in Figure 3-15 using the graphical dummy force method. Check your answer by the method of moments.

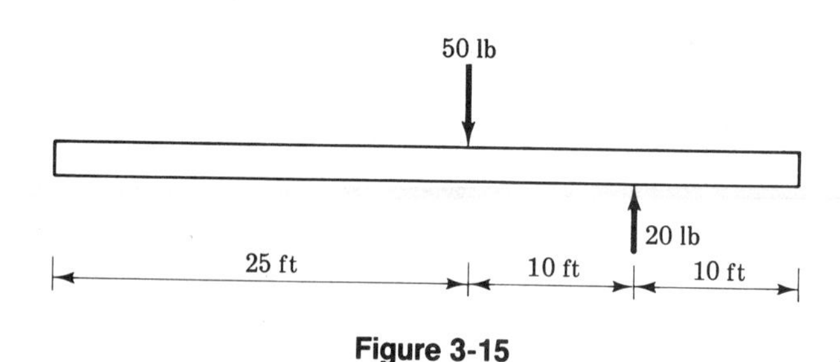

Figure 3-15

3.5 PARALLEL FORCES—GRAPHICAL PROPORTION METHOD

Another graphical method takes advantage of moment being in direct proportion to both force and moment arm. The technique for locating the resultant is quite simple, but it will not yield the magnitude of the resultant. The latter will have to be found by simple vector addition.

Example 3-6: Determine the magnitude, direction, and location of the resultant of the force system shown in Figure 3-11.

Solution: Trace the following steps in the graphical solution shown in Figure 3-16.

Step 1 Draw the body involved to some scale.

Step 2 Draw the parallel forces in to some scale *at each others points*

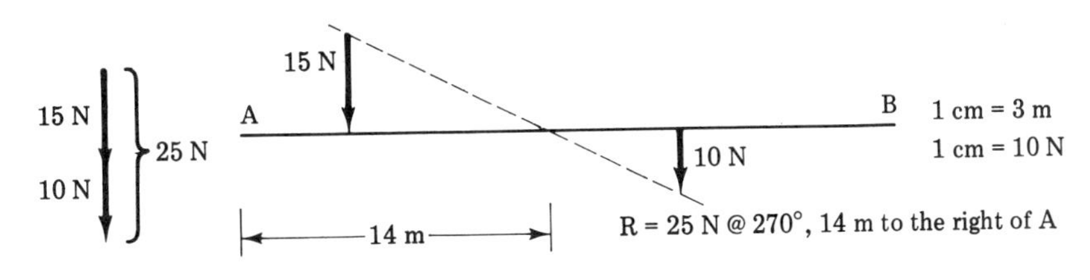

Figure 3-16

of application and with the *head of one vector at a point of application* and the *tail of the other vector at a point of application.*

Step 3 Draw a line from the free tail of the one vector to the free head of the other vector. Where this line crosses AB is the location of the resultant.

Step 4 Measure the distance from A to the location of the resultant.

Step 5 Draw the force diagram for the sum of the forces.

Step 6 Measure the magnitude of the resultant and note its direction. The complete answer is

25 N at 270° located 14 m to the right of A

Once again the description of the approach was much more time consuming than the approach. Particular care needs to be taken in executing Step 2.

Problem 3-6: Determine the magnitude, direction, and location of the resultant of the force system shown acting in Figure 3-17 using the graphical proportion method. Check your answer by the method of moments.

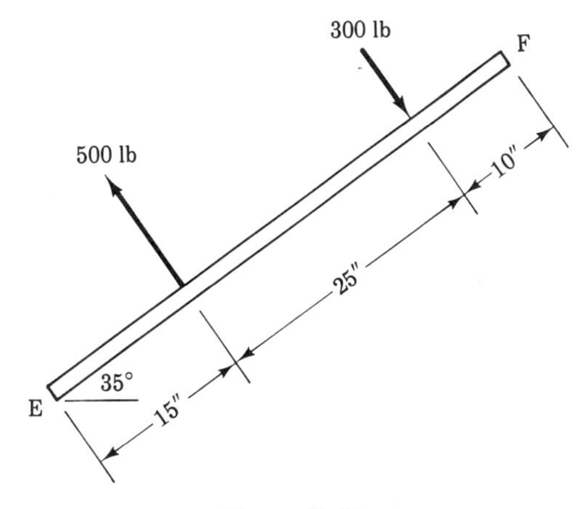

Figure 3-17

3.6 UNIFORMLY DISTRIBUTED LOADS

Until now, we have been dealing only with concentrated loads. These are loads that are applied at a point on a body which we have shown by use of a single arrow. If we examine the idea of a concentrated load closely, we realize that total concentration of force would mean that the force is applied to an area of zero magnitude. This is impossible, and only a few applications such as the use of needles, awls, ice picks, and nails even come close.

When we think of a force being applied to an area, we must consider the size of the area and the size and distribution of the force. The application of a force to an area is sometimes called a force field. Figure 3-18 shows two types of distributed loads, one that is uniformly applied and another that is nonuniformly applied.

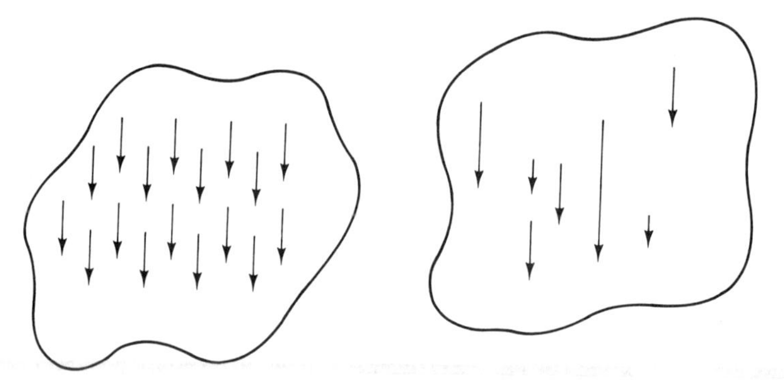

Figure 3-18. Uniform versus nonuniform loads.

In many cases, the uniformly applied load is applied over a relatively small, well-defined area and can be treated as a concentrated load. At the other extreme, some nonuniformly applied loads are so complex that extensive and involved analysis is required to reach a solution. Sometimes the nature of the problem determines how a load is to be considered. The weight of the flooring in an upper story of a building can quite safely be considered to be uniformly distributed. The weight of a full filing cabinet standing on that floor may basically be a uniformly distributed load. However, for the purpose of designing the support of the floor, the filing cabinet may be considered a concentrated load. Similarly a person standing on that floor may be considered a concentrated load on the floor. However, a shoe designer would consider the person's weight a nonuniformly distributed load on the shoe.

Many apparently complex problems can be analyzed readily if the nature of the distribution of the load is determined. The simplest case is one where a load is uniformly distributed over too large an area to consider it as a

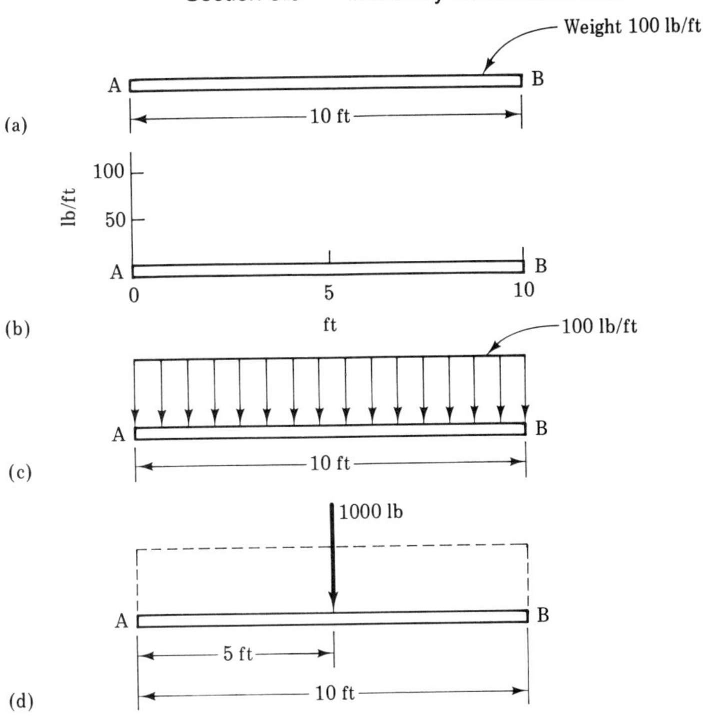

Figure 3-19. Representation of a uniformly distributed load.

concentrated load. In this chapter, we will consider uniformly distributed loads and combinations of them with concentrated loads. Some types of nonuniformly distributed loads will be dealt with in a later chapter.

A simple case of a uniformly distributed load is the weight of a beam. Figure 3-19 shows the development of the typical representation of a uniformly distributed load. First is shown a sketch of a beam that is 10 ft long and each foot of which weighs 100 lb. Next is shown a graphical representation of this information with the vertical axis scaled in lb/ft. The area, thus, becomes the total load, 1000 lb. This is sometimes shown as a series of parallel forces as shown next. For some types of analysis, particularly when taking moments, this uniformly distributed load can be represented by a single concentrated load acting through the center of the uniformly distributed load as shown last.

Example 3-7: Determine the magnitude, direction, and location of the resultant of the system of distributed loads shown in Figure 3-20a.

Solution:

Step 1 Recognize that there are three different distributed loads in the problem.

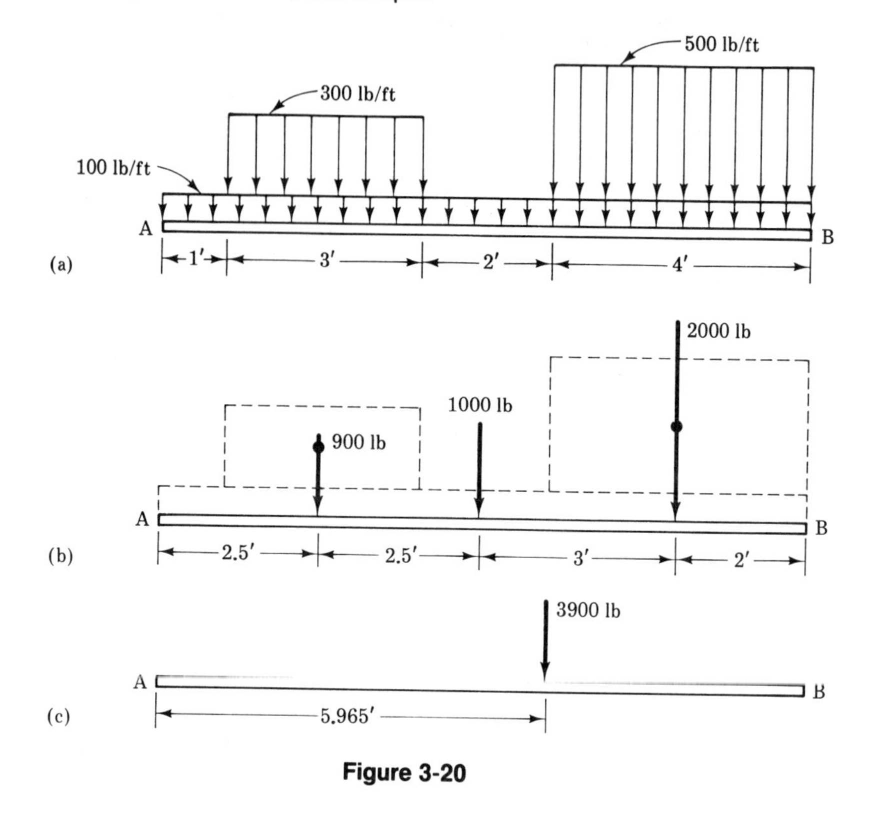

Figure 3-20

Step 2 Determine the magnitude of each load as follows:

$$F_1 = l_1 \frac{P_1}{l_1} = 3 \text{ ft } (300 \text{ lb/ft}) \qquad F_1 = 900 \text{ lb}$$

$$F_2 = l_2 \frac{P_2}{l_2} = 10 \text{ ft } (100 \text{ lb/ft}) \qquad F_2 = 1000 \text{ lb}$$

$$F_3 = l_3 \frac{P_3}{l_3} = 4 \text{ ft } (500 \text{ lb/ft}) \qquad F_3 = 2000 \text{ lb}$$

Step 3 Locate the equivalent concentrated loads at the centers of their respective distributed loads as shown in Figure 3-20b.

Step 4 Sum the force system as follows:

$$R = F_1 + F_2 + F_3 = 900 + 1000 + 2000 \text{ lb}$$

$$R = 3900 \text{ lb} \quad \text{at} \quad 270°$$

Step 5 Determine the location of the resultant by taking moments about A.

$$\Sigma M_A = F_1d_1 + F_2d_2 + F_3d_3$$

$$\Sigma M_A = -(900 \text{ lb})(2.5 \text{ ft}) - (1000 \text{ lb})(5 \text{ ft}) - (2000 \text{ lb})(8 \text{ ft})$$

$$\Sigma M_A = -23{,}250 \text{ lb ft} \circlearrowright$$

$$d_R = \frac{\Sigma M_A}{R} = \frac{23{,}250 \text{ lb ft}}{3900 \text{ lb}} \qquad d_R = 5.96 \text{ ft}$$

Since the moment is clockwise and R is directed downward, it must be located to the right of A. The complete answer is $R = 3900$ lb at 270°, located 5.96 ft to the right of A.

Problem 3-7: Determine the magnitude, direction, and location of the resultant of the force system shown in Figure 3-21.

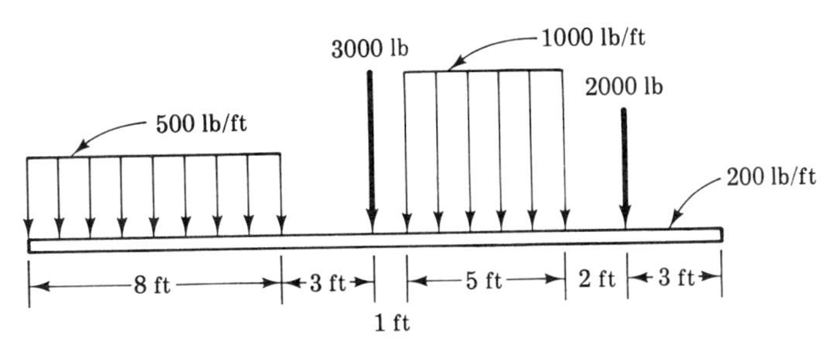

Figure 3-21

3.7 COUPLES

Up to now, we have discussed a variety of parallel force systems. One special type has been left to this section and is called a *couple.* A couple is a system of two forces that are *equal in magnitude, opposite in direction,* and *not collinear.* Since they are opposite in direction but not collinear, they must be parallel. Further, since they are opposite in direction and equal in magnitude, their resultant force is zero. Thus, a couple does not tend to produce linear motion but only tends to produce rotation. In the examples shown in Figure 3-22, (a) and (c) are couples, (b), (d), and (e) are not. The systems shown in (g) and (i) contain couples while those shown in (f), (h), and (j) do not.

In (g) the system is complex enough that the couple present may not be readily apparent. However, 2 N of the 4-N force does make a couple with the 2-N force. Couples in a complex system may be recognized and treated as such as long as the effect of the remaining force(s) is not ignored.

As indicated before, a couple produces no resultant force, only moment. Simple examples of a couple in daily life include the turning of an

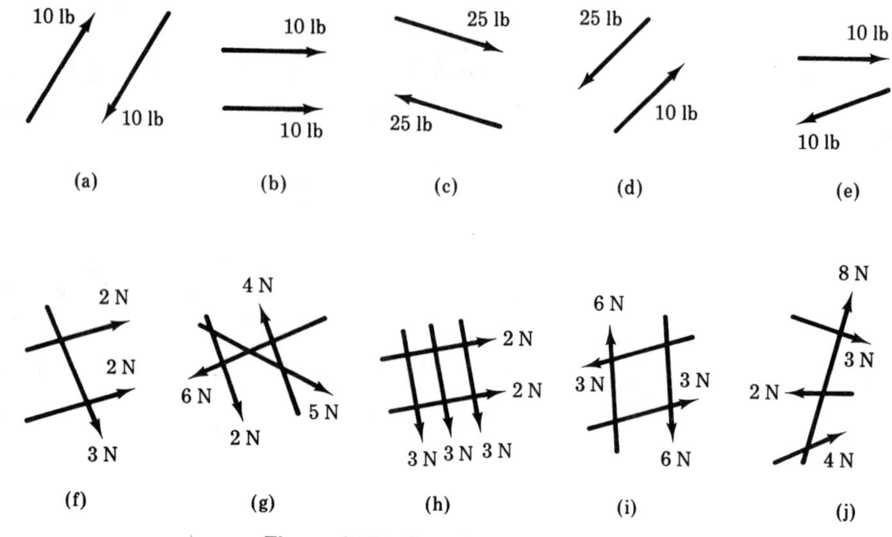

Figure 3-22. Couples and noncouples.

automobile steering wheel with both hands, and the removal of a jar top with one hand. The magnitude of the moment resulting from a couple can be determined by multiplying *one* of the two forces in the couple by the perpendicular distance between them. Thus, in Figure 3-23, we could determine the moment of the 10-lb couple from $M = Fd$, where $F = 10$ lb and $d = 3$ in. It would appear that no moment center is involved, but, in reality, with a couple any moment center will do. Let us calculate M about four different moment centers as well as by the equation noted above.

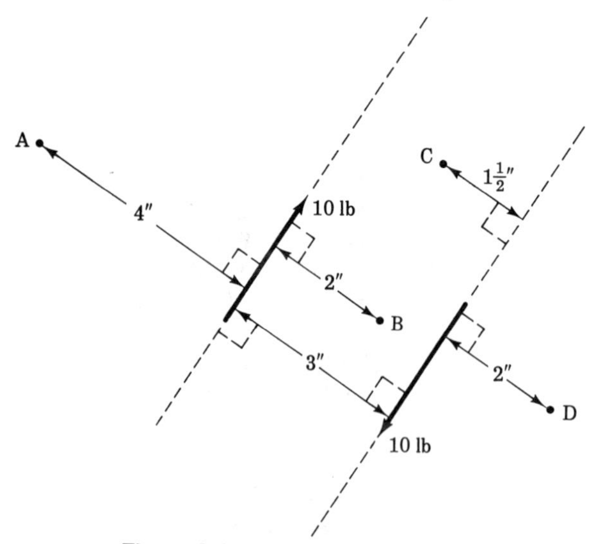

Figure 3-23. Moment of a couple.

$$M = Fd = -(10 \text{ lb})(3 \text{ in.}) \qquad M = -30 \text{ lb in.} \circlearrowright$$

$$M_A = F_1 d_{1A} + F_2 d_{2A} = +(10 \text{ lb})(4 \text{ in.}) - (10 \text{ lb})(7 \text{ in.})$$
$$M = -30 \text{ lb in.} \circlearrowright$$

$$M_B = F_1 d_{1B} + F_2 d_{2B} = +(10 \text{ lb})(2 \text{ in.}) - (10 \text{ lb})(1 \text{ in.})$$
$$M = -30 \text{ lb in.} \circlearrowright$$

$$M_C = F_1 d_{1C} + F_2 d_{2C} = -(10 \text{ lb})(1\tfrac{1}{2} \text{ in.}) - (10 \text{ lb})(1\tfrac{1}{2} \text{ in.})$$
$$M = -30 \text{ lb in.} \circlearrowright$$

$$M_D = F_1 d_{1D} + F_2 d_{2D} = -(10 \text{ lb})(5 \text{ in.}) + (10 \text{ lb})(2 \text{ in.})$$
$$M = -30 \text{ lb in.} \circlearrowright$$

In some cases, more than one couple may be acting on a body. As long as the couples are all coplanar, their moments may be added by simple vector algebra as shown in Figure 3-24.

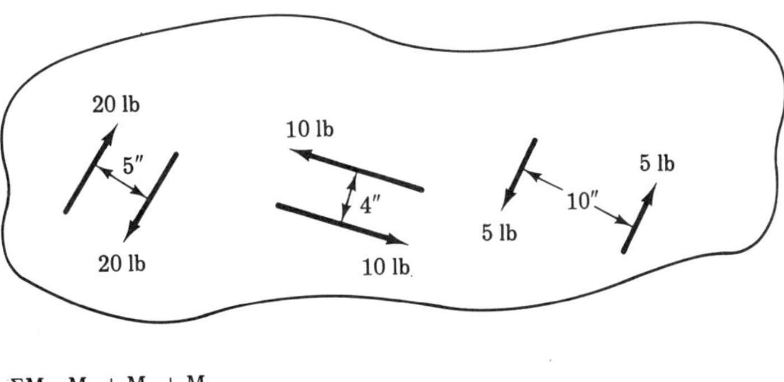

$\Sigma M = M_1 + M_2 + M_3$

$\Sigma M = -(20 \text{ lb})(5 \text{ in.}) + (10 \text{ lb})(4 \text{ in.}) + (5 \text{ lb})(10 \text{ in.})$

$\Sigma M = -10 \text{ lb} \cdot \text{in} \circlearrowright$

Figure 3-24. Summing moments of couples.

The properties of a couple exhibited in the discussion above permit a number of interesting operations to be performed with them. Each of the following operations may be performed without changing the effect of the couple on the body it is acting on.

1. A couple may be rotated or translated in its plane.
2. A couple may be moved to a parallel plane.
3. The distance between the forces may be changed as long as the magnitude of the forces is also changed so that the moment effect remains constant.

Example 3-8: Determine the magnitude and direction of the horizontal forces that must be applied at A and B to replace the existing force system shown in Figure 3-25.

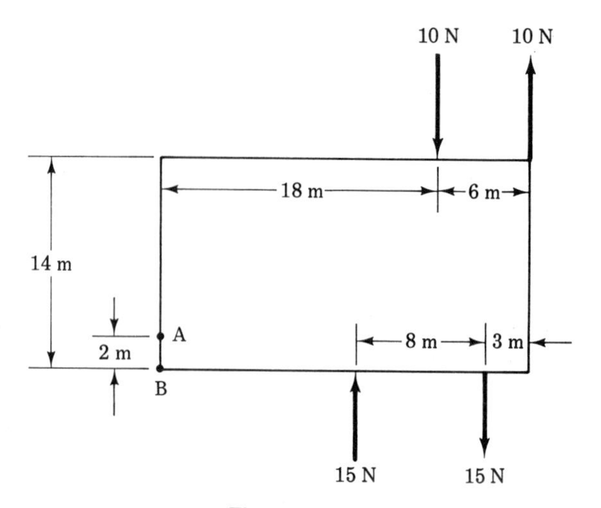

Figure 3-25

Solution:

Step 1 Recognize this as a problem involving solely couples.

Step 2 Calculate the total moment of the existing force system as follows:

$$\Sigma M = M_1 + M_2 = F_1 d_1 + F_2 d_2 = +(10\text{ N})(6\text{ m}) - (15\text{ N})(8\text{ m})$$

$$\Sigma M = -60\text{ N m} \circlearrowright$$

Step 3 If M_3, the moment of the couple at A and B, is to replace the existing force system, it must have no force effect and a moment effect equal to the total moment of the existing system.

$$M_3 = \Sigma M \qquad F_A = -F_B$$

Step 4 Calculate the magnitude of F_A and F_B.

$$F_A = -F_B = \frac{M_3}{AB} = \frac{\Sigma M}{AB} = \frac{60\text{ N m}}{2\text{ m}}$$

$$F_A = -\text{F}_\text{B} = 30\text{ N}$$

The direction of F_A and F_B must be determined by evaluating the direction of the moment they must cause. Thus for M_3 to be clockwise,

$$F_A = 30\text{ N} \quad \text{at} \quad 0^\circ \qquad F_B = 30\text{ N} \quad \text{at} \quad 180^\circ$$

Problem 3-8: Determine the total moment acting on the body shown in Figure 3-26 using couples.

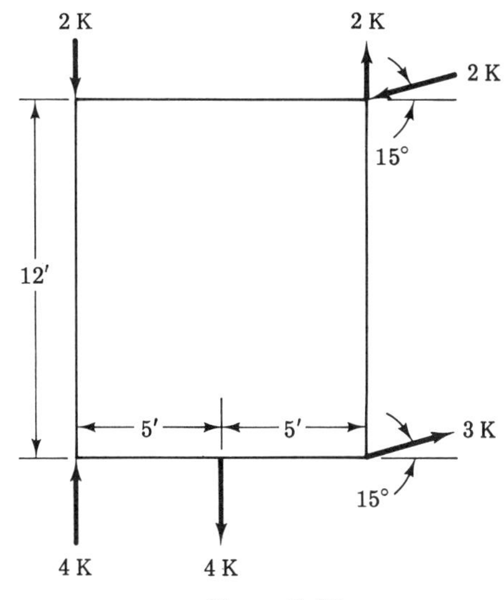

Figure 3-26

3.8 EQUILIBRIUM APPLICATIONS

Throughout this chapter, various concepts and methods have been evaluated, all pertaining to parallel force systems. Since the text is addressing static equilibrium, it is appropriate that we now examine some applications of these concepts and methods to equilibrium problems. Many of the applications will involve more than one of the concepts previously discussed.

Friction-Tipping versus Sliding

Many of us have experienced this problem. We are trying to slide something tall and thin along the floor and it starts to tip. At the cost of some discomfort or inconvenience, we lower the elevation at which we apply our force until the object slides without tipping. As shown in Figure 3-27, unless our force is applied exactly at and parallel to the friction interface, there is some tendency to tip. As soon as a force is applied above that level, it and the resulting friction force form a couple that attempts to rotate the body. In so doing, this couple shifts the normal reaction force forward of the force of gravity. Those two forces form another couple. If the friction force-applied force couple is greater than the gravity force-normal force couple tipping will occur.

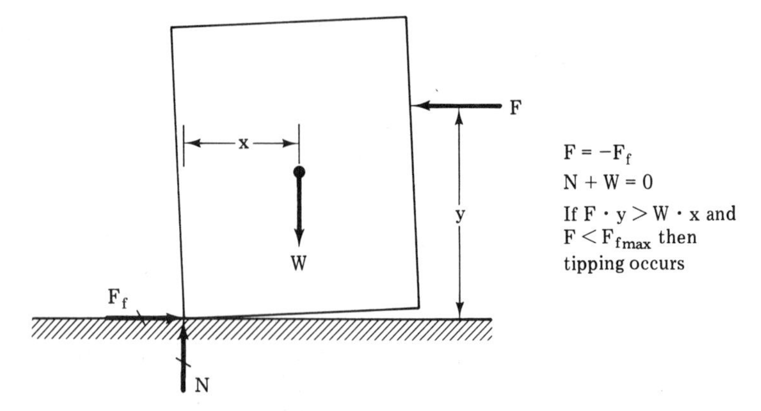

Figure 3-27. Tipping versus sliding.

Example 3-9: Determine the maximum height at which F can be applied and still cause sliding without tipping the box shown in Figure 3-28a.

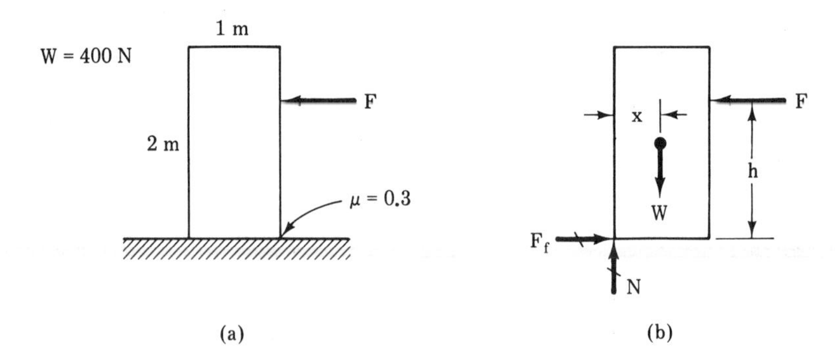

Figure 3-28

Solution:

Step 1 Recognize this as a couple equilibrium problem.

Step 2 Draw the free-body diagram as shown in Figure 3-28b assuming that gravity acts through the center of the box.

Step 3 Determine the force which will cause sliding as follows:

$$F = -\mathrm{F_f}\max = \mu\mathrm{N} = \mu\mathrm{W} = 0.3(400\ \mathrm{N}) \qquad F = 120\ \mathrm{N}\quad \text{at} \quad 180°$$

Step 4 The maximum height at which F can be applied and not cause tipping is the one at which the two couples are in equilibrium.

$$Wx + Fh = 0 \qquad -(400\ \mathrm{N})(\tfrac{1}{2}\ \mathrm{m}) + (120\ \mathrm{N})(h) = 0$$

$$h = 1.67\ \mathrm{m}$$

Problem 3-9: Determine if the box shown in Figure 3-29 will either tip or slide.

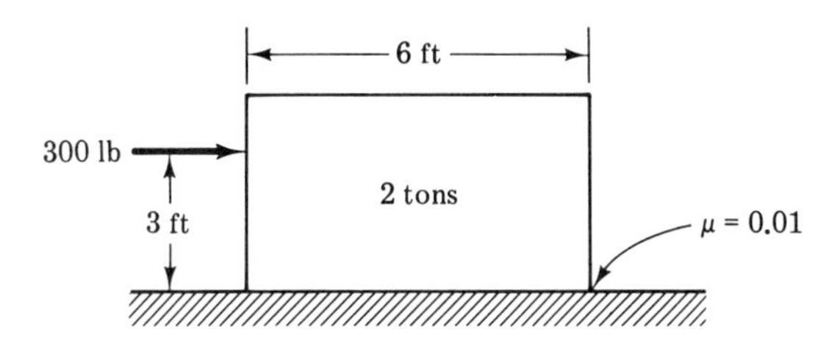

Figure 3-29

Friction-Rolling Resistance

Rolling friction, or rolling resistance as it is often called, is somewhat different from the friction due to the roughness of surfaces sliding on one another. Certainly, a rough-surfaced cylinder on a rough flat surface will experience minimal slippage. On the other hand a smoothly ground cylinder on a smoothly ground flat surface with lubrication will experience substantial slippage. However, in a rolling problem an additional effect must be considered. This, as illustrated in Figure 3-30, relates to the fact that either the cylinder or the surface may deform.

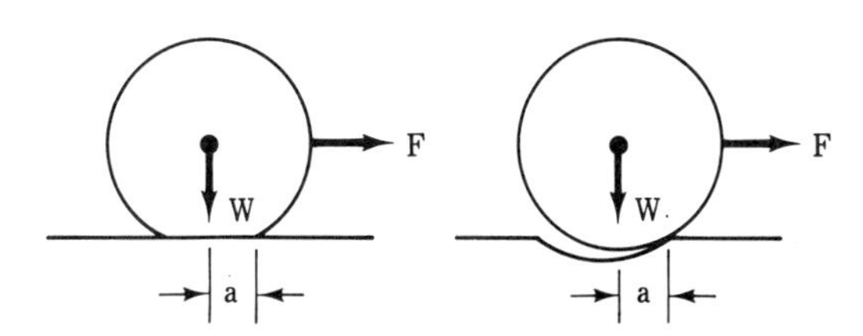

Figure 3-30. Rolling resistance.

When the cylinder is about to move the couple made by the applied force and the reaction to it equals the couple made by the force of gravity and the normal force. Any increase in the applied force will cause motion. Since the deformation is generally slight, the moment arm of F can be taken to be the radius of the cylinder. The moment arm of W, half the width of the deformation is generally designated as a, and known as the *coefficient of rolling resistance.* Appendix B, Table B-2, provides some typical values for this property.

Example 3-10: The coefficient of rolling resistance for a 60-cm steel drum on a steel shop floor is 0.008 cm. If the drum has a mass of 1000 kg, how much force must be applied parallel to the floor at the top of the drum to start it moving?

Solution:

Step 1 Draw the free-body diagram for the problem as shown in Figure 3-31.

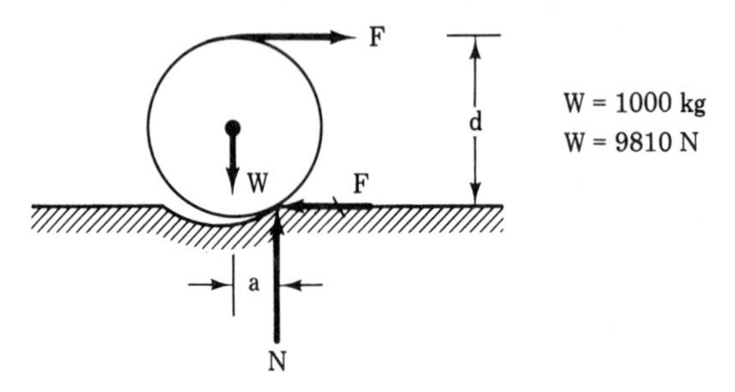

Figure 3-31

Step 2 Solve for F as follows:

$$Fd + Wa = 0 \qquad -(F)(60\text{ cm}) + (9810\text{ N})(0.008\text{ cm}) = 0$$

$$F = 1.31\text{ N}$$

Problem 3-10: A man is pulling horizontally on a 3-ft diameter steel lawn roller (through the center of the roller) with a force of 50 lb. If the lawn roller weighs 300 lb and does not move, what is the minimum value of the coefficient of rolling resistance between the roller and the lawn?

Friction-Belt Friction

The friction between a belt or rope and a cylindrical surface can be used to transmit power or to prevent (or at least slow down) motion. The tensions in the belt or rope approaching the cylindrical surface are described by the following equation and Figure 3-32:

$$T_1 = T_2 e^{\mu\beta}$$

where $e = 2.718$ (base of natural logarithms)

μ = coefficient of friction

β = angle of contact in radians

Figure 3-32. Belt friction.

In the equation above, T_1 is the larger of the two tensions. In most cases, such as a pulley driving a belt or a band brake slowing or stopping a drum, the higher tension is the one that opposes impending or actual motion. In the case of a driven pulley, however, the higher tension is in the same direction as impending or actual motion.

Example 3-11: The 18-in. pulley driving the belt in Figure 3-33a receives a counterclockwise torque of 500 lb ft from its shaft. The belt is essentially horizontal entering and leaving the pulley and the coefficient of friction between the belt and pulley is 0.4. Determine the tension in each belt and the reaction at pulley bearing.

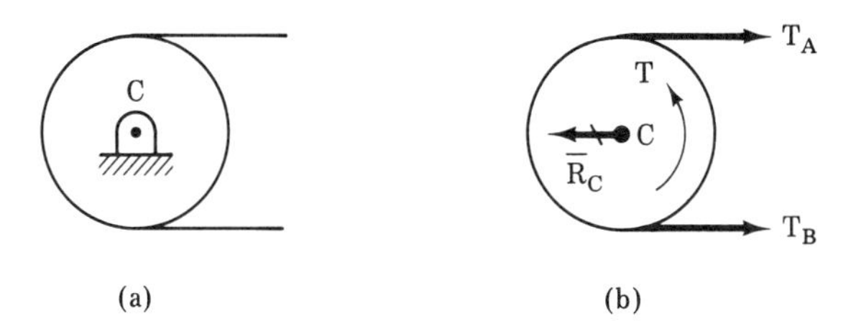

Figure 3-33

Solution:

Step 1 Draw the free-body diagram for the problem as shown in Figure 3-33b.

Step 2 Since the pulley is driving the belt, T_A is larger than T_B. Write the belt friction equation.

$$T_A = T_B e^{\mu\beta}$$

μ was given and β is π rad. However, we still have two unknowns in the equation, the two tensions.

Step 3 Search for another relationship containing T_A and T_B. Write the moment equilibrium equation.

$$T + T_A r + T_B r = 0 \quad (T \text{ is torque})$$

Step 4 Substitute knowns in both equations and then solve two equations by substitution as follows:

$$T_A = T_B e^{(0.4)(\pi)} \qquad T_A = 3.5T_B$$

$$+500 \text{ lb ft} - (T_A)(0.75 \text{ ft}) + (T_B)(0.75 \text{ ft})$$

$$T_A = 666.7 + T_B$$

Substituting: $666.7 + T_B = 3.5T_B$ $\quad T_B = 266.7$ lb
Substituting: $T_A = 666.7 - 148$ $\quad T_A = 936.4$ lb

Step 5 Write the force equilibrium equation.

$$\overline{R}_C + T_A + T_B = 0 \qquad \overline{R}_C = -(T_A + T_B)$$

$$\overline{R}_C = -(+933.4 + 266.7) \qquad \overline{R}_C = -1201 \text{ lb}$$

Problem 3-11: Determine the tension in either side of the brake band, the torque of the drum and the reactions at A and E for the band brake shown in Figure 3-34.

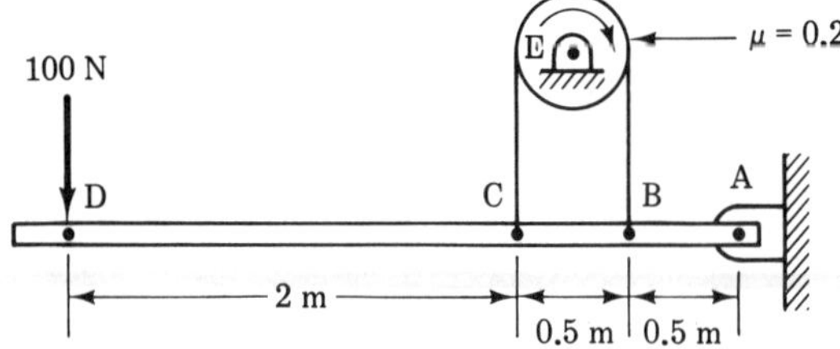

Figure 3-34

Simple Structural Equilibrium I

Some simply loaded structures form an equilibrium problem involving couples.

Example 3-12: Determine the reactions occurring at the supports of the structure shown in Figure 3-35a.

Solution:

Step 1 Draw the free-body diagram of the problem as shown in Figure 3-35b. It is based on examination of the structure which shows that the structure is pushing to the right on G so G must push back to the left. Then, for the structure to be in equilibrium, the reactions at F must be up and to the right.

Step 2 Recognize that the force system shown in Figure 3-35b consists of two couples.

Step 3 Write the vertical force equilibrium equation, and solve for $\overline{R}_{FY}$

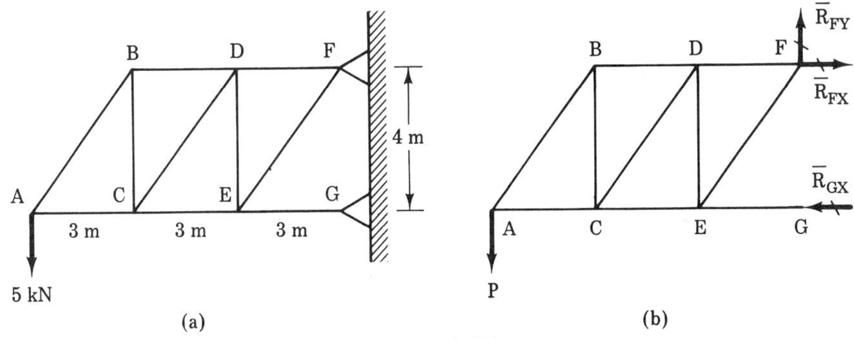

Figure 3-35

$$\Sigma F_Y = 0 \qquad P + \overline{R}_{FY} = 0 \qquad \overline{R}_{FY} = -P$$

$$\overline{R}_{FY} = -(-5 \text{ kN}) \qquad \overline{R}_{FY} = +5 \text{ kN} \quad \text{or} \quad \overline{R}_{FY} = 5 \text{ kN} \quad \text{at} \quad 90°$$

Step 4 Write the moment equilibrium equation and solve for $\overline{R}_{GX}$.

$$\Sigma M = 0 \qquad M_1 + M_2 = 0 \qquad +(5 \text{ kN})(9 \text{ m}) - (\overline{R}_{GX})(4 \text{ m}) = 0$$

$$\overline{R}_{GX} = 11.25 \text{ kN} \quad \text{at} \quad 180°$$

Step 5 Write the horizontal force equilibrium equation and solve for

$$\Sigma F_X = 0 \qquad \overline{R}_{FX} + \overline{R}_{GX} = 0 \qquad \overline{R}_{FX} = -\overline{R}_{GX}$$

$$\overline{R}_{FX} = -(-11.25 \text{ kN}) \qquad \overline{R}_{FX} = 11.25 \text{ kN} \quad \text{at} \quad 0°$$

Problem 3-12: Determine the reactions occurring at the supports of the structure shown in Figure 3-36.

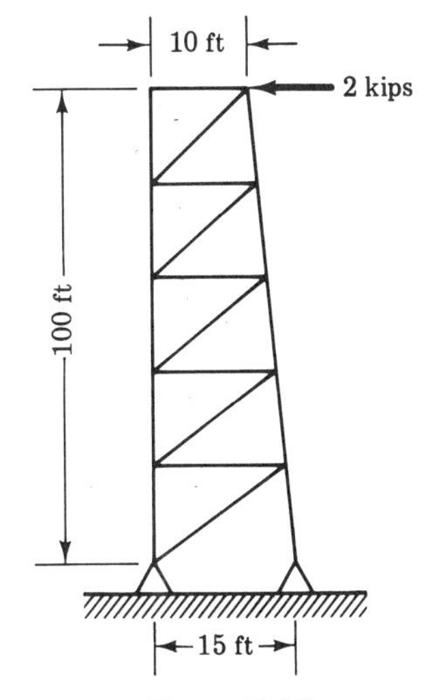

Figure 3-36

Simple Structural Equilibrium II

Some simply loaded structures form an equilibrium problem involving parallel forces.

Example 3-13: Determine the reactions occurring at the supports of the structure shown in Figure 3-37a.

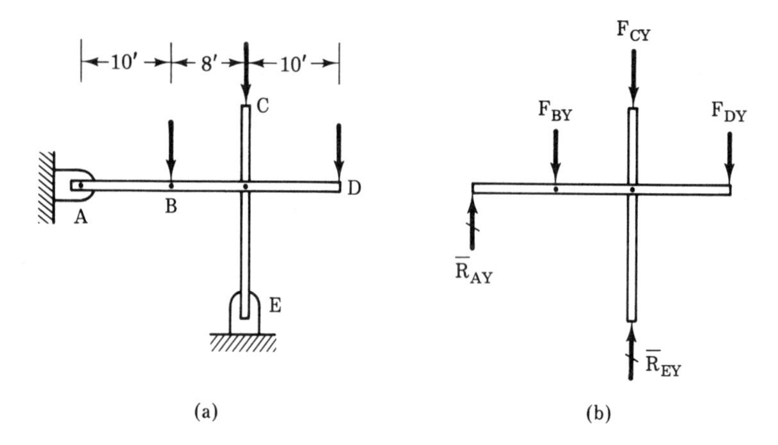

Figure 3-37

Solution:

Step 1 Draw the free-body diagram of the problem as shown in Figure 3-37b. It is based on an examination of the structure which suggests that no horizontal forces, acting or reacting, are involved.

Step 2 Recognize that the force system shown in Figure 3-37b consists entirely of parallel forces.

Step 3 Recognize that the vertical force equation would have two unknowns. Therefore, write the moment equation about either A or E, and solve.

$$\Sigma M_A = 0 \qquad \overline{R}_{AY} d_{AY} + F_{BY} d_{BY} + F_{CY} d_{CY} + F_{DY} d_{DY} + \overline{R}_{EY} d_{EY} = 0$$

$$\overline{R}_{AY}(0) - (1200 \text{ lb})(10 \text{ ft}) - (800 \text{ lb})(18 \text{ ft}) - (600 \text{ lb})(28 \text{ ft})$$

$$-(\overline{R}_{EY})(d_{EY}) = 0$$

$$\overline{R}_{EY} d_{EY} = +43{,}200 \text{ lb ft}$$

Since the moment of $\overline{R}_{EY}$ about A is counterclockwise, $\overline{R}_{EY}$ must be at 90° as indicated in the free-body diagram.

$$\overline{R}_{EY} = \frac{\overline{R}_{EY} d_{BY}}{d_{EY}} = \frac{43{,}200 \text{ lb ft}}{18 \text{ ft}}$$

$$R_{EY} = 2400 \text{ lb} \quad \text{at} \quad 90^\circ$$

Step 4 Write the vertical force equilibrium equation and solve

$$\Sigma F_Y = 0 \qquad \overline{R}_{AY} + F_B + F_C + F_D + \overline{R}_{EY} = 0$$

$$\overline{R}_{AY} = -(F_B + F_C + F_D + \overline{R}_{EY})$$

$$\overline{R}_{AY} = -(-1200 - 800 - 600 + 2400)$$

$$\overline{R}_{AY} = +200 \qquad \overline{R}_{AY} = 200 \text{ lb} \quad \text{at} \quad 90^\circ$$

Problem 3-13: Determine the reactions occurring at the supports of the structure shown in Figure 3-38.

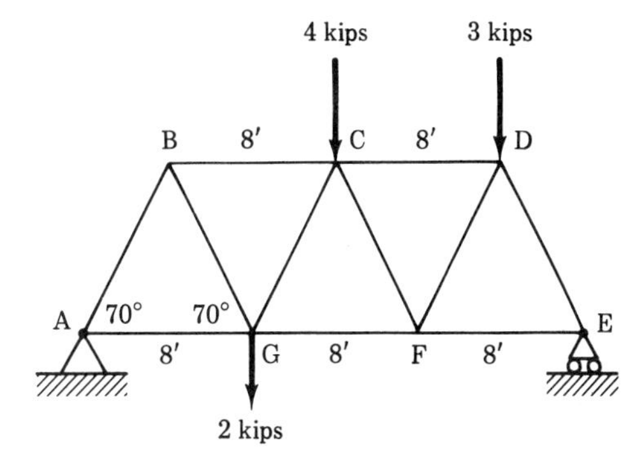

Figure 3-38

Pulley Systems

Many common pulley systems can be analyzed as parallel force systems with negligible error. They can be divided into two groups, single "rope" and multiple "rope." An example of each type follows.

Example 3-14: Determine the force required to lift a 200-kg load using the block and tackle shown in Figure 3-39a, and determine the reaction at A.

Solution:

Step 1 Neglect the angles of any ropes and consider the block and tackle a parallel force system.

Step 2 Recognize that there is only one rope involved. Assume no

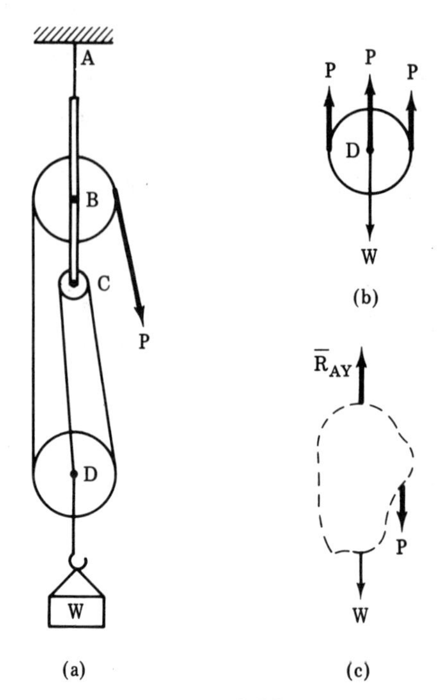

Figure 3-39

slippage of the rope on the pulleys and no friction at the pulley shafts. The tension in the rope then equals P throughout the rope.

Step 3 Draw the free-body diagram of pulley D as shown in Figure 3-39b.

Step 4 Write the vertical force equilibrium equation as follows and solve for P.

$$W + P + P + P = 0 \qquad P = -\frac{W}{3}$$

$$P = \frac{200 \text{ kg}(9.81 \text{ N/kg})}{3} \qquad P = 654 \text{ N}$$

This is often recognized as simply dividing the load by the number of supporting ropes (all of which are part of one rope).

Step 5 Draw the free-body diagram of the entire block and tackle as shown in 3-39c.

Step 6 Write the vertical force equilibrium equation as follows and solve for $\overline{R}_{AY}$.

$$\Sigma F_Y = 0 \qquad \overline{R}_{AY} + P + W = 0$$

$$\overline{R}_{AY} = -(P + W) \qquad \overline{R}_{AY} = -(-654 - 1962 \text{ N})$$

$$\overline{R}_{AY} = +2616 \text{ N} \quad \text{at} \quad 90°$$

Example 3-15: Determine the load that can be supported by the block and tackle shown in Figure 3-40a and determine the reactions at *A*, *B*, and *C*.

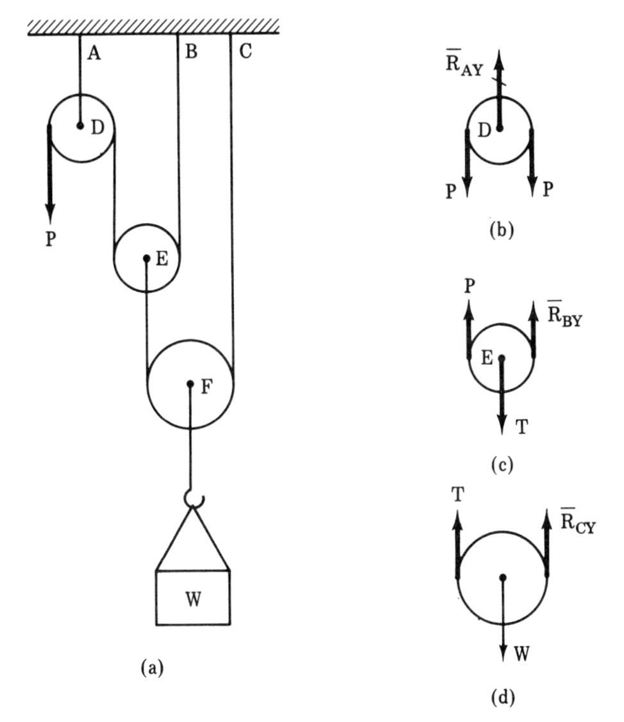

Figure 3-40

Solution:

Step 1 Neglect the angles of any ropes and consider the block and tackle a parallel force system.

Step 2 Recognize that there are two ropes involved. Assume no slippage of the ropes on pulleys and no friction at the pulley shafts. The tension on the rope being pulled on is equal to *P* throughout. The tension in the other rope will be constant throughout, but not necessarily equal to *P*.

Step 3 Draw the free-body diagram of pulley *D* as shown in 3-40b. Write the vertical force equilibrium equation and solve as follows:

$$F_Y = 0 \qquad \overline{R}_{AY} + P + P = 0$$

$$\overline{R}_{AY} = -2P = -2(-50 \text{ lb}) \qquad \overline{R}_{AY} = 100 \text{ lb} \quad \text{at} \quad 90°$$

Step 4 Draw the free-body diagram of pulley E as shown in 3-40c. Write the vertical force equilibrium equation and solve as follows:

$$F_Y = 0 \qquad T + P + \overline{R}_{BY} = 0 \qquad \overline{R}_{BY} = P$$

$$T = -2P = -2(+50 \text{ lb}) \qquad T = -100 \text{ lb}$$

$$T = 100 \text{ lb} \quad \text{at} \quad 270° \qquad \overline{R}_{BY} = 50 \text{ lb} \quad \text{at} \quad 90°$$

Step 5 Draw the free-body diagram of pulley F as shown in 3-40d. Write the vertical force equilibrium equation and solve as follows:

$$F_Y = 0 \qquad W + T + \overline{R}_{CY} = 0 \qquad T = \overline{R}_{CY}$$

$$W = -2T = -2(+100 \text{ lb}) \qquad W = -200 \text{ lb}$$

$$W = 200 \text{ lb} \quad \text{at} \quad 270° \qquad \overline{R}_{CY} = 100 \text{ lb} \quad \text{at} \quad 90°$$

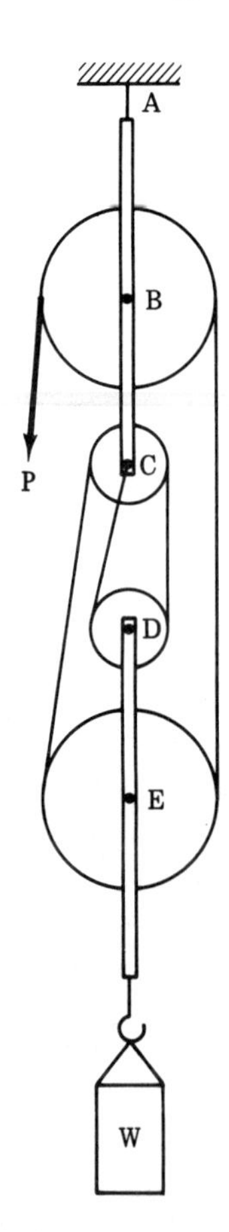

Figure 3-41

Problem 3-14: Determine the weight that can be lifted by applying a 60-lb force to the block and tackle shown in Figure 3-41.

Problem 3-15: Determine the force that must be applied to lift the weight shown in Figure 3-42.

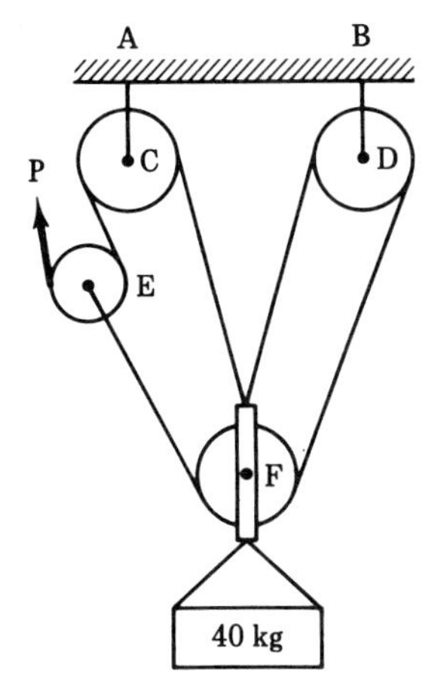

Figure 3-42

3.9 READINESS QUIZ

1. The distance used to determine a moment must be taken
   (a) perpendicular to the line of action of the force
   (b) parallel to the line of action of the force
   (c) perpendicular to the radius of rotation
   (d) parallel to the radius of rotation
   (e) none of the above

2. The moment of a force is a ________ quantity.
   (a) scalar
   (b) vector
   (c) coordinate
   (d) all of the above
   (e) none of the above

3. Appropriate units for moment would be
   (a) lb ft
   (b) lb in.
   (c) N m
   (d) all of the above
   (e) none of the above

4. The moment of force F about 0 in Figure 3-43 is

   (a) 20 in. lb clockwise
   (b) 80 in. lb clockwise
   (c) 20 ft. lb clockwise
   (d) 20 in. lb counterclockwise
   (e) none of the above

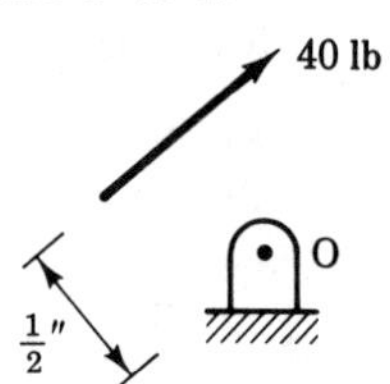

Figure 3-43

5. The total moment about A resulting from the forces shown in Figure 3-44 is

   (a) 150 in. lb clockwise
   (b) 50 in. lb counterclockwise
   (c) 150 in. lb counterclockwise
   (d) 100 in. lb clockwise
   (e) none of the above

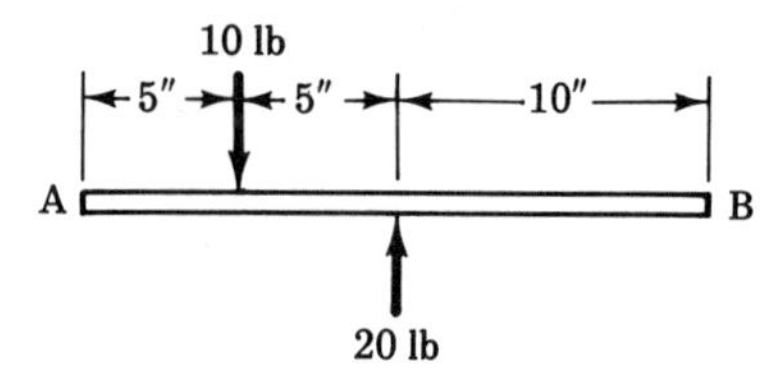

Figure 3-44

6. When a force has a moment it tends to cause

   (a) linear motion
   (b) rotation
   (c) no rotation
   (d) any of the above
   (e) none of the above

7. In actual applications, moment is often referred to as

   (a) sense
   (b) lever arm
   (c) torque
   (d) any of the above
   (e) none of the above

8. Forces that are coplanar yet can never meet are

   (a) collinear
   (b) concurrent
   (c) parallel
   (d) all of the above
   (e) none of the above

9. The resultant of the force system shown in Figure 3-45 will be located

(a) 10 ft to the right of A
(b) 4 ft to the right of A
(c) 6 ft to the right of A
(d) any of the above
(e) none of the above

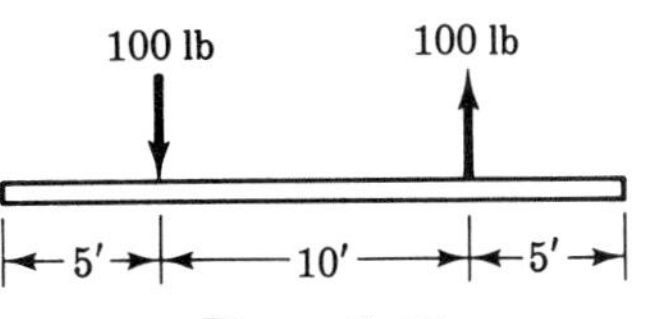

Figure 3-45

10. Two parallel forces must also be

    (a) coplanar
    (b) concurrent
    (c) collinear
    (d) all of the above
    (e) none of the above

11. The resultant of 100 N at 90° and 300 N at 270° is

    (a) 400 N at 0°
    (b) 200 N at 90°
    (c) 200 N at 270°
    (d) 400 N at 90°
    (e) 400 N at 270°

12. A 100-lb force is parallel to and in the same direction as a 200-lb force. Their resultant will lie

    (a) between them, closer to the 100-lb force
    (b) between them, closer to the 200-lb force
    (c) exactly halfway between
    (d) not between them, closer to the 100-lb force
    (e) not between them, closer to the 200-lb force

13. A 50-N force parallel to but opposite in direction to another 50-N force results in

    (a) moment but no force
    (b) force but no moment
    (c) force and moment
    (d) no force and no moment
    (e) any of the above

14. A couple can be moved __________ on a body without changing its effect on a body.

    (a) anywhere
    (b) to a plane perpendicular to the original plane
    (c) to a plane parallel to the original plane
    (d) any of the above
    (e) none of the above

15. The following include(s) a couple

(a) → 20 N / → 20 N (b) → 30 N / ← 10 N (c) ↓ 20 N / ← 20 N (d) all (e) none

16. A couple can be __________ on a body without changing its effect on a body

(a) turned 90° in the same plane
(b) moved to a parallel location in the same plane
(c) moved to a plane parallel to the original
(d) any of the above
(e) none of the above

17. The following is a couple:

(a) ↙ 20 N / 20 N ↗ (b) ↑↓ 10 N / 10 N (c) ↑↓ 30 N / 30 N
(d) all (e) none

18. The load acting on the beam shown in Figure 3-46 can be replaced by a concentrated load acting __________ to the right of A.

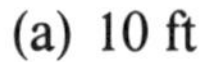

(a) 10 ft
(b) 5 ft
(c) 2.5 ft
(d) any of the above
(e) none of the above

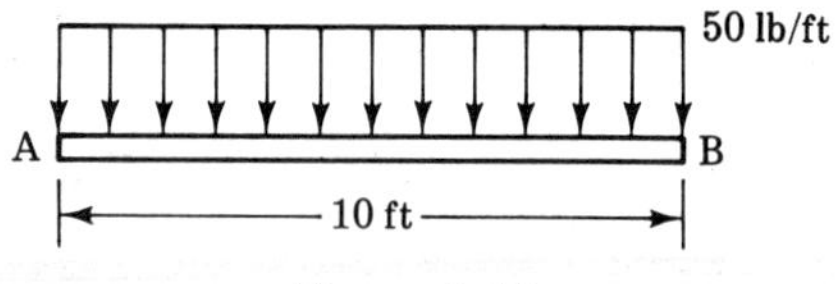

Figure 3-46

19. The total load shown acting on the beam in Figure 3-46 is

(a) 50 lb
(b) 5 lb
(c) 500 lb
(d) 250 lb
(e) none of the above

20. The load shown acting on the beam in Figure 3-46 is known as a

(a) concentrated load
(b) uniformly distributed load
(c) nonuniformly distributed load
(d) combination load
(e) none of the above

21. In a simple block and tackle, the applied force equals the load __________ the number of supporting ropes.

(a) multiplied by
(b) divided by

(c) raised to the power of
(d) multiplied by the log of
(e) none of the above

22. The parallel forces shown in Figure 3-47 could be replaced by a resultant of 5 lb up located

(a) to the left of F
(b) between F_1 and F_2
(c) to the right of F_2
(d) any of the above
(e) indeterminate

Figure 3-47

23. The resultant moment of the couples shown in Figure 3-48 is

(a) 50 lb in. clockwise
(b) 50 lb in. counterclockwise
(c) 100 lb in. clockwise
(d) 100 lb in. counterclockwise
(e) none of the above

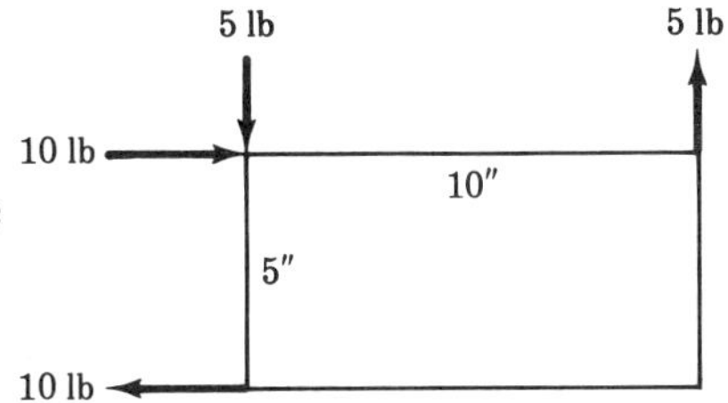

Figure 3-48

24. The block and tackle shown in Figure 3-49 is capable of lifting a load equal to

(a) P
(b) $2P$
(c) $3P$
(d) $4P$
(e) $5P$

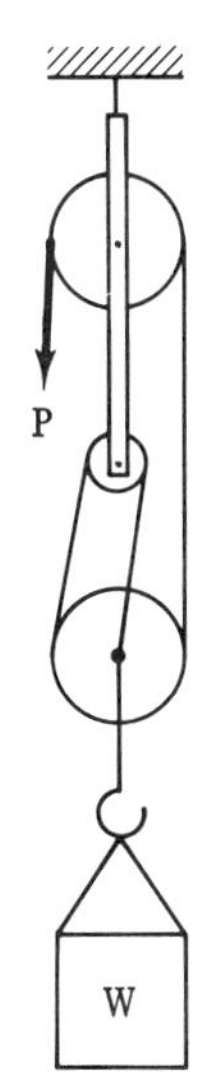

Figure 3-49

25. In the belt and pulley system shown in Figure 3-50

(a) $T_1 = T_2$
(b) $T_1 < T_2$
(c) $T_1 > T_2$
(d) any of the above
(e) indeterminate

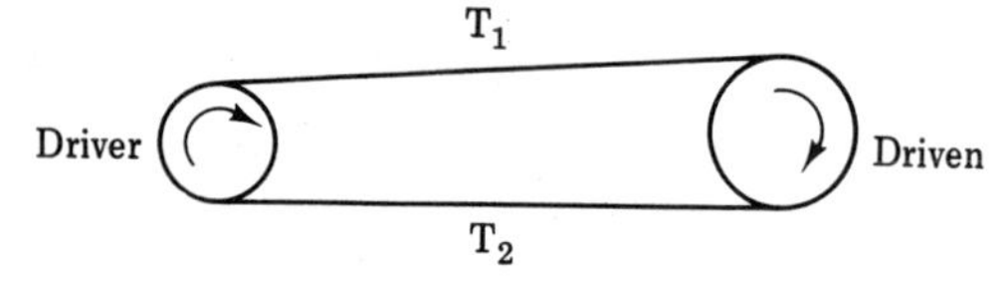

Figure 3-50

3.10 SUPPLEMENTARY PROBLEMS

3-16 The pressure above the piston shown in Figure 3-51 exerts a force of 2.5 kN on the piston. Determine the torque acting on the crankshaft.

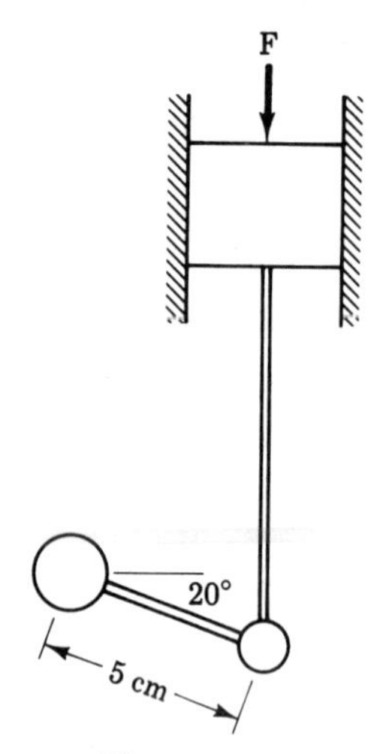

Figure 3-51

3-17 The bottle-capping machine partially shown in Figure 3-52 must exert a force of 15 lb on the cap. What force must be applied at A?

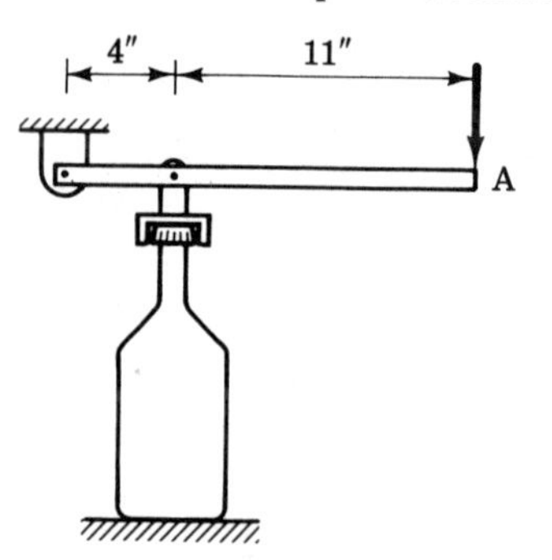

Figure 3-52

3-18 What are the reactions at A and C in Figure 3-53 if AB and BC both weigh 50 lb/ft?

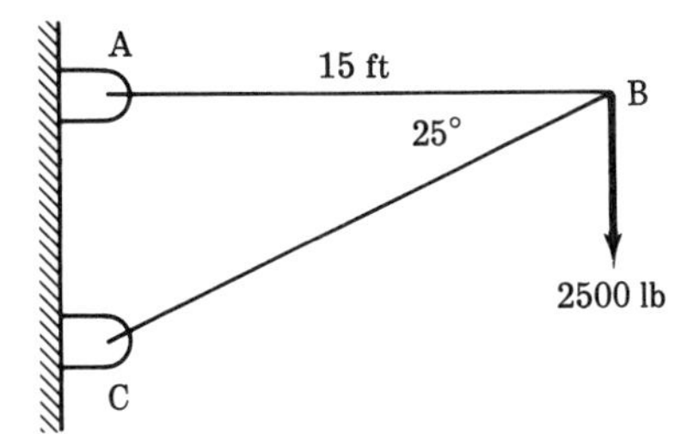

Figure 3-53

3-19 Determine the coefficient of friction between the block and the surface in Figure 3-54 if the block is just about to tip.

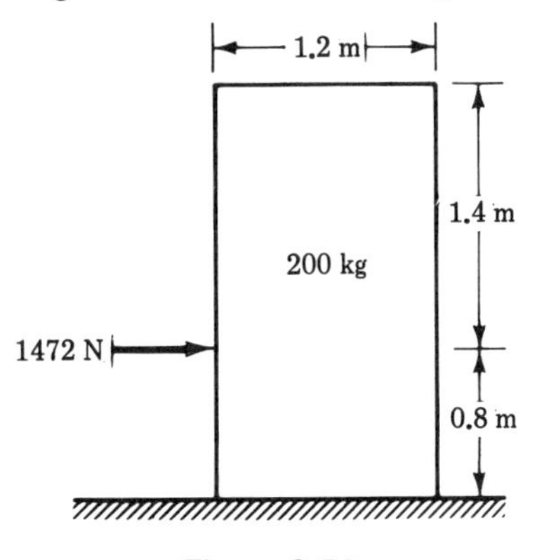

Figure 3-54

3-20 Determine the horizontal force required to pull a 100-car coal train if each loaded car weighs 30 tons. The wheels are 3.2 ft in diameter and the coefficient of rolling resistance is 0.009 in.

3-21 Determine the support reactions at A and B for the band brake shown in Figure 3-55 if the coefficient of friction between the band and the wheel is 0.25.

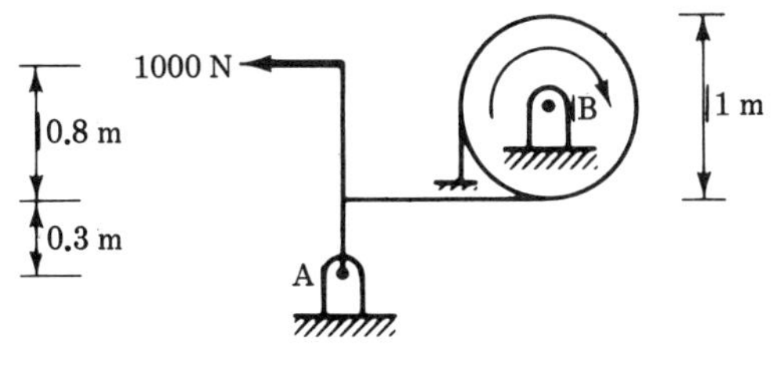

Figure 3-55

3-22 The differential chain hoist shown in Figure 3-56 must be capable of lifting 500 lb. What force must be applied to the chain?

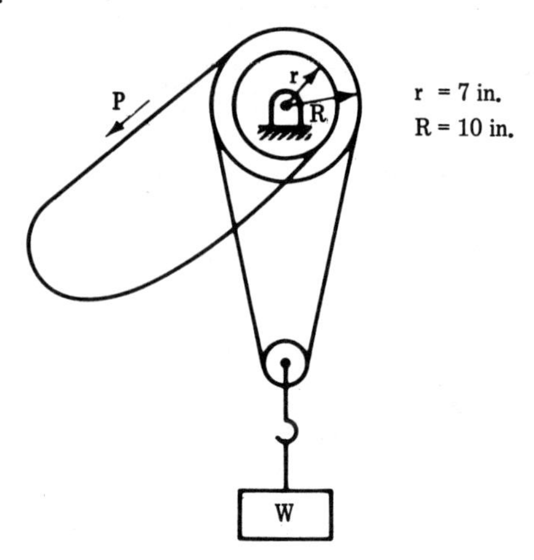

Figure 3-56

3-23 Determine the reactions at *A* and *B* for the beam shown in Figure 3-57.

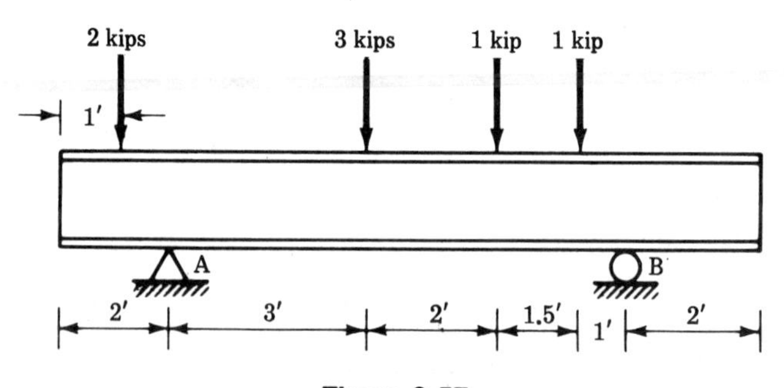

Figure 3-57

Chapter

4

Coplanar, Nonconcurrent Force Systems

4.1 OBJECTIVES

Upon completion of the work relating to this chapter you should be able to perform the following.

1. Given a coplanar, nonconcurrent force system, determine the magnitude, direction, and location of the resultant of the force system.
2. Given a coplanar, nonconcurrent force system acting on a supported body, determine the reactions at the supports.

4.2 COPLANAR, NONCONCURRENT FORCE SYSTEMS DEFINED

Throughout this text, we have been dealing with coplanar force systems. These systems consist of forces whose lines of action all lie in the same plane. Concurrent force systems consist of systems whose forces' lines of action intersect at a single point. In a coplanar, nonconcurrent force system, then, all the forces' lines of action must lie in a common plane; however, they must have more than one intersection, or in a special case, no intersections. This latter case involves parallel forces which we already studied in Chapter 3. Some more general cases of coplanar, nonconcurrent force systems are shown in Figure 4-1. Note that while some of these contain parallel forces, none consists solely of a single set of parallel forces.

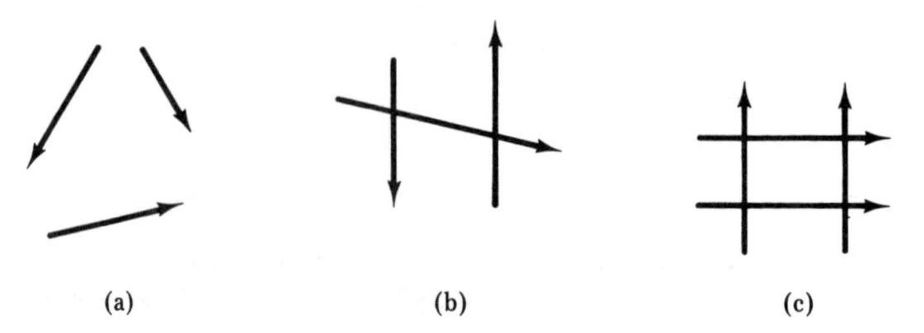

(a) (b) (c)

Figure 4-1. Coplanar, nonconcurrent force systems.

There are many such force systems that are not in equilibrium, meaning simply that acceleration is taking place. Since this course concentrates on static equilibrium, most of the work we do will be under those conditions.

4.3 DETERMINATION OF THE RESULTANT

As we found in Chapter 3 with parallel forces, the resultant of a system of nonconcurrent forces must be described in terms of not only magnitude and direction but also location. Recall that for concurrent forces the resultant was readily located because it had to pass through the point of concurrence. In nonconcurrent systems in general, we resort to one of the techniques used to locate the resultant of parallel forces. We recognize that not only must the resultant equal the vector sum of the forces in the system but the moment of the resultant must equal the vector sum of the moments of the individual forces in the system.

$$R = F = F_1 \leftrightarrow F_2 \leftrightarrow F_3 \cdots$$

$$M = \Sigma M = M_1 \leftrightarrow M_2 \leftrightarrow M_3 \cdots$$

$$Rd_R = F_1 d_1 \leftrightarrow F_2 d_2 \leftrightarrow F_3 d_3 \cdots$$

No subscript was used for the moment M of the resultant or for ΣM. This is because the subscript of the moment symbol is generally used to indicate the point about which moments are being taken. Note that the above equations are vector equations. *Never use force directions in a moment equation.* Use force directions only in ΣF_X and ΣF_Y equations. Use moment directions only in ΣM equations. In all other equations, use all absolute values. In those problems involving static equilibrium, ΣF and ΣM will both equal zero.

Example 4-1: Determine the magnitude, direction, and location with respect to 0 of the resultant of the force system shown in Figure 4-2a.

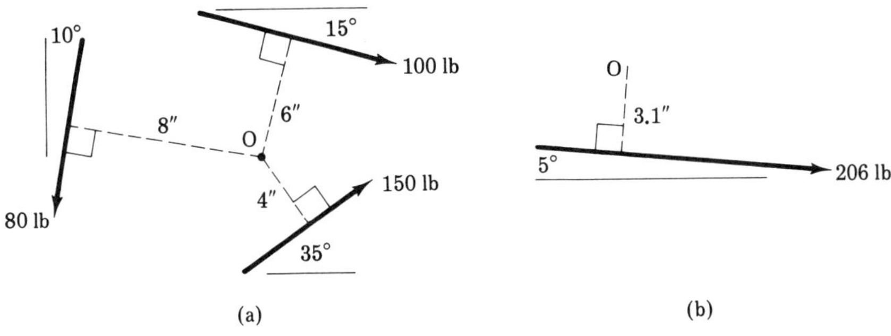

Figure 4-2

Solution:

Step 1 Obtain the magnitude and direction of the resultant by summing the forces in the same manner used for a system of concurrent forces. Any of the methods learned could be used. The method of components follows:

$$R_X = \Sigma F_X = F_{1X} + F_{2X} + F_{3X}$$

$$R_X = F_1 \cos \theta_1 + F_2 \cos \theta_2 + F_3 \cos \theta_3$$

$$R_X = (100 \text{ lb})\cos 345° + (150 \text{ lb})\cos 35° + (80 \text{ lb})\cos 260°$$

$$R_X = 96.6 + 122.91 - 13.91 \text{ lb} \qquad R_X = 205.6 \text{ lb}$$

$$R_Y = \Sigma F_Y = F_{1Y} + F_{2Y} + F_{3Y}$$

$$R_Y = F_1 \sin \theta_1 + F_2 \sin \theta_2 + F_3 \sin \theta_3$$

$$R_Y = (100 \text{ lb})\sin 345° + (150 \text{ lb})\sin 35° + (80 \text{ lb})\sin 260°$$

$$R_Y = -25.9 + 86.0 - 78.8 \text{ lb} \qquad R_Y = -18.71 \text{ lb}$$

$$R = \sqrt{R_X^2 + R_Y^2} \qquad R = \sqrt{(205.6)^2 + (-18.7)^2} \qquad R = 206 \text{ lb}$$

$$\theta_R = \arctan \frac{R_Y}{R_X} \qquad \theta_R = \arctan \frac{-18.7}{205.6} \qquad \theta_R = 355°$$

Locating the resultant involves knowing the perpendicular distance from the line of action of R to point 0 and knowing which side of 0 the resultant is on. The distance is the moment arm of the resultant and can be obtained by use of moments. The location of the resultant to one side or the other must be done by simultaneously evaluating the direction of the resultant and the direction of its moment.

$$\Sigma M_{0R} = M_{01} + M_{02} + M_{03} = F_1 d_1 + F_2 d_2 + F_3 d_3$$

$$\Sigma M_{0R} = -(100 \text{ lb})(6 \text{ in.}) + (150 \text{ lb})(4 \text{ in.}) + (80 \text{ lb})(8 \text{ in.})$$

$$\Sigma M_{0R} = +640 \text{ lb in.}$$

$$d_R = \frac{\Sigma M_{0R}}{R} \qquad d_R = \frac{640 \text{ lb in.}}{206 \text{ lb}} \qquad d_R = 3.1 \text{ in.}$$

Since R is to the right and slightly down and must provide counterclockwise moment, it must be located below 0 as shown in Figure 4-2b.

The total answer then is: $R = 206$ lb at 355° and a moment arm of 3.1 in., located below 0. If equilibrium is desired, there must be one or more reactions totaling 206 lb at 175° and a moment arm of 3.1 in. located below 0.

Sometimes it is desired to locate the resultant in terms of an X or Y intercept with respect to the moment center. With reference to Figure 4-2b:

$$X_R = -\frac{d_R}{\sin 5°} \qquad X_R = -\frac{3.1 \text{ in.}}{\sin 5°} \qquad X_R = 35.6 \text{ in. to left of } 0$$

$$Y_R = -\frac{d_R}{\cos 5°} \qquad Y_R = -\frac{3.1 \text{ in.}}{\cos 5°} \qquad Y_R = 3.11 \text{ in. below } 0$$

Problem 4-1: Determine the magnitude, direction, and location with respect to A of the resultant of the force system shown in Figure 4-3. Locate the resultant in terms of its X intercept.

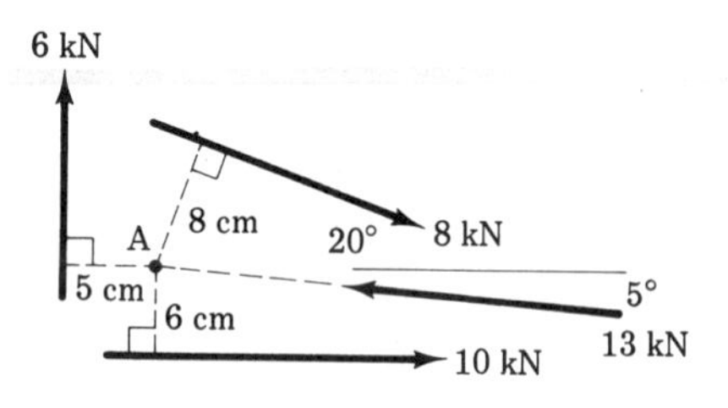

Figure 4-3

4.4 STRUCTURES IN EQUILIBRIUM

Many structures that are in equilibrium have a nonconcurrent force system acting on them. Such a system generally includes two support reactions. These may be two-force reactions as shown in the free-body diagram in Figure 4-4a or a force reaction and a moment reaction as shown in the free-body diagram in Figure 4-4b. Recall that earlier in the text we mentioned briefly a support that can sustain a moment acting on it, for example, a fencepost imbedded in the ground or a flagpole bolted to a concrete pad.

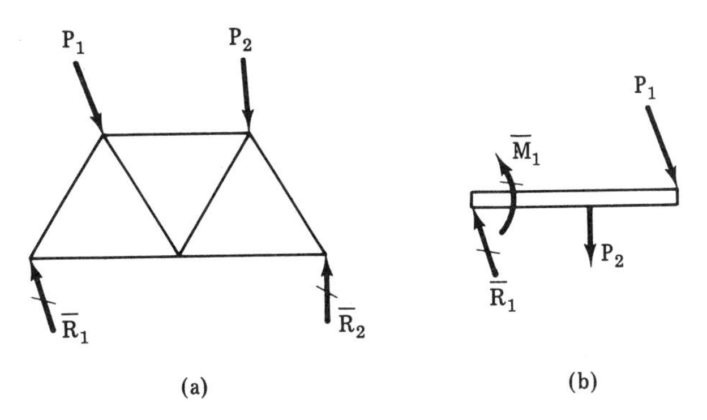

Figure 4-4. Nonconcurrent force system equilibrium.

Example 4-2: Determine the support reactions for the structure shown in Figure 4-5a.

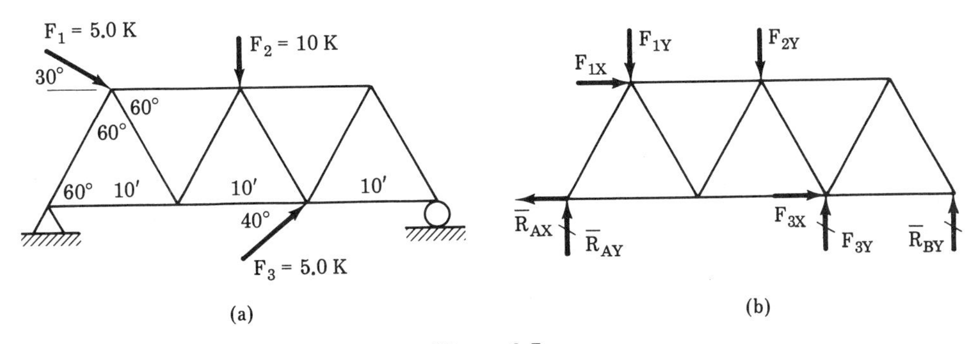

Figure 4-5

Solution:

Step 1 Make a free-body diagram of the truss as shown in Figure 4-5b. $\overline{R}_{BY}$ is the only reaction at B and is at 90° because it must act against the roller and must be perpendicular to the surface the roller is resting on. $\overline{R}_{AY}$ is at 90° because inspection of the structure suggests that this is necessary for $\Sigma M_B = 0$. $\overline{R}_{AX}$ is at 180° because all horizontal load components are at 0° and the only support that can cause $\Sigma F_X = 0$ is A.

Step 2 Write the two-force equilibrium equations as follows:

$$\Sigma F_X = 0 \qquad \overline{R}_{AX} + P_{1X} + P_{3X} = 0$$

$$\Sigma F_Y = 0 \qquad \overline{R}_{AY} + P_{1Y} + P_{2Y} + P_{3Y} + \overline{R}_{BY} = 0$$

Examination shows that the first contains only one unknown and is thus solvable. The second, however, contains two unknowns and cannot be solved. A moment equation, written about either A or B, will be needed to find one of the vertical components (K = kip):

$$\Sigma F_X = 0 \quad \overline{R}_{AX} + P_{1X} + P_{3X} = 0 \quad \overline{R}_{AX} = -(P_{1X} + P_{3X})$$

$$P_{1X} = P_1 \cos \theta_{P1} \quad P_{1X} = (5.0\text{ K})\cos 330° \quad P_{1X} = +4.33\text{ K}$$

$$P_{3X} = P_3 \cos \theta_{P3} \quad P_{3X} = (5.0\text{ K})\cos 40° \quad P_{3X} = +3.83\text{ K}$$

$$\overline{R}_{AX} = -(+4.33 + 3.83\text{ K}) \quad \overline{R}_{AX} = -8.16\text{ K}$$

$$\mathbf{\overline{R}_{AX} = 8.16\ K \quad at \quad 180°}$$

$$\overline{R}_{BY} d_{BY} = -(P_{1X} d_{1X} + P_{1Y} d_{1Y} + P_{2Y} d_{2Y} + P_{3Y} d_{3Y})$$

$$P_{1Y} = P_1 \sin \theta_{P1} \quad P_{1Y} = (5.0\text{ K})\sin 330° \quad P_{1Y} = -2.5\text{ K}$$

$$P_{3Y} = P_3 \sin \theta_{P3} \quad P_{3Y} = (5.0\text{ K})\sin 40° \quad P_{3Y} = +3.2\text{ K}$$

$$d_{1X} = (10\text{ ft})\sin 60° \quad d_{1X} = 8.66\text{ ft}$$

$$\overline{R}_{BY} d_{BY} = -[-(4.33)(8.66) - (2.5)(5) - (10)(15) + (3.2)(20)]$$

$$R_{BY} d_{BY} = +136\text{ K ft} \circlearrowleft$$

From the direction of $R_{BY} d_{BY}$ and the location of R_{BY}, it can be determined that R_{BY} must be +, as was assumed when making the free body.

$$\overline{R}_{BY} = \frac{\overline{R}_{BY} d_{BY}}{d_{BY}} \quad \overline{R}_{BY} = \frac{136\text{ K ft}}{30\text{ ft}}$$

$$\overline{R}_{BY} = 4.53\text{ K} \quad \text{or} \quad \mathbf{\overline{R}_{BY} = 4.53\ K \quad at \quad 90°}$$

$$\Sigma F_Y = 0 \quad \overline{R}_{AY} + P_{1Y} + P_{2Y} + P_{3Y} + \overline{R}_{BY} = 0$$

$$\overline{R}_{AY} = -(P_{1Y} + P_{2Y} + P_{3Y} + \overline{R}_{BY})$$

$$\overline{R}_{AY} = -(-2.5 - 10 - 3.2 + 4.53\text{ K})$$

$$\overline{R}_{AY} = +11.2\text{ K} \quad \text{or} \quad \mathbf{\overline{R}_{AY} = 11.2\ K \quad at \quad 90°}$$

Example 4-3: Determine the support reactions for the structure shown in Figure 4-6a.

Solution:

Step 1 Note the fixed support at the base (as opposed to a pin or roller). This support is capable of providing a moment reaction as well as a force reaction.

Step 2 Draw the free-body diagram for the structure as shown in Figure 4-6b.

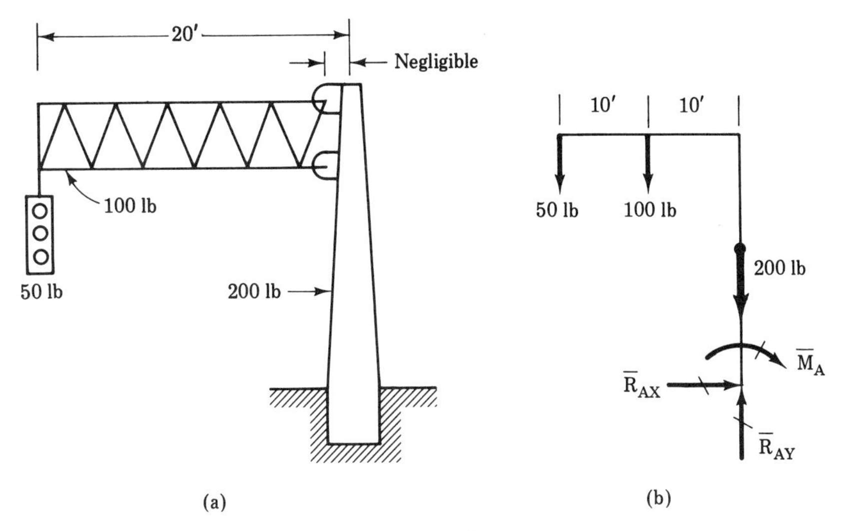

Figure 4-6

Step 3 Since a moment reaction is involved, it will be necessary to write a moment equilibrium equation as well as the two-force equilibrium equations.

$$\Sigma F_X = 0 \qquad \boldsymbol{R_{AX} = 0}$$

$$\Sigma F_Y = 0 \qquad P_1 + P_2 + P_3 + \overline{R}_{AY} = 0 \qquad \overline{R}_{AY} = -(P_1 + P_2 + P_3)$$

$$\overline{R}_{AY} = -(-50\text{ lb} - 100\text{ lb} - 200\text{ lb})$$

$$\overline{R}_{AY} = +350\text{ lb} \quad \text{or} \quad \overline{R}_{AY} = 350\text{ lb} \quad \text{at} \quad 90°$$

$$\Sigma M_A = 0 \qquad P_1d_1 + P_2d_2 + \overline{M}_A = 0 \qquad \overline{M}_A = -(P_1d_1 + P_2d_2)$$

$$\overline{M}_A = -+(50\text{ lb})(20\text{ ft}) + (100\text{ lb})(10\text{ ft}) \qquad \overline{M}_A = -2000\text{ lb ft}$$

Problem 4-2: Determine the support reactions for the structure shown in Figure 4-7.

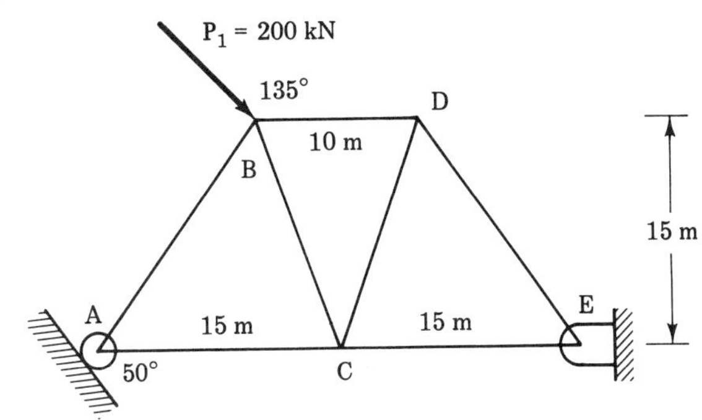

Figure 4-7

Problem 4-3: Determine the support reactions for the structure shown in Figure 4-8.

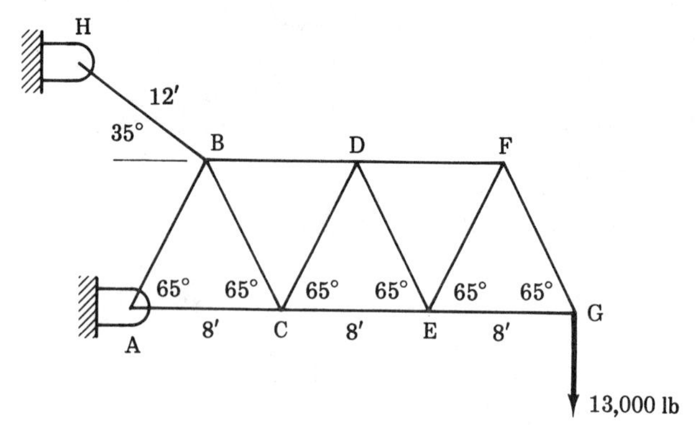

Figure 4-8

4.5 MECHANISMS IN EQUILIBRIUM

A great many mechanisms and machines have nonconcurrent force systems acting on them. Analysis in terms of static equilibrium is useful in itself when designing these devices. It also provides the basis for analysis in terms of dynamic equilibrium (not a part of this course).

Example 4-4: Determine the force required to counteract the 200-lb load on the lever shown in Figure 4-9a. Determine the reactions at the support.

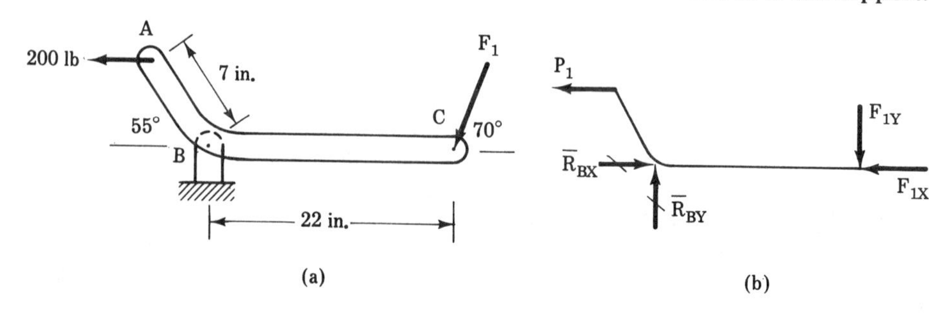

Figure 4-9

Solution:

Step 1 Draw the free-body diagram for the lever as shown in Figure 4-9b.

Step 2 Examination shows that there are two unknown horizontal components and two unknown vertical components. A moment equilibrium equation will thus be needed in addition to the two-force equilibrium equations.

$$\Sigma M_B = 0 \qquad Pd_P + F_Y D_{FY} = 0 \qquad F_Y d_{FY} = -Pd_P$$

$$d_P = (7 \text{ in.})\sin 55^\circ \qquad d_P = 5.73 \text{ in.}$$

$$F_Y d_{FY} = -(200 \text{ lb})(5.73 \text{ in.}) \qquad F_Y d_{FY} = -1147 \text{ lb in.}$$

$$F_Y = \frac{F_Y d_{FY}}{d_{FY}} \qquad F_Y = \frac{1147 \text{ lb in.}}{22 \text{ in.}} \qquad F_Y = -52 \text{ lb}$$

$$F = \frac{F_Y}{\sin \theta_F} \qquad F = \frac{-52 \text{ lb}}{\sin 250^\circ} \qquad \boldsymbol{F = 55.3 \text{ lb} \quad \text{at} \quad 250^\circ}$$

$$\Sigma F_Y = 0 \qquad \overline{R}_{BY} + F_Y = 0 \qquad \overline{R}_{BY} = -F_Y \qquad \overline{R}_{BY} = -(-52 \text{ lb})$$

$$\overline{R}_{BY} = +52 \text{ lb} \qquad \boldsymbol{\overline{R}_{BY} = 52 \text{ lb} \quad \text{at} \quad 90^\circ}$$

$$\Sigma F_X = 0 \qquad P + \overline{R}_{BX} + F_X = 0 \qquad F_X = F \cos \theta_F$$

$$F_X = (55.3 \text{ lb})\cos 250^\circ \qquad F_X = -18.9 \text{ lb}$$

$$\overline{R}_{BX} = -(P + F_X) \qquad \overline{R}_{BX} = -(-200 - 18.9 \text{ lb})$$

$$\overline{R}_{BX} = +219 \text{ lb} \qquad \boldsymbol{\overline{R}_{BX} = 219 \text{ lb} \quad \text{at} \quad 0^\circ}$$

Example 4-5: The bolt exerts a force of 3.0 kN on the pivot arm of the clamp shown in Figure 4-10a. Determine the total force exerted on the workpiece and the reactions at the pivot.

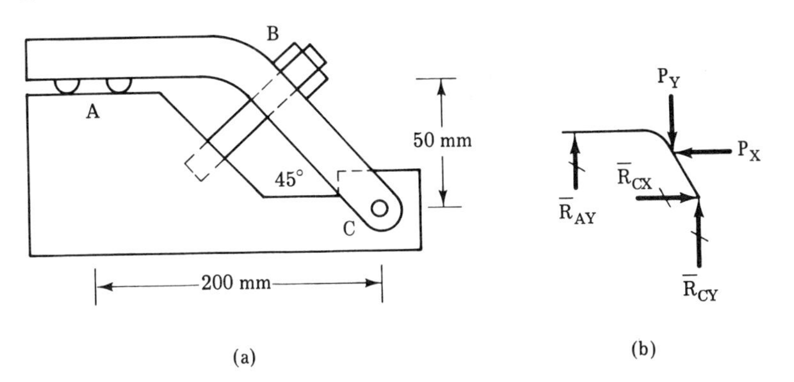

Figure 4-10

Solution:

Step 1 Draw the free-body diagram of the pivot arm as shown in Figure 4-10b.

Step 2

$$\Sigma F_X = 0 \qquad \overline{R}_{CX} + P_X = 0 \qquad \overline{R}_{CX} = -P_X \qquad P_X = P \cos \theta_P$$

$$P_X = (3.0 \text{ kN})\cos 225^\circ \qquad P_X = -2.12 \text{ kN}$$

$$\overline{\boldsymbol{R}}_{CX} = \mathbf{2.12\ kN} \quad \mathbf{at} \quad \mathbf{0°}$$

$$\Sigma M_C = 0 \qquad \overline{R}_{AY}d_{AY} + P_Y d_{PY} + P_X d_{PX} = 0$$

$$\overline{R}_{AY}d_{AY} = -(P_Y d_{PY} + P_X d_X)$$

$$\overline{R}_{AY}d_{AY} - [+(2.12\ \text{kN})(50\ \text{mm}) + (2.12\ \text{kN})(50\ \text{mm})]$$

$$R_{AY}d_{AY} = -212\ \text{kN mm}$$

$$\overline{R}_{AY} = \frac{\overline{R}_{AY}d_{AY}}{d_{AY}} \qquad \overline{R}_{AY} = \frac{212\ \text{kN mm}}{200\ \text{mm}}$$

$$\overline{\boldsymbol{R}}_{AY} = \mathbf{1.06\ kN} \quad \mathbf{at} \quad \mathbf{90°}$$

$$\Sigma F_Y = 0 \qquad \overline{R}_{AY} + P_Y + \overline{R}_{CY} = 0 \qquad \overline{R}_{CY} = -(\overline{R}_{AY} + P_Y)$$

$$\overline{R}_{CY} = -(+1.06 - 2.12\ \text{kN}) \qquad \overline{\boldsymbol{R}}_{CY} = \mathbf{1.06\ kN} \quad \mathbf{at} \quad \mathbf{90°}$$

Problem 4-4: Determine the tension being applied by the come-along shown in Figure 4-11. Determine the reactions at A and B.

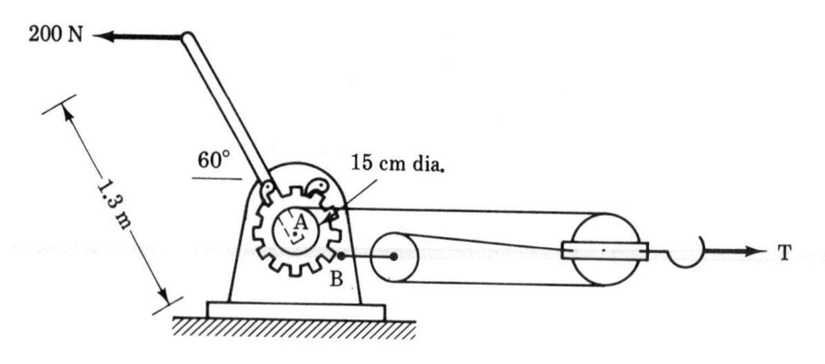

Figure 4-11

Problem 4-5: Determine the reactions at C and D for the hydraulically operated boom shown in Figure 4-12.

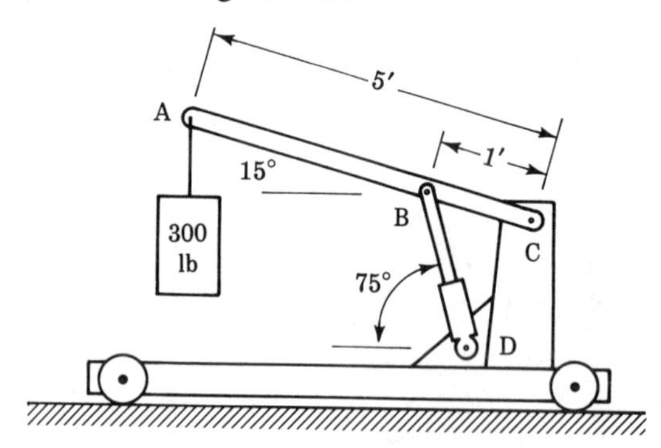

Figure 4-12

4.6 APPLICATIONS TO FRICTION

Some of the friction problems examined in earlier chapters become nonconcurrent, nonparallel force problems simply by changing the direction and/or location of just one of the forces. Other problems appear at first glance to be complex nonconcurrent force problems but can be solved by dividing the problem into more than one simple nonconcurrent or even concurrent force problems. The following examples illustrate a variety of such problems:

Example 4-6: Determine the least value of l required so that the pipe wrench shown in Figure 4-13a will not slip. Determine the force required to apply a torque of 150 N m to the pipe.

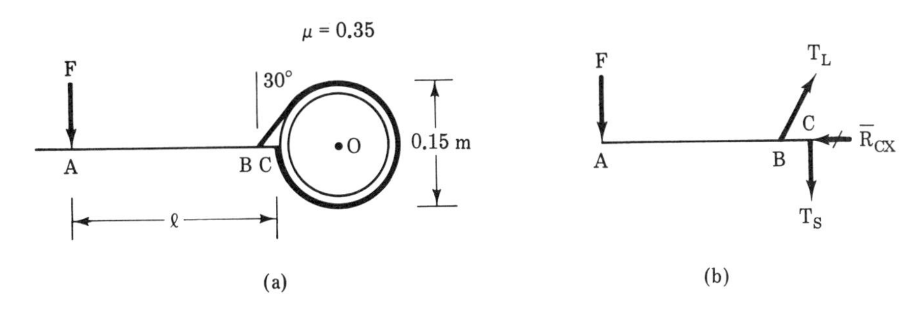

Figure 4-13

Solution:

Step 1 Recognize that determination of the least value of AC and the value of F corresponding to that value of AC are two separate problems.

Step 2 Recognize that the relationship of T_L and T_S must satisfy the belt friction equation and the equilibrium of the wrench lever. Draw the free-body diagram for ABC as shown in Figure 4-13b. Write the equations for these relationships as follows:

$$T_L = T_S e^{\mu\beta} \quad \text{or} \quad \left(\frac{T_L}{T_S}\right) \max = e^{\mu\beta}$$

(meaning that while an infinite number of combinations of T_L and T_S could provide adequate torque, any combination with a greater ratio of T_L to T_S will result in slippage)

$$\Sigma F_Y = 0 \qquad -T_S - F + T_{LY} = 0 \qquad T_{LY} = F + T_S$$

$$\Sigma M_C = 0 \qquad T_{LY}(BC) + F(AC) = 0$$

Step 3 By substitution, eliminate F, T_S, and T_L solving for AC.

$$\beta = 330° \frac{2\pi \text{ rad}}{360°} \qquad \beta = 5.76 \text{ rad}$$

$$T_S = T_L \div e^{\mu\beta} \qquad T_S = T_L \div e^{(0.35)(5.76)} \qquad T_S = 0.133T_L$$

Substitute for T_S in force equation

$$T_{LY} = F + T_S \qquad T_{LY} = F + 0.133T_L \qquad T_{LY} = T_L \sin\theta_{TL}$$

$$T_{LY} = T_L \sin 60° \qquad T_{LY} = 0.866T_L \qquad T_{LY} = T_L \sin\theta_{TL}$$

$$0.866T_L = F + 0.133T_L \qquad F = 0.733T_L$$

Substitute for F in moment equation

$$0.866T_L\,(BC) + 0.733T_L\,(AC) = 0$$

$$BC = \frac{0C}{\sin\theta_{TL}} - 0C \qquad BC = \frac{0.075 \text{ m}}{\sin 60°} - 0.075 \text{ m} \qquad BC = 0.0866 \text{ m}$$

$$0.866T_L(0.0866 \text{ m}) + 0.733T_L(AC) = 0 \qquad \mathbf{AC = 0.102\ m}$$

Step 4 Determine F from torque requirement

$$T = F(0A) \qquad F = \frac{T}{0A} \qquad F = \frac{150 \text{ N m}}{0.177 \text{ m}} \qquad \mathbf{F = 847\ N}$$

Step 5 Find $T_L T_S$. Check ratio.

$$T_L = \frac{F}{0.733} \qquad T_L = \frac{847 \text{ N}}{0.733} \qquad T_L = 1155 \text{ N}$$

$$T_S = T_{LY} - F \qquad T_S = 0.866(1155 \text{ N}) - 847 \qquad T_S = 153 \text{ N}$$

$$T_S = 0.133T_L \qquad T_S = 0.133(1155 \text{ N}) \qquad T_S = 154 \text{ N}$$

Check!

Problem 4-6: Answer the following questions pertaining to Example 4-6:

1. Keeping everything else, including AC, constant what happens if F is increased from the value found?
2. Keeping everything else, including AC, constant what happens if F is decreased from the value found?

3. Keeping everything else, including F, constant what happens if AC is increased from the value found?
4. Keeping everything else, including F, constant what happens if AC is decreased from the value found?

Example 4-7: Determine the force F required to just start the blocks shown in Figure 4-14a moving. Determine the tension in the rope between the blocks.

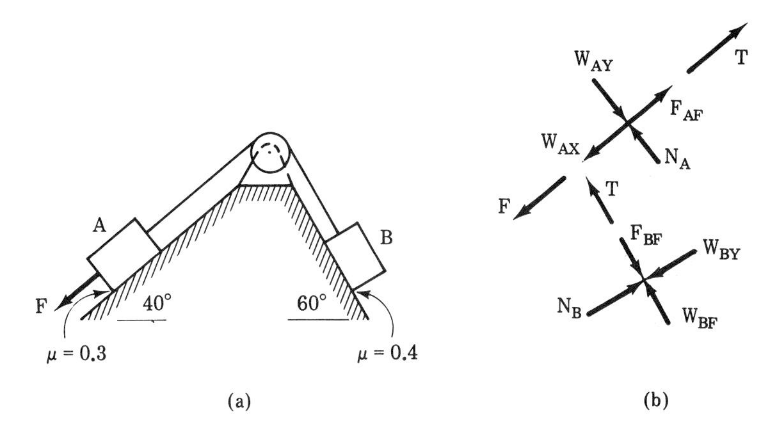

Figure 4-14

Solution:

Step 1 Draw complete free-body diagrams for each block as shown in Figure 4-14b.

Step 2 Recognize that the forces perpendicular to the slopes are in equilibrium by their nature and that the problem can be approached by considering only the forces parallel to the slopes.

Step 3 Recognize that once the readily determined forces parallel to the slopes have been found, the angles of the slopes no longer enter the problem. It becomes similar to a problem on a single horizontal surface.

Step 4 Consider the problem as one system making the tension an internal force and leaving the force F the only unknown.

Step 5 Write the equilibrium equation for the complete force system parallel to the slopes; determine the components of the weights and the friction forces; then solve for F.

$$F + W_{AX} + F_{AF} + W_{BX} + F_{BF} = 0$$

$$F = -(W_{AX} + F_{AF} + W_{BX} + F_{BF})$$

$$W_{AX} = W_A \sin \theta_A \qquad W_{AX} = 80 \text{ lb}(\sin 40°) \qquad W_{AX} = 51.4 \text{ lb}$$

$$W_{BX} = W_B \sin \theta_B \qquad W_{BX} = 70 \text{ lb}(\sin 60°) \qquad W_{BX} = 60.6 \text{ lb}$$

$$F_{AF} = \mu_A N_A = \mu_A W_{AY} = \mu_A \cos \theta_A W_A \qquad F_{AF} = 0.3(\cos 40°)(80 \text{ lb})$$

$$F_{AF} = 18.4 \text{ lb}$$

$$F_{BF} = \mu_B N_B = \mu_B W_{BY} = \mu_B \cos \theta_B W_B \qquad F_{BF} = 0.4(\cos 60°)(70 \text{ lb})$$

$$F_{BF} = 14 \text{ lb}$$

$$F = -(-51.4 + 18.4 + 60.6 + 14 \text{ lb}) \qquad \mathbf{F = -41.61 \text{ lb}}$$

Problem 4-7: Determine the maximum load that the weight W shown in Figure 4-15 can move up the incline at a constant velocity.

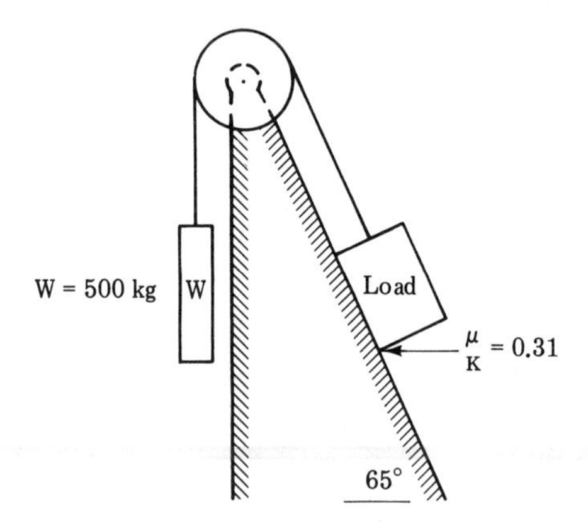

Figure 4-15

Example 4-8: Determine the force required to just start the block shown in Figure 4-16a sliding.

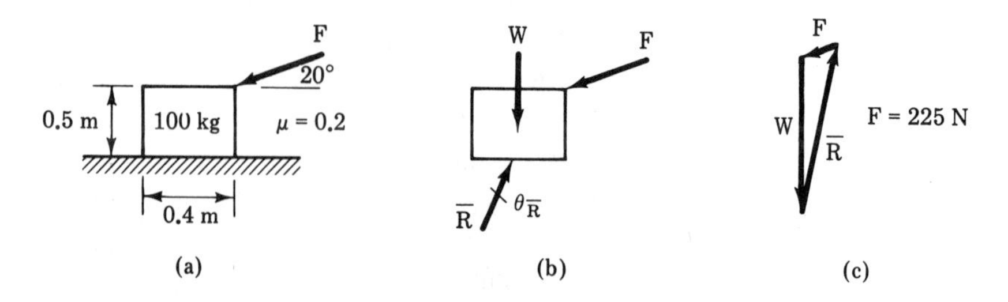

Figure 4-16

Solution:

Step 1 Draw the free-body diagram for the block as shown in Figure 4-16b.

Step 2 Recognize this as a nonconcurrent force problem requiring analysis of moment equilibrium as well as force equilibrium.

Step 3 Write the force and moment equations as follows and solve.

$$F \leftrightarrow W \leftrightarrow \overline{R} = 0$$

$$\theta_{\overline{R}} = 90° - \arctan \mu \qquad \theta_{\overline{R}} = 90° - \arctan 0.2 \qquad \theta_{\overline{R}} = 78.7°$$

Solve graphically as shown in Figure 4-16c.

Determine if tipping occurs by evaluating moments of F and W about the leading edge (0). For tipping not to occur $Wd_W > Fd_F$

$$dW = 0.2 \text{ m} \qquad dF = [0.5 \text{ m} - 0.4 \text{ m}(\tan 20°)]\cos 20° \qquad d_F = 0.33 \text{ m}$$

$$Wd_W = -(100 \text{ kg})(9.81 \text{ kN/kg})(0.2 \text{ m}) = -196 \text{ kN m}$$

$$Fd_F = +(225 \text{ kN})(0.333 \text{ m}) = +74.9 \text{ kN m}$$

The block will not tip. If desired, the moment equilibrium equation could be written to determine the exact location of $\overline{R}$ that achieves the moment equilibrium which we can see exists.

Problem 4-8: Determine the force required to just start the block shown in Figure 4-17 sliding.

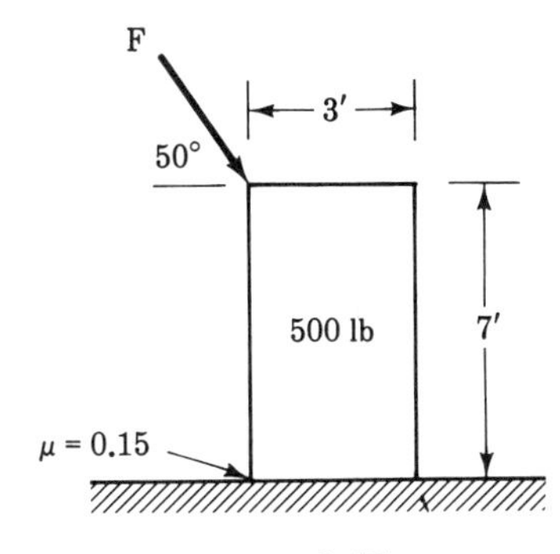

Figure 4-17

4.7 READINESS QUIZ

1. Of the force systems shown in Figure 4-18, the nonconcurrent one is

   (a)
   (b)
   (c)
   (d) all
   (e) none

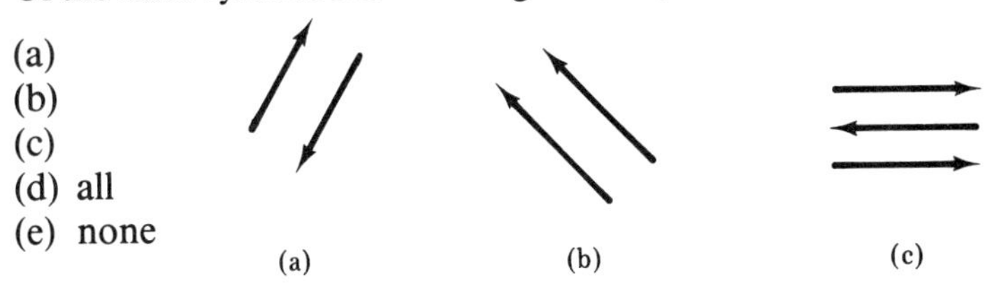

Figure 4-18

2. The total moment of a force system is equal to

   (a) the sum of the moments of all the forces in the system
   (b) the sum of the moments of the components of all the forces in the system
   (c) the resultant of the force system times its moment arm
   (d) all of the above
   (e) none of the above

3. Of the force systems shown in Figure 4-19, the nonconcurrent one is

   (a) (b) (c) (d) (e)

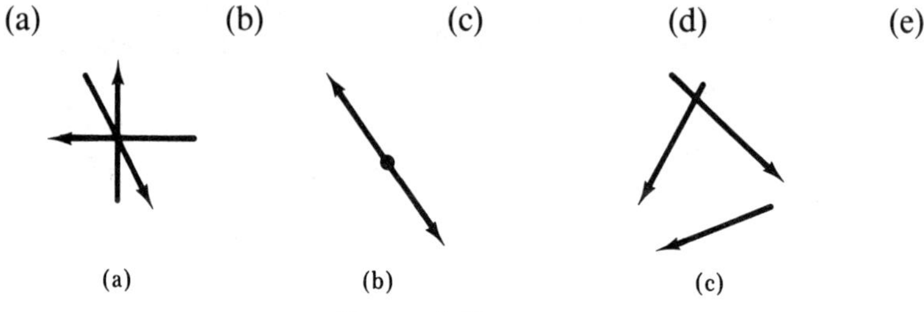

Figure 4-19

4. The parallel force system shown in Figure 4-20 has

   (a) moment but no resultant force
   (b) resultant force but no moment
   (c) resultant force and moment
   (d) neither resultant force nor moment
   (e) any of the above

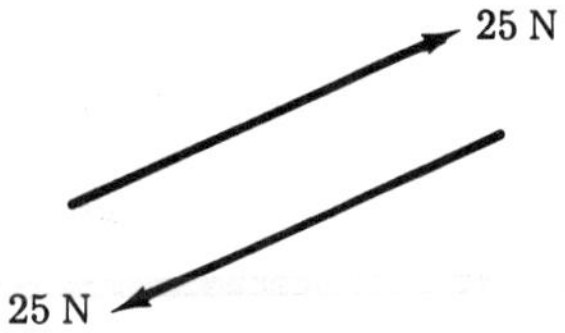

25 N

25 N

Figure 4-20

5. The reactions at the axle of the pulley shown in Figure 4-21 will be (assume static conditions and a friction-free bearing)

   (a) strictly horizontal
   (b) strictly vertical
   (c) vertical and horizontal
   (d) none of the above
   (e) insufficient information given

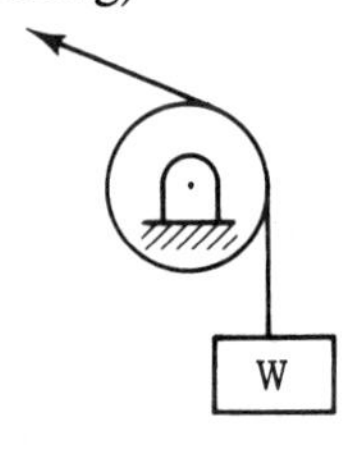

Figure 4-21

6. The reactions at A in Figure 4-22 are

   (a) left, up, counterclockwise
   (b) right, down, clockwise
   (c) left, down
   (d) any of the above
   (e) none of the above

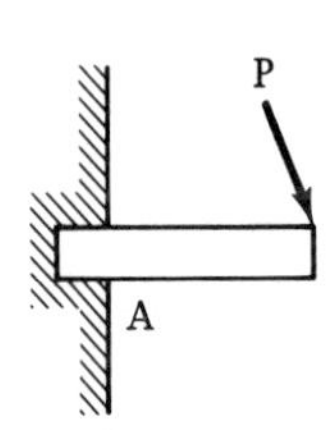

Figure 4-22

7. The reactions at B in Figure 4-23 are

(a) right
(b) left
(c) down
(d) any of the above
(e) none of the above

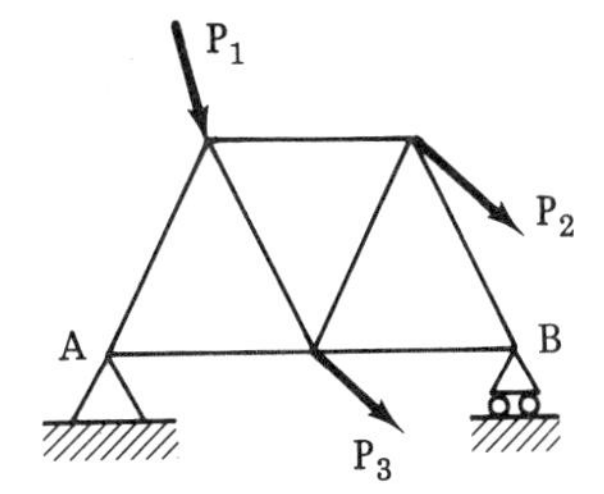

Figure 4-23

8. The reactions at A in Figure 4-24 are

(a) left
(b) right
(c) up and right
(d) down and left
(e) down and right

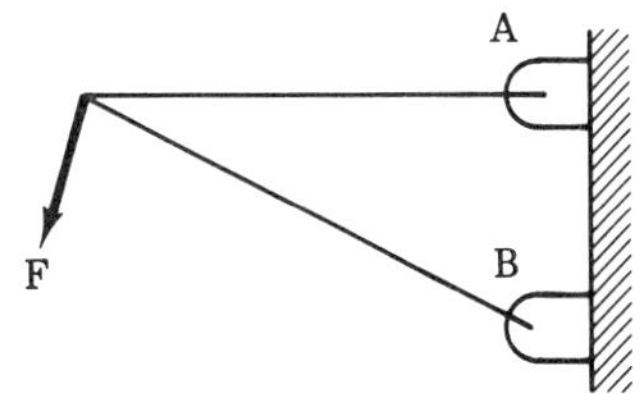

Figure 4-24

9. The reactions at B in Figure 4-24 above are

(a) left
(b) right
(c) up and left
(d) down and left
(e) down and right

10. The reactions at A for the truss shown in Figure 4-25 are

(a) down and left
(b) left
(c) up and right
(d) right
(e) up and left

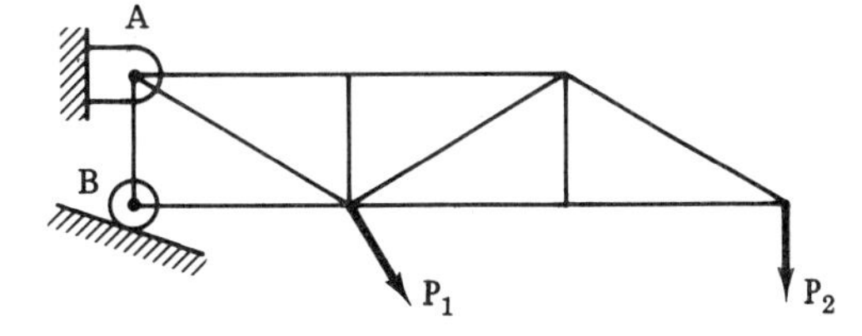

Figure 4-25

11. The reactions at B for the truss shown in Figure 4-25 above are

(a) down and left
(b) up and right
(c) up and left
(d) down and right
(e) none of the above

12. The idler pulley shown in Figure 4-26 has some frictional resistance in its bearing. For the direction of motion shown, the reaction at the shaft support will be

    (a) up
    (b) down
    (c) up and left
    (d) down and right
    (e) up and right

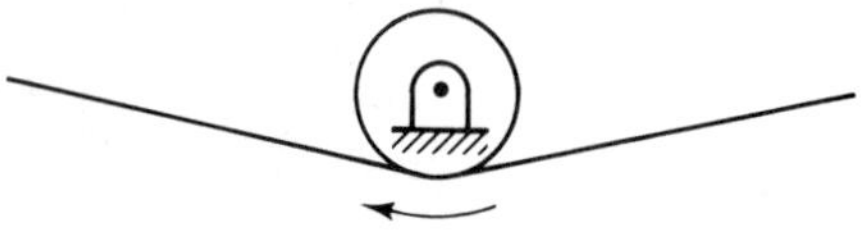

Figure 4-26

13. The weight of the cantilevered beam shown in Figure 4-27 *cannot* be considered negligible. The reaction(s) at A are

    (a) up
    (b) down
    (c) clockwise and up
    (d) clockwise and down
    (e) counterclockwise and up

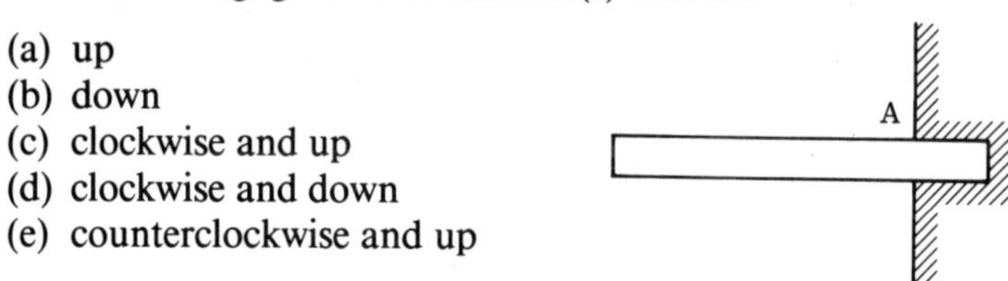

Figure 4-27

14. The relatively weightless frame shown in Figure 4-28 has a counterclockwise moment load applied to it. The reaction at B is

    (a) left
    (b) right
    (c) left and up
    (d) left and down
    (e) right and down

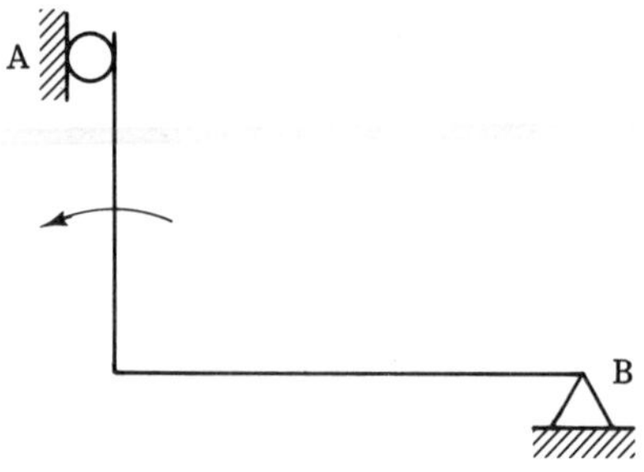

Figure 4-28

15. The reactions at A for the truss shown in Figure 4-29 are

    (a) up and left
    (b) up and right
    (c) down and left
    (d) down and right
    (e) any of the above

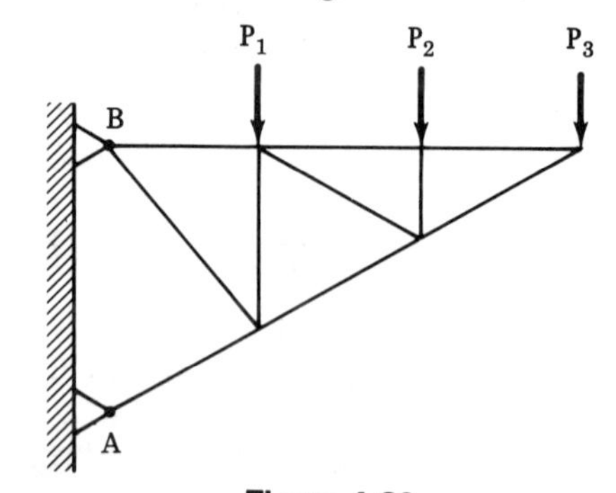

Figure 4-29

4.8 SUPPLEMENTARY PROBLEMS

4-9 Determine the reactions at A for the beam shown in Figure 4-30.

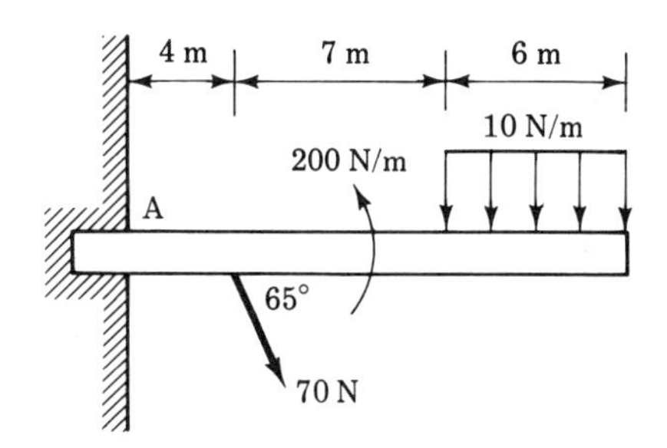

Figure 4-30

4-10 Determine the support reactions for the truss shown in Figure 4-31.

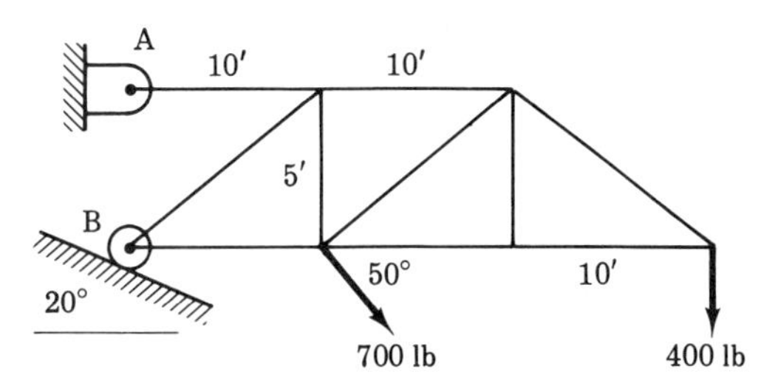

Figure 4-31

4-11 Determine the support reactions for the beam shown in Figure 4-32.

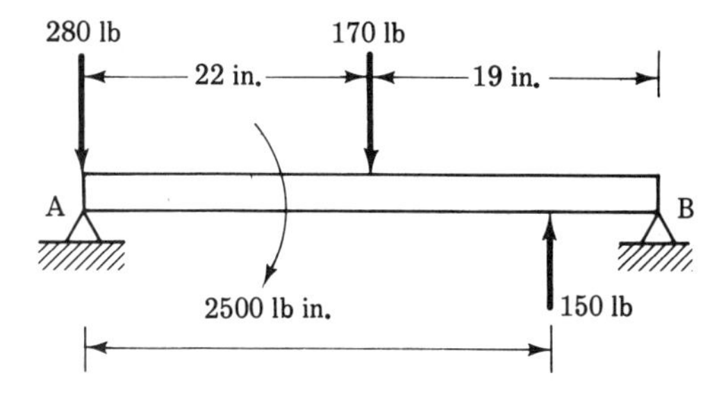

Figure 4-32

4-12 Determine the support reactions for the truss shown in Figure 4-33.

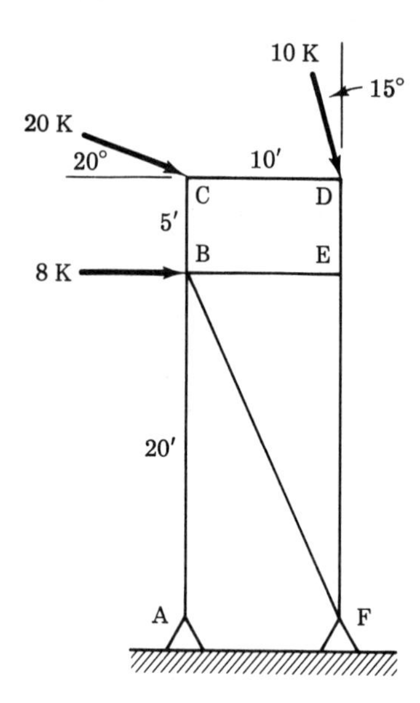

Figure 4-33

4-13 Three pulleys mounted on a single shaft are shown in end view in Figure 4-34. Determine the unknown belt tension and the support reactions at the shaft.

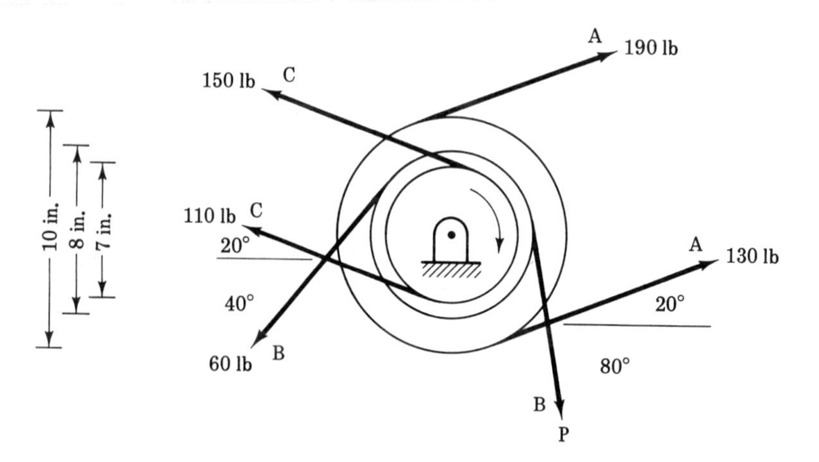

Figure 4-34

4-14 The bell crank shown in Figure 4-35 is in equilibrium. Determine the load P and the support reactions.

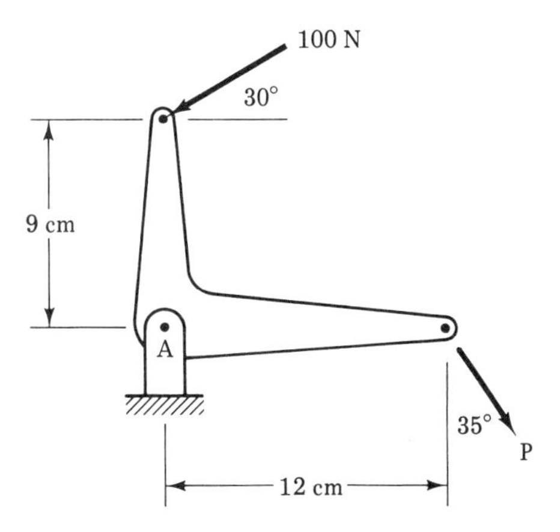

Figure 4-35

4-15 Determine the force required to just start the blocks shown in Figure 4-36 moving.

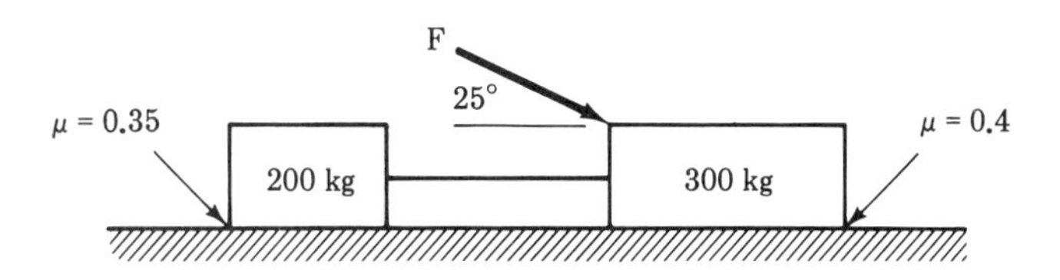

Figure 4-36

4-16 Determine the force required to just start the block shown in Figure 4-37 moving and check to see that it is in equilibrium.

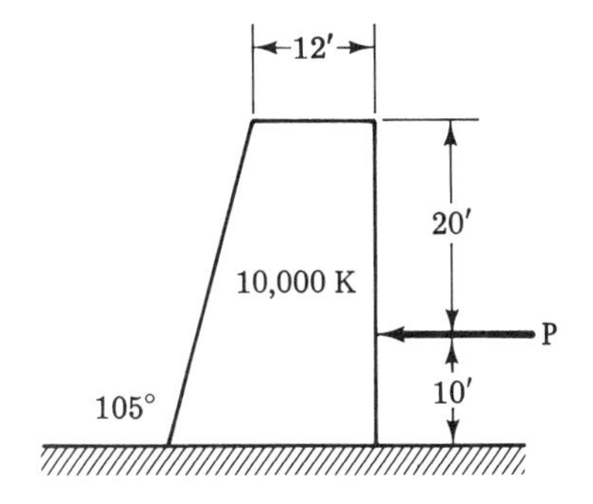

Figure 4-37

4-17 Determine the force required and its point of application to just impart 300 ft-lbs of braking torque on the wheel without slipping.

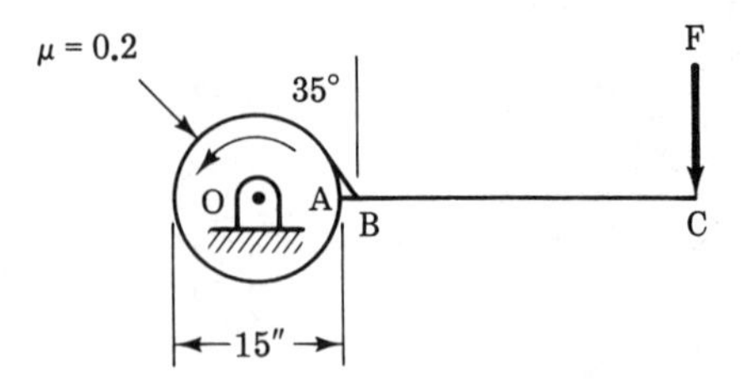

Figure 4-38

Chapter

5

Trusses

5.1 OBJECTIVES

Upon completion of the work relating to this chapter, you should be able to perform the following.

1. Given a list of properties, select those appropriate for a truss.
2. Given a loaded structure, determine whether or not it is a truss.
3. Given a simple truss and the loads acting on it, determine by inspection which members are nonload members, which members are in tension, and which members are in compression.
4. Given a truss and the loads acting on it, determine the force in any member using the method of sections.
5. Given a truss and the loads acting on it, determine the forces in all members using the combined graphical method.
6. Given a truss and the loads acting on it, determine the forces in all members using the method of joints.

5.2 NATURE OF TRUSSES

Simple Trusses Defined

In practice trusses are structures consisting of straight members made of wood or metal joined together in triangular groupings. They are used as

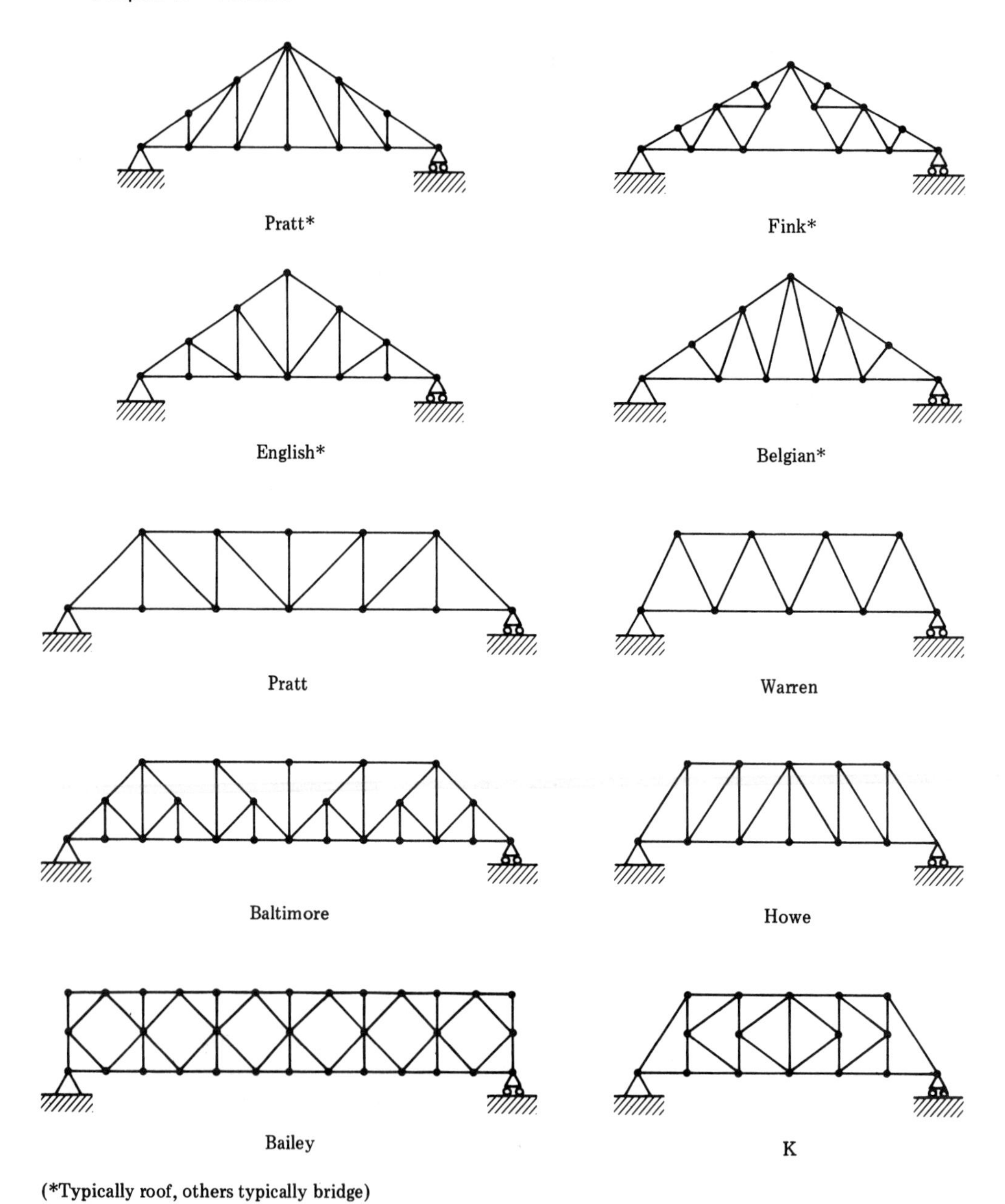

Figure 5-1. Common types of trusses.

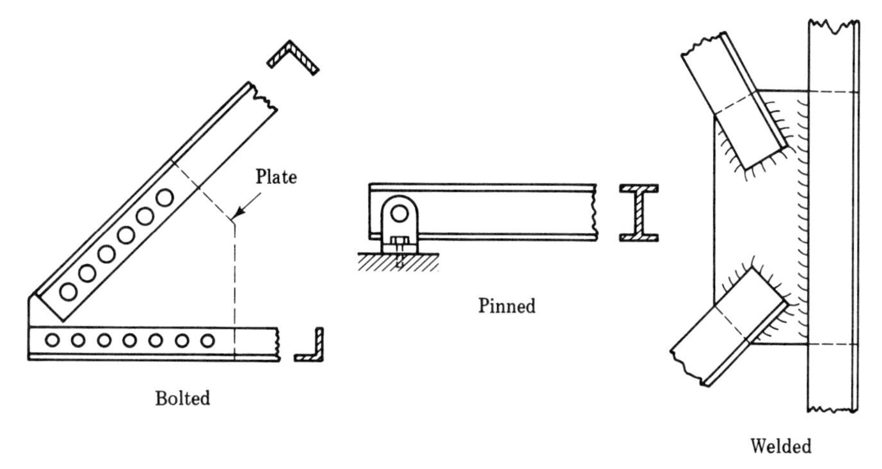

Figure 5-2. Joint connections.

bridges, roof supports, cranes, and a wide variety of structural and machine supports. Some of the more common types are shown in Figure 5-1. The members are generally connected by bolting, pinning, or welding as shown in Figure 5-2. Frequently an additional piece called a gusset is used to help bolt or weld together two members. You may see an occasional new structure, and frequently many older ones, where rivets were used to join members, but this method is rarely used today. Analysis of trusses can be greatly simplified by making a number of assumptions. Keep in mind that these assumptions do not seriously affect the results and corrections can be made later to compensate for them. These assumptions are:

1. All connections are made with frictionless pins.
2. All members are straight and are joined together to form triangles.
3. Each member is fastened only at its two ends.
4. All loads are applied only at connections (or joints as they are often called).
5. The weights of the members can be neglected since the load they put on the truss is minimal when compared to the external loads imposed on the truss.
6. The truss can be considered as a coplanar problem since most complex structures consist of two or more identical coplanar structures mounted in parallel.

Figure 5-3 shows some trusses and nontrusses based on some of the above assumptions.

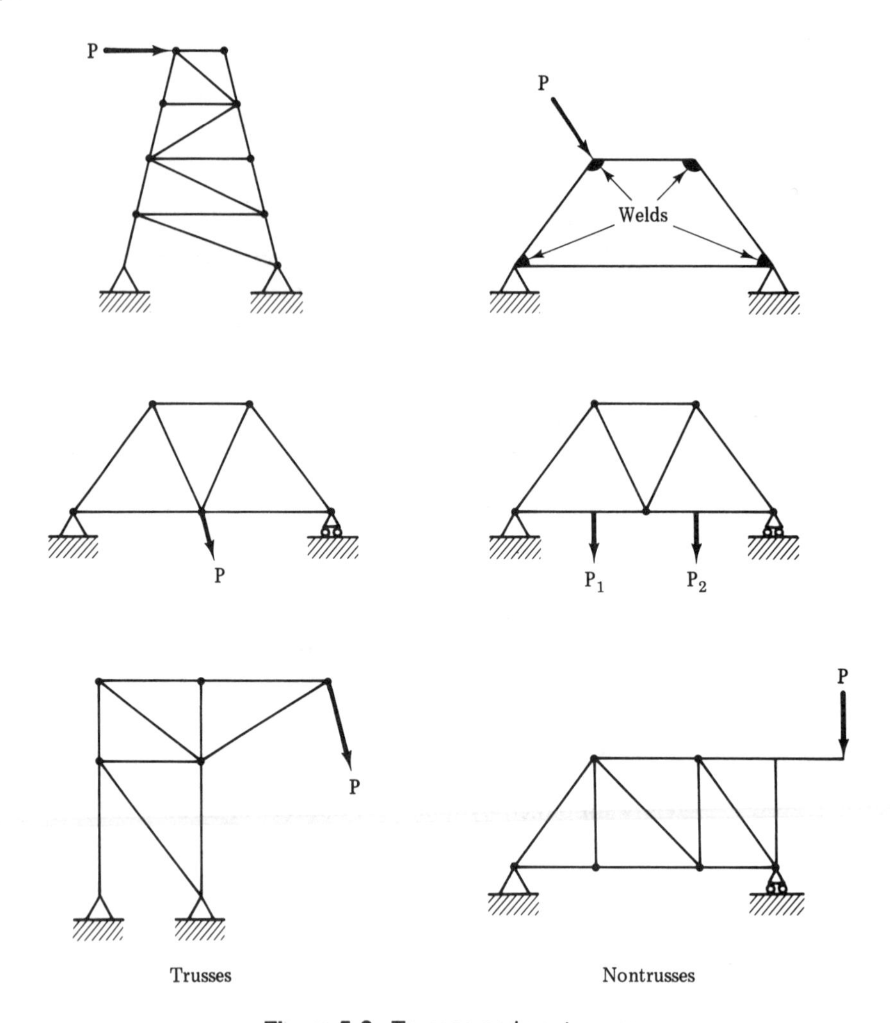

Figure 5-3. Trusses and nontrusses.

Tension Members and Compression Members

One of the effects of the above assumptions is that all members are simple two-force members. They are either in compression or tension. If the pins at a member's connecting points are pushing on the member as shown in Figure 5-4a, the member will be in compression. If the pins are pulling on the member, the member will be in tension as shown in Figure 5-4b. Figure 5-4 in essence shows the free-body diagrams for two members. A useful technique in truss analysis, regardless of the methods used, is to view the truss in terms of free-body diagrams of each joint. Doing this reduces the truss problem to a series of concurrent force problems. If all members of a

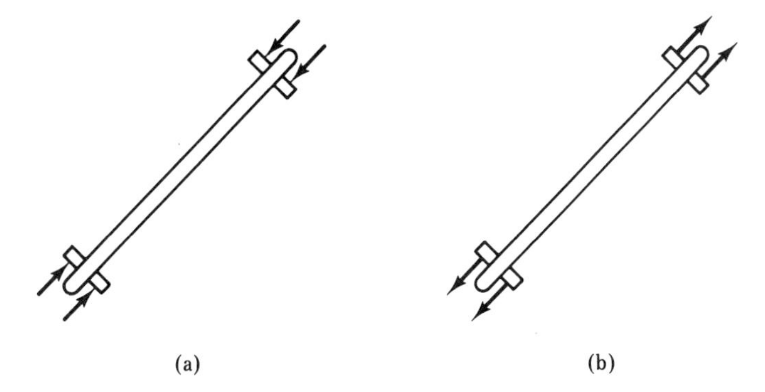

Figure 5-4. Tension and compression.

truss are either in compression or tension, then they are either pushing or pulling on each joint.

Example 5-1: Identify the nature of the force, tension or compression, in each member of the truss shown in Figure 5-5a.

Solution:

Step 1 Draw a free-body diagram of the truss as shown in Figure 5-5b.

Step 2 Recognize that the support reactions and load are such that the truss will tend to sag downward. This will tend to cause members *AB, BD,* and *DE* to be in compression, and members *AC* and *CE* to be in tension.

Step 3 Draw free-body diagrams of each joint applying the principle of concurrence and the method of components. Figure 5-5c shows the free-body diagrams of individual joints. As the following paragraphs show, the nature of any forces not already determined in Step 2 will result from a joint-by-joint examination and free-body construction.

A principal rule in analyzing truss joints is to deal only with those involving two or less unknowns. Starting off alphabetically at joint *A*, member *AB* is in compression. This is because its vertical component is the only force that can counter the vertically upward reaction at *A*. Thus the vertical equilibrium of *A* is maintained.

Horizontally, we cannot be absolutely sure about the force in *AC* although we've assumed it to be in tension. However, knowing that *AB* is in compression, permits working with joint *B*.

Since *AB* is in compression pushing upward on joint *B, BC* must be pulling downward in tension on *B* to achieve the vertical equilibrium of joint *B*. If *AB* is pushing to the right on *B* and *BC* is pulling to the right on *B*, then *BD* must be pushing to the left on *B* to maintain horizontal equilibrium of *B*. Since *BD* is pushing it must be in compression.

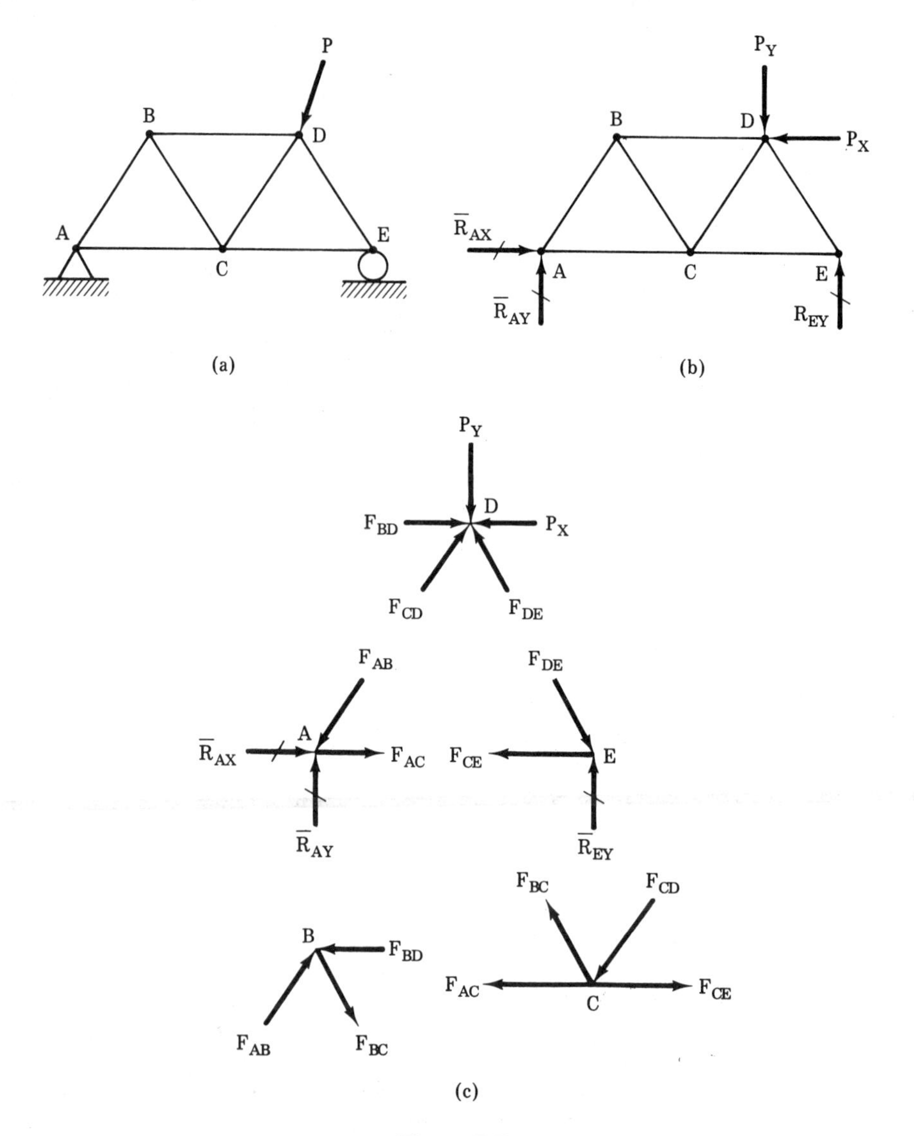

Figure 5-5

Knowing *BC*, we can partially solve at joint *C*, for if *BC* is in tension it is pulling up on joint *C*. Vertical equilibrium of *C* can only be achieved by *CD* pushing down on *C*, indicating that *CD* is in compression.

Moving to joint *E, DE* must be pushing downward in compression to counteract the vertical reaction of *E* and maintain *E* in vertical equilibrium. This means that *DE* is also pushing to the right requiring *CE* to pull to the left in tension to maintain the horizontal equilibrium of joint *E*.

In the process of doing joint *E*, we have also completed all the members connected to joint *D*. The only member still in question is *AC*, and we will have to stay with our original assumption that it is in tension. Occasionally, you may find in your solution that you erred in determining the tension or compression of a member. This is more than offset by almost always knowing the nature of the force in a member before introducing numbers to the problem.

Problem 5-1: Identify the nature of the force, tension or compression, in each member of the truss shown in Figure 5-6.

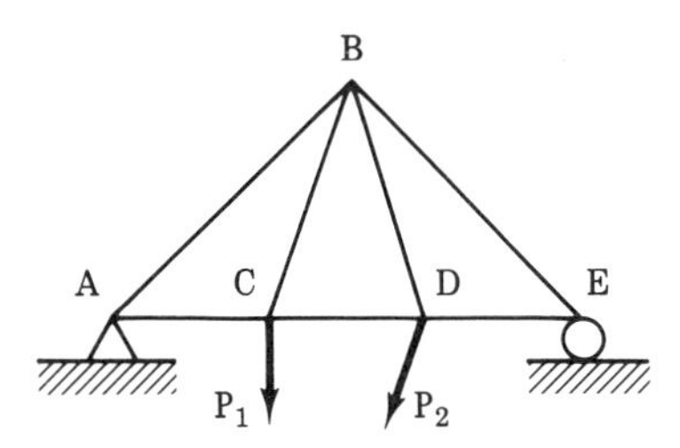

Figure 5-6

Rigidity and Zero-Force Members

Figure 5-7 shows a truss quite similar to the one shown in Figure 5-5. However, this new truss contains five additional members. When the truss is loaded as shown, these added members can be shown to have zero force in each of them and can be neglected for purposes of major-load analysis. Their purpose is to act as stiffeners for the structure. As you will see in a subsequent course, the addition of such stiffeners permits more economical design of the load-carrying members. Identifying zero-force members in a truss before starting analysis makes that analysis much simpler. Such identification requires inspection of each joint. Essentially, when three members are joined at a single joint and two of the members are collinear and there is no external load on the joint, the noncollinear member is a zero-force member. Several joints with zero-force members attached are shown in Figure 5-8. Most important, once a zero-force member has been identified by use of the

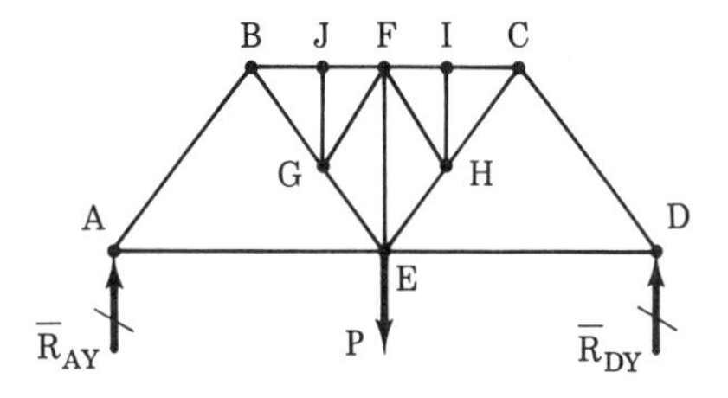

Figure 5-7. Truss with stiffeners.

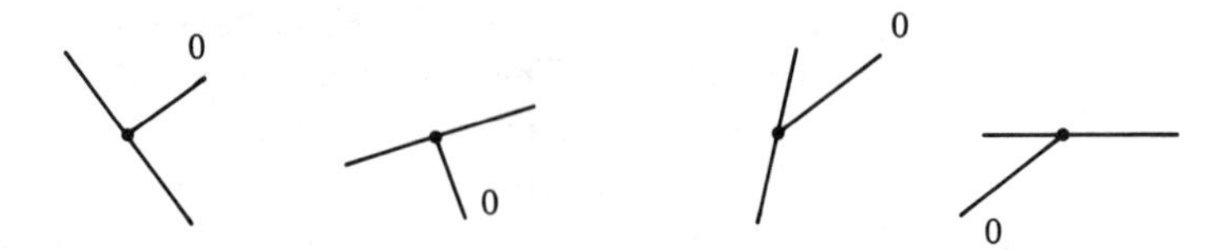

Figure 5-8. Joints with zero-force members.

above approach, either mentally or, better yet, with an eraser in a pencil sketch remove zero-force member. Then recheck the joints that the removed member had been attached to and apply the above technique again. Looking back at the truss in Figure 5-7, this technique would lead us to first remove *GJ* and *HI*, next remove *FG* and *FH*, and finally to remove *FE*. A final check of the truss reveals that it still meets the criteria for a truss as previously defined.

Problem 5-2: Identify all zero-force members in the truss shown in Figure 5-9.

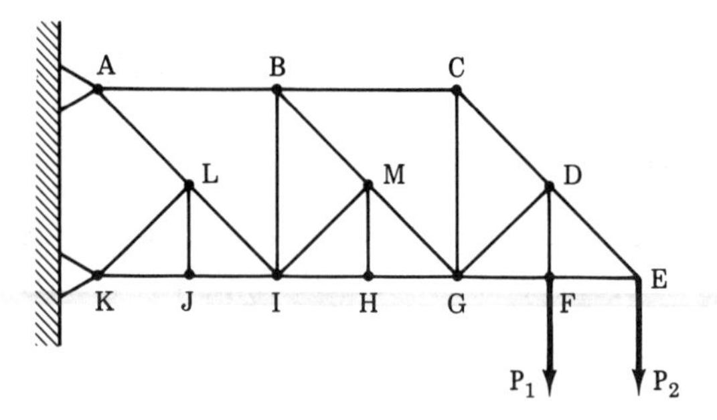

Figure 5-9

Problem 5-3: Identify all zero-force members in the truss shown in Figure 5-10.

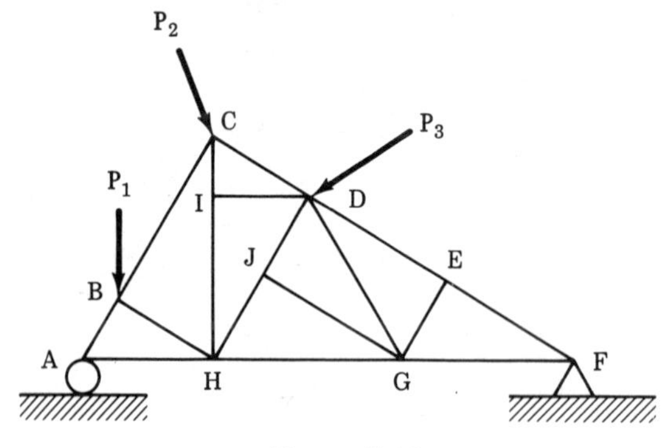

Figure 5-10

5.3 METHOD OF JOINTS: MATHEMATICALLY

One basic method that is used when it is necessary to determine the forces in all members of a truss is the method of joints, mathematically. This method entails the following steps:

1. Draw a free-body diagram of the truss.
2. Determine the magnitude and direction of all reactions.
3. Identify and delete any zero-force members.
4. Draw a free-body diagram and rough sketch of the force diagram for each joint. Sometimes the latter can be helpful in completing the former. There may, of course, be occasions when determination of direction must await specific analysis. Remember that a member force directed towards a joint is pushing on it and, therefore, that member is in compression. Similarly, a member force directed away from a joint is pulling on it and, therefore, that member is in tension. Recall, also, that if a member is pulling on a joint at its one end, it is pulling on the joint at its other end. The same is true in compression. Do not confuse free bodies of the joints with free bodies of the members. We are talking here about what the members are doing to the joints, not what the joints are doing to the members.
5. Each joint now consists of a separate simple concurrent force problem. Select any joint with two or less unknown forces on it as a starting point and proceed throughout the truss. When you reach the last joint, you will find that you have already determined all the forces on it. Do not stop. This is your check. These known forces must be in equilibrium. If they are not, an error exists in your work.

Example 5-2: Determine the forces acting in all members of the truss shown in Figure 5-11a using the method of joints mathematically.

Solution:

Step 1 Recognize this as a truss problem since all members are connected only at their ends, the structure consists of a series of triangles, and all loads and reactions are at joints.

Step 2 By inspection, identify members *BH* and *CE* as zero-force members.

Step 3 Draw the free-body diagram for the truss as shown in Figure 5-11b.

Step 4 Recognizing that there are two unknown vertical reactions and one unknown horizontal reaction, write and solve in the following order: the horizontal force equilibrium equation, the moment equilibrium equation about either point *A* or *C*, and the vertical force equilibrium equations:

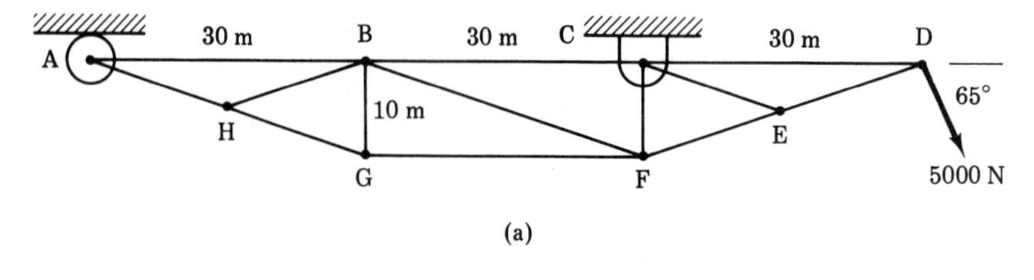

(a)

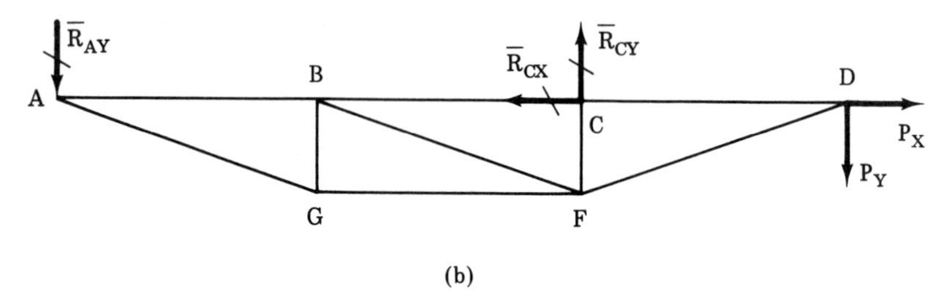

(b)

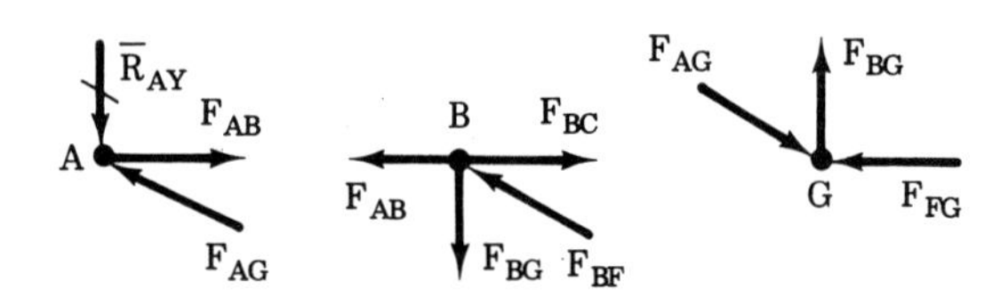

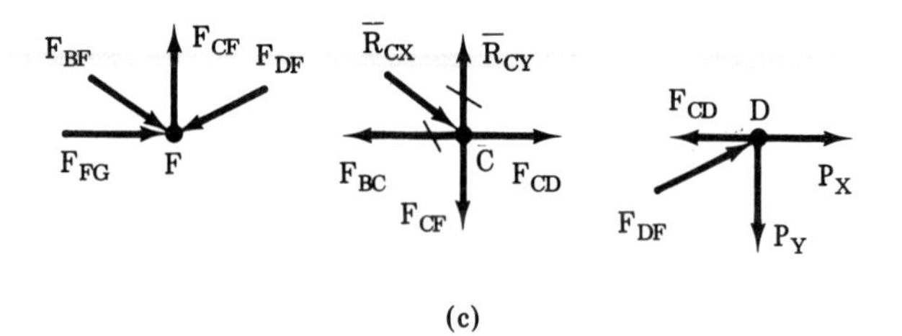

(c)

Figure 5-11

$$\Sigma F_X = 0 \qquad \overline{R}_{CX} + P_X = 0 \qquad \overline{R}_{CX} = -P_X$$

$$\overline{R}_{CX} = -(5000\text{ N})\cos 295°$$

$$\overline{R}_{CX} = -2113\text{ N} \quad \text{or} \quad \overline{R}_{CX} = 2113\text{ N} \quad \text{at} \quad 180°$$

$$\Sigma M_A = 0 \qquad \overline{R}_{CY} d_{CY} + P_Y d_{PY} = 0 \qquad \overline{R}_{CY} d_{CY} = -(P_Y d_{PY})$$

$$P_Y = (5000\text{ N})\sin 295° \qquad P_Y = -4532\text{ N}$$

$$\overline{R}_{CY} d_{CY} = -[-(4532\text{ N})(90\text{ m})] \qquad \overline{R}_{CY} d_{CY} = +407{,}838\text{ N m}\circlearrowleft$$

$$\overline{R}_{CY} = \frac{\overline{R}_{CY} d_{CY}}{d_{CY}} \qquad \overline{R}_{CY} = \frac{407{,}838\text{ N m}}{60\text{ m}} \qquad \overline{R}_{CY} = 6797\text{ N}$$

$$\overline{R}_{CY} = 6797 \text{ N} \quad \text{at} \quad 90^\circ$$

$$\Sigma F_Y = 0 \qquad \overline{R}_{AY} + \overline{R}_{CY} + P_Y = 0 \qquad \overline{R}_{AY} = -(R_{CY} + P_Y)$$

$$\overline{R}_{AY} = -(6797 - 4532 \text{ N}) \qquad \overline{R}_{AY} = -2265 \text{ N}$$

$$\overline{R}_{AY} = 2265 \text{ N} \quad \text{at} \quad 270^\circ$$

Note that many of the rules developed earlier were applied above, particularly those from Chapter 4 regarding appropriate use of force directions, moment directions, and absolute values.

Step 5 Identify all members as being either in tension or compression. Occasionally, it may not be possible to do this completely at this stage, but it will be helpful to identify as many as possible as shown in Figure 5-11b.

Step 6 Draw a free-body diagram of each joint as shown in Figure 5-11c.

Step 7 Treat each joint as a concurrent force problem and solve for the unknown forces in the members attached to that joint. Start with a joint with no more than two unknowns. Be sure to identify each force and indicate if it is in tension or compression. Although not shown below, you may find it useful to make a rough sketch of the force diagram even though you are solving each joint by method of components.

Joint *A*:

$$\Sigma F_Y = 0 \qquad R_{AY} + AG_Y = 0 \qquad AG_Y = -R_{AY} \qquad AG_Y = -(-2265 \text{ N})$$

$$AG_Y = +2265 \text{ N}$$

$$\theta_{AG} = \arctan \frac{+10 \text{ m}}{-30 \text{ m}} \qquad \theta_{AG} = 161.6^\circ$$

$$AG = \frac{AG_Y}{\sin \theta_{AG}} \qquad AG = \frac{2265 \text{ N}}{\sin 161.6^\circ} \qquad \boldsymbol{AG = 7200 \text{ N}(C)}$$

$$\Sigma F_X = 0 \qquad AB_X + AG_X = 0 \qquad AB_X = -(AG_X)$$

$$AG_X = AG \cos \theta_{AG} = (7200 \text{ N})\cos 161.6^\circ$$

$$AG_X = -6832 \text{ N} \qquad AB_X = -(-6832 \text{ N}) \qquad AB_X = +6832 \text{ N}$$

$$AB = AB_X \qquad \boldsymbol{AB = 6800 \text{ N}(T)}$$

Joint *G*:

$$\Sigma F_Y = 0 \qquad BG_Y + AG_Y = 0 \qquad BG_Y = -A_Y$$

$$BG_Y = -(-2265 \text{ N}) \qquad BG_Y = +2265 \text{ N}$$

$$BG = BG_Y \qquad \boldsymbol{BG = 2300 \text{ N}(T)}$$

$$\Sigma F_X = 0 \qquad AG_X + GF_X = 0 \qquad GF_X = -AG_X$$

$$GF_X = -(6832\text{ N}) \qquad GF_X = -6832\text{ N}$$

$$GF = GF_X \qquad \boldsymbol{GF = 6800\text{ N}(C)}$$

Joint *B*:

$$\Sigma F_Y = 0 \qquad BG_Y + BF_Y = 0 \qquad BF_Y = -BG_Y$$

$$BF_Y = -(-2265\text{ N}) \qquad BF_Y = +2265\text{ N}$$

$$\theta_{BF} = \theta_{AG} = 161.6°$$

$$BF = \frac{BF_Y}{\sin\theta_{BF}} \qquad BF = \frac{2265\text{ N}}{\sin 161.6°}$$

$$BF = +7176\text{ N} \qquad \boldsymbol{BF = 7200\text{ N}(C)}$$

$$\Sigma F_X = 0 \qquad AB_X + BC_X + BF_X = 0 \qquad BC_X = -(AB_X + BF_X)$$

$$BF_X = BF\cos\theta_{BF} \qquad BF_X = 7176\cos 161.6° \qquad BF_X = -6809\text{ N}$$

$$BC_X = -(-6832 - 6809\text{ N}) \qquad BC_X = +13{,}641\text{ N}$$

$$BC = BC_X \qquad \boldsymbol{BC = 13{,}600\text{ N}(T)}$$

Joint *C*:

$$\Sigma F_Y = 0 \qquad \overline{R}_{CY} + CF_Y = 0 \qquad CF_Y = -\overline{R}_{CY}$$

$$CF_Y = -(+6797\text{ N}) \qquad CF_Y = -6797\text{ N}$$

$$CF = CF_Y \qquad \boldsymbol{CF = 6800\text{ N}(T)}$$

$$\Sigma F_X = 0 \qquad BC_X + \overline{R}_{CX} + CD_X = 0 \qquad CD_X = -(BC_X + \overline{R}_{CX})$$

$$CD_X = -(-13{,}641 - 2113\text{ N}) \qquad CD_X = +15{,}754\text{ N}$$

$$CD = CD_X \qquad \boldsymbol{CD = 15{,}800\text{ N}(T)}$$

Joint *F*:

$$\Sigma F_Y = 0 \qquad BF_Y + CF_Y + DF_Y = 0 \qquad DF_Y = -(BF_Y + CF_Y)$$

$$DF_Y = -(-2265 + 6797\text{ N}) \qquad DF_Y = -4532\text{ N}$$

$$DF = \frac{DF_Y}{\sin\theta_{DF}} \qquad \theta_{DF} = \arctan\frac{-10\text{ m}}{-30\text{ m}} \qquad \theta_{DF} = 198.4°$$

$$DF = \frac{-4532\text{ N}}{\sin 198.4°} \qquad DF = 14{,}358\text{ N} \qquad \boldsymbol{DF = 14{,}400\text{ N}(C)}$$

Step 8 Although there is another joint left to work with, the forces in all members have been found. This is a result of every member being attached to two joints. However, the last joint serves a very useful purpose. Since all the forces acting on it are known, they must add up to zero if the above solutions are correct. The last joint is our check!

Joint *D*:

$$\Sigma F_Y = 0 \qquad DF_Y + P_Y = 0 \qquad 4532 - 4532\ \text{N} = 0$$

$$\theta_{DF} = \arctan \frac{10\ \text{m}}{30\ \text{m}} \qquad \theta_{DF} = 18.4^\circ$$

$$DF_X = DF \cos \theta_{DF} \qquad DF_X = (14{,}358\ \text{N}) \cos 18.4^\circ$$

$$DF_X = +13{,}624\ \text{N}$$

$$\Sigma F_X = 0 \qquad CD_X + DF_X + P_X = 0$$

$$-15{,}745 + 13{,}624 + 2113\ \text{N} = -17\ \text{N} \cong 0$$

Given the magnitudes of the forces involved, 17 N is a negligible error. Note that throughout the problem, values were carried through with 4 and 5 significant figures. Only the answers were rounded appropriately.

Example 5-3: Determine the forces acting in all members of the truss shown in Figure 5-12a using the method of joints mathematically.

Solution:

Step 1 Recognize this as a truss problem.

Step 2 By inspection, identify *BJ, BH, BF,* and *BD* as zero-force members.

Step 3 Draw the free-body diagram for the truss as shown in Figure 5-12b.

Step 4 Recognize that there are two unknown vertical reactions and two unknown horizontal reactions. Recognize also that the direction of the total reaction at *I* is known (perpendicular to the surface that the roller is on and against the roller). Write and solve in the following order: the moment equilibrium equation about point *A*, the horizontal force equilibrium equation, the vertical force equilibrium equation:

$$\Sigma M_A = 0 \qquad P_1 d_1 + P_2 d_2 + \overline{R}_I d_I = 0$$

d_I is not immediately apparent, but inspection of the given information indicates that the truss is made up of four congruent isosceles triangles with a 45° angle at their apexes and, therefore, 67.5° angles at their bases. Refer to Figure 5-12c for the following solution for d_I. $\overline{R}_I$ is directed at 60°, therefore,

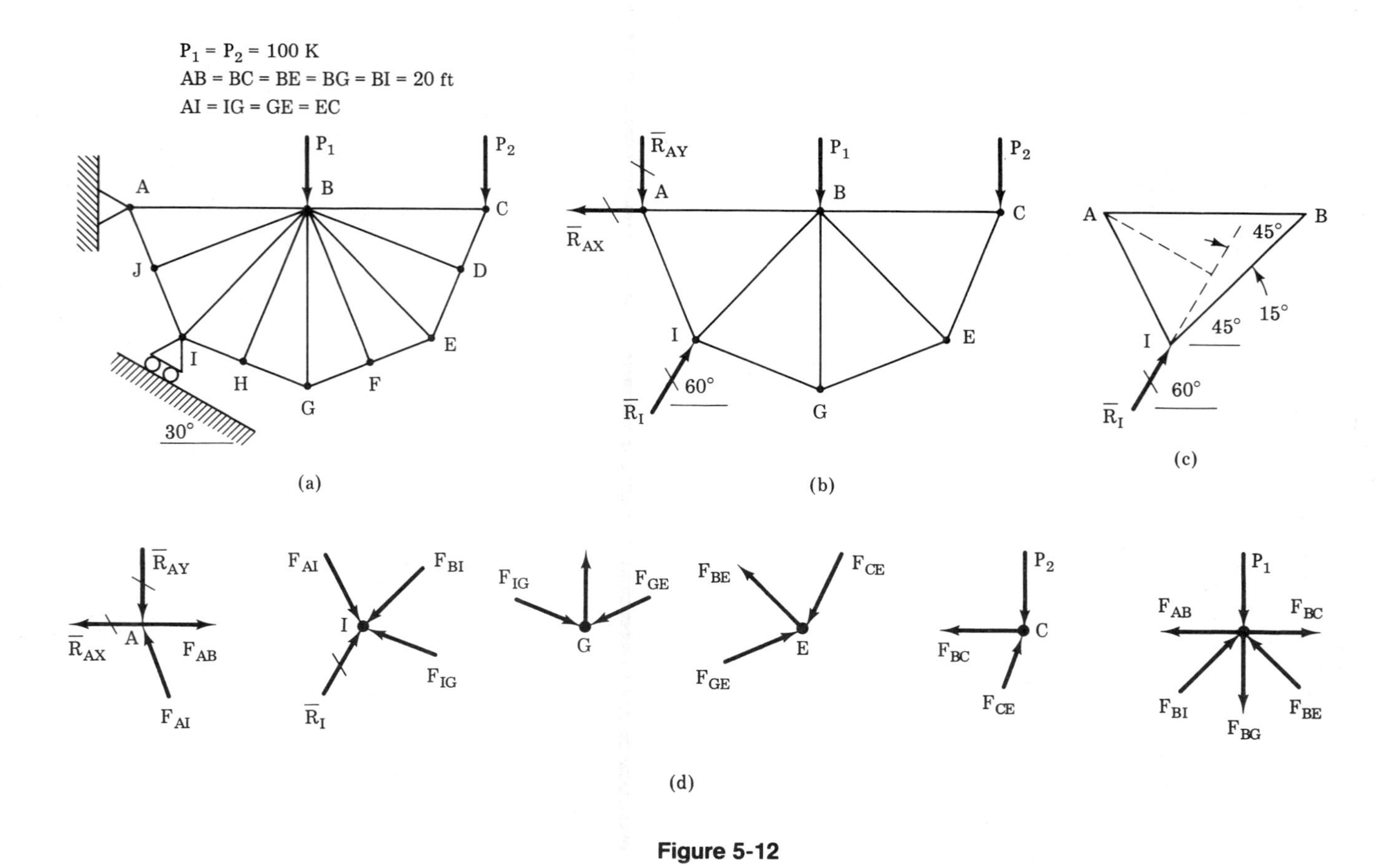

P1 = P2 = 100 K
AB = BC = BE = BG = BI = 20 ft
AI = IG = GE = EC
P1
P2
A
B
C
D
E
F
G
H
I
J
30°
(a)
RAY
RAX
60°
RI
(b)
45°
15°
60°
(c)
FAB
FAI
FBI
FIG
FGE
FBE
FCE
FBC
FBG
(d)

Figure 5-12

the angle between $\overline{R}_I$ and IB is 15°. The angle between $\overline{R}_I$ and AI is then 67.5° − 15° or 52.5°.

$$d_I = AI \sin 52.5°$$

Since $\frac{1}{2}AI$ and AB form a right triangle with a perpendicular drawn from B to AI

$$\tfrac{1}{2}AI = BI \cos 67.5° = (20 \text{ ft})\cos 67.5°$$

$$AI = 15.3 \text{ ft} \qquad d_I = 12.14 \text{ ft}$$

$$\overline{R}_I d_I = -(P_1 d_1 + P_2 d_2)$$

$$\overline{R}_I d_I = -[-(100 \text{ K})(20 \text{ ft}) - (100 \text{ K})(40 \text{ ft})]$$

$$\overline{R}_I d_I = +6000 \text{ K ft}\circlearrowleft$$

$$\overline{R}_I = \frac{\overline{R}_I d_I}{d_I} \qquad \overline{R}_I = \frac{6000 \text{ K ft}}{12.14 \text{ ft}} \qquad \overline{R}_I = 494 \text{ K}$$

In order to provide counterclockwise moment about A as indicated by $R_I d_I$ above, $\overline{R}_I$ must be directed at 60° (rather than 240°, the only other choice).

$$\overline{R}_{IX} = \overline{R}_I \cos \theta_{\overline{R}I} \qquad \overline{R}_{IX} = (494 \text{ K})\cos 60° \qquad \overline{R}_{IX} = +247 \text{ K}$$

$$\overline{R}_{IY} = \overline{R}_I \sin \theta_{\overline{R}I} \qquad \overline{R}_{IY} = +427.8 \text{ K}$$

$$\Sigma F_X = 0 \qquad \overline{R}_{AX} + \overline{R}_{IX} = 0 \qquad \overline{R}_{AX} = -\overline{R}_{IX}$$

$$\overline{R}_{AX} = -(+247 \text{ K}) \qquad \overline{R}_{AX} = -247 \text{ K}$$

$$\Sigma F_Y = 0 \qquad \overline{R}_{AY} + \overline{R}_{IY} + P_1 + P_2 = 0 \qquad \overline{R}_{AY} = -(\overline{R}_{IY} + P_1 + P_2)$$

$$\overline{R}_{AY} = -(427.8 - 100 - 100 \text{ K}) \qquad \overline{R}_{AY} = -227.8 \text{ K}$$

Step 5 Identify all members as being either in compression or tension. In this case, inspection suggests that *AI, IG, GE,* and *EC* would all be in compression while *AB* and *BC* are in tension. Inspection of joint G then suggests that *BG* is in tension. Following that with joint *E* suggests that *BE* is also in tension. Finally, examining joint *B* then suggests that *BI* is in compression. This may not all be readily apparent until you make the free-body diagram of each joint as indicated in Step 6.

Step 6 Draw a free-body diagram of each joint as shown in Figure 5-12d.

Step 7 Solve joint by joint as follows:

Joint A:

$$\Sigma F_Y = 0 \qquad \overline{R}_{AY} + AI_Y = 0 \qquad AI_Y = -\overline{R}_{AY} = -(-227.8\text{ K})$$

$$AI_Y = +227.8\text{ K} \qquad AI = \frac{AI_Y}{\sin\theta_{AI}} \qquad AI = \frac{227.8\text{ K}}{\sin\theta_{AI}} \qquad AI = \frac{227.8\text{ K}}{\sin 112.5°}$$

$$AI = 246.6\text{ K} \qquad \boldsymbol{AI = 250\text{ K}(C)}$$

$$\Sigma F_X = 0 \qquad \overline{R}_{AX} + AI_X + AB_X = 0 \qquad AB_X = -(\overline{R}_{AX} + AI_X)$$

$$AI_X = AI\cos\theta_{AI} \qquad AI_X = 246.6\text{ K}(\cos 112.5°)$$

$$AI_X = -94.37\text{ K} \qquad AB_X = -(-247 - 94.37\text{ K})$$

$$AB_X = +341.37\text{ K},\ AB = AB_X \qquad \boldsymbol{AB = 340\text{ K}(T)}$$

Joint I:

$$\Sigma F_Y = 0 \qquad AI_Y + \overline{R}_{IY} + IG_Y + BI_Y = 0$$

contains two unknowns.

$$\Sigma F_X = 0 \qquad AI_X + \overline{R}_{IX} + IG_X + BI_X = 0$$

also contains two unknowns, but if we write the unknowns as a function of their respective angles we find we have two equations for the same two unknowns.

$$\Sigma F_Y = -227.8 + 427.8\text{ K} + IG\sin\theta_{IG} + BI\sin\theta_{BI} = 0$$

$$\Sigma F_X = +94.37 + 247\text{ K} + IG\cos\theta_{IG} + BI\cos\theta_{BI} = 0$$

$$\Sigma F_Y = 200 + 0.3827\ IG - 0.7071\ BI = 0$$

$$\Sigma F_X = 341.37 - 0.9239\ IG - 0.7071\ BI = 0$$

Changing signs in the second equation and adding yields

$$-141.37 + 1.3066\ IG = 0 \qquad IG = 108.19 \qquad \boldsymbol{IG = 100\text{ K}(C)}$$

Substituting in ΣF_X yields

$$341.37 - 0.9239(108.19) - 0.7071\ BI = 0$$

$$BI = 341.41\text{ K} \qquad \boldsymbol{BI = 340\text{ K}(C)}$$

Joint *G*:

$$\Sigma F_X = 0 \qquad IG_X + GE_X = 0 \qquad GE_X = -IG_X$$

$$IG_X = IG \cos \theta_{IG} = +9239(108.19)$$

$$IG_X = +100 \text{ K} \qquad GE_X = -100 \text{ K} \qquad \theta_{GE} = 202.5°$$

$$GE = \frac{GE_X}{\cos \theta_{GE}} \qquad GE = \frac{-100 \text{ K}}{\cos 202.5°} \qquad GE = 108.2 \text{ K}$$

$$\boldsymbol{GE = 100 \text{ K}(C)}$$

$$\Sigma F_Y = 0 \qquad IG_Y + GE_Y + BG_Y = 0 \qquad BG_Y = -(IG_Y + GE_Y)$$

$$GE_Y = GE \sin \theta_{GE} \qquad GE_Y = (108.2 \text{ K}) \sin 202.5°$$

$$GE_Y = -41.4 \text{ K} \qquad IG_Y = IG \sin \theta_{IG}$$

$$IG_Y = 108.2 \sin 337.5° \qquad IG_Y = -41.4 \text{ K}$$

$$BG_Y = -(-41.4 - 41.4 \text{ K}) \qquad BG_Y = -82.8 \text{ K}$$

$$BG = BG_Y, \qquad \boldsymbol{BG = 80 \text{ K}(T)}$$

Joint *E*:

$$\Sigma F_Y = 0 \qquad GE_Y + BE_Y + EC_Y = 0$$

$$\Sigma F_X = 0 \qquad GE_X + BE_X + EC_X = 0$$

$$\Sigma F_Y = GE \sin \theta_{GE} + BE \sin \theta_{BE} + EC \sin \theta_{EC} = 0$$

$$\Sigma F_X = GE \cos \theta_{GE} + BE \cos \theta_{BE} + EC \sin 247.5° = 0$$

$$\Sigma F_X = +100 \text{ K} + BE \cos 135° + EC \cos 247.5° = 0$$

$$\Sigma F_Y = +41.4 \text{ K} + 0.707\ BE - 0.9239\ EC = 0$$

$$\Sigma F_X = +100 \text{ K} - 0.707\ BE - 0.3827\ EC = 0$$

Adding the two equations yields

$$141.4 \text{ K} - 1.3066\ EC = 0 \qquad EC = 108.2 \text{ K} \qquad \boldsymbol{EC = 110 \text{ K}(C)}$$

Substitutions in ΣF_X yields

$$100 \text{ K} - 0.707\ BE - 0.3827(108.2 \text{ K}) = 0$$

$$BE = 82.9 \text{ K} \qquad \boldsymbol{BE = 80 \text{ K}(T)}$$

Joint *C*:

$$F_X = 0 \qquad BC_X + EC_X = 0 \qquad BC_X = -EC_X$$

$$BC_X = -(EC \cos \theta_{EC}) \qquad BC_X = -108.2 \cos 67.5°$$

$$BC_X = -41.4 \text{ K} \qquad BC = BC_X \qquad \mathbf{BC = 40\ K(\mathit{T})}$$

Step 8 Check at joint *B*

$$F_Y = 0 \qquad P_1 + BI_Y + BG + BE_Y = 0$$

$$(100 + 341.4 \text{ K})\sin 45° - 82.8 \text{ K} + 82.9 \sin 315° = 0$$

$$0 = 0$$

$$F_X = 0 \qquad AB + BI_X + BE_X + BC = 0$$

$$(-341.4 + 341.4 \text{ K})\cos 45° + (82.9 \text{ K})\cos 315° + 41.4 \text{ K} = 0$$

$$0 = 0$$

Problem 5-4: Determine the forces acting in all members of the truss shown in Figure 5-13 using the method of joints mathematically.

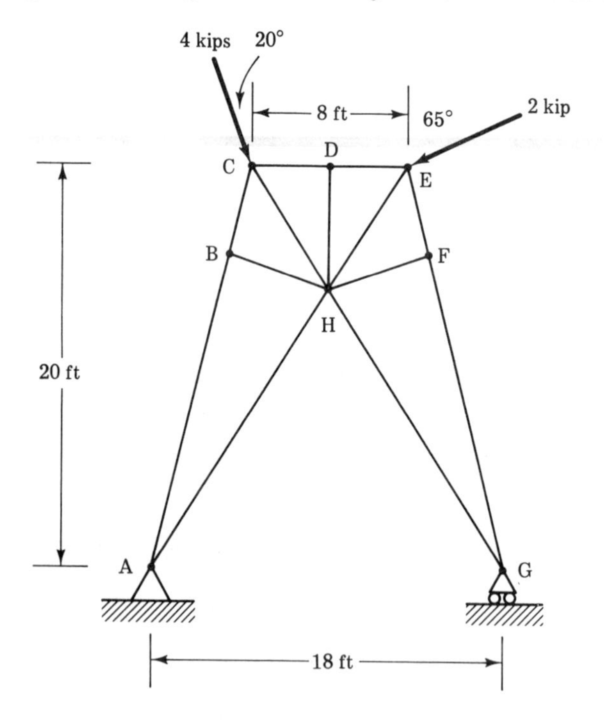

Figure 5-13

Problem 5-5: Determine the forces acting in all members of the truss shown in Figure 5-14 using the method of joints mathematically.

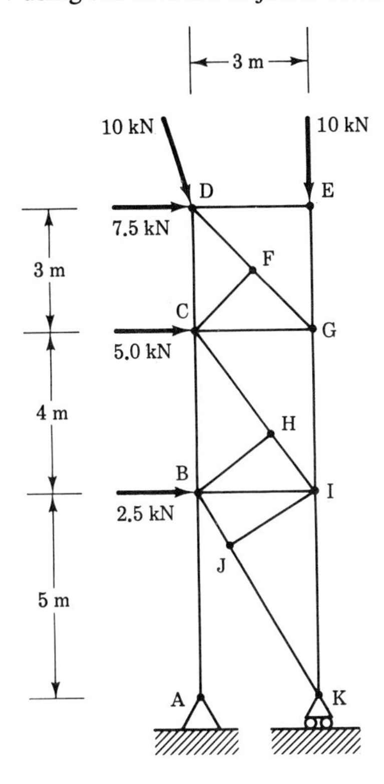

Figure 5-14

5.4 METHOD OF JOINTS: GRAPHICALLY

In Examples 5-2 and 5-3, the forces acting on each joint could have been found graphically instead of mathematically. This approach was examined in detail in Chapter 2 and utilized again in Chapter 4. However, we will now take it a step further to what is known as the combined graphical method. Instead of a separate force diagram for each joint, the diagrams will be, as the name of the method implies, combined. Instead of drawing the force in a member twice, once for each joint it is connected to, it will only be drawn once. This reduces the total work to be done and reduces the chance for error. Indeed, as you will see in the following examples, this method yields an automatic proof without any additional work. When the last unknown force in a member is found, one joint remains to be analyzed, just as it did when we worked mathematically. However, since we are working graphically and all forces have been found, when we get to the last joint, a force polygon indicating equilibrium for that joint should already exist.

Utilization of the combined graphical method requires learning one new technique. It is called *Bow's notation.* It involves numbering every space between external loads, and every internal space (triangle) of the truss. Every force then will be represented by a two-digit number.

The following examples will help to clarify the above procedure. Example 5-4 includes a detailed step-by-step examination of the process. Example 5-5 illustrates the way the process is usually applied.

Example 5-4: Determine the forces acting in all members of the truss shown in Figure 5-15a using the combined graphical method.

Solution:

Step 1 Recognize this as a truss problem.

Step 2 By inspection, identify *BJ, DI, EH, AG,* and *DC* as zero-force members. *DC* may not be immediately apparent, but removal of *DI* and subsequent examination of joint *D* will indicate *DC* to be a zero-force member.

Step 3 Draw the free-body diagram for the truss as shown in Figure 5-15b. Recognize that *AB* and *DE* are not *technically* a part of the truss but extensions of a reaction and a load respectively. *DE* is, therefore, in compression with a force of 3 K. *AB* is in tension with a force equal to $\overline{R}_{AY}$.

$$DE = 3\ \boldsymbol{K(c)}$$

Step 4 Recognizing that there are two unknown vertical reactions and one unknown horizontal reaction, write and solve in the following order: the horizontal force equilibrium equation about either point *A* or *F*, and the vertical force equilibrium equation.

$$\Sigma F_X = 0 \qquad P_1 + P_2 + \overline{R}_{FX} = 0 \qquad \overline{R}_{FX} = -(P_1 + P_2)$$

$$\overline{R}_{FX} = -(+4 + 2\ \text{K}) \qquad \overline{R}_{FX} = -6\ \text{K}$$

$$\Sigma M_F = 0 \qquad P_1 d_{P1} + P_2 d_{P2} + \overline{R}_{AY} d_{AY} = 0$$

$$\overline{R}_{AY} d_{AY} = -(P_1 d_{P1} + P_2 d_{P2})$$

$$\overline{R}_{AY} d_{AY} = -[-(4\ \text{K})(20\ \text{ft}) - (2\ \text{K})(25\ \text{ft})]$$

$$\overline{R}_{AY} d_{AY} = +130\ \text{K ft}\circlearrowleft \qquad \overline{R}_{AY} = \frac{\overline{R}_{AY} d_{AY}}{d_{AY}}$$

$$\overline{R}_{AY} = \frac{130\ \text{K ft}}{10\ \text{ft}} \qquad \overline{R}_{AY} = 13\ \text{K} \quad \text{at} \quad 270^\circ \quad \text{or} \quad -13\ \text{K}$$

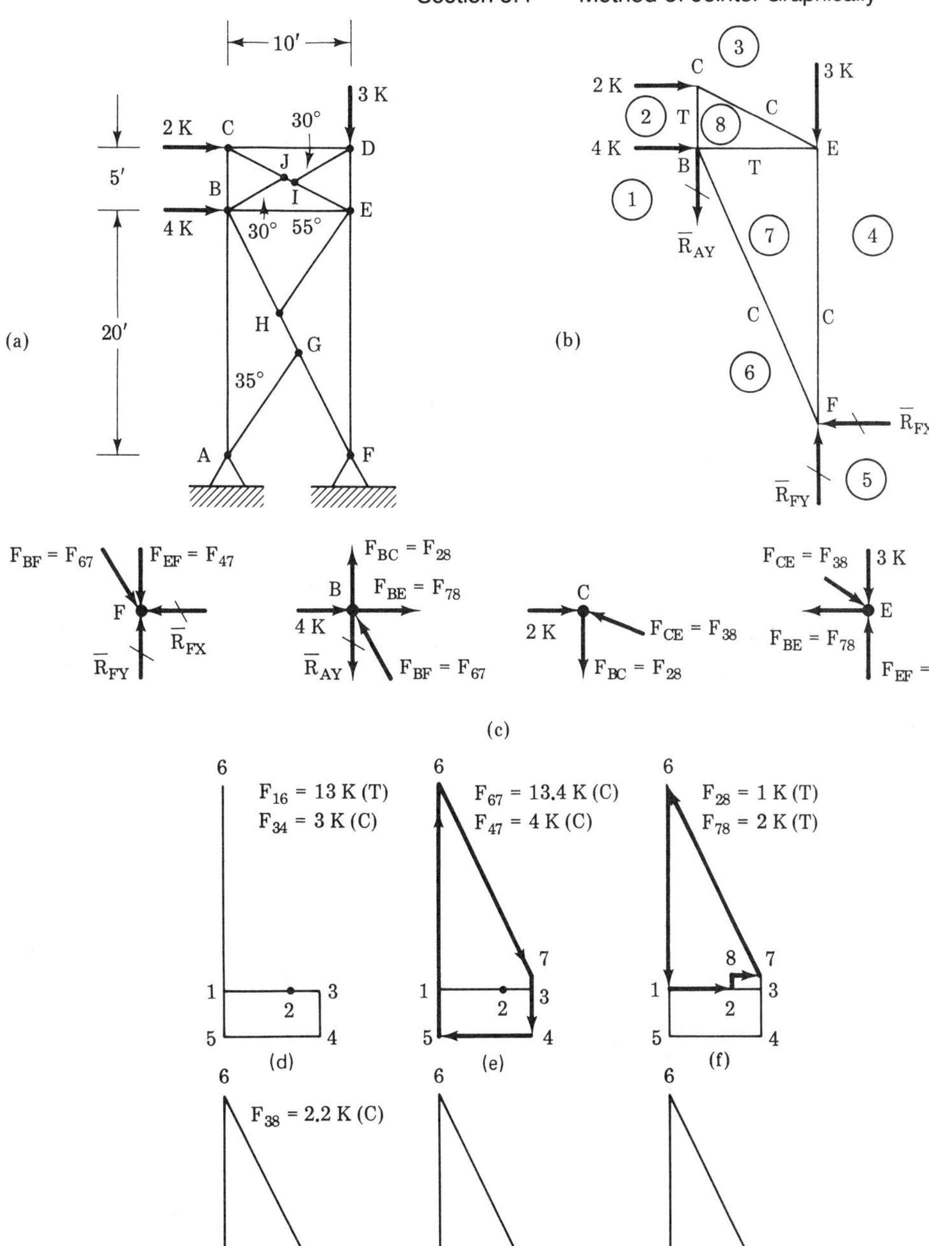

10′
3 K
2 K
C
30°
D
5′
J
B
I
E
4 K
30°
55°
(a)
20′
H
G
35°
A
F
(b)
3
C
2 K
3 K
2
T
8
C
4 K
E
B
T
1
R̄AY
7
4
C
C
6
F
R̄FX
R̄FY
5
FBF = F67
FEF = F47
F
R̄FX
R̄FY
FBC = F28
B
FBE = F78
4 K
R̄AY
FBF = F67
C
2 K
FCE = F38
FBC = F28
FCE = F38
3 K
E
FBE = F78
FEF = F47
(c)
F16 = 13 K (T)
F34 = 3 K (C)
(d)
F67 = 13.4 K (C)
F47 = 4 K (C)
(e)
F28 = 1 K (T)
F78 = 2 K (T)
(f)
F38 = 2.2 K (C)
(g)
(Check)
(h)
(Actual)
(i)

Figure 5-15

($\overline{R}_{AY}$ must be $-$ to provide the counterclockwise moment about F required by $\overline{R}_{AY} d_{AY}$).

$$AB = \overline{R}_{AY} \qquad \boldsymbol{AB = 13\ \mathrm{K}(T)}$$

$$F_Y = 0 \qquad \overline{R}_{AY} + P_3 + \overline{R}_{FY} = -(\overline{R}_{AY} + P_3)$$

$$\overline{R}_{FY} = -(-13 - 3\ \mathrm{K}) \qquad \overline{R}_{FY} = +16\ \mathrm{K}$$

Step 5 Identify all members as being either in tension or compression as shown in Figure 5-15b.

Step 6 Number the spaces between external forces and the triangular spaces formed by the truss members as shown in Figure 5-15b. Sometimes letters are used, but this is confusing if letters have already been used to label the joints.

Step 7 Draw a free-body diagram of each joint as shown in Figure 5-15c. As you become more experienced, you may be able to delete this step. At this time, however, you may wish to, in addition to the free-body diagram, draw a rough sketch of the force diagram for each joint.

Step 8 Examining Figure 5-15b, we can see that P_1 can now also be called force 12, P_2 force 23, and so on, around the outside of the truss. This becomes the basis for our graphical construction and solution. The first step is to draw to some scale the force diagram for the external forces acting on the truss. As we go from space 1 to space 2 on the free-body diagram, we encounter a horizontal force of +4 K. On the force diagram, we establish a point 1 and then, to scale, draw a horizontal force of +4 K as shown in Figure 5-15d. The end point of that line is labeled 2. We continue, discovering a +2 K horizontal force between spaces 2 and 3 on our free-body diagram and then drawing it in, to scale, on our force diagram. This process is continued until all the external forces are accounted for, in order, and the force diagram closes back on 1, as it should for a truss in equilibrium. The completed force diagram is shown in Figure 5-15d. *Normally, we would continue our graphical work on the same diagram.* In this example, only for the purpose of illustrating each joint solution, a series of diagrams will be made.

Step 9 Solve each joint graphically. The following solutions are illustrated in Figure 5-15e–h. Note that arrowheads are not normally used since directions change as one moves from joint to joint. Figure 5-15i contains all of the information shown in 5-15d–h and is all that would normally be drawn after drawing the free-body diagram of the truss.

Joint F: This joint has members 47 and 67 acting on it in addition to reactions 45 and 56. The reactions already exist on our force diagram from

Step 8. As shown in e, the line of action for force 4–7 is drawn in through point 4 (which is already on the diagram) at the same angle, to scale, as member 47. We do not know its other limit (point 7) at this time. Then the line of action for force 67 is drawn in, at the correct angle, through point 6. If 47 and 67 do not cross, we extend them until they do. The intersection of 47 and 67 is point 7 and the two previously unknown forces can be scaled from the force diagram.

Joint B: This joint has members 28 and 78 acting on it as well as reactions 61 and 12 and member 67 whose force was just found. The three known forces are already on our diagram. If we draw in 28 at the correct angle through point 2 and 78 at the correct angle through point 7, as shown in Figure 5-15f, point 8 will be determined. This permits scaling the force diagram for the values of 28 and 78.

Joint C: Only one unknown remains here, 38. Since points 3 and 8 already exist on the force diagram, it is only necessary to connect them as shown in g. As a check, use a protractor to see if the angle of 38 is correct. Scale the value of 38.

Step 10 Check the free-body diagram for joint *E*. Then examine the force diagram as shown in 5-15h to see if a closed force diagram exists for the forces 38, 34, 47, and 78. Since it does, all solutions are correct.

Example 5-5: Determine the forces acting in all members of the truss shown in Figure 5-16a using the combined graphical method.

Solution:

Step 1 Recognize this as a truss problem.

Step 2 By inspection determine that there are no zero-force members.

Step 3 Draw the free-body diagram for the truss as shown in Figure 5-16b. Although we will use components of P_1 in Step 4, it has been drawn in unresolved to avoid a minor confusion in subsequent steps.

Step 4 Recognizing that there are two unknown vertical reactions and one unknown horizontal reaction, write and solve in the following order: the horizontal force equilibrium equation, the moment equilibrium equation about either point *A* or *D*, and the vertical force equilibrium equation.

$$\Sigma F_X = 0 \qquad P_{1X} + \overline{R}_{AX} = 0 \qquad \overline{R}_{AX} = -P_{1X}$$

$$P_{1X} = P \cos \theta_{P1} \qquad P_{1X} = (40 \text{ kN})\cos 255° \qquad P_{1X} = -10.35 \text{ kN}$$

$$\overline{R}_{AX} = +10.35 \text{ kN}$$

$$\Sigma M_A = 0 \qquad P_{1Y}d_{1Y} + P_2 d_2 + \overline{R}_{DY}d_{DY}$$

$$\overline{R}_{DY}d_{DY} = -(P_{1Y}d_{1Y} + P_2 d_2) \qquad P_{1Y} = P_1 \sin \theta_{P1}$$

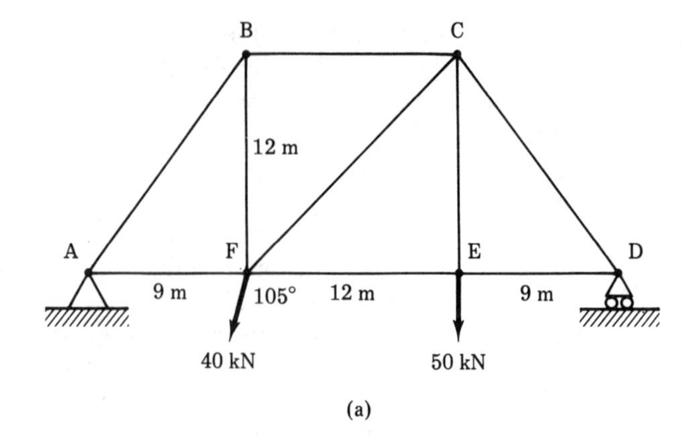

(a)

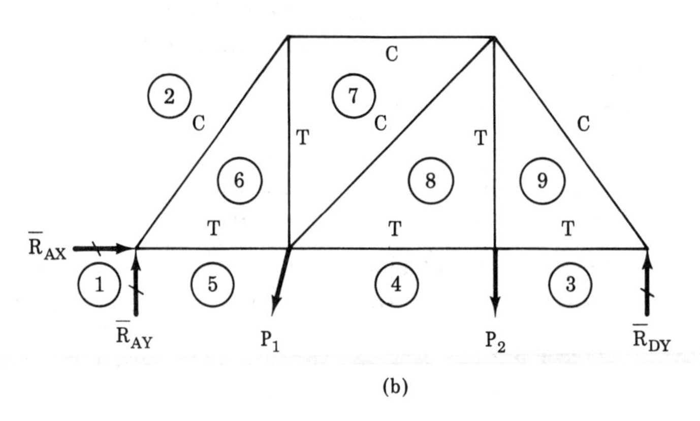

(b)

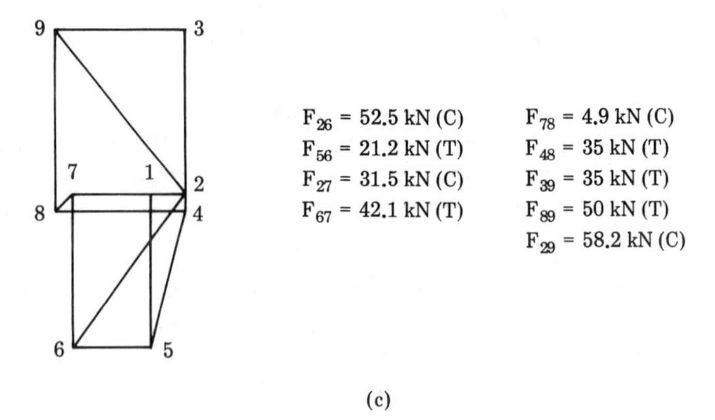

(c)

Figure 5-16

$$P_{1Y} = (40 \text{ kN})\sin 255° \qquad P_{1Y} = -38.64 \text{ kN}$$

$$\overline{R}_{DY}d_{DY} = -[-(38.64 \text{ kN})(9 \text{ m}) - (50 \text{ kN})(21 \text{ m})]$$

$$\overline{R}_{DY}d_{DY} = +1398 \text{ kN} \qquad \overline{R}_{DY} = \frac{\overline{R}_{DY}d_{DY}}{d_{DY}}$$

$$\overline{R}_{DY} = \frac{1398 \text{ kN}}{30 \text{ m}} \qquad \overline{R}_{DY} = +46.59 \text{ kN}$$

$$\Sigma F_Y = 0 \qquad R_{AY} + P_{1Y} + P_2 + R_{DY} = 0$$

$$R_{AY} = -(P_{1Y} + P_2 + R_{DY}) \qquad R_{AY} = -(-38.64 - 50 + 46.59 \text{ kN})$$

$$R_{AY} = +42.05 \text{ kN}$$

Step 5 Identify all members as being either in tension or compression as shown in Figure 5-16b.

Step 6 Number the spaces between external forces and the triangular spaces formed by the truss members as shown in Figure 5-16b.

Step 7 Draw, to scale, the force diagram for the external forces acting on the truss and then proceed to solve graphically each joint as shown in Figure 5-16c. The last joint done is the check.

Problem 5-6: Determine the forces acting in all members of the truss shown in Figure 5-17 using the combined graphical method.

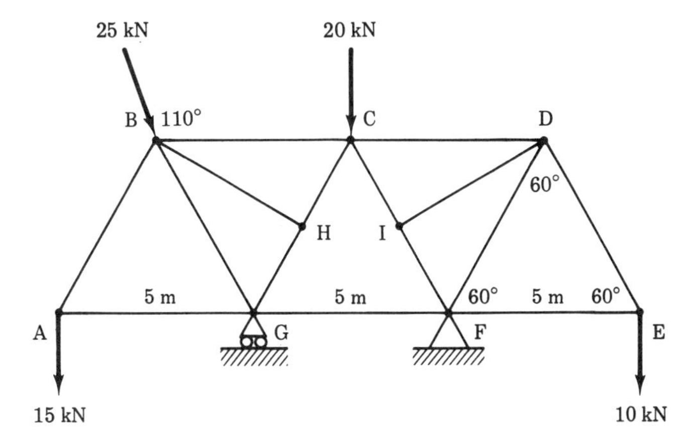

Figure 5-17

Problem 5-7: Determine the forces acting in all members of the truss shown in Figure 5-18 using the combined graphical method.

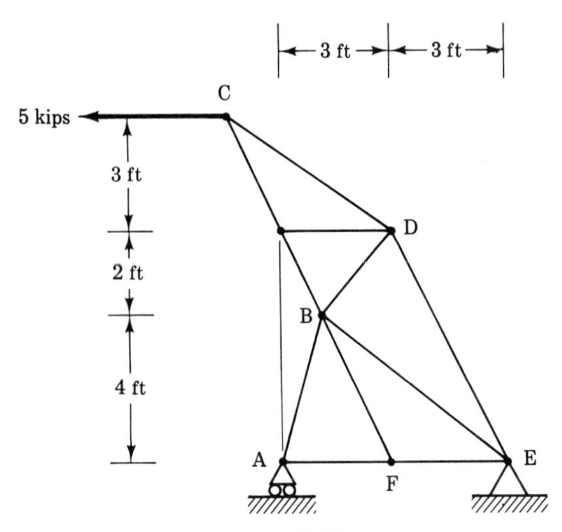

Figure 5-18

5.5 METHOD OF SECTIONS

There are occasions when the forces in only a few members of a truss are desired to be known. The technique called the method of sections provides a much faster method than the method of joints. The method of sections does not require finding any undesired forces and further does not require finding all reactions.

The technique consists of cutting the truss into two sections (hence the name of the method). The cut is made through those members whose forces are desired to be known. (Occasionally more than one cut may be required, but particularly on a truss with many members, this will still be faster than the method of joints.) One of the sections is selected for analysis. *Using the entire truss* the reaction(s) on that section are determined. Then, the section is treated as a separate body and solved for the unknown forces in the members cut.

Example 5-6: Determine the forces in members *BC, BG,* and *HG* in the truss shown in Figure 5-19a using the method of sections.

Solution:

Step 1 Recognize this as a truss problem.

Step 2 Recognize that since only a few forces are required to be known, the method of sections may be the best approach.

Step 3 By inspection, recognize that the truss contains no zero-force members.

Step 4 Draw the free-body diagram for the truss as shown in Figure 5-19b.

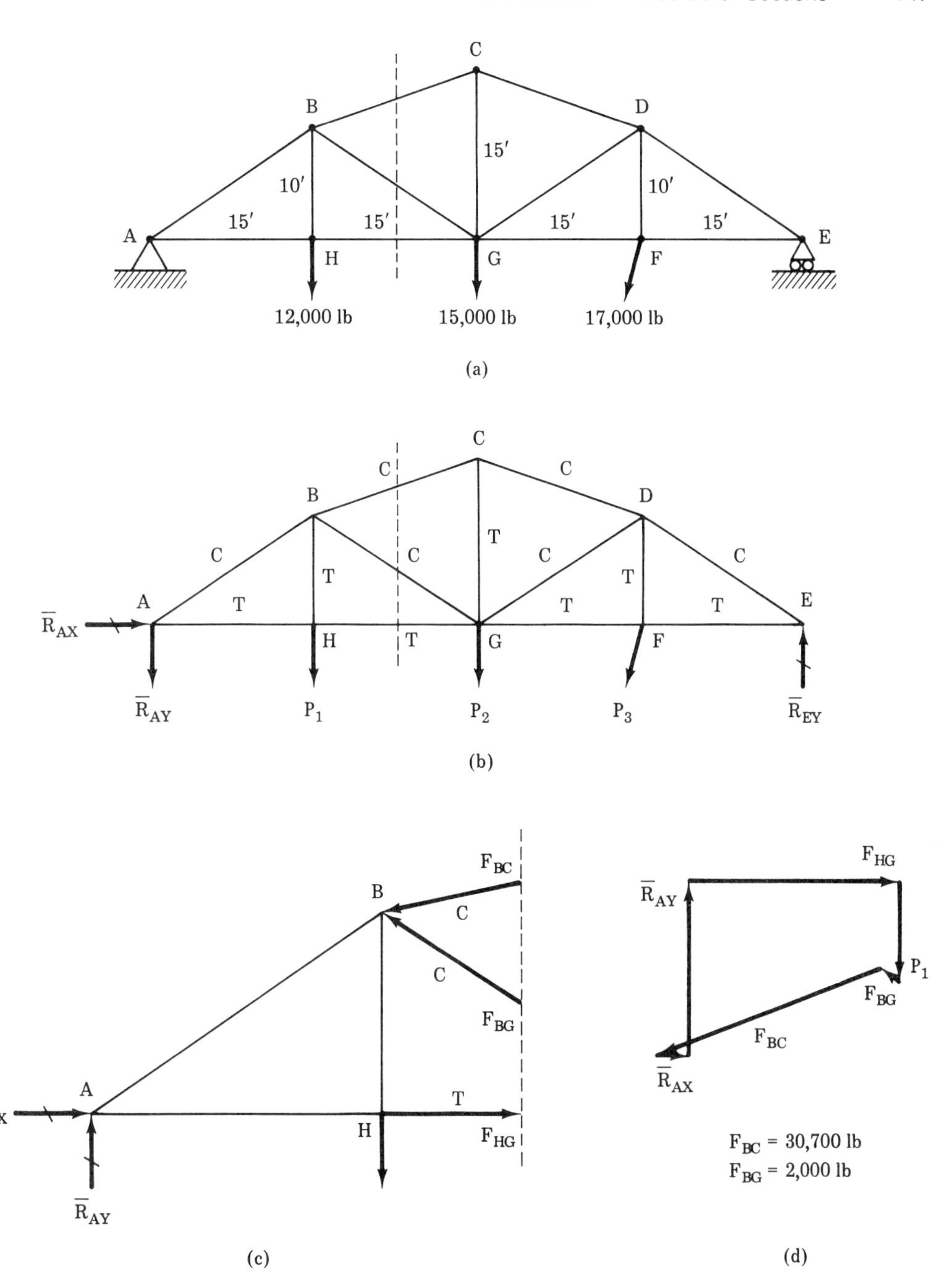
C
B
D
15′
10′
10′
A
15′
15′
15′
15′
E
H
G
F
12,000 lb
15,000 lb
17,000 lb
(a)
C
C
C
B
D
T
C
C
C
C
T
T
A
T
T
T
E
$\overline{R}_{AX}$
H
T
G
F
$\overline{R}_{AY}$
P_1
P_2
P_3
$\overline{R}_{EY}$
(b)
F_{BC}
B
C
C
F_{BG}
A
T
$\overline{R}_{AX}$
H
F_{HG}
$\overline{R}_{AY}$
(c)
F_{HG}
$\overline{R}_{AY}$
P_1
F_{BG}
F_{BC}
$\overline{R}_{AX}$
F_{BC} = 30,700 lb
F_{BG} = 2,000 lb
(d)

Figure 5-19

Step 5 Draw a cut line $a-a$ across the members whose forces are desired.

Step 6 Select either the left- or the right-hand section of the truss to work with.

Step 7 Having selected the left-hand half, determine the reactions at *A*. Since the truss contains one unknown horizontal reaction and two unknown vertical reactions, the horizontal force equilibrium equation and the moment equilibrium equation taken about *E* will be needed.

$$\Sigma F_X = 0 \qquad \overline{R}_{AX} + P_{3X} = 0 \qquad \overline{R}_{AX} = -P_{3X}$$

$$\overline{R}_{AX} = -P_3 \cos\theta_{P3} \qquad \overline{R}_{AX} = (-17{,}000 \text{ lb})\cos 255°$$

$$\overline{R}_{AX} = +4400 \text{ lb}$$

$$\Sigma M_E = 0 \qquad \overline{R}_{AY} d_{AY} + P_1 d_1 + P_2 d_2 + P_{3Y} d_{3Y} = 0$$

$$\overline{R}_{AY} d_{AY} = -(P_1 d_1 + P_2 d_2 + P_{3Y} d_{3Y})$$

$$P_{3Y} = P_3 \sin\theta_{P3} \qquad P_{3Y} = (17{,}000 \text{ lb})\sin 255° \qquad P_{3Y} = -16{,}421 \text{ lb}$$

$$\overline{R}_{AY} d_{AY} = -[+(12{,}000 \text{ lb})(45 \text{ ft}) + (15{,}000 \text{ lb})(30 \text{ ft}) + (16{,}421 \text{ lb})(15 \text{ ft})]$$

$$\overline{R}_{AY} d_{AY} = -1{,}236{,}315 \text{ lb ft} \qquad \overline{R}_{AY} = \frac{\overline{R}_{AY} d_{AY}}{d_{AY}}$$

$$\overline{R}_{AY} = \frac{1{,}236{,}315 \text{ lb ft}}{60 \text{ ft}} \qquad \overline{R}_{AY} = +20{,}605 \text{ lb}$$

Step 8 Make a free-body diagram of the left-hand section of the truss as shown in Figure 5-19c.

Step 9 Identify the forces in the three members in question as being either in tension or compression. *BG* may not be evident until the problem is partially solved.

Step 10 Examination of the free-body of the left-hand section shows that there are three unknown forces containing three unknown horizontal components and two unknown vertical components. Further examination shows that the force in *HG* can be found by taking moments about *B*. The forces in the other two members could be found by any one of three methods: method of components with simultaneous equations, law of sines (after the known forces are summed into one resultant force), and graphical.

$$\Sigma M_B = 0 \qquad \overline{R}_{AX} d_{AX} + \overline{R}_{AY} d_{AY} + HG d_{HG} = 0$$

$$HG d_{HG} = -(\overline{R}_{AX} d_{AX} + \overline{R}_{AY} d_{AY})$$

$$HG d_{HG} = -[+(4400 \text{ lb})(10 \text{ ft}) - (20{,}605 \text{ lb})(15 \text{ ft})]$$

$$HGd_{HG} = +265{,}075 \text{ lb ft}$$

$$HG = \frac{HGd_{HG}}{d_{HG}} \qquad HG = \frac{265{,}075 \text{ lb ft}}{10 \text{ ft}}$$

$$HG = +26{,}508 \text{ lb} \qquad \boldsymbol{HG = +26{,}500 \text{ lb}(T)}$$

BC and BG have been solved for graphically as shown in Figure 5-19d.

Problem 5-8: Determine the forces in the members of the truss shown in Figure 5-20 that are cut by line $a-a$ using the method of sections.

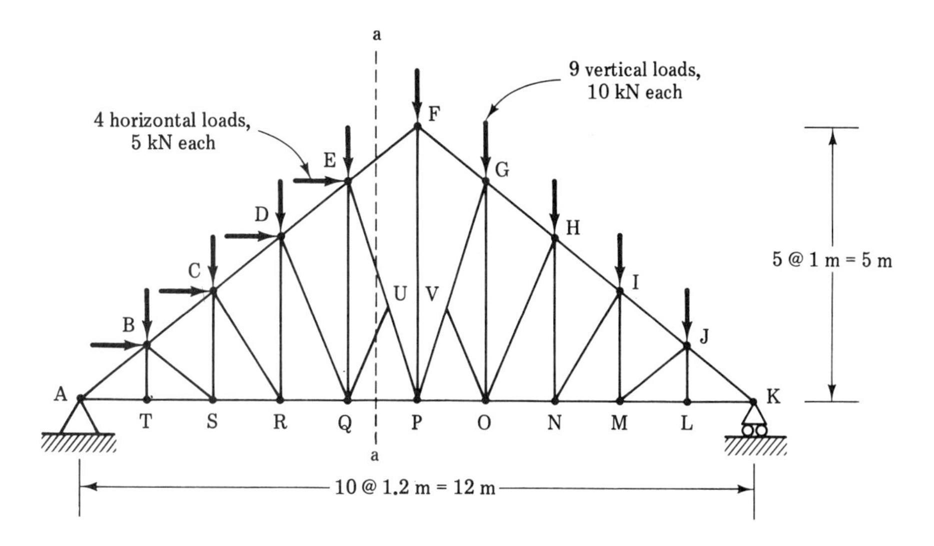

Figure 5-20

5.6 READINESS QUIZ

1. In the truss shown in Figure 5-21, the members in tension are

   (a) *AB, BC, CD,* and *BE*
   (b) *AB, BC, CD,* and *CE*
   (c) *AB, BC, CD,* and *BF*
   (d) *AB, BC, CD, BE,* and *BF*

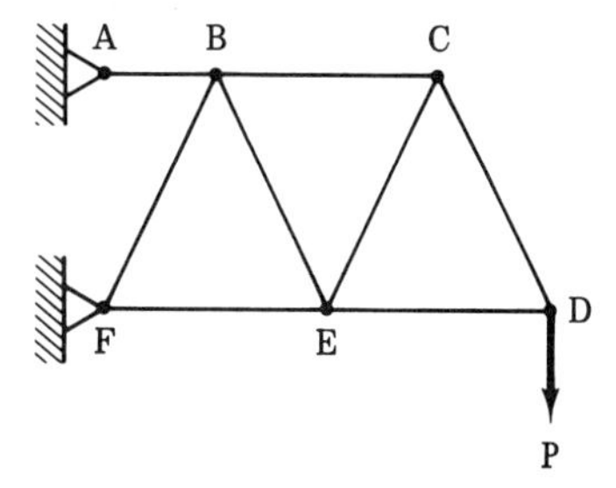

Figure 5-21

2. Before attempting an analysis by method of joints it is best to

   (a) determine all reactions
   (b) determine which members are in compression and which are in tension
   (c) determine which members carry no load
   (d) all of the above
   (e) none of the above

3. Simple analysis of a truss requires that

   (a) the weights of the individual members be considered negligible
   (b) the members be joined by smooth pins causing all joints to act as hinges
   (c) loads be applied only at joints
   (d) all of the above
   (e) none of the above

4. The members in the truss shown in Figure 5-22 that carry no load are

   (a) *BL, BK, CK, CJ, EJ, EI, FI,* and *FH*
   (b) *BL, BK, CK, FI, FH,* and *EI*
   (c) *BL, BK, CK, EI, FI,* and *FH*
   (d) *BL* and *FH*
   (e) none of the above

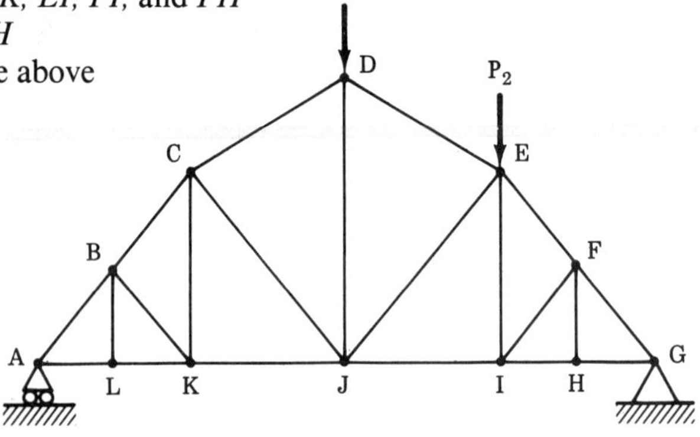

Figure 5-22

5. A member of a truss may be in

   (a) shear
   (b) torsion
   (c) bending
   (d) any of the above
   (e) none of the above

6. In the truss shown in Figure 5-23, the members in compression are

(a) *BC, BE, CE,* and *CD*
(b) *AB, BC, CD, AE,* and *ED*
(c) *AB, BC, CD,* and *CE*
(d) *AB, BE, CE,* and *CD*
(e) none of the above

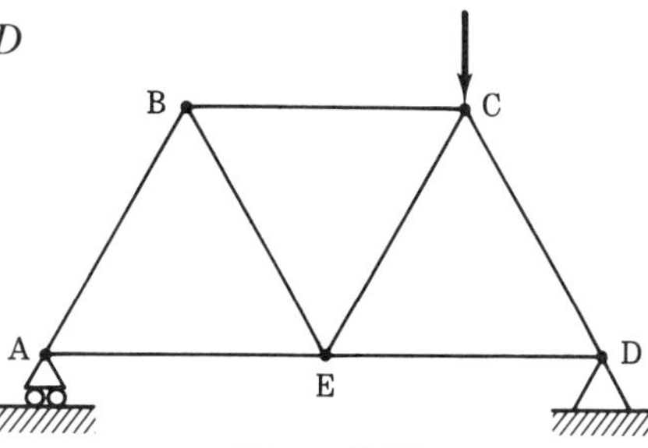

Figure 5-23

7. In the truss shown in Figure 5-24, the members that carry no load are

(a) *AF* and *FE*
(b) *BF, CF,* and *DF*
(c) *AF, CF,* and *FE*
(d) *AB, CF,* and *DE*
(e) none of the above

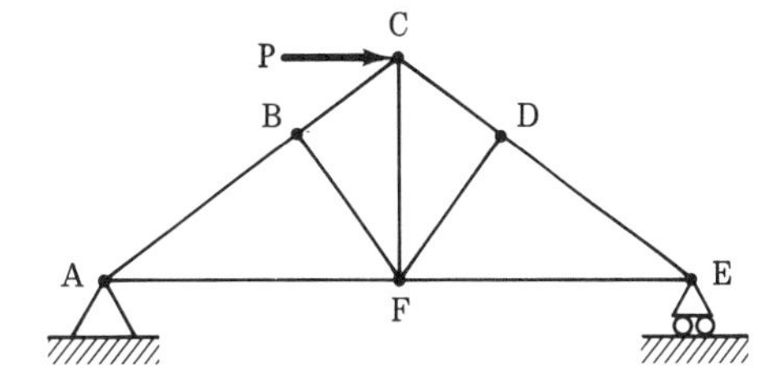

Figure 5-24

8. Simple analysis of a truss requires that every member be a

(a) one-force member
(b) two-force member
(c) three-force member
(d) any of the above
(e) none of the above

9. Member *DF* in the truss shown in Figure 5-25

(a) carries no load
(b) is in tension
(c) is in compression
(d) any of the above
(e) none of the above

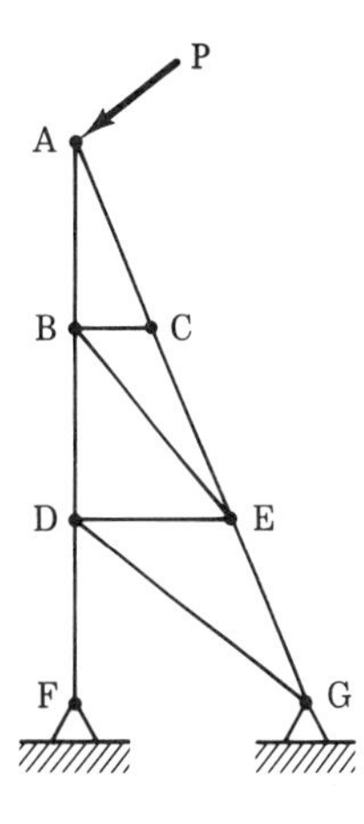

Figure 5-25

10. Member AC in the truss shown in Figure 5-26

    (a) carries no load
    (b) is in tension
    (c) is in compression
    (d) any of the above
    (e) none of the above

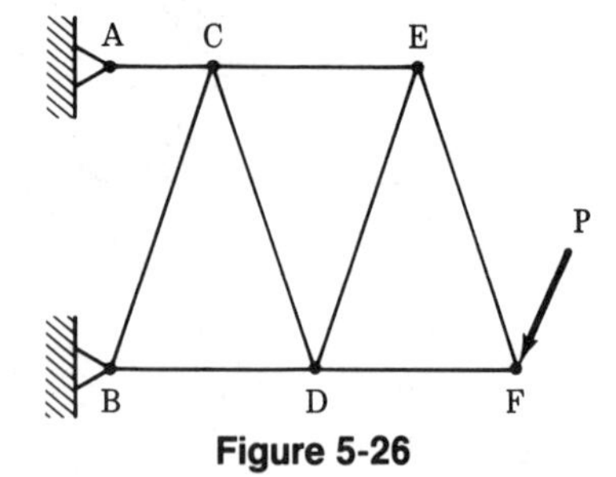

Figure 5-26

11. Using the method of joints mathematically

    (a) consists of a series of concurrent force problems
    (b) need not involve any non-right-triangle trigonometry
    (c) includes a check that proves the solution to be correct
    (d) all of the above
    (e) none of the above

12. The method of sections

    (a) requires some graphical work
    (b) is used only when the forces in all members of a truss are desired to be known
    (c) requires no use of moments
    (d) all of the above
    (e) none of the above

13. The combined graphical method

    (a) does not require finding all the reactions
    (b) requires free-body diagrams of individual joints
    (c) requires drawing each force to scale only once
    (d) all of the above
    (e) none of the above

14. Of the structures shown in Figure 5-27 ________ is a truss.

    (a) (b) (c) (d) all (e) none

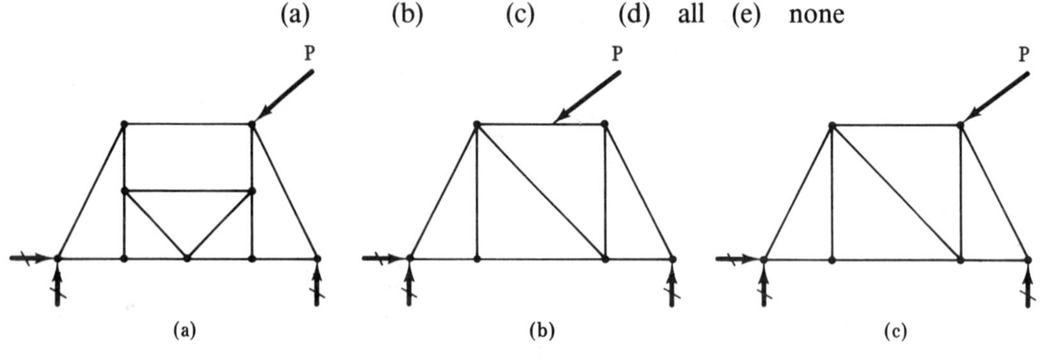

Figure 5-27

15. Of the structures shown in Figure 5-28 ________ is a truss.
    (a) (b) (c) (d) all (e) none

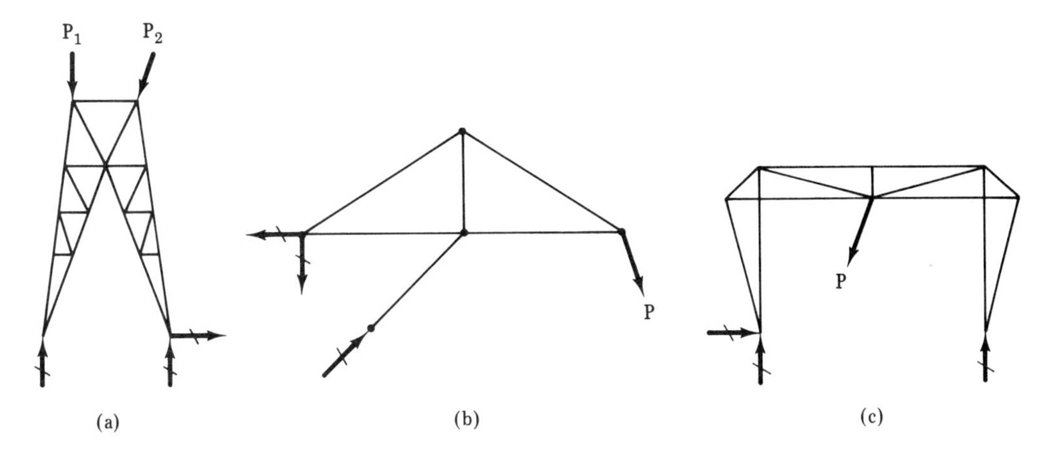

Figure 5-28

5.7 SUPPLEMENTARY PROBLEMS

5-9 Determine the forces acting in all members of the truss shown in Figure 5-29 using the method of joints mathematically.

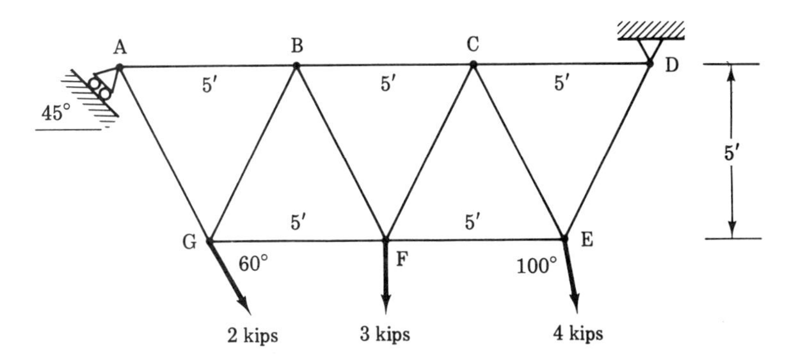

Figure 5-29

5-10 Determine the forces acting in all members of the truss shown in Figure 5-30 using the method of joints mathematically.

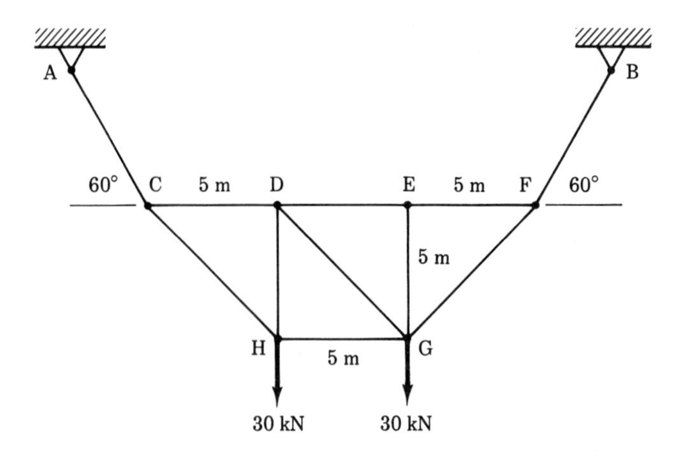

Figure 5-30

5-11 Determine the forces acting in all members of the truss shown in Figure 5-31 using the combined graphical method.

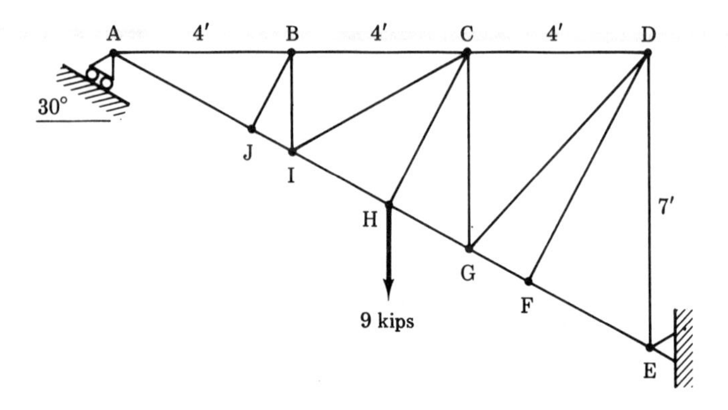

Figure 5-31

5-12 Determine the forces acting in all members of the truss shown in Figure 5-32 using the combined graphical method.

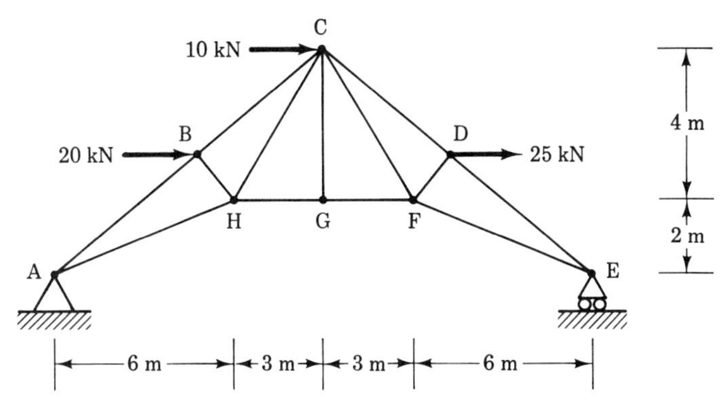

Figure 5-32

5-13 Determine the forces acting in all members of the truss shown in Figure 5-33 using the combined graphical method.

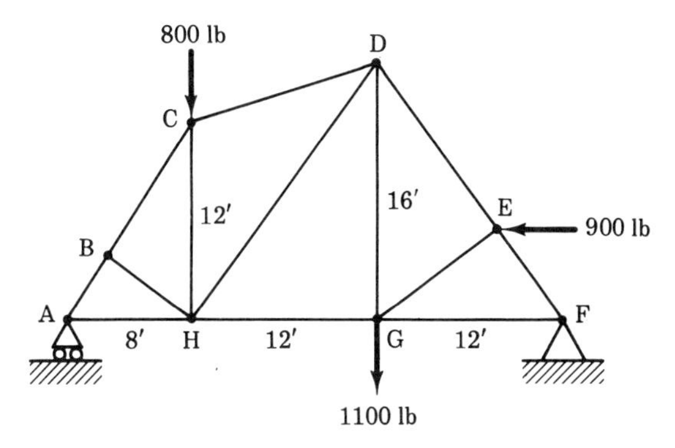

Figure 5-33

5-14 Determine the forces acting in all members cut by line $a-a$ in the truss shown in Figure 5-34 using the method of sections.

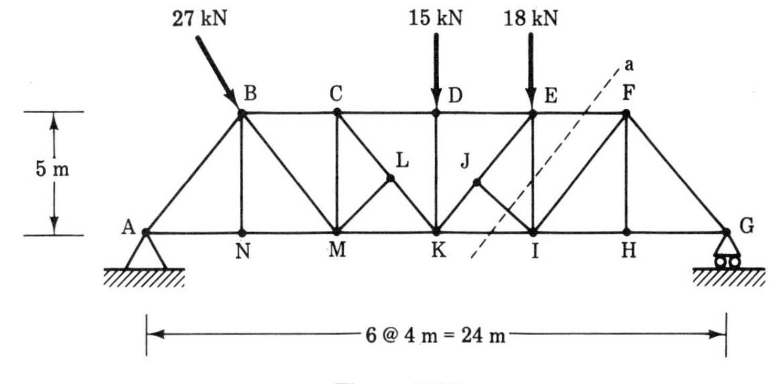

Figure 5-34

5-15 Determine the forces acting in all members cut by line $a-a$ in the truss shown in Figure 5-35 using the method of sections.

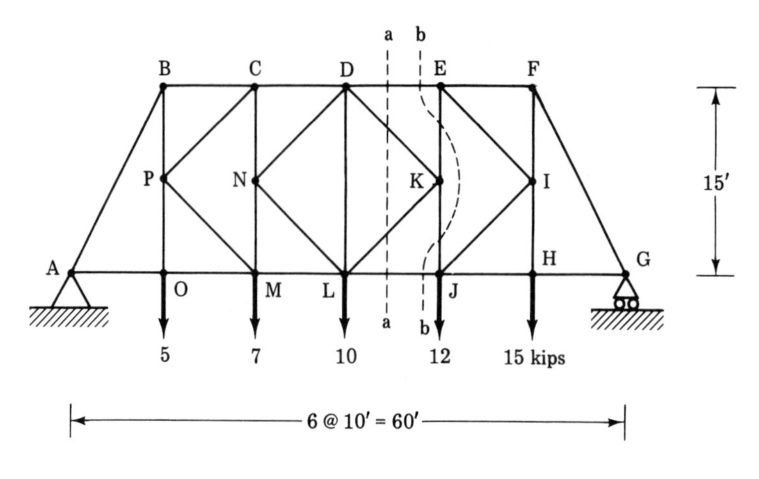

Hint: Find F_{DE} first by cutting through on b–b.

Figure 5-35

Chapter 6

Frames

6.1 OBJECTIVES

Upon completion of the work relating to this chapter, you should be able to perform the following.

1. Given a series of structures with loads and supports, identify those that are frames.
2. Identify zero-force and two-force members in a frame.
3. Determine all the forces acting on each individual member of a frame using the method of members.

6.2 NATURE OF FRAMES

Defining a frame can best be accomplished by comparing it to a truss. A frame is different in at least one of the following three ways:

1. At least one member extends through more than two joints.
2. At least one load is not located at a joint.
3. At least one member is not straight.

The effect of any one of the above criteria is that the member affected is no longer a simple two-force member. In addition to being subjected to tension or compression, the member is also subjected to bending.

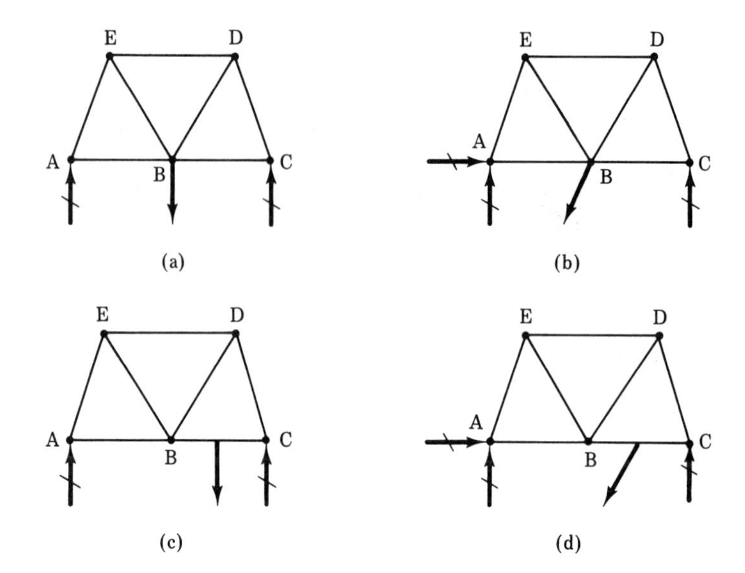

Figure 6-1. Frames versus trusses, effect of load.

In Figure 6-1a, a simply loaded truss is shown. In 6-1b, the load has been angled with the sole result that a horizontal reaction now occurs at *A*. These are situations that you are already familiar with from the preceding chapter. However, in 6-1c the load is no longer at a joint and *BC* is now subjected to bending as well as tension. In 6-1d, *BC* is also subjected to bending while the left-hand portion is subjected to compression and the right-hand portion to tension.

It is important to note from the above that only one member of a structure not being a two-force member makes the structure a frame. However, the fact that, in this particular frame, all of the remaining members are two-force members does make complete analysis easier than for most frames.

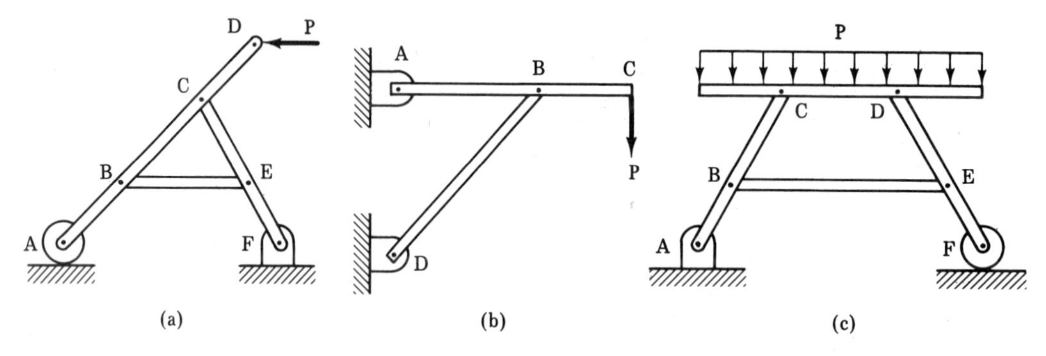

Figure 6-2. Members extending through pin joints.

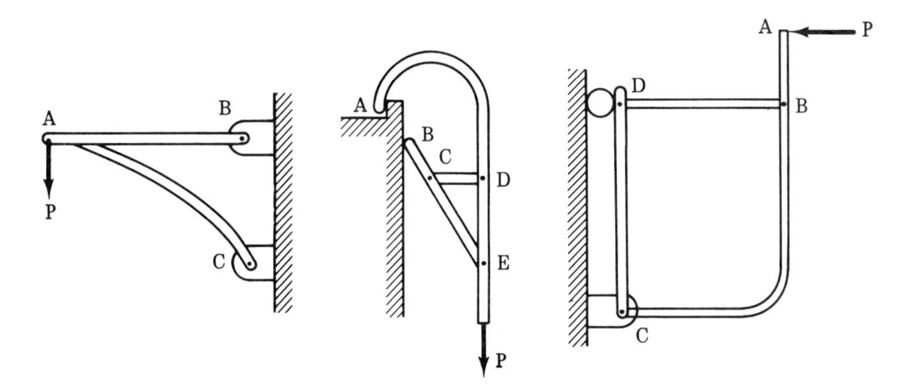

Figure 6-3. Curved members.

Figure 6-2 shows three examples of frames that have at least one member extending through more than two joints. Close inspection shows that each contains a simple two-force member, *BE* in 6-2a, *BD* in 6-2b, and *BE* in 6-2c. This may not appear to be too earthshaking, but it proves to be an invaluable help in frame analysis.

Figure 6-3 shows three examples of frames that contain a curved member. Once again, in each case, existence of a simple two-force member simplifies analysis.

6.3 SOLUTION OF FRAMES: METHOD OF MEMBERS

The process of solving for the forces acting in a frame is often called the method of members. This is because the heart of the process involves drawing a free-body diagram of each member. The basic steps to the process follow:

Step 1 Recognize that the problem involves a frame rather than a truss.

Step 2 By inspection, identify and remove any zero-force members.

Step 3 By inspection, identify any two-force members and categorize them as being in either tension or compression.

Step 4 By inspection, identify any joints that connect only two-force members. The method of joints can be applied at these.

Step 5 Draw the free-body diagram for the frame.

Step 6 Determine all reactions for the frame.

Step 7 Draw a free-body diagram for each member of the frame.

Step 8 Determine the forces acting on each member. Note that it may not be possible to complete the work on any one member at the start. That is, you may start on one member and find some of the forces acting on it, hit a dead end, work on another member, and eventually feed back information to the first member to complete its solution.

Step 9 Upon reaching the last member you will find that all the forces acting on it are already known. It, therefore, serves as a check since all these known forces must add up to zero and all the moments of these forces must add up to zero. The latter is extremely important since frame members are prone to including couples in their force systems.

6.4 APPLICATIONS

The following examples and problems provide some typical frame problems. Section 6.6 at the end of the chapter provides further variety.

Example 6-1: Determine the forces acting on each member of the frame shown in Figure 6-4a.

Solution:

Step 1 The structure is definitely a frame since it has a load located at some point other than a joint.

Step 2 *CG* is a zero-force member since it connects only two joints, has no load imposed between the joints, and is unopposed at *G*. At first glance, it appears that *CG* could cause *BGE*, which extends through three joints, to bend. However, further examination of the frame shows that *BGE* could be two members, *BG* and *GE* without changing the functioning of the frame.

Step 3 Elimination of *CG* makes *BGE* a simple two-force member in tension.

Step 4 There are no joints with only two-force members attached to them.

Step 5 Draw the free-body diagram for the entire frame as shown in Figure 6-4b.

Step 6 Determine all reactions for the frame as follows:

$$\Sigma M_A = 0 \qquad Pd_P + R_{FY}d_{FY} = 0 \qquad R_{FY}d_{FY} = -(Pd_P)$$

$$d_P = AD \qquad AD^2 = (4)^2 + (1.333)^2 \qquad d_P = 4.22 \text{ m}$$

$$R_{FY}d_{FY} = -[-(40 \text{ N})(4.22 \text{ m})] \quad R_{FY}d_{FY} = +168.7 \text{ N m} \circlearrowleft$$

$$R_{FY} = \frac{R_{FY}d_{FY}}{d_{FY}} \qquad R_{FY} = \frac{168.7 \text{ N m}}{2 \text{ m}} \qquad R_{FY} = +84.35 \text{ N}$$

$$\theta_P = \arctan\frac{-1.333}{+4} \qquad \theta_P = 346°$$

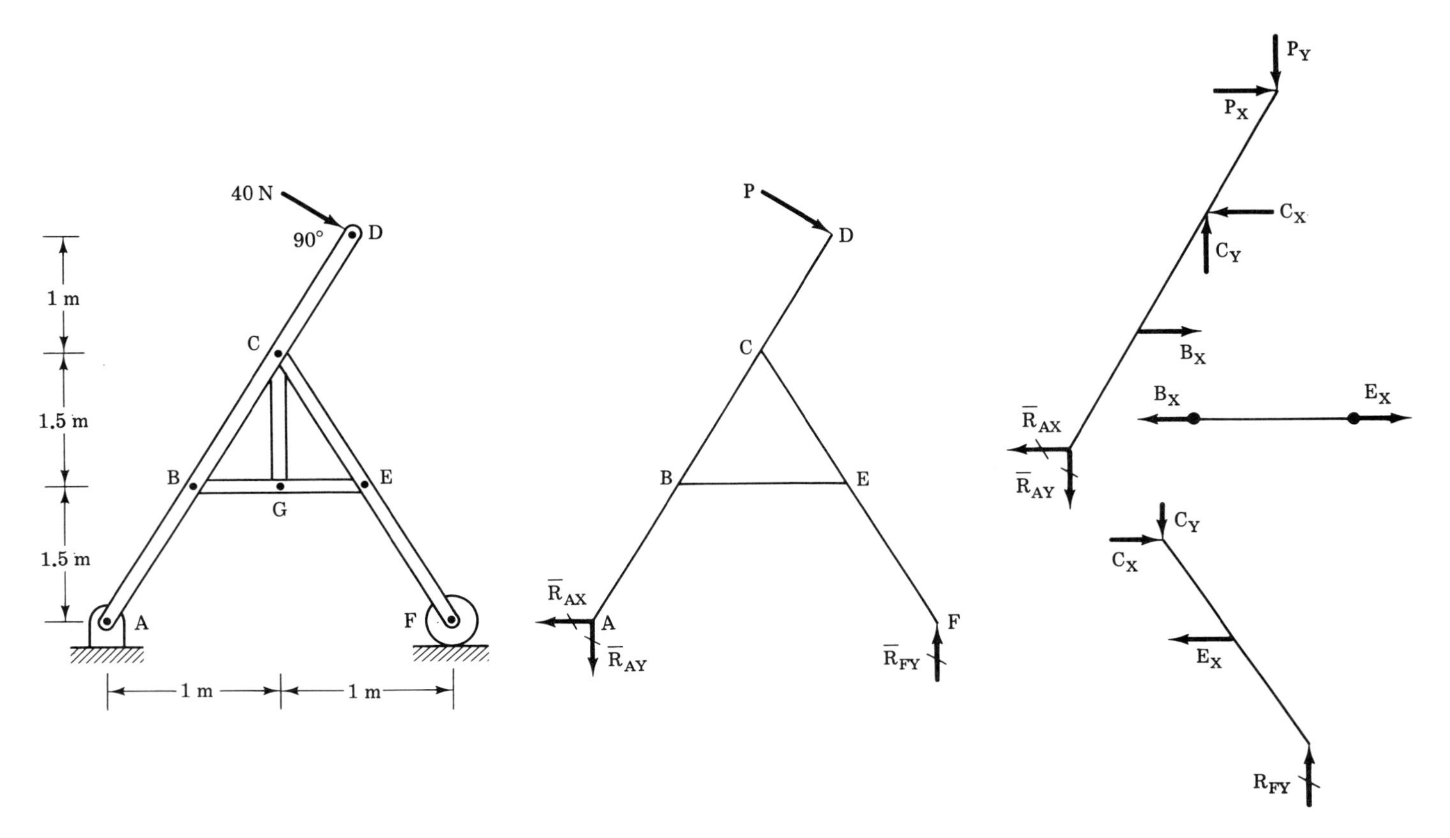
40 N
90°
D
C
B
G
E
A
F
1 m
1.5 m
1.5 m
1 m
1 m
P
$\overline{R}_{AX}$
$\overline{R}_{AY}$
$\overline{R}_{FY}$
P_Y
P_X
C_X
C_Y
B_X
B_X
E_X
$\overline{R}_{AX}$
$\overline{R}_{AY}$
C_Y
C_X
E_X
R_{FY}

Figure 6-4

$$P_X = P\cos\theta_P \qquad P_X = (40\text{ N})\cos 346° \qquad P_X = 38.8\text{ N}$$

$$P_Y = P\sin\theta_P \qquad P_Y = (40\text{ N})\sin 346° \qquad P_Y = -9.68\text{ N}$$

$$\Sigma F_X = 0 \qquad R_{AX} + P_X = 0 \qquad R_{AX} = -P_X \qquad R_{AX} = -38.8\text{ N}$$

$$\Sigma F_Y = 0 \qquad R_{AY} + P_Y + R_{FY} = 0 \qquad R_{AY} = -(P_Y + R_{FY})$$

$$R_{AY} = -(-9.68 + 84.35\text{ N}) \qquad R_{AY} = -74.67\text{ N}$$

Step 7 Draw a free-body diagram of each member of the frame as shown in Figure 6-4c.

Step 8 Determine the forces acting on each member as follows:
Member *ABCD*:

$$\Sigma M_C = 0 \qquad R_{AX}d_{AX} + R_{AY}d_{AY} + B_X d_{BX} + P_X d_X + P_Y d_Y = 0$$

$$B_X d_X = -(R_A d_{AX} + R_{AY}d_{AY} + P_X d_X + P_Y d_Y)$$

$$B_X d_X = -[-(38.8\text{ N})(3\text{ m}) + (74.6\text{ N}) \times (1\text{ m}) - (38.8\text{ N})(1\text{ m}) - (9.68\text{ N})(0.33\text{ m})]$$

$$B_X d_X = +83.72\text{ N m}↺$$

$$B_X = \frac{B_X d_X}{d_X} \qquad B_X = \frac{83.72\text{ N m}}{1.5\text{ m}} \qquad B_X = +55.8\text{ N}$$

$$\Sigma F_X = 0 \qquad P_X + C_X + B_X + R_{AX} = 0 \qquad C_X = -(P_X + B_X + R_{AX})$$

$$C_X = -(38.8 + 55.8 - 38.8\text{ N}) \qquad C_X = -55.8\text{ N}$$

$$\Sigma F_Y = 0 \qquad P_Y + C_Y + R_{AY} = 0 \qquad C_Y = -(P_Y + R_{AY})$$

$$C_Y = -(-9.68 - 74.67\text{ N}) \qquad C_Y = +84.4\text{ N}$$

Member *BE*:
A simple two-force member in 55.8 N tension. Pulls horizontally to left on *E* of *CEF*.

Member *BEF*:
By transfer from *ABCD* and *BE*, all forces acting on this member are known. Our equilibrium equations should provide a check on our solution.

$$\Sigma F_X = 0 \qquad C_X + E_X = 0 \qquad +55.8 - 55.8\text{ N} = 0$$

$$\Sigma F_Y = 0 \qquad C_Y + R_{FY} = 0 \qquad -84.4 + 84.4\text{ N} = 0$$

These forces consist of two couples:

$$\Sigma M = 0 \qquad M_1 + M_2 = 0$$

$$-(55.8\text{ N})(1.5\text{ m}) + (84.4\text{ N})(1\text{ m}) = 0$$

$$-83.7 + 84.4\text{ N} = 0,\ \text{less than } \tfrac{1}{2}\%\ \text{error}$$

Problem 6-1: Determine the forces acting on each member of the frame shown in Figure 6-5.

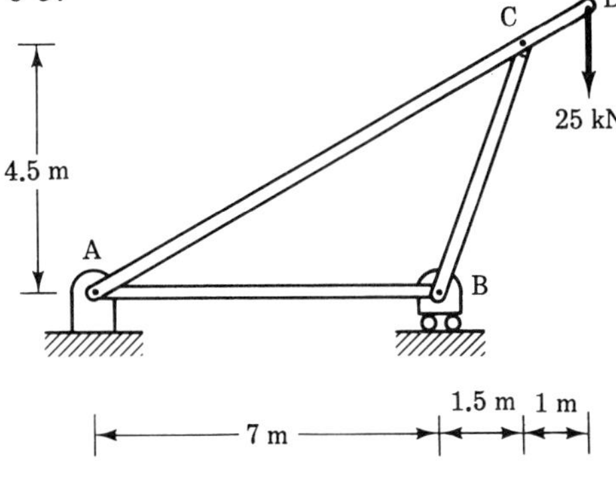

Figure 6-5

Example 6-2: Determine the forces acting on each member of the frame shown in Figure 6-6a.

Solution:

Step 1 The structure is definitely a frame since it consists of a curved member and two members that extend through three joints.

Step 2 There are no zero-force members.

Step 3 There are no two-force members.

Step 4 Given Step 3, there's no need to look for joints with only two-force members attached.

Step 5 Draw the free-body diagram of the entire frame as shown in Figure 6-6b.

Step 6 Determine all reactions for the frame as follows:

$$P_X = P\cos\theta_P \qquad P_X = (1500\text{ lb})\cos 235° \qquad P_X = -860\text{ lb}$$

$$P_Y = P\sin\theta_P \qquad P_Y = (1500\text{ lb})\sin 235° \qquad P_Y = -1229\text{ lb}$$

$$\Sigma F_X = 0 \qquad R_{AX} + P_X = 0 \qquad R_{AX} = -P_X \qquad R_{AX} = +860\text{ lb}$$

$$\Sigma M_A = 0 \qquad R_{EX}d_{EX} + P_Y d_Y = 0 \qquad R_{EX}d_{EX} = -P_Y d_Y$$

$$R_{EX}d_{EX} = -[-(1229\text{ lb})(5.5\text{ ft})] \qquad R_{EX}d_{EX} = +6760\text{ lb ft}↺$$

$$R_{EX} = \frac{R_{EX}d_{EX}}{d_{EX}} \qquad R_{EY} = \frac{6760\text{ lb ft}}{4\text{ ft}} \qquad R_{EY} = +1690\text{ lb}$$

$$\Sigma F_Y = 0 \qquad R_{AY} + R_{EY} + P_Y = 0 \qquad R_{AY} = -(R_{EY} + P_Y)$$

$$R_{AY} = -(+1690 - 1229\text{ lb}) \qquad R_{AY} = -461\text{ lb}$$

Step 7 Draw a free-body diagram of each member of the frame as shown in Figure 6-6c.

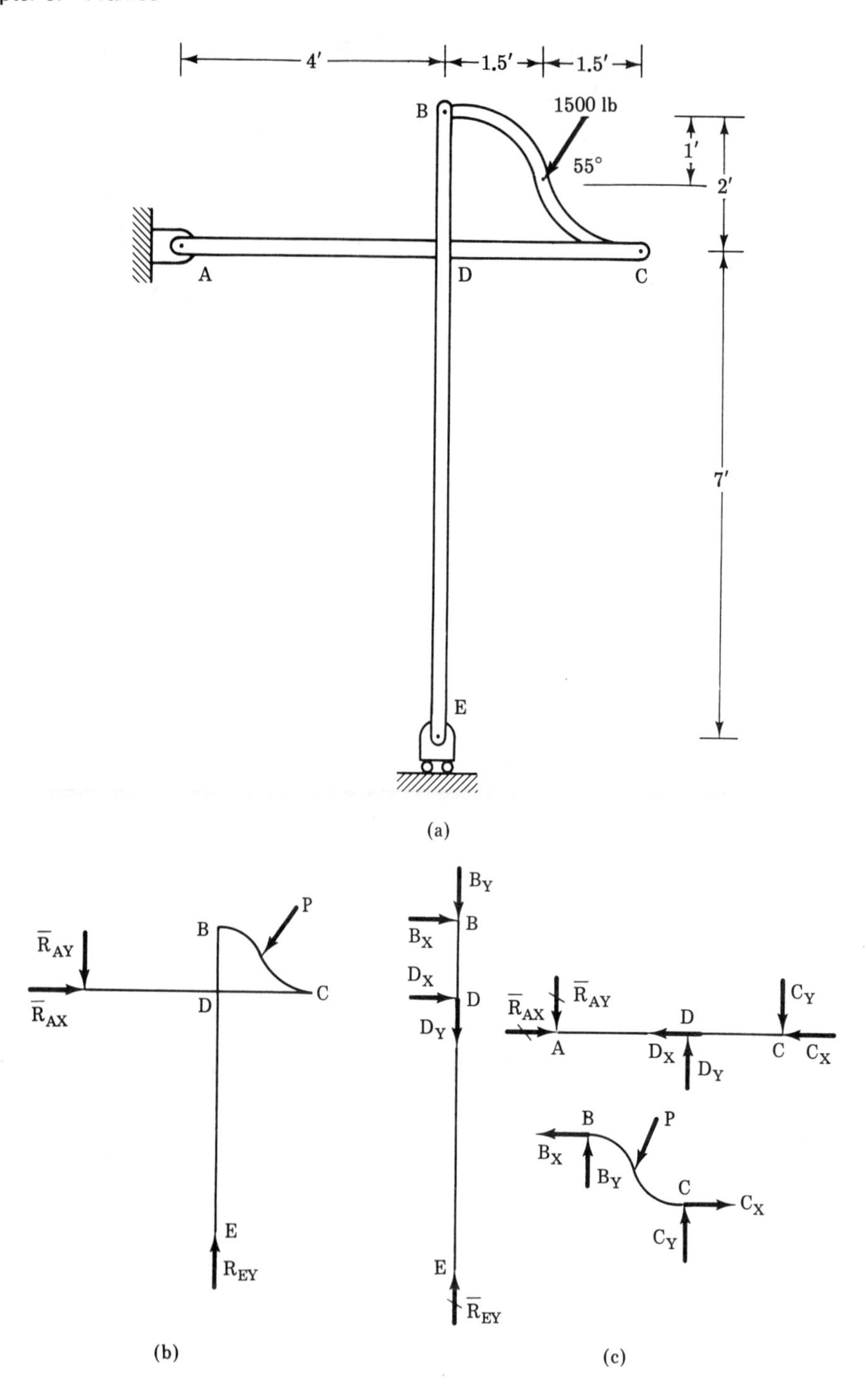

4′
1.5′
1.5′
1500 lb
55°
1′
2′
7′
A
B
C
D
E
(a)
P
$\overline{R}_{AY}$
$\overline{R}_{AX}$
R_{EY}
(b)
B_Y
B_X
D_X
D_Y
C_Y
C_X
$\overline{R}_{EY}$
(c)

Figure 6-6

Step 8 Determine the forces acting on each member as follows:
Member *ADC*:

$$\Sigma M_D = 0 \qquad R_{AY}d_{AY} + C_Y d_{CY} = 0 \qquad C_Y d_{CY} = -R_{AY}d_{AY}$$

$$C_Y d_{CY} = -[+(461\text{ lb})(4\text{ ft})] \qquad C_Y d_Y = -1844\text{ lb ft}$$

$$C_Y = \frac{C_Y d_{CY}}{d_{CY}} \qquad C_Y = \frac{1844\text{ lb ft}}{3\text{ ft}} \qquad C_Y = -615\text{ lb}$$

$$\Sigma F_Y = 0 \qquad R_{AY} + D_Y + C_Y = 0 \qquad D_Y = -(R_{AY} + C_Y)$$

$$D_Y = -(-461 - 615\text{ lb}) \qquad D_Y = +1076\text{ lb}$$

No further work can be done on *ADC* at this time.
Member *BDE*:

$$\Sigma F_Y = 0 \qquad B_Y + D_Y + E_Y = 0 \qquad B_Y = -(D_Y + E_Y)$$

$$B_Y = -(-1076 + 1690\text{ lb}) \qquad B_Y = -614\text{ lb}$$

$$\Sigma M_D = 0 \qquad B_X d_X = 0 \qquad B_X = 0$$

$$\Sigma F_X = 0 \qquad B_X + D_X = 0 \qquad D_X = -B_X \qquad D_X = 0$$

Back to *ADC*:

$$\Sigma F_X = 0 \qquad R_{AX} + D_X + C_X = 0 \qquad C_X = -(R_{AX} + D_X)$$

$$C_X = -(+800\text{ lb} + 0) \qquad C_X = -860\text{ lb}$$

Member *BC*: Check

$$F_X = 0 \qquad B_X + P_X + C_X = 0 \qquad 0 - 860 + 860\text{ lb} = 0$$

$$F_Y = 0 \qquad B_Y + P_Y + C_Y = 0 \qquad +614\text{ lb} - 1229\text{ lb} + 615\text{ lb} = 0$$

Problem 6-2: Determine the forces acting on each member of the frame shown in Figure 6-7.

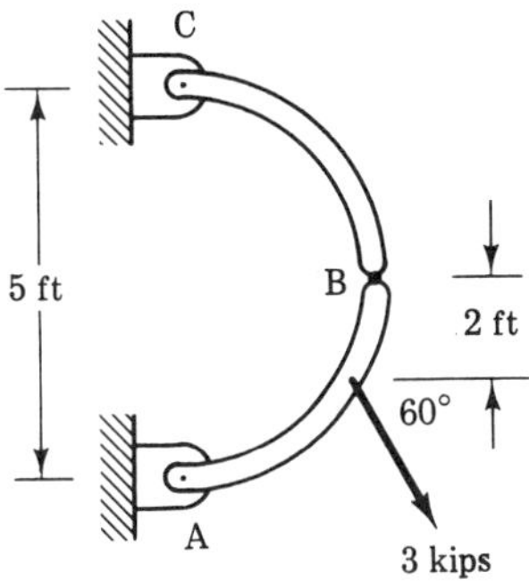

Figure 6-7

Example 6-3: Determine the forces acting on each member of the frame shown in Figure 6-8a.

Solution:

Step 1 The structure is definitely a frame since it consists of three members, one of which has loads not at a joint, and two of which extend through more than two joints.

Step 2 There are no zero-force members.

Step 3 There are no two-force members.

Step 4 Given Step 3, there's no need to look for joints with only two-force members attached.

Step 5 Combine loads as follows:

BC by proportion is 3.5 ft long.

The 100-kN/m load is therefore spread over 1.5 m.

$$P_T = P_1 + P_2 + P_3 \qquad P_T = (-100 \text{ kN})(1.5 \text{ m}) + (-50 \text{ kN/m})\ (2 \text{ m}) + (-200 \text{ kN}),\ P_T = -450 \text{ kN}$$

$$\Sigma M_B = P_T d_T \qquad P_1 d_1 + P_2 d_2 + P_3 d_3 = P_T d_T$$

$$P_T d_T = -(150 \text{ kN})(0.75 \text{ m}) - (100 \text{ kN})(2.5 \text{ m}) - (200 \text{ kN})(3.5 \text{ m})$$

$$P_T d_T = -1062.5 \text{ kN m} \circlearrowright \qquad d_T = \frac{P_T d_T}{P_T} \qquad d_T = \frac{1062.5 \text{ kN m}}{450 \text{ kN}}$$

$$d_T = 2.36 \text{ m} \quad \text{or} \quad P_T \text{ is located 2.36 m to right of } B$$

Step 6 Draw the free-body diagram of the entire frame as shown in Figure 6-8b.

Step 7 Determine all reactions for the frame as follows:

$$\Sigma F_Y = 0 \qquad R_{AY} + P_T = 0 \qquad R_{AY} = -P_T \qquad R_{AY} = +450 \text{ kN}$$

$$\Sigma M_E = 0 \qquad R_{AX} d_{AX} + P_T d_{PT} = 0 \qquad R_{AX} d_{AX} = -P_T d_{PT}$$

$$R_{AX} d_{AX} = -[-(450 \text{ kN})(2.86 \text{ m})] \qquad R_{AX} d_{AX} = +1287 \text{ kN m} \circlearrowleft$$

$$R_{AX} = \frac{R_{AX} d_{AX}}{d_{AX}} \qquad R_{AX} = \frac{1287 \text{ kN m}}{8 \text{ m}} \qquad R_{AX} = -161 \text{ kN}$$

$$\Sigma F_X = 0 \qquad R_{AX} + R_{EX} = 0 \qquad R_{EX} = -R_{AX} \qquad R_{EX} = +161 \text{ kN}$$

Step 8 Draw a free-body diagram of each member of the frame as shown in Figure 6-8c.

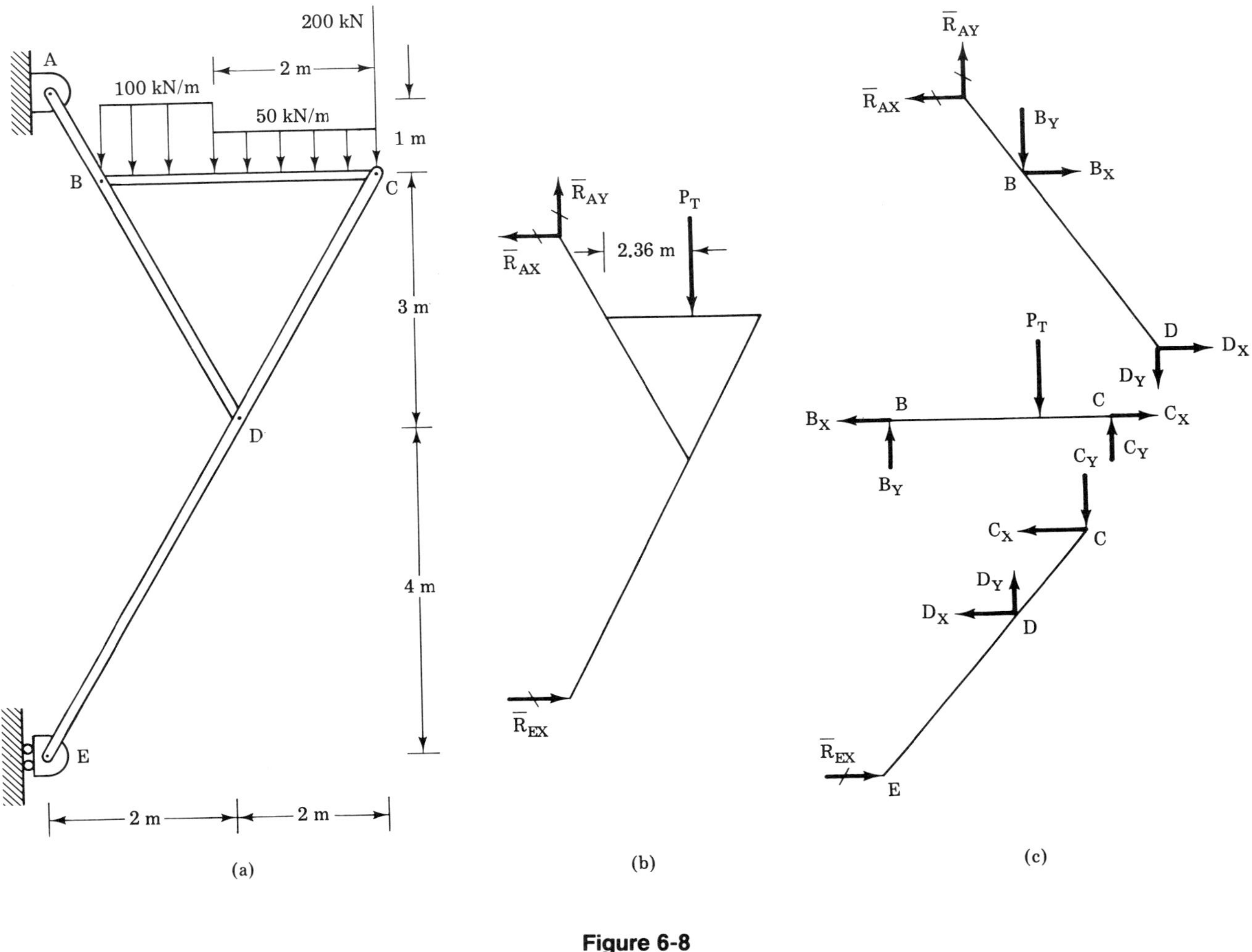

200 kN
A
100 kN/m
2 m
50 kN/m
1 m
B
C
3 m
D
4 m
E
2 m
2 m
(a)
R̄AY
R̄AX
PT
2.36 m
R̄EX
(b)
R̄AY
R̄AX
BY
BX
B
D
DX
DY
PT
BX
B
C
CX
CY
CY
BY
CX
C
DY
DX
D
R̄EX
E
(c)

Figure 6-8

Step 9 Determine the forces acting on each member of the frame as follows:

Member *BC*:

$$\Sigma M_B = 0 \qquad P_T d_{PT} + C_Y d_{CY} = 0 \qquad C_Y d_{CY} = -P_T d_{PT}$$

$$C_Y d_{CY} = -[-(450 \text{ kN})(2.36 \text{ m})] \qquad C_Y d_{CY} = +1062 \text{ kN m}\circlearrowleft$$

$$C_Y = \frac{C_Y d_{CY}}{d_{CY}} \qquad C_Y = \frac{1062 \text{ kN m}}{3.5 \text{ m}} \qquad C_Y = +303.4 \text{ kN}$$

$$\Sigma F_Y = 0 \qquad B_Y + P_T + C_Y = 0 \qquad B_Y = -(P_T + C_Y)$$

$$B_Y = -(-450 \text{ kN} + 303.4 \text{ kN}) \qquad B_Y = +146.6 \text{ kN}$$

Member *ABD*:

$$\Sigma M_D = 0 \qquad R_{AX} d_{AX} + R_{AY} d_{AY} + B_Y d_Y + B_X d_X = 0$$

$$B_X d_X = -(R_{AX} d_{AX} + R_{AY} d_{AY} + B_Y d_Y)$$

$$B_X d_X = -[+(161 \text{ kN})(4 \text{ m}) - (450 \text{ kN})(2 \text{ m}) + (146.6 \text{ kN})(1.5 \text{ m})]$$

$$B_X d_X = -36.1 \text{ kN m} \qquad B_X = \frac{B_X d_X}{d_X} \qquad B_X = \frac{36.1 \text{ kN m}}{3 \text{ m}}$$

$$B_X = +12 \text{ kN}$$

$$\Sigma F_X = 0 \qquad R_{AX} + B_X + D_X = 0 \qquad D_X = -(R_{AX} + B_X)$$

$$D_X = -(-161 + 12 \text{ kN}) \qquad D_X = +149 \text{ kN}$$

$$\Sigma F_Y = 0 \qquad R_{AY} + B_Y + D_Y = 0 \qquad D_Y = -(R_{AY} + B_Y)$$

Member *BC*:

$$\Sigma F_X = 0 \qquad B_X C_X = 0 \qquad C_X = -B_X \qquad C_X = +12 \text{ kN}$$

Member *CDE*: Check

$$\Sigma F_X = 0 \qquad C_X + D_X + R_{EX} = 0 \qquad -12 = 149 + 161 \text{ kN} = 0$$

$$\Sigma F_Y = 0 \qquad C_Y + D_Y = 0 \qquad -303.4 + 303.4 \text{ kN} = 0$$

$$\Sigma M_E = 0 \qquad C_Y d_{CY} + C_X d_{CX} + D_Y d_{DY} + D_X d_{DX} = 0$$

$$-(303.4 \text{ kN})(4 \text{ m}) + (12)(7 \text{ m}) + (303.4 \text{ kN})(2 \text{ m}) + (149 \text{ kN})(4 \text{ m}) = 0$$

$$-10.8 \text{ kN m} \cong 0$$

Problem 6-3: Determine the forces acting on each member of the frame shown in Figure 6-9.

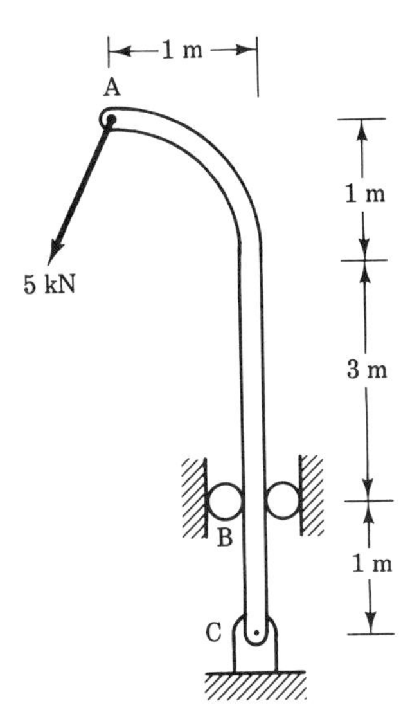

Figure 6-9

6.5 READINESS QUIZ

1. Simple analysis of a frame requires that

   (a) the weights of the individual members be considered negligible
   (b) the members be joined by smooth pins, causing all joints to act as hinges
   (c) loads be applied only at joints
   (d) all of the above
   (e) none of the above

2. Force analysis of a frame will require a free body of

   (a) a section of the frame
   (b) one or more joints of the frame
   (c) one or more members of the frame
   (d) all of the above
   (e) none of the above

3. In the frame shown in Figure 6-10 the members that are in pure tension are

(a) AC and BE
(b) AC and AF
(c) BE and AF
(d) AC only
(e) none of the above

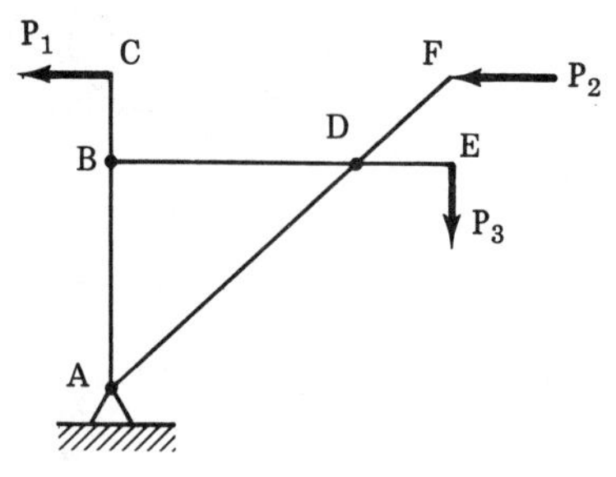

Figure 6-10

4. A member of a frame can be in

(a) tension
(b) compression
(c) bending
(d) any of the above
(e) none of the above

5. Before making a force analysis of a frame, it is necessary to

(a) make a free-body diagram of the frame
(b) determine all external forces acting on the frame
(c) make a free-body diagram of each member of the frame
(d) all of the above
(e) none of the above

6. The structure shown in Figure 6-11

(a) contains only two-force members
(b) contains no zero-force members
(c) is a truss
(d) all of the above
(e) none of the above

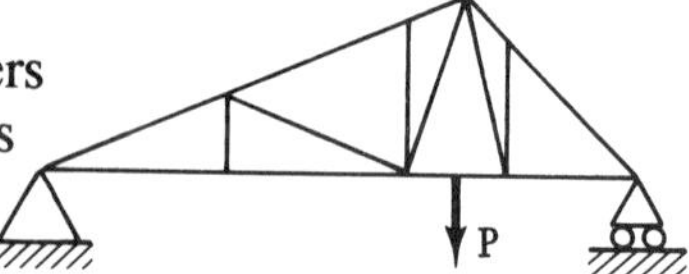

Figure 6-11

7. The structure shown in Figure 6-12 contains ________ two-force member(s).

(a) no
(b) one
(c) two
(d) three
(e) four

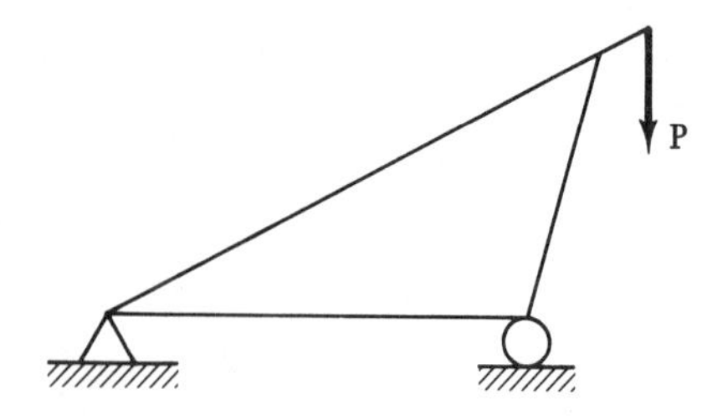

Figure 6-12

8. __________ of the structures shown in Figure 6-13 is a frame.

(a) None
(b) One
(c) Two
(d) Three
(e) Four

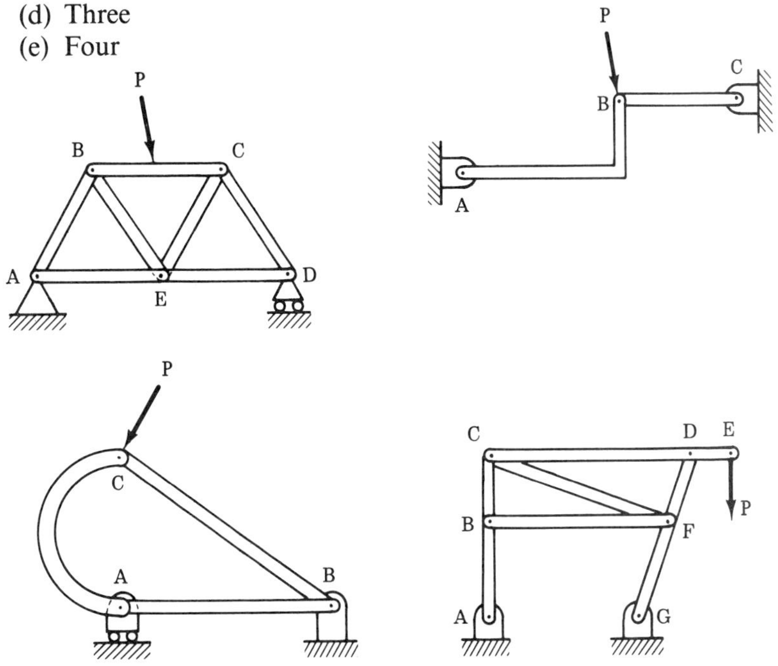

Figure 6-13

9. In the frame shown in Figure 6-14, member *BC* is in

(a) tension
(b) compression
(c) bending
(d) torsion
(e) none of the above

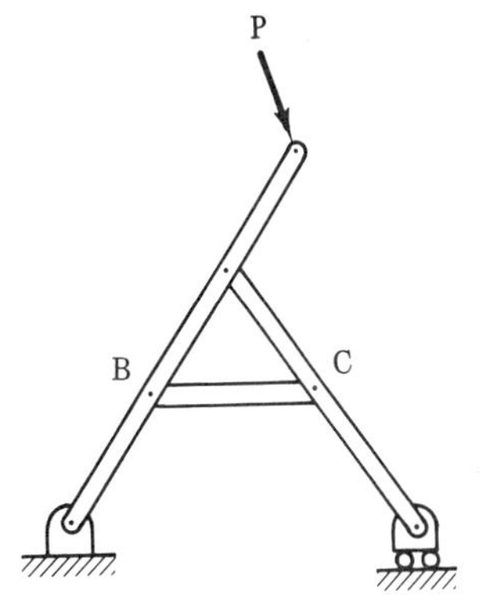

Figure 6-14

10. In the frame shown in Figure 6-15, there are ________ members subjected to bending.

(a) zero
(b) one
(c) two
(d) three
(e) four

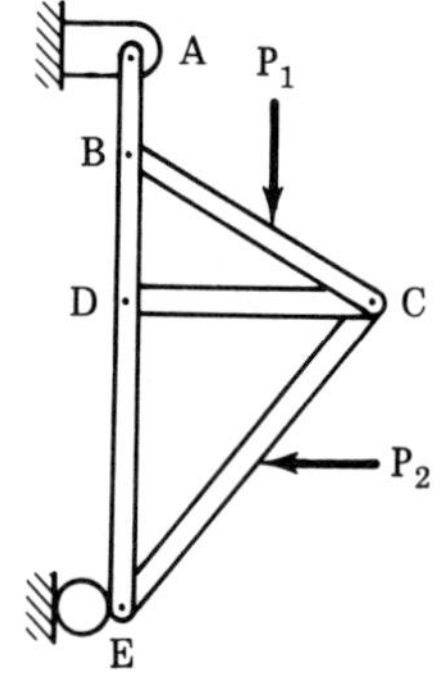

Figure 6-15

6.6 SUPPLEMENTARY PROBLEMS

In each of the following problems, determine the forces acting on each member of the frame.

6-4

6-5

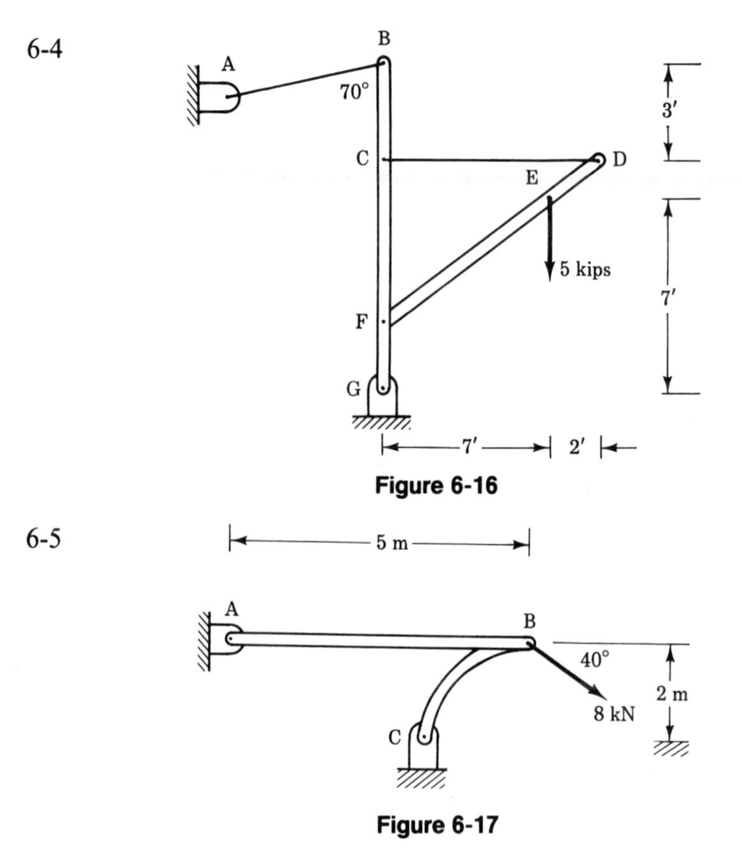

Figure 6-16

Figure 6-17

6-6

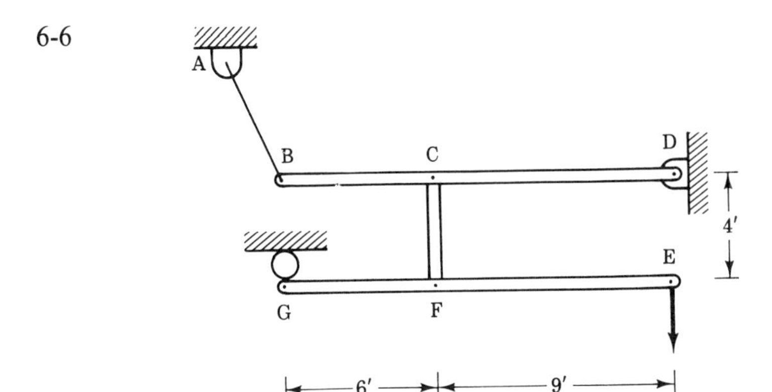

Figure 6-18

6-7

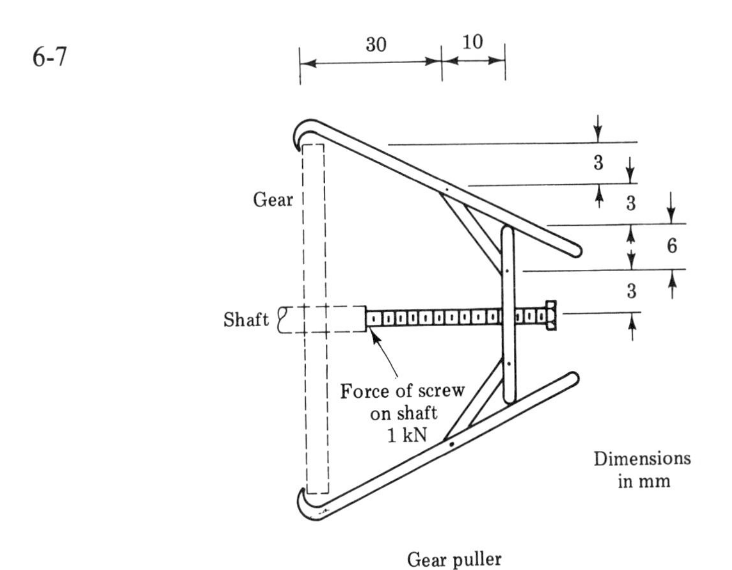

Gear puller

Figure 6-19

6-8

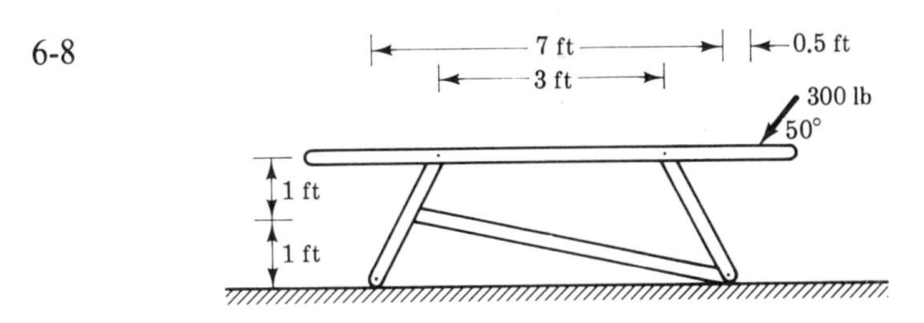

Hint: Consider friction

Figure 6-20

6-9

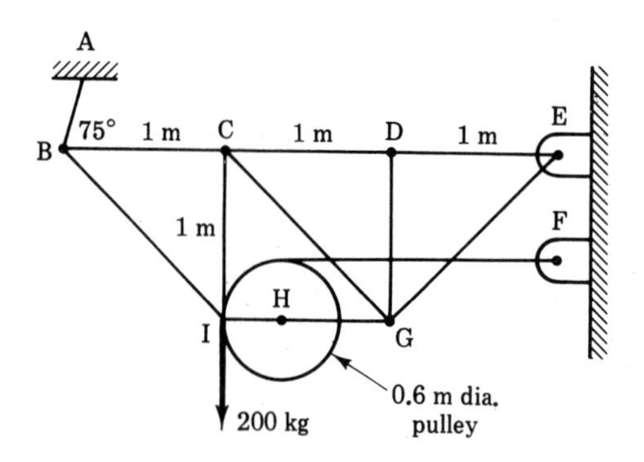

Figure 6-21

6-10

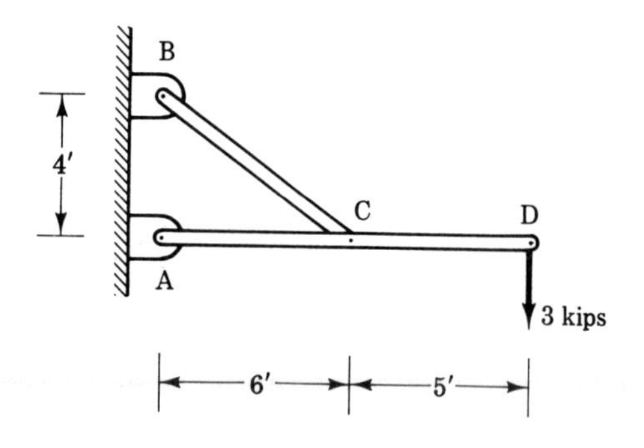

Figure 6-22

Chapter 7

Cables

7.1 OBJECTIVES

Upon completion of the work relating to this chapter, you should be able to perform the following.

1. Given a series of symmetrically distributed loads acting on a cable determine the maximum tension in the cable.
2. Given a series of unsymmetrically distributed loads acting on a cable determine the maximum tension in the cable.
3. Given the type of cable and the loading on it determine the size required.

7.2 NATURE OF CABLES

Cables are structural members that have a very large length-to-diameter ratio. This, in combination with the types of materials used for cables, makes them extremely flexible. Cables, therefore, cannot develop compression and so cannot develop bending either. A cable could be used for any two-force tension member in the structures and mechanisms you have already studied. We are able to consider those members as rigid (even if they are cables) only because they are simple two-force members. The instability

Figure 7-1. Single load on cable.

or lack of rigidity of a cable is evident in Figure 7-1, where the cable has changed shape as a result of the tightrope walker's weight being added to the weight of the cable. This is typical of a cable. If several loads are applied to a cable, it takes a string polygon shape, as shown in Figure 7-2, that is a function of the magnitude, direction, and location of the loads.

In working with cables, we shall make several assumptions that will simplify our work, yet do not affect our results appreciably.

1. The change in length and, therefore, shape of a cable resulting from the tensile forces developed in it can be considered negligible compared to the effect of the loads.

2. The two support points for a cable are at the same elevation. There can be exceptions to this, but this simplifying assumption covers the vast majority of the problems you are likely to encounter.

3. The horizontal distance between support points and the amount of sag permitted are known. These are reasonable assumptions since they are usually physical limitations of the specific problem.

4. Generally, the weight of the cable is negligible compared to the other forces acting on it.

Finally, let us consider two important points dealing with the nature and application of cables. First, the general description of a cable developed above would permit it to be made of many materials and in any cross-sectional shape. In reality, cables are circular in cross section, generally steel, and almost always of stranded construction. The latter feature is for purposes of structural efficiency and economy and is dealt with in depth in most machine design texts and references. Second, most of us are familiar with the use of cables in lifting applications including elevators, bridge supports, and, perhaps, ski lifts. However, extensive applications occur in electrical power transmission, roof design, and materials handling equipment.

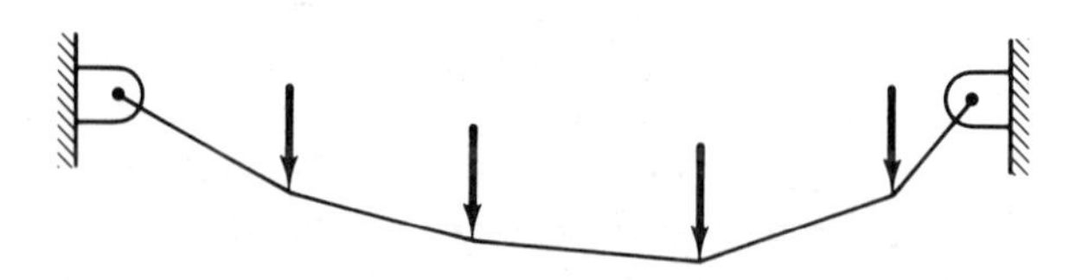

Figure 7-2. Multiloaded cable.

7.3 SYMMETRICALLY LOADED CABLES

Symmetrical loading of a cable may be of any one of the three types shown in Figures 7-3a–c or a combination of types as shown in Figure 7-3d. The only requirement for symmetry is that the cable can be divided into a left-hand half and a right-hand half with their loadings mirror images of one another.

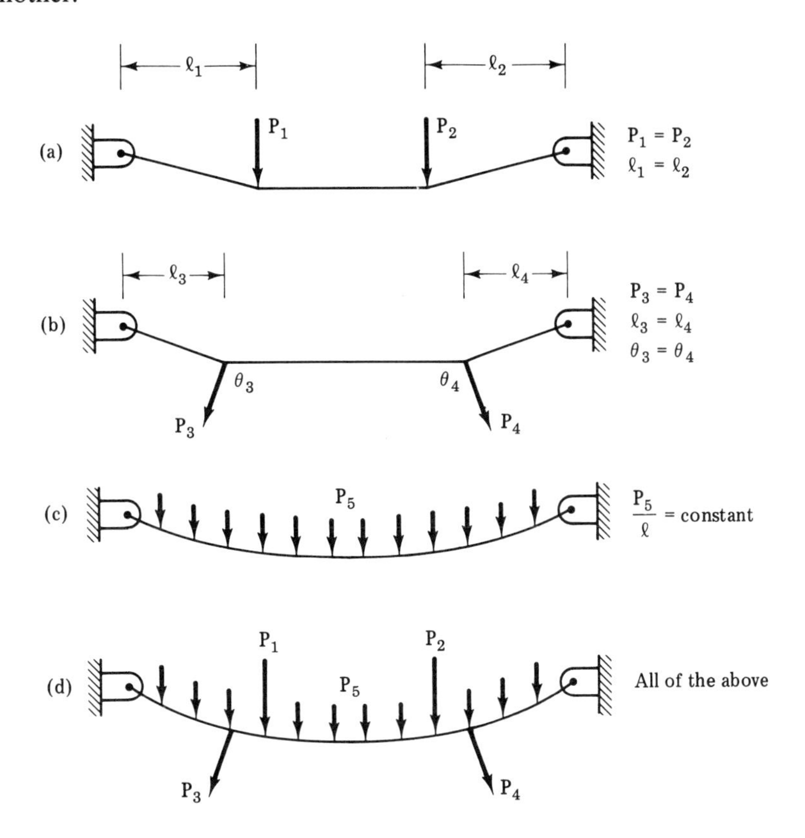

Figure 7-3. Symmetrically loaded cables.

In this section, we shall first deal with concentrated symmetrical loads and then uniformly distributed loads.

Let us examine the cable shown in Figure 7-4a. Its supports are at the same elevation; it has a span l and a sag f. Every load on the left-hand half has an equal load at a similar location on the right-hand half. It therefore fits the definition of a symmetrically loaded cable and is in accord with two of the assumptions made earlier. Let us assume that the other two assumptions also hold and proceed with an analysis of the cable.

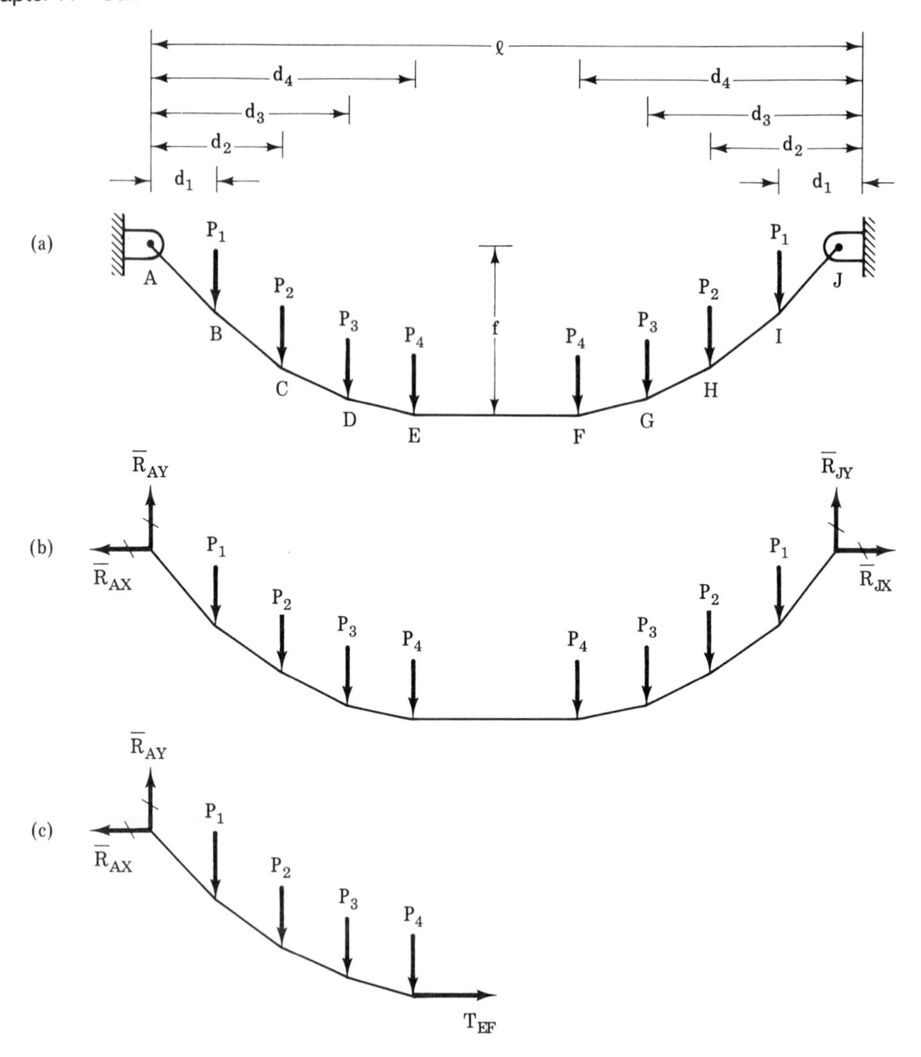

Figure 7-4. Cable analysis, concentrated loads.

First, let us draw a free-body diagram of the entire cable as shown in Figure 7-4b. Inspection shows that $\overline{R}_{AY} = \overline{R}_{JY} = \frac{1}{2}\Sigma P$. This could also be attained mathematically by taking moments about either A or J and then summing the vertical forces.

It is apparent from $\Sigma F_X = 0$ that $\overline{R}_{AX} = -\overline{R}_{JX}$. However, neither can be obtained by further analysis of the free-body diagram of the entire cable. The method of sections approach used with trusses can be applied here. In Figure 7-4c is shown the free-body diagram of the left-hand half of the cable. Taking moments about A yields

$$\Sigma M_A = 0 \qquad P_1 d_1 + P_2 d_2 + P_3 d_3 + P_4 d_4 + T_{EF} f = 0$$

$$T_{EF} = \frac{1}{f}\,(P_1 d_1 + P_2 d_2 + P_3 d_3 + P_4 d_4)$$

The above is algebraically correct since the moments about A of all the loads are clockwise while the moment about A of T_{EF} is counterclockwise. Since $R_{AX} = -TEF$ and $R_{JX} = -R_{AX}$, all the external forces acting on the cable are now known. The magnitude and direction of the tension in each portion of the cable can now be determined by joint analysis using any one of the three methods previously learned for analyzing concurrent force problems.

It is worthwhile to note that the horizontal thrust of the cable is inversely proportional to the sag.

Example 7-1: Determine the support reaction and the tension in each section of the cable shown in Figure 7-5a. Make a scaled sketch showing the displacement of the cable.

Solution:

Step 1 Recognize this as a symmetrically loaded cable.

Step 2 Draw the free-body diagram of the entire cable as shown in Figure 7-5b.

Step 3 Determine R_{AY} and R_{HY} by writing and solving the equations for $\Sigma M_A = 0$ and $\Sigma F_Y = 0$.

$$\Sigma M_A = 0 \qquad P_1 d_1 + P_2 d_2 + P_3 d_3 + P_4 d_4 + P_5 d_5 + P_6 d_6 + \overline{R}_{HY} l = 0$$

$$\overline{R}_{HY} l = -[-(5\text{ MN})(10\text{ m}) - (7\text{ MN})(22\text{ m}) - (3\text{ MN})(37\text{ m}) - (3\text{ MN})(42\text{ m}) - (7\text{ MN})(57\text{ M}) - (5\text{ MN})(69\text{ m})]$$

$$R_{HY} l = +1185\text{ MN m} \quad \overline{R}_{HY} = \frac{\overline{R}_{HY} l}{l} = \frac{1185\text{ MN m}}{79\text{ m}} \quad \mathbf{\overline{R}_{HY} = +15\text{ MN}}$$

$$\Sigma F_Y = 0 \qquad R_{AY} + P_1 + P_2 + P_3 + P_4 + P_5 + P_6 + \overline{R}_{HY} = 0$$

$$\overline{R}_{AY} = -(-5 - 7 - 3 - 3 - 7 - 5 + 15\text{ MN})$$

$$\mathbf{\overline{R}_{AY} = +15\text{ MN}}$$

Step 4 Draw the free-body diagram for the left-hand half of the cable as shown in Figure 7-5c.

Step 5 Determine R_{AX} by writing and solving the equations for $\Sigma M_A = 0$ and $\Sigma F_X = 0$ for the left-hand half of the cable.

$$\Sigma M_A = 0 \qquad P_1 d_1 + P_2 d_2 + P_3 d_3 + T_{DE} f = 0$$

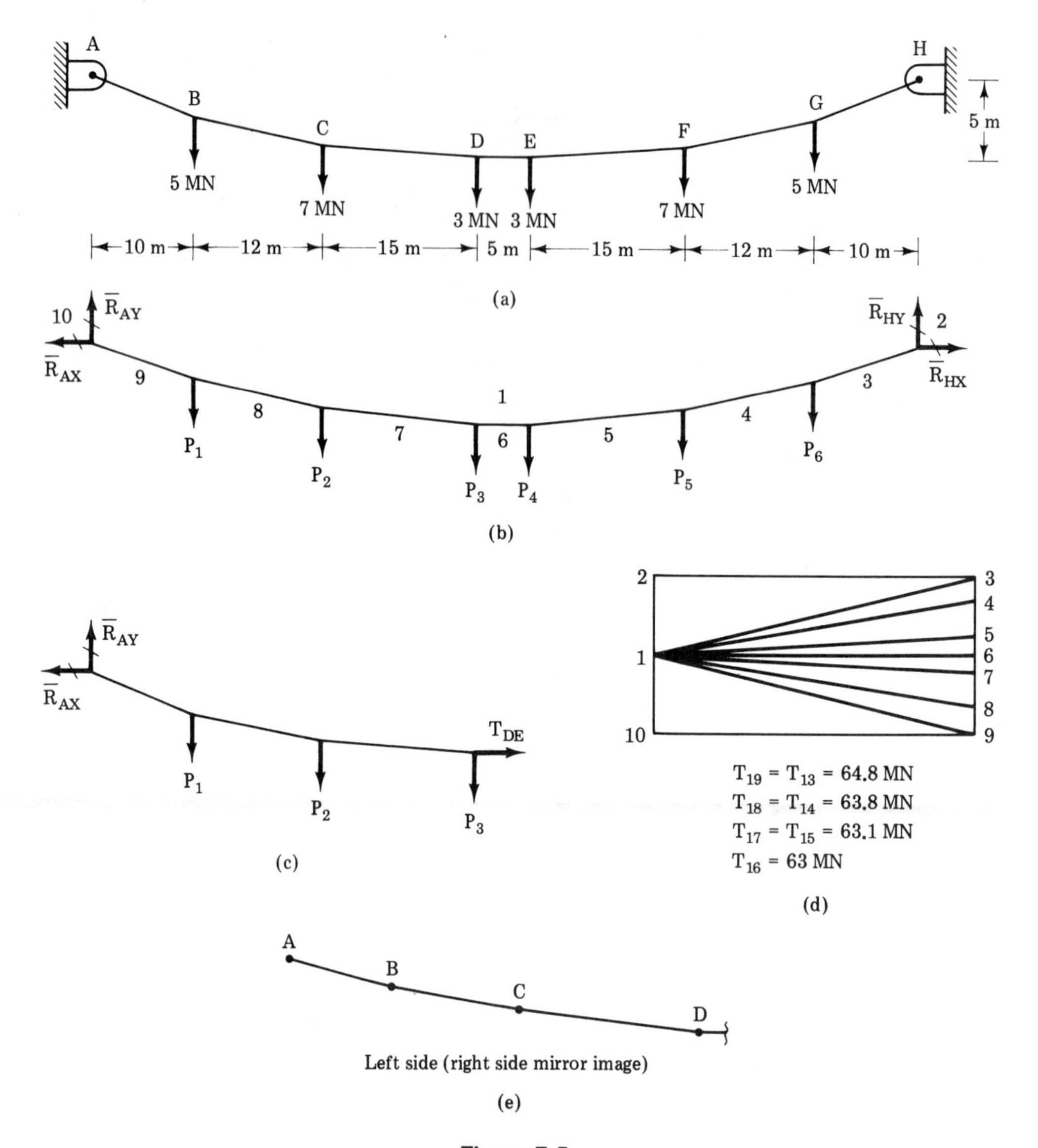

Figure 7-5

$$T_{DE}f = -[-(5\text{ MN})(10\text{ m}) - (7\text{ MN})(22\text{ m}) - (3\text{ MN})(37\text{ m})]$$

$$T_{DE}f = +315\text{ MN m} \qquad T_{DE} = \frac{T_{DE}f}{f} = \frac{315\text{ MN m}}{5\text{ m}} \qquad T_{DE} = +63\text{ MN}$$

$$\Sigma F_X = 0 \qquad \overline{R}_{AX} + T_{DE} = 0 \qquad \overline{R}_{AX} = -T_{DE} \qquad \overline{R}_{AX} = -(+63\text{ MN})$$

$$\boldsymbol{\overline{R}_{AX} = -63\textbf{ MN}}$$

Step 6 Determine R_{HX} by writing and solving the equation for $\Sigma F_X = 0$ for the entire cable.

$$\Sigma F_X = 0 \qquad \overline{R}_{AX} + \overline{R}_{HX} = 0 \qquad \overline{R}_{HX} = -\overline{R}_{AX} \qquad \boldsymbol{\overline{R}_{HX} = +63\ \mathrm{MN}}$$

Step 7 Apply Bow's notation to the cable free-body diagram as shown in Figure 7-5b.

Step 8 Determine the tensions in all sections of the cable by the combined graphical method as shown in Figure 7-5d.

Step 9 Make a scale drawing of the cable as displaced by the loads as shown in Figure 7-5e.

Problem 7-1: Rework Example 7-1 using a sag of 1 m.

Problem 7-2: Determine the support reactions and the tension in each section of the cable shown in Figure 7-6. Make a scaled sketch showing the displacement of the cable.

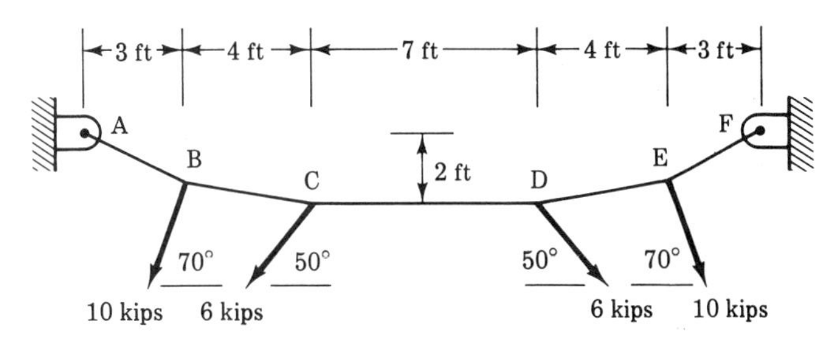

Figure 7-6

As the number of loads acting on a cable is increased, the string polygon of displacement begins to approach a smooth curve. If all the loads are equal and relatively closely spaced, the total load becomes essentially a uniformly distributed load as shown in Figure 7-7a.

Referring to Figure 7-7a, the same type of analysis performed for concentrated loads can be used here. Taking moments about A for the entire cable yields

$$\Sigma M_A = 0 \qquad P\frac{l}{2} + \overline{R}_{BY} l = 0 \qquad \overline{R}_{BY} = -\frac{1}{2}P$$

Summing vertical forces yields

$$\Sigma F_Y = 0 \qquad \overline{R}_{AY} + P + \overline{R}_{BY} = 0 \qquad \overline{R}_{AY} = -(P + \overline{R}_{BY})$$

$$\overline{R}_{AY} = -\frac{P}{2}$$

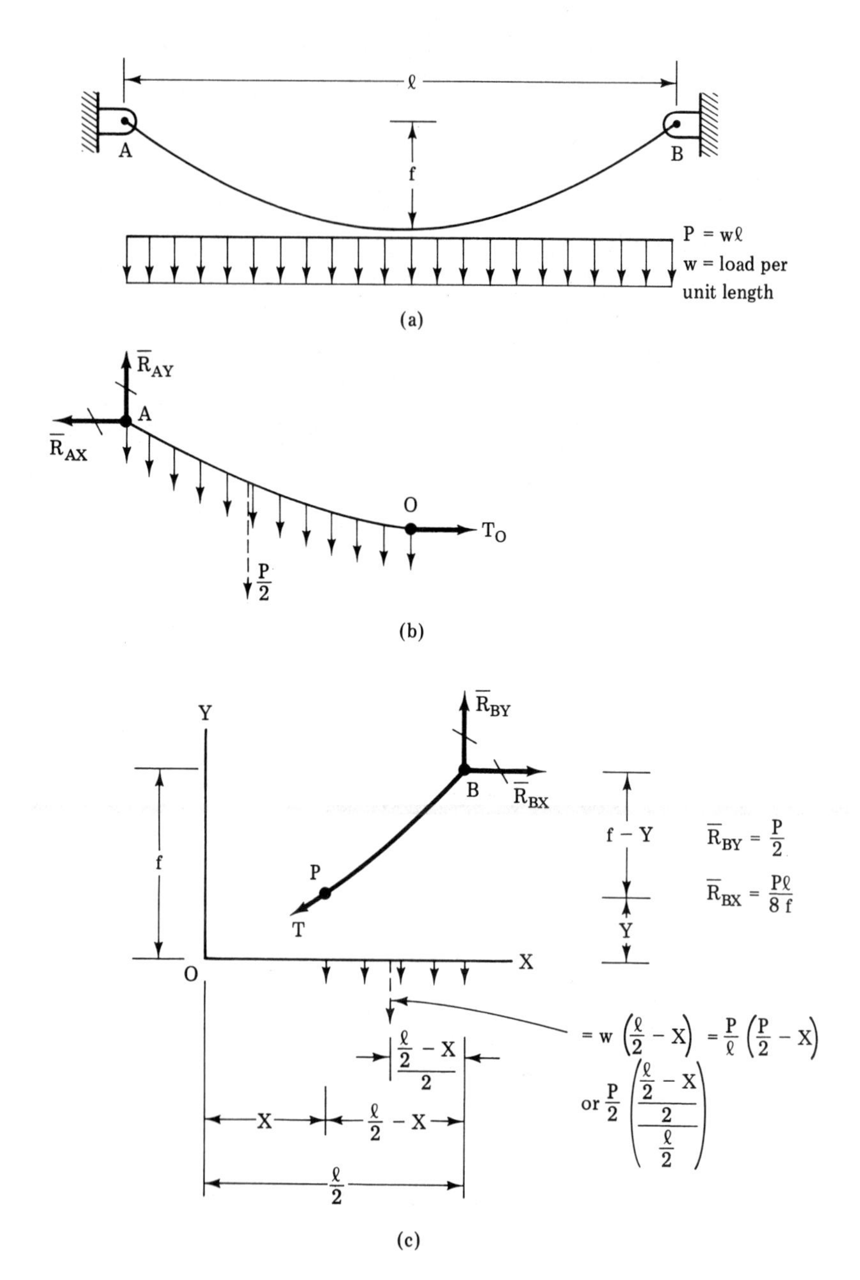

Figure 7-7. Cable analysis, uniformly distributed load.

Drawing a free-body diagram for the left-hand half of the cable as shown in Figure 7-7b and taking moments about A yields

$$\Sigma M_A = 0 \qquad \left(\frac{P}{2}\right)\left(\frac{l}{4}\right) + T_0 f = 0 \qquad T_0 = \frac{Pl}{8f}$$

Summing the horizontal forces yields

$$\Sigma F_X = 0 \qquad T_0 + \overline{R}_{AX} = 0 \qquad \overline{R}_{AX} = -T_0$$

Summing the horizontal forces for the entire cable yields

$$\overline{R}_{AX} + \overline{R}_{BX} = 0 \qquad \overline{R}_{BX} = -\overline{R}_{AX}$$

The maximum value for the cable tension occurs at the supports, as before

$$T = \sqrt{R_{AX}^2 + R_{AY}^2} = \sqrt{\frac{(R_{AX})^2}{(R_{AX})^2}\,(R_{AX}^2 + R_{AY}^2)}$$

$$T = R_{AX}\sqrt{1 + \left(\frac{R_{AY}}{R_{AX}}\right)^2} = R_{AX}\sqrt{1 + \left(\frac{P}{2} \div \frac{Pl}{8f}\right)^2}$$

$$T = R_{AX}\sqrt{1 + 16\left(\frac{f}{l}\right)^2}$$

Since the ratio of f to l is generally 0.1 or less, it can be seen that the maximum tension tends to be less than 10% more than the horizontal thrust.

The shape of the curve of the cable formed by the uniformly distributed load can be shown to be a quadratic parabola. Referring to Figure 7-7c, consider point P with ordinate y and abscissa x. Taking moments about P [only that portion of the load distributed over $(l/2) - x$ should be used] yields

$$+\frac{P}{2}\left(\frac{l}{2} - x\right) - \left(\frac{Pl}{8f}\right)(f - y) - \frac{P}{2}\left(\frac{\frac{l}{2} - x}{l/2}\right)\left(\frac{\frac{l}{2} - x}{2}\right) = 0$$

$$\left(\frac{l}{2} - x\right) - \frac{l}{4f}(f - y) - \frac{\left(\frac{l}{2} - x\right)^2}{l} = 0$$

$$\frac{l^2}{2} - x - \frac{l^2}{4} + \frac{l^2 y}{4f} - \frac{l^2}{4} + xl - x^2 = 0$$

$$y = 4f\left(\frac{x}{l}\right)^2$$

Thus, for any cable with uniformly distributed load and known length and sag, a specific parabola describes its deformation.

Example 7-2: Determine the support reactions, the maximum cable tension, and the displacement 7 ft to the right of *A* for the cable shown in Figure 7-8a.

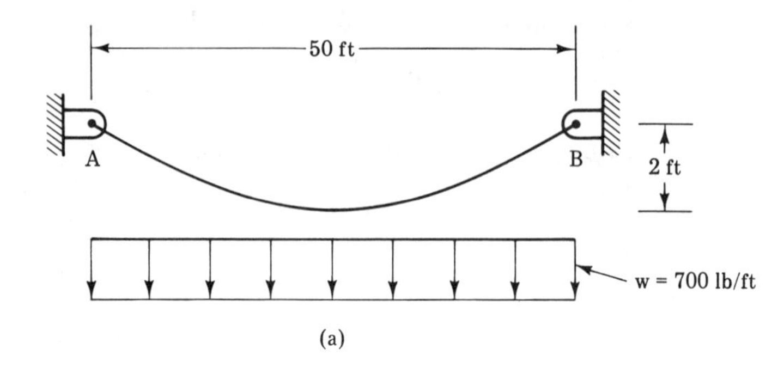

(a)

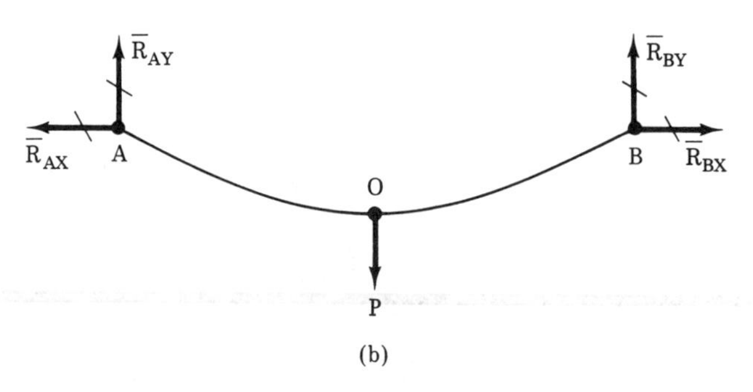

(b)

Figure 7-8

Solution:

Step 1 Draw a free-body diagram of the entire cable.

Step 2 Determine the total load, R_{AY} and R_{BY}, as follows:

$$P = wl = 700 \text{ lb/ft } (50 \text{ ft}) \qquad P = -3500 \text{ lb}$$

$$R_{AY} = R_{BY} = -\frac{P}{2} \qquad \boldsymbol{R_{AY} = R_{BY} = +1750 \text{ lb}}$$

$$R_{AX} = -\frac{Pl}{8f} = -\frac{3500 \text{ lb}(50 \text{ ft})}{8(2 \text{ ft})} \qquad \boldsymbol{R_{AX} = -10{,}940 \text{ lb}}$$

$$R_{BX} = -R_{AX} \qquad \boldsymbol{R_{BX} = +10{,}940 \text{ lb}}$$

Step 3 Determine the maximum tension as follows:

$$T = R_{AX}\sqrt{1 + 16\,(f/l)^2} \qquad T = 10{,}940 \text{ lb}\sqrt{1 + 16(2/50)^2}$$

$$\boldsymbol{T = 11{,}080 \text{ lb}}$$

Step 4 Determine the displacement at 8 ft to the right of A. At that point $x = -17$ ft

$$y = 4f\left(\frac{x}{l}\right)^2 = 4(2 \text{ ft})\left(\frac{17 \text{ ft}}{50 \text{ ft}}\right)^2 \qquad \boldsymbol{y = 0.925 \text{ ft}}$$

Problem 7.3: Determine the support reactions, the maximum cable tension, and the displacement 4 m to the left of B for the cable shown in Figure 7-9.

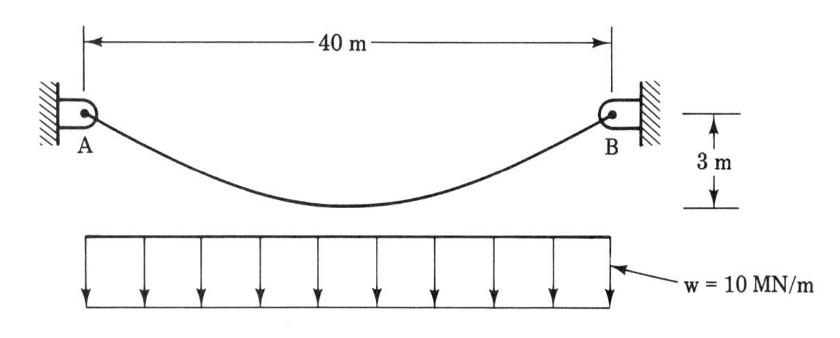

Figure 7-9

Problem 7-4: Determine the support reactions, the maximum cable tension, and the displacement 5 ft to the right of A for the cable shown in Figure 7-10.

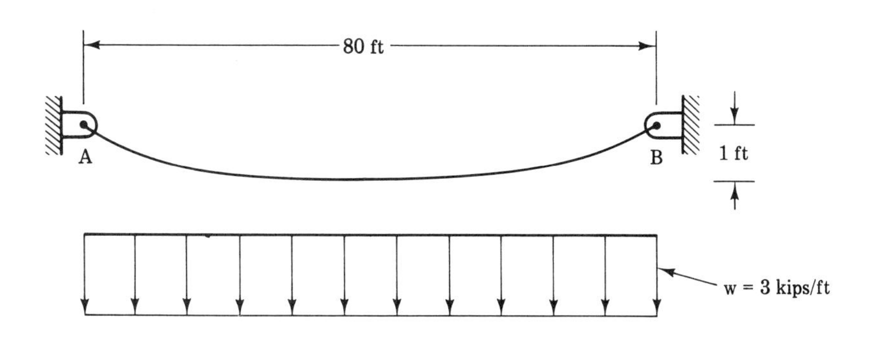

Figure 7-10

7.4 UNSYMMETRICALLY LOADED CABLES

Unsymmetrically loaded cables may exhibit a substantial lateral shift of points on the cable upon loading. Further, the point of maximum sag is not always apparent by inspection. However, it is reasonable to assume that the point(s) of load application remain constant and that the direction(s) of load application remain constant. Given these two assumptions, the techniques of the preceding section can be applied.

Example 7-3: Determine the support reactions and the tension in each section of the cable shown in Figure 7-11a. Make a scaled sketch showing the displacement of the cable.

Solution:

Step 1 Draw the free-body diagram for the entire cable as shown in Figure 7-11b.

Step 2 Since the direction, magnitude, and line of action for each load is known, the vertical reactions can be found by writing the equations for $\Sigma M_A = 0$ and $\Sigma F_Y = 0$ (K = kip):

$$\Sigma M_A = 0 \qquad P_1 d_1 + P_{2X} d_{2X} + P_{2Y} d_{2Y} + \overline{R}_{BY} d_{BY} = 0$$

$$P_{2X} = P_2 \cos\theta_{P2} \qquad P_{2X} = (11\text{ K})\cos 300° \qquad P_{2X} = 5.5\text{ K}$$

$$P_{2Y} = P_2 \sin\theta_{P2} \qquad P_{2Y} = (11\text{ K})\sin 300° \qquad P_{2Y} = -9.53\text{ K}$$

$$\overline{R}_{BY} d_{BY} = -[-(8\text{ K})(10\text{ ft}) + (5.5\text{ K})(2\text{ ft}) - (9.53\text{ K})(20\text{ ft})] = +260\text{ K ft}$$

$$\overline{R}_{BY} = \frac{\overline{R}_{BY} d_{BY}}{d_{BY}} = \frac{260\text{ K ft}}{35\text{ ft}} \qquad \boldsymbol{\overline{R}_{BY} = +7.43\text{ K}}$$

$$\Sigma F_Y = 0 \qquad \overline{R}_{AY} + P_1 + P_{2Y} + \overline{R}_{BY} = 0$$

$$\overline{R}_{AY} = -(P_1 + P_{2Y} + R_{BY})$$

$$= -(-8 - 9.53 + 7.43\text{ K}) \qquad \boldsymbol{\overline{R}_{AY} = +10.1\text{ K}}$$

Step 3 Examining P_1 and P_{2Y}, it can be seen that maximum sag will occur at P_2. From this, the angle of DB can be found. Analyzing the forces acting at B indicates that the direction of all forces and the magnitude of one of them are known. $\overline{R}_{BX}$ can thus be found and by $\Sigma F_X = 0$, $\overline{R}_{AX}$ can also be found.

$$\theta_{DB} = \arctan\frac{-2}{-15} \qquad \theta_{DB} = 187.6°$$

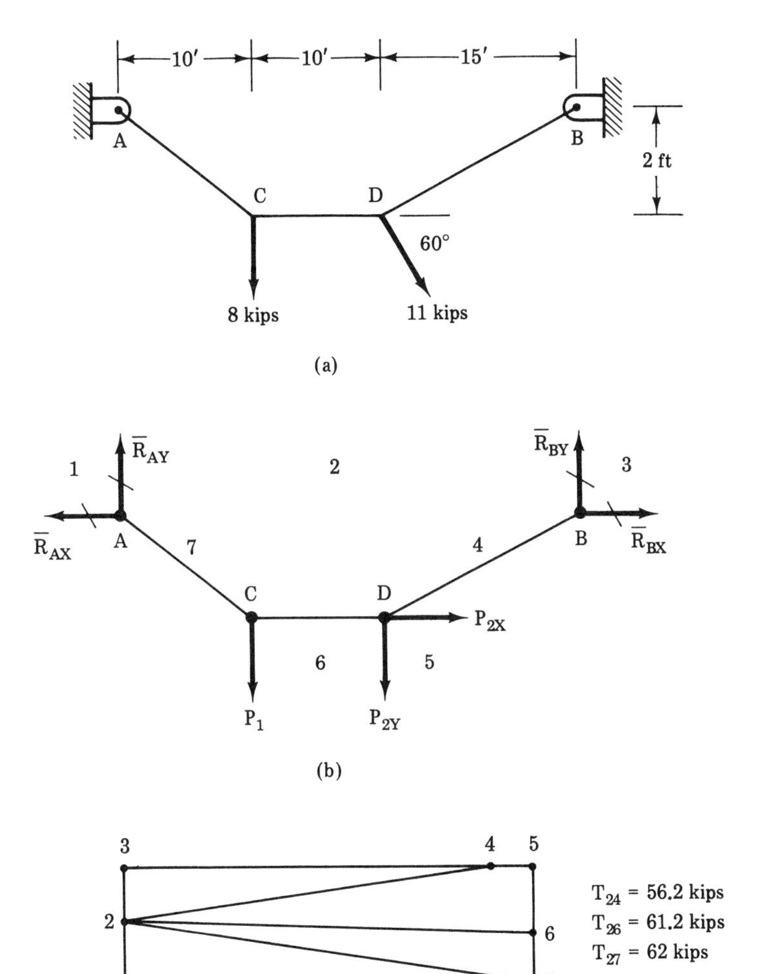
10′
10′
15′
A
B
2 ft
C
D
60°
8 kips
11 kips
(a)
$\bar{R}_{AY}$
1
2
$\bar{R}_{BY}$
3
$\bar{R}_{AX}$
A
7
4
B
$\bar{R}_{BX}$
C
D
P_{2X}
6
5
P_1
P_{2Y}
(b)
3
4
5
2
6
1
7
T_{24} = 56.2 kips
T_{26} = 61.2 kips
T_{27} = 62 kips
(c)

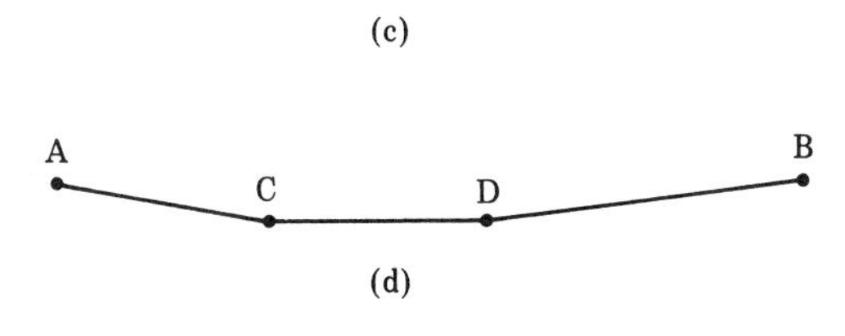
A
B
C
D
(d)

Figure 7-11

for joint *B*:

$$\Sigma F_Y = 0 \qquad \overline{R}_{BY} + F_{DBY} = 0 \qquad F_{DBY} = -\overline{R}_{BY} = -7.43\text{ K}$$

$$\Sigma F_X = 0 \qquad \overline{R}_{BX} + F_{DBX} = 0 \qquad F_{DBX} = \frac{F_{DBY}}{\tan\theta_{DB}}$$

$$F_{DBX} = \frac{-7.43\text{ K}}{\tan 187.6^\circ} \qquad F_{DBX} = -55.7\text{ K} \qquad \boldsymbol{\overline{R}_{BX} = +55.7\text{ K}}$$

for entire cable:

$$\Sigma F_X = 0 \qquad \overline{R}_{AX} + P_{2X} + \overline{R}_{BX} = 0 \qquad \overline{R}_{AX} = -(P_{2X} + \overline{R}_{BX})$$

$$P_{2X} = P_2 \sin\theta_{P2} \qquad P_{2X} = (11\text{ K})\cos 300 \qquad P_{2X} = +5.5\text{ K}$$

$$R_{AX} = -(5.5 + 55.7\text{ K}) \qquad \boldsymbol{\overline{R}_{AX} = -61.2\text{ K}}$$

Step 4 Apply Bow's notation to the free-body diagram as shown in Figure 7-11b. Solution for cable tensions can be obtained by the combined graphical method as shown in Figure 7-11c.

Step 5 Finally, a scaled drawing of the cable displacement can now be made as shown in Figure 7-11d.

Problem 7-5: Determine the support reactions and the tension in each section of the cable shown in Figure 7-12. Make a scaled sketch showing the displacement of the cable.

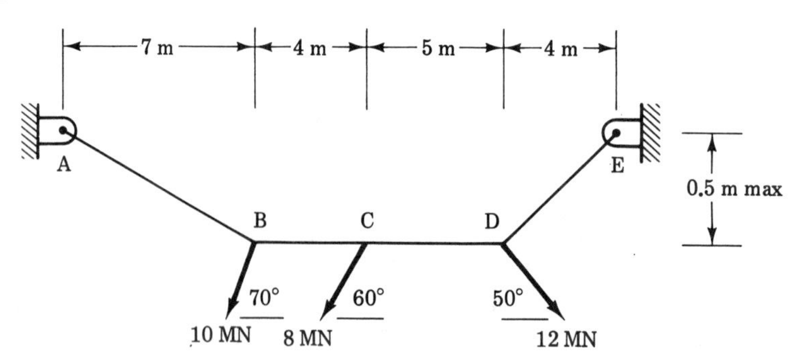

Figure 7-12

7.5 CABLE DESIGN

Cable design is a matter of

1. a choice of material,

2. selection of a factor of safety (dependent on the application),
3. selection of a solid or stranded construction (flexibility for shipment and installation, and even on-site assembly for large sizes, favor stranded construction for most applications), and
4. sizing of diameter and number of strands.

The two basic relationships required for cable design are thus:

$$\text{Factor of Safety} = \frac{\text{ultimate strength}}{\text{working strength}}$$

$$FS = \frac{\sigma_u}{\sigma_w}$$

$$\text{Working Strength} = \frac{\text{maximum tension}}{\text{cross-sectional area}}$$

$$\sigma_w = \frac{T}{A}$$

In application, these can be combined and solved for A to yield

$$A = \frac{T \cdot FS}{\sigma_u}$$

Since we are dealing with circular cross sections where $A = 0.7854d^2$, the above can be further refined to

$$d = \sqrt{\frac{1.273T \cdot FS}{\sigma_u}}$$

When working with a stranded cable of n strands, the total area is equal to nA, A being the cross-sectional area of one strand. The above equation then becomes

$$d = \sqrt{\frac{1.273T \cdot FS}{n\sigma_u}}$$

Example 7-4: Using steel with an ultimate strength of 85,000 psi and applying a factor of safety of 5, determine for a cable tension of 10,000 lb:

1. the diameter required for a single strand,
2. the strand diameter if 20 strands are used, and
3. the number of strands required if a strand diameter of $\frac{1}{4}$ inch is used.

Step 1 In answer to 1

$$d = \sqrt{\frac{1.273T \cdot FS}{\sigma_u}}$$

$$d = \sqrt{\frac{1.273(10{,}000 \text{ lb})(5)}{85{,}000 \text{ lb/in.}^2}}$$

$d = 0.865$ in. (next largest commercial size)

Step 2 In answer to 2

$$d = \sqrt{\frac{1.273T \cdot FS}{n\sigma_u}}$$

$$d = \sqrt{\frac{1.273(10{,}000 \text{ lb})(5)}{20(85{,}000 \text{ lb/in.}^2)}}$$

$d = 0.193$ in. (next largest commercial size)

Step 3 In answer to 3

$$n = \frac{1.273T \cdot FS}{d^2\sigma_u}$$

$$n = \frac{1.273\ (10{,}000 \text{ lb})(5)}{(0.25 \text{ in.})^2(85{,}000 \text{ lb/in.}^2)}$$

$$n = 11.98 \qquad n = 12$$

Problem 7-6: Determine the number of 25-mm diameter steel strands ($\sigma_u = 400$ MPa) required for the cable in Example 7-1 if a factor of safety of 6 is specified.

7.6 READINESS QUIZ

1. The tension in a cable is a function of its

   (a) span
   (b) sag
   (c) loading
   (d) all of the above
   (e) none of the above

2. If the load on a cable is uniformly distributed, the cable

   (a) remains straight

(b) takes the form of a parabola
(c) takes the form of a hyperbola
(d) all of the above
(e) none of the above

3. The diameter of a stranded cable is __________ the diameter of a solid cable of equal strength.

   (a) equal to
   (b) larger than
   (c) less than
   (d) any of the above
   (e) none of the above

4. Maximum cable tension is always __________ either of the horizontal reactions at the supports.

   (a) larger than
   (b) smaller than
   (c) equal to
   (d) any of the above
   (e) insufficient information

5. In working with an unsymmetrically loaded cable, one difficulty encountered is that

   (a) the direction of a load may not be clear
   (b) the location of a load may not be clear
   (c) the point of maximum sag may not be immediately evident
   (d) all of the above
   (e) none of the above

6. In designing cables, the following may be safely neglected:

   (a) deformation of cable due to tensile stress
   (b) displacement of cable due to loads
   (c) total permissible sag
   (d) all of the above
   (e) none of the above

7. In order to design a cable with no sag

   (a) a material with extremely high strength-to-weight ratio must be used
   (b) loads must be symmetrical
   (c) loads must be uniformly distributed
   (d) all of the above
   (e) none of the above

8. The vertical support reaction at the end of a cable ________ the horizontal support reaction.

   (a) is always equal to
   (b) can be equal to
   (c) is always smaller than
   (d) is never larger than
   (e) none of the above

9. The weight of the cable itself with no load applied but with sag f (assume supports at equal elevation) is always ________ distributed.

   (a) symmetrically
   (b) unsymmetrically
   (c) uniformly
   (d) any of the above
   (e) none of the above

10. The less sag a cable of known span and loading has, the

   (a) greater the maximum tension
   (b) smaller the thrust of the cable
   (c) smaller the vertical reactions at the supports
   (d) all of the above
   (e) none of the above

7.7 SUPPLEMENTARY PROBLEMS

7-7 Determine the support reactions and the tension in each section of the cable shown in Figure 7-13. Make a scaled sketch showing the displacement of the cable.

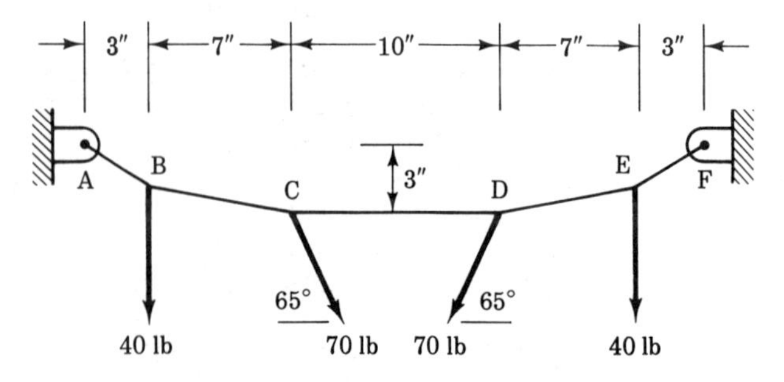

Figure 7-13

7-8 Determine the support reactions, the maximum cable tension, and the displacement 3 m to the left of B for the cable shown in Figure 7-14.

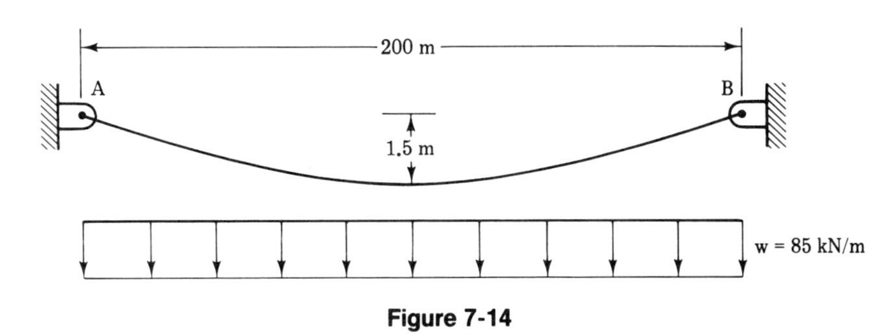

Figure 7-14

7-9 Determine the support reactions and the tension in each section of the cable shown in Figure 7-15. Make a scaled sketch showing the displacement of the cable.

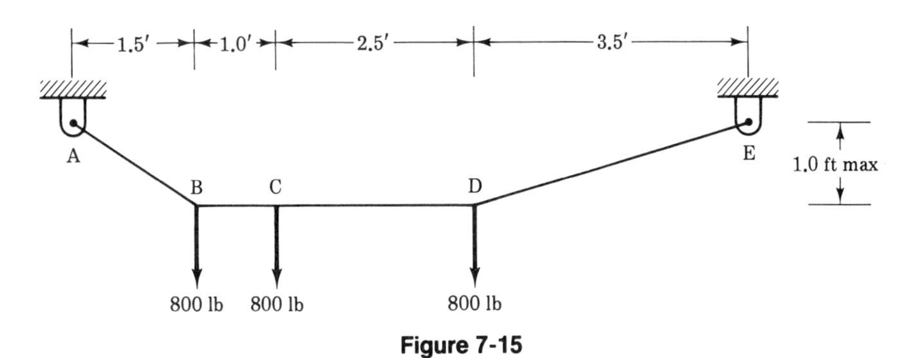

Figure 7-15

7-10 Using a bronze with an ultimate strength of 52,000 psi and applying a factor of safety of 2.5 determine the minimum diameter permissible for a single-strand cable for Problem 7-7.

7-11 Determine the minimum number of 40-mm strands of an aluminum alloy with an ultimate strength of 200 MPa required in Problem 7-8 if a factor of safety of 4 is to be used.

7-12 Determine the strand diameter required in Problem 7-9 if 15 strands of steel with an ultimate strength of 95,000 psi are to be used with a factor of safety of 12.

Chapter

8

Noncoplanar Force Systems

8.1 OBJECTIVES

Upon completion of the work relating to this chapter, you should be able to perform the following.

1. Given a noncoplanar force, resolve it into components lying along each axis of a given three-axis cartesian coordinate system.
2. Given forces lying along each of three axes of a cartesian coordinate system, resolve them into a single noncoplanar force.
3. Given two or more noncoplanar forces, resolve them into a single resultant.
4. Given a noncoplanar system of forces acting on a structure, mechanism, or machine, determine any unknown support reactions, forces in members, and torques in members.

8.2 NONCOPLANAR FORCE SYSTEMS DEFINED

There are many three-dimensional force systems that can be separated into two or more coplanar force systems and solved using the techniques covered in the first seven chapters. Some of the applications examined in this chapter will be of that nature. However, many applications will involve noncoplanar force systems that are parallel (but neither uniformly nor even

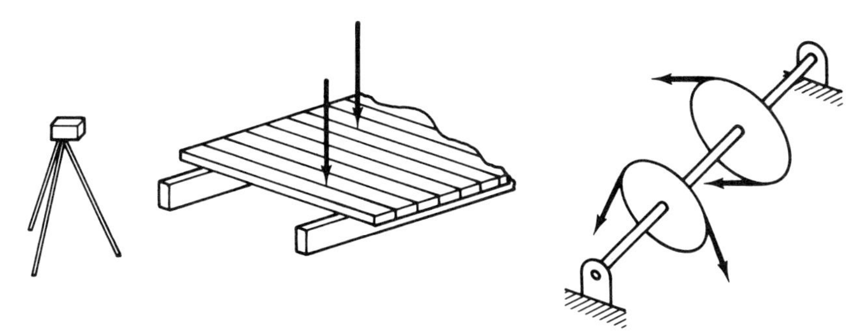

Figure 8-1. Noncoplanar force systems.

symmetrically distributed), concurrent, or nonconcurrent, that cannot be reduced to a number of simpler systems for solutions.

There are enough examples in our lives to convince us that noncoplanar systems are common enough to warrant our consideration. The tripod as used for a camera or a surveying instrument is basically a noncoplanar, concurrent force problem. A building floor with its symmetrical loads of beams, uniform loads of flooring but highly unsymmetrical loading of fixtures and equipment is a noncoplanar, parallel force problem. A shaft with two supports and two or more pulleys, even though there is an orientation to the shaft axis is a noncoplanar, nonconcurrent, nonparallel force problem. These examples are illustrated in Figure 8-1. Combinations of such systems, including many forces, exist in real life. The techniques of this chapter can be applied to problems much more complex than the ones we shall consider.

8.3 VECTORS IN SPACE

Working with vectors in three-dimensional space requires that our use of a cartesian or rectangular coordinate system be extended to three axes as shown in Figure 8-2a. The y and x axes are oriented as before. The third, or z axis, can be envisioned as being directed toward (+) and away (−) from us, perpendicular to both the x and the y axes.

In Figure 8-2b, a three-dimensional vector and its three components has been drawn onto the coordinate system. Envision this as creating a volume, a rectangular parallelepiped, probably like the room you are in. The three-dimensional vector is a straight line from one corner of the room at floor level to the diagonally opposite corner of the room at ceiling level. If you start at corner 0, there is no other straight path to another corner that cuts through space. Line $0E$ can be called R, a space vector. $0A$, $0B$, and $0C$ or respectively the three components of the space vector R_X, R_Y, and R_Z.

Note that every possible pair of these four vectors form a plane (as you

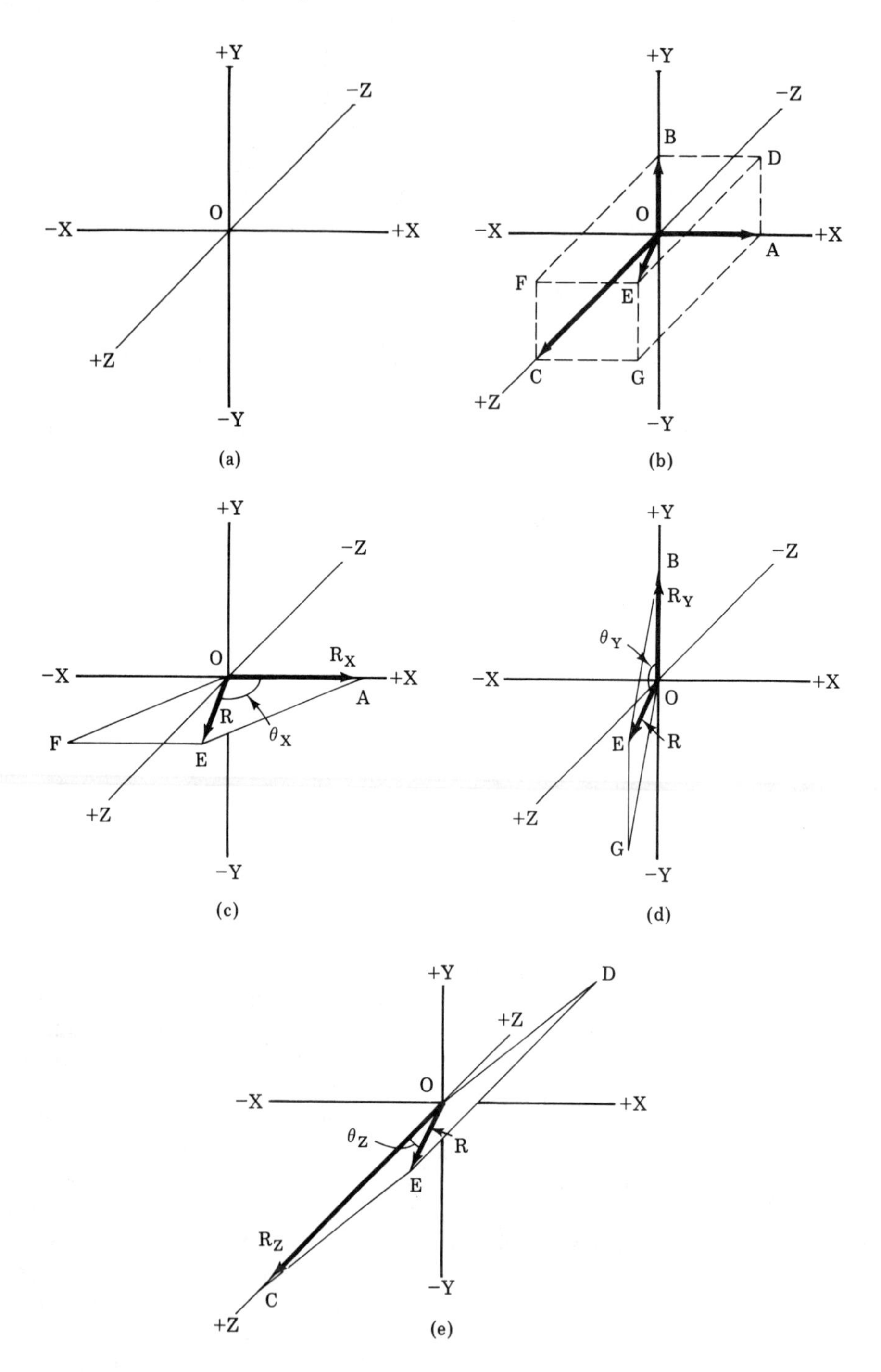

Figure 8-2. Space vectors.

would expect of two intersecting straight lines). A number of useful relationships can be derived from these planes.

First, neglecting the y axis for the time being, let us add the vectors R_X and R_Z to form a resultant in the XZ plane.

$$R_{XZ} = \sqrt{R_X^2 + R_Z^2}$$

This vector lies along $0G$. However, $0G$ lies in the vertical plane $0BEG$, which also contains the component R_Y. Since R_Y is, by definition, perpendicular to R_X and R_Z, it is also perpendicular to plane XZ and vector R_{XZ}. Therefore, R_Y and R_{XZ} being perpendicular to each other in the plane $0BEG$ can be added by Pythagoras' theorem to form the total resultant or space force R:

$$R = \sqrt{R_Y^2 + R_{XZ}^2}$$

Since $R_{XZ}^2 = R_X^2 + R_Z^2$

$$R = \sqrt{R_X^2 + R_Y^2 + R_Z^2}$$

If this requires proof, get a friend and a long tape measure (or a string and a meter stick) and measure the dimensions involved in any appropriate room. After all, vectors are vectors whether they involve distance, velocity, acceleration, force, or anything else that has direction as well as magnitude.

Second, note that in each plane formed by the space force and one of its components (e.g., plane $0AEF$, R_X and R) a specific relationship exists, namely the angle between the space force and its component (in this case θ_X). The other leg of the triangle thus formed is always the vector sum of the other two components. The cosines of these angles are called direction cosines, leading us to the following relationships:

$$R_X = R \cos \theta_X \qquad \cos \theta_X = \frac{R_X}{R}$$

$$R_Y = R \cos \theta_Y \qquad \cos \theta_Y = \frac{R_Y}{R}$$

$$R_Z = R \cos \theta_Z \qquad \cos \theta_Z = \frac{R_Z}{R}$$

If any two of the angles are known, the third angle is automatically specified. This not only appears logical physically but can be proven mathematically as follows:

$$R^2 = R_X^2 + R_Y^2 + R_Z^2$$

$$R^2 = R^2\cos^2\theta_X + R^2\cos^2\theta_Y + R^2\cos^2\theta_Z$$

$$1 = \cos^2\theta_X + \cos^2\theta_Y + \cos^2\theta_Z$$

Specifying two of these angles specifies their cosines thus specifying the cosine and, therefore, the magnitude of the third. These angles are always taken as acute angles. Since R always has an absolute value, the sign of the cosine of an angle always indicates the direction of the component associated with that angle. Thus the indication of the direction of all three cosines immediately fixes which of the eight possible spaces the space force is located in.

Example 8-1: Determine the resultant of $F_X = +100$ N, $F_Y = -150$ N, and $F_Z = -50$ N. Provide a sketch.

Solution:

Step 1 Sum up the components using Pythagoras' theorem.

$$R = \sqrt{R_X^2 + R_Y^2 + R_Z^2}$$

$$R = \sqrt{(100\text{ N})^2 + (-150\text{ N})^2 + (-50\text{ N})^2}$$

$$R = 187\text{ N}$$

Step 2 Establish the direction of R by finding the appropriate angles and the corresponding direction cosines.

$$\cos\theta_X = \frac{R_X}{R} = \frac{100}{187} \qquad \cos\theta_X = +0.535 \qquad \theta_X = 57.7°$$

$$\cos\theta_Y = \frac{R_Y}{R} = \frac{-150}{187} \qquad \cos\theta_Y = -0.802 \qquad \theta_Y = 36.7°$$

$$\cos\theta_Z = \frac{R_X}{R} = \frac{-50}{187} \qquad \cos\theta_Z = -0.267 \qquad \theta_Z = 74.5°$$

If you wish to check this step, see if the sum of the squares of the cosines of the three angles equals one. (It does!)

Step 3 Make a sketch by drawing a coordinate system like the one shown in Figure 8-2a. Then lay out the three vectors, each along its appropriate axis and in the correct direction. Draw the parallel lines to construct the "box." Then sketch in R and label the angles appropriately. The completed sketch is shown in Figure 8-3.

Problem 8-1: Determine the resultant of $F_X = -500$ lb, $F_Y = -200$ lb, and $F_X = -300$ lb. Provide a sketch.

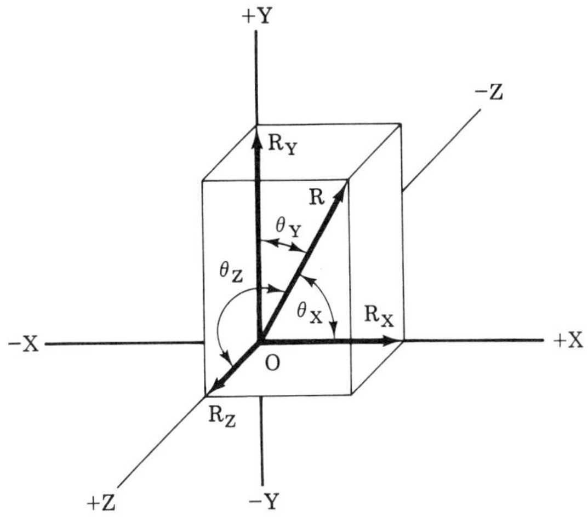

Figure 8-3

Example 8-2: Determine the components of the space force shown in Figure 8-4.

Solution:

Step 1 Determine θ_Z

$$\cos^2\theta_X + \cos^2\theta_Y + \cos^2\theta_Z = 1$$

$$\cos\theta_Z = \sqrt{1 - \cos^2\theta_X - \cos^2\theta_Y}$$

$$\theta_Z = \arccos\sqrt{1 - \cos^2 15^\circ - \cos^2 80^\circ}$$

$$\theta_Z = 14.2^\circ$$

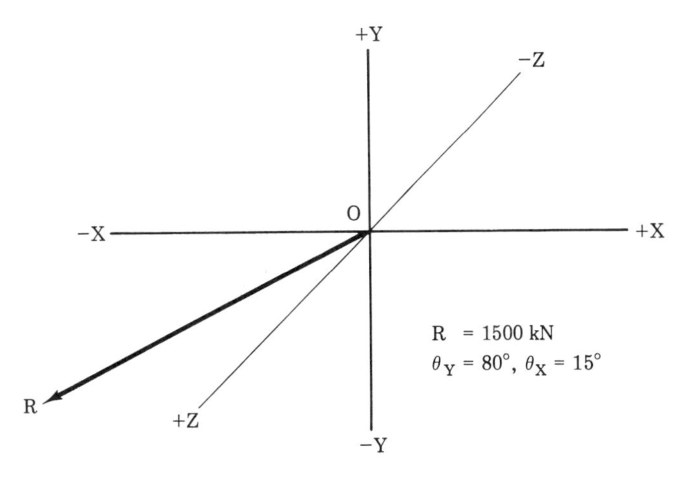

Figure 8-4

Step 2 Determine the components using direction cosines. The sign of the cosine must be the same as the sign of the direction of the component which can be determined by inspection from Figure 8-4.

$$\cos \theta_X = -\cos 15° = -0.966$$

$$\cos \theta_Y = -\cos 80° = -0.174$$

$$\cos \theta_Z = +\cos 14.2° = +0.969$$

$$F_X = R \cos \theta_X = 1500 \text{ kN}(-0.966) \qquad \boldsymbol{F_X = -1449 \text{ kN}}$$

$$F_Y = R \cos \theta_Y = 1500 \text{ kN}(-0.174) \qquad \boldsymbol{F_Y = -260 \text{ kN}}$$

$$F_Z = R \cos \theta_Z = 1500 \text{ kN}(+0.969) \qquad \boldsymbol{F_Z = +1454 \text{ kN}}$$

Problem 8-2: Determine the components of the space force shown in Figure 8-5.

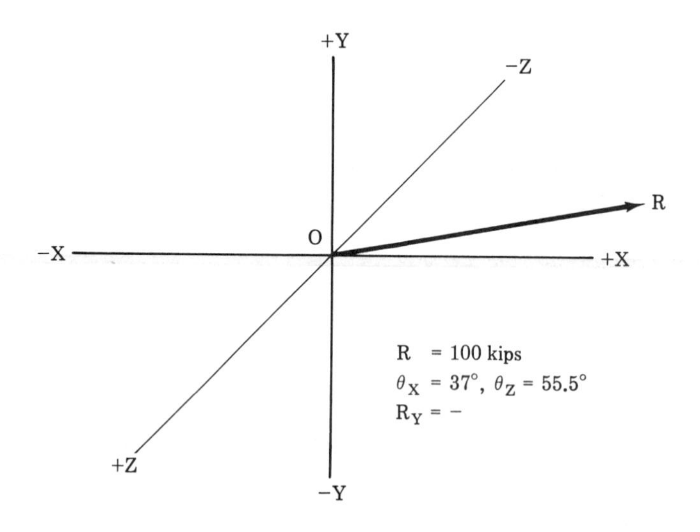

Figure 8-5

Inevitably, we will have to deal with a system of space forces rather than a single space force. Our previous work with the method of components together with our work on direction cosines can be combined to work such problems including those involving equilibrium.

Example 8-3: Determine the resultant of the space forces shown in Figure 8-6.

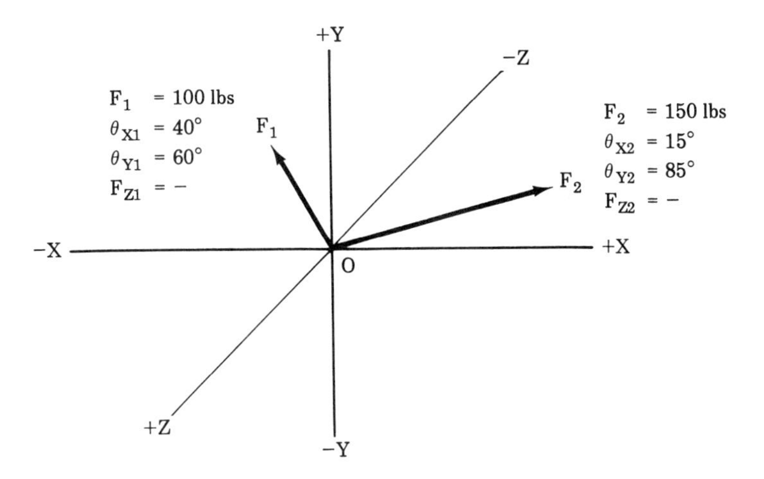

Figure 8-6

Solution:

Step 1 Determine θ_{Z1} and θ_{Z2}

$$\cos^2\theta_{X1} + \cos^2\theta_{Y1} + \cos^2\theta_{Z1} = 1$$

$$\cos\,\theta_{Z1} = \sqrt{1 - \cos^2\theta_{X1} - \cos^2\theta_{Y1}}$$

$$\theta_{Z1} = \arccos\sqrt{1 - \cos^2 40^\circ - \cos^2 60^\circ}$$

$$\theta_{Z1} = 66.2^\circ$$

$$\cos^2\theta_{X2} + \cos^2\theta_{Y2} + \cos\theta_{Z2} = 1$$

$$\cos\theta_{Z2} = \sqrt{1 - \cos^2\theta_{X2} - \cos^2\theta_{Y2}}$$

$$\theta_{Z2} = \arccos\sqrt{1 - \cos^2 15^\circ - \cos^2 85^\circ}$$

$$\theta_{Z2} = 75.9^\circ$$

Step 2 Determine the components using direction cosines.

$$\cos\,\theta_{X1} = -\cos\,40^\circ \qquad \cos\,\theta_{X1} = -0.766$$

$$\cos\,\theta_{Y1} = +\cos\,60^\circ \qquad \cos\,\theta_{Y1} = +0.500$$

$$\cos\,\theta_{Z1} = -\cos\,66.2^\circ \qquad \cos\,\theta_{Z1} = -0.404$$

$$\cos\,\theta_{X2} = +\cos\,15^\circ \qquad \cos\,\theta_{X2} = +0.969$$

$$\cos\,\theta_{Y2} = +\cos\,85^\circ \qquad \cos\,\theta_{Y2} = +0.87$$

$$\cos\,\theta_{Z2} = -\cos\,75.9^\circ \qquad \cos\,\theta_{Z2} = -0.244$$

$$F_{X1} = F_1 \cos\theta_{X1} = 100\text{ lb}(-0.766) \qquad F_{X1} = -76.6\text{ lb}$$
$$F_{Y1} = F_1 \cos\theta_{Y1} = 100\text{ lb}(+0.500) \qquad F_{Y1} = +50.0\text{ lb}$$
$$F_{Z1} = F_1 \cos\theta_{Z1} = 100\text{ lb}(-0.404) \qquad F_{Z1} = -40.4\text{ lb}$$
$$F_{X2} = F_2 \cos\theta_{X2} = 150\text{ lb}(+0.969) \qquad F_{X2} = +145.4\text{ lb}$$
$$F_{Y2} = F_2 \cos\theta_{Y2} = 150\text{ lb}(+0.087) \qquad F_{Y2} = +13.1\text{ lb}$$
$$F_{Z2} = F_2 \cos\theta_{Z2} = 150\text{ lb}(-0.244) \qquad F_{Z2} = -36.6\text{ lb}$$

Step 3 Determine the components of the resultant and the resultant itself.

$$R_X = \Sigma F_X = F_{X1} + F_{X2} = -76.6 + 145.5 \qquad R_X = +68.8\text{ lb}$$
$$R_Y = \Sigma F_Y = F_{Y1} + F_{Y2} = +50.0 + 13.1 \qquad R_Y = +63.1\text{ lb}$$
$$R_Z = \Sigma F_Z = F_{Z1} + F_{Z2} = -40.4 - 36.6 \qquad R_Z = -77\text{ lb}$$
$$R = \sqrt{R_X^2 + R_Y^2 + R_Z^2}$$
$$R = \sqrt{(68.8)^2 + (63.1)^2 + (-77)^2} \qquad \mathbf{R = 121\text{ lb}}$$
$$\theta_{XR} = \arccos\frac{R_X}{R} = \arccos\frac{68.8}{121} \qquad \boldsymbol{\theta_{XR} = 55.3^\circ}$$
$$\theta_{YR} = \arccos\frac{R_Y}{R} = \arccos\frac{63.1}{121} \qquad \boldsymbol{\theta_{YR} = 58.6^\circ}$$
$$\theta_{ZR} = \arccos\frac{R_Z}{R} = \arccos\frac{-77}{121} \qquad \boldsymbol{\theta_{ZR} = 50.5^\circ}$$

The above was somewhat lengthy and steps can be combined to shorten the process now that its details are clear.

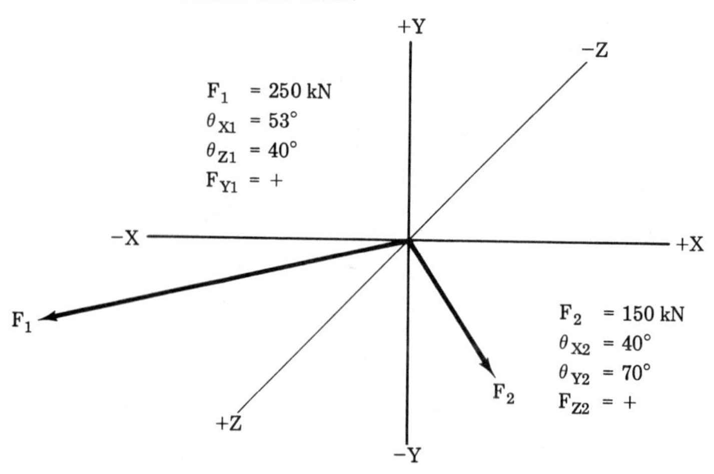

Figure 8-7

Problem 8-3: Determine the resultant of the space forces shown in Figure 8-7.

As previously indicated, there are many types of three-dimensional problems other than the concurrent force system type dealt with above. One of these is the noncoplanar, parallel force system. As with the coplanar force system, we shall have to resort to the use of moment equations as well as force equations to obtain a solution.

Example 8-4: The floor shown in Figure 8-8a weighs 50 lb/ft². In addition, there are concentrated loads located as shown. Determine the magnitude and location of the resultant of those loads.

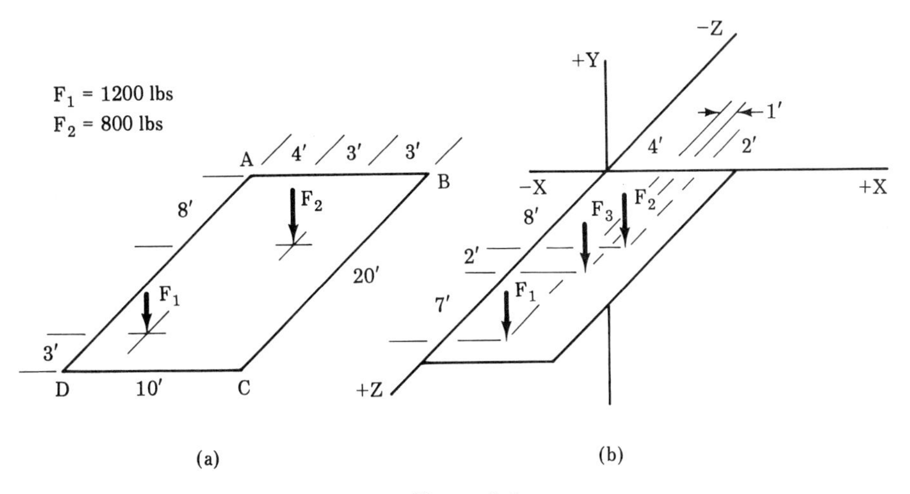

Figure 8-8

Solution:

Step 1 Recognize this as a problem involving noncoplanar, parallel forces. (Do not be confused by the fact that the points of load application are all essentially on one plane. Coplanar means that the *lines of action* are all in one plane.

Step 2 Apply a cartesian coordinate system putting all the forces parallel to the Y axis, the 10-ft sides parallel to the X axis, and the 20-ft sides parallel to the Z axis as shown in Figure 8-8b.

Step 3 Determine the magnitude and location of the concentrated load equivalent to the weight of the floor system.

$$F_3 = A\left(\frac{W}{A}\right) = (10 \text{ ft})(20 \text{ ft})\left(\frac{50 \text{ lb}}{\text{ft}^2}\right) \qquad F_3 = 10{,}000 \text{ lb}$$

There is no reason to assume that the floor is not uniform; therefore, the concentrated load would be located at $X = 5$ ft, $Z = 10$ ft as shown in Figure 8-8b.

Step 4 Determine the resultant by vector addition.

$$R = R_Y = \Sigma F_Y = F_{1Y} + F_{2Y} + F_{3Y} = F_1 + F_2 + F_3$$

$$R_Y = (-1200 \text{ lb}) + (-800 \text{ lb}) + (-10{,}000 \text{ lb})$$

$$\boldsymbol{R_Y = -12{,}000 \text{ lb}}$$

Step 5 Locate the resultant by use of moments. In this case, this must be done twice; once about the X axis and once about the Z axis to establish d_{RX}, the distance from the Z axis, and d_{RZ}, the distance from the X axis, both in the plane $ABCD$. Taking moments about the Z axis,

$$\Sigma M_Z = F_{1Y}(d_{1X}) + F_{2Y}(d_{2X}) + F_{3Y}(d_{3X})$$

$$\Sigma M_Z = -(1200 \text{ lb})(4 \text{ ft}) - (800 \text{ lb})(7 \text{ ft}) - (10{,}000 \text{ lb})(5 \text{ ft})$$

(Moment directions are usually taken looking from the positive end of the axis towards the negative end of the axis.)

$$\Sigma M_Z = -60{,}400 \text{ lb ft} \curvearrowright$$

$$\Sigma M_Z = R_Y d_{RX} \qquad d_{RX} = \frac{\Sigma M_Z}{R_Y} = \frac{60{,}400 \text{ lb ft}}{12{,}000 \text{ lb}}$$

$$\boldsymbol{d_{RX} = +5.033 \text{ ft}}$$

(Direction of d_{RX} was determined by examining the directions of both R_Y and M_Z.)

$$\Sigma M_X = F_{1Y}(d_{1Z}) + F_{2Y}(d_{2Z}) + F_{3Y}(d_{3Z})$$

$$\Sigma M_X = +(1200 \text{ lb})(17 \text{ ft}) + (800 \text{ lb})(8 \text{ ft}) + (10{,}000 \text{ lb})(10 \text{ ft})$$

$$\Sigma M_X = +126{,}800 \text{ lb ft} \curvearrowleft$$

$$\Sigma M_X = R_Y d_{RZ} \qquad d_{RZ} = \frac{\Sigma M_X}{R_Y} = \frac{126{,}800 \text{ lb ft}}{12{,}000 \text{ lb}}$$

$$\boldsymbol{d_{RZ} = +10.57 \text{ ft}}$$

Problem 8-4: Determine the magnitude and location of the resultant of the force system shown in Figure 8.9.

Finally, the most difficult type of problem, is the noncoplanar, noncon-current, nonparallel force system. This may require summing forces in all directions as well as taking moments about all three axes.

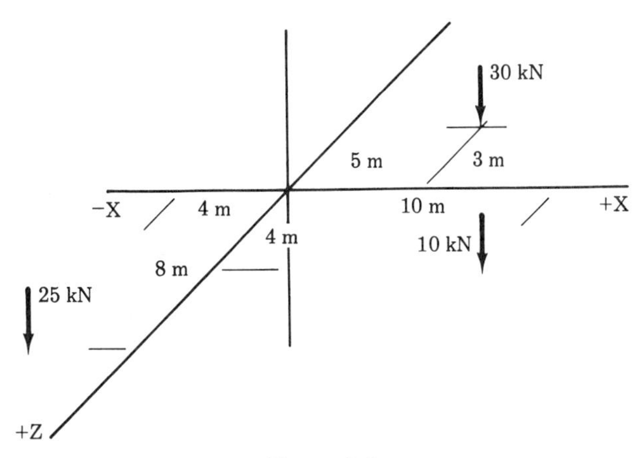

Figure 8-9

Example 8-5: Determine the magnitude, direction, and location of the resultant of the forces acting on the block shown in Figure 8-10.

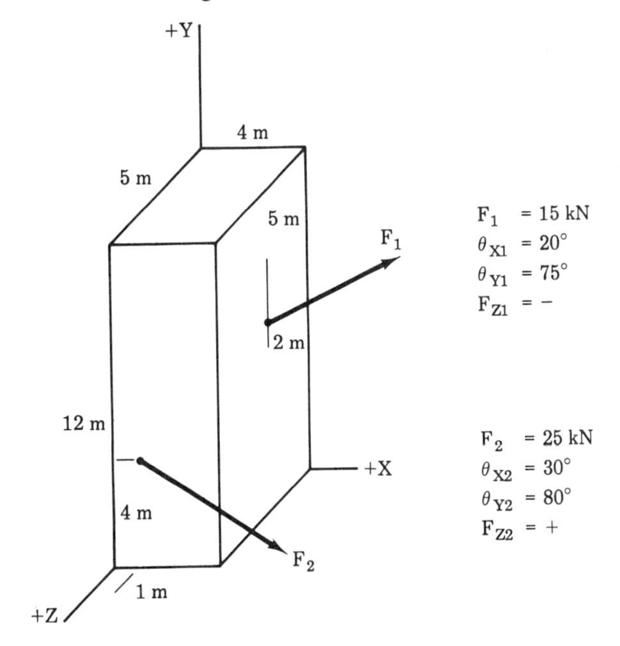

Figure 8-10

Solution:

Step 1 Determine the components of the forces and of the resultant, and the magnitude and direction of the resultant.

$$\theta_{Z1} = \arccos\sqrt{1 - \cos^2\theta_{X1} - \cos^2\theta_{Y1}}$$

$$\theta_{Z1} = \arccos\sqrt{1 - \cos^2 20° - \cos^2 75°}$$

$$\theta_{Z1} = 77.1°$$

$$\theta_{Z2} = \arccos\sqrt{1 - \cos^2\theta_{XZ} - \cos^2\theta_{YZ}}$$

$$\theta_{Z2} = \arccos\sqrt{1 - \cos^2 30° - \cos^2 80°}$$

$$\theta_{Z2} = 87.2°$$

$$R_X = F_{1X} + F_{2X} = F_1\cos\theta_{X1} + F_2\cos\theta_{XZ}$$

$$R_X = 15\text{ kN}(+\cos 20°) + 25\text{ kN}(+\cos 30°)$$

$$R_X = +35.7\text{ kN}$$

$$R_Y = F_{1Y} + F_{2Y} = F_1\cos\theta_{Y1} + F_2\cos\theta_{Y2}$$

$$R_Y = 15\text{ kN}(+\cos 75°) + 25\text{ kN}(-\cos 80°)$$

$$R_Y = -0.46\text{ kN}$$

$$R_Z = F_{1Z} + F_{2Z} = F_1\cos\theta_{Z1} + F_2\cos\theta_{Z2}$$

$$R_Z = 15\text{ kN}(-\cos 77.1°) + 25\text{ kN}(+\cos 87.2°)$$

$$R_Z = -2.13\text{ kN}$$

$$R = \sqrt{R_X^2 + R_Y^2 + R_Z^2} = \sqrt{(35.7)^2 + (0.46)^2 + (2.13)^2} \qquad \boldsymbol{R = 35.8\text{ kN}}$$

$$\theta_{RX} = \arccos\frac{R_X}{R} = \arccos\frac{35.7}{35.8} \qquad \boldsymbol{\theta_{RX} = 4.3°}$$

$$\theta_{RY} = \arccos\frac{R_Y}{R} = \arccos\frac{0.46}{35.8} \qquad \boldsymbol{\theta_{RY} = 89.3°}$$

$$\theta_{RZ} = \arccos\frac{R_Z}{R} = \arccos\frac{2.13}{35.8} \qquad \boldsymbol{\theta_{RY} = 86.6°}$$

Step 3 Determine the location of the resultant. This will require taking moments about each axis. Since two sets of components have moment about each axis (e.g., both X and Y components have moments about the Z axis) we will end up with three equations each containing two different pairs of the three unknowns.

$$\Sigma M_X = F_{1Z}d_{1Y} + F_{1Y}d_{1X} + F_{2Z}d_{2Y} + F_{2Y}d_{2X} = R_Z d_{RY} + R_Y d_{RX}$$

$$-(15\text{ kN})\cos 77.1°(7) - (15\text{ kN})\cos 75°(4) + (25\text{ kN})\cos 87.2°(4) + (25\text{ kN})\cos 80°(5) = R_Z d_{RY} + R_Y d_{RX}$$

$$(1) \qquad R_Z d_{RY} + R_Y d_{RX} = -12.6 \text{ kNm}$$

$$\Sigma M_Y = F_{1Z} d_{1X} + F_{1X} d_{1Z} + F_{2Z} d_{2X} + F_{2X} d_{2Z} = R_Z d_{RX} + R_X d_{RZ}$$

$$+(15 \text{ kN})\cos 77.1°(4) + (15 \text{ kN})\cos 20°(2) - (25 \text{ kN})\cos 87.2°(1) + (25 \text{ kN})\cos 30°(5) = R_Z d_{RX} + R_X d_{RZ}$$

$$(2) \qquad R_Z d_{RX} + R_X d_{RZ} = +148.6 \text{ kNm}$$

$$\Sigma M_Z = F_{1X} d_{1Y} + F_{2X} d_{2Y} + F_{1Y} d_{1X} + F_{2Y} d_{2X} = R_X d_{RY} + R_Y d_{RX}$$

$$-(15 \text{ kN})\cos 20°(7) - (25 \text{ kN})\cos 30°(4) + (15 \cos 75°)(4) - (25 \text{ kN})\cos 80°(1) = R_X d_{RY} + R_Y d_{RX}$$

$$(3) \qquad R_X d_{RY} + R_Y d_{RX} = -174 \text{ kNm}$$

Note that when the force components were entered into the moments equation as $F \cos \theta_F$, no indication of force direction was made. Also, moment direction for each known moment was indicated. We now have three equations for three unknowns; let us go one step further and indicate directions for the unknown moments. We will also enter the absolute values for the components of the resultant.

$$(1) \qquad -(2.13 d_{RY}) + (0.46 d_{RX}) = -12.6$$

$$(2) \qquad -(2.13 d_{RX}) + (35.7 d_{RZ}) = 148.6$$

$$(3) \qquad -(35.7 d_{RY}) - (0.46 d_{RX}) = -174$$

Solving the three equations above either simultaneously or by substitution yields:

$$\mathbf{d_{RX} = 4.6 \text{ m} \qquad d_{RY} = 6.9 \text{ m} \qquad d_{RZ} = 4.4 \text{ m}}$$

Actually, the above represent one point on the line of action of R. However, it could be proven, for example, that if coordinates further back along R were chosen, the negative moment of R_Y about the Z axis would be decreasing as d_{RX} decreased; but the increase in the negative moment of R_X about the axis would be increasing equally as d_{RY} increased. The moment of R about the Z axis would thus stay constant. The same could be proven for the other two axes.

The above was certainly very complex and required careful attention to follow. Fortunately, most of our practical problems involve equilibrium conditions where all the points of force application are known. Also, some force components may be eliminated by the nature of a support. The next section of this chapter will discuss equilibrium in three dimensions, and the last two sections will provide applications.

8.4 EQUILIBRIUM IN THREE DIMENSIONS

The conditions for equilibrium in three dimensions are quite similar to those for two dimensions. In two dimensions, equilibrium exists if the sum of all forces is zero and if the sum of the moments of these forces taken about any point is zero. In this general statement, the only change needed for three-dimensional equilibrium would be to talk of moments about an axis rather than a point.

It is generally convenient to work with components of the forces. The equations for equilibrium become

$$\Sigma F_X = 0 \qquad \Sigma M_X = 0$$

$$\Sigma F_Y = 0 \qquad \Sigma M_Y = 0$$

$$\Sigma F_Z = 0 \qquad \Sigma M_Z = 0$$

In the moment equations, X, Y, and Z refer to the axes about which moments are taken. These are still very broad statements. In two-dimensional equilibrium, we could take moments about any point in the XY plane. In three-dimensional equilibrium, those points become Z axes. Similarly, a point in the XZ plane becomes a Y axis and a point in the YZ plane becomes an X axis.

This leads us to two approaches to solving three-dimensional problems. Both will be illustrated in subsequent sections. One approach is to make the free-body diagram three-dimensional, possibly isometric. This approach was used with the space forces in the previous section. It does require some visualization. The second approach involves taking three views of the problem and dealing with one at a time; the XY plane and moments about the Y axis, and the YZ plane and moments about the X axis. As stated before, we use a conventional cartesian coordinate system and view each plane and associated axis from the positive end of the axis (i.e., looking toward the negative end). The views then become, in the same order as above, the front, top, and the right side view.

Supports were mentioned above, and the nature of various types in terms of the reactions they can provide must be considered just as they were in two dimensions. The simple pin joint of two dimensions now becomes a choice between a simple hinge which can support moment reactions about two axes and a ball joint which cannot support any moment reactions. The simple cylindrical roller of two dimensions becomes a choice between a simple cylindrical roller with friction resistance in one direction and rolling resistance in another direction and a spherical roller that has only rolling resistance in any direction in the plane it rests on. Actions and reactions for a variety of supports in three dimensions are shown in Figure 8-11.

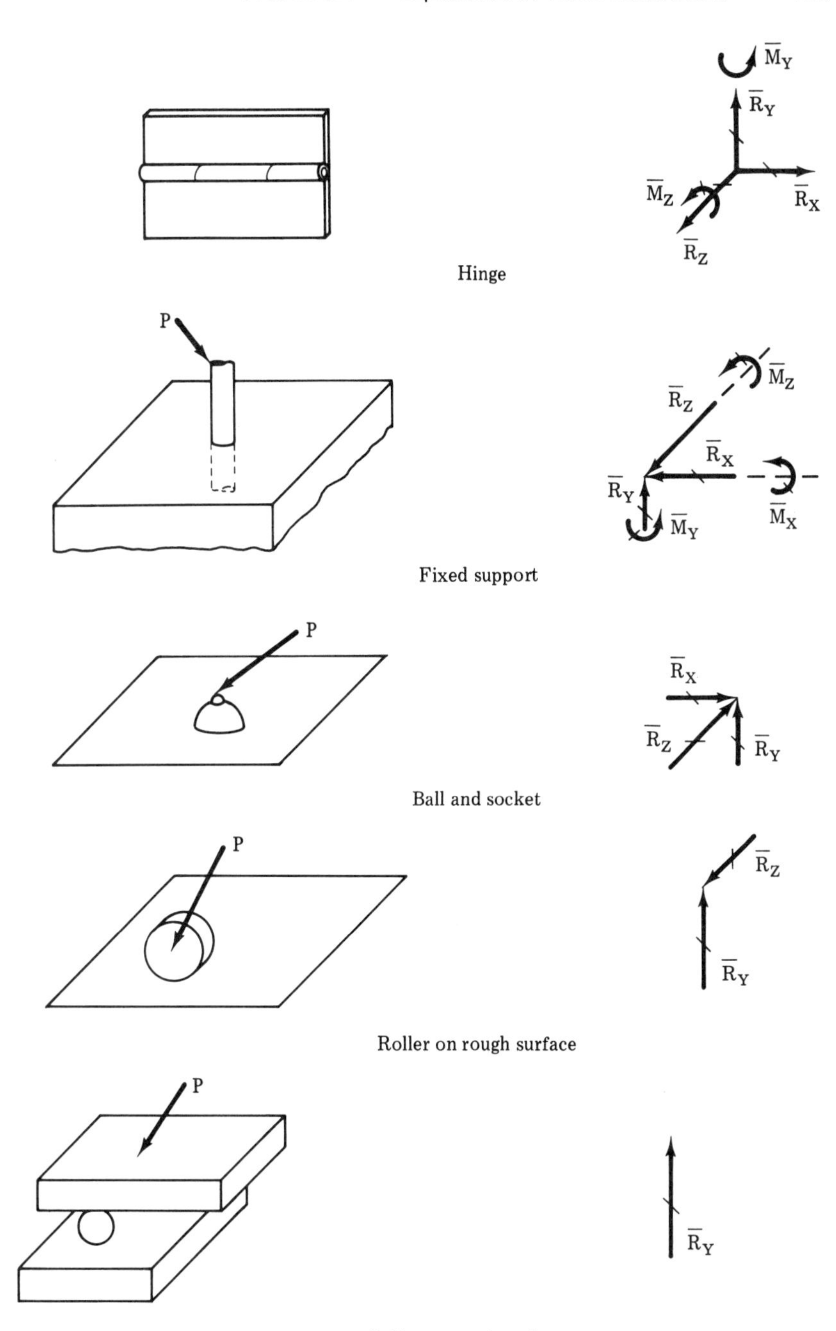

Figure 8-11. Support reactions in three dimensions.

8.5 APPLICATIONS TO STRUCTURES

One of the simplest types of three-dimensional equilibrium problems consists of a system of parallel noncoplanar loads and reactions. Many floor loadings in residential, commercial, and industrial buildings are of this nature.

Example 8-6: Determine the reactions to the loads on the floor shown in Figure 8-12a.

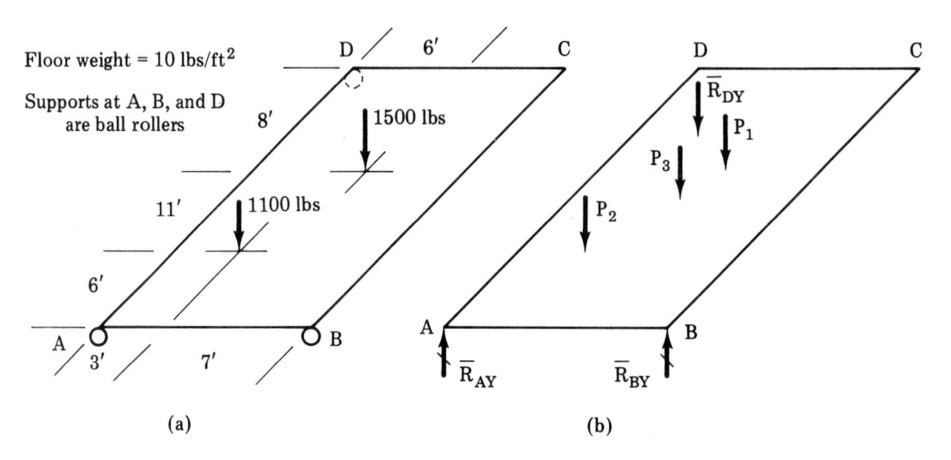

Figure 8-12

Solution:

Step 1 Recognize this as a simple parallel, noncoplanar force system since all the loads are straight down and all the reactions must be straight up (only reaction possible at a ball roller).

Step 2 Draw the free-body diagram as shown in Figure 8-12b. Note the inclusion of a concentrated load equivalent to the floor weight and located at the centroid of the rectangular floor.

Step 3 It is quite apparent that while $\Sigma F_Y = 0$, this approach will not lead to a solution initially. Moments *about axes* need to be taken.

$$\Sigma M_{AD} = 0 \qquad -(P_1 d_1) - (P_2 d_2) - (P_3 d_3) + (R_B d_{RB}) = 0$$

$$-(1500 \text{ lb})(4 \text{ ft}) - (1100 \text{ lb})(3 \text{ ft}) - (2500 \text{ lb})(5 \text{ ft}) + 10R_B = 0$$

$$\boldsymbol{R_B = +2180 \text{ lb}}$$

Step 4

$$\Sigma M_{AB} = 0 \qquad -(P_1 d_1) - (P_2 d_2) - (P_3 d_3) + (R_D d_{AD}) = 0$$

$$-(1500\text{ lb})(17\text{ ft}) - (1100\text{ lb})(6\text{ ft}) - (2500\text{ lb})(12.5\text{ ft}) + 25R_D = 0$$

$$\boldsymbol{R_D = +2534\text{ lb}}$$

Step 5

$$\Sigma F_Y = 0 \qquad P_1 + P_2 + P_3 + R_A + R_B + R_D = 0$$

$$R_A = -(P_1 + P_2 + P_3 + R_B + R_D)$$

$$R_A = -(-1500 - 1100 - 2500 + 2180 + 2534\text{ lb})$$

$$\boldsymbol{R_A = +386\text{ lb}}$$

Step 6 If desired, in lieu of Step 5, R_A could be found by taking moments about axis *BD*. This could also be done as a check. *Note:* In Steps 4 and 5, some shortcuts have been taken in that the direction of the moment of the unknown force is recognized and then the direction of the force is recognized. If you feel more comfortable including the missing steps, be sure to do so. Further, if P_1 and/or P_2 had been only a little further to the right and/or back, R_A would have been negative. The floor would have been unstable since the roller at *A* is not capable of pulling down.

Problem 8-5: Determine the support reactions for the floor shown in Figure 8-13.

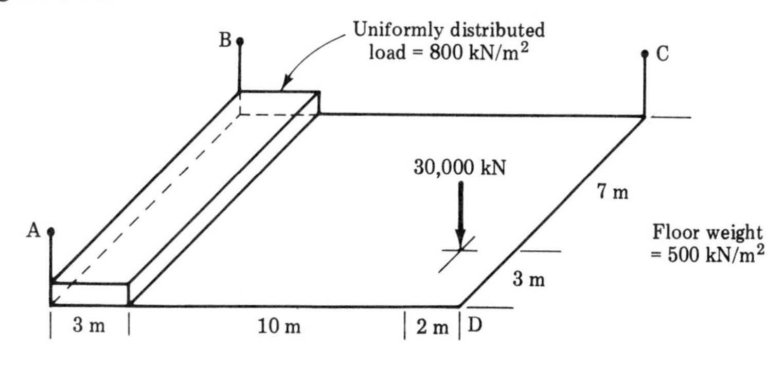

Figure 8-13

Example 8-7: The billboard shown in Figure 8-14a weighs 1000 kg. It also has a wind load on it of 200 kN/m^2 at an angle of 40°. Determine the support reactions.

Step 1 Recognize that this is truly a three-dimensional problem with force components lying along all three axes.

Step 2 Assume that the loads are equally divided between the supports, and draw the free-body diagram as shown in Figure 8-14b.

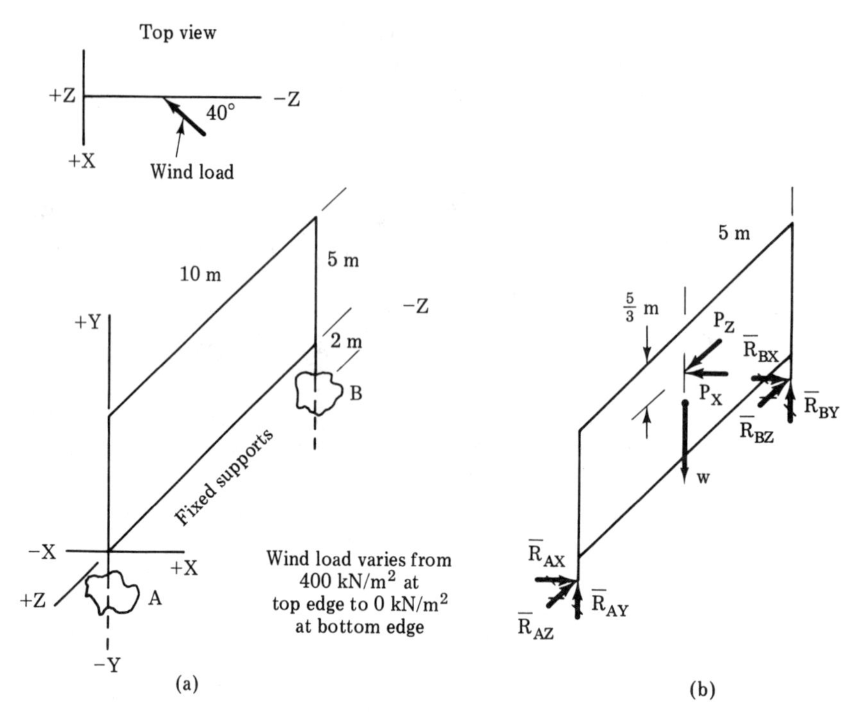

Figure 8-14

Step 3 Determine the wind load.

$$P = wA = (200 \text{ kN/m}^2)(50 \text{ m}^2)$$

$P = 10{,}000$ kN located at the centroid of the face, 2.5 m down from the top and 5 m from either end.

$$P_X = P \cos \theta_P \qquad P_X = (10{,}000 \text{ kN})\cos 230° \qquad P_X = -6428 \text{ kN}$$

Similarly, $P_Z = 7660$ kN. (Full reaction to this assumes 100% drag, which is unrealistically high; but for simplicity, we'll work with it.)

Step 4 Determine the force reactions in the Z direction.

$$\Sigma F_Z = 0 \qquad P_Z + \overline{R}_{AZ} + \overline{R}_{BZ} = 0 \qquad \overline{R}_{AZ} = \overline{R}_{BZ}$$

$$2R_{AZ} = -P_Z \qquad Z\overline{R}_{AZ} = -(+7660 \text{ kN})$$

$$\boldsymbol{\overline{R}_{AZ} = -3830 \text{ kN} \qquad \overline{R}_{BZ} = -3830 \text{ kN}}$$

Step 5 Determine the force reactions in the X direction.

$$\Sigma F_X = 0 \qquad P_X + R_{AX} + R_{BX} = 0 \qquad R_{AX} = R_{BX}$$

$$\overline{R}_{AX} + \overline{R}_{BX} = -P_X \qquad \overline{R}_{AX} + \overline{R}_{BX} = -(-6428 \text{ kN}) = +6428 \text{ kN}$$

$$\boldsymbol{\overline{R}_{AX} = +3214 \text{ kN} \qquad \overline{R}_{BX} = +3214 \text{ kN}}$$

Step 6 Determine the force reactions in the Y direction.

$$W = 1000 \text{ kg} = 9810 \text{ kN}$$

$$F_Y = 0 \qquad W + \overline{R}_{AY} + \overline{R}_{BY} = 0 \qquad \overline{R}_{AY} = \overline{R}_{BY}$$

$$\overline{R}_{AY} + R_{BY} = -W \qquad \overline{R}_{AY} + \overline{R}_{BY} = -(-9810 \text{ kN}) = +9810 \text{ kN}$$

$$\boldsymbol{\overline{R}_{AY} = +4905 \text{ kN} \qquad \overline{R}_{BY} = +4905 \text{ kN}}$$

Step 7 Taking moments about the Y axis for each support will show that

$$\boldsymbol{\overline{M}_{AY} = 0 \qquad \overline{M}_{BY} = 0}$$

Step 8 Determine the moment reaction about the Z axis for each support.

$$M_Z = 0 \qquad \overline{M}_{AZ} + \overline{M}_{BZ} + P_X d_{PX} = 0 \qquad \overline{M}_{AZ} = \overline{M}_{BZ}$$

$$\overline{M}_{AZ} + \overline{M}_{BZ} = -(P_X d_{PX}) = -[+(6428 \text{ kN})(4.5 \text{ m})] = -28{,}926 \text{ kN m} \circlearrowright$$

$$\boldsymbol{\overline{M}_{AZ} = -14{,}463 \text{ kN m} \circlearrowright \; = \overline{M}_{BZ}}$$

Step 9 Taking moments about the Y axis for each support will show that

$$\boldsymbol{\overline{M}_{AY} = 0 \qquad \overline{M}_{BY} = 0}$$

Problem 8-6: Determine the support reactions for the platform shown in Figure 8-15.

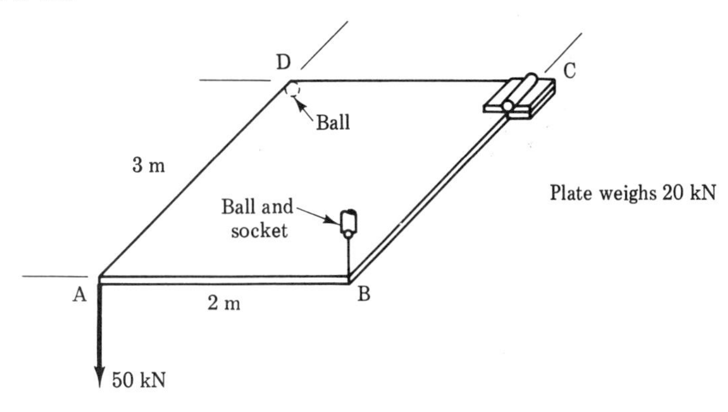

Figure 8-15

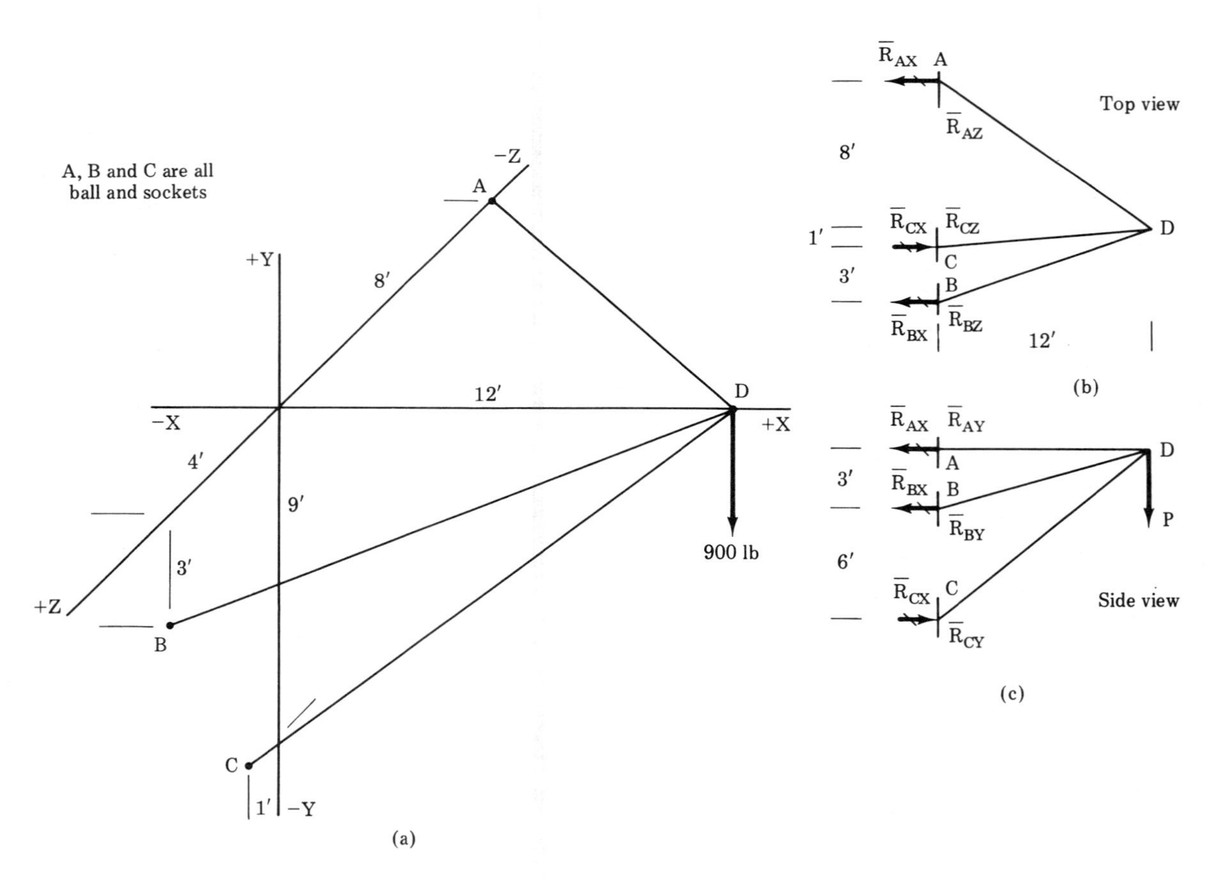
A, B and C are all
ball and sockets
+Y
−Y
−X
+X
+Z
−Z
A
B
C
D
8′
12′
4′
9′
3′
1′
900 lb
(a)
$\overline{R}_{AX}$
$\overline{R}_{AZ}$
$\overline{R}_{CX}$
$\overline{R}_{CZ}$
$\overline{R}_{BX}$
$\overline{R}_{BZ}$
Top view
8′
1′
3′
12′
(b)
$\overline{R}_{AX}$
$\overline{R}_{AY}$
$\overline{R}_{BX}$
$\overline{R}_{BY}$
$\overline{R}_{CX}$
$\overline{R}_{CY}$
P
3′
6′
Side view
(c)

Figure 8-16

Example 8-8: Determine the forces acting in each member of the truss shown in Figure 8-16a, as well as all support reactions.

Step 1 Examine the supports and recognize that since they are all ball joints; they can all support force in any direction but no moments at all.

Step 2 Recognize that all members are simple two-force members; *AD* and *BD* in tension, *CD* in compression.

Step 3 Draw the free-body diagrams of the top view and the left-side view, as shown in Figures 8-16b and 8-16c.

Step 4 Based on our previous work with direction cosines, if we find the lengths of the three members, we can find the direction cosines and thus the relationships among the components of each space force involved. Since all the members are simple two-force members, finding the total reaction is the same as finding the force in the member.

$$R_{AX} = R_A \cos \theta_{AX} \qquad R_{AX} = F_{AD} \frac{AD_X}{AD}$$

$$AD = \sqrt{12^2 + 0^2 + 8^2} \qquad AD = 14.4 \text{ ft}$$

$$R_{AX} = F_{AD}\left(\frac{12}{14.4}\right) \qquad R_{AX} = 0.833 F_{AD}$$

$$R_{AY} = R_A \cos \theta_{AY} \qquad R_{AY} = F_{AD} \frac{AD_Y}{AD}$$

$$R_{AY} = F_{AD}\left(\frac{0}{14.4}\right) \qquad \mathbf{R_{AY} = 0}$$

$$R_{AZ} = R_A \cos \theta_{AZ} \qquad R_{AZ} = F_{AD} \frac{AD_Z}{AD}$$

$$R_{AZ} = F_{AD}\left(\frac{8}{14.4}\right) \qquad R_{AZ} = 0.556 F_{AD}$$

$$R_{BX} = R_B \cos \theta_{BX} \qquad R_{BX} = F_{BD} \frac{BD_X}{BD}$$

$$BD = \sqrt{12^2 + 3^2 + 4^2} \qquad BD = 13.0 \text{ ft}$$

$$R_{BX} = F_{BD}\left(\frac{12}{13.0}\right) \qquad R_{BX} = 0.923 F_{BD}$$

$$R_{BY} = R_B \cos \theta_{BY} \qquad R_{BY} = F_{BD} \frac{BD_Y}{BD}$$

$$R_{BY} = F_{BD}\left(\frac{3}{13.0}\right) \qquad R_{BY} = 0.231 F_{BD}$$

$$R_{BZ} = R_B \cos \theta_{BZ} \qquad R_{BZ} = F_{BZ} \frac{BD_Z}{BD}$$

$$R_{BZ} = F_{BZ} \left(\frac{4}{13.0} \right) \qquad R_{BZ} = 0.308 F_{BD}$$

$$R_{CX} = R_C \cos \theta_{CX} \qquad R_{CX} = F_{CD} \frac{CD_X}{CD}$$

$$CD = \sqrt{12^2 + 9^2 + 1^2} \qquad CD = 15.0 \text{ ft}$$

$$R_{CX} = F_{CD} \left(\frac{12}{15.0} \right) \qquad R_{CX} = 0.80 F_{CD}$$

$$R_{CY} = R_C \cos \theta_{CY} \qquad R_{CY} = F_{CD} \frac{CD_Y}{CD}$$

$$R_{CY} = F_{CD} \left(\frac{9}{15.0} \right) \qquad R_{CY} = 0.60 F_{CD}$$

$$R_{CZ} = R_C \cos \theta_{CZ} \qquad R_{CZ} = F_{CD} \frac{CD_Z}{CD}$$

$$R_{CZ} = F_{CD} \left(\frac{1}{15.0} \right) \qquad R_{CZ} = 0.067 F_{CD}$$

Step 5 Write two moment equations, one about C in the top view, and one about C in the side view.

$$\text{(top)} \qquad \Sigma M_C = 0 \qquad R_{AX} d_{AX} + R_{BX} d_{BX} = 0$$

$$9R_{AX} + 3R_{BX} = 0 \qquad R_{BX} = 3R_{AX}$$

$$\text{(side)} \qquad \Sigma M_C = 0 \qquad R_{AX} d_{AX} + R_{BX} d_{BX} + P_X d_X = 0$$

$$+(9R_{AX}) + (6R_{BX}) - (12)(900) = 0$$

Substitutions (from top)

$$+(9R_{AX}) + (18R_{AX}) - 10{,}800 = 0 \qquad \boldsymbol{R_{AX} = -400 \text{ lb}}$$

$$\text{Since } R_{AX} = 0.833 F_{AD} \qquad F_{AD} = \frac{R_{AX}}{0.833}$$

$$F_{AD} = \frac{400}{0.833} \qquad \boldsymbol{F_{AD} = 480 \text{ lb}(T)}$$

$$R_{AZ} = 0.556F_{AD} = 0.556(480) \qquad \boldsymbol{R_{AZ} = -268\ \text{lb}}$$

$$R_{BX} = 3R_{AX} = 3(400) \qquad \boldsymbol{R_{BX} = -1200\ \text{lb}}$$

$$\text{Since } R_{BX} = 0.923F_{BD} \qquad F_{BD} = \frac{R_{BX}}{0.923}$$

$$F_{BD} = \frac{1200}{0.923} \qquad \boldsymbol{F_{BD} = 1300\ \text{lb}(T)}$$

$$R_{BY} + 0.231F_{BD} \qquad F_{BD} = 0.231(1300) \qquad \boldsymbol{R_{BY} = -300\ \text{lb}}$$

$$R_{BZ} = 0.308F_{BD} = 0.308(1300) \qquad \boldsymbol{R_{BZ} = +400\ \text{lb}}$$

Step 6 Write the equilibrium equation for F_X for either view.

$$\Sigma F_X = 0 \qquad R_{AX} + R_{BX} + R_{CX} = 0$$

$$R_{CX} = -(R_{AX} + R_{BX}) \qquad R_{CX} = -(-400 - 1200) \qquad \boldsymbol{R_{CX} = +1600\ \text{lb}}$$

$$\text{Since } R_{CX} = 0.80F_{CD}F_{CD} = \frac{R_{CX}}{0.80}$$

$$F_{CD} = \frac{1600}{0.80} \qquad \boldsymbol{F_{CD} = 2000\ \text{lb}(C)}$$

$$R_{CY} = 0.60F_{CD} = 0.60(2000) \qquad \boldsymbol{R_{CY} = +1200\ \text{lb}}$$

$$F_{CZ} = 0.067F_{CD} = 0.067(2000) \qquad \boldsymbol{R_{CZ} = -134\ \text{lb}}$$

Step 7 Check by summing in X, Y, and Z directions

$$\Sigma F_X = 0 \qquad R_{AX} + R_{BX} + R_{CX} + P_X = 0$$

$$-400 - 1200 + 1600 + 0 = 0$$

$$\Sigma F_Y = 0 \qquad R_{AY} + R_{BY} + R_{CY} + P_Y = 0$$

$$0 - 300 + 1200 - 900 = 0$$

$$\Sigma F_Z = 0 \qquad R_{AZ} + R_{BZ} + R_{CZ} + P_Z = 0$$

$$-267 + 400 - 134 + 0 \cong 0$$

Problem 8-7: Determine the forces acting in each member of the tripod shown in Figure 8-17 as well as all support reactions. If all legs are resting on horizontal surfaces (X–Z planes) and the coefficient of friction is 0.2, will the tripod slip?

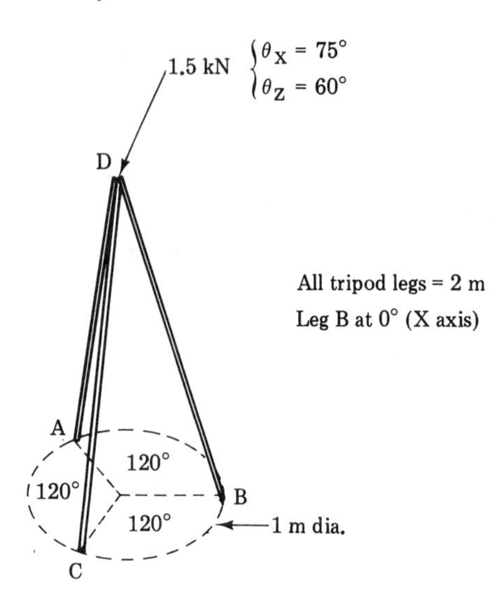

Figure 8-17

Example 8-9: The damper shown in Figure 8-18a is held in place by two springs with $k = 2$ kN/mm and a free length of 0.8 m. Determine the force in cable *AE* and all support reactions when the damper is lifted to a position 30° below the horizontal.

Step 1 Draw the free-body diagram of the damper as shown in Figure 8-18b.

Step 2 Recognize that with a little work the spring forces acting on *C* and *D* can be found as follows: Since the side of the damper is 3 m and the spring post is 1 m, the distance from *C* (or *D*) to the *X* axis through B is

$$\sqrt{3^2 + 1^2} = 3.16 \text{ m}$$

The angle between the 3.16-m line and the 3-m line is arctan $\frac{1}{3} = 18.4°$. Since the side of the damper makes a 30° angle with the horizontal, the 3.16-m line makes an angle of 11.6° with the horizontal or an angle of 78.4° with the vertical. This angle is the included angle between two known lines, the 3.16-m line and 3-m side of the damper opening. The side opposite the 78.4° angle is the extended spring length.

$$\ell\text{ext} = \sqrt{(3)^2 + (3.16)^2 - 2(3)(3.16)\cos 78.4°}$$

$$\ell\text{ext} = 3.9 \text{ m}$$

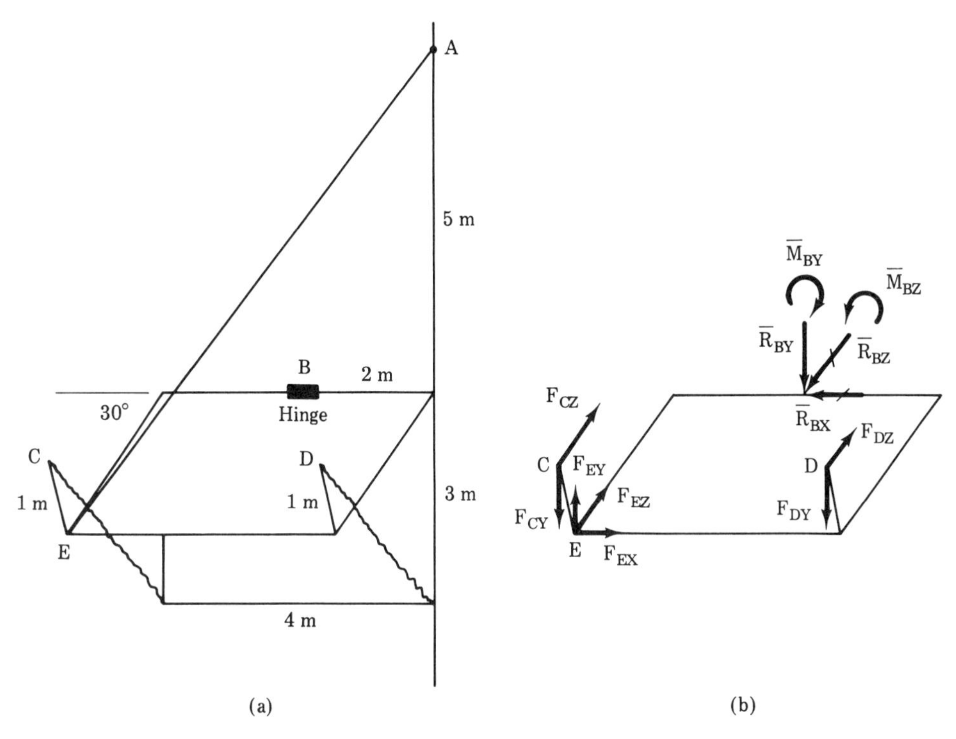

Figure 8-18

The spring force then equals

$$F_S = k\Delta X = 2000 \text{ kN/m}(3.9 - 0.8 \text{ m}) = 6200 \text{ kN}$$

Step 3 The spring forces have no X component and their Y and Z components can be found as follows:

$$F_{CZ} = F_{DZ} = -F_S \cos\theta_{FS} = -(6200 \text{ kN})(\cos 11.6°)$$

$$F_{CZ} = F_{DZ} = 6073 \text{ kN}$$

$$F_{CY} = F_{DY} = -F_S \sin\theta_{FS} = -(6200 \text{ kN})(\sin 11.6°)$$

$$F_{CY} = F_{DY} = 1247 \text{ kN}$$

Step 4 Equilibrium equations can now be written for ΣF_X, ΣF_Y, ΣF_Z, ΣM_{BX}, ΣM_{BY}, and ΣM_{BZ}. Inspection of these possibilities (or actually writing them and evaluating them one by one) results in the selection of M_{BX} as the only immediately feasible approach. Since the hinge at B turns about the X axis, there is no moment reaction and

$$\Sigma M_{BX}=0 \qquad F_{CY}d_{CY}+F_{CZ}d_{CZ}+F_{DY}d_{DY}+F_{DZ}d_{DZ}+F_{EY}d_{EY}+F_{EZ}d_{EZ}=0$$

This appears to leave two unknowns, but these are related by the direction of AE.

$$AE=\sqrt{(4)^2+(5+3\sin 30°)^2+(3\cos 30°)^2}$$

$$AE=8.06 \text{ m}$$

$$\cos\theta_X=\frac{4}{8.06}=0.496$$

$$\cos\theta_Y=\frac{5+3\sin 30°}{8.06}=0.806$$

$$\cos\theta_Z=\frac{3\cos 30°}{8.06}=0.322$$

$$F_{EX}=F_E\cos\theta_X \qquad F_{EY}=F_E\cos\theta_Y \qquad F_{EZ}=F_E\cos\theta_Z$$

$$\text{then } F_E=\frac{F_{EY}}{\cos\theta_Y}=\frac{F_{EZ}}{\cos\theta_Z}$$

$$\text{and } F_{EY}=\frac{\cos\theta_Y}{\cos\theta_Z}(F_{EZ})=\frac{0.806}{0.322}F_{EZ} \qquad F_{EY}=2.5F_{EZ}$$

Substitution of this in our moment equation will then yield F_{EZ}. Checking on moment arms

$$d_{CY}=d_{DY}=3.16\cos 11.6°=3.1 \text{ m}$$
$$d_{CZ}=d_{DZ}=3.16\sin 11.6°=0.635 \text{ m}$$

$$d_{EY}=3\cos 30°=2.6 \text{ m}$$

$$d_{EZ}=3\sin 30°=1.5 \text{ m}$$

$$-(2.5F_{EZ})(2.6 \text{ m})+(F_{EZ})(1.5 \text{ m})=-[+(1247 \text{ Kc})(3.1 \text{ m})+(6073 \text{ kN}) \times (0.635 \text{ m})+(1247 \text{ kN})(3.1 \text{ m})+(6073 \text{ kN})(0.635)]$$

$$F_{EZ}=3089 \text{ kN}$$

$$F_{EY}=2.5F_{EZ} \qquad F_{EY}=7722 \text{ kN}$$

$$F_E=\frac{F_{EY}}{\cos\theta_Y}=\frac{7722 \text{ kN}}{0.806}=9635 \text{ kN}$$

Force in cable $AE=9653$ kN

$$F_{EX}=F_E\cos\theta_X=9653(0.496)=4788 \text{ kN}$$

Step 5 Summing in X direction

$$\Sigma F_X = 0 \qquad F_{EX} + \overline{R}_{BX} = 0 \qquad \overline{R}_{BX} = -F_{EX} = -4788 \text{ kN}$$

$$\boldsymbol{R_{BX} = 4788 \text{ kN}}$$

Step 6 Summing in Y direction

$$\Sigma F_Y = 0 \qquad \overline{R}_{BY} + F_{CY} + F_{DY} + F_{EY} = 0 \qquad \overline{R}_{BY} = -(F_{CY} + F_{DY} + F_{EY})$$

$$\overline{R}_{BY} = -(-1247 - 1247 + 7722 \text{ kN}) \qquad \boldsymbol{\overline{R}_{BY} = -5228 \text{ kN}}$$

Step 7 Summing in Z direction

$$\Sigma F_Z = 0 \qquad \overline{R}_{BZ} + F_{CZ} + F_{DZ} + F_{EZ} = 0 \qquad \overline{R}_{BZ} = -(F_{CZ} + F_{DZ} + F_{EZ})$$

$$\overline{R}_{BZ} = -(-6073 - 6073 - 3089 \text{ kN}) \qquad \overline{R}_{BZ} = +15{,}235 \text{ kN}$$

Step 8 Sum moments about Y axis through B

$$\Sigma M_{BY} = 0 \qquad \overline{M}_{BY} + F_{CZ}d_{CZ} + F_{DZ}d_{DZ} + F_{EZ}d_{EZ} + F_{EX}d_{EX} = 0$$

$$\overline{M}_{BY} = -(F_{CZ}d_{CZ} + F_{DZ}d_{DZ} + F_{EZ}d_{EZ} + F_{EX}d_{EX})$$

$$\overline{M}_{BY} = -[-(6073 \text{ kN})(2 \text{ m}) + (6073 \text{ kN})(2 \text{ m}) - (3089)(2 \text{ m}) + (4788 \text{ kN})(3 \cos 30^\circ \text{ m})]$$

$$\boldsymbol{\overline{M}_{BY} = -6262 \text{ kN m}} \circlearrowright$$

Step 9 Sum moments about Z axis through B:

$$\Sigma M_{BZ} = 0 \qquad \overline{M}_{BZ} + F_{CY}d_{CY} + F_{DY}d_{DY} + F_{EX}d_{EX} + F_{EY}d_{EY}$$

$$\overline{M}_{BZ} = -(F_{CY}d_{CY} + F_{DY}d_{DY} + F_{EX}d_{EX} + F_{EY}d_{EY})$$

$$\overline{M}_{BZ} = -[+(1247 \text{ kN})(2 \text{ m}) - (1247 \text{ kN})(2 \text{ m}) + (4788 \text{ kN})(3 \sin 30^\circ) - (7772 \text{ kN})(2 \text{ m})]$$

$$\boldsymbol{\overline{M}_{BZ} = +8262 \text{ kN m}} \circlearrowleft$$

Problem 8-8: Determine the support reactions for the boom shown in Figure 8-19.

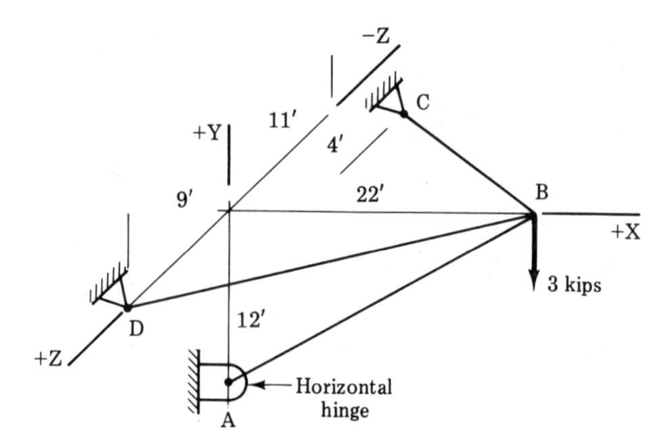

Figure 8-19

8.6 APPLICATIONS TO SHAFTS

At first glance, a single shaft may appear to be a relatively simple problem. However, as can be seen in Figure 8-20, shaft problems may vary from a simple case where all the forces and reactions lie parallel to one another to problems where the forces merely lie in parallel planes to true three-dimensional problems with force components along all three axes.

A further complication provides a glimpse into some aspects of machine design and beam design. If we examine a simple journal bearing, ball bearing, or roller bearing as shown in Figure 8-21a, b, and c, respectively, we note that they all must be considered fixed supports. Therefore, any force loads on a shaft supported by such bearings will place a moment load on those bearings. Only a bearing such as the uniball shown in Figure 8-21d can be considered a simple support. In our work in this section, all bearings will be considered as simple supports. In a later chapter, we will deal with beams (a shaft is simply a special type of beam) supported by one or two fixed supports. It should be noted that if there is much of an axial (parallel to the

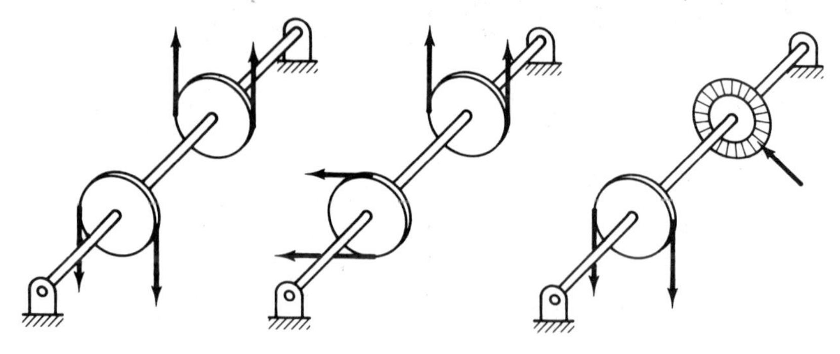

Figure 8-20. Shaft loadings.

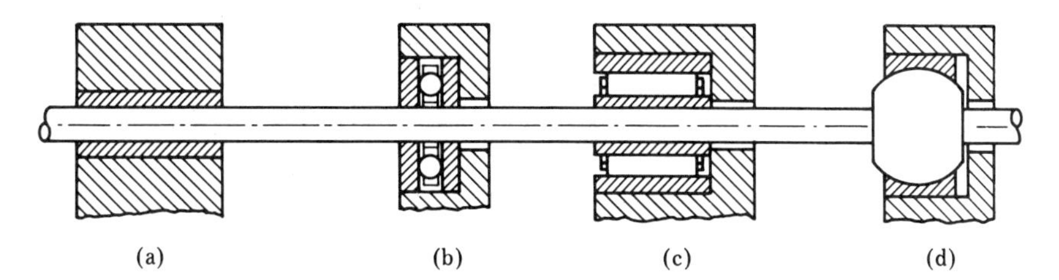

Figure 8-21. Shaft bearings.

centerline of the shaft) load placed on the shaft, at least one bearing must be selected to take this load.

Example 8-10: Pulley B drives the shaft shown in Figure 8-22. Power is taken off at pulleys C and D. Determine the tension T and the bearing reactions at A and E. Assume that the bearings are frictionless and act as simple supports.

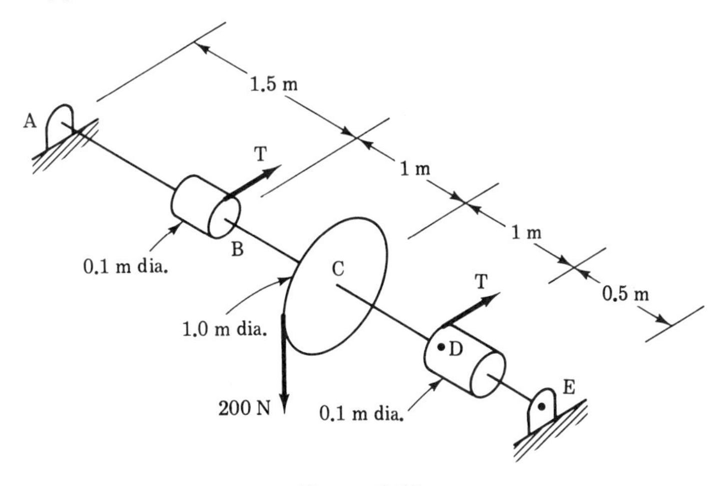

Figure 8-22

Solution:

Step 1 Draw the free-body diagram. With no loads in the axial (x) direction, there will be no reactions at A or E in the x direction. Since all loads in the Z direction are positive, the reactions in the Z direction at both A and E will be negative. Reactions in the Y direction are a little harder to evaluate, but calculations will bear out that both are negative. Since the bearings are friction free and act as simple supports, there are no moment reactions at either A or E.

Step 2 Determine the missing tension by summing moments about the X axis.

$$\Sigma M_X = 0 \qquad M_B + M_C + M_D = 0$$

Since the outputs are clockwise, M_B must be counterclockwise making T greater than 50 lb.

$$+(T - 50\text{ lb})(2.5\text{ in.}) - (240\text{ lb})(6\text{ in.}) - (100\text{ lb})(5\text{ in.}) = 0$$

$$\boldsymbol{T = 726\text{ lb}}$$

Step 3 Find $\overline{R}_{EZ}$ by summing moments about A in the XZ plane.

$$\Sigma M_A = 0 \qquad F_{BZ}d_{BZ} + \overline{R}_{EZ}d_{EZ} \qquad \overline{R}_{EZ} = F_{BZ}\left(\frac{d_{BZ}}{d_{EZ}}\right)$$

Since there are only two moments involved, they must be in opposite directions. Because of the location of the moment center, the two related forces must also be in opposite directions.

$$\overline{R}_{EZ} = 776\text{ lb}\,\frac{2\text{ ft}}{11\text{ ft}} \qquad \boldsymbol{\overline{R}_{EZ} = -141\text{ lb}}$$

Note that the sum of the forces acting on B was used.

Step 4 Find $\overline{R}_{AZ}$ by summing forces in the Z direction.

$$\Sigma F_Z = 0 \qquad \overline{R}_{AZ} + F_{BZ} + \overline{R}_{EZ} = 0 \qquad \overline{R}_{AZ} = -(F_{BZ} + \overline{R}_{EZ})$$

$$\overline{R}_{AZ} = -(+776 - 141\text{ lb}) \qquad \boldsymbol{\overline{R}_{AZ} = -635\text{ lb}}$$

Step 5 Find $\overline{R}_{EY}$ by summing moments about A in the XY plane.

$$\Sigma M_A = 0 \qquad F_{CY}d_{CY} + F_{DY}d_{DY} + \overline{R}_{EY}d_{EY} = 0$$

$$\overline{R}_{EY}d_{EY} = -(F_{CY}d_{CY} + F_{DY}d_{DY})$$

$$\overline{R}_{EY}d_{EY} = -[+(360\text{ lb})(5\text{ ft}) - (200\text{ lb})(7\text{ ft})]$$

$$\overline{R}_{EY}d_{EY} = -400\text{ lb ft} \circlearrowright$$

Therefore $\overline{R}_{EY}$ is

$$\overline{R}_{EY} = \frac{\overline{R}_{EY}d_{EY}}{\overline{R}_{EY}} = \frac{400\text{ lb ft}}{11\text{ ft}} \qquad \boldsymbol{\overline{R}_{EY} = -36\text{ lb}}$$

Step 6 Find $\overline{R}_{AY}$ by summing forces in the Y direction.

$$\Sigma F_Y = 0 \qquad \overline{R}_{AY} + F_{CY} + F_{DY} + \overline{R}_{EY} = 0$$

$$\overline{R}_{AY} = -(F_{CY} + F_{DY} + \overline{R}_{EY})$$

$$\overline{R}_{AY} = -(+360 - 200 - 361 \text{ lb}) \qquad \boldsymbol{\overline{R}_{AY} = -124 \text{ lb}}$$

Problem 8-9: For the double winch shown in Figure 8-23 determine the force in each cable and the bearing reactions. Assume that the bearings are friction free and act as simple supports.

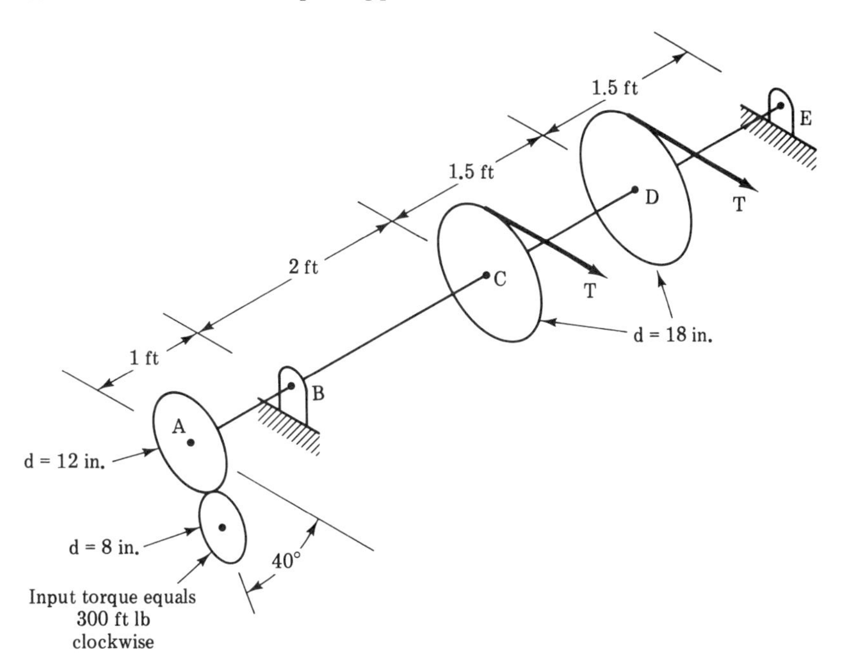

Figure 8-23

Example 8-11: The crank shown in Figure 8-24a has a smooth bearing at *B* and a socket at *E*. Determine the force required at *A* and the bearing reactions if the torque required at *C* is 10 ft lb counterclockwise and the torque required at *D* is 15 ft lb counterclockwise. Assume that the bearings are frictionless, that *B* acts as a simple support, and that the coefficient of belt friction is 0.3 for *C* and 0.25 for *D*.

Step 1 Draw the free-body diagram as shown in Figure 8-24b.

Step 2 Determine the belt tensions for pulley *C*.

$$M_C = (T_{CL} - T_{CS})(r_C) \qquad T_{CL} = T_{CS}e^{\mu\beta}$$

$$T_{CL} = T_{CS}e^{(0.3)(\pi)} \qquad T_{CL} = 2.566T_{CS}$$

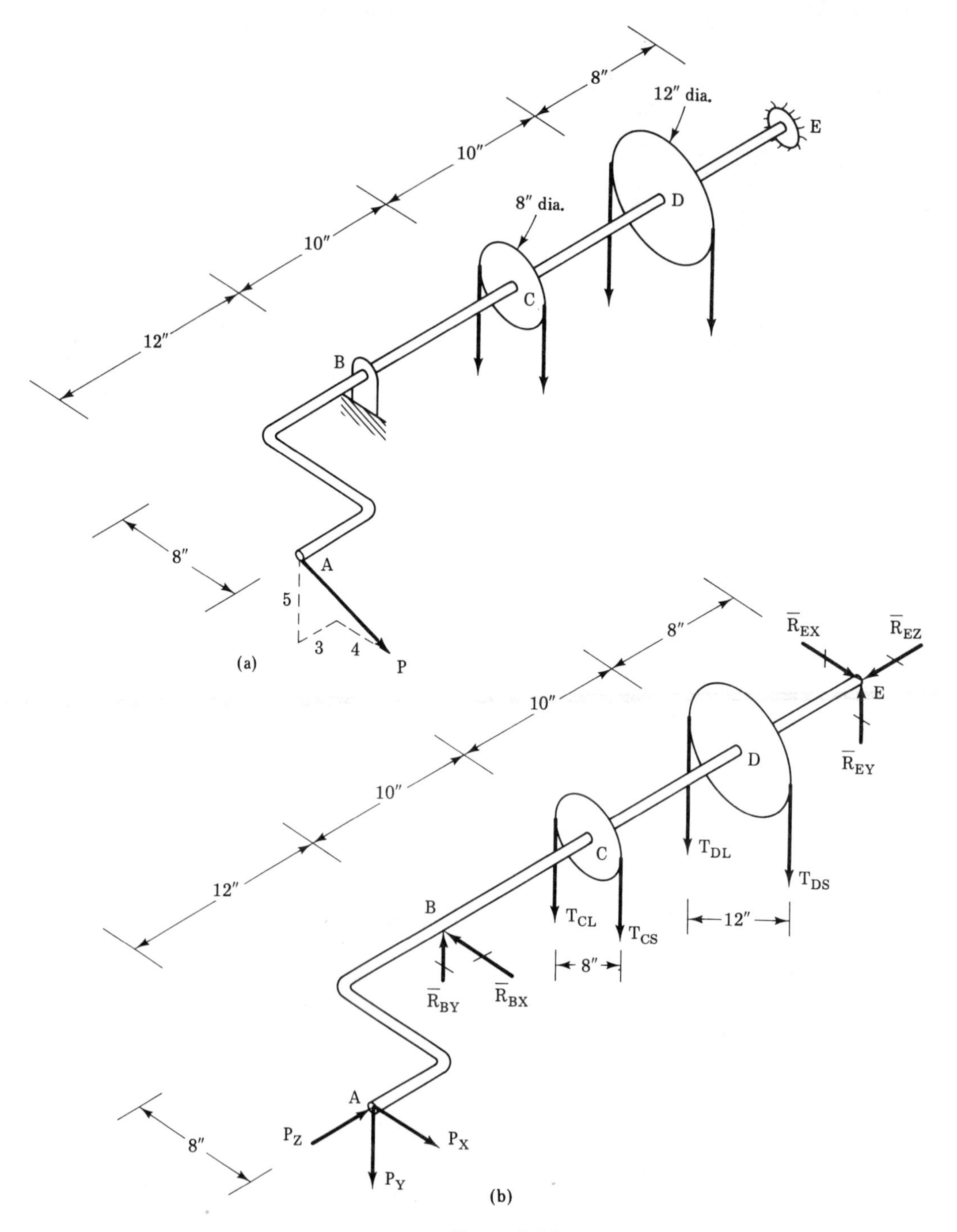
8″
12″ dia.
E
10″
8″ dia.
D
10″
C
12″
B
8″
A
5
3
4
P
(a)
8″
R̄EX
R̄EZ
E
10″
D
R̄EY
10″
C
TDL
12″
TDS
B
TCL
TCS
12″
8″
R̄BY
R̄BX
A
8″
PZ
PX
PY
(b)

Figure 8-24

$$M_C = (2.566T_{CS} - T_{CS})(r_C) \qquad T_{CS} = \frac{M_C}{1.566r_C}$$

$$T_{CS} = \frac{10 \text{ ft lb}}{1.566\left(\dfrac{2}{12}\text{ ft}\right)} \qquad T_{CS} = 38.3 \text{ lb}$$

$$T_{CL} = 2.566T_{CS} = 2.566(38.31 \text{ lb}) \qquad T_{CL} = 98.3 \text{ lb}$$

Step 3 Determine the belt tensions for pulley D.

$$M_D = (T_{DL} - T_{DS})(r_D) \qquad T_{DL} = T_{DS}e^{\mu\beta}$$

$$T_{DL} = T_{DS}e^{(0.25)(\pi)} \qquad T_{DL} = 2.193T_{CS}$$

$$M_D = (2.193T_{DS} - T_{DS})(r_D) \qquad T_{DS} = \frac{M_D}{1.193r_D}$$

$$T_{DS} = \frac{15 \text{ ft lb}}{1.193\left(\dfrac{3}{12}\text{ ft}\right)} \qquad T_{DS} = 50.3 \text{ lb}$$

$$T_{DL} = 2.193T_{DS} = 2.193(50.3 \text{ lb}) \qquad T_{DL} = 110.3 \text{ lb}$$

Step 4 Resolve the force on the crank into components. From the directional components given, the proportional length of the diagonal can be determined.

$$l_P = \sqrt{5^2 + 4^2 + 3^2} = 7.07$$

$$\cos\theta_X = \frac{4}{7.07} = 0.566$$

$$\cos\theta_Y = \frac{5}{7.07} = 0.707$$

$$\cos\theta_Z = \frac{3}{7.07} = 0.424$$

$$P_X = P\cos\theta_X \qquad P_Y = P\cos\theta_Y \qquad P_Z\cos\theta_Z$$

Step 5 Take moments about the Z axis to determine P_Y

$$\Sigma M_X = 0 \qquad P_Yd_Y + M_C + M_D = 0$$

$$P_Yd_Y = -(M_C + M_D) = -(+10 + 15 \text{ ft lb})$$

$$P_Yd_Y = -25 \text{ ft lb}$$

making P_Y negative.

$$P_Y = \frac{P_Y d_Y}{d_Y} = \frac{25 \text{ ft lb}}{\frac{8}{12} \text{ ft}} \qquad P_Y = 37.5 \text{ lb}$$

Step 6 Determine P, P_X, and P_Z by cosines.

$$P = \frac{P_Y}{\cos \theta_Y} = \frac{37.5 \text{ lb}}{0.707} \qquad \boldsymbol{P = 53 \text{ lb}}$$

$$P_X = P \cos \theta_X = 53 \text{ lb}(0.566) \qquad \boldsymbol{P_X = 30 \text{ lb}}$$

$$P_Z = P \cos \theta_Z = 53 \text{ lb}(0.424) \qquad \boldsymbol{P_Z = -22.5 \text{ lb}}$$

Step 7 Determine $\overline{R}_{EZ}$ by summing forces in the Z direction.

$$\Sigma F_Z = 0 \qquad P_Z + \overline{R}_{EZ} = 0 \qquad \overline{R}_{EZ} = -P_Z \qquad \boldsymbol{\overline{R}_{EZ} = +22.5 \text{ lb}}$$

Step 8 Find $\overline{R}_{BY}$ by taking moments about E in the YZ plane.

$$P_Y d_{PY} + \overline{R}_{BY} d_{BY} + F_{CY} d_{CY} + F_{DY} d_{DY} = 0$$

$$\overline{R}_{BY} d_{BY} = -(P_Y d_{PY} + F_{CY} d_{CY} + F_{DY} d_{DY})$$

$$\overline{R}_{BY} d_{BY} = -\left[+(37.5 \text{ lb})\left(\frac{40}{12} \text{ ft}\right) + (136.6 \text{ lb})\left(\frac{18}{12} \text{ ft}\right) + (160.6 \text{ lb})\left(\frac{8}{12} \text{ ft}\right)\right]$$

$$\overline{R}_{BY} d_{BY} = -437 \text{ ft lb} \circlearrowright$$

Thus $\overline{R}_{BY}$ is positive.

$$\overline{R}_{BY} = \frac{\overline{R}_{BY} d_{BY}}{d_{BY}} = \frac{437 \text{ ft lb}}{\frac{28}{12} \text{ ft}} \qquad \boldsymbol{\overline{R}_{BY} = +187 \text{ lb}}$$

Step 9 Find $\overline{R}_{EY}$ by summing forces in the Y direction.

$$\Sigma F_Y = 0 \qquad P_Y + \overline{R}_{BY} + F_{CY} + F_{DY} + \overline{R}_{EY} = 0$$

$$\overline{R}_{EY} = -(P_Y + \overline{R}_{BY} + F_{CY} + F_{DY})$$

$$\overline{R}_{EY} = -(-37.5 + 187 - 136.6 - 160.6 \text{ lb})$$

$$\boldsymbol{\overline{R}_{EY} = +147.7 \text{ lb}}$$

Step 10 Find $\overline{R}_{BX}$ by taking moments about E in the XZ plane.

$$\Sigma M_E = 0 \qquad P_X d_{PX} + P_Z d_{PZ} + \overline{R}_{BX} d_{BX} = 0$$

$$\overline{R}_{BX} d_{BX} = -(P_X d_{PX} + P_Z d_{PZ})$$

$$\overline{R}_{BX} d_{BX} = -\left[+(30 \text{ lb})\left(\frac{40}{12} \text{ ft}\right) + (22.5 \text{ lb})\left(\frac{8}{12} \text{ ft}\right)\right]$$

$$\overline{R}_{BX} d_{BX} = -1215 \text{ ft lb} \circlearrowright$$

Thus $\overline{R}_{BX}$ is negative.

$$\overline{R}_{BX} = \frac{\overline{R}_{BX} d_{BX}}{d_{BX}} = \frac{1215 \text{ ft lb}}{\frac{28}{12} \text{ ft}} \qquad \boldsymbol{\overline{R}_{BX} = -521 \text{ lb}}$$

Step 11 Find $\overline{R}_{EX}$ by summing forces in the X direction.

$$\Sigma F_X = 0 \qquad P_X + \overline{R}_{BX} + \overline{R}_{EX} = 0 \qquad \overline{R}_{EX} = -(P_X + \overline{R}_{BX})$$

$$\overline{R}_{EX} = -(+30 - 521 \text{ lb}) \qquad \boldsymbol{\overline{R}_{EX} = +491 \text{ lb}}$$

Problem 8-10: A single-cylinder engine is shown in Figure 8-25. The connecting rod is 10 in. long and the crankshaft has a 4-in. offset. Determine the larger belt tension and all bearing reactions (assume bearings are simple supports). What is the minimum permissible value for the coefficient of friction for the belt and pulley? If this is too large, what do you suggest?

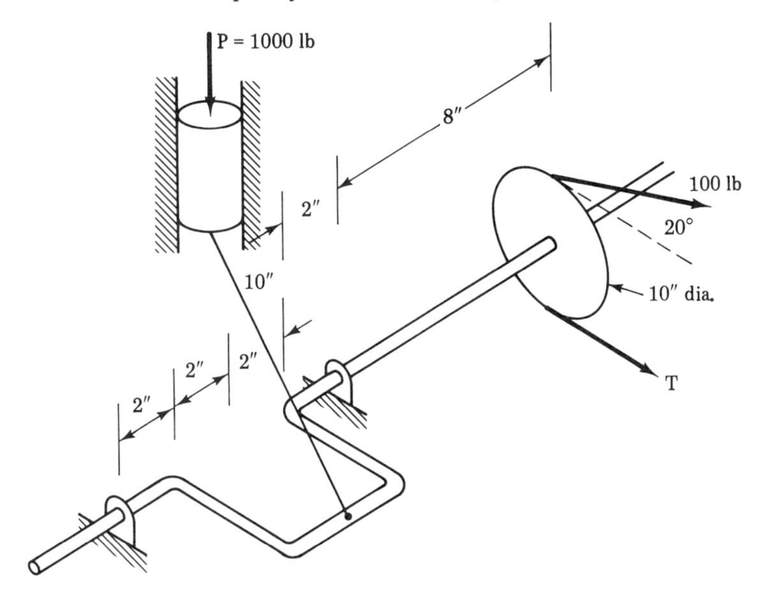

Figure 8-25

8.7 APPLICATIONS TO FRICTION

Many of the aspects of friction discussed earlier in this text could be three-dimensional simply by the directional orientation of the problem. Problems of this nature would simply involve application of the three-coordinate system to concepts and applications we are already familiar with. However, one particular combination is unique enough to warrant special attention. This special combination is the screw thread, used to fasten materials together and to transmit power. Some of the more common types of threads are shown in Figure 8-26. The remainder of this section will concentrate on square threads. Any good text on machine design or appropriate handbook may be consulted regarding other types of threads. The concepts are the same.

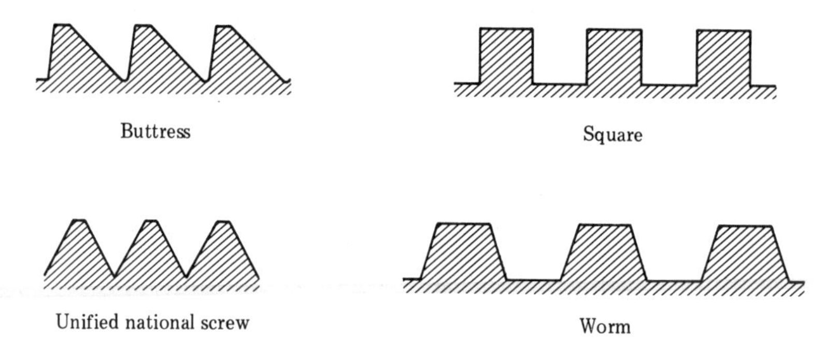

Figure 8-26. Screw threads.

The thread of a screw is an incline. If the axis of the screw is placed in a vertical position, one can think of the thread as a circular staircase winding around a cylinder. As one walks up or down the incline one also walks around the surface of the cylinder. Most of us have used a bolt or screw in combination with a nut to fasten something. Sometimes the nut is entirely stationary and may even be a threaded hole in a large object. As the screw is turned it has not only angular motion but also axial motion. The reverse may be true as in the case of a lug nut holding a wheel on a car; the nut has all the motion. The other two possibilities are where one of the components has only angular motion and the other only axial motion and vice versa. The same thinking can be applied to power screws.

Some basic nomenclature is helpful in working with threads. Reference should be made to Figure 8-27 during this discussion. First, the *horizontal* distance traveled in the process of moving up or down the incline through one complete revolution or 360 *horizontal* degrees is a circumference, πd.

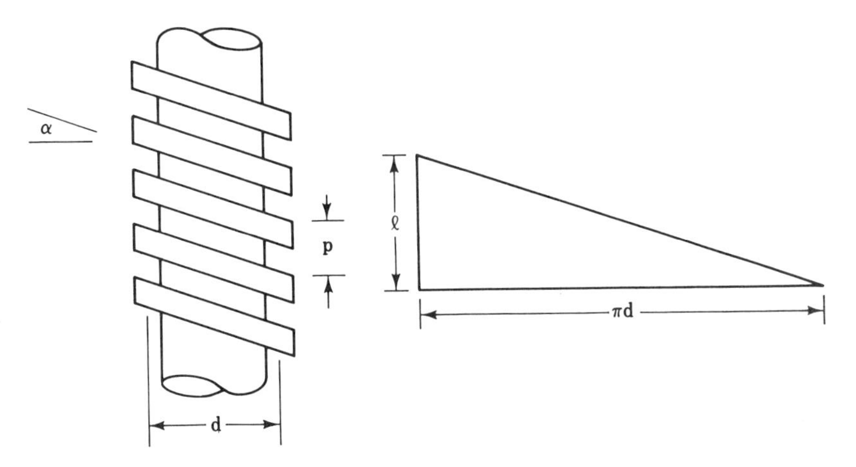

Figure 8-27. Square screw thread.

The diameter used is the mean or average diameter. This can be expressed as the outside diameter minus the thread depth or thickness (these are essentially the same for a square thread). The vertical or axial distance traveled is called the *lead.* For a *single thread,* this is the same as the *pitch.* The pitch is the axial distance between identical points on two adjacent threads. There are screws with multiple threads, most commonly two or three. This is analogous to having more than one staircase, starting equally spaced around the base of a circular tower, winding their way around it and ending equally spaced at the top of the tower. The lead would be the pitch times the number of threads. Multiple threading, thus, increases the ratio of axial motion to angular motion by a factor equal to the number of threads. The winding of the thread about the base cylinder is frequently called a helix. The angle whose tangent is the ratio of the lead to the mean circumference is known as the helix angle.

Continuing the analogy of the incline, let us examine the relationship between axial force and the torque required to achieve and maintain that force. Reference should be made to Figure 8-28 during this discussion.

Recall that, in dealing with friction, the area of contact is irrelevant. Thus the length of the screw thread path (πd times the number of turns in contact) and the width of the path (depth of the thread) do not enter into this discussion. They are important, of course, in terms of the ability of the thread to withstand the forces applied to it. However, that is a separate issue that will be dealt with later in this text in terms of the strength of materials.

The relationship of the load P, the applied force F_S, and the reaction R is shown in Figure 8-28. Theoretically, each of these forces can be broken into components parallel to and perpendicular to the incline. The perpendicular component of F_S and the perpendicular component of P are countered by the normal force. The parallel component of F_S counters the

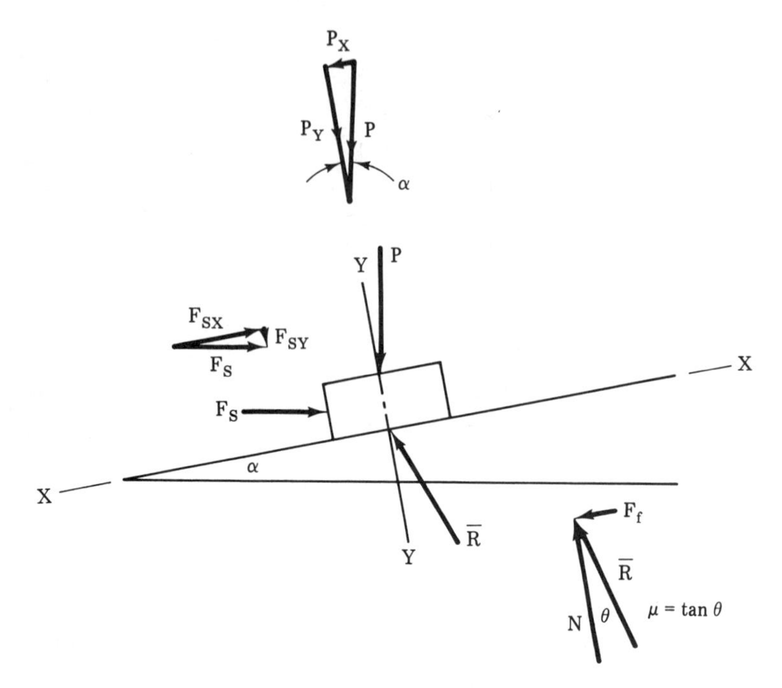

Figure 8-28. Torque and axial force.

parallel component of P and the friction force. The normal force and the friction force make up the reaction.

Example 8-12: The bench vise shown in Figure 8-29a has a screw with a 25-mm diameter (outside) square threads (single) with a 3-mm pitch. Determine the force required at the end of the lever that will result in 1 kN of force between the vise jaws. The coefficient of friction is 0.19. Neglect the friction at A and B.

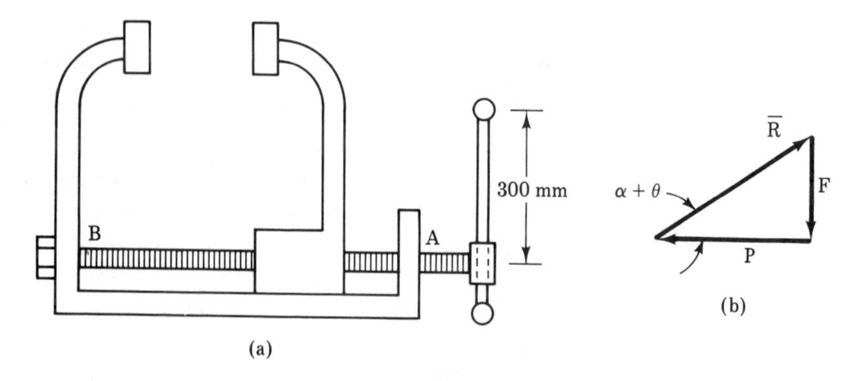

Figure 8-29

Solution:

Step 1 Assume that the moment effect of the vise jaw force is taken up at the base of the movable jaw and, therefore, has a negligible effect on the screw. The axial load on the screw is then 1 kN.

Step 2 Draw the generalized free-body diagram for a screw thread as shown in Figure 8-29b.

Step 3 Determine the force applied to the screw thread as follows: Draw the force polygon for the problem as shown in Figure 8-29c.

$$\tan\alpha = \frac{l}{\pi d} \qquad \alpha = \arctan\frac{l}{\pi d}$$

$$d = 25 - 1.5 \text{ mm (thickness} = \tfrac{1}{2}\text{ pitch)}$$

$$d = 23.5 \text{ mm} \qquad \alpha = \arctan\frac{3 \text{ mm}}{\pi(23.5 \text{ mm})} \qquad \alpha = 2.33°$$

$$\theta = \arctan\mu = \arctan 0.19 \quad \theta = 10.76°$$

$$\tan(\alpha + \theta) = \frac{F_s}{P} \qquad F_s = \text{P}\tan(\alpha + \theta)$$

$$F_s = (1 \text{ kN})\tan 13.09° \qquad F_s = 232.5 \text{ N}$$

Step 4 Determine the force needed at the lever by summing moments about the screw axis.

$$\Sigma M = 0 \qquad F_s r_s + F_L d_{FL} = 0 \qquad F_L = \frac{r_s}{d_{FL}} F_s$$

$$F_L = \frac{11.75 \text{ mm}}{300 \text{ mm}}(232.5 \text{ N}) \qquad \boldsymbol{F_L = 9.11 \text{ N}}$$

Note that this problem gives an indication of the tremendous mechanical advantage resulting from the combination of the lever and the screw.

$$MA = \frac{F \text{ out}}{F \text{ in}} = \frac{1000 \text{ N}}{9.11 \text{ N}} \qquad MA = 110$$

Problem 8-11: A house jack shown in Figure 8-30 has a double square thread, 1 in. in outside diameter, with 6 threads per in. The coefficient of friction for the threads is 0.1. What force is required at the jack handle 10 in. from the screw axis to lift a load of 5 ton? If the jack handle is released under load, will the load stay up? Explain.

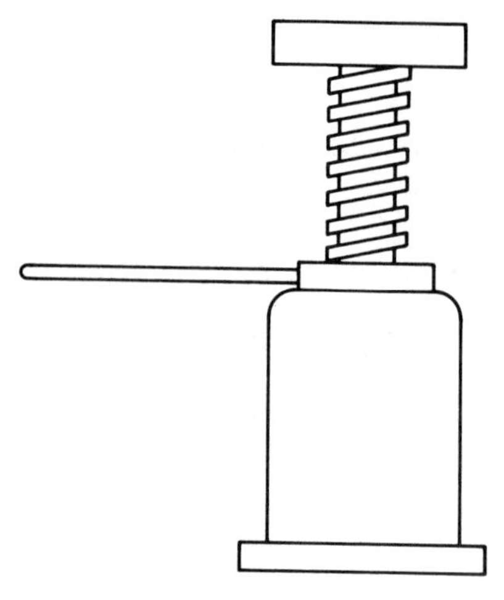

Figure 8-30

8.8 READINESS QUIZ

1. Three-dimensional equilibrium requires that

   (a) the sum of the forces in all directions add up to zero
   (b) the sum of the moments about any axis add up to zero
   (c) the sum of the moments about any point in any plane add up to zero
   (d) all of the above
   (e) none of the above

2. Each rectangular component of a space force is related to that force by the __________ of the angle between the component and the force.

   (a) sine
   (b) cosine
   (c) tangent
   (d) all of the above
   (e) none of the above

3. An example of a noncoplanar force system is

   (a) a piano standing on a stage
   (b) wind acting on one side of a building
   (c) a piston, connecting rod, and crankshaft combination in an engine
   (d) all of the above
   (e) none of the above

4. Pythagoras' theorem expanded to three dimensions states that

   (a) $F_X^2 + F_Y^2 + F_Z^2 = 0$
   (b) $F_X + F_Y + F_Z = 0$
   (c) $F^2 = F_X^2 + F_Y^2 + F_Z^2$
   (d) $F = F_X + F_Y + F_Z$
   (e) none of the above

5. Problems dealing with single shafts may consist of

   (a) parallel forces in space
   (b) nonparallel forces in parallel planes
   (c) forces in all three dimensions
   (d) all of the above
   (e) none of the above

6. The pitch of a thread is always equal to the

   (a) lead
   (b) axial distance from a point on one thread to an identical point on the next thread
   (c) thread thickness
   (d) all of the above
   (e) none of the above

7. If the applied torque on a screw under axial load is removed, motion will reverse if the angle of friction is __________ the helix angle.

   (a) smaller than
   (b) greater than
   (c) equal to
   (d) parallel to
   (e) perpendicular to

8. If a space force has components of $F_X = 2$ lb, $F_Y - 3$ lb, and $F_Z = 6$ lb, the force has a magnitude of

   (a) 7 kN
   (b) 11 lb
   (c) 7 lb
   (d) 11 kN
   (e) none of the above

9. A double-threaded screw has a lead equal to __________ times its pitch.

(a) 4
(b) $\frac{1}{2}$
(c) $\sqrt{2}$
(d) any of the above
(e) none of the above

10. The mechanical advantage of a screw is generally

(a) small
(b) negative
(c) large
(d) any of the above
(e) none of the above

8.9 SUPPLEMENTARY PROBLEMS

8-12 Determine the resultant of $F_X = +200$ kN, $F_Y = -600$ kN, $F_Z = -300$ kN. Provide a sketch.

8-13 Determine the components of the space force shown in Figure 8-31.

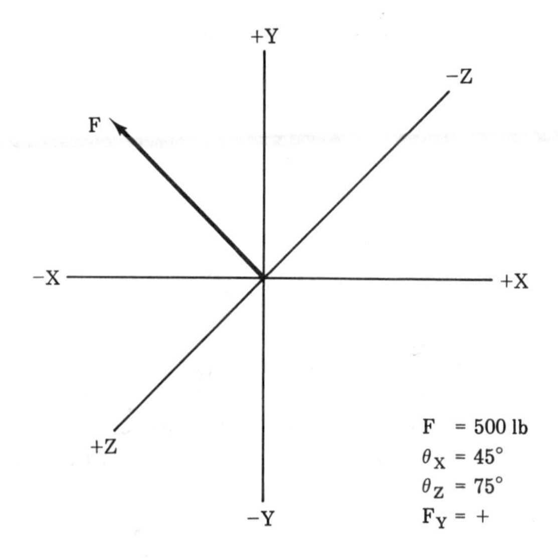

Figure 8-31

8-14 Determine the resultant of the space forces shown in Figure 8-32.

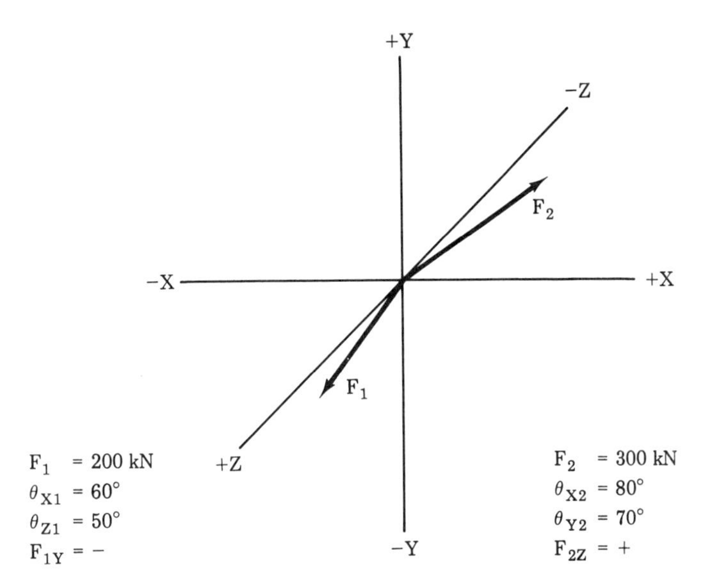

Figure 8-32

8-15 Determine the magnitude and location of the resultant of the deck weight (20 lb/ft²) and the vehicle weights for the ferry deck shown in Figure 8-33.

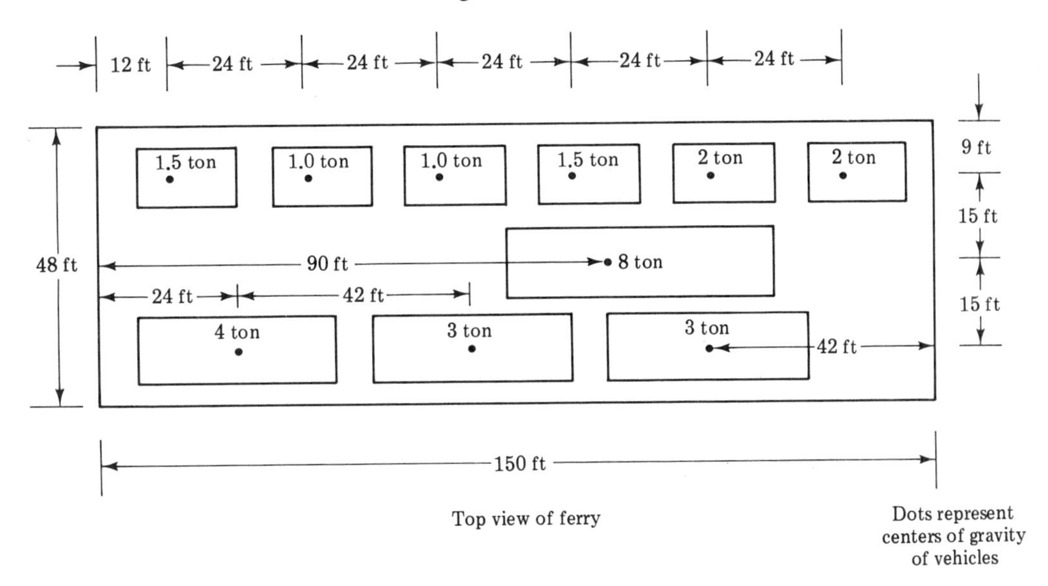

Figure 8-33

8-16 The load shown in Figure 8-34 is raised by a combination of the two hydraulic cylinders and the winch. Assume that all bearings and pivots are friction free. Assume that bearing *A* is smooth, allowing axial motion. Determine the support reactions at *A*, *B*, *D*, *F*, and *G*, and the pin reactions at *C*, *E*, and *H*.

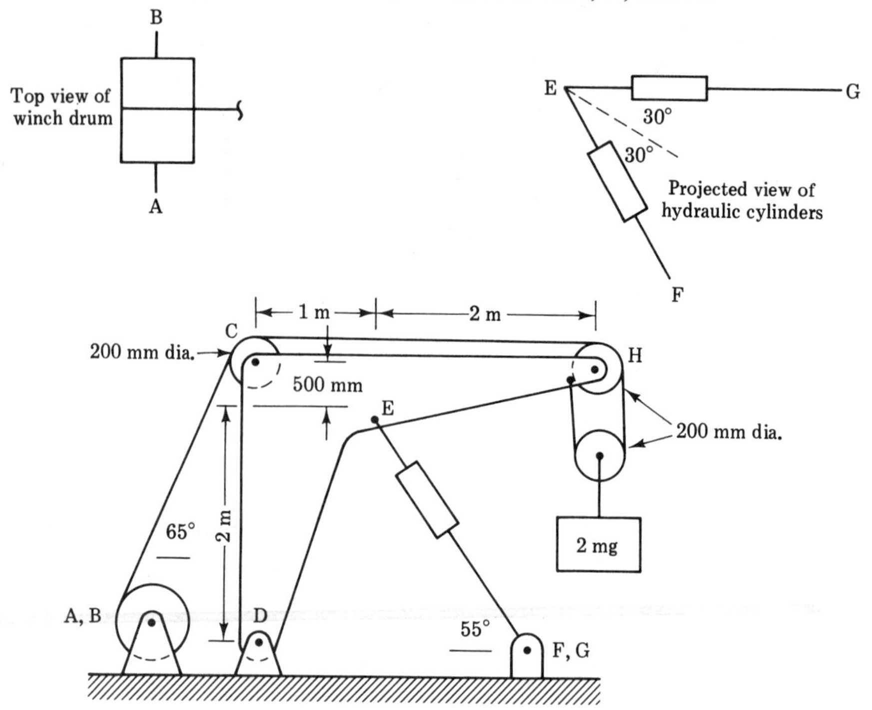

Figure 8-34

8-17 What axial load can be achieved in a triple-threaded square thread screw with a thread thickness of $\frac{1}{4}$ in., a mean diameter of 1.75 in. and a coefficient of friction of 0.2 if a force of 40 lb is applied (perpendicular to the plane made by the screw axis and the wrench) 2 ft from the screw axis? What is the mechanical advantage?

Chapter

9

Centers of Gravity, Centroids, and Area Moments of Inertia

9.1 OBJECTIVES

Upon completion of the work relating to this chapter, you should be able to perform the following.

1. For the terms listed below:
   (a) Select from several definitions the correct one for any given term.
   (b) Given a definition, select the correct term from a given list of terms.

   center of gravity
   centroid
   centroidal moment of inertia
   inertia
   moment of inertia
   polar moment of inertia
   radius of gyration

2. Determine the location of the centroid of any complex plane figure (including cross sections of composite structural members) with respect to any known point.
3. Determine the moment of inertia of an irregular area about any given axis.
4. Determine the polar moment of inertia of an irregular area about any given axis.

5. Determine the radius of gyration of an irregular area measured from any given axis.

9.2 CENTERS OF GRAVITY

The *center of gravity* of any object or group of objects can be simply defined as the point through which the force of gravity acts on the mass of the object or group of objects. This phenomena attains great personal significance when we try to balance ourselves on a single ice skate or a bicycle. In order to maintain our balance, the line of action of our weight (force of gravity) must pass directly through our support (skate) or a line drawn through our supports (bicycle tire contacts with ground).

The complexity of such problems varies greatly. They may be very simple collinear problems, coplanar, or even noncoplanar. However, if we limit ourselves to problems involving relatively short distances on the earth's (or any other large body's) surface, all forces of gravity can be considered, for practical purposes, to be parallel. Consider that the earth's radius is roughly 20×10^6 ft. Then a distance of 1000 ft subtends an angle of less than 0.003°, a difference in direction which we can usually ignore. Thus, all problems of this nature will be parallel force problems. Problems involving the center of gravity of an object can be further complicated by the fact that a relatively simple object may be a combination of two materials that are quite dissimilar in their mass densities. We shall proceed to examine a variety of such problems from the simple to the complex.

Example 9-1: A cube of iron (490 lb/ft³) is to be balanced as shown in Figure 9-1a. What force must be exerted on the edge to result in equilibrium?

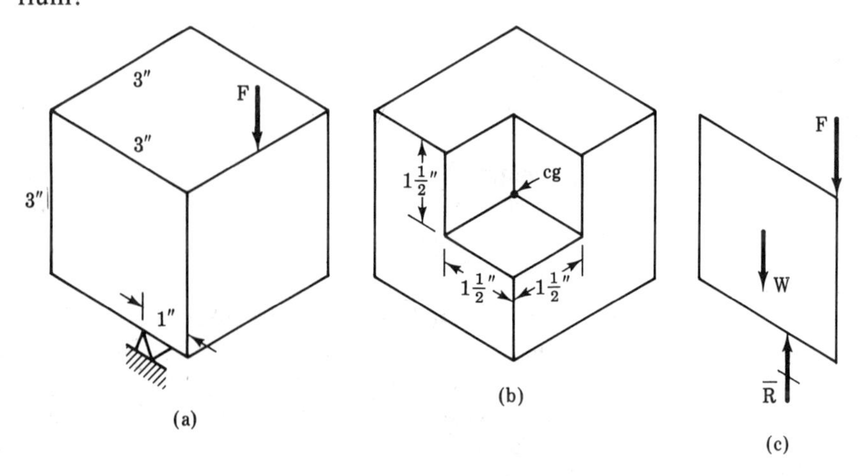

Figure 9-1

Solution:

Step 1 Recognize that the center of gravity of a homogeneous cube will be located at one-half the length of an edge from each face as shown in Figure 9-1b.

Step 2 Draw a free-body diagram looking at the front face of the cube as shown in Figure 9-1c.

Step 3 Write the force and moment equilibrium equations and solve for F as follows:

$$W + F + \overline{R} = 0$$

Since there are two unknowns this will not immediately yield a solution. However, a value for $\overline{R}$ was not requested so it may be convenient to write a moment equation about the point of application of $\overline{R}$.

$$Wd_W + Fd_F = 0$$

Since there are only two moments, their directions must be opposite and are obvious from the free-body diagram. The equation can be rewritten as

$$F = \frac{Wd_W}{d_F}$$

$$W = \gamma V = \left(490 \frac{\text{lb}}{\text{ft}^3}\right)\left(\frac{1 \text{ ft}^3}{1728 \text{ in.}^3}\right)(3 \text{ in.})^3 \qquad W = 7.66 \text{ lb}$$

$$F = \frac{7.66 \text{ lb } (0.5 \text{ in.})}{1.0 \text{ in.}} \qquad \mathbf{F = 3.83 \text{ lb}}$$

Example 9-2: Determine the center of gravity of the block of wood ($\rho = 800 \text{ kg/m}^3$) shown in Figure 9-2a.

Solution:

Step 1 Recognize that the block can be divided into two simple rectangular parallelepipeds either along an XY plane or a ZY plane as shown in Figure 9-2b. The location of the center of gravity of each of these can be readily determined since it is at a distance of half an edge from each face.

Step 2 Using the first division illustrated and selecting as an origin for our coordinate system the back lower left-hand corner, draw the free-body diagram of the block as shown in Figure 9-2c. There is no support shown, however, if we wished to balance this block on a point on its bottom surface, that point would have to lie directly below the center of gravity for the total block.

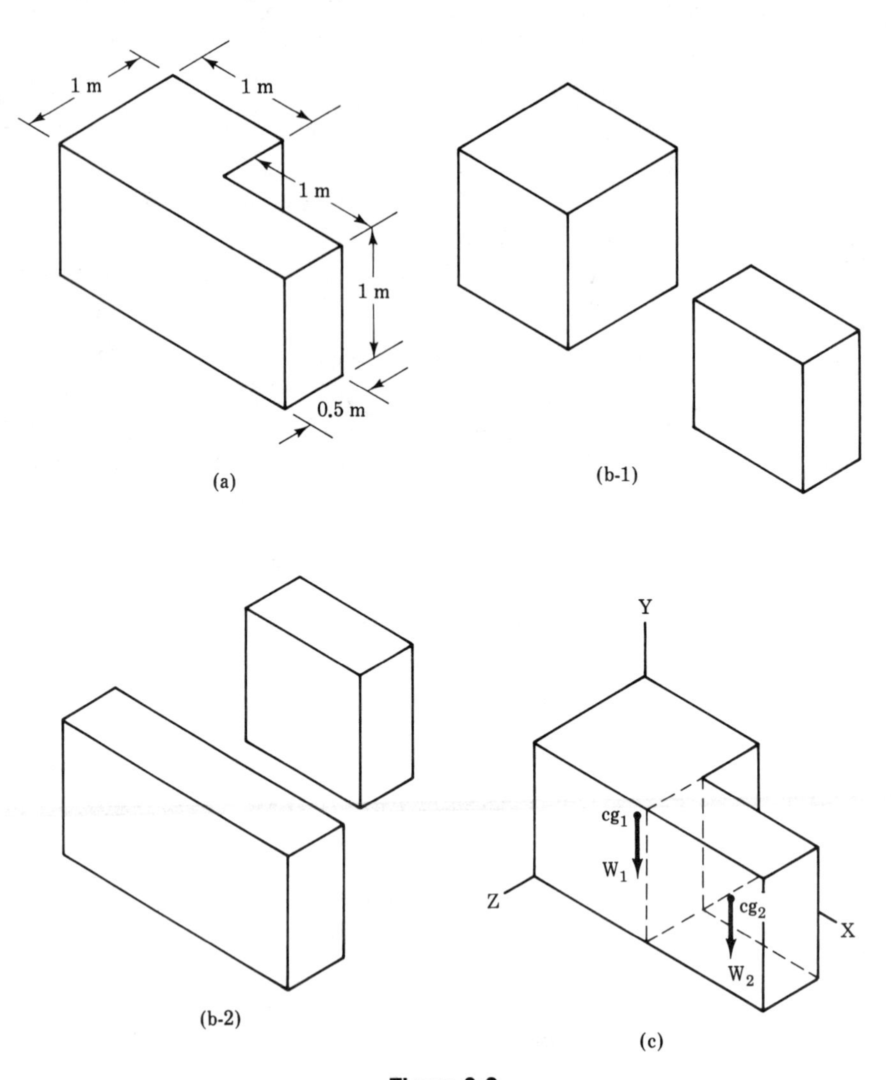

Figure 9-2

Step 3 The center of gravity of the left-hand portion has coordinates of $X_1 = +0.5$ m, $Y_1 = +0.5$ m, and $Z_1 = +0.5$ m. The center of gravity of the right-hand portion has coordinates of $X_2 = +1.5$ m, $Y_2 = +0.5$ m, and $Z_2 = +0.75$ m. The coordinates of the center of gravity for the total block must lie between these two points (again, envision balancing this block), closer to W_1 than W_2 since W_1 is the larger.

Step 4 Since the total weight is the sum of the individual weights, and weights are forces, we are simply dealing with the location of the resultant of two parallel forces. This can be done by any of the methods learned earlier.

The total weight can be readily determined and the location determined by moments about axes as follows:

$$m_T = m_1 + m_2 = \rho v_1 + \rho v_2$$

$$m_T = (800\ \text{kg/m}^3)(1\ \text{m}^3) + (800\ \text{kg/m}^3)(1\ \text{m}^2)(0.5\ \text{m})$$

$$m_T = 800 + 400\ \text{kg}$$

$$m_T = 1200\ \text{kg}$$

(This could be converted to N or kN but there is really no need to in this case since ultimately we are only interested in distances.)

$$\Sigma M_Z = m_1 x_1 + m_2 x_2 = m_T x_T$$

$$x_T = \frac{m_1 X_1 + m_2 X_2}{m_T}$$

$$x_T = \frac{800(0.5) + (400)(1.5)}{1200} \qquad \boldsymbol{x_T = 0.833\ \text{m}}$$

$$\Sigma M_X = m_1 Z_1 + m_2 Z_2 = m_T Z_T$$

$$Z_T = \frac{m_1 Z_1 + m_2 Z_2}{m_T}$$

$$Z_T = \frac{800(0.5) + 400(0.75)}{1200} \qquad \boldsymbol{Z_T = 0.583\ \text{m}}$$

Since the block is homogeneous and has a constant height, $Y_1 = Y_2 = Y_T$

$$\boldsymbol{Y_T = 0.5\ \text{m}}$$

The usefulness of this last dimension may not be immediately apparent, but consider turning the block 90° about either the X or the Z axis. We shall make use of this technique in some subsequent examples.

Example 9-3: Determine the center of gravity of the tank of water ($\gamma = 0.0361\ \text{lb/in.}^3$) shown in Figure 9-3a (neglect the tank walls and bottom).

Solution:

Step 1 Divide the tank into two sections, the low flat rectangular parallelepiped and the upright cylinder, W_1 and W_2 respectively.

Step 2 Select the origin of the coordinate system at the upper right-hand back corner of the parallelepiped (the choice is arbitrary or at your convenience—in this case it was chosen to illustrate the feasibility of using

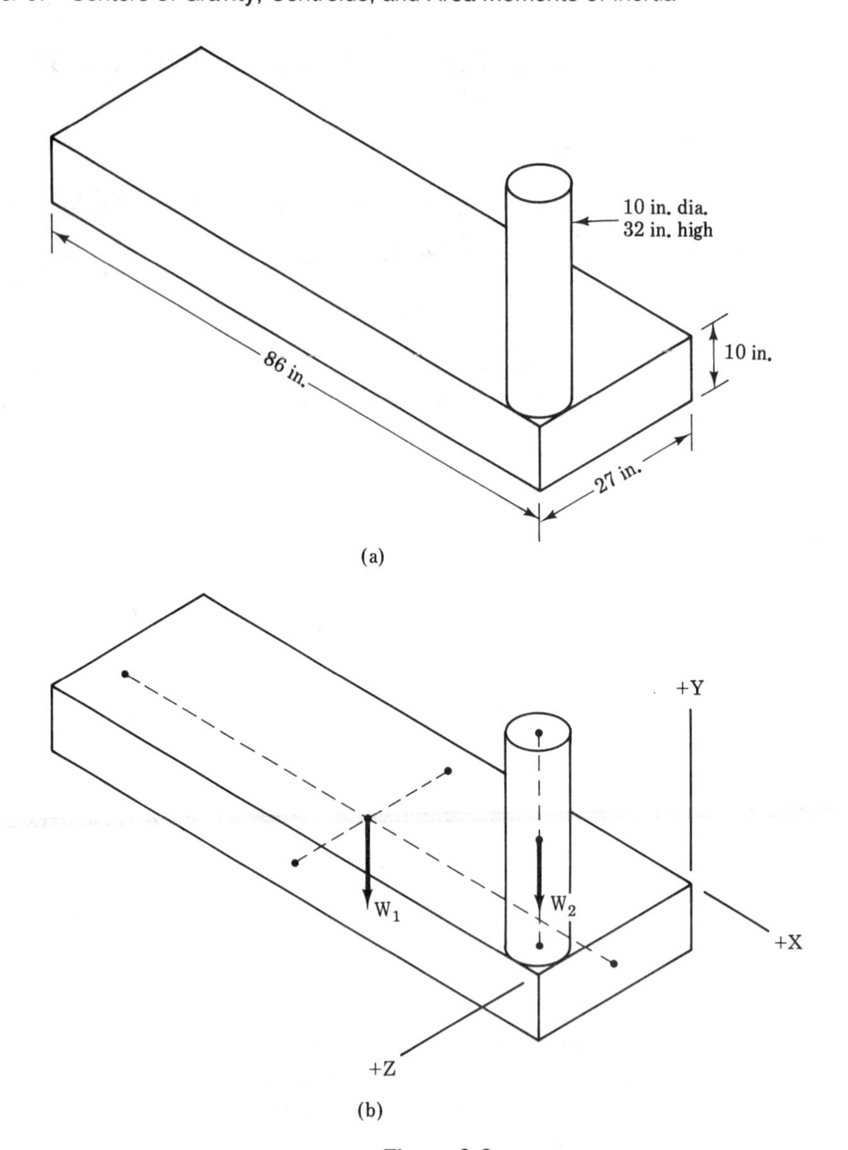

Figure 9-3

negative coordinates). Since the center of gravity of a cylinder is at the center of its circular cross section and half way up its height, we can say that $X_1 = -43$ in., $Y_1 = -5$ in., $Z = +13.5$ in., $X_2 = -5$ in., $Y_2 = +16$ in., and $Z_2 = +22$ in.

Step 3 Determine the weights

$$W_1 = \gamma V_1 = (0.0361 \text{ lb/in.}^3)(86 \text{ in.})(10 \text{ in.})(27 \text{ in.})$$

$$W_1 = 838 \text{ lb}$$

$$W_2 = \gamma V_2 = \gamma \pi r^2 h = (0.0361 \text{ lb/in.}^3)(\pi)(5 \text{ in.}^2)(32 \text{ in.})$$

$$W_2 = 90.7 \text{ lb}$$

$$W_T = W_1 + W_2 = 838 + 90.7 + 90.7 \text{ lb} \qquad W_T = 928.7 \text{ lb}$$

Step 4 Take moments about each axis to determine X_T, Y_T, and Z_T. By inspection, it can be estimated that X_T will be slightly less than X_1, Y_T will be appreciably less than Y_1, and Z_T will be slightly greater than Z_1.

$$\Sigma M_X = W_1 Z_1 + W_2 Z_2 = W_T Z_T$$

$$Z_T = \frac{W_1 Z_1 + W_2 Z_2}{W_2}$$

$$Z_T = \frac{838(13.5) + 90.7(22)}{928.7} \qquad \boldsymbol{Z_T = +4.3 \text{ in.}}$$

$$\Sigma M_Z = W_1 X_1 + W_2 X_2 = W_T X_T$$

$$X_T = \frac{W_1 X_1 + W_2 X_2}{W_T}$$

$$X_T = \frac{838(-43) + 90.7(-5)}{928.7} \qquad \boldsymbol{X_T = -39.3 \text{ in.}}$$

To obtain Y_T, we can either turn the object about either the X axis or the Z axis. Or, we can simply turn the effect of gravity temporarily and point the weight forces perpendicular to either the XY or the ZY plane. Then

$$\Sigma M_X = \Sigma M_Z = W_1 Y_1 + W_2 Y_2 = W_T Y_T$$

$$Y_T = \frac{W_1 Y_1 + W_2 Y_2}{W_T}$$

$$Y_T = \frac{838(-5) + 90.7(16)}{928.7} \qquad \boldsymbol{Y_T = -2.95 \text{ in.}}$$

It may have occurred to you that the specific weights of the materials are really irrelevant in all of the above problems. You are absolutely right! All of the calculations could have been made using volumes. In that case we would have said that we had located the *centroid* of the total volume. Since, in each example, the material was homogeneous, the locations of the centroid and the center of gravity are identical. However, there are many times when an object is composed of two or more materials and the centroid and center of gravity are identical only by chance.

Example 9-4: Determine the location of both the centroid and the center of gravity of the arm shown in Figure 9-4a.

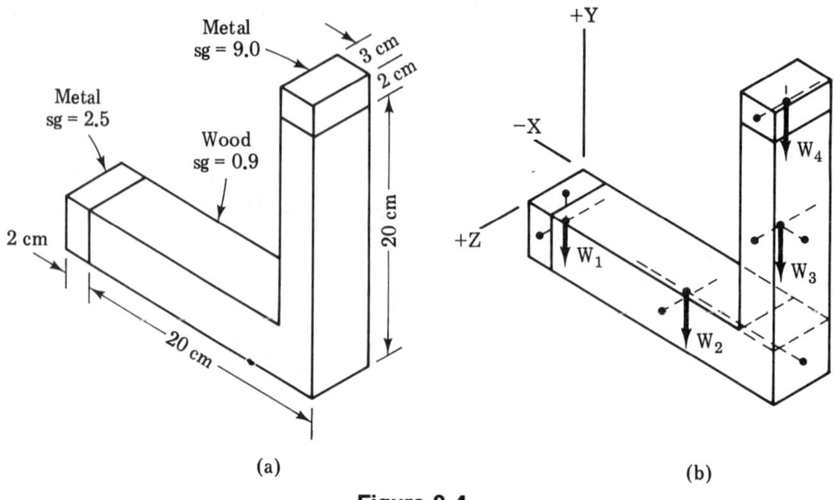

Figure 9-4

Solution:

Step 1 Recognize the need to separate the object into four rectangular parallelepipeds.

Step 2 Select an origin for the coordinate system at the top back left corner of the lower portion of the arm as shown in Figure 9-4b.

Step 3 Make a free-body diagram as shown in Figure 9-4b.

Step 4 Recognize the need for two sets of calculations: One for the centroid, based on volume, and another based on weight or mass for the center of gravity.

Step 5 Determine coordinates of centroid/center of gravity for each homogeneous part of the arm.

$$X_1 = +1 \text{ cm} \qquad Y_1 = -2 \text{ cm} \qquad Z_1 = +2.5 \text{ cm}$$

$$X_2 = +12 \text{ cm} \qquad Y_2 = -2 \text{ cm} \qquad Z_2 = +2.5 \text{ cm}$$

$$X_3 = +20.5 \text{ cm} \qquad Y_3 = +8 \text{ cm} \qquad Z_3 = 2.5 \text{ cm}$$

$$X_4 = +20.5 \text{ cm} \qquad Y_4 = +17 \text{ cm} \qquad Z_4 = +2.5 \text{ cm}$$

Step 6 Determine all volumes.

$$V_1 = 4 \times 5 \times 2 \qquad V_1 = 40 \text{ cm}^3$$

$$V_2 = 4 \times 5 \times 20 \qquad V_2 = 400 \text{ cm}^3$$

$$V_3 = 16 \times 5 \times 3 \qquad V_3 = 240 \text{ cm}^3$$

$$V_4 = 2 \times 5 \times 3 \qquad V_4 = 30 \text{ cm}^3$$

$$V_T = 710 \text{ cm}^3$$

Step 7 Determine all masses. Since *sg* for water is 1 and ρ for water is 1 gr/cm³, each *sg* can be read as ρ with those units.

$$m_1 = \rho_1 V_1 = 2.5(40) \qquad m_1 = 100 \text{ gr}$$
$$m_2 = \rho_2 V_2 = 0.9(400) \qquad m_2 = 360 \text{ gr}$$
$$m_3 = \rho_3 V_3 = 0.9(240) \qquad m_3 = 216 \text{ gr}$$
$$m_4 = \rho_4 V_4 = 0.0(30) \qquad m_4 = 270 \text{ gr}$$
$$m_T = 946 \text{ gr}$$

Step 8 Determine location of centroid.

$$x_T = \frac{V_1X_1 + V_2X_2 + V_3X_3 + V_4X_4}{V_T}$$

$$x_T = \frac{(40)(1) + (400)(12) + (240)(20.5) + (30)(20.5)}{710}$$

$$\mathbf{x_T = +14.6 \text{ cm}}$$

$$y_T = \frac{(40)(-2) + (400)(-2) + (240)(+8) + (30)(+17)}{710}$$

$$\mathbf{y_T = +2.2 \text{ cm}}$$

$$Z_T = \frac{(40)(2.5) + (400)(2.5) + (240)(2.5) + (30)(2.5)}{710}$$

$$\mathbf{Z_T = +2.5 \text{ cm}}$$

Step 9 Determine location of center of gravity.

$$x_T = \frac{m_1x_1 + m_2x_2 + m_3x_3 + m_4x_4}{m_T}$$

$$x_T = \frac{(100)(1) + (360)(12) + (216)(20.5) + (270)(20.5)}{946}$$

$$\mathbf{x_T = 15.2 \text{ cm}}$$

$$y_T = \frac{(100)(-2) + (360)(-2) + (210)(+8) + (270)(+17)}{946}$$

$$\mathbf{y_T = 5.7 \text{ cm}}$$

$$Z_T = \frac{(100)(2.5) + (360)(2.5) + (216)(2.5) + (270)(2.5)}{946}$$

$$\mathbf{Z_T = 2.5 \text{ cm}}$$

In comparing the locations of the centroid and the center of gravity, several things should be noted. First, the metal end caps had little effect on the x coordinate because of the relatively long moment arms of the wood portions. In the y direction, however, the combination of the relatively long moment arm and the high density of the one cap had a decided effect. Finally, in the Z direction, the results could have been attained by inspection, because the dimension was constant and at any and every point the material was homogeneous in the Z direction. It is worth noting that neither fell on the arm, i.e., they were located in space. This is not unusual for a curved or hollow object. After all, going back to the idea of a single balancing point lying below the center of gravity: How would you balance an inflated inner tube on one finger if the tube's largest circumference is in a plane parallel to the ground?

9.3 DEFINING THE CENTROID OF AN AREA

The centroid of the volume of an object has already been discussed in the previous section. If we look at the simple block shown in Figure 9-5, we can readily see the location of the centroid. If we make the 3-in. depth of the

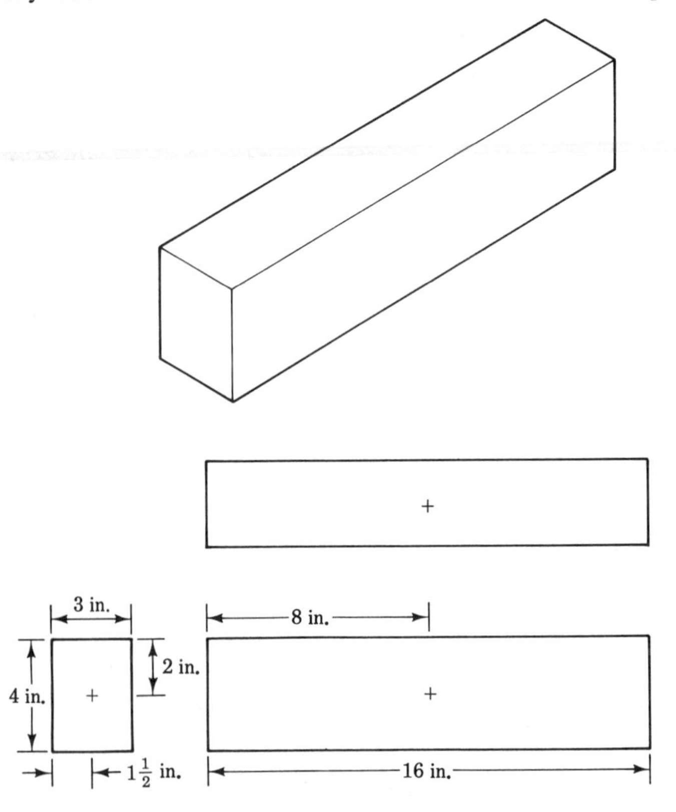

Figure 9-5. Centroid of a volume.

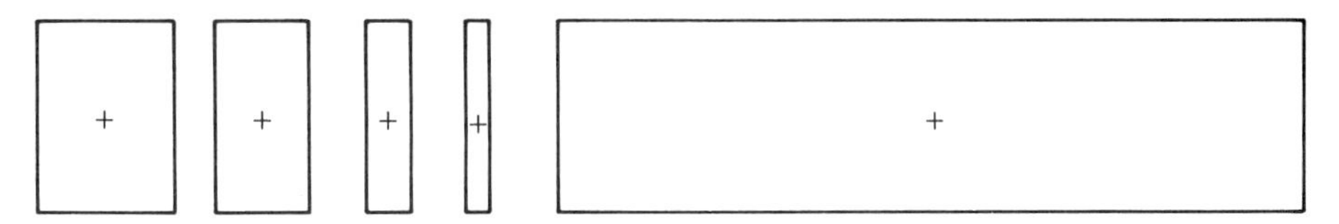

Figure 9-6. Centroid of an area.

block thinner and thinner, the location of the centroid in that direction becomes shallower and shallower, always half the total depth. However, as shown in Figure 9-6, the location of the centroid in the other two directions remains constant. If the depth is reduced to zero, we are left with a plane figure, an area. This area still has bidirectional properties including a properly located centroid. It sometimes helps to continue to envision such areas as a thin sheet of dense material with the centroid being the single point upon which we can balance the sheet. The usefulness of the centroid will become more apparent in the next section as well as in later chapters of this text.

In the previous section, the location of the centroid of a volume was accepted somewhat intuitively. The concept of a "balance point" was inherent in our calculations but was never examined very thoroughly.

At first glance it would appear that a straight line drawn through the centroid of an area simply divides the area in half. If we examine the area from our previous figures and divide it vertically, then horizontally, and finally diagonally, as shown in Figure 9-7, we can check this out.

The areas of the two halves for each split are computed and compared below.

		Area	
	Left		Right
Vertically divided	4×8	32	4×8
	Upper		Lower
Horizontally divided	2×16	32	2×16
	Upper left		Lower right
Diagonally divided	$\frac{1}{2}(4 \times 16)$	32	$\frac{1}{2}(4 \times 16)$

2 in.
2 in.
8 in.
8 in.

Figure 9-7. Centroid of an area described.

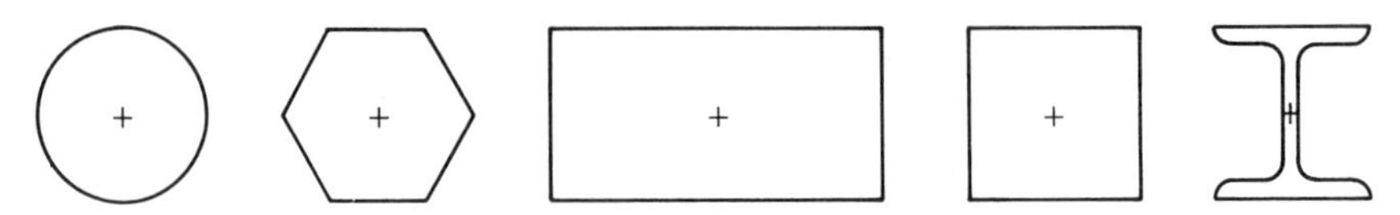

Figure 9-8. Centroids of symmetrical areas.

The same could be shown for any line drawn through the centroid of that rectangle. Indeed, it would be true for any symmetrically shaped areas such as those shown in Figure 9-8.

Let us now look at a totally nonsymmetrical area such as the right triangle shown in Figure 9-9a. Half the area would be 27. Since we are creating similar triangles with a height-to-base ratio of 4 : 3, the smaller triangles, would each have an area of $\frac{1}{2}b$ (4b/3) equal to 27. Their bases would then be 6.36 and their heights 8.48. This would locate the centroid at a point 3.52 above the base of the initial triangle and 2.64 to the right of the altitude of the initial triangle as shown in Figure 9-9b.

However, if we go back to our concept of balancing, we would find that the triangle would not balance on the point we have just found. Why not? Because balancing does not involve simply equal weights, masses, volumes, or areas. It involves the *moments* of those quantities. Recall the childhood seesaw. It was not sufficient for two children to be of equal weight for the seesaw to balance—they also had to be at equal distances from the fulcrum. Or, if they were of unequal weight, their distance from the fulcrum had to be in inverse proportion to their weight—the heavier one closer to the fulcrum, the lighter one further away from the fulcrum. The centroid then is really the fulcrum point.

From the above discussion, it becomes evident that the centroid is the point about which the moments of the areas involved equal zero. Further, those moments can be taken about any axis through the centroid. The reason it was so simple when dealing with totally symmetrical areas was that

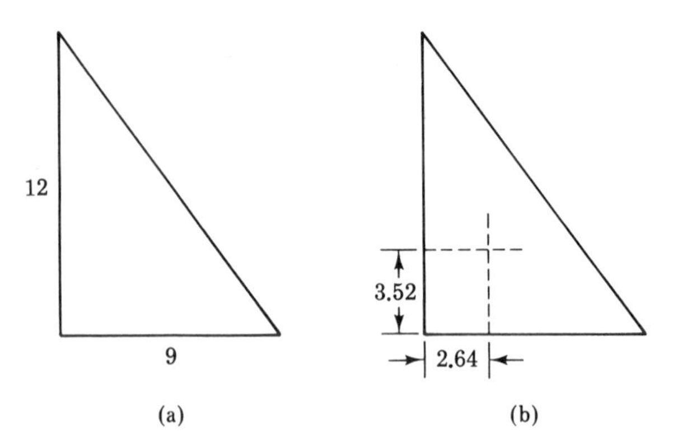

Figure 9-9. Centroid of a triangle.

the areas were not just equal on both sides of any axis drawn through the centroid but the areas were also uniformly distributed in terms of their distances from that axis.

9.4 LOCATING THE CENTROID OF A REGULAR AREA

Determining the location of the centroid of any area, including one that is asymmetrical can be accomplished by applying a concept learned in working with forces. Recall that not only must the resultant of a force system equal the sum of the forces in order to replace the system but that the resultant must also be located in such a manner that the moment of the resultant equals the sum of the moments of the forces. If we replace the word "resultant" with "total area" and the word "force" with "area," then we can write

$$\Sigma M = A_T d_T = A_1 d_1 + A_2 d_2 + A_3 d_3 + \cdots$$

Appropriate subscripts can be used to indicate the axis about which moments are being taken and x, y, and z can be used as needed in place of d above. If the axis about which moments are being taken passes through the centroid then the sum of the moments will equal zero.

This approach can be used in two ways. First, it can be used to locate the centroid of any asymmetrical area. Second, it can be used to locate the centroid of any complex area.

Let us first correctly locate the centroid of the triangle we examined in Figure 9-9. It has been reproduced in Figure 9-10a. In Figure 9-10b, x and y

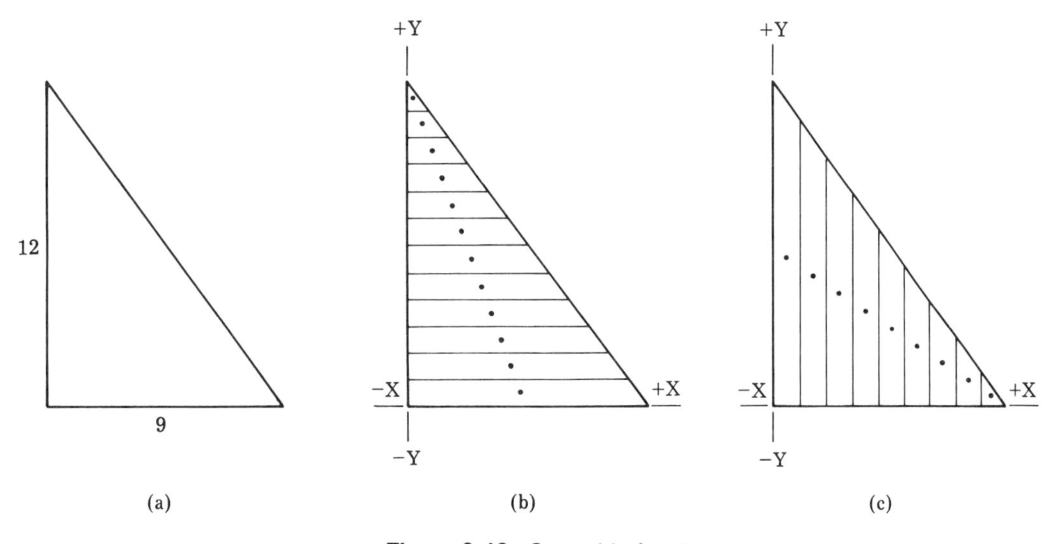

Figure 9-10. Centroid of a triangle determined.

axes have been added and the triangle divided into 1-in. horizontal strips. In Figure 9-10c, x and y axes have been added and the triangle divided into 1-in. vertical strips. We shall now proceed to correctly locate the centroid of the triangle by use of moments.

If we treat each trapezoid as a rectangle with a length equal to the average length of the trapezoid then the centroid of each will fall at 0.5 in., 1.5 in., and so on. The horizontal strip at the base will have an average length of 8.625, the next 7.875, and so on. The vertical strip at the altitude will have an average height of 11.33 in., the next 10 in., and so on.

$$\Sigma M_x = A_T y_T + A_1 y_1 + A_2 y_2 + A_3 y_3 + \cdots$$

A	y	M
8.625	0.5	4.3125
7.875	1.5	11.8125
7.125	2.5	17.8125
6.375	3.5	22.3125
5.625	4.5	25.3125
4.875	5.5	26.8125
4.125	6.5	26.8125
3.375	7.5	25.3125
2.625	8.5	22.3125
1.875	9.5	17.8125
1.125	10.5	11.8125
0.375	11.5	4.3125
54.0		216.75

$$\bar{y}_T = \frac{\Sigma M_x}{A_T} = \frac{216.75}{54} \qquad \bar{y}_T = 4.014 \text{ in.}$$

$$\Sigma M_y = A_T x_T = A_1 x_1 + A_2 x_2 + A_3 x_3 + \cdots$$

This can be treated in a similar manner to ΣM_x to arrive at

$$\bar{x}_T = 3.019 \text{ in.}$$

If the methods of calculus were used to do the above calculations with an infinitely large number of infinitely thin strips, the precise answers of 4 and 3 would have been obtained. Thus it can be said that the centroid of any triangle is located at a distance of $\frac{1}{3}$ of any altitude above the base associated with that altitude. Appendix B-3 locates the centroids of a number of common shapes that have been determined by the method just illustrated.

Note that the centroid always needs to be located with respect to some set of coordinates. In general, some combination of axes of symmetry and a defining edge are used. Figure 9-11 illustrates these. In 9-11a the cross section of the *I* beam is symmetrical in both directions so the axes of symmetry have been used as references. In 9-11b, the isosceles triangle is symmetrical only about the *y* axis, so that axis of symmetry and the base of the triangle have been used as references. In 9-11c, the quarter circle is asymmetrical so the axes running through the center of curvature have been used as references. Note that a bar has been placed above *x* and *y* to denote centroid location. This is a common symbol used to denote "average."

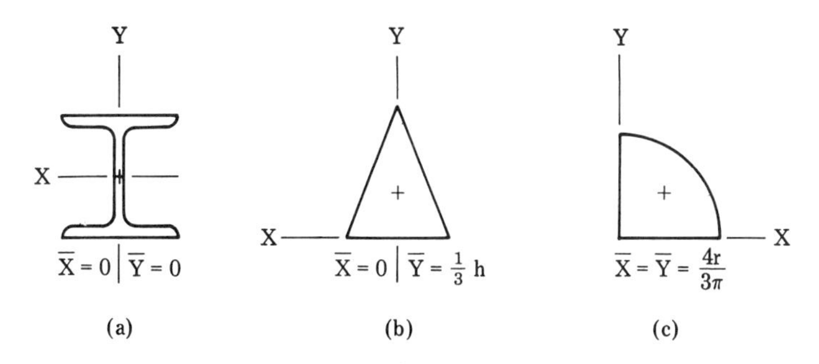

Figure 9-11. Symmetrical and asymmetrical areas.

9.5 LOCATING THE CENTROID OF A COMPLEX AREA

As has already been suggested, the same techniques just used to locate the centroid of a regular area can be employed in finding the centroid of a complex area. The following steps must be followed:

1. Identify the axes from which the location of the centroid is to be measured—these may be given or else must be assumed and so noted.
2. Identify and label all the regular areas that make up the complex area.
3. Determine the size of each regular area and the total area.
4. Locate the centroid of each area on the area itself.
5. Locate the centroid of each area with respect to the axes identified in 1 above. This needs to be done with care but is rather like dimensioning when drafting.
6. Write and apply the moment equations to determine $\bar{x}_T$ and $\bar{y}_T$.

Example 9-5: Locate the centroid of the area shown in Figure 9-12a with respect to the given axes.

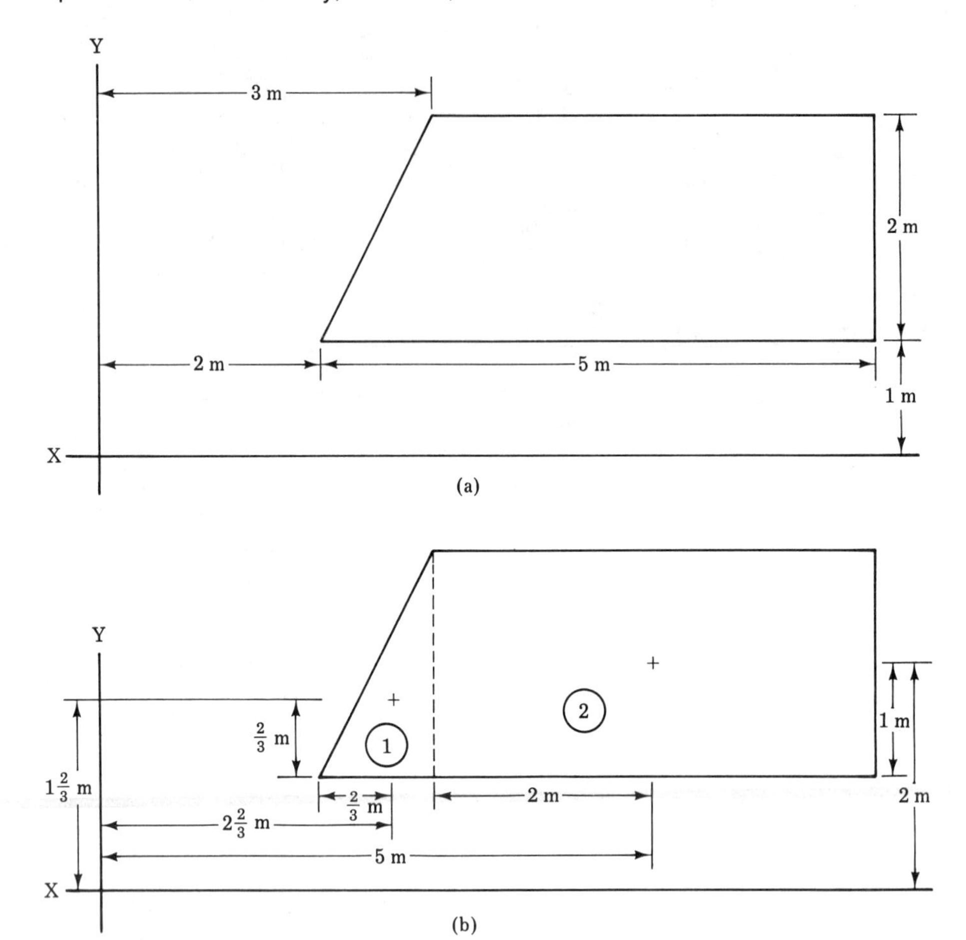

Figure 9-12

Solution:

Step 1 The axes are given.

Step 2 The area can be divided into a triangle, labeled 1, and a rectangle labeled 2.

Step 3

$$A_1 = \tfrac{1}{2}bh = \tfrac{1}{2}(1\text{ m})(2\text{ m}) \qquad A_1 = 1\text{ m}^2$$

$$A_2 = bh = (4\text{ m})(2\text{ m}) \qquad A_2 = 8\text{ m}^2$$

$$A_T = A_1 + A_2 = 1 + 8\text{ m}^2 \qquad A_T = 9\text{ m}^2$$

Step 4 The centroids of each area have been located in Figure 9-12b.

Step 5 The location of the centroid of each area has been dimensioned with respect to the given axes in Figure 9-12b.

Step 6 $\bar{x}_T$ can be estimated as being slightly smaller than $\bar{x}_2$. $\bar{y}_T$ can be estimated as being slightly smaller than $\bar{y}_2$.

$$\bar{x}_T = \frac{A_1\bar{x}_1 + A_2\bar{x}_2}{A_T}$$

$$\bar{x}_T = \frac{(1)(2.67) + (8)(5)}{9} \qquad \bar{x}_T = 4.74 \text{ m}$$

$$\bar{y}_T = \frac{A_1\bar{y}_1 + A_2\bar{y}_2}{A_T}$$

$$\bar{y}_T = \frac{(1)(1.67) + (8)(2)}{9} \qquad \bar{y}_T = 1.96 \text{ m}$$

These figures appear quite realistic in the light of our initial estimate.

Problem 9-1: Locate the centroid of the area shown in Figure 9-13 with respect to the given axes.

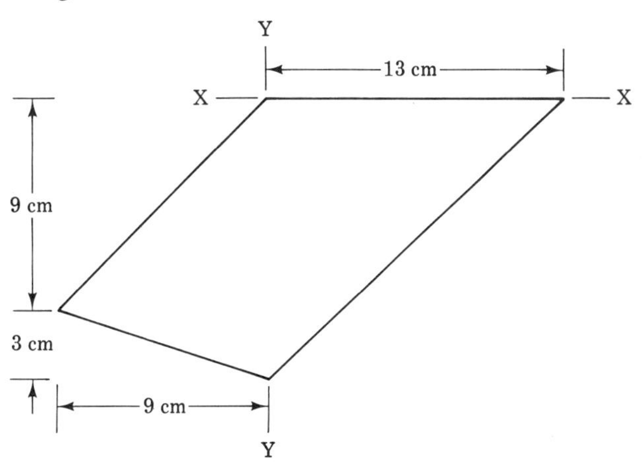

Figure 9-13

Example 9-6: Determine the location of the centroid of the area shown in Figure 9-14a with respect to the given axes.

Solution:

Step 1 The axes are given.

Step 2 The area can be divided into a rectangle (1), a quarter circle (2), and a triangle (3).

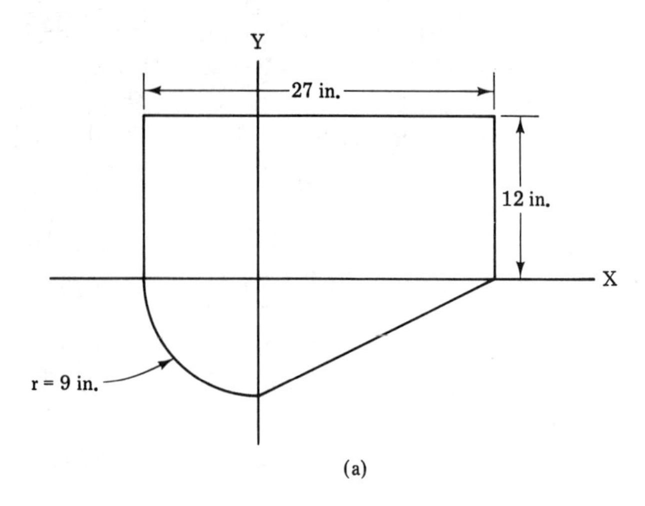

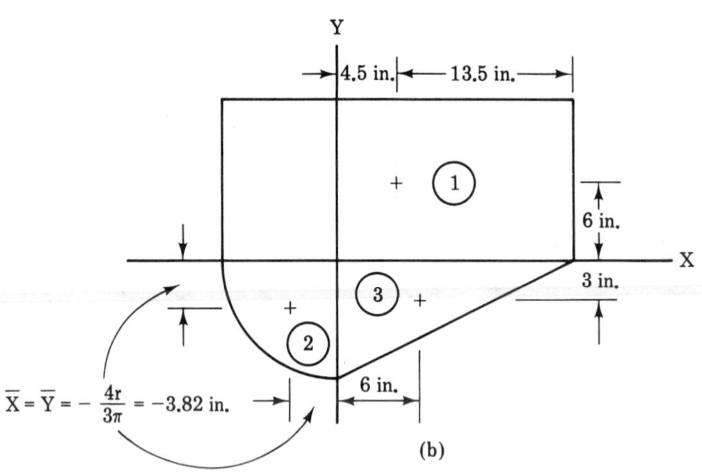

Figure 9-14

Step 3

$$A_1 = bh = (27)(12) \qquad A_1 = 324 \text{ in.}^2$$

$$A_2 = \tfrac{1}{4}\pi \qquad A_2 = 63.6 \text{ in.}^2$$

$$\tfrac{1}{4}\pi r^2 = \tfrac{1}{4}\pi(9)^2$$

$$A_3 = \tfrac{1}{2}bh = \tfrac{1}{2}(18)(9) \qquad A_3 = 81 \text{ in.}^2$$

$$A_T = A_1 + A_2 + A_3 \qquad A_T = 468.6 \text{ in.}^2$$

Step 4 The centroids of each area have been located in Figure 9-14b.

Step 5 The location of the centroid of each area has been dimensioned with respect to the given axes in Figure 9-14b (the orientation of this problem is such that this step was largely completed in Step 4).

Step 6 $\bar{x}_T$ can be estimated as being somewhat less than $\bar{x}_1$, while $\bar{y}_T$ can be estimated as being substantially less than $\bar{y}_1$ but still positive.

$$\bar{x}_T = \frac{A_1\bar{x}_1 + A_2\bar{x}_2 + A_3\bar{x}_3}{A_T}$$

$$\bar{x}_T = \frac{(324)(4.5) + (63.6)(-3.82) + (81)(6)}{468.6}$$

$$\mathbf{\bar{x}_T = 3.63\ in.}$$

$$\bar{y}_T = \frac{A_1\bar{y}_1 + A_2\bar{y}_2 + A_3\bar{y}_3}{A_T}$$

$$\bar{y}_T = \frac{(324)(6) + (63.6)(-3.82) + (81)(-3)}{468.6}$$

$$\mathbf{\bar{y}_T = 3.11\ in.}$$

These figures fit in well with our estimates. *Note the need to use negative moment arms in this problem.*

Problem 9-2: Determine the location of the centroid of the area shown in Figure 9-15 with respect to the given axes.

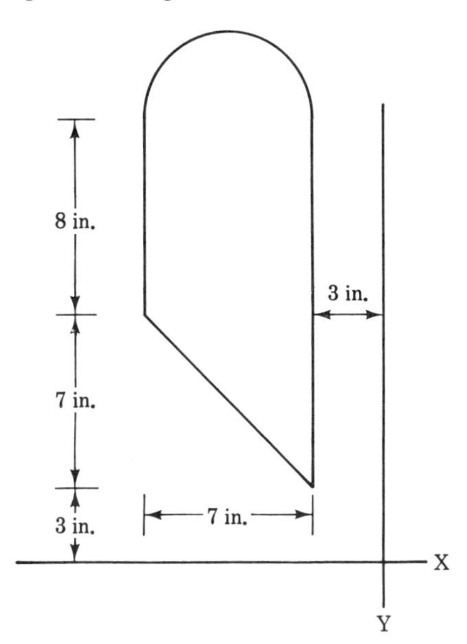

Figure 9-15

Example 9-7: Determine the location of the centroid of the area shown in Figure 9-16a with respect to the given axes.

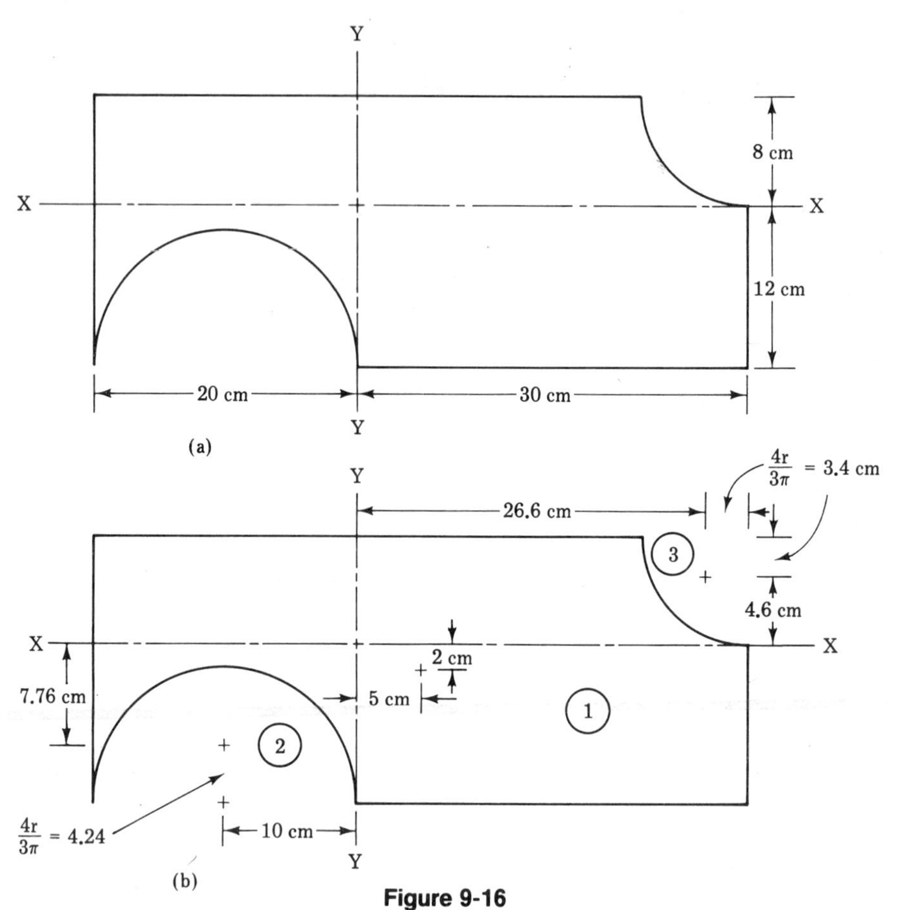

Figure 9-16

Solution:

Step 1 The axes are given.

Step 2 The area can be divided into a rectangle (1) with a semicircle (2) removed and a quarter-circle (3) removed. (Any area removed must be treated as a negative both in summing the total area and in computing the total moment.)

Step 3

$$A_1 = bh = (50)(20) \qquad A_1 = 1000 \text{ cm}^2$$

$$A_2 = -\tfrac{1}{2}\pi r^2 = -\tfrac{1}{2}\pi(10)^2 \qquad A_2 = -157 \text{ cm}^2$$

$$A_3 = -\tfrac{1}{4}\pi r^2 = -\tfrac{1}{4}\pi(8)^2 \qquad A_3 = -50.3 \text{ cm}^2$$

$$A_T = A_1 + A_2 + A_3 \qquad A_T = 792.7 \text{ cm}^2$$

Step 4 The centroid of each area has been located in Figure 9-16b.

Step 5 The location of the centroid of each area has been dimensioned with respect to the given axes in Figure 9-16b.

Step 6 $\bar{x}_T$ can be estimated as being slightly more positive than $\bar{x}_1$ while $\bar{y}_T$ can be estimated as being slightly less negative than $\bar{y}_1$.

$$\bar{x}_T = \frac{A_1\bar{x}_1 + A_2\bar{x}_2 + A_3\bar{x}_3}{A_T}$$

$$\bar{x}_T = \frac{(1000)(5) + (-157)(-10) + (-50.3)(26.6)}{792.7}$$

$$\mathbf{\bar{x}_T = 6.6\ cm}$$

$$\bar{Y}_T = \frac{A_1\bar{Y}_1 + A_2\bar{Y}_2 + A_3\bar{Y}_3}{A_T}$$

$$\bar{Y}_T = \frac{(1000)(-2) + (-157)(-7.76) + (-50.3)(4.6)}{792.7}$$

$$\mathbf{\bar{Y}_T = -1.28\ cm}$$

Problem 9-3: Determine the location of the centroid of the area shown in Figure 9-17 with respect to the given axes.

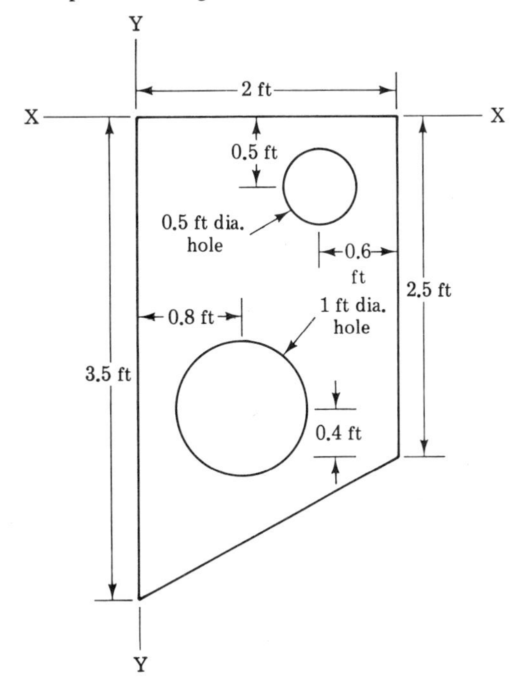

Figure 9-17

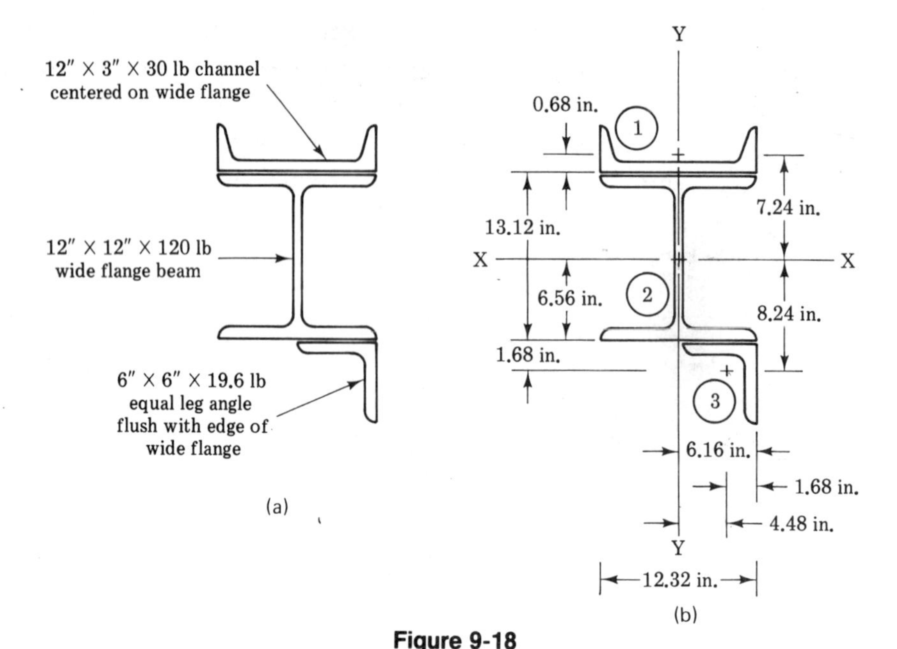

Figure 9-18

Example 9-8: Determine the location of the centroid of the cross section of the composite structural member shown in Figure 9-18a.

Solution: Before proceeding with the solution, it is worth noting that this type of problem is even simpler than the preceding ones. This is because the areas and centroid locations for many standard structural steel and aluminum shapes can be found in Appendix B-4 through B-7. However, a *word of caution: the given, or nominal dimensions of a structural member are not always the actual dimensions.* The actual dimensions must also be taken from a table.

Step 1 Since the wide flange beam is not only the largest member but also the only symmetrical one, the axes running through its centroid have been chosen as reference axes.

Step 2 As shown in Figure 9-18b, the channel has been labeled 1, the wide flange 2, and the angle 3.

Step 3 The areas for each member as extracted from the appropriate table are:

A_1	8.79 in.^{2}
A_2	35.31 in.^{2}
A_3	5.75 in.^{2}
A_T	49.85 in.^{2}

Step 4 The location of the centroid for each member as well as any dimensions needed were taken from the appropriate table and are shown in Figure 9-18b.

Step 5 The location of each individual centroid has been dimensioned to the reference axes in Figure 9-18b.

Step 6 $\bar{x}_T$ can be estimated to be somewhat to the right of the y axis while $\bar{y}_T$ can be estimated to be slightly above the x axis.

$$\bar{x}_T = \frac{\bar{x}_1 A_1 + \bar{x}_2 A_2 + \bar{x}_3 A_3}{A_T}$$

$$\bar{x}_T = \frac{(0)(8.79) + (0)(35.31) + (4.48)(5.75)}{49.85}$$

$$\bar{x}_T = +0.52 \text{ in.}$$

$$\bar{y}_T = \frac{\bar{y}_1 A_1 + \bar{y}_2 A_2 + \bar{y}_3 A_3}{A_T}$$

$$\bar{y}_T = \frac{(7.24)(8.79) + (0)(35.31) + (-8.24)(5.75)}{49.85}$$

$$\bar{y}_T = +0.33 \text{ in.}$$

Problem 9-4: Determine the location of the centroid of the cross section of the composite structural member shown in Figure 9-19.

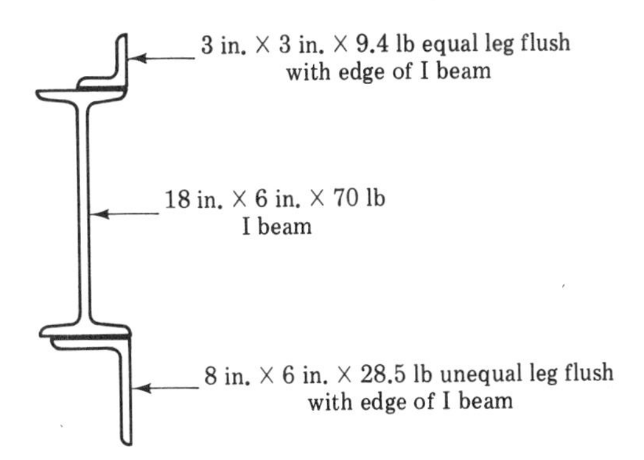

Figure 9-19

9.6 AREA MOMENT OF INERTIA DEFINED

The moment of inertia of an area is a specific property of an area. As might be expected from its name, this property is directional in nature, being a function of the axis about which moments are taken. It is sometimes called

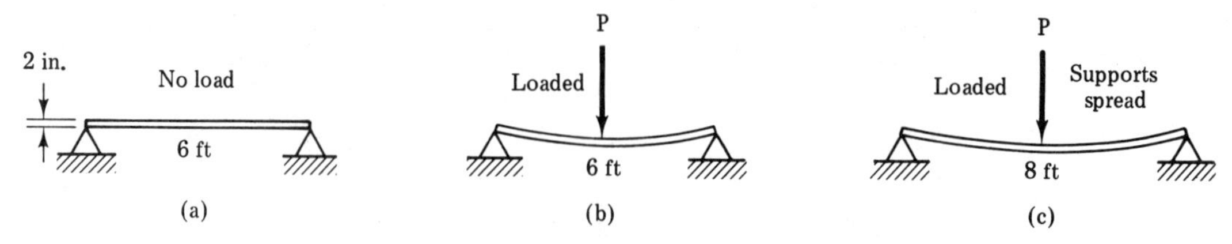

Figure 9-20. Nature of area moment of inertia.

the second moment of the area because it is a function of the square of the moment arm rather than just the moment arm. It is used in the study and design of beams and columns that are subjected to bending and/or torsion.

Much of the above will become more apparent as this chapter progresses. Subsequent chapters make substantial application of this property. At this time, however, let us examine the nature of the moment of inertia in more detail.

Figure 9-20 shows a 12 × 2 in. wood beam lying with its 12-in. dimension horizontal and its 2-in. dimension vertical and supported at two points. In Figure 9-20a the beam lies essentially flat (in reality, gravity acting on its own mass will cause it to sag slightly). In 9-20b the beam has a downward load applied causing it to sag or deflect further. In 9-20c the same load has been applied but the supports are further apart causing the deflection to be greater.

The above should seem reasonable from experience. It would also appear reasonable that if the beam were made thicker, say 3 × 12 in., or wider, say 2 × 18 in., it would be stronger and would deflect less. Similarly, if we changed material from wood to steel the strength would also increase and deflection would decrease. The converse of these statements would also be true, that is, reducing size or substituting a weaker material would result in a weaker beam and greater deflection.

Let us return to the 12 × 2 in. wood beam and examine a different phenomenon from those above. In Figure 9-21, this beam is shown in two positions, flat and on edge. All other variables are being held constant: material, cross-sectional shape, cross-sectional area, load, location of load,

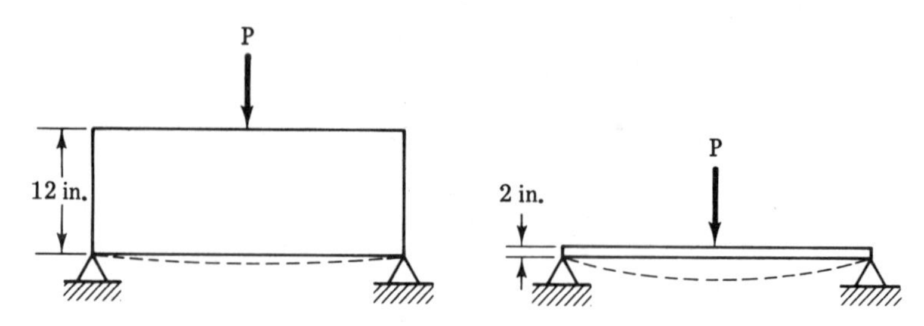

Figure 9-21. Area moment of inertia and orientation.

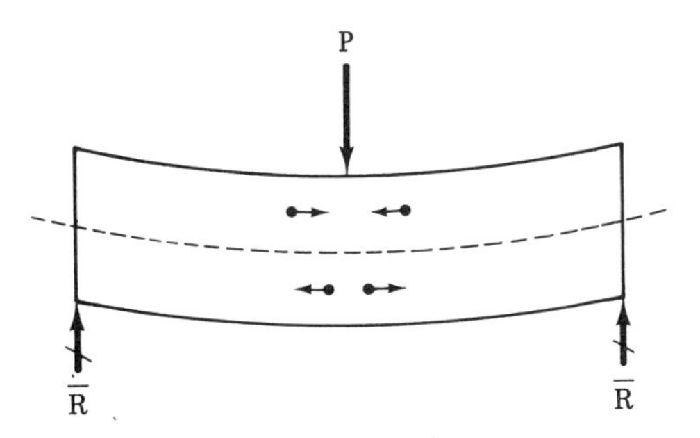

Figure 9-22. The neutral axis.

distance between supports. Yet the deflection of the upright beam is substantially less than that of the flat-lying beam. You may say "of course" to this observation, but there must be a way to explain the difference. Further, the explanation must be in terms of the orientation of the beam to the load and reactions. The explanation lies in the property known as the area moment of inertia.

In the case of a beam in bending, it can be shown that a neutral axis runs the length of the beam through the centroid of the cross section of the beam. Such a beam is shown in Figure 9-22. While there is no stress in the fibers at the neutral axis, those above the axis are in compression while those below the axis are in tension. Also, since the magnitude of the tension or compression at any point is in direct proportion to the perpendicular distance from the neutral axis, it follows that adjacent horizontal layers of fibers are under different magnitudes of tensile or compressive stress. This means that in addition to tensile or compressive stresses there are also horizontal shear stresses occurring as adjacent layers deform in different amounts and try to slide on each other. The resistance of a beam to bending thus becomes a function not only of the strength of the individual fibers but also their distance from the neutral axis. Actually, it can be shown both theoretically and experimentally that this resistance to bending is in direct proportion to the square of the distance. In general form, this is expressed as

$$I = Ar^2$$

It can be seen from the above that for a rectangular beam the moment of inertia will be different depending on the orientation of the load to the beam. Another possibility exists—the beam could be in torsion. This is extremely common in machine shafts, which are simply a special type of beam. Instead of, or sometimes in addition to, a bending load, the beam is subjected to torsion which is simply shear stress in a circular path around the neutral axis. Figure 9-23 shows a shaft under torsion with an imaginary cut made through the shaft to show the shearing forces. Once again, the

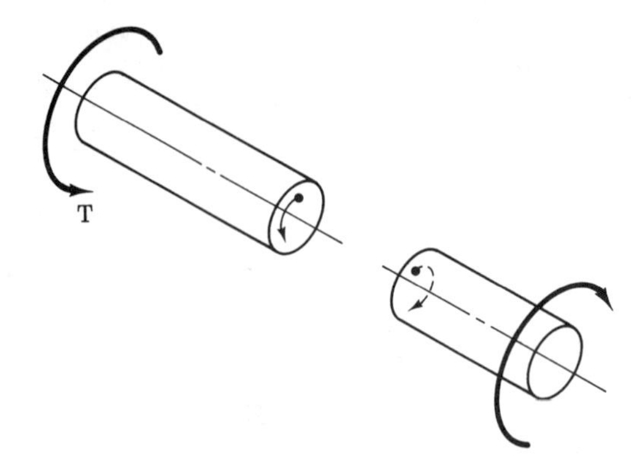

Figure 9-23. Area moment of inertia and torsion.

resistance to deflection is a function of the distance squared. The distance is measured from the lateral or longitudinal axis of the shaft which is perpendicular to the cross-sectional area of the shaft. This form of the area moment of inertia is known as the polar moment of inertia.

The reference axis may be anywhere and may or may not pass through the area involved. However, there are two reasons for examining the area moment of inertia with respect to a centroidal axis. First of all, many problems are of a nature where this is the desired value of the moment of inertia. Second, if we know the centroidal moment of inertia, as it is commonly called, then a very simple technique permits us to determine the moment of inertia about any other axis. In the next section we shall consider the determination of the centroidal moment of inertia for simple areas. In the subsequent section we shall consider the determination of the centroidal moment of inertia of an area about any axis and in the following section of this chapter the moment of inertia of a complex area.

9.7 DETERMINING THE CENTROIDAL AREA MOMENT OF INERTIA OF A SIMPLE AREA

Let us once again examine our 2×12 in. beam standing on edge as shown in Figure 9-24a. Earlier in this chapter, we discovered that the centroid of the cross-sectional area could be easily determined. Calculating the value of the cross-sectional area is also quite easy. In the previous section, the area moment of inertia was generally defined as area times radius squared. However, what radius should be used? Certainly not 6 in. for only the very top and bottom fibers are at that radius. A radius of 3 in. would seem to be a good average until we realize that we are working with the radius squared not just the radius, and 3 squared is certainly not the average of 6 squared and 0 squared. The answer is to use a technique applied earlier in this chapter. As shown in Figure 9-24b, we divide the cross section of the beam into horizontal strips. These must be narrow enough that the error

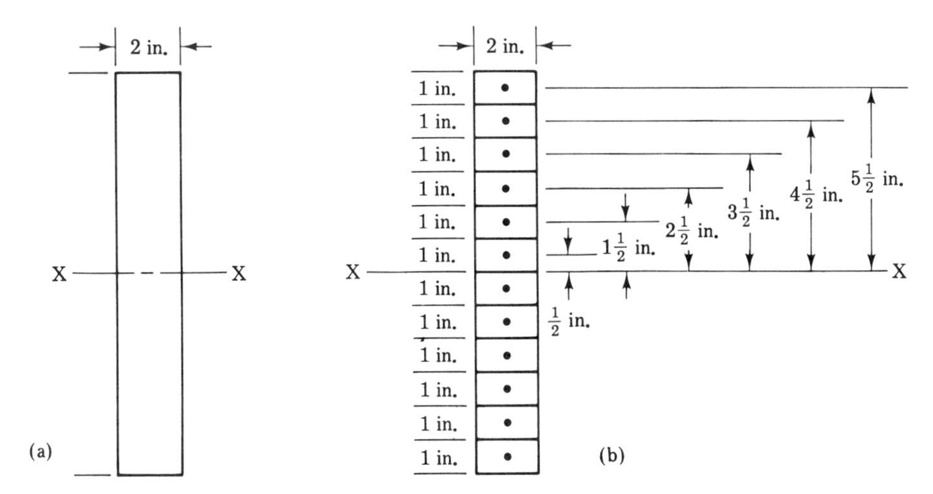

Figure 9-24. Determining area moment of inertia.

resulting from using a strip's average radius for the entire strip is negligible. In our case we are using 1-in.-high strips. The total moment of inertia will be the sum of all the individual moments of inertia. This is an application of an important principle that will be applied through the remainder of this chapter: *The moment of inertia of an area with respect to an axis is the sum of the moments of inertia of all the elements that comprise that area taken with respect to the same area.*

Applying this principle to the beam in Figure 9-24,

$$\bar{I}_x = \Sigma I_x = \Sigma A r^2$$

(Note the bar over the I to indicate that this is a centroidal moment of inertia and the use of the subscript x to indicate the reference axis about which moments are being taken.)

$$\begin{aligned}
(2\text{ in.})(1\text{ in.})(5.5\text{ in.}^2) &= 60.5\text{ in.}^4\\
(2\text{ in.})(1\text{ in.})(4.5\text{ in.}^2) &= 40.5\text{ in.}^4\\
(2\text{ in.})(1\text{ in.})(3.5\text{ in.}^2) &= 24.5\text{ in.}^4\\
(2\text{ in.})(1\text{ in.})(2.5\text{ in.}^2) &= 12.5\text{ in.}^4\\
(2\text{ in.})(1\text{ in.})(1.5\text{ in.}^2) &= 4.5\text{ in.}^4\\
(2\text{ in.})(1\text{ in.})(0.5\text{ in.}^2) &= 0.5\text{ in.}^4\\
(2\text{ in.})(1\text{ in.})(-0.5\text{ in.}^2) &= 0.5\text{ in.}^4\\
(2\text{ in.})(1\text{ in.})(-1.5\text{ in.}^2) &= 4.5\text{ in.}^4\\
(2\text{ in.})(1\text{ in.})(-2.5\text{ in.}^2) &= 12.5\text{ in.}^4\\
(2\text{ in.})(1\text{ in.})(-3.5\text{ in.}^2) &= 24.5\text{ in.}^4\\
(2\text{ in.})(1\text{ in.})(-4.5\text{ in.}^2) &= 40.5\text{ in.}^4\\
(2\text{ in.})(1\text{ in.})(-5.5\text{ in.}^2) &= \underline{60.5\text{ in.}^4}\\
&\ 286\ \ \text{in.}^4
\end{aligned}$$

The application of calculus to the above would permit the use of an infinitely large number of areas of infinitely small height. It would result in a precisely accurate answer of 288 $in.^4$ Our error in using 1-in.-high strips is, therefore, less than 1%. More important, the use of calculus permits the derivation of a formula for the centroidal moment of inertia for a rectangle that states

$$\bar{I}_x = \frac{1}{12} bh^3$$

We would, of course, have been quite accurate in deriving that formula from the algebraic technique just used.

Since we earlier considered the same beam laid flat, it is interesting to note that application of the formula to the flat-laid boom yields an $\bar{I}$ equal to 8 $in.^4$ Thus, the beam has substantially less resistance to bending when laid flat than when standing on edge (8:288 or 1:36).

Through the above technique and with the application of calculus formulas for the centroidal moments of inertia of many common areas and values for the centroidal moments of inertia for cross sections of all common structural shapes have been determined. Many of these are found in Appendices B-3 and B-4 through B-7.

Problem 9-5: Determine the centroidal moment of inertia about the x axis of the area shown in Figure 9-25 by approximation using $\frac{1}{2}$-m strips. Check by use of the formula from the Appendix.

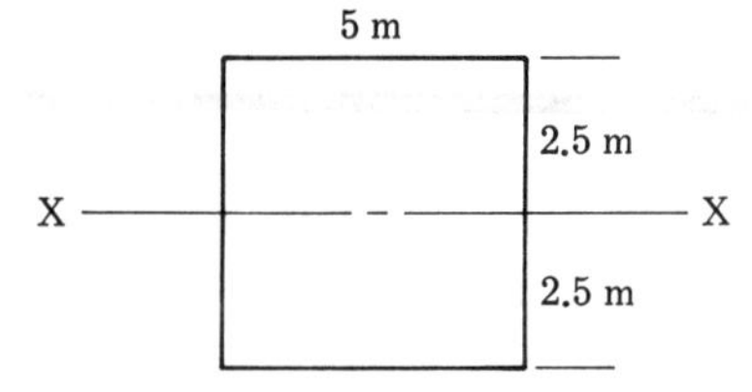

Figure 9-25

Problem 9-6: Determine the centroidal moment of inertia about both the x and the y axes of the area shown in Figure 9-26 by approximation using $\frac{1}{2}$-ft. strips. Check by use of the formula from the Appendix.

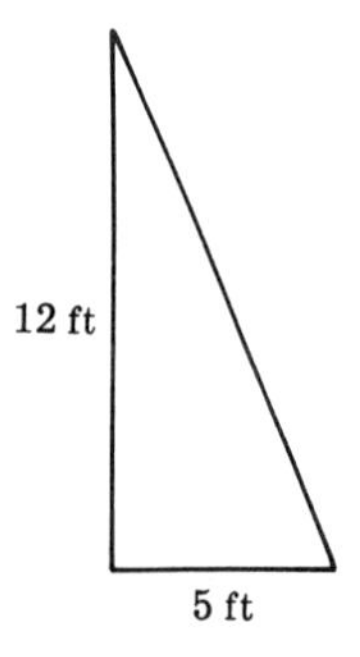

Figure 9-26

9.8 DETERMINING THE CENTROIDAL AREA MOMENT OF INERTIA OF AN AREA ABOUT ANY AXIS

Returning to the 2 × 12 in. beam on edge, suppose that the reference axis does not pass through the centroid. The same principle still applies, that is, $I = \Sigma Ar^2$. Let us consider the x axis as passing through a point 2 in. below the bottom edge of the beam as shown in Figure 9-27. Then

$$
\begin{aligned}
(2\text{ in.})(1\text{ in.})(13.5\text{ in.}^2) &= 364.5\text{ in.}^4\\
(2\text{ in.})(1\text{ in.})(12.5\text{ in.}^2) &= 312.5\text{ in.}^4\\
(2\text{ in.})(1\text{ in.})(11.5\text{ in.}^2) &= 264.5\text{ in.}^4\\
(2\text{ in.})(1\text{ in.})(10.5\text{ in.}^2) &= 220.5\text{ in.}^4\\
(2\text{ in.})(1\text{ in.})(9.5\text{ in.}^2) &= 180.5\text{ in.}^4\\
(2\text{ in.})(1\text{ in.})(8.5\text{ in.}^2) &= 144.5\text{ in.}^4\\
(2\text{ in.})(1\text{ in.})(7.5\text{ in.}^2) &= 112.5\text{ in.}^4\\
(2\text{ in.})(1\text{ in.})(6.5\text{ in.}^2) &= 84.5\text{ in.}^4\\
(2\text{ in.})(1\text{ in.})(5.5\text{ in.}^2) &= 60.5\text{ in.}^4\\
(2\text{ in.})(1\text{ in.})(4.5\text{ in.}^2) &= 40.5\text{ in.}^4\\
(2\text{ in.})(1\text{ in.})(3.5\text{ in.}^2) &= 24.5\text{ in.}^4\\
(2\text{ in.})(1\text{ in.})(2.5\text{ in.}^2) &= \underline{12.5\text{ in.}^4}\\
& 1822.5\text{ in.}^4
\end{aligned}
$$

The same results could have been achieved much more simply using the *parallel axis theorem: The moment of inertia of any area about any axis is the sum of the centroidal moment of inertia of the area and the product of the area and the square of the distance from the centroidal axis to the parallel axis.*

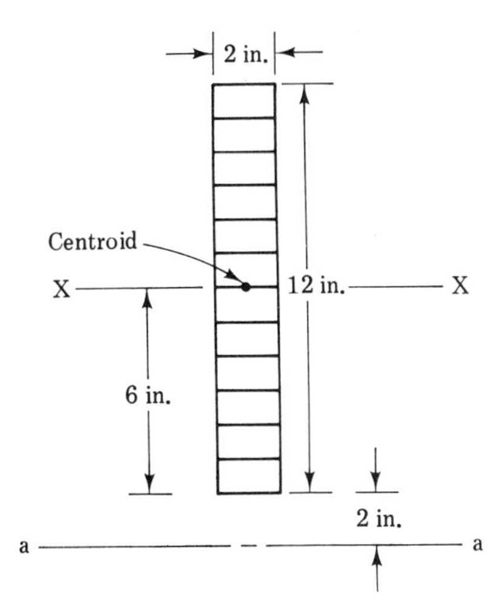

Figure 9-27. Parallel-axis theorem.

$$I_{a-a} = \bar{I}_x + Ad^2 = \frac{1}{12} bh^3 + Ad^2$$

$$I_{a-a} = 288 \text{ in.}^4 + (2 \text{ in.})(12 \text{ in.})(8 \text{ in.}^2)$$

$$I_{a-a} = 1824 \text{ in.}^4$$

Once again, the approximation which illustrated the principle, provided an accurate answer (0.08% error) but once the principle is grasped, the parallel axis theorem provides a much quicker method (based, of course, upon the same principle).

Problem 9-7: Determine the moments of inertia about the given axes of the area shown in Figure 9-28 by approximation. Check by use of the appropriate formula from the Appendix.

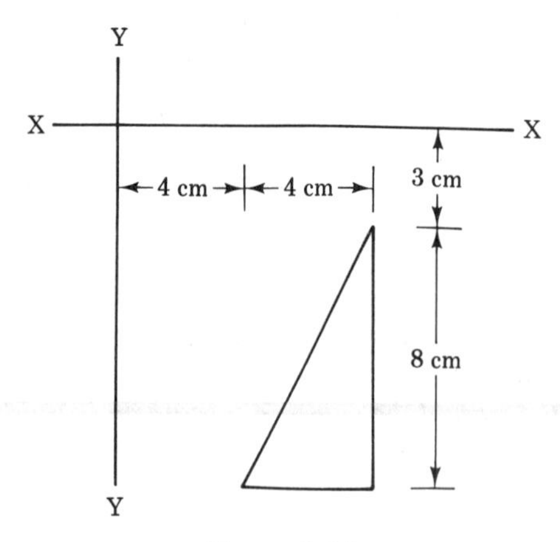

Figure 9-28

9.9 DETERMINING THE AREA MOMENT OF INERTIA OF A COMPLEX AREA

Combining the principle of $I = \Sigma Ar^2$ and the parallel axis theorem permits us to readily determine the area moment of inertia of a complex area.

Example 9-9: Determine the area moment of inertia of the complex area shown in Figure 9-29a with respect to the given axes.

Solution:

Step 1 Divide the complex area into regular areas triangle (1), rectangle (2), circular hole (3), and semicircle (4).

Step 2 Determine the value of the four areas.

$$A_1 = \tfrac{1}{2}bh_1 = \tfrac{1}{2}(3)(6) \qquad A_1 = 9 \text{ cm}^2$$

$$A_2 = bh_2 = (10)(6) \qquad A_2 = 60 \text{ cm}^2$$

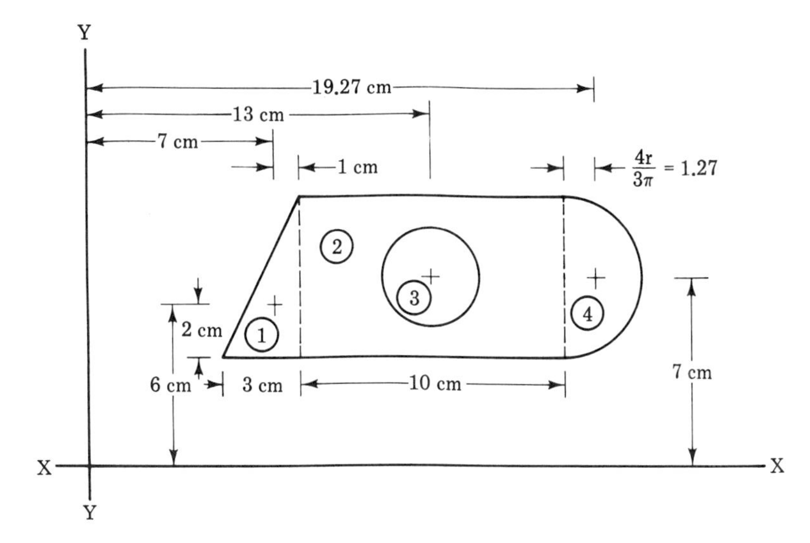

Figure 9-29

$$A_3 = -\pi r^2 = -\pi(2)^2 \qquad A_3 = -12.57 \text{ cm}^2$$

$$A_4 = \tfrac{1}{2}\pi r^2 = \tfrac{1}{2}\pi(3)^2 \qquad A_4 = 14.14 \text{ cm}^2$$

$$A_T = 70.57 \text{ cm}^2$$

Step 3 Determine the location of the centroid of each area as shown in Figure 9-29.

Step 4 Determine the location of the centroid of each area with respect to the given axes as shown in Figure 9-29.

Step 5 Determine the centroidal moment of inertia of each area using the formulas from the Appendix.

$$\bar{I}_{1x} = \frac{1}{36} bh^3 = \frac{1}{36}(3)(6)^3 \qquad \bar{I}_{1x} = 18 \text{ cm}^4$$

$$\bar{I}_{2x} = \frac{1}{12} bh^3 = \frac{1}{12}(10)(6)^3 \qquad \bar{I}_{2x} = 180 \text{ cm}^4$$

$$\bar{I}_{3x} = \frac{-\pi}{4} r^4 = \frac{-\pi}{4}(2)^4 \qquad \bar{I}_{3x} = -12.57 \text{ cm}^4$$

$$\bar{I}_{4x} = \frac{\pi}{8} r^4 = \frac{\pi}{8}(3)^4 \qquad \bar{I}_{4x} = 31.8 \text{ cm}^4$$

$$\bar{I}_{1y} = \frac{1}{36} hb^3 = \frac{1}{36}(6)(3)^3 \qquad \bar{I}_{1y} = 4.5 \text{ cm}^2$$

$$\bar{I}_{2y} = \frac{1}{12}hb^3 = \frac{1}{12}(6)(10)^3 \qquad \bar{I}_{2y} = 500 \text{ cm}^2$$

$$\bar{I}_{3y} = \bar{I}_{3x} \qquad \bar{I}_{3y} = -12.57 \text{ cm}^2$$

$$\bar{I}_{4y} = 0.11r^4 = 0.11(3)^4 \qquad \bar{I}_{4y} = 8.91 \text{ cm}^2$$

Step 6 Determine the area moments of inertia by applying the parallel axis theorem.

$$I_{Tx} = \bar{I}_{1x} + A_1y_1^2 + \bar{I}_{2x} + A_2y_2^2 + \bar{I}_{3x} + A_3y_3^2 + \bar{I}_{4x} + A_4y_4^2$$

$$I_{Tx} = 18 + 9(6)^2 + 180 + (60)(7)^2 + (-12.57) + (-12.57)(7)^2 + 31.8 + (14.14)(7)^2 \qquad \mathbf{I_{Tx} = 3558 \text{ cm}^4}$$

$$I_{Ty} = \bar{I}_{1y} + A_1x_1^2 + \bar{I}_{2y} + A_2x_2^2 + \bar{I}_{3y} + A_3x_3^2 + \bar{I}_{4y} + A_4x_4^2$$

$$I_{Ty} = 4.5 + (9)(7)^2 + 500 + (60)(13)^2 + (-12.57)(13)^2 + 8.91 + (14.14)(19.27)^2 \qquad \mathbf{I_{Ty} = 14220 \text{ cm}^4}$$

Unquestionably, there are two major sources of error in these computations. One is the failure to be meticulously careful. The second is to fail to recognize that in determining moments of inertia about the y axis one is working with x moment arms and vice versa.

Note that the same results could have been obtained by determining the centroid of the complex area, determining the centroidal moment of inertia and then transferring. While correct, the method would require more steps with an increased chance for error.

Problem 9-8: Determine the area moment of inertia of the complex area shown in Figure 9-30.

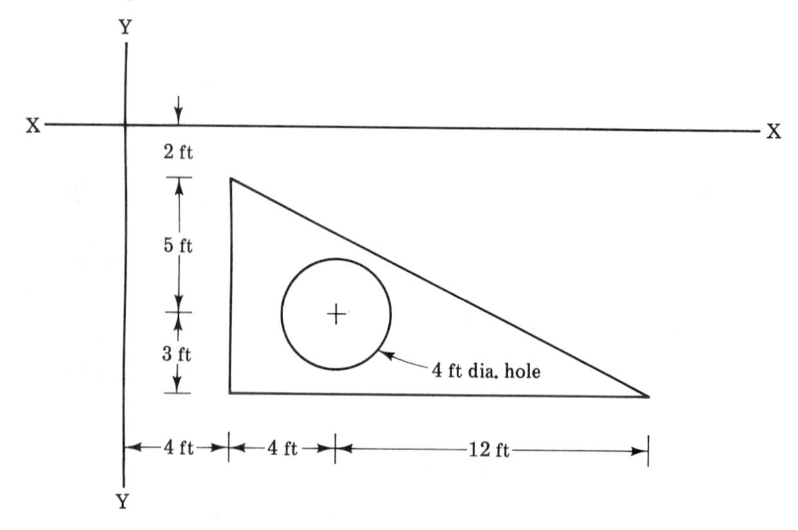

Figure 9-30

Problem 9-9: Determine the area moment of inertia of the cross section of the composite structural member shown in Figure 9-31 with respect to the given axes.

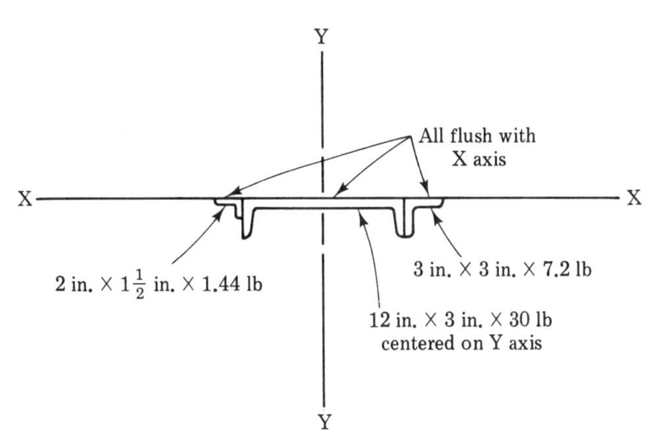

Figure 9-31

9.10 POLAR MOMENT OF INERTIA

Up to now, the moment of inertia has been considered with respect to some horizontal or vertical axis *coplanar* with the area being considered. However, earlier, brief mention was made of the moment of inertia of the cross-sectional area of a beam or column in torsion. The radii for determining this property are measured in the plane of the area from the z axis. This appears in the plane as the origin, that is, the intersection of the x and y axis. Then $r^2 = x^2 + y^2$.

One might assume that this moment of inertia would be designated as I_z. However, to make a point of the fact that this moment of inertia is different from the others (being taken about an axis perpendicular to, rather than in the plane, of the area), it is designated J and called the *polar moment of inertia.* Since we often have a need to calculate I_x and I_y, the polar moment of inertia can be simply calculated from the equation at left below rather than the formula given in the Appendix and the equation at right below.

$$J = I_x + I_y \qquad J = \bar{J} + Ar^2$$

Example 9-10: Determine the polar moment inertia of the area shown in Figure 9-29 with respect to an axis passing through the given origin.

$$J_T = I_{Tx} + I_{Ty} \qquad J = 3558 + 14{,}220 \qquad J = 17{,}780 \text{ cm}^4$$

Problem 9-10: Determine the polar moment of inertia of the area shown in Figure 9-30 with respect to an axis passing through the given origin.

Problem 9-11: Determine the polar moment of inertia of the area shown in Figure 9-31 with respect to an axis passing through the given origin.

9.11 RADIUS OF GYRATION

Another property of an area that is frequently encountered is the *radius of gyration.* Going back to the original equation, $I = Ar^2$, and rearranging it so that $r = \sqrt{I/A}$ we can recognize that there could be many areas that have different shapes but are equal in total area and arranged in such a way that they have equal moments of inertia. Others may not be equal in area but may still have the same proportion of I to A. These all have a common r. This r derived from I and A is called the radius of gyration and designated k. It is an indication of commonality among areas that may appear rather different otherwise. One way that it may be envisioned is to consider it as the distance from the axis we are taking moments about to an area that is equal in both area and moment of inertia to the original area but spread out in an infinitely thin line parallel to the axis. Figure 9-32 illustrates this for our 2 × 12 in. beam located 2 in. above some horizontal axis. Since algebraic methods do not permit us to work with an infinitely thin line, let us make it relatively thin, say $\frac{1}{4}$ in. as shown.

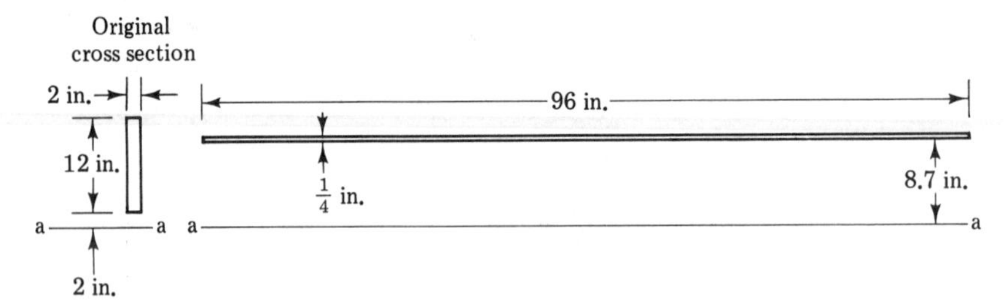

Figure 9-32. Radius of gyration.

We had previously determined that $I_{a-a} = 1824$ in.^{4} and that $A = 24$ in. Then

$$k_{a-a} = \frac{I}{A} = \frac{1824 \text{ in.}^4}{24 \text{ in.}^2} \qquad k_{a-a} = 8.717 \text{ in.}$$

Applying the parallel axis theorem to the area spread out at an average distance from the axis equal to the radius of gyration:

$$I_{a-a} = \bar{I}_x + A_y^2 = \tfrac{1}{12}\, bh^3 + Ak^2$$

$$I_{a-a} = \tfrac{1}{12}(96 \text{ in.})(\tfrac{1}{4} \text{ in.})^3 + (24 \text{ in.}^2)(8.717 \text{ in.})^2$$

$$I_{a-a} = 1823.79 \text{ in.}^4$$

Spreading the area out even thinner would illustrate the concept even more accurately but this hardly seems necessary since the error encountered above is only 0.01%.

Example 9-11: Determine the radius of gyration about all three axes for the area shown in Figure 9-29.

$$k_x = \sqrt{\frac{I_{Tx}}{A}} = \sqrt{\frac{3558 \text{ cm}^4}{70.57 \text{ cm}^2}} \qquad \boldsymbol{k_x = 7.1 \text{ cm}}$$

$$k_y = \sqrt{\frac{I_{Ty}}{A}} = \sqrt{\frac{14{,}220 \text{ cm}^4}{70.57 \text{ cm}^2}} \qquad \boldsymbol{k_y = 14.2 \text{ cm}}$$

$$k_0 = \sqrt{\frac{J}{A}} = \sqrt{\frac{17{,}780 \text{ cm}^4}{70.57 \text{ cm}^2}} \qquad \boldsymbol{k_0 = 15.9 \text{ cm}}$$

Note the use of the subscript 0 to designate the radius of gyration about the Z axis passing through the origin.

Example 9-12: Locate the centroid of a 3 × 2 in. rectangular shaft that is to have the same radii of gyration as the round shaft shown in Figure 9-33. (Assume rectangular shaft edges are parallel to the given axes with the 3-in. side horizontal.

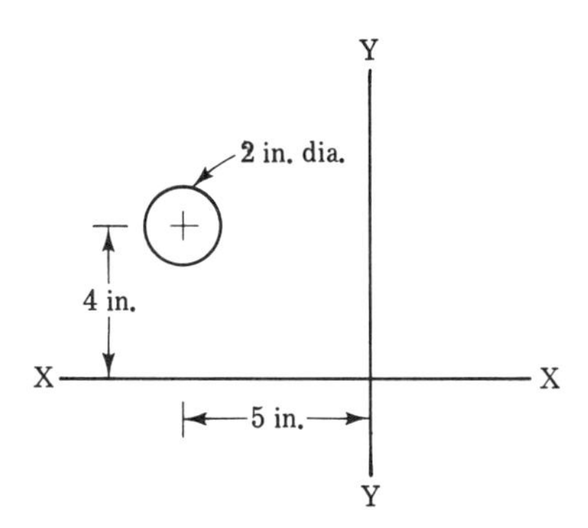

Figure 9-33

Step 1 The centroid of the shaft is at the geometric center of the shaft.

Step 2 (circle)

$$A = \pi r^2 \qquad A = \pi \text{ in.}^2$$

Step 3 (circle)

$$\bar{I}_x = \bar{I}_y = \frac{\pi}{4} r^4 \qquad \bar{I}_x = \bar{I}_y = \frac{\pi}{4} \text{ in.}^4$$

Step 4 (circle)

$$I_{xx} = \bar{I}_x + A_y^2 = \frac{\pi}{4} + \pi(4)^2 \qquad I_{xx} = 51.05 \text{ in.}^4$$

$$I_{yy} = \bar{I}_y + A_x^2 = \frac{\pi}{4} + \pi(5)^2 \qquad I_{yy} = 79.33 \text{ in.}^4$$

Step 5 (circle)

$$k_{xx} = \sqrt{\frac{I_{xx}}{A}} = \sqrt{\frac{51.05}{\pi}} \qquad k_{xx} = 4.03 \text{ in.}$$

$$k_{yy} = \sqrt{\frac{I_{yy}}{A}} = \sqrt{\frac{79.33}{\pi}} \qquad k_{yy} = 5.02 \text{ in.}$$

According to the stated problem, these radii of gyration will be the same for the rectangle.

Step 6 (rectangle)

$$A = bh = (3 \text{ in.})(2 \text{ in.}) \qquad A = 6 \text{ in.}^2$$

Step 7 (rectangle)

$$\bar{I}_x = \tfrac{1}{12}\, bh^3 = \tfrac{1}{12}\,(3 \text{ in.})(2 \text{ in.}^3) \qquad \bar{I}_x = 1 \text{ in.}^4$$

$$\bar{I}_y = \tfrac{1}{12}\, hb^3 = \tfrac{1}{12}\,(2 \text{ in.})(3 \text{ in.}^3) \qquad \bar{I}_y = 1.5 \text{ in.}^4$$

Step 8 (rectangle)

$$I_{xx} = k_{xx}^2 A = (4.03 \text{ in.}^2)(6 \text{ in.}^2) \qquad I_{xx} = 97.45 \text{ in.}^4$$

$$I_{yy} = k_{yy}^2 A = (5.02 \text{ in.}^2)(6 \text{ in.}^2) \qquad I_{yy} = 151.2 \text{ in.}^4$$

Step 9 (rectangle)

$$I_{xx} = \bar{I}_x + A^2 \qquad y = \sqrt{\frac{I_{xx} - \bar{I}_x}{A}}$$

$$y = \frac{97.45 - 1}{6} \qquad y = 9.86 \text{ in.}$$

$$I_{yy} = \bar{I}_y + A^2 \qquad x = \sqrt{\frac{I_{yy} - \bar{I}_y}{A}}$$

$$x = \frac{151.2 - 1.5}{6} \qquad x = 12.29 \text{ in.}$$

Problem 9-12: Determine k_x, k_y, and k_0 for the area shown in Figure 9-30.

Problem 9-13: Determine k_x, k_y, and k_0 for the area shown in Figure 9-31.

9.12 READINESS QUIZ

1. The y coordinate for the centroid of the semicircular area shown in Figure 9-34 will be equal to

   (a) one-half the radius
   (b) more than one-half the radius
   (c) less than one-half the radius
   (d) zero
   (e) none of the above

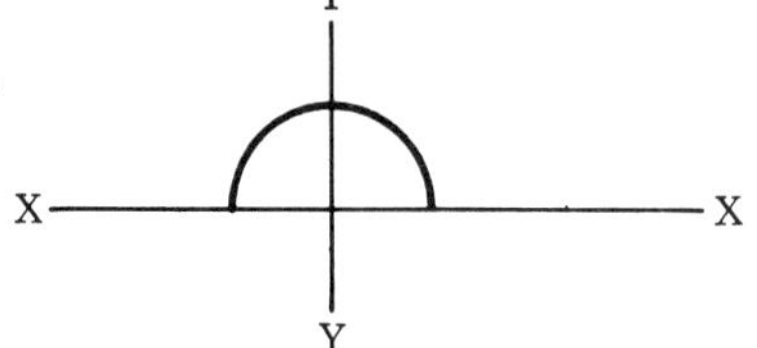

Figure 9-34

2. In the figure shown in Figure 9-35, a steel plate has been bonded to an aluminum plate of equal area and thickness. The centroid will be located

   (a) to the right of AB
   (b) to the left of AB
   (c) on AB
   (d) insufficient information given
   (e) none of the above

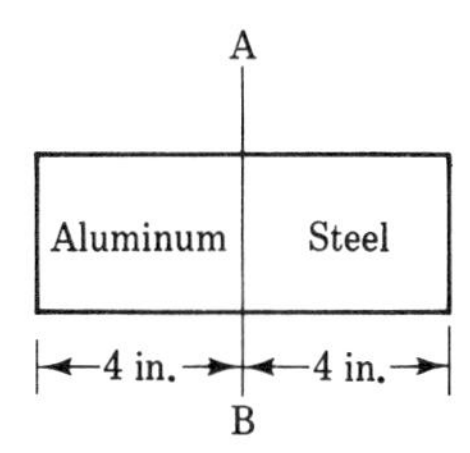

Figure 9-35

3. The center of gravity of an object

   (a) is always the geometric center of the object
   (b) is the point through which the force of gravity acts irrespective of the position of the object
   (c) cannot be determined experimentally
   (d) all of the above
   (e) none of the above

4. For the area shown in Figure 9-36, the X coordinate for the centroid will be located

(a) on line AB
(b) to the right of line AB
(c) to the left of line AB
(d) insufficient information given
(e) none of the above

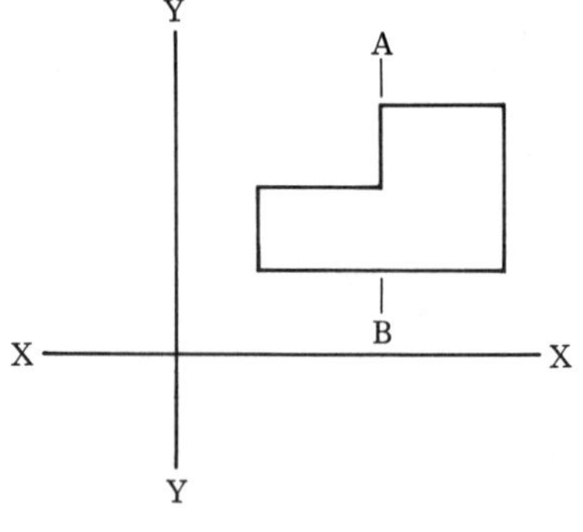

Figure 9-36

5. For the area shown in Figure 9-37, the location of the centroid with respect to the center of the square will be

(a) to the right and up
(b) to the right and down
(c) to the left and up
(d) to the left and down
(e) none of the above

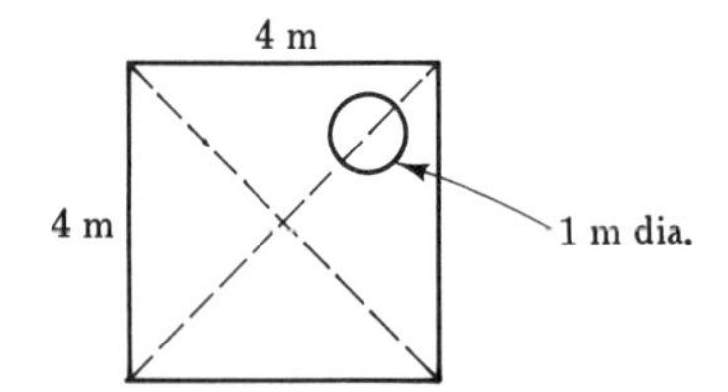

Figure 9-37

6. The center of gravity for the two particles shown in Figure 9-38 will be located

(a) 5 in. to left of 0
(b) 8 in. to left of 0
(c) 9 in. to left of 0
(d) 2 in. to right of 0
(e) none of the above

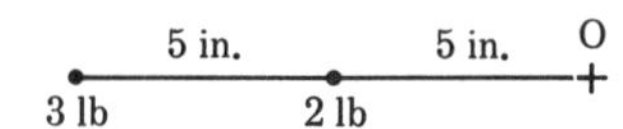

Figure 9-38

7. A centroid of a figure is always the

(a) same as the center of gravity
(b) arithmetic center of the figure
(c) geometric center of the figure
(d) center of symmetry
(e) none of the above

8. The X coordinate for the centroid of the quarter circle shown in Figure 9-39 will be

(a) equal to $\frac{1}{2}r$
(b) less than $\frac{1}{2}r$
(c) more than $\frac{1}{2}r$
(d) zero
(e) none of the above

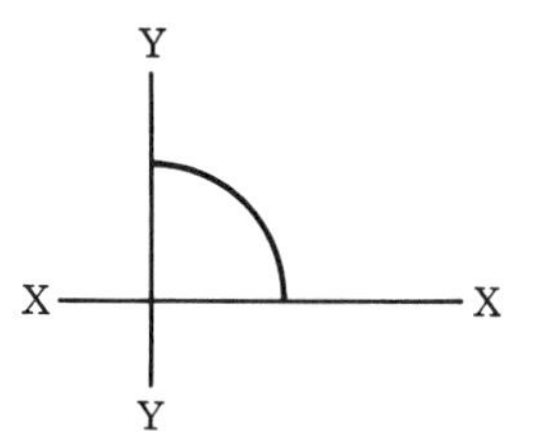

Figure 9-39

9. The center of gravity for the two aluminum spheres of equal density shown in Figure 9-40 will be located

(a) 12 in. below 0
(b) 14 in. below 0
(c) 15 in. below 0
(d) 18 in. below 0
(e) none of the above

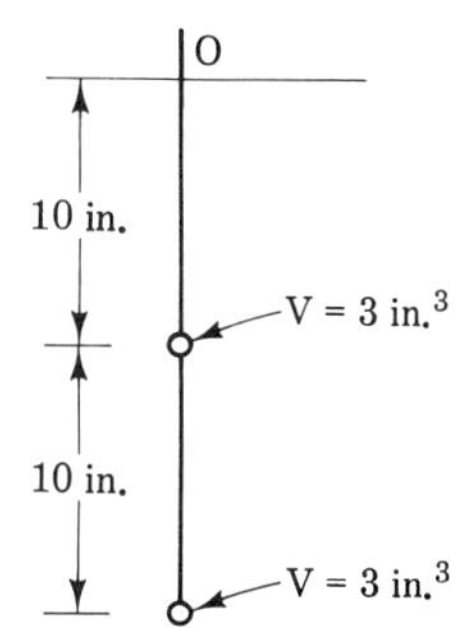

Figure 9-40

10. Centroids relate to distribution of

(a) area
(b) mass
(c) weight
(d) any of the above
(e) none of the above

11. Moment of inertia must be determined relative to some

(a) point
(b) axis
(c) plane
(d) any one of the above
(e) none of the above

12. In determining the moment of inertia of an area by approximation, greater accuracy is attained if

(a) the area is relatively large
(b) the area is relatively small
(c) the number of increments taken is large
(d) the number of increments taken is small
(e) none of the above

13. The radius of gyration for an area is

    (a) equal to the physical radius of that area
    (b) equal to the square root of the average of the squared radii for all parts of the area
    (c) equal to the square root of the physical radius
    (d) equal to the square of the physical radius
    (e) none of the above

14. The parallel axis theorem deals with

    (a) solely moment of inertia of an area
    (b) transfer of any moment of inertia to an axis other than the centroidal
    (c) solely moment of inertia of a body
    (d) radii of gyration
    (e) none of the above

15. For the area shown in Figure 9-41

    (a) $\bar{I}_x$ is equal to $\bar{I}_y$
    (b) $\bar{I}_x$ is smaller than $\bar{I}_y$
    (c) $\bar{I}_x$ is greater than $\bar{I}_y$
    (d) $\bar{I}_x$ is directly proportional to $\bar{I}_y$
    (e) none of the above

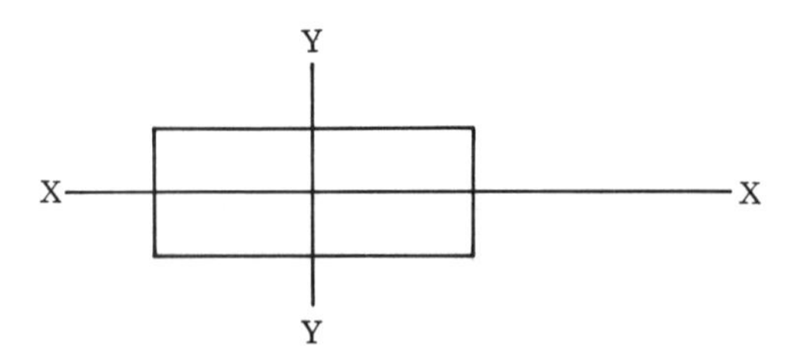

Figure 9-41

16. The polar moment of inertia J is generally taken with respect to

    (a) the Z axis
    (b) the X axis
    (c) the Y axis
    (d) the XY plane
    (e) none of the above

17. For the equal areas shown in Figure 9-42

    (a) $\bar{I}_{ay} = \bar{I}_{by}$

(b) $\bar{I}_{ay}$ is smaller than $\bar{I}_{by}$
(c) $\bar{I}_{ay}$ is greater than $\bar{I}_{by}$
(d) $\bar{I}_{ay}$ is inversely proportional to $\bar{I}_{by}$
(e) none of the above

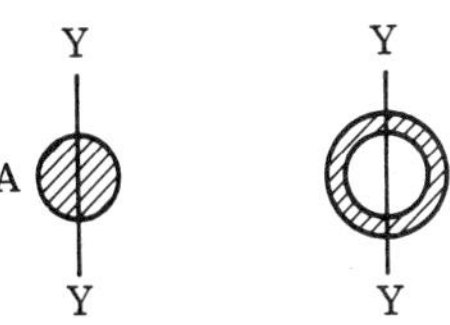

Figure 9-42

18. The moment of inertia is useful in the study of
    (a) beams
    (b) angular motion
    (c) vibrations
    (d) all of the above
    (e) none of the above

19. The moment of inertia of an area
    (a) helps determine the resistance to bending for a beam
    (b) can be determined from the size of the area and its distribution
    (c) can be determined from the size of the area and its radius of gyration
    (d) all of the above
    (e) none of the above

20. For the area shown in Figure 9-43
    (a) $I_{bb} = 4I_{yy}$
    (b) $I_{aa} = 2I_{bb}$
    (c) $2I_{aa} = I_{bb}$
    (d) $I_{aa} = 4I_{bb}$
    (e) none of the above

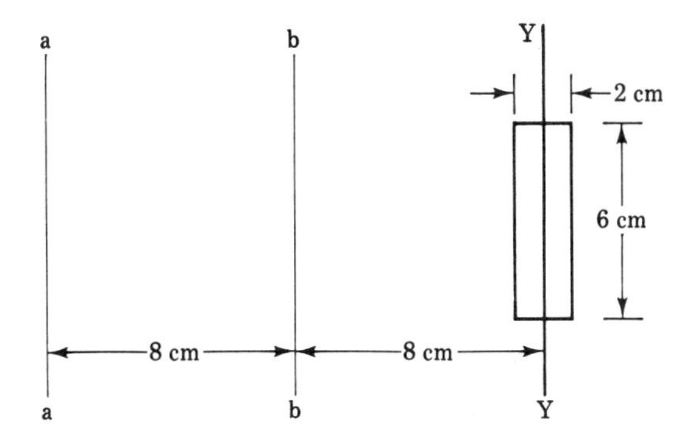

Figure 9-43

9.13 SUPPLEMENTARY PROBLEMS

9-14 Locate the centroid of the area shown in Figure 9-44 with respect to the given axes.

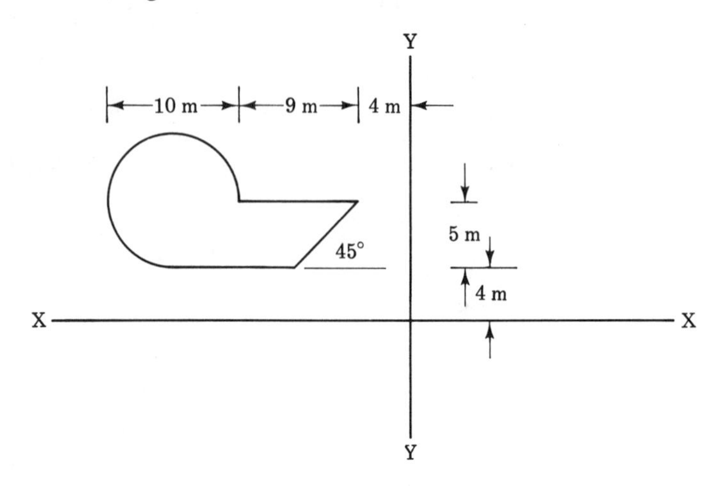

Figure 9-44

9-15 Locate the centroid of the area shown in Figure 9-45 with respect to the given axes.

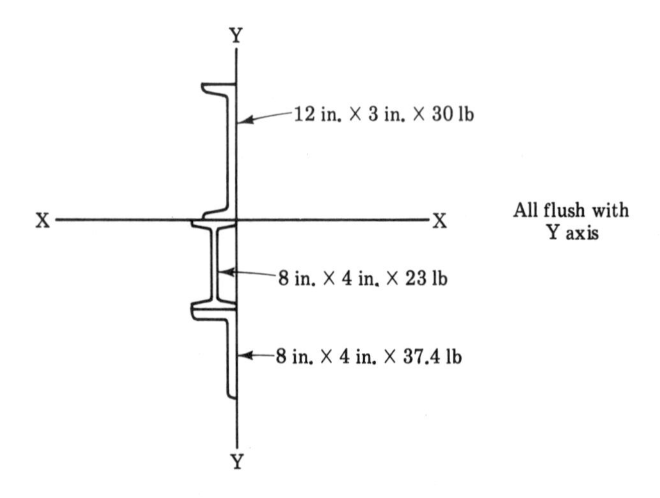

Figure 9-45

9-16 Determine the centroidal moments of inertia about the x and y axes as well as the polar moment of inertia of the area shown in Figure 9-46. Determine k_x, k_y, and k_0.

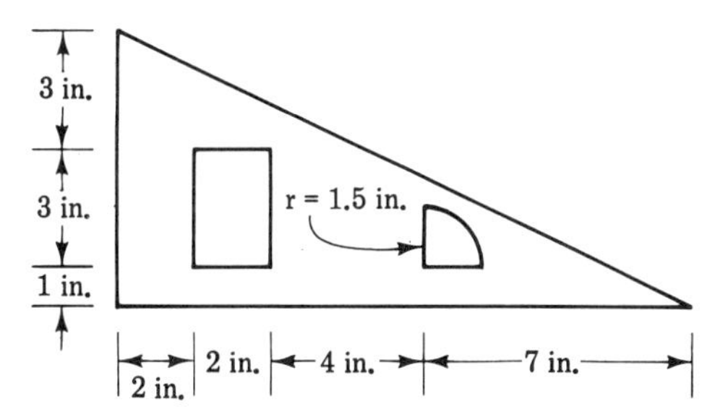

Figure 9-46

9-17 Determine the centroidal moments of inertia about the x and y axes as well as the polar moment of inertia of the area shown in Figure 9-47. Determine k_x, k_y, and k_0.

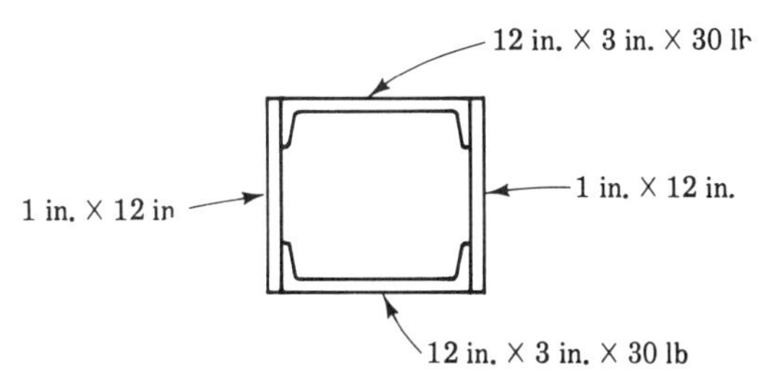

Figure 9-47

9-18 Determine the moments of inertia of the area shown in Figure 9-48 about all three axes through 0. Determine k_x, k_y, and k_0.

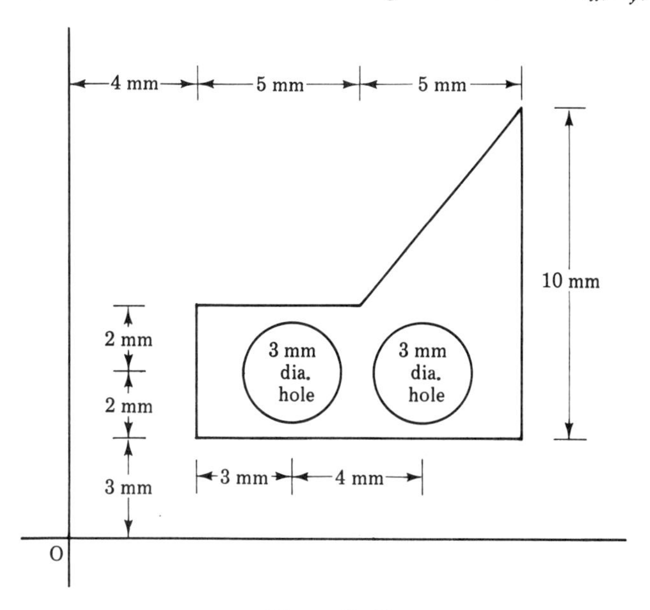

Figure 9-48

9-19 Determine the moments of inertia of the area shown in Figure 9-49 about all three axes through 0. Determine k_x, k_y, and k_0.

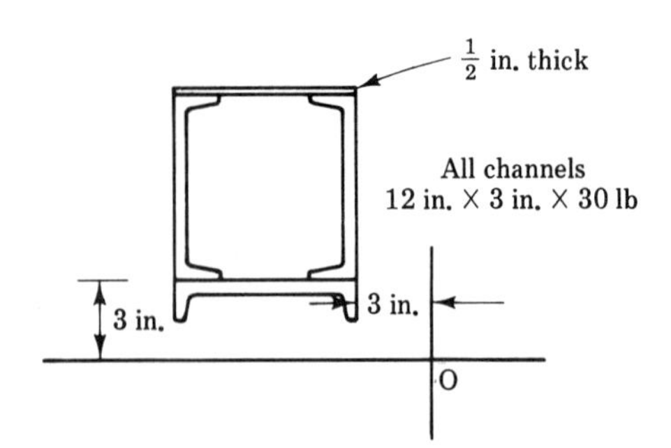

Figure 9-49

9-20 Determine the moments of inertia of the area shown in Figure 9-50 about all three axes through 0. Determine k_x, k_y, and k_0.

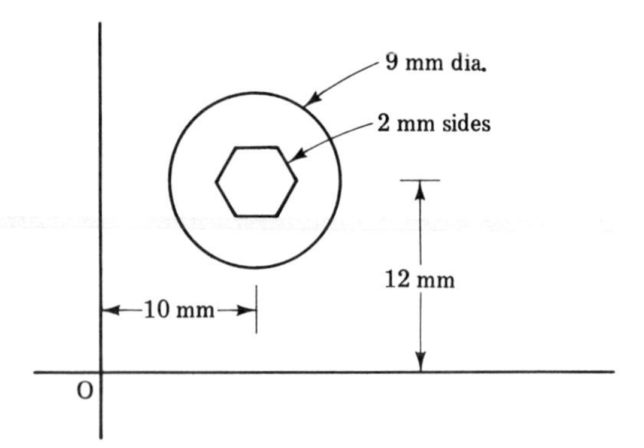

Figure 9-50

Chapter

10

Shear Forces and Bending Moments in Beams

10.1 OBJECTIVES

Upon completion of the work relating to this chapter, you should be able to perform the following.

1. For the terms listed below:
   (a) Select from several definitions the correct one for any given term.
   (b) Given a definition, select the correct term from a given list of terms.

   bending movement
   cantilever beam
   continuous beam
   fixed beam
   overhanging beam
   propped beam
   shear force
   simple beam
   statically determinate
   statically indeterminate

2. Determine the shearing force and bending moment of a statically determinate beam given the nature, size, and location of the loads on the beam.
3. Draw the shear and bending moment diagrams for a statically determinate beam given the nature, size, and location of the loads on the beam.

4. Utilize the relationships among loads, shear forces, and bending moments to solve statically determinate beam problems.

10.2 INTRODUCTION

A beam is a part or member of a machine or structure which has loads applied to it in a direction perpendicular to its longitudinal axis. Loads can, of course, also be applied in a direction parallel to the longitudinal axis of such a member. Such loads are called axial loads and the member is known as a column. A later chapter deals with columns while another deals with combined loads.

Ultimately, the purpose of the knowledge and skills gained in completing this chapter will permit you to find the critical section of a beam where the maximum stresses occur.

10.3 TYPES OF BEAMS

Beams are made with many different cross-sectional shapes and of many different materials. These are variables with which we must eventually concern ourselves. At this point, however, we are still analyzing the effect of

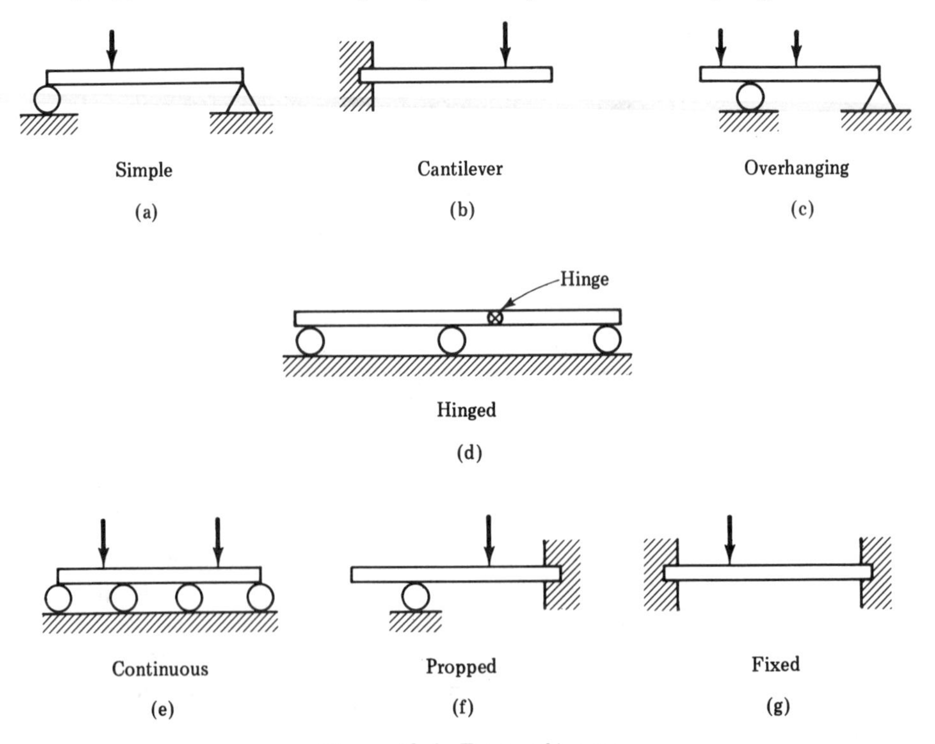

Figure 10-1. Types of beams.

loads on the beam which will then lead us to select an appropriate shape and material.

The major differentiation that needs to be made at this point is between statically determinate and statically indeterminate. As the names might suggest to you, the former can be worked on with the methods and techniques learned in statics while the latter will require some new techniques. These new techniques are discussed in the last chapter of this text.

Several types of beams are illustrated in Figure 10-1. The simple (a), cantilever (b), overhanging (c), and hinged (d) are all statically determinate. The continuous (e), propped (f), and fixed (g) are all statically indeterminate. The only type that is load dependent is the overhanging beam (c) where it is assumed that at least part of the load is on the actual overhang (to the right of the roller). To assume otherwise would reduce the overhanging beam to a simple beam (a).

10.4 LOADS ON BEAMS

The types of loads we will deal with have, for the most part, already been discussed earlier in this text. Illustrated in Figure 10-2 are concentrated

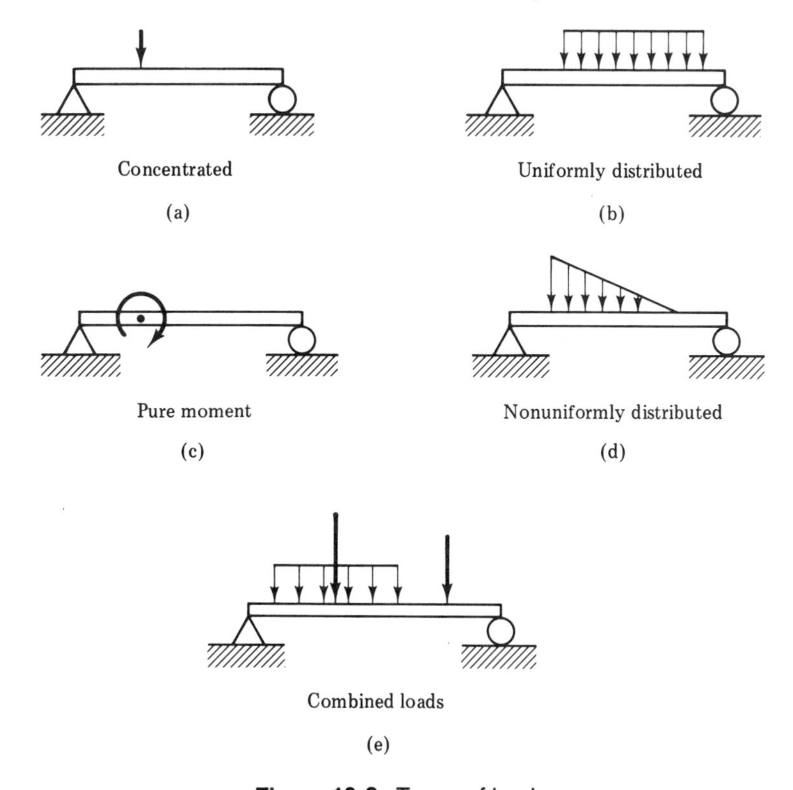

Figure 10-2. Types of loads.

load (a), uniformly distributed load (b), pure-moment load (c), nonuniformly distributed load (d), and combined loads (e). Of these, only pure-moment loads and nonuniformly distributed loads may be unfamiliar to you.

Pure-moment load occurs, for example, when a couple is applied to an axis running perpendicular to the longitudinal axis of the beam. As seen in Figure 10-3 this does result in forces as well as moment acting on the beam.

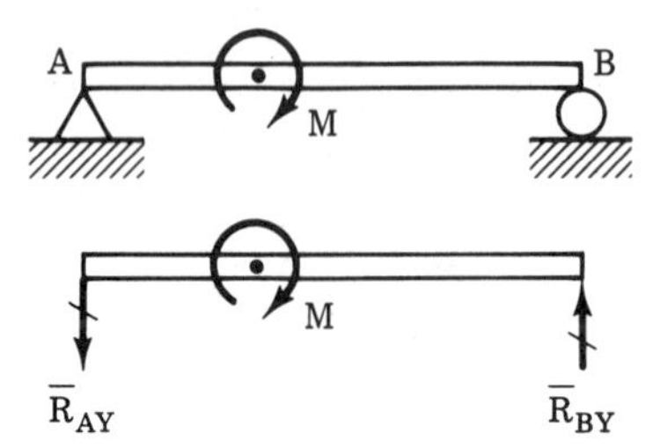

Figure 10-3. Pure-moment load.

Nonuniformly distributed load can be dealt with in the same manner as uniformly distributed load. Recall that for a distributed load, the load is determined by calculating the area under the load curve *($P = wl$)*. Also, for the purposes of taking moments, the uniformly distributed load can be considered as a concentrated load acting through the centroid of the area under the load curve. The same can be said for a nonuniformly distributed load. The only difference lies in the degree of complexity in calculating the area and locating the centroid. If complex enough, the methods of calculus must be used; however, the vast majority of problems can be solved using the methods of algebra as shown in Example 10-1.

Example 10-1: Determine the magnitude and location (with respect to *A*) of the equivalent concentrated load for the load shown acting on the beam in Figure 10-4a.

Solution:

Step 1 Recognize that the load is a nonuniform one in the shape of a trapezoid (or a rectangle and triangle combined).

Step 2 Calculate the total load as follows:

$$P = \Delta l \left(\frac{w_{\text{min}} + w_{\text{max}}}{2} \right)$$

$$P = 4 \text{ ft} \left(\frac{100 + 200 \text{ lb/ft}}{2} \right) \qquad P = 600 \text{ lb}$$

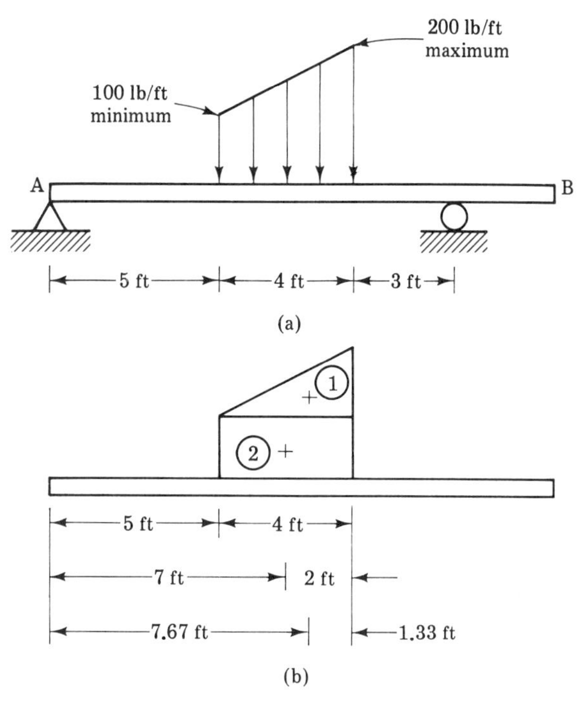

Figure 10-4

Step 3 Sketch the centroid problem as shown in Figure 10-4b. Label the triangle 1 and the rectangle 2.

Step 4 Since the centroid of the rectangle is half way across it, it is located 7 ft to the right of A. Since the centroid of the triangle is located one-third of the length of its base from its altitude, it is located 7.67 ft to the right of A.

Step 5 Calculate the location of the centroid of the trapezoid as follows.

$$x_T = \frac{A_1x_1 + A_2x_2}{A_T}$$

$$A_1 = \tfrac{1}{2}b_1h_1 = \tfrac{1}{2}(4)(100) = 100$$

$$A_2 = b_2h_2 = (4)(100) = 400$$

$$x_T = \frac{100(7.67) + (400)(7)}{600} \qquad x_T = 7.22 \text{ ft}$$

The distance could, of course, have been computed to any desired point other than A.

10.5 SHEAR FORCE

Whenever forces are acting radially (perpendicular to the longitudinal axis) on a beam a tendency exists to shear the beam in the direction of the forces. This is evident in the clevis joint shown in Figure 10-5a. The pin connecting the parts is the beam in this case. If the parts are machined to close tolerances, pure shear results and there is a tendency to cut or shear (slide the molecules) along planes perpendicular to the shaft at lines $a-a$ and $b-b$. If we make the shaft longer, leaving the one force in the middle and moving the others out to the ends as shown in Figure 10-5c we have modified the problem. There is now a tendency to bend the shaft as well as shear it. We will leave the bending aspect to the next section of this chapter. For now, let us examine the shear effects in this elongated beam. At all points

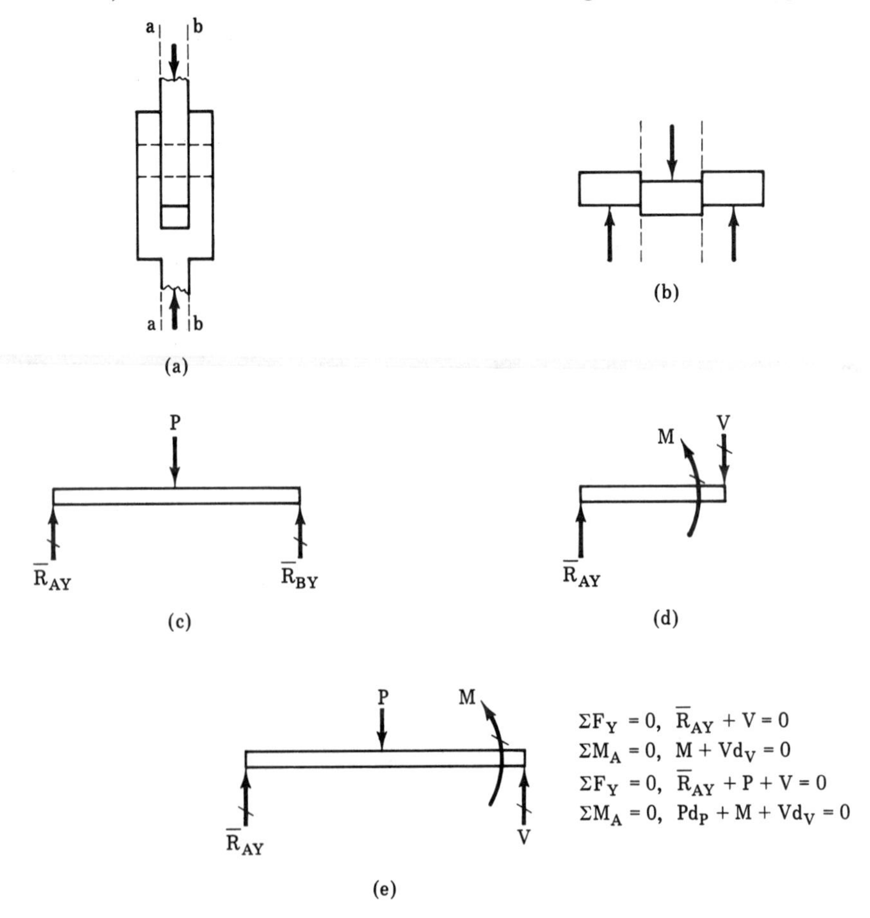

Figure 10-5. Shear force.

between the left-hand end and the load P there exists a shearing force equal to $\overline{R}_{Ay}$, while at all points between the load and the right-hand end there exists a shearing force equal to $\overline{R}_{By}$. This may not be immediately evident but can be shown by taking moments of parts of the beam as shown in Figure 10-5d and 10-5e.

In Figure 10-5d, a cut has been made any place between $\overline{R}_{Ay}$ and P and the left-hand piece is isolated. The shearing force V is the internal force required for force equilibrium with $\overline{R}_{Ay}$ to occur. An internal bending moment reaction is necessary to establish moment equilibrium with the couple formed by $\overline{R}_{Ay}$ and V. In Figure 10-5e a cut has been made anyplace between P and $\overline{R}_{By}$ and the left-hand piece is isolated. Once again, some internal shearing force V is required for force equilibrium with $\overline{R}_{Ay}$ and P. Also, for moment equilibrium, an internal bending moment reaction is necessary in combination with the moments of the forces present.

We shall deal in more detail with the internal bending moments in the next section of this chapter. For now, let us look at the shearing force in more detail. The beam shown in Figure 10-5c–e is a simply supported and a very simply loaded beam. It is quite easy to see that if we solve either or both of the force equations, *the shear force at any point is equal and opposite to the sum of the forces lying to the left of that point.* Based upon the principle of force equilibrium, this statement is true no matter how complex the support and loading arrangement of a beam may be. It is important to note that *a downward shear force is considered to be positive while an upward shear force is considered to be negative,* contrary to the way we usually denote forces. This can be readily explained by the fact that at any given point in the beam there must actually exist equal and opposite shear forces in equilibrium, as shown in Figure 10-6. One may be considered the resultant of the forces to the left of the point while the other is the reaction to those forces. We dealt above with the reactive shear to prove by force equilibrium that such a force must really exist. However, it is really the active shear force that we are interested in, hence, the change in direction sign. A plot of the shear force along the length of the beam gives us a picture of what is happening to the entire beam and is called the *shear force diagram* or often simply the *shear diagram.*

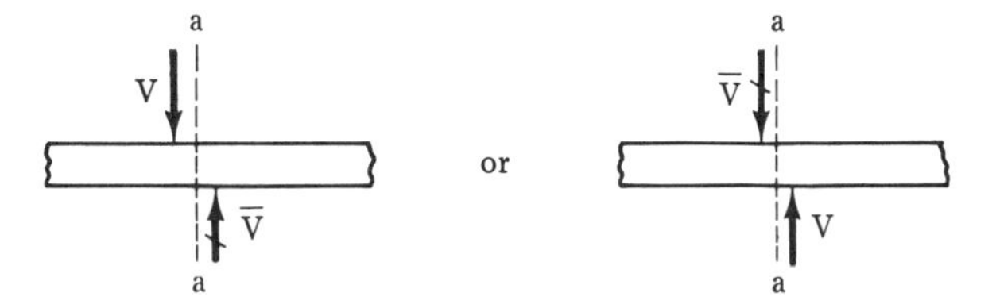

Figure 10-6. Shear direction.

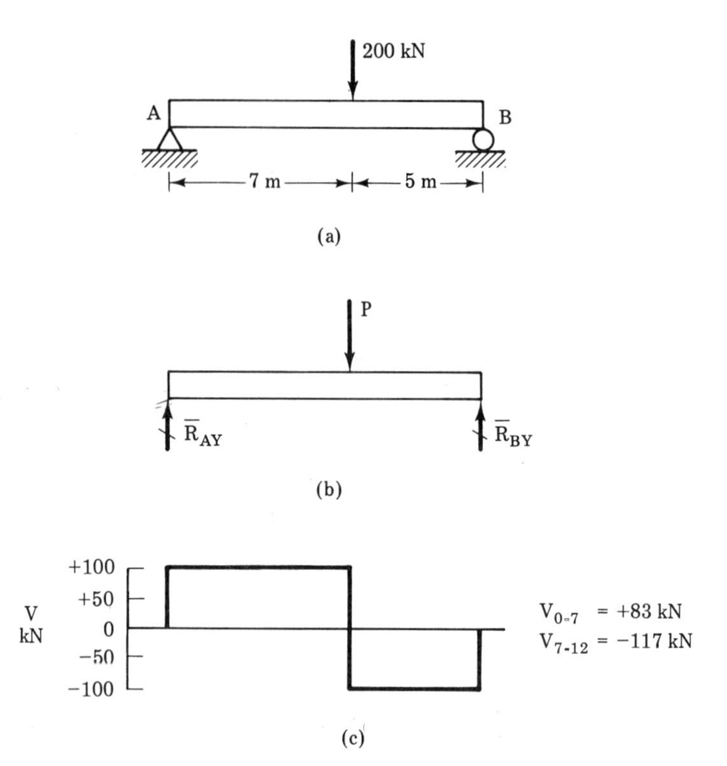

Figure 10-7

Example 10-2: Determine the shear force at all points in the beam shown in Figure 10-7a and draw the shear diagram representing these forces.

Solution:

Step 1 Draw the free-body diagram of the beam as shown in Figure 10-7b.

Step 2 Determine the reactions as follows:

$$\Sigma M_A = 0 \qquad Pd_p + \overline{R}_{BY} d_{BY} = 0$$

$$\overline{R}_{BY} = \frac{Pd_p}{d_{BY}} = \frac{200 \text{ kN } (7 \text{ m})}{12 \text{ m}} \qquad \overline{R}_{BY} = 117 \text{ kN}$$

$$\Sigma F_Y = 0 \qquad \overline{R}_{AY} + P + \overline{R}_{BY} = 0 \qquad \overline{R}_{AY} = -(P + \overline{R}_{BY})$$

$$\overline{R}_{AY} = -(-200 + 117 \text{ kN}) \qquad \overline{R}_{AY} = +83 \text{ kN}$$

Step 3 Recall that the shear force acting in the beam at any point is the sum of the forces lying to the left of that point.

Step 4 Draw the shear diagram as shown in Figure 10-7c recognizing that the shear force at the left end rises instantaneously to a value equal to $\overline{R}_{AY}$, then stays at that value as we move to the right until another force acts on the beam. That would be the load P which instantaneously changes the shear force to a negative value equal to $\overline{R}_{AY} + P$. This new value continues until the next force, $\overline{R}_{BY}$, is applied to the beam completing thc diagram.

Problem 10-1: Determine the shear force at all points in the beam shown in Figure 10-8 and draw the shear diagram representing these forces.

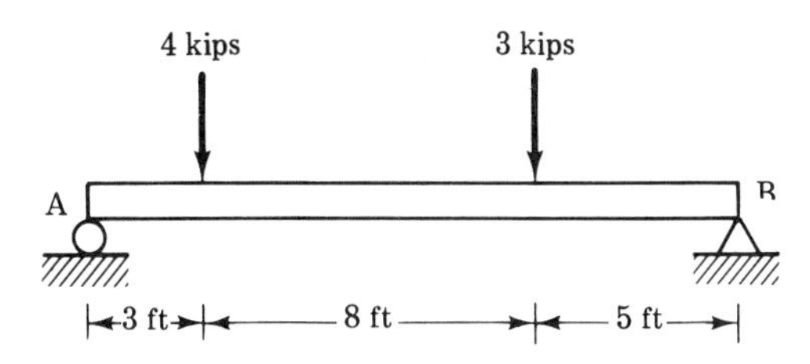

Figure 10-8

10.6 BENDING MOMENT

Only in very specific force applications such as the one shown earlier in Figure 10-5a does pure shear occur. In a beam of any length l, there are internal bending moment reactions as well as shear force reactions. The reality of this became evident in our elongated beam in Figures 10-5c–e.

As might be expected from our earlier discussion, moment equilibrium applied to any section of the beam will include the internal bending moment reaction at the point where the beam was sectioned. Thus, *the moment reaction at any point in the beam* is equal to the negative *sum of the moments of all the forces acting on the beam to the left of the point* (the point in question being the moment center). *Moment reaction tending to bend the beam counterclockwise is considered positive while moment reaction tending to bend the beam counterclockwise is considered negative.* Unlike shear where the *acting* force direction is used to analyze the beam, the *reacting* moment direction is used. The usefulness of this mathematically will be illustrated shortly. In terms of one of our ultimate goals of determining the point of maximum stress in the beam, the directions themselves are not really significant, only the absolute values. Later in this text we will examine the deflection or actual deformation of the beam due to the loads placed upon it. For that purpose directions will need to be considered.

A plot of the bending moment along the length of the beam gives us a picture of what is happening to the entire beam and is called the moment diagram.

Example 10-3: Determine the bending moment reaction at all points in the beam shown in Figure 10-9a and draw the moment diagram. (Note that this is the same problem for which we found the shear forces in Example 10-2.)

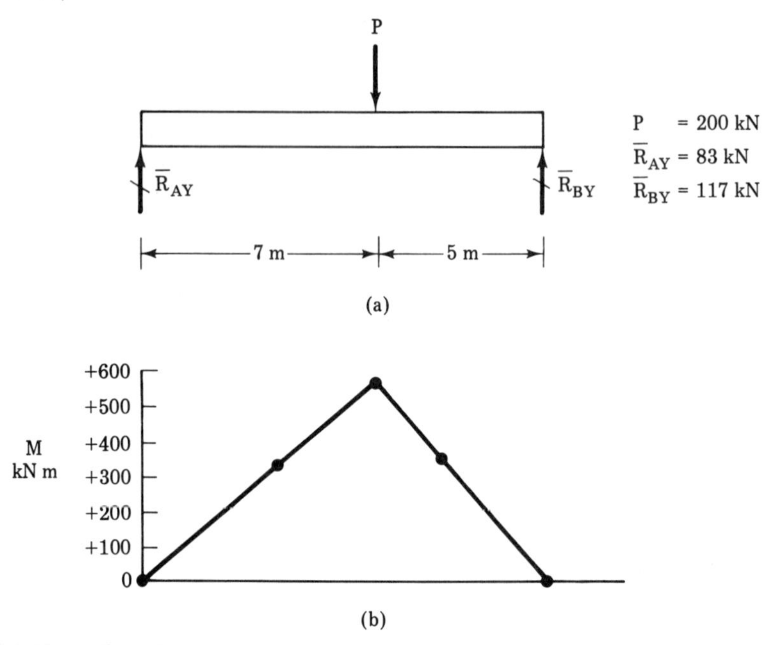

Figure 10-9

Solution:

Step 1 Recall that the negative sum of the moments of the forces to the left of the point in question (taken about the point in question) is equal to the moment reaction at that point.

Step 2 Moving from left to right in this problem, the internal bending moment at any point from $\overline{R}_{Ay}$ to P is $M = -(\overline{R}_{Ay} l)$, l being the distance of the point from A. This is a simple linear relationship, so we really only need values from M at A and at P. However, as a check we will calculate M for one point in between as well.

$$M = -(\overline{R}_{AY} l) \qquad M_0 = -(83)(0) = 0$$

$$M_4 = -[-(83)(4)] = +332 \text{ kN m}$$

$$M_7 = -[-(83)(7)] = +581 \text{ kN m}$$

Note that the moment $\overline{R}_{Ay} l$ was itself clockwise and thus negative and that M was the opposite or negative of $\overline{R}_{Ay} l$ and thus turned out positive.

Step 3 Moving from left to right in this problem, the internal bending moment at any point from P to $\overline{R}_{By}$ is $M = -(\overline{R}_{Ay}l + Pd_p)$, being the distance of the point from A, and d_p being the distance from load P to the point. Again, this is a linear relationship and a beginning and end point would suffice.

$$M = -(\overline{R}_{Ay}l + Pd_p) \quad \text{at } M_7 \quad d_p = 0 \quad M_7 = 581 \text{ kN m}$$

$$M_9 = -[-(83)(9) + (200)(2)] \quad M_9 = +347 \text{ kN m}$$

$$M_{12} = -[-(83)(12) + (200)(5)] \quad M_{12} \cong 0$$

Step 4 Draw the moment diagram based on the above calculations as shown in Figure 10-9b.

Problem 10-2: Determine the bending moment reaction at all points in the beam shown in Figure 10-8 and draw the moment diagram.

10.7 RELATIONSHIPS BETWEEN LOAD, SHEAR, AND MOMENT

It should already be apparent to you from the work you have just done that load, shear, and moment are unquestionably interrelated. Every value for shear and every value for moment were calculated from the load and reaction data. However, we can go one step further and clarify and organize these relationships to the point that they can be defined in very specific mathematical terms.

In Figure 10-10, the figures for Examples 10-2 and 10-3 have been reproduced. Let us examine them for some mathematical clues.

Comparing the free-body diagram with the shear diagram we note that when a force is concentrated ($w = P/l = P/0 = \infty$) the slope of the shear curve is also infinite. Also, when there is no change in force ($w = P/l = 0/l = 0$) the slope of the shear curve is also zero. This would lead us to assume that *the slope of the shear curve at any point equals the force per unit length at that point.*

Comparing the shear diagram with the moment diagram we note that at any point the slope of the moment curve is constant when the value of the shear curve is constant. Further, we note that *the slope of the moment curve at any point equals the shear value at that point.*

Further investigation reveals that *the area under the shear curve between any two points equals the change in moment between those two points.* The same relationship holds between load and shear, that is, *the area under the load curve between any two points equals the change in shear between those two points.*

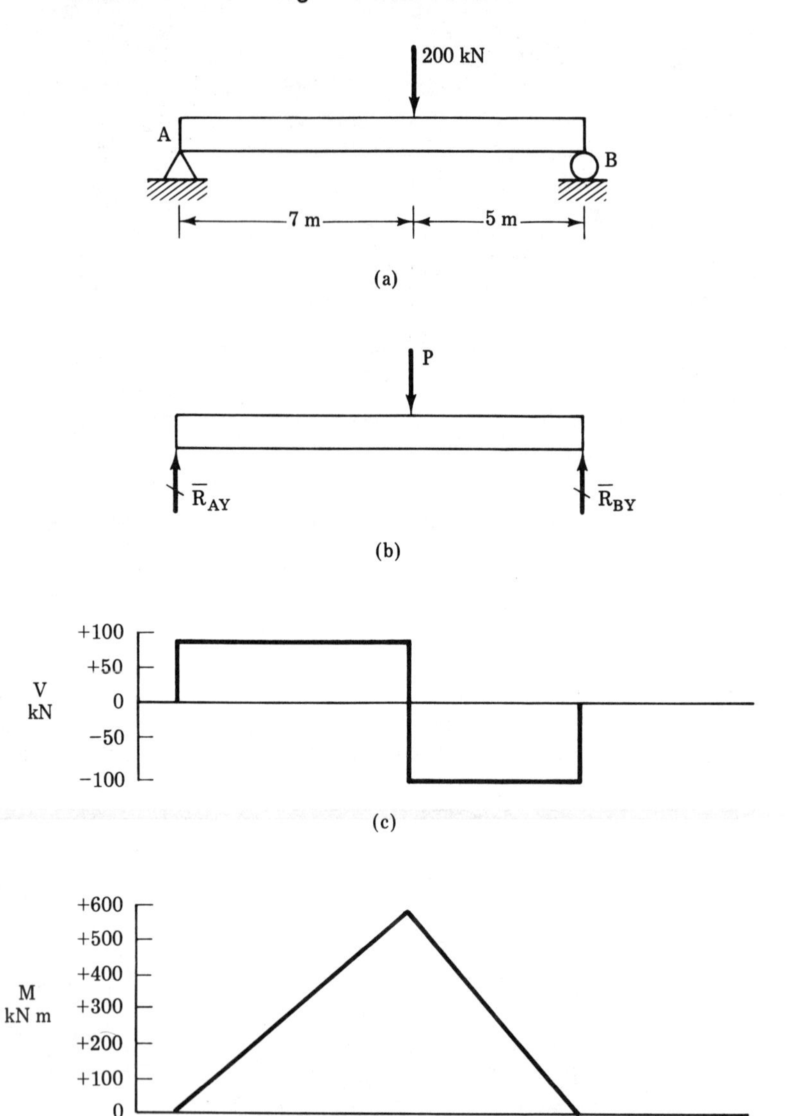

Figure 10-10. Relationships between load, shear, and moment.

In summation

$$\frac{\Delta V}{\Delta l} = w \qquad \Delta l(w) = \Delta V$$

$$\frac{\Delta M}{\Delta l} = V \qquad \Delta l(V) = \Delta M$$

Explicit in the above is a direct relationship between load and moment which you have already experienced in your earlier calculations. These relationships combined with the use of calculus permit one to evaluate the two unknowns if one of the three variables (load, shear, and moment) are defined, no matter how complex that variable may be. However, you will shortly see that algebraic methods can handle a wide range of problems.

Let us look at one more example in some detail and then proceed to the next section which applies these concepts to a variety of combinations of loads and supports.

Example 10-4: Determine both by calculation and graphically the shear diagram and the moment diagram for the beam shown in Figure 10-11a.

Solution:

Step 1 Draw the free-body diagram of the beam as shown in Figure 10-11b. Note the equivalent concentrated load passing through the centroid of the load area.

Step 2 Determine the total load and the reactions as follows:

$$P = wl = 150 \text{ lb/ft}(15 \text{ ft}) \qquad P = 2250 \text{ lb}$$

$$\Sigma M_A = 0 \qquad Pd_p + \overline{R}_{By} d_{By} = 0$$

$$\overline{R}_{By} = \frac{Pd_p}{d_{By}} = \frac{2250 \text{ lb}(7.5 \text{ ft})}{9 \text{ ft}} \qquad \overline{R}_{By} = 1875 \text{ lb}$$

$$\Sigma F_y = 0 \qquad \overline{R}_{Ay} + P + \overline{R}_{By} = 0 \qquad \overline{R}_{Ay} = -(P + \overline{R}_{By})$$

$$\overline{R}_{Ay} = -(-2250 + 1875 \text{ lb}) \qquad \overline{R}_{Ay} = +375 \text{ lb}$$

Step 3 Recall that the shear force acting in the beam is the negative sum of the forces lying to the left of the point. Thus, at $l = 0$, $V = -\overline{R}_{Ay} = -375$ lb. As we move to the right between $\overline{R}_{Ay}$ and $\overline{R}_{By}$ the distributed load acts with a load w for every unit of length. The shear force at any point between those two points can be calculated from $V = -(\overline{R}_{Ay} + wl) = -375 \text{ lb} + 150l$. Thus $V_1 = -225$ lb, $V_2 = -75$ lb, $V_3 = +75$ lb, and so on, until $V_9 = +975$ lb. At 9 ft, there is an instantaneous change in V as $\overline{R}_{By} = +1875$ lb is applied giving us a second value for $V_9 = -(\overline{R}_{Ay} + wl + \overline{R}_{By}) = [-375 + 150(9) - 1875 \text{ lb}] = -900$ lb. The same general equation will hold from $\overline{R}_{By}$ to the end of the beam. Thus $V_{10} = -750$ lb and so on until $V_{15} = 0$.

Step 4 From the results obtained in Step 3, plot the shear diagram (remembering to reverse all directions) as shown in Figure 10-11c.

Step 5 Examine Figures 10-11b and 10-11c and note that the change in shear at A is instantaneous meaning that the slope of the shear is infinite. This follows from the fact that the shear slope equals the rate of force

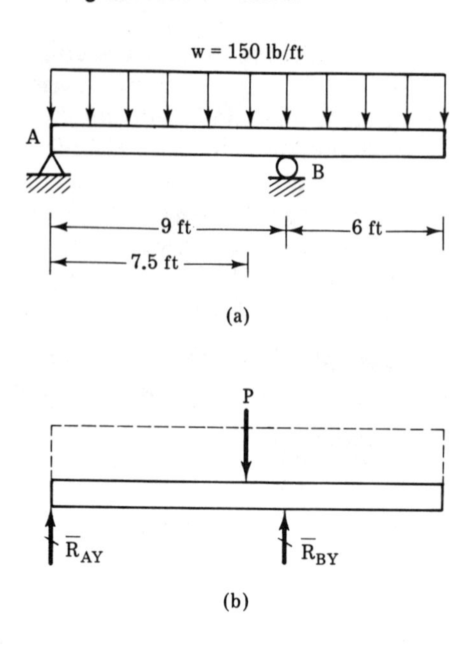
w = 150 lb/ft
A
B
9 ft
6 ft
7.5 ft
(a)
P
$\overline{R}_{AY}$
$\overline{R}_{BY}$
(b)

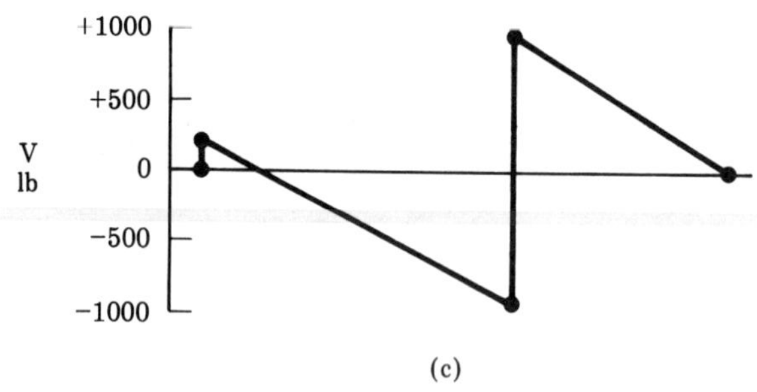
+1000
+500
V
lb
0
−500
−1000
(c)

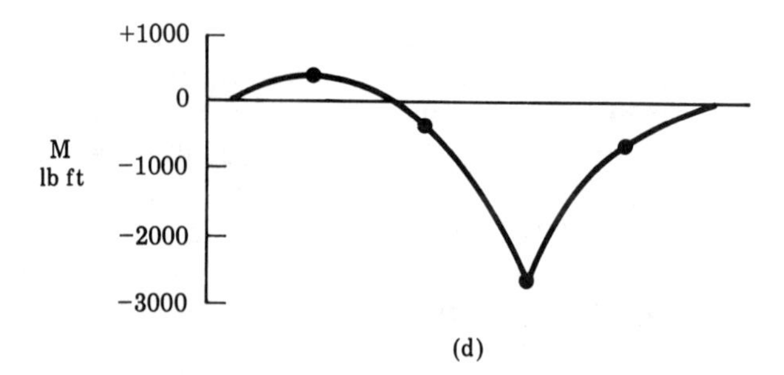
+1000
0
M
lb ft
−1000
−2000
−3000
(d)

Figure 10-11

application which in this case equals $P/l = \overline{R}_{Ay}/l = 375/0 = \infty$. Between A and B, the rate of force application $w = -150$ lb/ft and the shear slope is -150 lb/ft. At B, there is an instantaneous force application again and thus a section of the shear curve at that point has an infinite slope. The remainder of the shear curve has a slope of $w = -150$ lb/ft. Further, let us select any two points, let us say $l = 4$ ft and $l = 7$ ft and examine the relationship between area under the load curve and the change in shear. The area under the load curve $= w\Delta l = -150$ lb/ft(3 ft) $= -450$ lb. The change in shear $= V_7 - V_4 = -675 - (-225) = -450$ lb.

Step 6 Recall that the negative sum of the moments of the forces to the left of the point in question (moments taken about that point) is equal to the moment reaction at that point. Moving from left to right in this problem between points A and B, the moment reaction is the negative sum of the moment of $\overline{R}_{Ay}$ (or $-\overline{R}_{Ay}l$, l being the distance from A to the point in question) and the moment of that part of the distributed load between A and the point in question (or $+wl \cdot \frac{1}{2}l$). Thus, $M = -[-(\overline{R}_{Ay})(l) + \frac{1}{2}wl^2]$. This can be rewritten as $M = +375l - 75l^2$. At $l = 0$, $M = 0$. There will be a maximum M where $V = 0 = -(\overline{R}_{Ay} + wl)$ or $-[+375 + (-150)l] = 0$, that is, where $l = 2.5$ ft. At that point, $M_{2.5} = +469$ lb ft. Arbitrarily, for another point, $M_6 = -450$ lb ft at $l = 6$ ft. At $l = 9$ ft, M will reach a minimum because once again $V = 0$ (at least instantaneously). $M_9 = -2700$ lb ft. Between $l = 9$ and $l = 15$ the moment equation will be $M = -[-(\overline{R}_{Ay})(l) + \frac{1}{2}wl^2 - (\overline{R}_{By})(l - 9)]$ or substituting in known values, $M = +375l - 75l^2 + 1875l - 16{,}875$ or $M = 2250l - 75l^2 - 16{,}875$. As a check, we know that M should equal zero at $l = 15$ ft and if we substitute we find that it indeed is zero. Since this last portion of the moment curve represents a second degree curve we should calculate at least one more point between $l = 9$ ft and $l = 15$ ft. At $l = 12$ ft, $M_{12} = -675$ lb ft. Naturally, the more points we calculate, the more accurate our moment diagram will be. If desired the additional point where $M = 0$ can be calculated from the first moment equation.

Step 7 From the results obtained in Step 6, plot the moment diagram (leaving the directions as found) as shown in Figure 10-11d.

Step 8 Examine Figures 10-11c and 10-11d and note that at all points where $V = 0$, $\Delta M/\Delta l = 0$. Where V is negative $\Delta M/\Delta l$ is negative. Except at specific points, we can only approximate using algebra, but between any two points, the area under the shear curve should equal the change in moment.

$$\left(\frac{V_{2.5} + V_0}{2}\right)(l_{2.5} - l) = \left(\frac{0 + 375}{2}\right)(2.5) = +469 \text{ lb ft.}$$

Checking the moment diagram, $M_{2.5} - M_0 = +469$ lb ft.

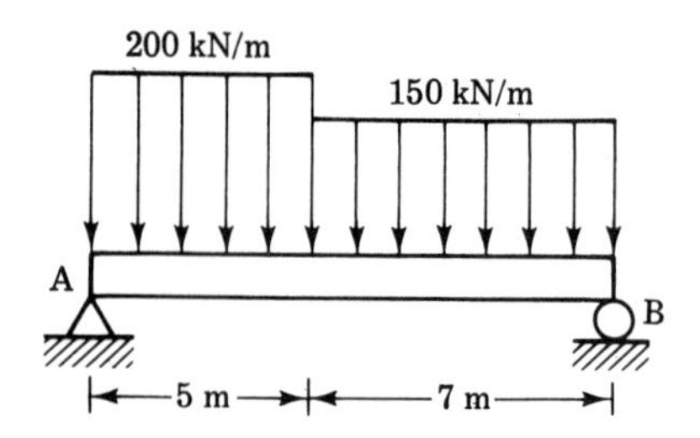

Figure 10-12

Problem 10-3: Determine both by calculation and graphically the shear diagram and the moment diagram for the beam shown in Figure 10-12.

10.8 ADDITIONAL APPLICATIONS

The above examples were purposely somewhat lengthy in explanation. The following examples provide some insight into various combinations of loads and supports for a beam. For each example a comparable problem for practice has been provided. The examples have been done in relatively short form the way you might do them.

Example 10-5: Determine by calculation the shear diagram and the moment diagram for the beam shown in Figure 10-13a.

Solution:

Step 1 Draw the free-body diagram of the beam as shown in Figure 10-13b.

Step 2 Determine the reactions as follows:

$$\Sigma F_y = 0 \qquad \overline{R}_{Ay} + P = 0 \qquad \overline{R}_{Ay} = -P \qquad \overline{R}_{Ay} = +700\text{ kN}$$

$$\Sigma M_A = 0 \qquad \overline{M}_A + Pd_p = 0 \qquad \overline{M}_A = -(Pd_p)$$

$$\overline{M}_A = -[-(700\text{ kN})(4\text{ m})] \qquad \overline{M}_A = +2800\text{ kN m}$$

Step 3 Determine shear at all points on the beam as follows:

$$V_{0-4} = -\overline{R}_{Ay} \qquad V_0 = V_1 = V_2 = V_3 = V_4 = -700\text{ kN}$$

Step 4 Plot the shear diagram as shown in Figure 10-13c.

Step 5 Determine the moment at every point in the beam as follows:

$$M_{0-4} = -(\overline{M}_A + \overline{R}_{Ay} l)$$

$$M_{0-4} = -[2800 - (700)l] = 700l - 2800$$

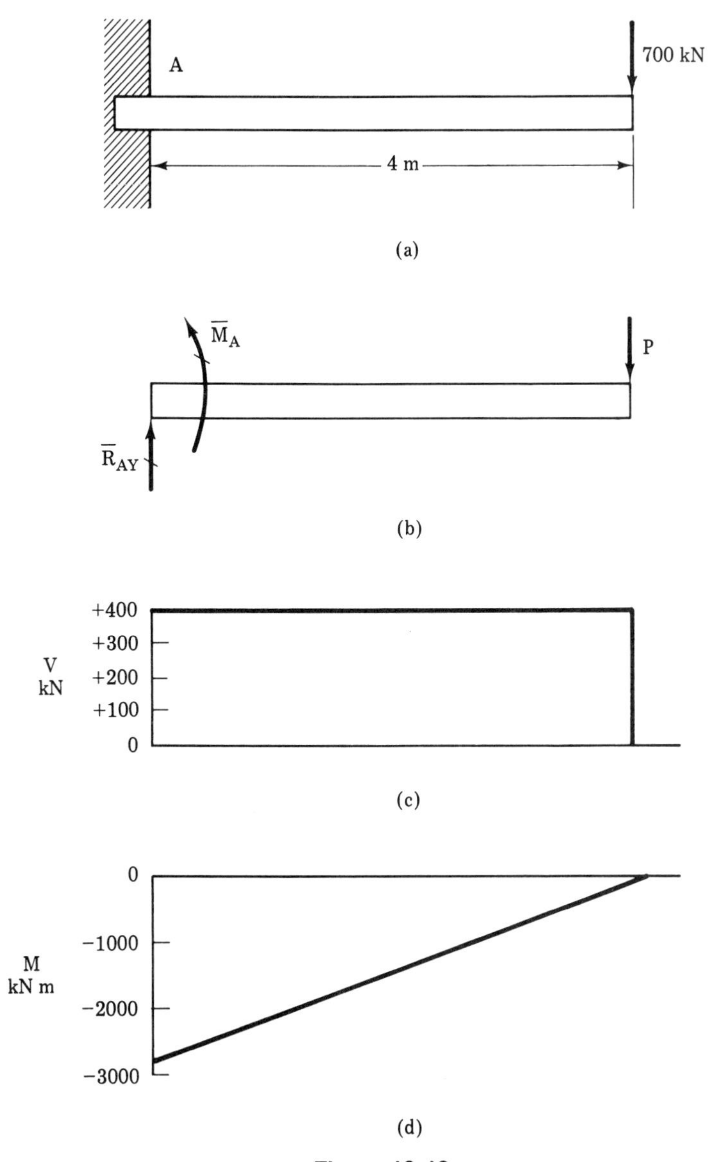

Figure 10-13

$$M_0 = -2800 \text{ kN m} \qquad M_1 = -2100 \text{ kN m}$$

$$M_2 = -1400 \text{ kN m} \qquad M_3 = -700 \text{ kN m} \qquad M_4 = 0$$

Step 6 Plot the moment diagram as shown in Figure 10-13d.

Problem 10-4: Determine by calculation the shear diagram and the moment diagram for the beam shown in Figure 10-14.

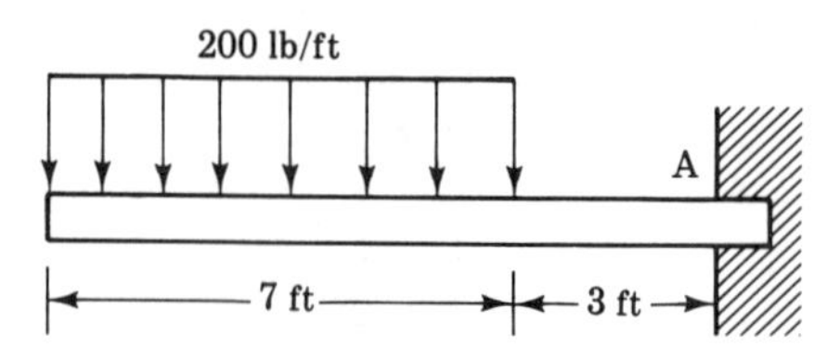

Figure 10-14

Example 10-6: Determine the shear diagram and the moment diagram for the beam shown in Figure 10-15a.

Solution:

Step 1 Draw the free-body diagram of the beam as shown in Figure 10-15b.

Step 2 Determine the total distributed load and the reactions (K = kip):

$$P_1 = wl = 2(10) = 20 \text{ K} \quad \text{at} \quad l = 5 \text{ ft}$$

$$\Sigma M_A = 0 \qquad P_1 d_1 + P_2 d_2 + \overline{R}_{By} d_{By} = 0 \qquad \overline{R}_{BY} d_{BY} = -[P_1 d_1 + P_2 d_2]$$

$$\overline{R}_{By} d_{By} = -[-(20)(2) - (5)(4)] = +60 \text{ K ft}$$

$$\overline{R}_{By} = \frac{\overline{R}_{By} d_{By}}{d_{By}} = \frac{60}{12} \qquad \overline{R}_{By} = +5 \text{ K}$$

$$\Sigma F_y = 0 \qquad \overline{R}_{Ay} + P_1 + P_2 + \overline{R}_{By} = 0$$

$$\overline{R}_{Ay} = -(P_1 + P_2 + \overline{R}_{By}) = -(-20 - 5 + 5)$$

$$\overline{R}_{Ay} = +20 \text{ K}$$

Step 3 Draw the shear diagram from the free-body diagram as shown in Figure 10-15c. Using the distributed load, not its equivalent concentrated load, the shear force starts at 0, drops 2 K/ft for 3 ft, instantaneously rises 20 K, then continues to drop 2 K/ft for another 4 ft, then instantaneously drops 5 K. It then drops 2 K/ft for another 3 ft, and finally stays constant for 5 ft and then rises 5 K to 0. This can be read directly from the free-body diagram and plotted step by step to form the shear diagram.

Step 4 Draw the moment diagram from the shear diagram as shown in Figure 10-15d. First, note that the slope of the moment curve is zero at $l = 0, 3, 7.5$, and 15. Calculate the moment at $l = 1, 2, 3, 4, 6, 7, 7.5, 8, 9$ and 10. This is done by calculating the area under the shear curve to the left of the point in question. The reason for the large number of points is that the moment curve is primarily second degree (except the last 5 ft).

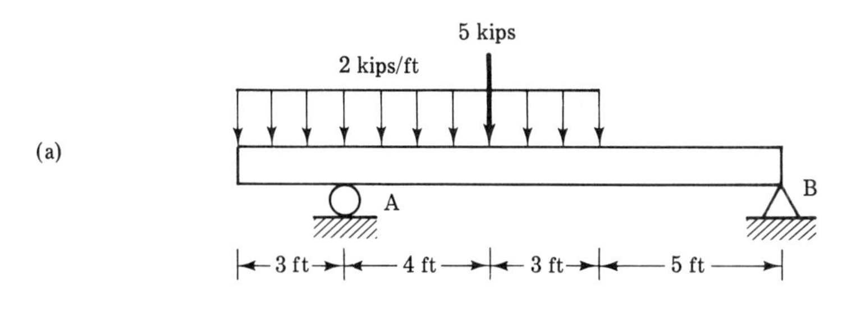

(a)
5 kips
2 kips/ft
A
B
3 ft
4 ft
3 ft
5 ft

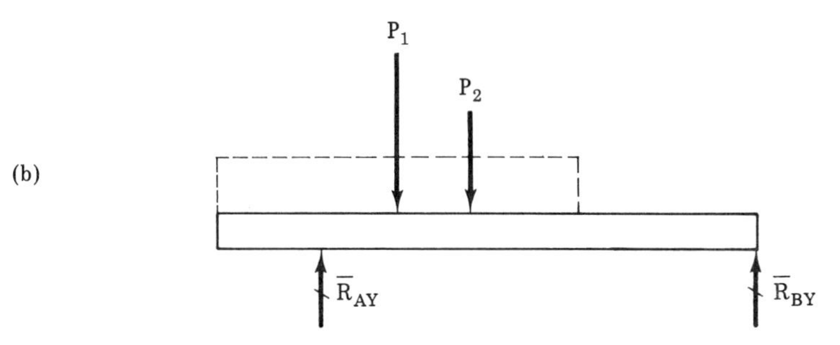

(b)
P1
P2
R̄AY
R̄BY

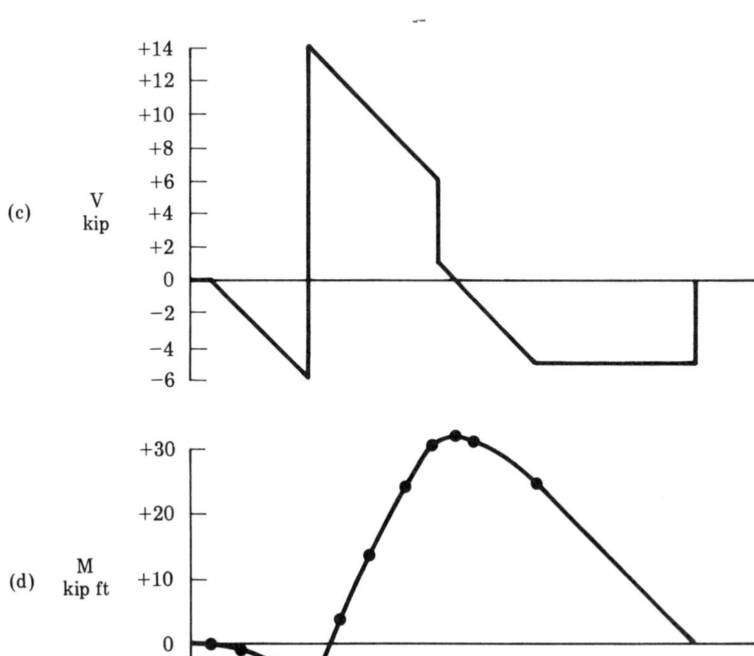

(c)
V
kip
+14
+12
+10
+8
+6
+4
+2
0
−2
−4
−6
(d)
M
kip ft
+30
+20
+10
0
−10

Figure 10-15

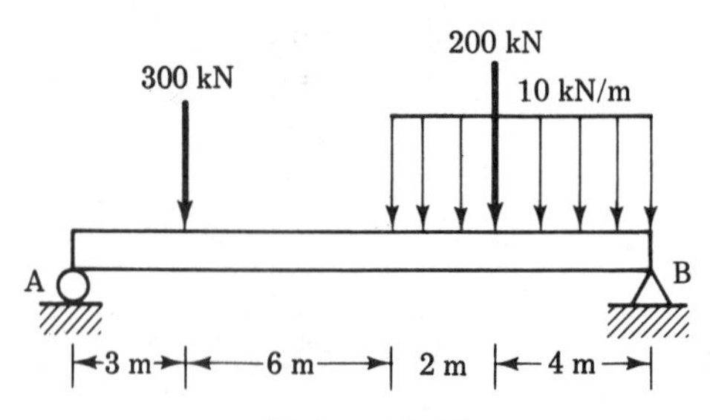

Figure 10-16

Problem 10-5: Determine the shear diagram and the moment diagram for the beam shown in Figure 10-16.

Example 10-7: Determine the shear diagram and the moment diagram for the beam shown in Figure 10-17a.

Solution:

Step 1 Draw the free-body diagram of the beam as shown in Figure 10-17b.

Step 2 Determine the total distributed load and the reactions

$$P_2 = wl = 20(5) = 100 \text{ kN} \quad \text{at} \quad l = 5.5 \text{ m}$$

$$\Sigma M_A = 0 \qquad P_1 d_1 + P_2 d_2 + P_3 d_3 + P_4 d_4 + \overline{R}_{By} d_{By}$$

$$\overline{R}_{By} d_{By} = -[P_1 d_1 + P_2 d_2 + P_3 d_3 + P_4 d_4 + \overline{R}_{By} d_{By}]$$

$$\overline{R}_{By} d_{By} = -[-(100)(0) - (100)(5.5) - (200)(10) - (50)(14)]$$

$$\overline{R}_{By} d_{By} = +3240 \text{ kN m}$$

$$\overline{R}_{By} = \frac{\overline{R}_{By} d_{By}}{d_{By}} = \frac{3250}{14} \qquad \overline{R}_{By} = +232 \text{ kN}$$

$$\Sigma F_y = 0 \qquad \overline{R}_{Ay} + P_1 + P_2 + P_3 + P_4 + \overline{R}_{By} = 0$$

$$\overline{R}_{Ay} = -(P_1 + P_2 + P_3 + P_4 + \overline{R}_{By}) = -(-100 - 100 - 200 - 50 + 232)$$

$$\overline{R}_{Ay} = +218 \text{ kN}$$

Step 3 Draw the shear diagram from the free-body diagram as shown in Figure 10-17c. At A, the *net* shearing force is 118 kN up. It stays at that value until $l = 3$ m, where it starts to drop 20 kN/m and continues to do so until $l = 8$ m. It then remains constant 18 kN up until $l = 10$ m where an instantaneous drop of 200 kN to 182 kN down takes place. It stays at that value until $l = 14$ m where a net instantaneous force of 182 kN up completes the diagram.

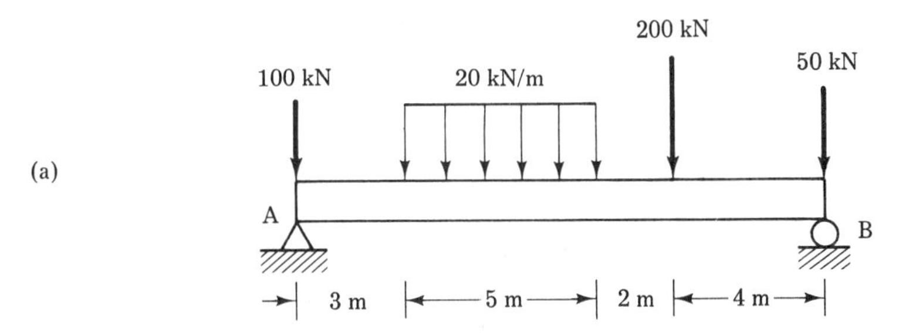

(a)
100 kN
20 kN/m
200 kN
50 kN
A
B
3 m
5 m
2 m
4 m

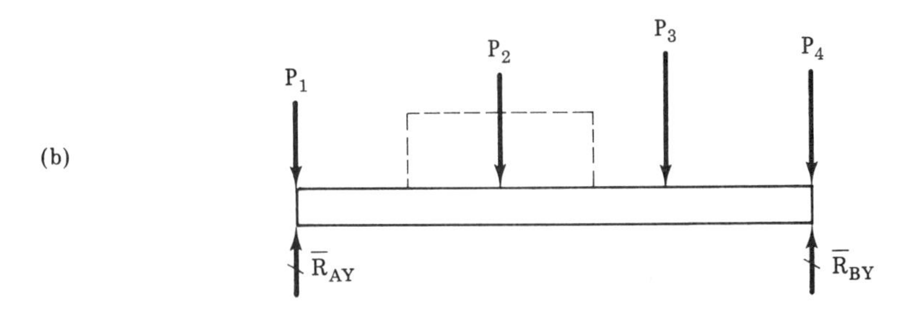

(b)
P_1
P_2
P_3
P_4
$\overline{R}_{AY}$
$\overline{R}_{BY}$

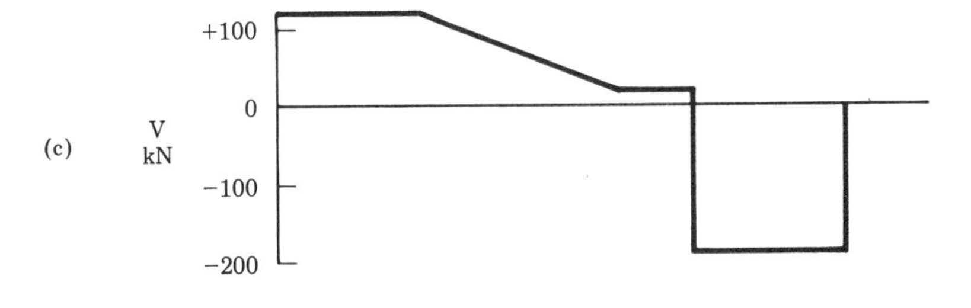

(c)
V
kN
+100
0
−100
−200

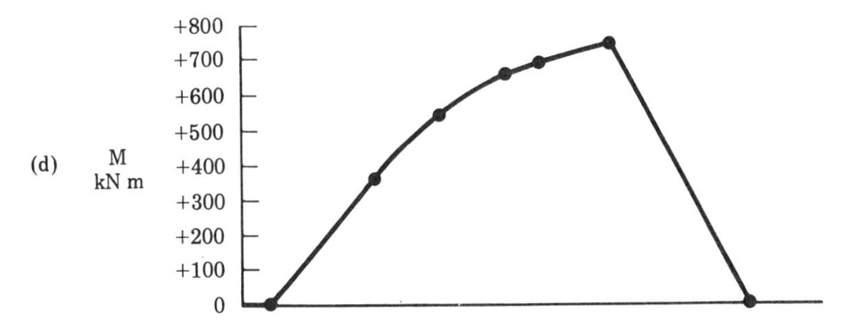

(d)
M
kN m
+800
+700
+600
+500
+400
+300
+200
+100
0

Figure 10-17

Step 4 Draw the moment diagram from the shear diagram as shown in Figure 10-17d. It starts at 0 at A and since the value of the shear is constant for 3 m, so is the slope of the moment curve. The value of the moment at $l = 3$ m is the area under the shear curve to the left of that point or $118(3) = 354$ kN m. The value of the shear then drops uniformly for 5 m so the slope of the moment curve while still positive is decreasing. Since this is a second-degree curve, two values between $l = 3$ m and $l = 8$ m must be calculated. This amounts to adding the area of a trapezoid under the shear curve to the area of the rectangle already determined.

$$M_5 = 354 + \tfrac{1}{2}(118 + 78)(2) = +550 \text{ kN m}$$

$$M_7 = 354 + \tfrac{1}{2}(118 + 38)(4) = +666 \text{ kN m}$$

At $l = 8$, the added area is still a trapezoid so $M_8 = 354 + \tfrac{1}{2}(118 + 18)(5) = +694$ kN m. For the next 2 m the shear is constant so the slope of the moment curve is constant (still positive) and the additional moment to arrive at M_{10} will be the rectangular area under that portion of the shear curve, $M_{10} = +694 + (18)(2) = +730$ kN m. At $l = 10$ m, there is an instantaneous change in shear to a negative value. Thus at $l = 10$ m, the slope of the moment curve has a value of zero for an instant as it switches from its old positive slope to its new negative slope. For the next 4 m, shear is a constant so the moment curve has a constant slope. The value for the moment at $l = 14$ m can be determined by adding the area of the rectangle under the shear curve to M_{10}.

$$M_{14} = +730 - (182)(4) \cong 0.$$

Problem 10-6: Determine the shear diagram and the moment diagram for the beam shown in Figure 10-18.

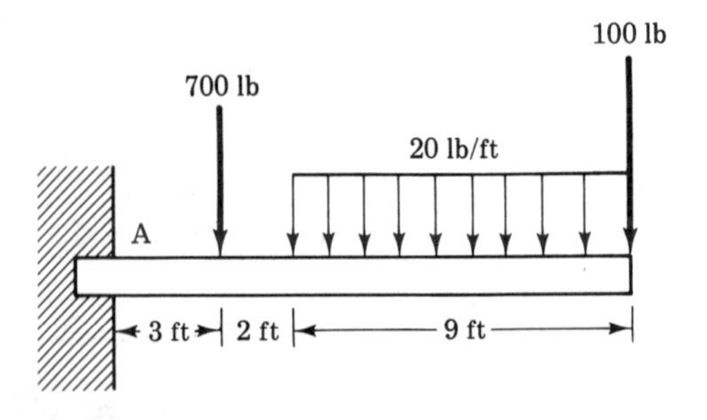

Figure 10-18

Example 10-8: Determine the shear diagram and the moment diagram for the beam shown in Figure 10-19a.

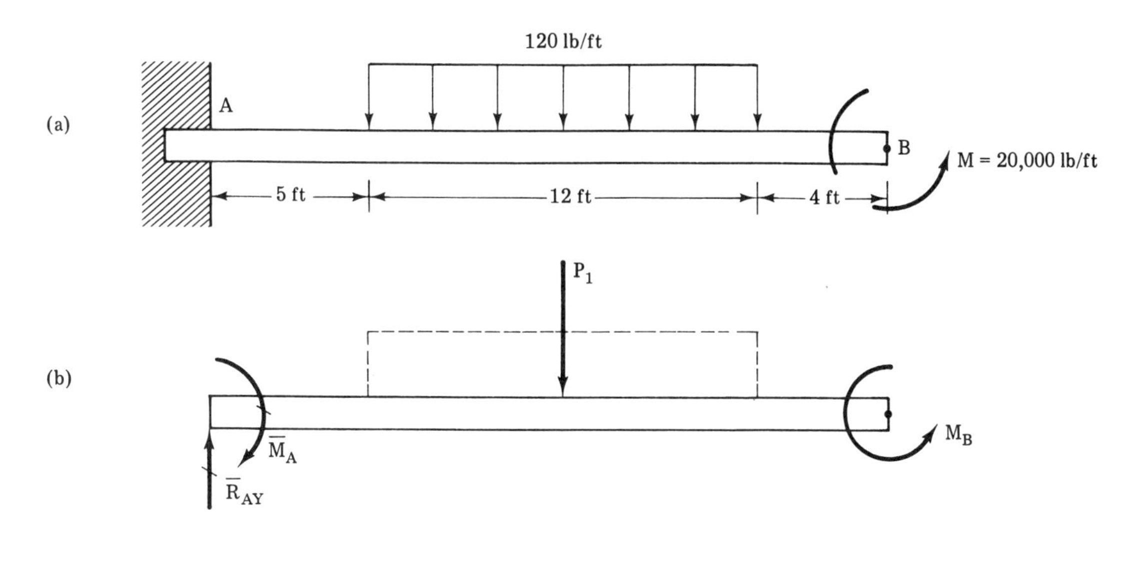

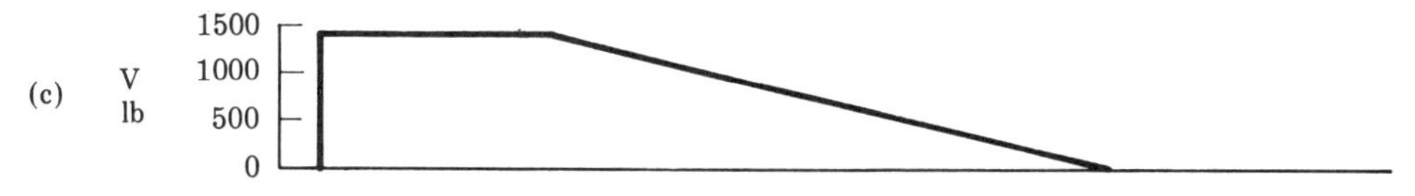

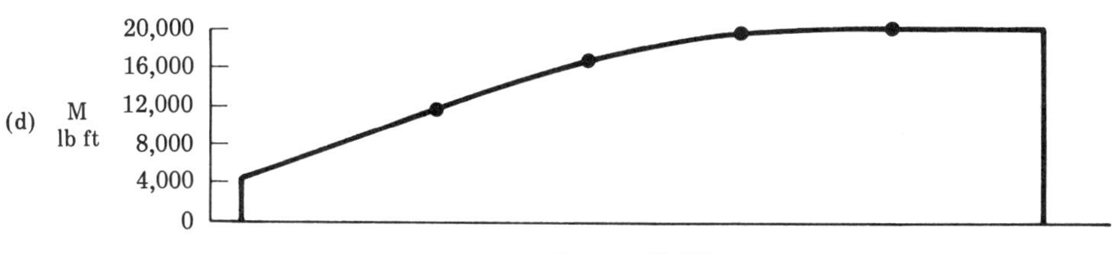

Figure 10-19

Solution:

Step 1 Draw the free-body diagram of the beam as shown in Figure 10-19b.

Step 2 Determine the total distributed load and the reactions.

$$P_1 = wl = 120(12) = 1440 \text{ lb} \quad \text{at} \quad l = 11 \text{ ft}$$

$$\Sigma M_A = 0 \qquad \overline{M}_A + P_1 d_1 + M_B = 0$$

$$\overline{M}_A = -[P_1 d_1 + M_B] = -[-(1440)(11) + 20{,}000]$$

$$\overline{M}_A = -4160 \text{ lb ft}$$

$$\Sigma F_y = 0 \qquad \overline{R}_{Ay} + P_1 = 0 \qquad \overline{R}_{Ay} = -P_1$$
$$\overline{R}_{Ay} = -(-1440) \qquad \overline{R}_{Ay} = +1440 \text{ lb}$$

Step 3 Draw the shear diagram from the free-body diagram as shown in Figure 10-19c. The shear at A is 1440 lb up and continues at this value until $l = 5$ ft. At that point, the shear starts to drop at a constant rate of 120 lb/ft until it reaches a value of zero at $l = 17$ ft. It continues at a value of zero to the end of the beam.

Step 4 Draw the moment diagram of the beam from the free-body diagram and the shear diagram. At $l = 0$, a moment of 4160 lb ft exists. This seems strange at first since we just calculated $\overline{M}_A = -4160$ but recall that the internal moment at any point is the *negative* of the moments to the left of that point. For the next 5 m the slope of the moment curve is constant and positive (equal to V). The moment at $l = 5$ is equal to the initial moment plus the change in moment represented by the area of the rectangle above the shear curve.

$$M_5 = +4160 + (1440)(5) + 11{,}360 \text{ lb ft}$$

The slope of the moment curve then decreases uniformly over the next 12 ft. Thus, it will have a second degree curve.

$$M_9 = +11{,}360 + \tfrac{1}{2}(1440 + 960)(4) = +16{,}660 \text{ lb ft}$$
$$M_{13} = +11{,}360 + \tfrac{1}{2}(1440 + 480)(8) = +19{,}040 \text{ lb ft}$$
$$M_{17} = +11{,}360 + \tfrac{1}{2}(1440)(12) = +20{,}000 \text{ lb ft}$$

The moment then stays constant until at $l = 21$ ft, $M_{21} = +20{,}000 - (+20{,}000) = 0$.

Problem 10-7: Determine the shear diagram and the moment diagram for the beam shown in Figure 10-20.

No further illustration of a hinged beam will be made since once the initial step of separating it into two beams is done, analysis is similar to the

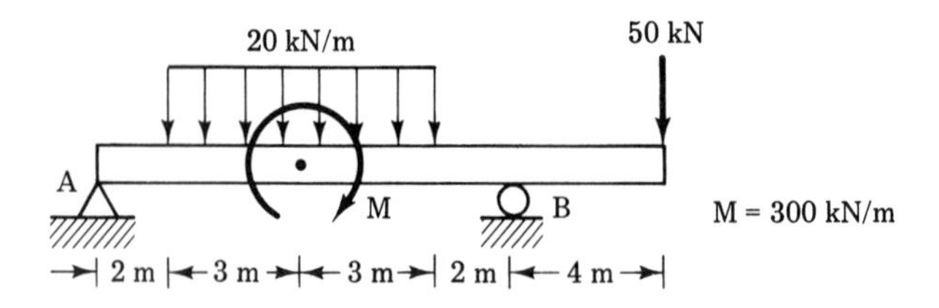

Figure 10-20

above for each part. Moment resulting from a triangularly distributed load can only be approximated by the area method since the top edge of the trapezoidal area formed under the shear curve is actually a second-degree curve. However, you could revert to the calculation of the negative sum of the moments of the forces which while somewhat tedious will give you precise answers. If you wish to try there is such a problem in the "Supplementary Problems" at the end of the chapter.

10.9 READINESS QUIZ

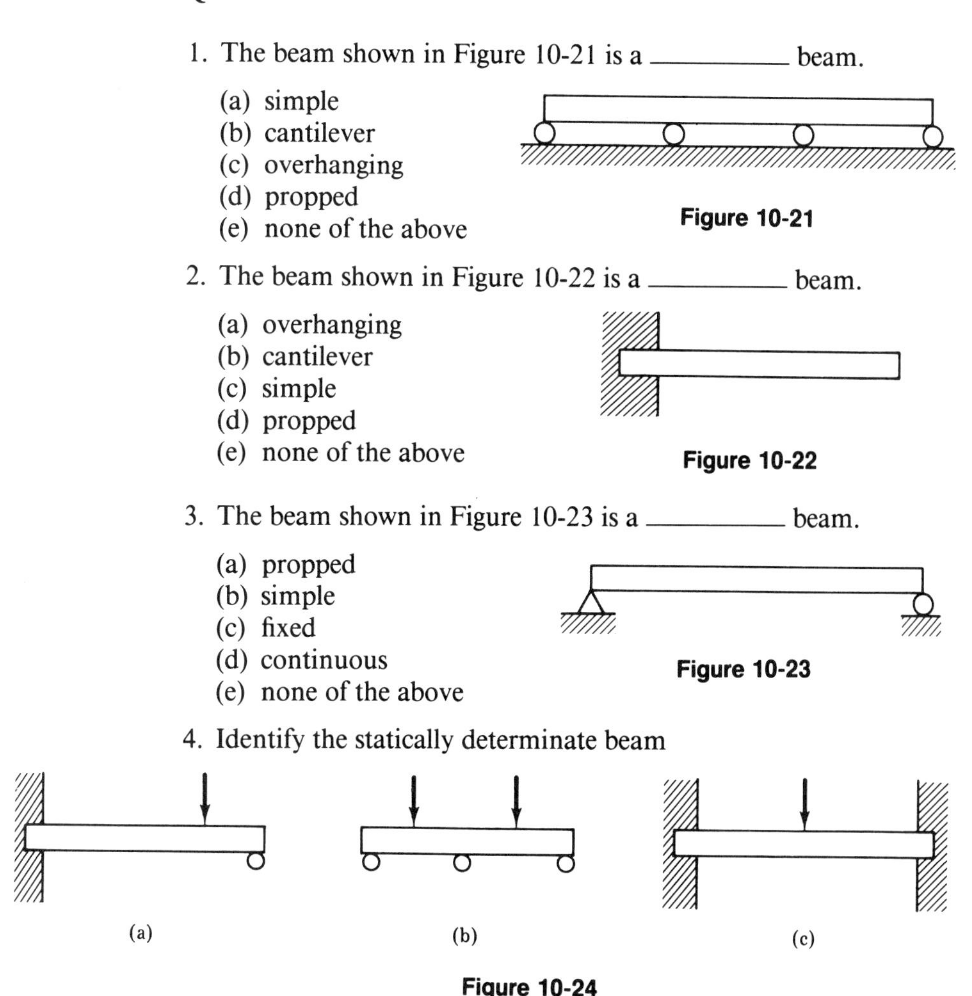

1. The beam shown in Figure 10-21 is a ________ beam.

   (a) simple
   (b) cantilever
   (c) overhanging
   (d) propped
   (e) none of the above

Figure 10-21

2. The beam shown in Figure 10-22 is a ________ beam.

   (a) overhanging
   (b) cantilever
   (c) simple
   (d) propped
   (e) none of the above

Figure 10-22

3. The beam shown in Figure 10-23 is a ________ beam.

   (a) propped
   (b) simple
   (c) fixed
   (d) continuous
   (e) none of the above

Figure 10-23

4. Identify the statically determinate beam

Figure 10-24

   (d) all of the above (e) none of the above

5. The bending moment at $a-a$ in the beam shown in Figure 10-25 is

   (a) 100 lb ft
   (b) 200 lb ft
   (c) 300 lb ft
   (d) 400 lb ft
   (e) none of the above

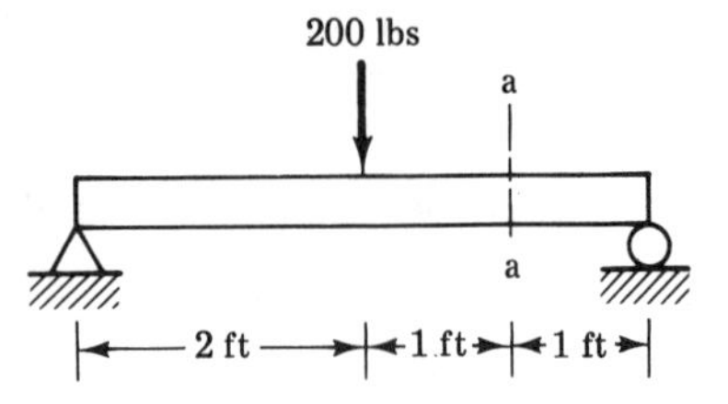

Figure 10-25

6. The shear force at $a-a$ in the beam shown in Figure 10-25 is

   (a) 100 lb
   (b) 200 lb
   (c) 300 lb
   (d) 400 lb
   (e) none of the above

7. The shear diagram corresponding to the free-body diagram shown in Figure 10-26 is

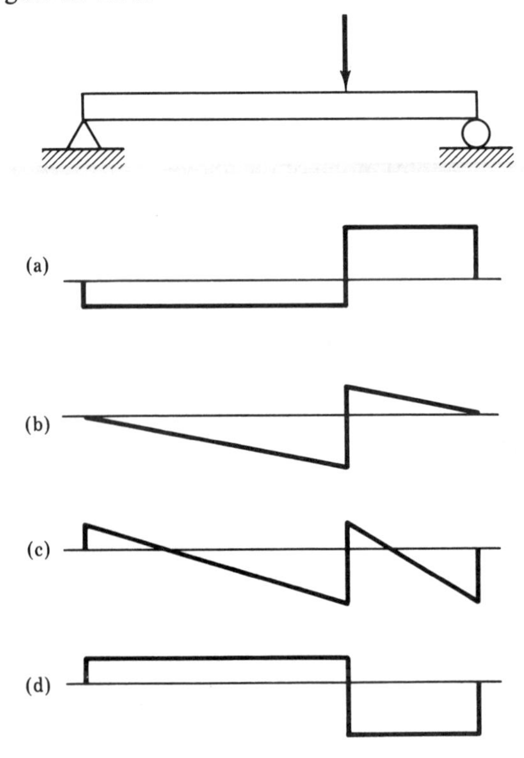

(e) None of the above

Figure 10-26

8. The moment diagram corresponding to the free-body diagram shown in Figure 10-27 is

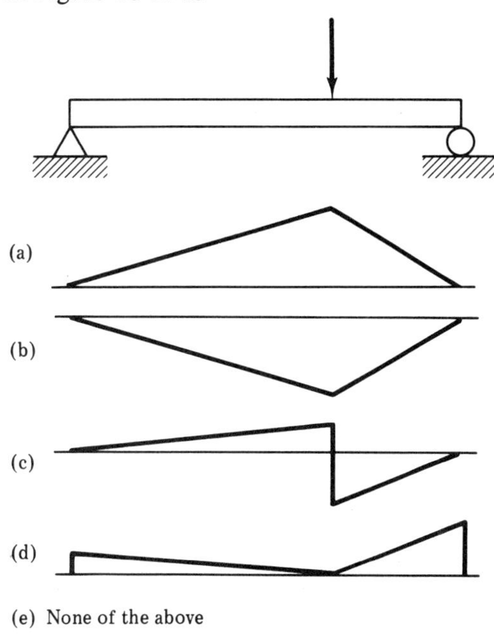

Figure 10-27

9. The moment diagram corresponding to the shear diagram shown in Figure 10-28 is

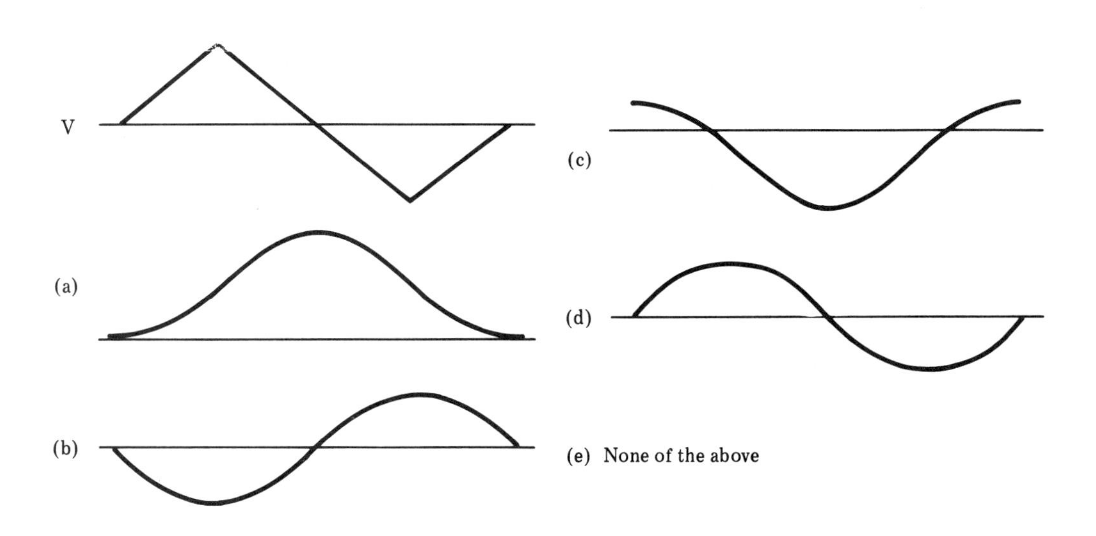

Figure 10-28

10. The shear diagram corresponding to the moment diagram shown in Figure 10-29 is

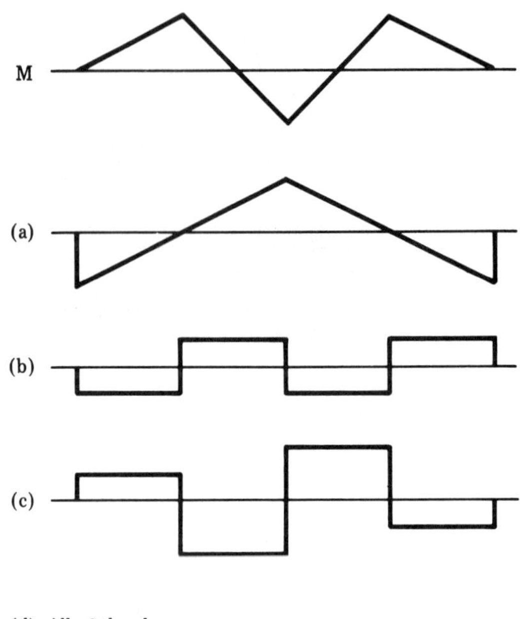

(d) All of the above

(e) None of the above

Figure 10-29

10.10 SUPPLEMENTARY PROBLEMS

10-8 Determine by calculation the shear force at all points in the beam shown in Figure 10-30 and draw the shear diagram representing these forces.

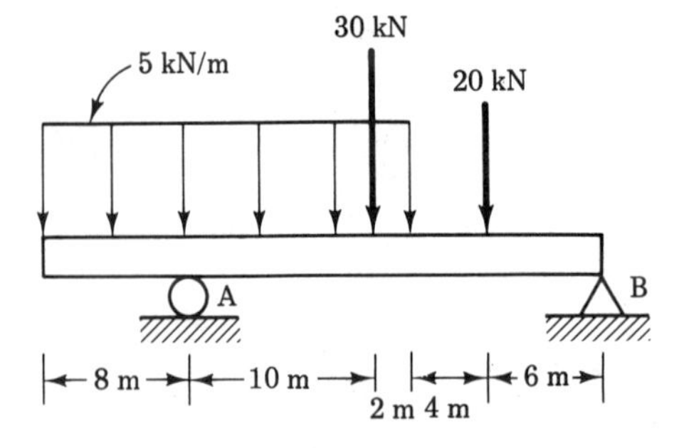

Figure 10-30

10-9 Determine by calculation the bending moment reaction at all points in the beam shown in Figure 10-31 and draw the moment diagram.

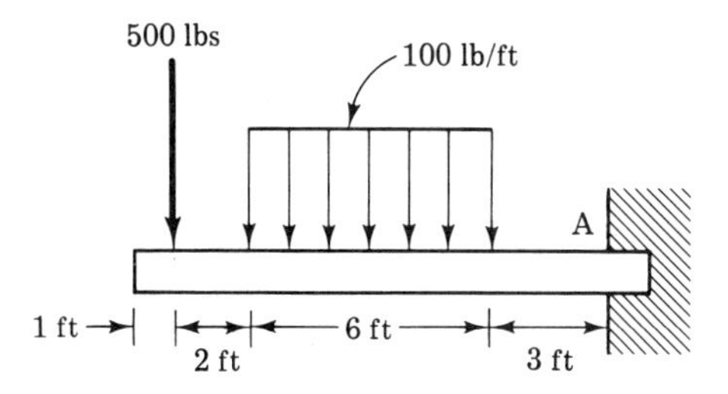

Figure 10-31

10-10 Determine the shear diagram and the moment diagram for the beam shown in Figure 10-32.

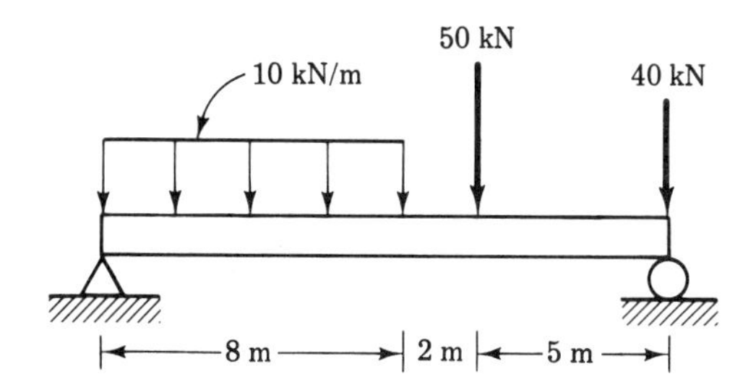

Figure 10-32

10-11 Determine the shear diagram and the moment diagram for the beam shown in Figure 10-33.

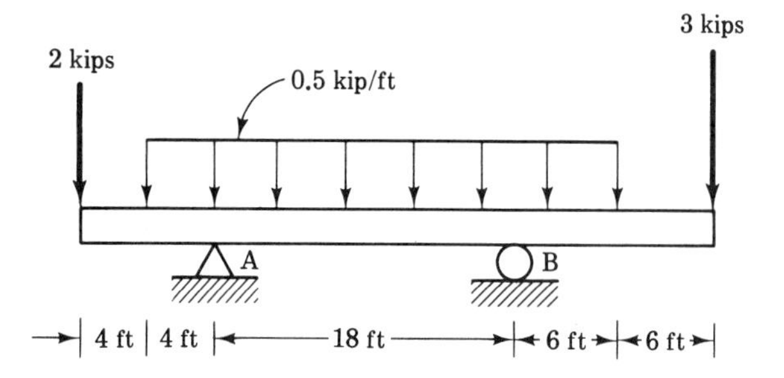

Figure 10-33

10-12 Determine the shear diagram and the moment diagram for the beam shown in Figure 10-34.

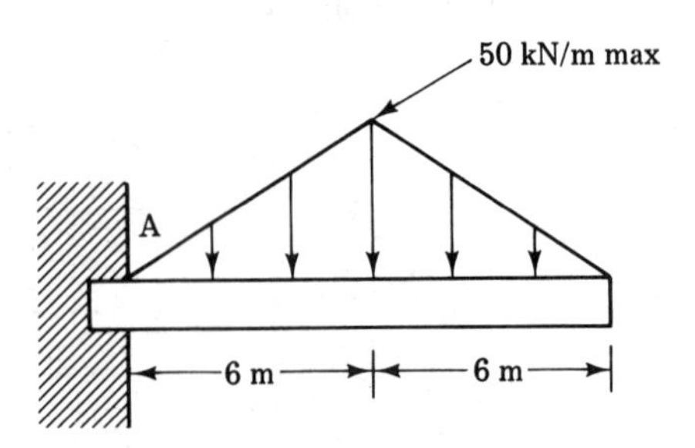

Figure 10-34

Chapter

11

Stress and Strain

11.1 OBJECTIVES

Upon completion of the work relating to this chapter, you should be able to perform the following.

1. For the terms listed below:
   (a) Select from several definitions the correct one for any given term.
   (b) Given a definition, select the correct term from a given list of terms.

axial strain	modulus of elasticity	strain
deformation	offset method	stress
elastic limit	plastic range	ultimate strength
elastic range	Poisson's ratio	yield point
elongation	proportional limit	Young's modulus
lateral strain	rupture strength	

2. Given the data from a standard tensile test
   (a) determine all stresses and strains.
   (b) plot the stress-strain diagram.
   (c) determine the modulus of elasticity.

(d) identify all significant points and areas of the stress-strain diagram.

3. Determine the deformation of a body perpendicular to the line of action of the applied forces.
4. Determine the stresses and forces in a restrained body resulting from temperature change.
5. Determine stresses and strains in all components of a multimaterial member under load.

11.2 STRESS DEFINED

When any type of matter is at rest, the molecules that it consists of are in an equilibrium position, neither too close nor too far away. The exact distance is a function of the particular molecules and helps determine some of the properties of the material. Whenever outside forces attempt to disturb this equilibrium, the molecular structure resists any change. This resistance is known as *stress.* This can be used as a very general term measured in units of force. However, usually the term stress is used to mean *unit stress* measured in units of force per unit area such as pounds per square inch (psi), kips per square inch (ksi), or newtons per square meter (Pa). (It is worth noting that except under either natural or artificial zero-gravity conditions, even materials "at rest" are under stress due to the force of gravity.)

There are three basic types of stress depending upon the way in which the external forces are attempting to disturb the molecular equilibrium. One of these types is *tensile stress* which is the resistance to being pulled apart or trying to increase the distance between molecules in the direction of the applied and reacting forces. Examples include the cables in a suspension bridge, the ropes in a child's swing, and a simple clothesline.

Directly opposite to tensile stress is *compressive stress.* As this infers, it is the resistance to being pushed together or trying to decrease the distance between molecules in the direction of the applied and reacting forces. Examples include the foundations of buildings, bases of machines, legs on a chair, and for that matter, any stationary object that has gravitational force acting on it.

The third type of basic stress is perhaps a little more difficult to visualize. This is *shear stress,* the resistance of molecules to being slid relative to one another. It differs from the other two basic stresses in that the cross-sectional area being stressed is parallel to the lines of action of the applied and reacting forces. Examples include a sheet of metal having a hole punched in it, scissors cutting a piece of paper, and the action on the bolt holding the two parts of the scissors together. Figure 11-1 shows some examples before and after stressing.

The details in each case show the change in molecular orientation;

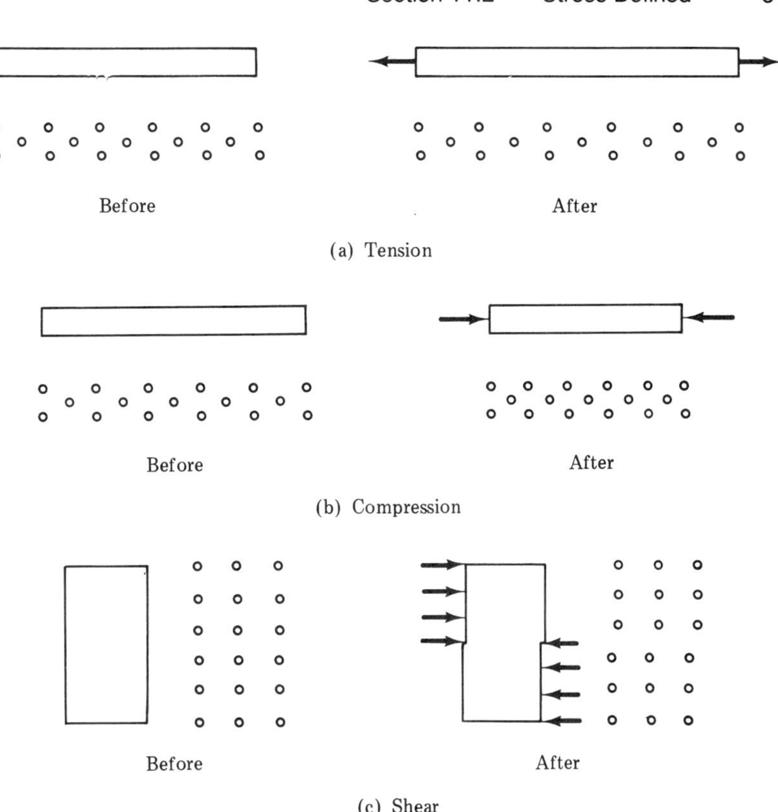

Figure 11-1. Types of stress.

spread apart further in tension, closer together in compression, and a shift in layers in shear.

In later chapters variations on the basic stresses (such as torsion, a form of shear) and combinations of the basic stresses (as occur in bending, a common phenomena) will be discussed.

The concept of a unit stress or stress per unit area requires some attention. Any solid material can support any size load if the load is spread out over a sufficient area. A paper cup will be crushed if a person attempts to stand on it. It can be argued that substituting steel to make the same size cup will solve the problem. However, it can also be argued that redesigning the paper cup with much thicker walls will also solve the problem. Ultimately, it is neither the load nor the size of the member it is applied to that solves the problem, but a combination of the two. Further, regarding the choice between two materials, it is the bonding force that is at issue. More force is required to separate (in any manner) the molecules of some materials (such as steel) than others (such as paper). In addition, other properties of the materials must be considered—the paper cup won't rust but a carbon steel cup would. Last but not least, economics must be considered. The material and manufacturing costs may be greater for a thick-walled paper cup than a thin-walled steel cup.

Back to stress, the distances and forces between molecules are very small quantities. Therefore, we deal with more practical quantities such as pounds, kips, newtons, and kilonewtons for force and square inches, square feet, and square meters for area. As Figure 11-2 shows, it is the cross-sectional area of the member under stress that determines how many pairs of molecules are being pulled apart, pushed together, or slid apart. The greater the area, the more pairs there are to resist the applied force. If the area is held constant, then the greater the total applied force, the greater the force acting between each pair of molecules.

Thus, the unit stress in a material, generally simply called stress, can be expressed as:

$$\sigma = \frac{F}{A}$$

where σ (the Greek letter sigma) stands for stress, F for force, and A for area. Units for stress are thus lb/in.^{2} (psi), kips/in.^{2} (ksi), N/m^2 (Pa), and kN/m^2 (kPa).

If the idea of unit stress is not clear to you, try the following simple experiments. First, lay one of your hands flat on a table, palm up. Gradually, one after another, pile books on your hand. The area to which the force is being applied is constant but the applied force is increasing. You can certainly feel the increased stress with increased load. Second, put only the first book back on your hand momentarily. Remove it and place a dull pencil vertically on your hand, point down. Place the same book on top of the pencil, balancing it but letting the pencil support it. Since the pencil has relatively little mass, the applied force is essentially the same as when the book was directly on your hand. However, you can certainly feel the higher level of stress when using the pencil and are probably glad it doesn't have a good sharp point! In this case, the force was kept constant but the area was

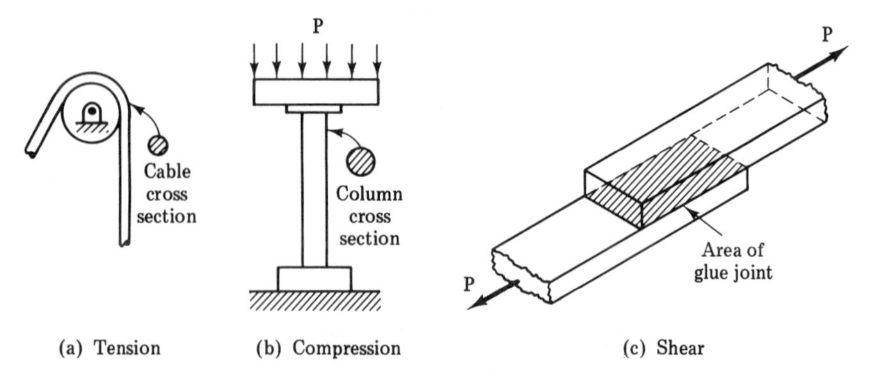

Figure 11-2. Unit stress.

reduced, thus increasing stress. Let us look at some examples and problems that will illustrate stress further.

Example 11-1: The cylindrical steel rod shown in Figure 11-3a has a specific weight of 495 lb/ft³. What is the stress in the rod neglecting its weight? including its weight?

Solution:

Step 1 Draw the free-body diagram as shown in Figure 11-3b.

Step 2 Sum the vertical forces to determine $\overline{R}_{AY}$ (K = kip):

$$\Sigma F_Y = 0 \qquad P + \overline{R}_{AY} = -P \qquad \overline{R}_{AY} = -(-20 \text{ Kip})$$

$$\overline{R}_{AY} = +20 \text{ Kip}$$

Step 3 Since P and $\overline{R}_{AY}$ are directed toward the rod, they are attempting to push it together. The rod is under compressive stress or "in compression."

Step 4 Determine the magnitude of the stress as follows:

$$\sigma = \frac{F}{A} = \frac{F}{0.7854(d)^2} \qquad \frac{20 \text{ Kip}}{0.7854(1.5)^2} \qquad \sigma = 11.32 \text{ ksi}$$

Step 5 The weight of the rod can be determined from

$$w = \gamma V = \gamma \pi r^2 h$$

$$w = 495 \text{ lb/ft}^3\ (\pi)\left(\frac{0.75}{12}\text{ ft}^2\right)(10 \text{ ft}) \qquad w = 81 \text{ lb}$$

$$w = 0.081 \text{ Kip}$$

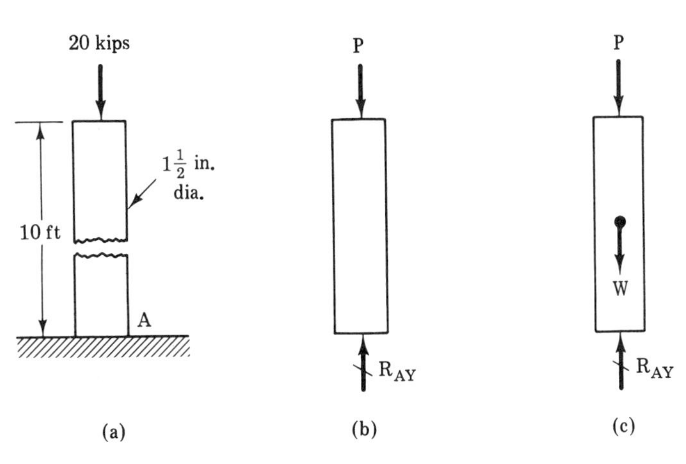

Figure 11-3

Step 6 Draw the free-body diagram as shown in Figure 11-3c.

Step 7 The weight acts through the center of gravity of the rod. However, at the very top of the rod there is no effect of the weight. Therefore, the stress will be the same as calculated in Step 4. However, the full effect of the weight is felt at the base of the rod. Summing vertical forces:

$$\Sigma F_Y = 0 \qquad P + W + \overline{R}_{AY} = 0 \qquad \overline{R}_{AY} = -(P + W)$$

$$\overline{R}_{AY} = -(-20 - 0.081 \text{ Kip}) \qquad \overline{R}_{AY} = +20.081 \text{ Kip}$$

Step 8 The stress at the bottom of the rod can be calculated by

$$\sigma = \frac{F}{A} = \frac{\overline{R}_{AY}}{A} = \frac{20.081 \text{ Kip}}{0.7854(1.5 \text{ in.}^2)}, \qquad \boldsymbol{\sigma = 11.36 \text{ ksi}}$$

Note that the answer obtained in Step 8 is only about one-third of one percent greater than the answer obtained in Step 4. This small difference is the basis for our generally ignoring the effects of the weights of structural or machine members.

Problem 11-1: A 200-kN weight is suspended from a rod as shown in Figure 11-4. Determine the stress in both the upper and lower portions of the rod.

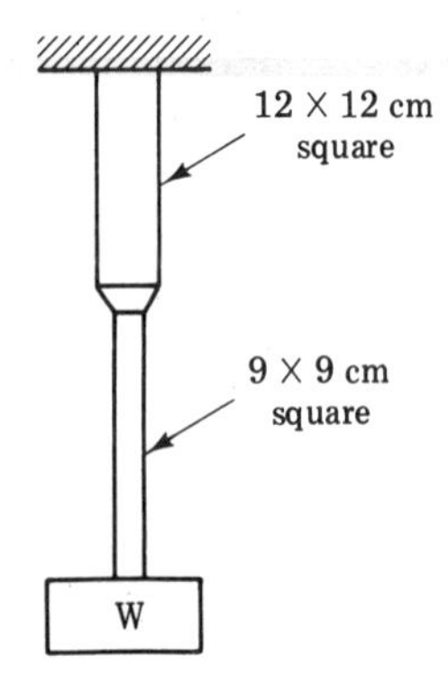

Figure 11-4

Example 11-2: Determine the shear stress in the glue joint shown in Figure 11-5 and the tensile stress in each of the pieces.

Solution:

Step 1 Identify the cross-sectional areas that are in tension and shear as follows:

$$A_A = (0.02 \text{ m})(0.01 \text{ m}) \qquad A_A = 0.0002 \text{ m}^2 \text{ in tension}$$

$$A_B = (0.02 \text{ m})(0.015 \text{ m}) \qquad A_B = 0.0003 \text{ m}^2 \text{ in tension}$$

$A_S = (0.02 \text{ m})(0.03 \text{ m})$ $\qquad A_S = 0.0006 \text{ m}^2$ in shear

Step 2 Recognize that the forces causing the two tensions and the shear are all equal.

Step 3 Determine the three stresses as follows:

$$\sigma_A = \frac{F}{A_A} = \frac{6 \text{ kN}}{0.0002 \text{ m}^2} \qquad \boldsymbol{\sigma_A = 30 \text{ MPa tension}}$$

$$\sigma_B = \frac{F}{A_B} = \frac{6 \text{ kN}}{0.0003 \text{ m}^2} \qquad \boldsymbol{\sigma_B = 20 \text{ MPa tension}}$$

$$\sigma_S = \frac{F}{A_S} = \frac{6 \text{ kN}}{0.0006 \text{ m}^2} \qquad \boldsymbol{\sigma_S = 10 \text{ MPa shear}}$$

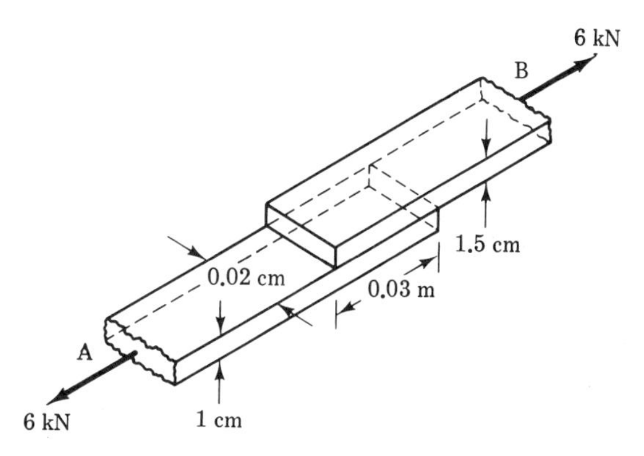

Figure 11-5

Problem 11-2: Determine the compressive stress in both sections of the column shown in Figure 11-6 and the shear stress in each bolt.

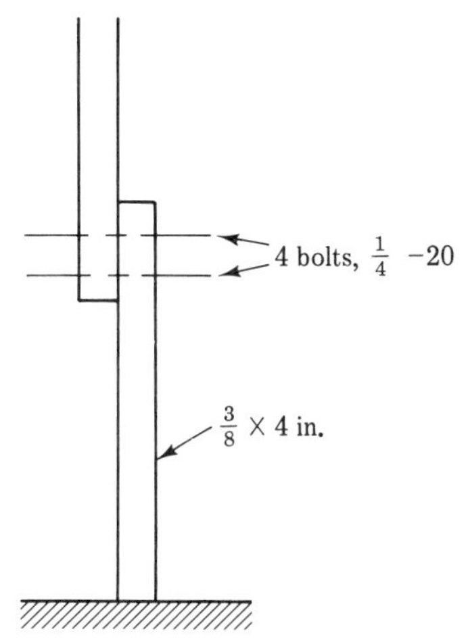

Figure 11-6

11.3 STRAIN DEFINED

When materials are under the effect of forces they change shape. This deformation is quite obvious when we stretch a rubber band. It is much less obvious, indeed unnoticeable, when we turn a cube of steel from one side to another. The new vertical dimension quickly reduces slightly as the force of gravity takes effect. The fact that the apparent effects are very different in magnitude is irrelevant. The nature of the effect is exactly the same: when equal and opposite forces are applied to a material it deforms or changes shape.

Measuring *deformation* can be very crude or very exacting. If you have a large wide rubber band, you can mark some length on it, say 3 in. or perhaps 76 mm, using a ballpoint pen. If you now stretch the rubber band and measure the distance between the two marks you may find it to be as much as 3.5 in. or perhaps 89 mm long. This change in length is sometimes designated as ΔL and often simply δ (delta).

The relevance of the deformation of 0.5 in. in the above case is much more significant than a deformation, or change of length, of 0.5 in. on an original length of 30 in. This relationship, the change in length proportioned to the original length, is known as *strain*. It is designated by the Greek letter ϵ (epsilon) and thus

$$\epsilon = \frac{\delta}{L}$$

where ϵ is strain, δ is total deformation, and L is original length. Strain is sometimes called unit deformation. As we look at the equation above, we can see why dividing total deformation by total original length will yield the amount each unit length was deformed. Thus the units for strain will be of the nature of in./in., ft/ft, mm/mm, or m/m.

Example 11-3: Determine the strain imposed on a 4-ft long, 2-in. diameter bar when a force of 7 Kip compresses it 2 in.

Step 1 Recognize that strain is a function of original length and total deformation. Thus the cross-sectional area and the force involved are not relevant to the stated problem.

Step 2 Determine strain as follows:

$$\epsilon = \frac{\delta}{L} = \frac{2 \text{ in.}}{48 \text{ in.}} \qquad \boldsymbol{\epsilon = 0.042 \textbf{ in./in.}}$$

Problem 11-3: A 200-m long cable is stretched 40 cm. Determine the strain in the cable.

Let us examine our shapes more closely now. As you stretch the rubber band, if you look carefully you will notice that not only does it get longer, but its cross section becomes narrower and thinner. These changes in dimensions are certainly a deformation. Further, we can also determine the original dimensions. Strain, by our earlier definition, has thus occurred in a direction perpendicular or transverse to the line of action of the applied forces. The transverse strain is related to the longitudinal strain (collinear with the line of action of the applied forces) previously identified. This relationship is a property of each material and was discovered by the French mathematician, Poisson (poy-sahn′). Known as Poisson's ratio, it is stated as

$$\mu = \frac{\epsilon_t}{\epsilon_l}$$

where μ is Poisson's ratio, ϵ_t is transverse strain, and ϵ_l is longitudinal strain. Some typical values for μ are given in Table 11-1.

Example 11-4: A 7-ft long wire 0.03 in in diameter made of copper is stretched 1 in. What is its new diameter?

Solution:

Step 1 Determine the longitudinal strain as follows:

$$\epsilon_l = \frac{\delta_l}{L} = \frac{1 \text{ in.}}{84 \text{ in.}} \qquad \epsilon_l = 1.19 \times 10^{-2} \text{ in./in.}$$

Step 2 Determine the transverse strain as follows:

$$\mu = \frac{\epsilon_t}{\epsilon_l} \qquad \epsilon_t = \mu\epsilon_l = 0.355\ (1.19 \times 10^{-2} \text{ in./in.})$$

$$\epsilon_t = 4.23 \times 10^{-3} \text{ in./in.}$$

Table 11-1
Poisson's Ratio

Material	μ
Aluminum alloys	0.330
Brass	0.340
Bronze	0.350
Cast iron	0.270
Concrete	0.200
Copper	0.355
Steels, carbon	0.288
Steels, stainless	0.305

Step 3 Determine the transverse deformation as follows:

$$\epsilon_t = \frac{\delta_t}{D_0} \qquad \delta_t = \epsilon_t D = 4.23 \times 10^{-3} \text{ in./in. } (0.03 \text{ in.})$$

$$\delta_t = 1.27 \times 10^{-4} \text{ in.}$$

Step 4 Determine the new diameter. Since it is being stretched, it will reduce in diameter. Thus

$$D_f = D_0 - \delta_t = 0.03 \text{ in.} - 1.27 \times 10^{-4} \text{ in.}$$

$$\mathbf{D_f = 0.0299 \text{ in.}}$$

Problem 11-4: A cube of steel 0.04 m on each side is compressed until its transverse dimensions each increase by 0.001 m. Determine the new longitudinal dimension.

11.4 RELATIONSHIP OF STRESS AND STRAIN

Stress and strain have been discussed in previous sections without relating them to each other. However, as we look again at our stretching of the rubber band it becomes apparent that no deformation takes place unless a force is applied. Further, the greater the applied force the greater the deformation. Thus deformation is a function of applied force.

$$\delta = f(F)$$

Earlier in this chapter we examined first the relationship between force and area, and next the relationship between original length and deformation. These relationships were stated as

$$\sigma = \frac{F}{A} \quad \text{and} \quad \epsilon = \frac{\delta}{L}$$

Returning to the rubber band, it can be seen from the above that if we use a thicker or wider rubber band, its cross-sectional area will increase and more force must be applied to achieve the same level of stress. On the other hand, if the cross section of the rubber band is kept constant and a constant force is applied, the longer the section of the length we examine the greater the total deformation will be. This leads to combining the previous equations as follows

$$\delta = f(F) \qquad \frac{\delta}{L} = f\left(\frac{F}{A}\right) \qquad \epsilon = f(\sigma)$$

This relationship is fundamental to all materials and is generally depicted graphically in what is known as a *stress-strain diagram.* Such a diagram can be established for any material by carefully conducted testing. Examination of such a diagram can tell us a great deal about the nature of a material. Figure 11-7 shows a typical stress-strain diagram.

The early or low-strain portion of the curve has special significance. For reasons that will quickly become apparent, any items that are to be used over and over again must not be stressed beyond this early range.

In the mid-seventeenth century, the English scientist Robert Hooke first observed that under initial stress the relationship between stress and strain was constant for a given material. It varied, however, from material to material. This may seem obvious to you, for you know instinctively that if the rubber band discussed earlier were made of steel with the same dimensions, you would be hard put to strain the steel much even though you may apply a great deal more force, and thus stress, to the steel than you did to the rubber band. Because of this discovery, the relationship is often called "Hooke's law."

Early in the nineteenth century, another Englishman, Robert Young, was able to determine the constant of proportionality that defined this

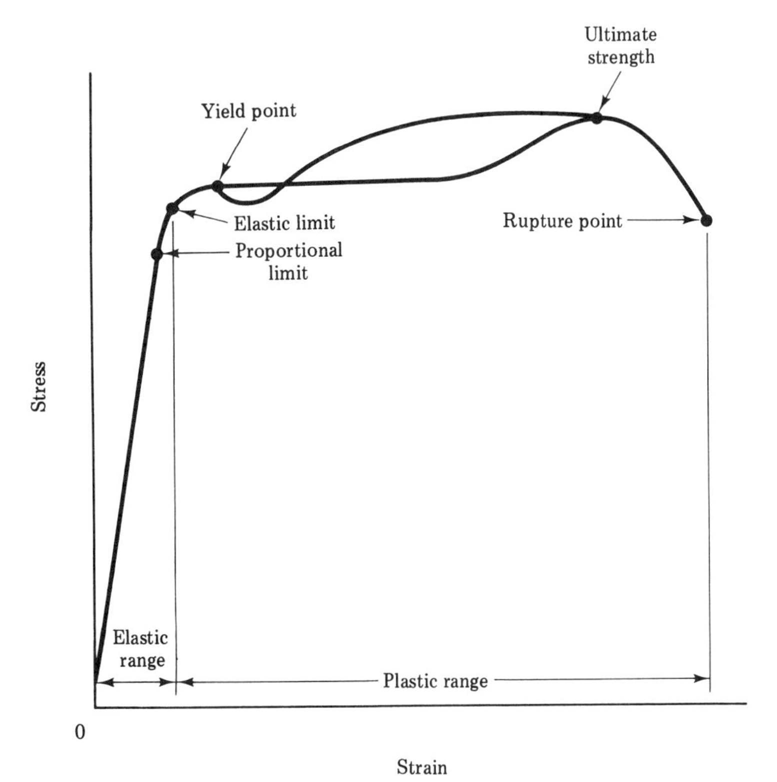

Figure 11-7. Stress-strain diagram.

straight-line relationship. Generally known as the modulus of elasticity, it is often called "Young's modulus" and denoted by the letter E. Thus

$$\sigma = E\epsilon \qquad E = \frac{\sigma}{\epsilon} \qquad E = \frac{F/A}{\delta/L} \qquad E = \frac{FL}{A\delta}$$

This relationship is also quite useful when expressed in terms of the deformation

$$\delta = \frac{FL}{EA} \qquad \delta = \sigma\frac{L}{E}$$

Since strain is really a dimensionless proportion, the modulus of elasticity has the same units as stress, force units divided by units of area, for example, psi or kPa. As shown in Figure 11-7, the straight-line relationship extends to a point called the *proportional limit.* The stress may be increased or decreased at will (below the stress level at the proportional limit) and the strain will follow proportionately. If any stress higher than that at the proportional limit is applied, the additional change in strain will not be proportional to the additional stress. In other words, above the proportional limit, Hooke's law and Young's modulus no longer apply. Typical values of E for many materials are found in the Appendices.

A slight digression is necessary here for you may have noticed that in Figure 11-7, stress is not zero when strain is zero. This is the result of testing procedures set forth by ASTM (American Society for Testing and Materials). These procedures require that a small initial load be placed on a specimen prior to taking any deformation readings. This is done to assure that the testing machine is firmly gripping the specimen and thus that none of the deformation recorded is slippage.

The next significant point on the stress-strain diagram is the *elastic limit.* Load may be applied and then relieved time and again if the stress is kept below that indicated at the elastic limit and the material will always return to its original shape when the load returns to zero. We need to know where this limit is for a material so that we can stay safely within the *elastic range* when a component of a machine or structure is under stress. The application of sufficient load to induce a stress above the level indicated for the elastic limit will result in permanent deformation. The atoms of the material have then been sufficiently rearranged to prevent their attractive forces from pulling them back into their original position. Old bonds have been broken and new bonds formed. It should be noted that for many materials, the proportional limit and the elastic limit are one.

As more load is applied to a material and stress increases, many materials evidence a phenomenon known as *yield.* The nature of yield is a flattening of the stress-strain curve, small increments of stress change causing large increments of strain change. The point where this begins to occur as stress increases is known as the *yield point*. Testing will clearly indicate this point for some materials. It has sufficient design significance that, for other materials, the yield point is determined mathematically. The procedure is called the 0.2% offset method. It consists of constructing a line parallel to the proportional section of the stress-strain diagram and passing through a point of zero stress and 0.002 strain. Where this line crosses the stress-strain curve is the accepted location of the yield point as shown in Figure 11-8.

As more load is applied stress increases to a maximum, known as the *ultimate strength*. Past this point, lateral strain becomes significant and the *rupture* or breaking point soon follows.

The strain range from the elastic limit to the rupture point is known as the *plastic range*. This is extremely important in many manufacturing processes, for in drawing out wire from rod or forging a crankshaft, for example, the elastic limit must be exceeded in order to reshape the material. However, the process must stay well short of the rupture point to assure that even small localized failures or cracks do not occur.

Example 11-5: An aluminum bar 1.5 m long with a rectangular cross section of 5 by 10 cm is subjected to a tensile force of 5000 kN along its longest axis. The modulus of elasticity for the aluminum is 70×10^6 kPa

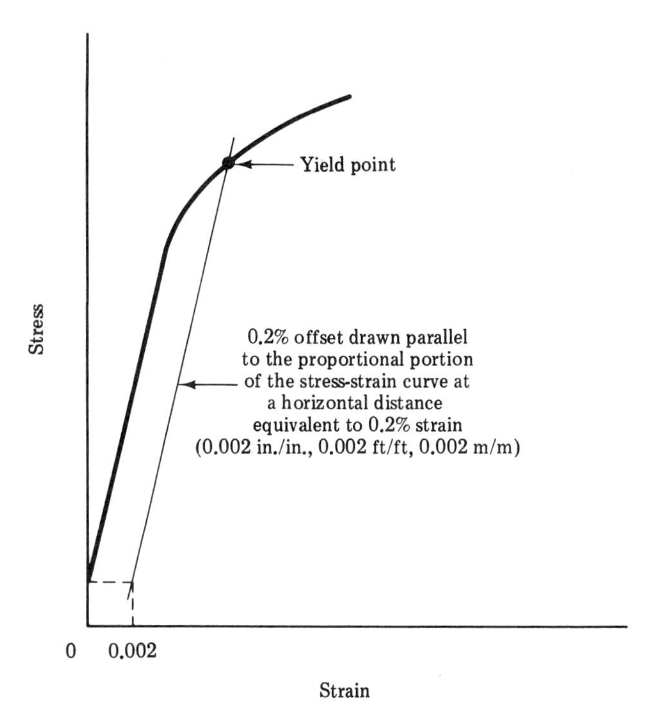

Figure 11-8. Yield point-offset method.

and Poisson's ratio is 0.33. Determine the deformation in all three directions.

Solution:

Step 1 Recognize that sufficient information is presented to determine deformation along the longest axis from $E = \sigma/\epsilon$. Once δ_l is determined, Poisson's ratio will permit us to find δ_w and δ_h for the cross section.

Step 2 Determine δ_l as follows:

$$E = \frac{\sigma}{\epsilon} = \frac{F/A}{\delta_l/L_l} = \frac{FL_l}{A_l}$$

$$\delta_l = \frac{FL_0}{AE} = \frac{5000 \text{ kN } (1.5 \text{ m})}{(0.05 \times 0.10 \text{ m})(70 \times 10^6 \text{ kPa})}$$

$$\delta_l = 0.0214 \text{ m} \qquad \boldsymbol{\delta_l = 21.4 \text{ mm}}$$

Step 3 Determine δ_w as follows:

$$\epsilon_w = \mu\epsilon_l \qquad \frac{\delta_w}{L_w} = \mu\frac{\delta_l}{L_l} \qquad \delta_w = \mu\delta_l\frac{L_w}{L_l}$$

$$\delta_w = 0.33(0.0214 \text{ m})\frac{.05 \text{ m}}{1.5 \text{ m}} \qquad \boldsymbol{\delta_w = -0.235 \text{ mm}}$$

Step 4 Determine δ_h as follows:

$$\epsilon_h = \mu\epsilon_l \qquad \frac{\delta_h}{L_h} = \mu\frac{\delta_l}{L_l} \qquad \delta_h = \mu\delta_l\frac{L_h}{L}$$

$$\delta_h = 0.33(0.0214 \text{ m})\frac{0.1 \text{ m}}{1.5 \text{ m}} \qquad \delta_h = -0.471 \text{ mm}$$

Note the minus signs. As the bar is being stretched, its cross section will contract.

Problem 11-5: A solid stainless steel post 4 ft long and 2 in. in diameter is subjected to a compressive load of 60 kip along its long axis. E is 28×10^6 psi and $\mu = 0.305$. Determine its new length and new diameter.

11.5 TEMPERATURE EFFECTS

Materials change their dimensions as temperature changes. This is generally stated in a one-dimensional sense as

$$\delta = \alpha l(\Delta t)$$

Table 11-2
Coefficients of Linear Thermal Expansion

Material	$\times 10^{-6}$ Per °F	$\times 10^{-6}$ Per °C
Aluminum	12.8	23.1
Brass	10.4	18.8
Brick	3.4	6.1
Bronze	10.1	18.1
Cast iron (gray)	5.9	10.6
Concrete	5.5	9.9
Copper	9.3	16.8
Fir (parallel to grain)	2.1	3.7
Magnesium alloys	16.	29.
Nickel	7.0	12.6
Pine	3.0	5.4
Steel, carbon	6.5	11.7
Steel, stainless	9.6	17.3

where δ is deformation as previously defined, l is original length, Δt is the change in temperature, and α is the *coefficient of thermal expansion* which is expressed in units of length divided by length times temperature, for example, in./in. °F. This deformation is for a material that is unrestrained, or free to expand or contract. Table 11-2 lists values of α for a number of materials.

Example 11-6: The dimensions shown for the two electrical contacts in Figure 11-9 are at 100 °F. At what temperature will they make contact?

Solution:

Step 1 Recognize that an increase in temperature is needed to bring the two contacts together.

Step 2 Recognize that the total deformation required is the sum of the individual deformations

$$\delta_t = \delta_b + \delta_c$$

Step 3 Write the equations for the individual deformations.

$$\delta_b = \alpha_b l_b(\Delta t) \qquad \delta_c = \alpha_c l_c(\Delta t)$$

Step 4 Recognize that Δt will be the same for both materials and can be written as $t_f - t_i$. Combine the above equations and solve for t_f.

$$\delta_t = \alpha_b l_b (t_f - t_i) + \alpha_c l_c (t_f - t_i)$$

$$\delta_t = (t_f - t_i)(\alpha_b l_b + \alpha_c l_c)$$

$$t_f = t_i + \frac{\delta_t}{\alpha_b l_b + \alpha_c l_c}$$

$$t_f = 100\ ^\circ\text{F} + \frac{0.005\ \text{in.}}{(10.4 \times 10^{-6}\ \text{in./in.}\ ^\circ\text{F})(4\ \text{in.}) + (9.2 \times 10^{-6}\ \text{in./in.}\ ^\circ\text{F})(6\ \text{in.})}$$

$$\mathbf{t_f = 152\ ^\circ F}$$

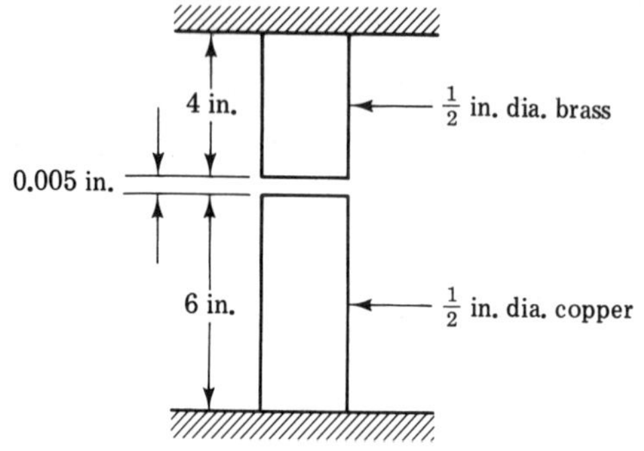

Figure 11-9

Problem 11-6: The load on the steel column shown in Figure 11-10 is free to move while the base under the concrete column is not. How much will the load be lowered with a drop in temperature of 150 °C?

Frequently problems arise involving the combination of deformation resulting from load and deformation resulting from temperature change. In such cases, total deformation is expressed as

$$\delta_{\text{total}} = \delta_{\text{stress}} + \delta_{\text{temp}}$$

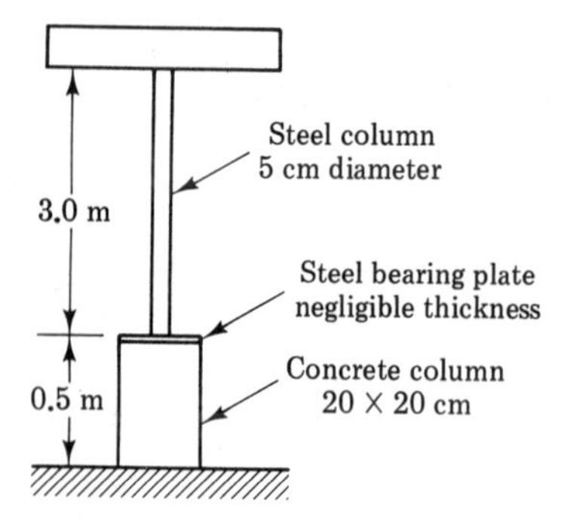

Figure 11-10

or total deformation equals the sum of the deformation due to load and the deformation due to temperature change.

Example 11-7: The two rods shown in Figure 11-11 fit snugly at the stated dimensions at 0 °C. What will be the new length of each rod if the walls are fixed and the temperature rises to 200 °C?

Solution:

Step 1 Recognize that no change in total length will take place. Total deformation is zero. Therefore the potential free thermal expansion is offset by the compression resulting from the fixed walls.

$$\delta_{\text{total}} = \delta_{\text{stress}} + \delta_{\text{temp}} = 0$$

Since one deformation is an expansion and the other a contraction

$$\delta_{\text{stress}} = -\delta_{\text{temp}}$$

Step 2 Recognize that the resultant stresses in the two rods must be proportional to their area since *the resultant compressive force is constant throughout both rods.*

$$F = \sigma_M A_M = \sigma_{CI} A_{CI}$$

Step 3 Recognize that if a definite proportion exists between the stresses then a definite proportion must exist between the deformations due to stress. Moreover, we already have a relationship for these deformations since their sum is equal and opposite to the "thwarted" thermal expansion

$$\delta_{\substack{\text{stress}\\\text{total}}} = \delta_{\substack{\text{stress}\\\text{Monel}}} + \delta_{\substack{\text{stress}\\CI}} = -\delta_{\substack{\text{temp}\\\text{total}}} = -\delta_{\substack{\text{temp}\\\text{Monel}}} - \delta_{\substack{\text{temp}\\CI}}$$

Step 4 Determine the potential total thermal deformation as follows:

$$\delta_{\substack{\text{temp}\\\text{total}}} = \alpha_M l_M \Delta t_M + \alpha_{CI} l_{CI} \Delta t_{CI}$$

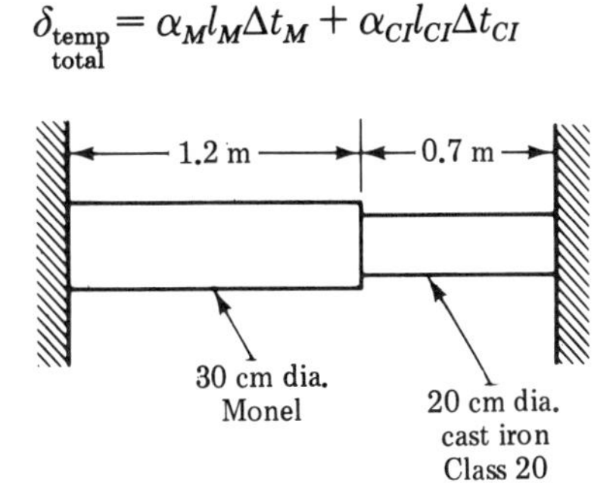

Figure 11-11

$$\delta_{\substack{\text{temp}\\\text{total}}} = (14.0 \times 10^{-6}\ \text{m/m °C})(1.2\ \text{m})(200\ \text{°C}) + (11.3 \times 10^{-6}\ \text{m/m °C})(0.7\ \text{m})(200\ \text{°C})$$

$$\delta_{\substack{\text{temp}\\\text{total}}} = 3360 \times 10^{-6} + 1582 \times 10^{-6}\ \text{m}$$

$$\delta_{\substack{\text{temp}\\\text{total}}} = +4942 \times 10^{-6}\ \text{m} = -\delta_{\substack{\text{stress}\\\text{total}}}$$

Step 5 Determine the compressive force as follows:

$$\delta_{\substack{\text{stress}\\\text{total}}} = \delta_M + \delta_{CI} = \frac{FL_M}{A_M E_M} + \frac{FL_{CI}}{A_{CI} E_{CI}}$$

$$F = \frac{\delta_{\text{stress total}}}{\dfrac{L_M}{A_M E_M} + \dfrac{L_{CI}}{A_{CI} E_{CI}}}$$

$$F = \frac{4.942 \times 10^{-3}\ \text{m}}{\dfrac{1.2\ \text{m}}{0.7854(0.3\ \text{m})^2(180 \times 10^9\ \text{N/m}^2)} + \dfrac{0.7\ \text{m}}{0.7854(0.2\ \text{m})^2(75 \times 10^9\ \text{N/m}^2)}}$$

$$F = 12.63 \times 10^6\ \text{N}$$

Step 6 Determine the individual deformations due to stress

$$\delta_M = \frac{F_M L_M}{A_M E_M} = \frac{(12.63 \times 10^6\ \text{N})(1.20336\ \text{m})}{.7854(0.2\ \text{m})^2\ (180 \times 10^9\ \text{N/m}^2)}$$

$$\delta_M = 0.00119\ \text{m}$$

$$\delta_{CI} = \frac{F_{CI} L_{CI}}{A_{CI} E_{CI}} = \frac{(12.63 \times 10^6\ \text{N})(0.70158\ \text{m})}{0.7854(0.2\ \text{m})^2\ (75 \times 10^9\ \text{N/m}^2)}$$

$$\delta_M = 0.00376\ \text{m}$$

Note that the thermally expanded lengths were used to calculate the deformation due to compression.

Step 7 Determine the new lengths:

$$L_{fM} = L_{iM} + \delta_{\text{temp}} + \delta_{\text{stress}}$$

$$L_{fM} = 1.2 + 0.00336 - 0.00119\ \text{m} \qquad \boldsymbol{L_{fM} = 1.202\ \text{m}}$$

$$L_{fCI} = L_{iCI} + \delta_{\text{temp}} + \delta_{\text{stress}}$$

$$L_{fCI} = 0.7 + 0.00158 - 0.00376\ \text{m} \qquad \boldsymbol{L_{fCI} = 0.698\ \text{m}}$$

The reason the cast iron shortened and the Monel lengthened is that the cast iron is less resistant to deformation due to stress.

Problem 11-7: At a temperature of 150 °F, a 1020 steel shaft and a 302 stainless steel wheel have the dimensions shown in Figure 11-12. Since the shaft is larger, it must be chilled to permit assembly. To what temperature must the shaft be chilled to permit 0.001 in radial clearance during assembly if the wheel will be at a minimum of 50 °F during assembly?

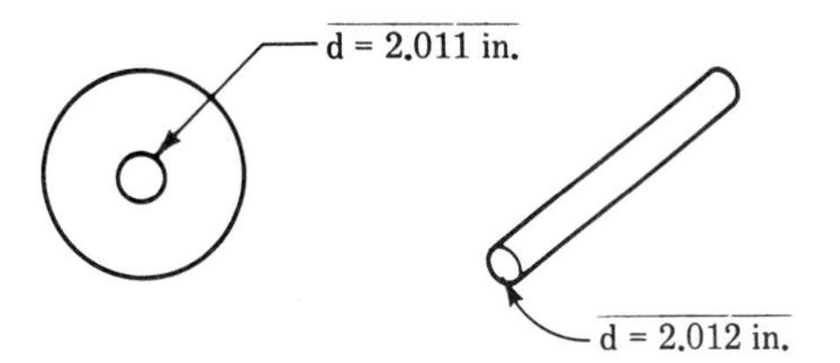

Figure 11-12

11.6 MULTIMATERIAL MEMBERS

Up to this point, we have considered only members or components fabricated from one material. However, members are frequently made of more than one material, for example, floors, columns, and beams made of steel-reinforced concrete.

The way in which the materials are combined must be related to the direction of the load in order to solve a problem. Figure 11-13 shows three possible arrangements.

If the series arrangement looks familiar, it should, since we already dealt with it in Example 11-7. The following summary of the nature of these combinations will be useful in solving subsequent problems.

$$\textit{Series: } F_1 = F_2 \qquad \delta_{\text{total}} = \delta_1 + \delta_2$$

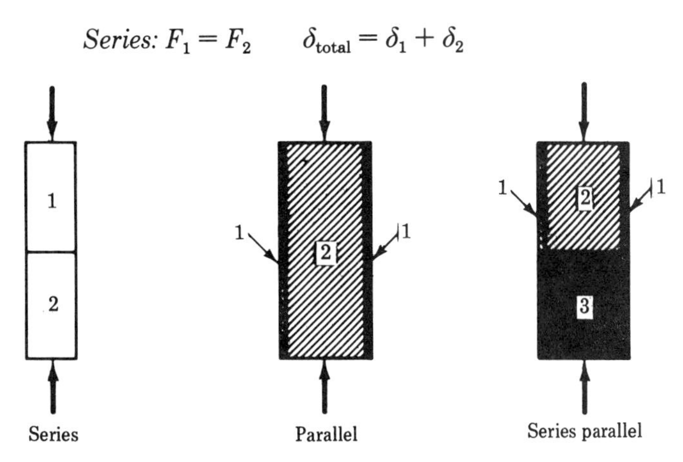

Figure 11-13. Multimaterial members.

$$\textit{Parallel: } F_{\text{total}} = F_1 + F_2 = A_1\sigma_1 + A_2\sigma_2$$

$$\delta_1 = \delta_2 \qquad L_1 = L_2 \qquad \epsilon_1 = \epsilon_2 \qquad \frac{\sigma_1}{E_1} = \frac{\sigma_2}{E_2}$$

$$\textit{Series Parallel: } F_3 = F_1 + F_2 \qquad \delta_{\text{total}} = \delta_1 + \delta_3 = \delta_2 + \delta_3$$

The fact that each combination is shown in compression is irrelevant. The same relationships would hold true for tension.

Example 11-8: Determine the total deformation for the rod shown in Figure 11-14 if the dimensions given are for a no-load condition.

Solution:

Step 1 Recognize that the materials are in series with respect to the load.

Step 2 Since $F_{Al} = F_{br}$ and $\delta_{\text{total}} = \delta_{Al} + \delta_{br}$ and $E = \sigma/\epsilon$, $\sigma = F/A$ and $\epsilon = \delta/L$, δ_{total} can be found as follows:

$$\delta_{\text{total}} = \delta_{Al} + \delta_{br}, \qquad \delta_{\text{total}} = \epsilon_{Al}L_{Al} + \epsilon_{br}L_{br}$$

$$\delta_{\text{total}} = \frac{\sigma_{Al}L_{Al}}{E_{Al}} + \frac{\sigma_{br}L_{br}}{E_{br}}$$

$$\delta_{\text{total}} = \frac{FL_{Al}}{A_{Al}E_{Al}} + \frac{FL_{br}}{A_{br}E_{br}}$$

$$\delta_{\text{total}} = \frac{(40{,}000\text{ lb})(22\text{ in.})}{0.7854(3\text{ in.}^2)(10.4 \times 10^6\text{ lb/in.}^2)} + \frac{(40{,}000\text{ lb})(13\text{ in.})}{(2\text{ in.}^2)(15 \times 10^6\text{ lb/in.}^2)}$$

$$\boldsymbol{\delta_{\text{total}} = 0.021\text{ in.}}$$

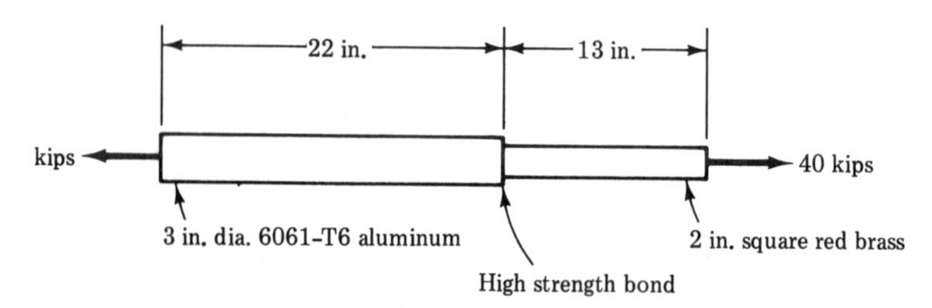

Figure 11-14

Problem 11-8: What load is required to cause a total deformation of −0.5 mm in the column shown in Figure 11-15?

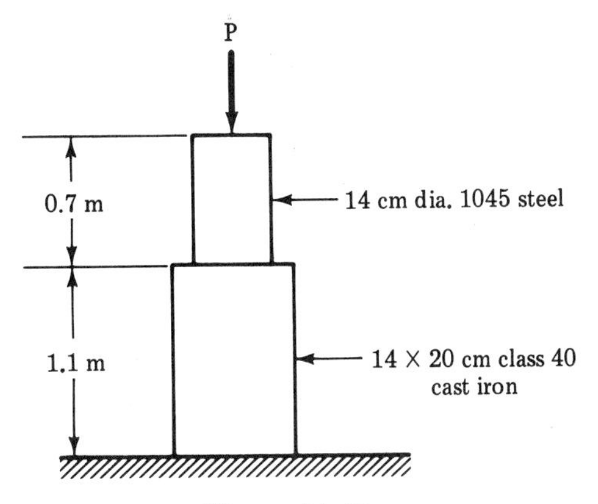

Figure 11-15

Example 11-9: Determine the stress in both materials in the member shown in Figure 11-16, as well as the deformation.

Solution:

Step 1 Recognize that the materials are in parallel with respect to the load.

Step 2 Since $\delta_f = \delta_s$ and $L_f = L_s$, $\epsilon_1 = \epsilon_2$ and thus $\sigma_f/E_f = \sigma_s/E_s$.

Step 3 Also, $F_{\text{total}} = F_f + F_s = A_f\sigma_f + A_s\sigma_s$

Step 4 By substitution, determine σ_f

$$\sigma_s = \frac{E_s}{E_f}\sigma_f \qquad F_{\text{total}} = A_f\sigma_f + A_s\sigma_f\frac{E_s}{E_f}$$

$$F_{\text{total}} = \sigma_f\left(A_f + A_s\frac{E_s}{E_f}\right) \qquad \sigma_f = \frac{F_{\text{total}}}{A_f + A_s\dfrac{E_s}{E_f}}$$

$$\sigma_f = \frac{50\text{ kN}}{(0.5\text{ m})(0.16\text{ m}) + 2(0.005\text{ m})(.16\text{ m})\left(\dfrac{200 \times 10^9\text{ Pa}}{8.3 \times 10^9\text{ Pa}}\right)}$$

$$\boldsymbol{\sigma_f = 5.96\ \text{MPa}}$$

Step 5 Determine σ_s as follows:

$$\sigma_s = \frac{E_s}{E_f}\sigma_f \qquad \sigma_s = \frac{200 \times 10^9\text{ Pa}}{8.3 \times 10^9\text{ Pa}} \qquad (5.96\text{ MPa})$$

$$\boldsymbol{\sigma_s = 144\ \text{MPa}}$$

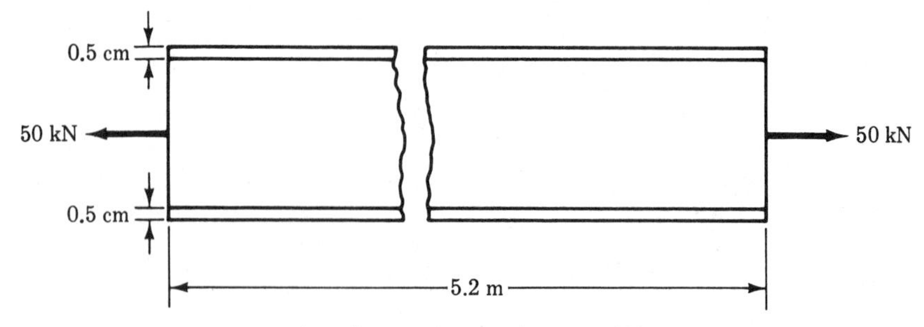

Figure 11-16

Step 6 Determine δ_f, δ_s as follows:

$$\delta_f = \delta_s = \epsilon_f L_f = \epsilon_s L_s = \frac{\sigma_f L_f}{E_f} = \frac{\sigma_s L_s}{E_s}$$

$$\delta_f = \frac{5.96 \text{ MPa}(5.2 \text{ m})}{8.3 \times 10^9 \text{ Pa}} \qquad \boldsymbol{\delta_f = 0.00373 \text{ m}}$$

$$\text{check: } \delta_s = \frac{144 \text{ MPa}(5.2 \text{ m})}{200 \times 10^9 \text{ Pa}} \qquad \boldsymbol{\delta_s = 0.00374 \text{ m}}$$

The minor difference of much less than 1% is undoubtedly due to rounding variations and is negligible.

Problem 11-9: What minimum thickness (in inches) of 6061-T6 aluminum needs to be added to both sides of the spruce column shown in Figure 11-17 to limit the stress in the spruce to a maximum of 750 psi? What will the deformation be?

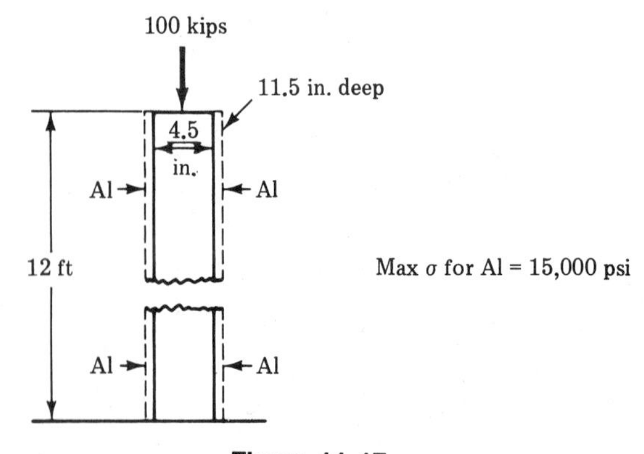

Figure 11-17

Problem 11-10: Determine the stress in the wood, concrete, and steel for the column and base shown in Figure 11-18. Determine the total deformation.

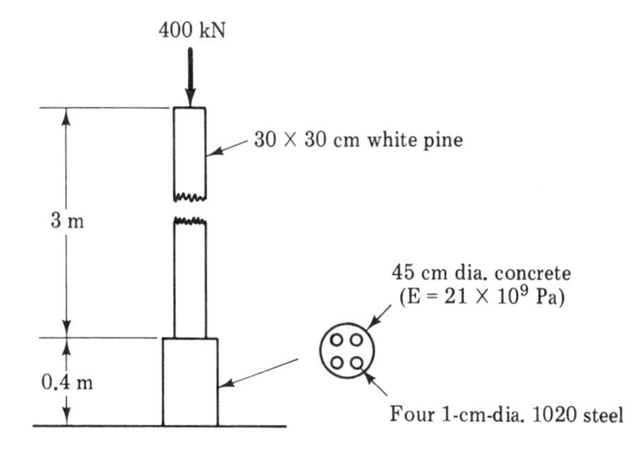

Figure 11-18

11.7 READINESS QUIZ

1. As the compressive load on a column increases
   (a) stress increases
   (b) strain increases
   (c) deformation increases
   (d) all of the above
   (e) none of the above
2. Strain is simply
   (a) the force between two molecules
   (b) unit stress
   (c) unit deformation
   (d) all of the above
   (e) none of the above
3. If a rod is fixed at both ends heating it will not result in any
   (a) overall deformation
   (b) stress
   (c) strain
   (d) all of the above
   (e) none of the above
4. Under a given load, substituting a stronger material of the same size will always result in
   (a) less stress
   (b) less strain
   (c) less deformation
   (d) all of the above
   (e) none of the above

5. The elastic modulus of a material is a ratio of

   (a) strain over stress
   (b) stress over strain
   (c) deformation over original length
   (d) force over area
   (e) none of the above

6. The proportional limit for a material is always ________ the elastic limit.

   (a) equal to
   (b) greater than
   (c) less than
   (d) none of the above
   (e) insufficient information given

7. Poisson's ratio relates

   (a) lateral strain to stress
   (b) axial strain to stress
   (c) lateral stress to axial stress
   (d) lateral strain to axial strain
   (e) none of the above

8. If two materials in a member are in parallel with respect to the load then

   (a) $F_1 = F_2$
   (b) $\delta_1 = \delta_2$
   (c) $\sigma_1 = \sigma_2$
   (d) all of the above
   (e) none of the above

9. The elastic range of a material is always that range of strain

   (a) between zero and the breaking point
   (b) throughout which stress over strain is a constant
   (c) that includes all strains that a material can be subjected to and still return to its original shape upon unloading
   (d) all of the above
   (e) none of the above

10. Young's modulus is another name for

    (a) Poisson's ratio
    (b) modulus of elasticity
    (c) modulus of rigidity
    (d) thermal expansion coefficient
    (e) none of the above

11. The offset method is a mathematical tool used to find

(a) rupture strength
(b) ultimate strength
(c) yield strength
(d) all of the above
(e) none of the above

12. Stress will occur as a result of ________ load.

(a) tensile
(b) compressive
(c) shear
(d) any of the above
(e) none of the above

13. If the cross-sectional area perpendicular to a load in a member is increased, ________ will also increase.

(a) stress
(b) strain
(c) deformation
(d) all of the above
(e) none of the above

14. The yield point of some materials is noticeable on the stress-strain curve by the fact that

(a) strain increases with no increase in stress
(b) strain becomes proportional to stress
(c) stress over strain increases noticeably
(d) all of the above
(e) none of the above

15. If two materials in a member are in series with respect to the load then the following is always true:

(a) $\delta_1 = \delta_2$
(b) $\epsilon_1 = \epsilon_2$
(c) $\sigma_1 = \sigma_2$
(d) all of the above
(e) none of the above

16. The modulus of elasticity of cast iron is ten times that of pine. Therefore, we can say that cast iron is ________ as strong as pine.

(a) ten times
(b) one hundred times
(c) one-tenth

(d) one-hundredth
(e) none of the above

17. Poisson's ratio for a material is 0.33. If a 10-in. long, 1-in. diameter member is under tension along its long axis

(a) lateral strain is 0.33 in.
(b) lateral strain is 3.3 in.
(c) lateral strain is 0.33 times axial strain
(d) lateral strain is 3.3 times axial strain
(e) none of the above

18. The plastic range of the stress-strain diagram provides

(a) an indication of the workability of a material when we attempt to permanently change its shape
(b) the strength of the material
(c) the elastic modulus of the material
(d) all of the above
(e) none of the above

19. Dividing change in stress by change in strain can give us

(a) the proportional limit
(b) the modulus of elasticity
(c) Poisson's ratio
(d) the elastic limit
(e) none of the above

20. An axe applies a load of 25 lb in the plane of the smallest cross section of a $2 \times 2 \times 20$ in. piece of wood. The resultant shear stress is

(a) 6.25 psi
(b) 12.5 psi
(c) 1.25 psi
(d) 0.625 psi
(e) none of the above

11.8 SUPPLEMENTARY PROBLEMS

11-11 A 1-in. diameter punch is to punch holes in $\frac{1}{8}$-in.-thick plate. This requires a force of 35 Kip. What are the compressive stress in the punch and the shear stress in the plate?

11-12 A steel surveyor's tape is 50 m long, 0.8 cm wide, and 0.05 cm thick when fully supported at 20 °C with a 50-N pull on it. What are its dimensions at 40 °C with a 150-N pull?

11-13 Determine the tensile stress and the lateral strain in each rod and the shear stress in the connecting pin shown in Figure 11-19.

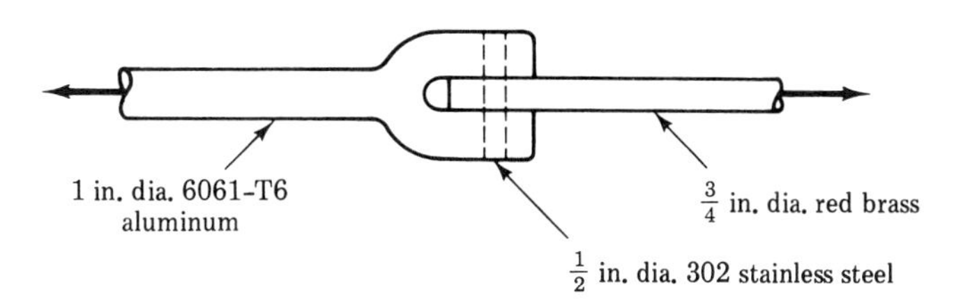

Figure 11-19

11-14 An electrical conductor consists of 6061 aluminum with an outside diameter of $\frac{1}{2}$ in. and a core of $\frac{1}{8}$-in.-diameter 1020 steel. Under a tensile load of 5000 lb, what is the stress in the aluminum? In the steel?

11-15 The dimensions given in Figure 11-20 are at 10 °C. At 40 °C, what are the shear stress in the 1045 steel, the shear stress in the yellow brass, the compressive stress in the 1095 steel, and the lateral strain in the 1095 steel?

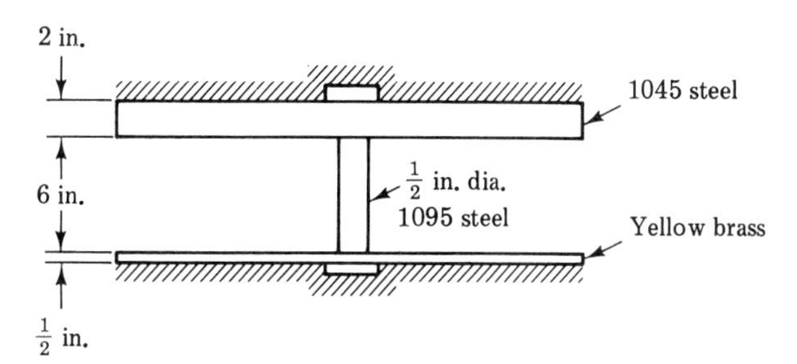

Figure 11-20

Chapter 12

Nature of Materials

12.1 OBJECTIVES

Upon completion of the work relating to this chapter, you should be able to perform the following.

1. For the words listed below:

   (a) Select from several definitions the correct one for a given word.
   (b) Given a definition, select the correct word from a choice of words.

allowable stress	factor of safety	plasticity
alloy	fatigue	stiffness
brittleness	hardness	toughness
creep	malleability	ultimate stress
ductility	modulus of elasticity	working stress
elasticity	modulus of rigidity	yield stress

2. Identify the factors involved in the selection of materials.
3. Identify the properties of wood, stone, brick, concrete, metals, and plastics.
4. Apply the factor of safety to problems.

12.2 PROPERTIES OF MATERIALS

Cost is inevitably a major determinant in selecting a material. However, it cannot be considered until we have narrowed down the choices by matching the properties of a material to the application. While a study of materials is beyond the scope and intent of this text, the following discussion should provide at least a basic foundation and illustrate the need for further study.

Elasticity

In a previous chapter we discussed the elastic range and the elastic limit of a material. These are illustrated again in Figure 12-1. The slope of the elastic portion of the stress-strain curve is known as the modulus of elasticity, E. Recall that a material may be loaded and unloaded time and again with no permanent change in its unloaded dimension as long as it is not loaded above its elastic limit. A high elastic limit is generally quite desirable since the higher it is the less material we need for a given load. The components of a structure or machine must never be permanently deformed under load. Generally a high modulus of elasticity is also desirable. In order to maintain dimensions and clearances within an acceptable range, the smaller the change in strain for a given change in stress, the better. It is worth noting that E is the same for a material in either tension or compression. The property of a material that relates to E is often called *stiffness*. An exception to the general desire for a high degree of stiffness should be kept in mind. Materials with a low modulus of elasticity exhibit a relatively high degree of strain for a given change in stress. This would be highly desirable if we are seeking a material that will absorb impact or shock loads (assuming substantial dimensional variations are acceptable). This is sometimes considered the inverse of stiffness and called *resilience*.

Plasticity

The plastic range of the stress-strain diagram was also previously discussed and is illustrated in Figure 12-1. The plasticity of a material is a measure of how great this range is. It is extremely important in the reshaping of a material by forging (hammering), bending, stretching, rolling, drawing (pulling through a die), and extruding (pushing through a die). As might be expected, the plasticity of a material varies with temperature and sometimes materials are hot worked rather than cold worked. Hot working is done at a relatively high temperature but, of course, below the melting point of a material. Cold working is done at temperatures in the neighborhood of what we normally consider room temperature. If the process of working on the

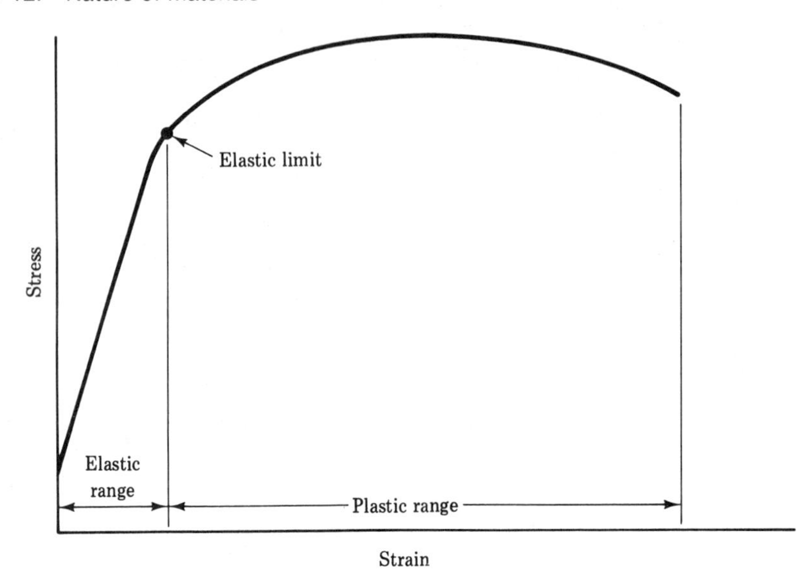

Figure 12-1. Typical stress-strain diagram.

material involves compressive forces, as in forging or rolling, the plasticity of the material is more commonly referred to as *malleability.* On the other hand, if tensile forces are applied to work the material as in drawing or stretching, then the plasticity of the material is commonly called *ductility.* Since the plastic range for a material may be rather different in tension than in compression, it may be either more or less ductile than it is malleable. This is further complicated by the fact that the nature of the working, hot or cold, tensile or compressive, affects the final properties of the material. Once again, this is an area best left to a materials text and course.

Toughness

Our discussion has generally assumed slowly applied or constant-load situations. However, many machine components experience very rapidly applied loads often called impact loads. An air hammer used to break up concrete and an internal combustion engine with its high pressure explosions are two common examples of this.

There are tests, known as impact tests, that break a test specimen and measure the energy expended in breaking the specimen. These values provide us with a convenient means of comparing the toughness of materials.

The above testing does not really define toughness. This can be better accomplished by examining the stress-strain diagrams for several materials.

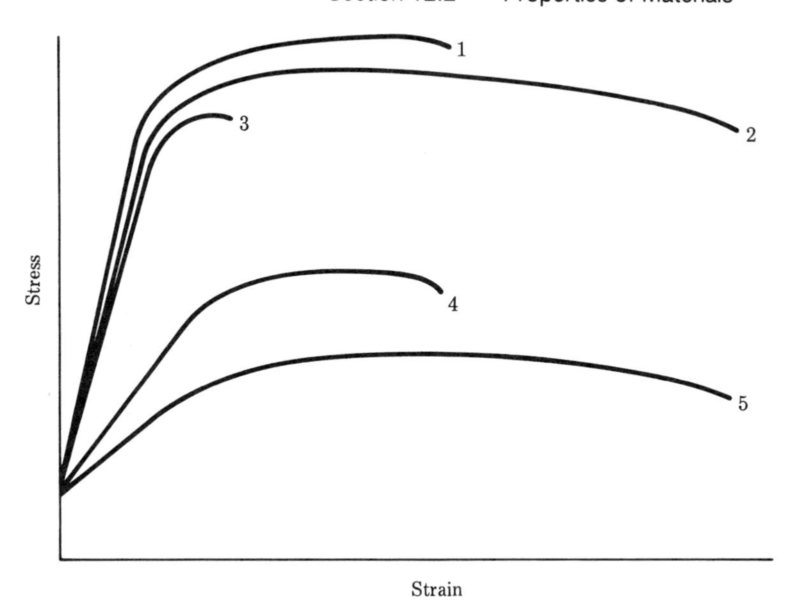

Figure 12-2. Comparative stress-strain diagram.

Figure 12-2 shows the stress-strain diagrams for five different materials plotted together. Careful examination shows that E is fairly similar for materials 1, 2, and 3. Materials 4 and 5 have fairly similar values for E but at a lower value than for 1, 2, and 3. The elastic ranges vary somewhat among the materials, but the plastic ranges and strengths vary greatly. Thus materials 4 and 1 have similar plastic ranges but greatly different strengths. However, the strengths of 1 and 2 are quite similar. Material 3 is quite different from the others in that it has virtually no plastic range. Such a material is considered *brittle,* and while it can be machined successfully, it cannot be reshaped by deforming it with force. Ordinary cast iron is a good example of this.

What then of toughness? Let us consider the source of the stress-strain diagram, a test where force is applied through a distance. The amount of force varies as the distance changes, but the fact remains that we could easily plot the data as a force versus distance. The area under the curve would be the work done (or energy expended) on the specimen to break it. Thus, the total area under the stress-strain curve is also a measure of the energy expended in breaking a specimen and, therefore, a measure of the toughness of a material. Referring to Figure 12-2, it would appear that material 2 is the toughest, material 1 about half as tough as 2, material 4 about one-quarter as tough as 2, and material 5 about a third as tough as 2. Material 3, while rather high in strength and having a relatively high modulus of elasticity, has a very low toughness, perhaps one-eighth that of 2.

Rigidity

Earlier, stiffness was described as the ability to resist deformation under tensile or compressive loads (in the elastic range) and was thus related to the modulus of elasticity, E. We have also discussed a third means of loading a material, shear, where the load is attempting to slide layers of molecules along one another. The resistance to shear deformation is known as rigidity and is measured by the shear modulus, or *modulus of rigidity,* designated as G. The phenomenon of shear is found in some pin or bolt connections such as the pin connecting your watchband to your watch case—any pulling on the band tends to cut through or shear the connecting pin. Punching operations as simple as a paper punch are another example. Shear also occurs with circular direction. When a torque is put on a shaft, there is a tendency for layers of molecules to slide on one another with circular motion.

A torsion test performed on a material can be used to determine G. It is interesting to note that for a given material E and G, while not the same, do have a definite relationship which can be expressed as follows:

$$G = \frac{E}{2(1 + \mu)}$$

μ being Poisson's ratio which we discussed earlier. If you review that section in conjunction with the above you will note an interesting phenomenon. Materials that exhibit little necking down or thickening under tension or compression have a relatively high modulus of elasticity. This is particularly common among brittle materials. Note from the above equation, though, that G can never be greater than one-half E.

Hardness

Hardness is generally defined as the ability of a material to resist wear or penetration. Since wear is a rubbing or abrasion resulting from forces applied parallel to a surface, those forces are shear forces. Penetration, on the other hand, involves compression and some shear. One significant outcome of the above is that hardness of a material is directly related to its strength. Another outcome is that hardness (and indirectly strength) can be tested by a penetration-type hardness test. In this test, a known force pushes a very hard material with a specific shape into the material being tested. The depth (or sometimes the diameter) of the indentation is used as a measure of hardness.

Fatigue

Many machine components are repeatedly loaded and unloaded many times during that lifetime. This raising and lowering of the stress on the material results in a change in its properties. Thus the design strength used for a material under such conditions would not be the same design strength used under static load conditions. A specialized piece of equipment, appropriately called a fatigue tester, is used to evaluate this property. It takes a standard size specimen, stresses it to a stress level below the elastic limit and then relieves the stress. It repeats this very rapidly until the specimen breaks, counting the number of cycles required to break it.

Creep

Creep is a phenomenon that, in a sense, is opposite to impact. It is not that the load is applied very slowly, but that, once it is applied (below the elastic limit) deformation is not a constant with constant load. Instead, over a long period of time deformation very slowly increases. This effect is much more marked as temperature rises. Gas turbine blades are an excellent example of an application where creep must be evaluated. As might be expected, materials vary in both their creep strength and the way in which they react to temperature change. The equipment that tests for this property is known simply as a creep tester.

Additional Properties

There are, of course, many other properties that a designer must take into account in selecting a material. Just a few of these are electrical and thermal conductivity, weight, corrosion resistance, color, weldability, and castability. However, the ones discussed above are those of primary concern in this text along with the general consideration of strength discussed earlier.

12.3 TYPES OF MATERIALS

Much of the foregoing discussion referenced metals, and for good reason; they make up both a large and significant portion of the materials used. However, their use was certainly preceded by the use of stone and wood, followed in time by brick and later concrete. In the last half century the use of plastics has mushroomed. The fact that this group of materials is often maligned is generally the result of improper selection.

Stone

One of the two oldest materials used by man, even the selection of stone required a judgment that comes from experience. Some types are suitable for structures, whether to carry loads or as decoration. Others are suitable for tools but the selection varies depending on whether one wishes to pound, cut, or grind. While the use of stone as a tool material is very limited, its use as a structural material in buildings, roads, and bridges is still extensive, as support, decoration, or as an ingredient for concrete.

Wood

Wood, the other "old" material also still sees extensive use. It has the disadvantage that it is not inert (unaffected by the environment) and must be protected. It is frequently an economical alternative and in the long run is a renewable resource when properly managed. Its orientation in use must be considered since its properties are substantially different when measured across the grain as opposed to with the grain. It can be combined by nailing, bolting, or gluing into a wide variety of structural shapes, some of which are even ornamental as well as functional. It can be made both resistant to decay and fire resistant by chemical treatment. Impregnation with a monomer and subsequent exposure to gamma radiation can increase its strength substantially.

Brick and Refractories

Early in history people observed the sun making a natural building material other than stone and wood—sun-baked clay bricks. They learned to fashion these into more desirable shapes, discovered, probably by accident, that baking these in a fire produced stronger brick than sun baking. Eventually, they discovered that some clays made better bricks than others. Others learned to shape the clay into pots, urns, and other useful objects before firing or baking them. Today, this ancient art is a major industry providing common inexpensive brick, high-temperature refractory bricks for furnaces, common and fine dishes, as well as specialized ceramics used in gas turbines and rockets. They can be generally categorized as being relatively poor in shear and tension but excellent in compression. They tend to be poor in impact; most of them are brittle.

Concrete

Frequently one hears people speak of a cement wall or walk or patio. This is generally incorrect for cement is a chemical that reacts with water to make a hard brittle material after curing. Its properties and cost are such that it is rarely used alone in quantity. Instead, the cement and water are mixed with sand and frequently also gravel to make concrete. Like brick, concrete is strong in compression but poor in tension and shear. However, concrete is often reinforced with steel mesh and bars to improve its properties while retaining its economy. This approach is even expanded to make prestressed concrete beams. In these, steel rods in the bottom portion of the beam are put under tension while the concrete is poured around them and cures. After the concrete cures, the rods are released and place the concrete in compression. As a load on the beam causes it to bend, the lower portion of the beam attempts to stretch, putting the concrete in tension. However, in the prestressed concrete beam, this will not occur until the compressive stress in the concrete has been relieved. See Figure 12-3.

Plastics

Plastics have often been maligned as a cheap substitute for metal or wood. This is unfortunate because many plastics are relatively inexpensive, easily formed, and have desirable physical and chemical properties. As can happen with any type of material, selection of the wrong plastic for an application or misuse of a component made of plastic can result in failure. In the initial decades of the use of plastics, incorrect applications were frequent as a result of ignorance.

The physical properties of plastics can be improved by the use of a wide variety of fillers, the best known of which are probably glass fibers. Their general tendency to brittleness at low temperatures, breakdown in sunlight, and a relatively high coefficient of thermal expansion must be contrasted with their formability, machineability, and thermal and electrical insulating properties as well as excellent individual properties of specific plastics when making a selection.

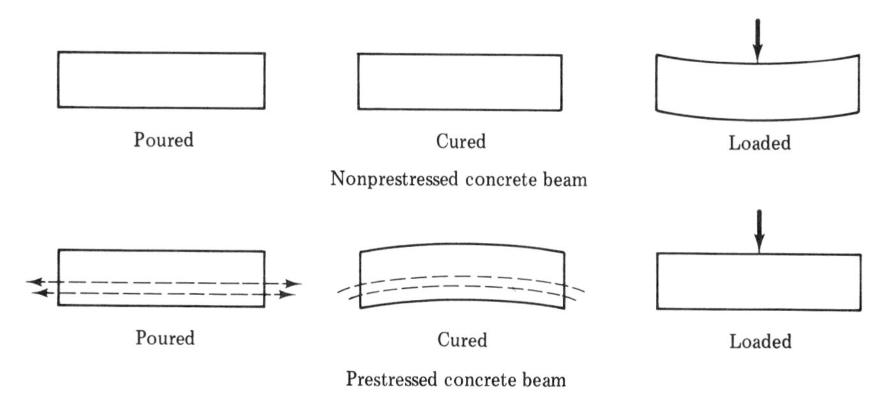

Figure 12-3. Prestressing of concrete beams.

Metals

Metals have made a significant enough impact on civilization that two of the ages or periods in our development took their names from metals—the Bronze Age and the Iron Age. As soon as we speak of bronze we move into the world of alloys, combinations of two or more metals. Most bronzes are alloys of copper and zinc. There are very few applications for pure (unalloyed) metals. Even jewelry and coinage made from precious metals are alloyed with other materials to give them added strength and durability. Generally speaking, metals have a higher strength-to-weight ratio than other materials. However, that ratio varies greatly among the many alloys available. Metals are also known for their high thermal and electrical conductivities.

Combining metals into a wide variety of alloys provides a great variability in the properties available. However, these properties can be further altered by the way in which they are processed—forging, rolling, drawing, and even machining. Some metals can even have their properties altered by heating and cooling them in very specific ways. All of these processes, alloying, forming, heat treating, affect the nature and size of the crystals or grains that make up the structure of the metal. It is the nature and size of these crystals that give a metal its unique properties.

12.4 ALLOWABLE STRESS AND FACTOR OF SAFETY

The *allowable stress* is also known by a number of other names, among them design stress, working stress, and safe working stress. By any name, it is the maximum stress which the material is to be subjected to in a specific application.

The allowable stress for a material is not a single value but varies with the type of loading and the conditions under which the load is applied.

The *factor of safety* is a value for a material that can be determined in two different ways. First, it can be determined as a ratio of the ultimate stress to the allowable stress. Second, it can be determined as a ratio of yield stress to allowable stress. Thus, the factor of safety based on ultimate stress is

$$FS_u = \frac{\text{ultimate stress}}{\text{allowable stress}} = \frac{\sigma_u}{\sigma_a}$$

while the factor of safety based on yield stress is

$$FS_y = \frac{\text{yield stress}}{\text{allowable stress}} = \frac{\sigma_y}{\sigma_a}$$

Under no terms should the factor of safety be interpreted as an indication of the amount of stress a material *really* can take as compared to the allowable stress! Some typical values for allowable stresses and factors of safety are found in the Appendices.

12.5 SELECTION OF MATERIALS

Our discussion of materials in this chapter has already started to give us some insight into the process of material selection. It rapidly becomes apparent that this selection process involves a meeting of the properties of a material and the working conditions which must be met. There may be several materials that are suitable but one will undoubtedly have the lowest cost. This seems to be very straightforward, a problem that a computer could be programmed to handle. Unquestionably, a computer could store and sort a great deal of information much more effectively than a person could. However, many considerations currently require the use of a human mind.

Costs are a big factor. By the time the material is actually purchased, will a deterioration of relations with a foreign country, civil war in a foreign country, or a strike in one's own country change the price and perhaps even the availability of the material? Materials research is continuous; last year's best solution may not be this year's best solution. Perhaps no one material can currently meet the requirements. What requirement is it best to compromise? What will be the effect of any compromise on the cost of maintenance or the frequency of replacement? Sometimes the quality of a property is relative. After all, while the casing of a gas-fired hot-air furnace should certainly not be flammable, it does not need the heat resistance that the burners require. Light weight may be worth additional cost for certain components and sometimes entire pieces of equipment, particularly mobile and transport equipment. In other cases, a heavier weight may be desirable to provide stability or absorb vibration. Obviously, experience becomes a very important factor in material selection.

12.6 READINESS QUIZ

1. The hardness of a material is an indication of its

   (a) toughness
   (b) plasticity
   (c) strength
   (d) all of the above
   (e) none of the above

2. The factor of safety for a material is its

   (a) yield stress ÷ ultimate stress
   (b) ultimate stress ÷ yield stress
   (c) allowable stress ÷ ultimate stress
   (d) allowable stress ÷ yield stress
   (e) none of the above.

3. The degree to which strain extends beyond the elastic range is an indication of a material's

   (a) plasticity
   (b) ductility
   (c) malleability
   (d) any of the above
   (e) none of the above

4. The area under the stress-strain curve is an indication of a material's

   (a) strength
   (b) toughness
   (c) elasticity
   (d) all of the above
   (e) none of the above

5. The properties of a metal may be altered by

   (a) alloying
   (b) heat treatment
   (c) working
   (d) any of the above
   (e) none of the above

6. The resistance of a material to shear is known as its

   (a) rigidity
   (b) plasticity
   (c) elasticity
   (d) all of the above
   (e) none of the above

7. The modulus of rigidity and modulus of elasticity are related by

   (a) a torsion test
   (b) Poisson's ratio
   (c) the factor of safety
   (d) all of the above
   (e) none of the above

8. The tendency of a material to change its properties under load over a long period of time is known as

   (a) fatigue
   (b) creep
   (c) brittleness
   (d) all of the above
   (e) none of the above

9. The tendency of a material to change its properties under repeated loading and unloading is known as

   (a) fatigue
   (b) creep
   (c) stiffness
   (d) all of the above
   (e) none of the above

10. __________ always have different strength properties in different directions

    (a) Metals
    (b) Plastics
    (c) Woods
    (d) All of the above
    (e) None of the above

11. A mixture of sand, water, and cement hardens to become

    (a) an alloy
    (b) a laminate
    (c) concrete
    (d) all of the above
    (e) none of the above

12. Metals in general are known for their

    (a) strength-to-weight ratio
    (b) conductivity
    (c) crystalline structure
    (d) all of the above
    (e) none of the above

13. Concrete is known for its __________ strength.

    (a) shear
    (b) compressive
    (c) tensile

(d) all of the above
(e) none of the above

14. Prestressed concrete has had __________ added to it to improve its properties.

    (a) metal chips
    (b) glass fibers
    (c) gravel
    (d) all of the above
    (e) none of the above

15. Plastics are generally known for their high

    (a) strength
    (b) formability
    (c) heat resistance
    (d) all of the above
    (e) none of the above

Chapter

13

General Stress

13.1 OBJECTIVES

Upon completion of the work relating to this chapter, you should be able to perform the following.

1. For the words listed below:
   (a) Select from several definitions the correct one for any given word.
   (b) Given a definition, select the correct word from a choice of words.

   bearing stress
   modulus of resiliency
   resiliency
   strain energy
   stress concentration

2. Given a simple two-force system acting on a body in either tension or compression, determine the resulting shear stress acting along any plane cutting through the line of action of the two-force system.
3. Given a simple two-force system acting on a body, make design decisions based on the strain energy concept.
4. Given a simple two-force system acting on a body apply the strain energy concept to impact loading.

5. Given a simple two-force system acting on a body, make the appropriate corrections to stress-strain calculations to account for the effect of stress concentrations caused by steps, notches, and holes in the body.

13.2 BASIC STRESSES REVIEWED

In an earlier chapter, we discussed the three basic stresses, tension, compression, and shear. Before delving into relationships among these stresses, let us make a quick review of each on its own as well as a new one, bearing stress. These are all illustrated in Figure 13-1.

Tensile Stress

Examples of simple tensile stress occur all around. Some of the members in both simple building and bridge trusses are in simple tension. Any cable is always in tension unless there is no load on it. Any bolt fastening two or more parts together is in tension as soon as it is tightened up.

Compressive Stress

Compressive stress is probably the most commonly encountered. This is because the force of gravity acting on the mass of a body causes a compressive stress on all but the top layer of molecules. The exception to this occurs when a body is suspended from one or more points. Thus a tree standing on a lawn tends to be compressed by the force of gravity while a blanket airing on a clothesline tends to be stretched by the force of gravity. Compressive forces, of course, are often man made as in the case of presses of many kinds (forging, ironing, printing, and wine to mention a few).

Shear Stress

Shear stress is somewhat different than either tensile or compressive stress. In tension, the load tends to increase the distance between molecules in the direction of the load. In compression, the load tends to decrease the distance between molecules in the direction of the load. In both these cases, the cross-sectional area (a measure of the number of molecular pairs) that the load is acting on is measured perpendicular to the direction of the load. However, in shear, the effect of the load is to slide the molecules in layers in the direction of the load. Thus, the cross-sectional area that the load is applied to lies parallel to the direction of the load.

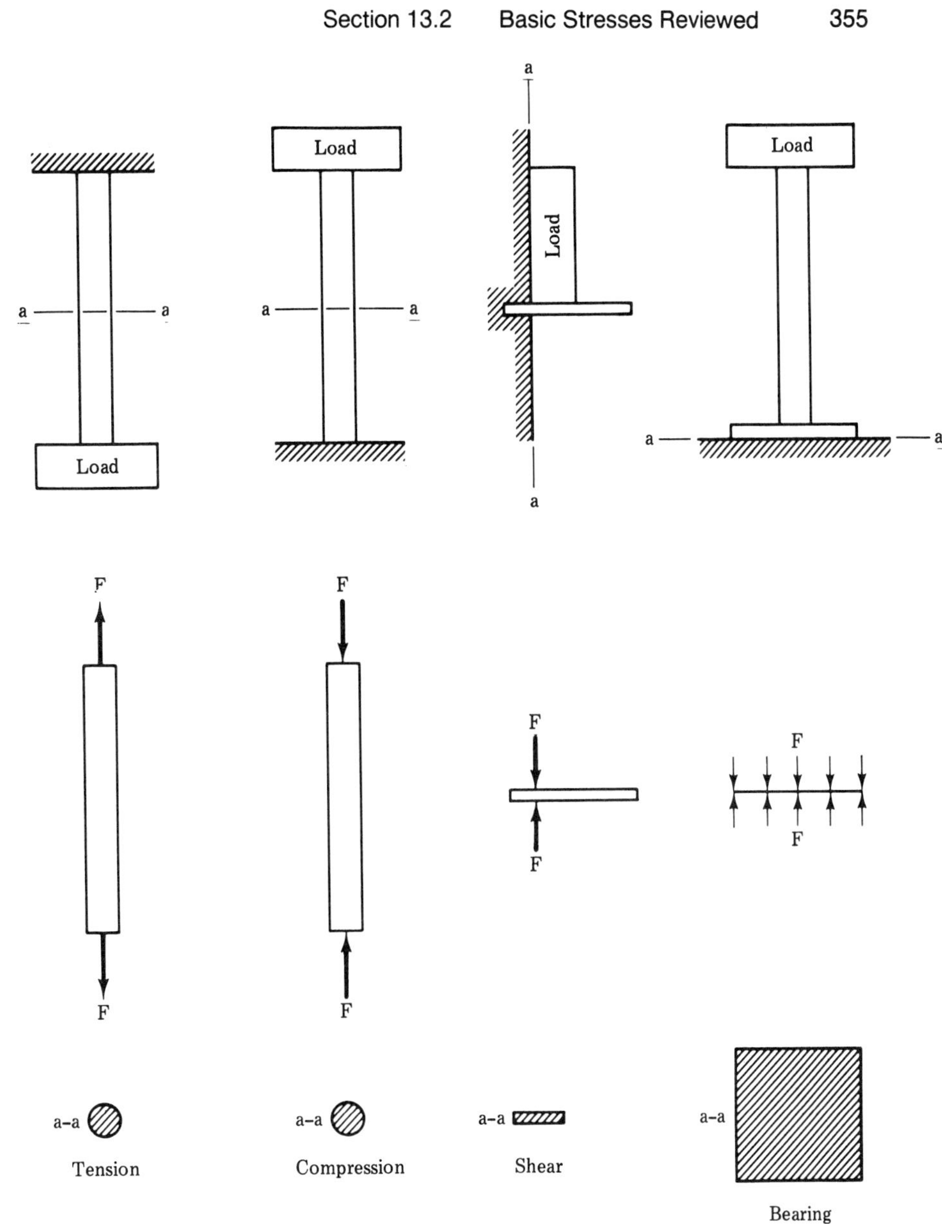

Figure 13-1. Basic stresses illustrated.

Bearing Stress

Bearing stress can be considered a type of compressive stress. However, rather than being an internal reaction, it is a surface reaction where two bodies are pressing against each other. Examples are the feet of your chair acting on the floor (and the floor reacting) and a loaded shaft pressing on a bearing (and the bearing pressing back).

Example 13-1: The column shown in Figure 13-2a is 4 in. in diameter. The cross bar passing through it is 1 in. in diameter. Determine the shear stress in the cross bar, the bearing stress at the bar-column, the compressive stress in the column, and the bearing stress at the floor.

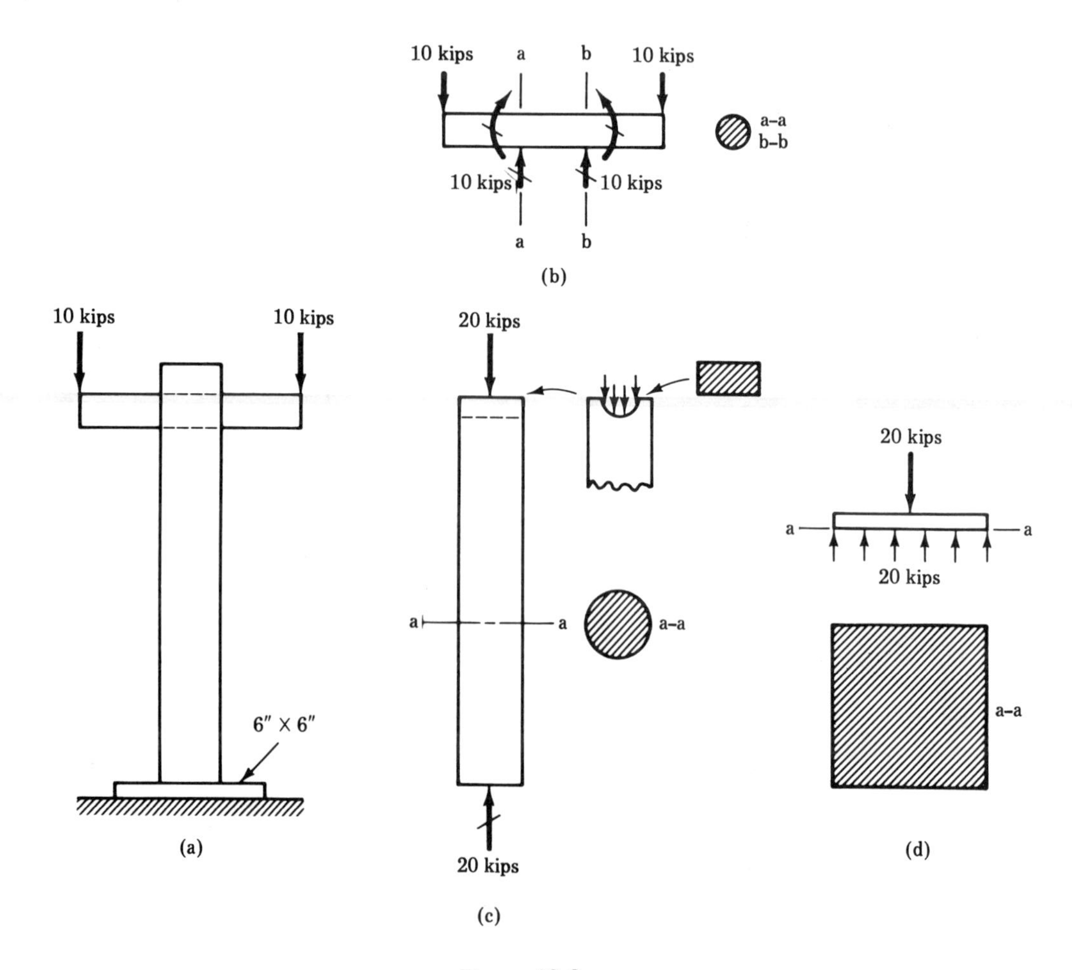

Figure 13-2

Solution:

Step 1 Draw the free-body diagram of the cross bar, as shown in Figure 13-2b. Except for the portion hidden within the column, all portions of the bar experience a 10-K shearing force. The cross-sectional area of the bar parallel to the shearing force can readily be found from its diameter and then the shear stress can be found.

$$A_s = 0.7854 d_{\text{bar}}^2 = 0.7854(1 \text{ in.}^2) \qquad A_s = 0.07854 \text{ in.}^2$$

$$\tau = \frac{P}{A} = \frac{10{,}000 \text{ lb}}{0.7854 \text{ in.}^2} \qquad \boldsymbol{\tau = 12{,}700 \text{ psi}}$$

Step 2 Draw the free-body diagram of the column as shown in Figure 13-2c. It is important to recognize here that the bearing area through which the bar is pushing on the column is the *projected* area (as we look from above or below) of the semicylindrical surface. Actually, because of the 2-in. *radius* of the column, the average length of the projected area is slightly less than 4 in. However, we will consider this small error as negligible to simplify our calculation. The bearing area bearing stress can then be determined as follows:

$$A_b = d_{\text{col}} d_{\text{bar}} \qquad A_b = (4 \text{ in.})(1 \text{ in.}) \qquad A_b = 4 \text{ in.}^2$$

$$\sigma_b = \frac{P}{A_b} = \frac{20{,}000 \text{ lb}}{4 \text{ in.}^2} \qquad \boldsymbol{\sigma_b = 5000 \text{ psi}}$$

Step 3 Using the same free-body diagram (13-2c) recognize that the cross-sectional area of the column perpendicular to the load can be found from its diameter. Compressive stress can then be easily found.

$$A_c = 0.7854 d^2{}_{\text{col}} \qquad A_c = 0.7854(4 \text{ in.}^2) \qquad A_c = 12.45 \text{ in.}^2$$

$$\sigma_c = \frac{P}{A} = \frac{20{,}000 \text{ lb}}{12.45 \text{ in.}^2} \qquad \boldsymbol{\sigma_c = 1610 \text{ psi}}$$

Step 4 Draw a free-body diagram of the bearing plate-floor interface as shown in Figure 13-2d. The horizontal cross section of the plate is the area through which the load is transmitted to the floor.

$$A_b = s^2 = 6 \text{ in.}^2 \qquad A_b = 36 \text{ in.}^2$$

$$\sigma_b = \frac{P}{A} = \frac{20{,}000 \text{ lb}}{36 \text{ in.}^2} \qquad \boldsymbol{\sigma_b = 560 \text{ psi}}$$

Note You may have wondered about our ignoring the moment reactions in the bar. The bar is acting as a beam and in a subsequent chapter we will investigate the stresses resulting from bending in a beam. This will not negate the work we have just done but add to it. Thus, this part of the problem is not finished. The total stress will be greater than that just determined.

Problem 13-1: Determine the tensile stress on each component of the clevis shown in Figure 13-3. Also determine the shear stress in the pin and the bearing stresses at both the center and the ends of the pin.

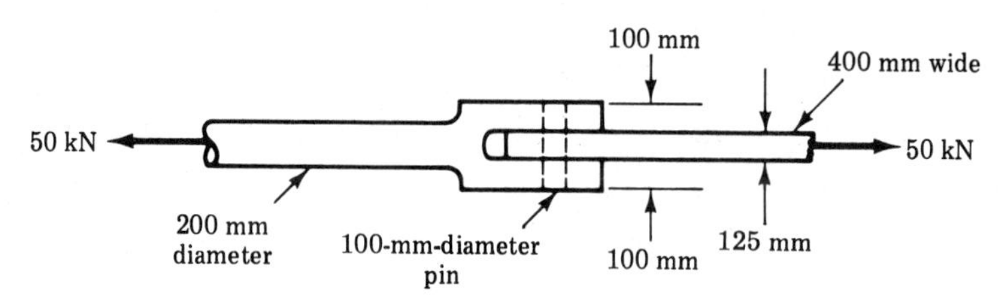

Figure 13-3

13.3 SHEAR RESULTING FROM SIMPLE TENSION OR COMPRESSION

Much of the discussion up to now has concentrated on simple tension and compression acting on a material. Thus, we have spoken of molecules being pulled further apart or pushed closer together. More specifically, we spoke of this change in length as deformation, and the change in unit length as strain. However, we also noted that as a material stretches under axial tension it also gets thinner in the transverse direction (perpendicular to the tension axis). In the case of compression, of course, there would be a thickening of material in the transverse direction. Motion is obviously taking place at some angle as tension or compression increases or decreases.

Indeed, if we examine test specimens, cylindrical in cross section that have been tested to destruction, we find the fracture surface to be cone shaped. As we will see in the following analysis, this is not because the component of the basic stress perpendicular to the cone surface is greater than the basic stress (it is actually less). Rather, it is because the molecular bonds failed in shear, sliding along the cone surface.

Let us examine a simple piece of material with a cross-sectional area A and a tensile force P as shown in Figure 13-4a. This causes a tensile stress σ_t which, if great enough, will cause the material to part or fail at some transverse plane $a-a$. Perhaps some imperfections or a slightly smaller A caused that to be the weakest plane.

However, we have already noted that many materials fail along planes

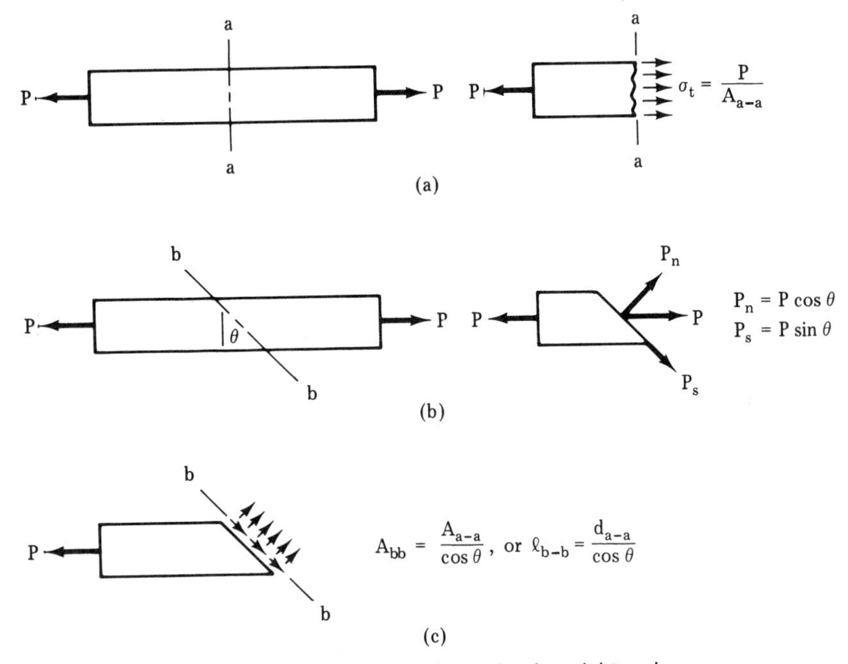

Figure 13-4. Shear stress from simple axial tension.

other than one that is perpendicular to the axis. Let us consider the stresses in the material at some plane $b-b$ at some angle θ, as shown in 13-4b. The forces acting at that plane can then be analyzed as a tensile force acting normal to the plane (trying to pull the material apart in that direction) and a shear force parallel to the plane (trying to slide the molecules along that plane).

Taking the analysis one step further, let us recognize that the area along plane $b-b$ is substantially larger than that along plane $a-a$. Their relationship is a function of angle θ as shown in Figure 13-4c. The normal tensile stress can then be determined as follows:

$$\sigma_n = \frac{P_n}{A_{b-b}} = \frac{P\cos\theta}{\dfrac{A_{a-a}}{\cos\theta}} \qquad \sigma_n = \frac{P}{A_{a-a}}\cos^2\theta$$

The shear stress can be determined by

$$\tau = \frac{P_s}{A_{b-b}} = \frac{P\sin\theta}{\dfrac{A_{a-a}}{\cos\theta}} \qquad \tau = \frac{P}{A_{a-a}}\sin\theta\cos\theta$$

Since we can go back to our trigonometric identities and find that

$$\sin\theta\cos\theta = \tfrac{1}{2}\sin 2\theta$$

the shear stress can also be written as

$$\tau = \frac{P}{2A_{a-a}}\sin 2\theta$$

Examining these two equations closely, we can see that the normal stress is at a minimum when $\theta = 90°$, and at a maximum when $\theta = 0°$. Also, the shear stress is at a maximum when $\theta = 45°$ and at a minimum at both 0° and 90°.

Thus, when materials fail in shear, rather than in tension or compression (as might be expected when the basic stress is tension or compression) we should expect a fracture angle of 45°. Furthermore, if we consider that at 45°, $\frac{1}{2}\sin 2\theta$ equals 0.5, it becomes evident that for a material under tension or compression to fail in shear, its ultimate strength in shear must be less than half its ultimate strength in tension or compression.

Example 13-2: A 50-mm-diameter bar of material is subjected to an axial compressive force of 3 kN. What is the shear stress in a plane at an angle of 35° from the transverse plane?

Solution:

Step 1 Determine the area in shear. Recognize that any plane cutting through a cylinder other than the axial or transverse forms an ellipse. The one axis of the ellipse is the diameter of the cylindrical bar, 50 mm. The other axis will be $d \div \cos\theta$ or 50 mm ÷ cos 35° = 61 mm.

$$A = \pi ab = \pi\left(\frac{50 \text{ mm}}{2}\right)\left(\frac{61 \text{ mm}}{2}\right) \qquad A = 2395 \text{ mm}^2 = 2.395 \times 10^{-3} \text{ m}^2$$

Step 2 Determine the shear force.

$$P_s = P\sin\theta = 3 \text{ kN}(\sin 35°) \qquad P_s = 1.72 \text{ kN}$$

Step 3 Determine the shear stress.

$$\tau = \frac{P_s}{A} = \frac{1.72 \text{ kN}}{2.395 \times 10^{-3} \text{ m}^2} \qquad \boldsymbol{\tau = 720 \text{ kPa}}$$

Problem 13-2: A part with a rectangular cross section of $\frac{1}{2} \times 1\frac{1}{2}$ in. is to be made of a material with an ultimate compressive strength of 100,000 psi and an ultimate shear strength of 40,000 psi. Using an ultimate factor of safety of 5, what is the maximum axial compressive load that can be safely applied to the part?

13.4 STRAIN ENERGY

Throughout our discussions, we have considered points, members, mechanisms, and structures under static conditions. In our earlier discussion of deformation, we realized perhaps that work must be done in order for deformation to occur; however, that aspect of our problems was, at least temporarily, sidestepped. Reality forces us to consider how we move from one static condition to another and how the nature of our material affects this change.

In Figure 13-5a, a load P in the shape of a collar is located on a vertical rod. It rests on two supports A and B so that it exerts no force on the flange at the bottom of the rod. When the supports are retracted, the weight acts on the rod and a deformation of the rod occurs. The work done by the weight is force times distance divided by 2 as in a simple linear spring. Within the elastic range of a material, the force displacement relationship is exactly like that of a spring. We have already seen this in our earlier discussions of stress and strain.

$$U = \tfrac{1}{2}FS \qquad U = \tfrac{1}{2}PS$$

Since

$$\delta = \frac{PL}{AE} \qquad U = \tfrac{1}{2}\frac{P^2L}{AE} = \frac{P^2LA}{A^2 2E}$$

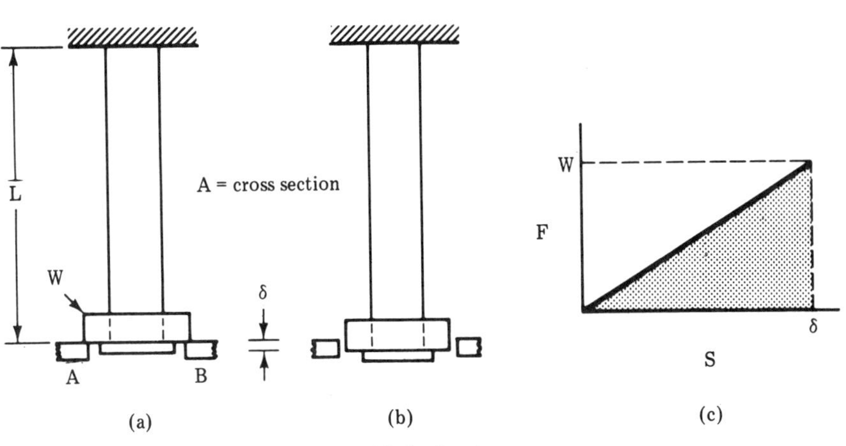

Figure 13-5. Strain energy.

The reason for the last algebraic step is that $P/A = \sigma$ and so

$$U = \sigma^2 \frac{LA}{2E}$$

The work done on the rod is stored in the rod as *strain energy.* The maximum possible storage of energy that can take place occurs when the stress level at the elastic limit is reached. Any attempt to store additional energy would result in some permanent deformation and the additional energy is irretrievable.

$$U_{max} = \sigma_e{}^2 \frac{LA}{2E}$$

where σ_e is the stress level at the elastic limit.

A material's ability to absorb energy is called its *resiliency.* The modulus of resiliency is the amount of energy a material can absorb *per unit volume* when stressed to its elastic limit. Since volume can be expressed as LA, the modulus of resiliency E_R can be expressed as

$$E_R = \frac{\sigma_e{}^2 \frac{LA}{2E}}{LA} \qquad E_R = \frac{\sigma_e{}^2}{2E}$$

$$\text{Since } E = \frac{\sigma}{\epsilon} \qquad E_R = \frac{\sigma_e{}^2}{2(\sigma/\epsilon)} \qquad E_R = \frac{1}{2}\sigma_e \epsilon$$

where ϵ is strain at the elastic limit. Thus, the modulus of resiliency is the area under the stress-strain curve for the elastic range.

Example 13-3: Two post designs need to be evaluated in terms of their ability to absorb energy. Both are steel with an E of 29,000,000 psi and a σ_e of 40,000 psi. The dimensions are given in Figure 13-6.

Solution:

Step 1 Recognize that for each post the difference in cross section must be dealt with and that in each case the smallest cross section will be at σ_e. Use l for larger cross section and s for smaller cross section.

Step 2 Determine all areas

$$A_{as} = S_s^2 = 1 \text{ in.}^2 \qquad A_{as} = 1 \text{ in.}^2$$

$$A_{al} = S_l^2 = 3 \text{ in.}^2 \qquad A_{al} = 9 \text{ in.}^2$$

$$A_{bs} = 0.7854 d_s^2 = 0.7854(1 \text{ in.}^2) \qquad A_{bs} = 0.7854 \text{ in.}^2$$

$$A_b = 0.7854 d_l^2 = 0.7854(3 \text{ in.}^2) \qquad A_b = 7.07 \text{ in.}^2$$

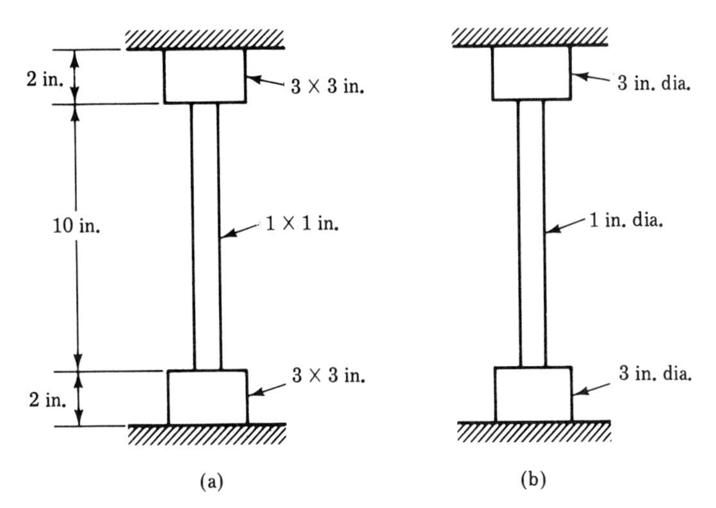

Figure 13-6

Step 3 Determine all lengths

$$L_{al} = 2 \text{ in.} = L_{bl}$$
$$L_{as} = 10 \text{ in.} = L_{bs}$$

Step 4 Determine the maximum load in each post

$$P_a = A_{as}\sigma_e = (1 \text{ in.}^2)(40{,}000 \text{ psi}) \qquad P_a = 40{,}000 \text{ lb}$$
$$P_b = A_{bs}\sigma_e = (0.7854 \text{ in.}^2)(40{,}000 \text{ psi}) \qquad P_b = 31{,}416 \text{ lb}$$

Step 5 Determine the stress in the larger cross section of each post.

$$\sigma_{al} = \frac{P_a}{A_{al}} = \frac{40{,}000 \text{ lb}}{9 \text{ in.}^2} \qquad \sigma_{al} = 4444 \text{ psi}$$
$$\sigma_{bl} = \frac{P_b}{A_{bl}} = \frac{31{,}416 \text{ lb}}{7.07 \text{ in.}^2} \qquad \sigma_{bl} = 4444 \text{ psi}$$

Step 6 Determine the maximum strain energy absorbed for each post.

$$U_{a_{max}} = \frac{\sigma_e^2 L_{as} A_{as}}{2E} + \frac{\sigma^2 L_{al} A_{al}}{2E}$$

$$U_{a_{max}} = \frac{(40{,}000 \text{ psi}^2)(10 \text{ in.})(1 \text{ in.}^2)}{2(29{,}000{,}000 \text{ psi})} + \frac{(4444 \text{ psi}^2)(2 \text{ in.})(9 \text{ in.}^2)}{2(29{,}000{,}000 \text{ psi})}$$

$$\mathbf{U_{a_{max}} = 282 \text{ in. lb}}$$

$$U_{b_{max}} = \frac{\sigma_e^2 L_{bs} A_{bs}}{2E} + \frac{\sigma^2 L_{bl} A_{bl}}{2E}$$

$$U_{b_{max}} = \frac{(40{,}000 \text{ psi}^2)(10 \text{ in.})(0.7854 \text{ in.}^2)}{2(29{,}000{,}000 \text{ psi})}$$

$$+ \frac{(4444 \text{ psi}^2)(2 \text{ in.})(7.07 \text{ in.}^2)}{2(29{,}000{,}000 \text{ psi})}$$

$$\mathbf{U_{b_{max}} = 221 \text{ in. lb}}$$

Note The above may appear deceptively simple but that is a result of some commonalities of dimensions and proportions. The larger mass does not always have the greater energy absorbtive ability.

Problem 13-3: The two-cylinder designs shown in Figure 13-7 need to be evaluated in terms of their ability to absorb energy under tensile load. Both are steel with an E of 200 GPa and a σ_e of 275 MPa. Compare their abilities to absorb energy.

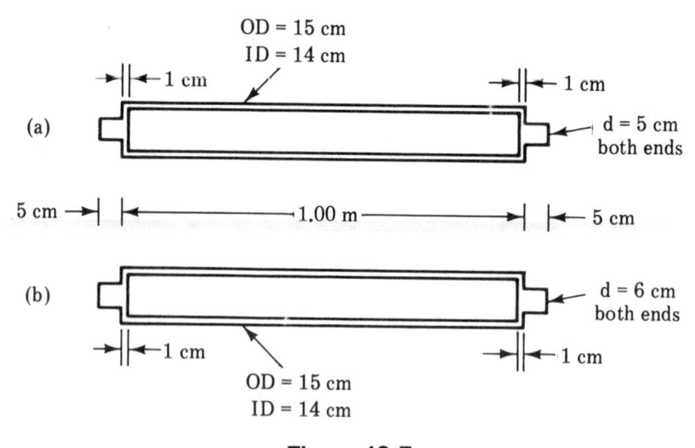

Figure 13-7

13.5 DYNAMIC LOADS

In Section 13.4, we evaluated the ability of a material to absorb and store energy. The energy exchange involved the displacement of a load, that displacement being simply the deformation of a component as the result of the relatively statically applied load (see Figure 13-5). However, if the load was moving relatively rapidly, the absorption of its kinetic energy will have a substantially greater effect than the simple static load. In Figure 13-8, we see the same flanged rod and collar as in Figure 13-5. However, this time the collar is initially at the top of the rod. When it is dropped, its initial effect will

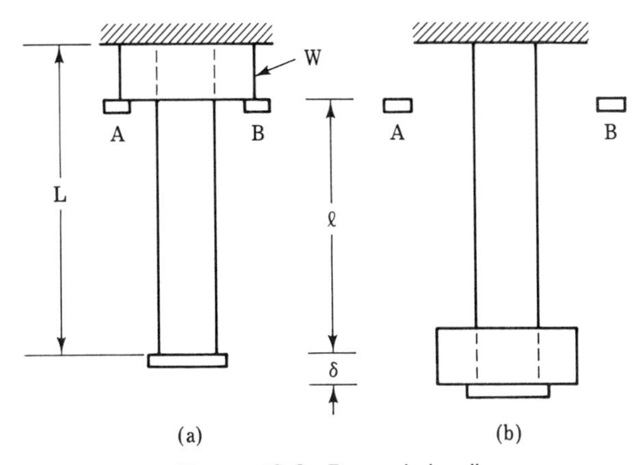

Figure 13-8. Dynamic loading.

be to convert the initial potential energy to kinetic energy which upon impact is converted to strain energy in the rod. There will be a tendency for "bounce," that is, the flanged rod under tension will push up on the collar which will bounce up on impact, then come down and reimpact. This will continue until finally the static condition shown in Figure 13-5 is reached. This period of oscillation involves the study of vibrations, which is beyond the scope of this text. Our concern is the maximum energy that must be absorbed upon the initial impact.

Once again, the weight initially held in place by supports A and B is released by retracting those supports. This time, however, the weight drops the length of the rod (less the thickness of the weight) plus the distance the rod is deformed. The change in potential energy therefore is

$$\Delta PE = W(l + \delta)$$

This energy is converted to strain energy

$$U = \frac{\sigma^2 LA}{2E}$$

Setting the above energies equal to each other and substituting for δ from $E = \sigma/\epsilon = \sigma/(\delta/L)$ yields

$$W\left(l + \frac{\sigma L}{E}\right) = \frac{\sigma^2 LA}{2E}$$

The next step depends on what unknown we wish to solve for. This might be the maximum weight the collar can be without exceeding the stress at the elastic limit. It might be the actual stress resulting from a given weight collar.

Or, it might be a very basic design problem: what cross section of column is needed to prevent exceeding the stress at the elastic limit for a given weight.

Example 13-4: Given a design similar to that shown in Figure 13-8, if the weight is 250 N and with a thickness of 10 cm and the rod is 1.00 m long (not including the flange thickness) what minimum diameter must the aluminum ($E = 73$ GPa, and $e = 300$ MPa) rod be? What would determine the minimum flange thickness?

Solution:

Step 1 Recognize that l is equal to L less the thickness of the weight.

Step 2 Since the rod is circular in cross section, the area A will be equal to $0.7854d^2$.

Step 3 Substituting from Step 2 into the equation just developed for dynamic loading yields

$$W\left(l + \frac{\sigma L}{E}\right) = \frac{\sigma^2 L(0.7854d^2)}{2E}$$

Step 4 Solving for d and using σ_e yields

$$d = \left[\frac{2.546EW}{\sigma_e^2 L}\left(l + \frac{\sigma_e L}{E}\right)\right]^{0.5}$$

Step 5 Substituting knowns

$$d = \left[\frac{2.546(7.3 \times 10^{10}\text{ Pa})(250\text{ N})}{(3.0 \times 10^8\text{ Pa}^2)(1\text{ m})}\left(0.9\text{ m} + \frac{3.0 \times 10^8\text{ Pa}(1\text{ m})}{7.3 \times 10^{10}\text{ PA}}\right)\right]^{0.5}$$

$$\mathbf{d = 0.0216\ m \quad or \quad 21.6\ mm}$$

Step 6 The thickness of the flange can be determined by recognizing that the strain energy is transferred from the flange to the rod by shear forces acting along a plane that is the surface of a cylinder with the diameter of the rod and a length equal to the thickness of the flange. An equation can then be developed based upon the shear modulus and the shear stress at the elastic limit.

Problem 13-4: The 10-lb disk slides on the rod as shown in Figure 13-9, striking the flanged end at a velocity of 10 fps. The material is steel ($E = 29 \times 10^6$ psi). What is the stress in the rod upon impact? (Recall that kinetic energy is equal to $\frac{1}{2}mv^2$.)

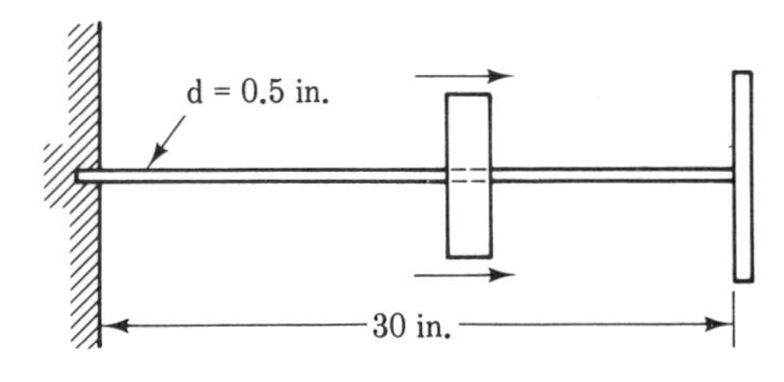

Figure 13-9

13.6 STRESS CONCENTRATION

Until now, we have considered, for the most part, components with a constant cross section. In those cases where a change in cross section occurred (as in the preceding two sections) we gave no special thought to any effect of the change in cross section. As illustrated in Figure 13-10a, forces flow in a uniform direction and with uniform distribution when the cross section of a component is constant. However, when the component is notched as in 13-10b, the flow of forces is disturbed just as the flow of water in a stream with restrictions is disturbed. The disturbance is greatest near the notches; the lines of force gathering closer together or concentrating. This raises the stress near the notches to a level higher than would be expected by simply dividing the total force by the reduced cross section at the notches. The maximum actual stress is expressed as

$$\sigma_{\max} = K\sigma_{\text{avg}}$$

where the average stress is based upon the area at the minimum cross section and K is a stress concentration factor determined by experimentation. In such experiments it is actually strain that is measured by the use of strain gauges and stress is calculated by multiplying the strain by E, the modulus of

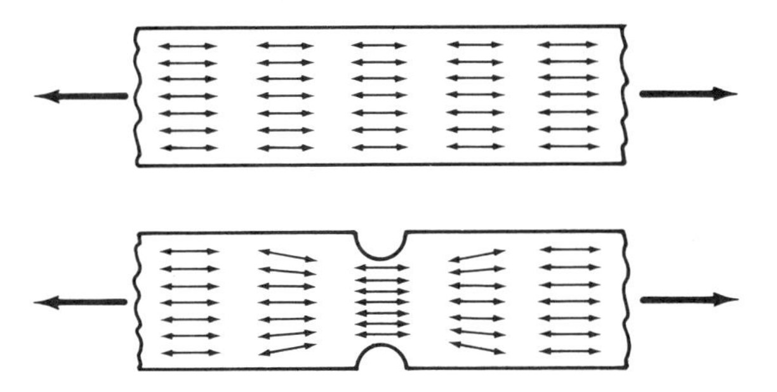

Figure 13-10. Concentration of forces and stresses.

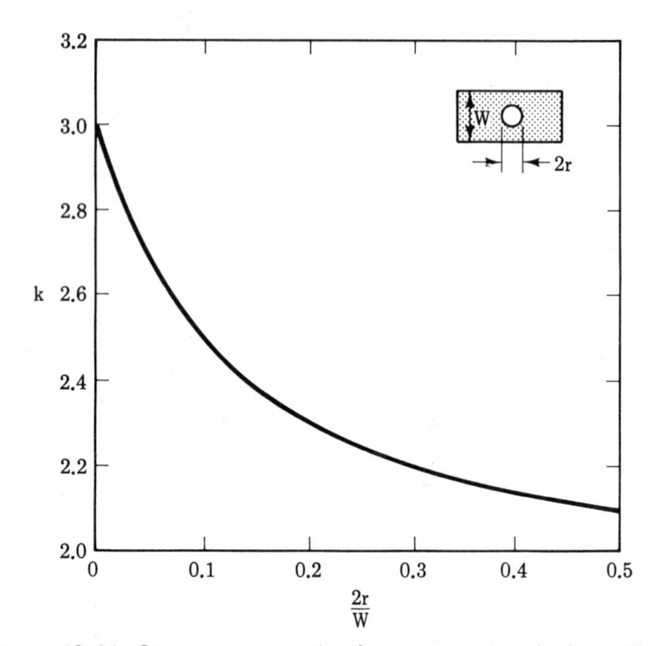

Figure 13-11. Stress concentration factor—rectangular bar with hole.

elasticity. Some results of this experimentation are in Figures 13-11 through 13-15.*

Example 13-5: Determine the maximum axial load that can be carried by the bar shown in Figure 13-16 if the tensile stress cannot exceed 240 MPa.

Solution:

Step 1 Recognize that two solutions must be determined, one for each point of stress concentration. The maximum permissible load will be the lower of the two solutions.

Step 2 Label the groove 1 and the step 2 to clarify the solution.

Step 3 Analyzing the groove, refer to Figure 13-15. Determine h_1/r_1 and r_1/d_1 as follows, then establish K from Figure 13-15.

$$h_1 = \tfrac{1}{2}(D_1 - d_1) = \tfrac{1}{2}(0.05 - 0.03\text{ m}) \qquad h_1 = 0.01\text{ m}$$

$$\frac{h_1}{r_1} = \frac{0.01\text{ m}}{0.005\text{ m}} \qquad \frac{h_1}{r_1} = 2$$

$$\frac{r_1}{d_1} = \frac{0.005\text{ m}}{0.03\text{ m}} \qquad \frac{r_1}{d_1} = 0.167$$

$$K_1 = 1.76$$

* These figures are from Levinson, Irving J. *Mechanics of Materials,* 2nd ed. Englewood Cliffs, NJ: Prentice-Hall, Inc., 1970.

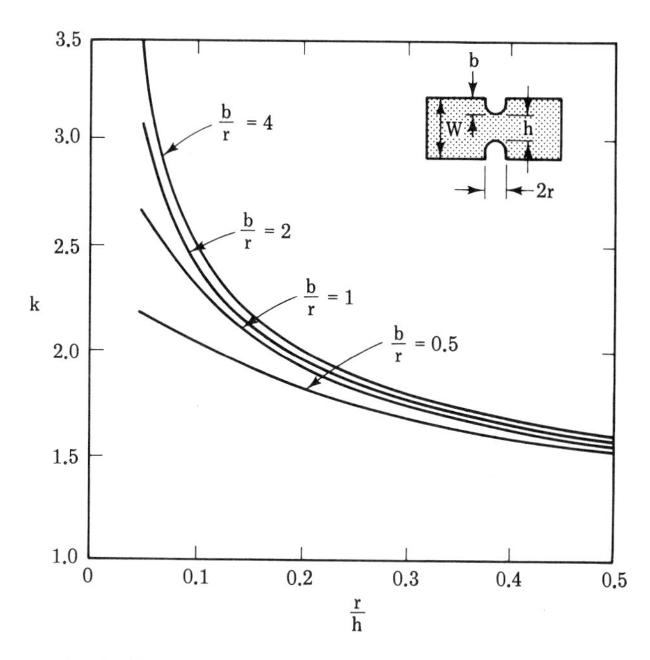

Figure 13-12. Stress concentration factor—notched rectangular bar.

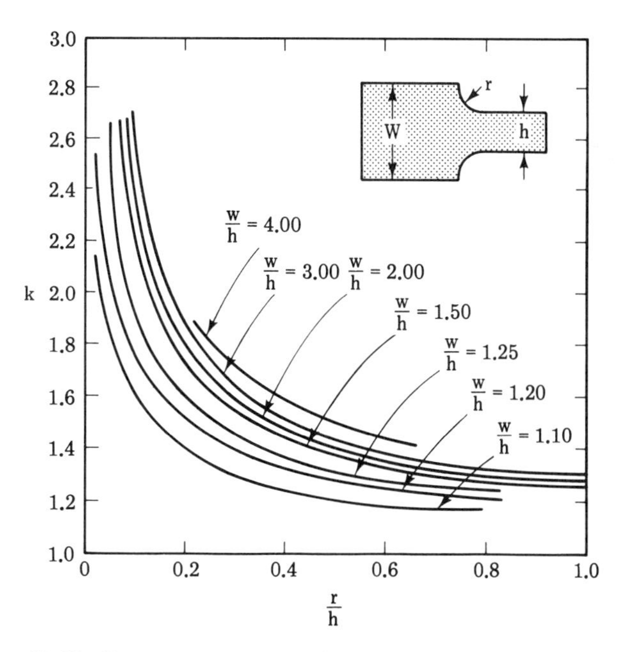

Figure 13-13. Stress concentration factor—stepped rectangular bar.

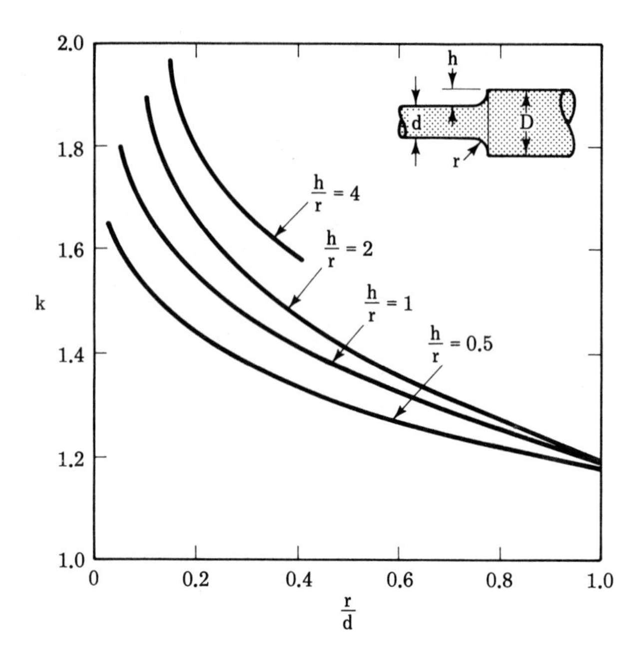

Figure 13-14. Stress concentration factor—stepped round bar.

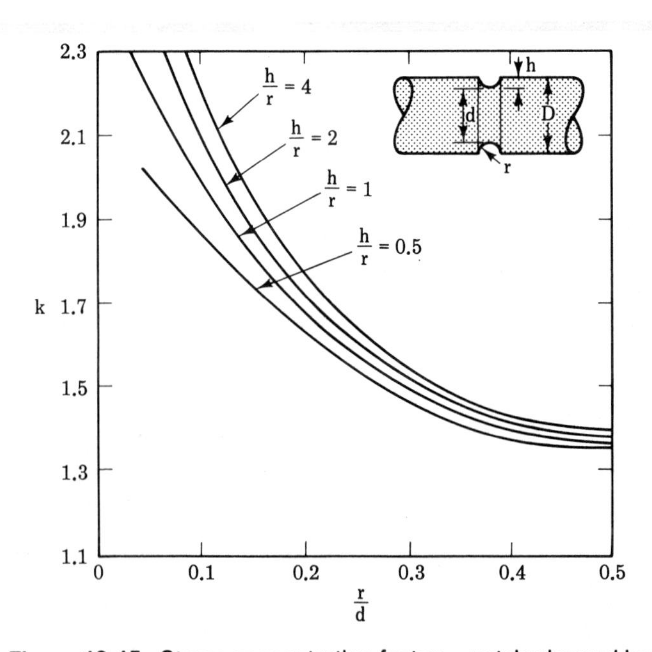

Figure 13-15. Stress concentration factor—notched round bar.

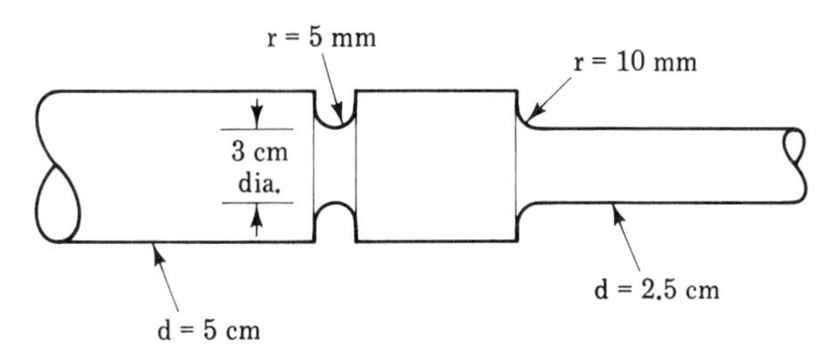

Figure 13-16

Step 4 Determine the maximum load at the groove as follows:

$$\sigma_{1\text{max}} = K_1 \sigma_{1\text{avg}} = K_1 \frac{P_{1\text{max}}}{A_1}$$

$$P_{1\text{max}} = \frac{\sigma_{1\text{max}} A_1}{K_1} = \frac{240 \text{ MPa}(0.7854)(0.03 \text{ m}^2)}{1.76}$$

$$\boldsymbol{P_{1\text{max}} = 96.4 \text{ kN}}$$

Step 5 Analyzing the step, refer to Figure 13-14. Determine h_2, h_2/r_2, and r_2/d_2 as follows, then establish K from Figure 13-14.

$$h_2 = \tfrac{1}{2}(D_2 - d_1) = \tfrac{1}{2}(0.05 - 0.025 \text{ m}) \qquad h_2 = 0.0125 \text{ m}$$

$$\frac{h_2}{r_2} = \frac{0.0125 \text{ m}}{0.01 \text{ m}} \qquad \frac{h_2}{r_2} = 1.25$$

$$\frac{r_2}{d_2} = \frac{0.010 \text{ m}}{0.025 \text{ m}} \qquad \frac{r_2}{d_2} = 0.4$$

$$K_2 = 1.42$$

Step 6 Determine the maximum load at the step as follows:

$$\sigma_{2\text{max}} = K_2 \sigma_{2\text{avg}} = K_2 \frac{P_{2\text{max}}}{A_2}$$

$$P_{2\text{max}} = \frac{\sigma_{2\text{max}} A_2}{K_2} = \frac{240 \text{ MPa}(0.7854)(0.025 \text{ m}^2)}{1.42}$$

$$\boldsymbol{P_{2\text{max}} = 83 \text{ kN}}$$

Step 7 Since the maximum load at the step, 83 kN, is the lower of the two allowable loads determined, it will have to be the maximum axial load that the bar may be subjected to.

Problem 13-5: Determine the maximum stress occurring in the bar shown in Figure 13-17.

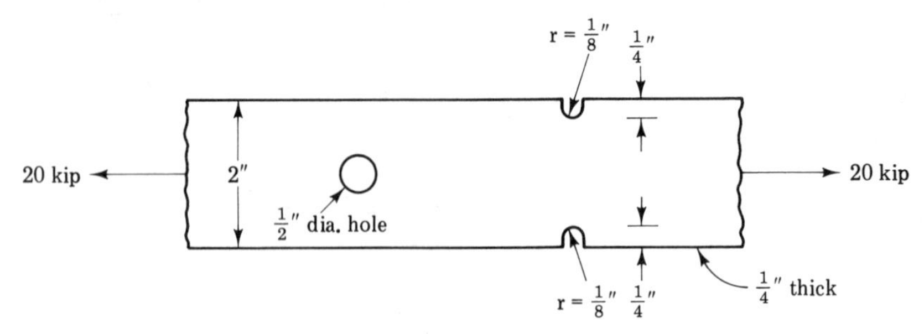

Figure 13-17

13.7 READINESS QUIZ

1. When compressing a body, force times displacement yields

   (a) deformation
   (b) stress
   (c) strain energy
   (d) all of the above
   (e) none of the above

2. Radial notches in an axially loaded member will result in ________ at a section through the notches.

   (a) increased stress
   (b) increased strain
   (c) stress concentration
   (d) all of the above
   (e) none of the above

3. When a body is subjected to a simple tensile load, ________ occurs.

   (a) torsion stress
   (b) shear stress
   (c) compressive stress
   (d) all of the above
   (e) none of the above

4. A 5-lb load striking (at some velocity) a fixed object exerts a load ________ 5 lb.

   (a) equal to
   (b) greater than
   (c) less than

(d) all of the above
(e) none of the above

5. The ability of a material to absorb energy is known as its
   (a) elasticity
   (b) plasticity
   (c) ductility
   (d) any of the above
   (e) none of the above

6. The modulus of resiliency is expressed in units of
   (a) force divided by area
   (b) length divided by length
   (c) force times area
   (d) area divided by force
   (e) energy divided by volume

7. If a hole is drilled in a body and a tensile load is applied perpendicular to the axis of the hole, the concentration of stress will be greatest
   (a) near the hole
   (b) at the edge of the body nearest the hole
   (c) at the point(s) in the body most distant from the hole
   (d) any of the above
   (e) none of the above

8. The resiliency of a material is a measure of its ability to
   (a) absorb impact
   (b) resist stress
   (c) absorb energy
   (d) all of the above
   (e) none of the above

9. The effect of a radially loaded shaft on its supports is known as
   (a) shear stress
   (b) tensile stress
   (c) bearing stress
   (d) all of the above
   (e) none of the above

10. The greatest shear stress resulting from a simple tensile load occurs in planes
    (a) perpendicular to the load direction
    (b) at 45° to the load direction
    (c) parallel to the load direction
    (d) any of the above
    (e) none of the above

13.8 SUPPLEMENTARY PROBLEMS

13-6 The shaft shown in Figure 13-18 is 5 cm in diameter. Determine the direct shear stress in the shaft, the bearing stress at each support, the compressive stress in each support and the bearing stress on the floor.

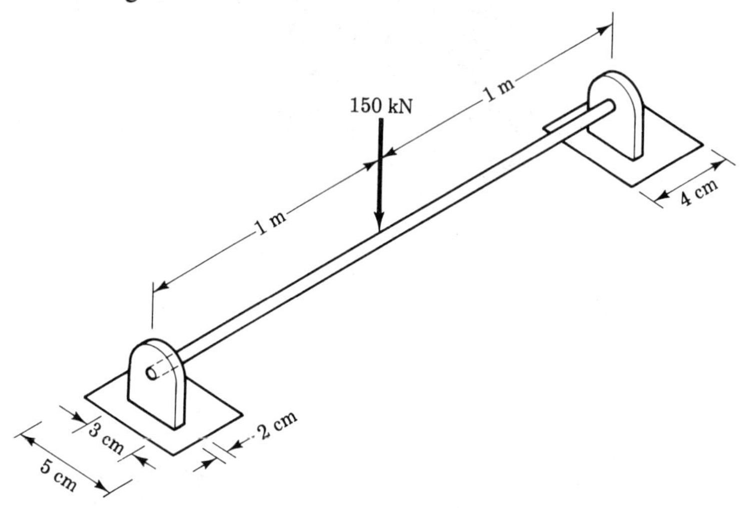

Figure 13-18

13-7 A 30-cm diameter shaft with an axial tensile loading is made of a material with an ultimate strength of 750 MPa and an ultimate shear strength of 300 MPa. Using an ultimate factor of safety of

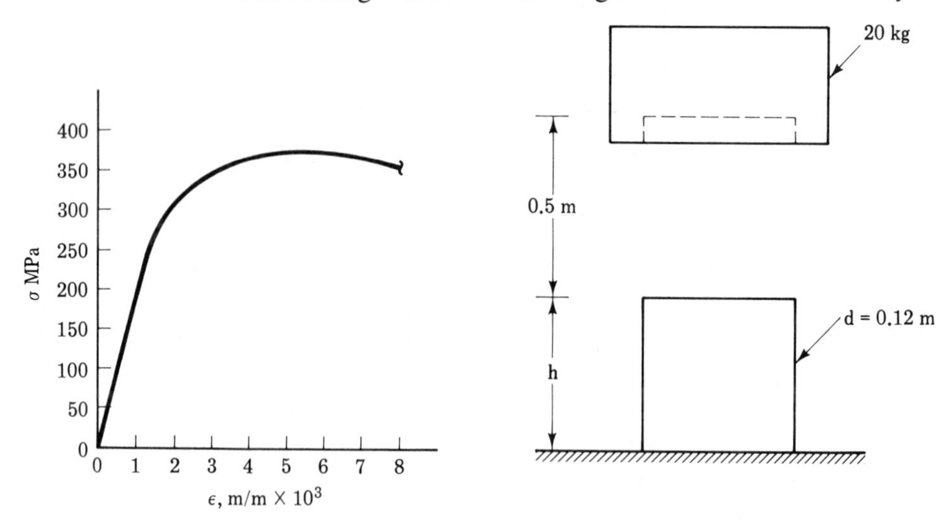

Figure 13-19

4, what is the maximum axial tensile load that can be safely applied to the shaft?

13-8 Figure 13-19 shows a drop forge hammer poised above the anvil and the stress-strain diagram for the anvil material. How high must the anvil be if its elastic limit is not to be exceeded when the hammer is dropped with no work on the anvil?

13-9 Determine the thickness required for the bar shown in Figure 13-20 if the maximum stress is to be 28,000 psi.

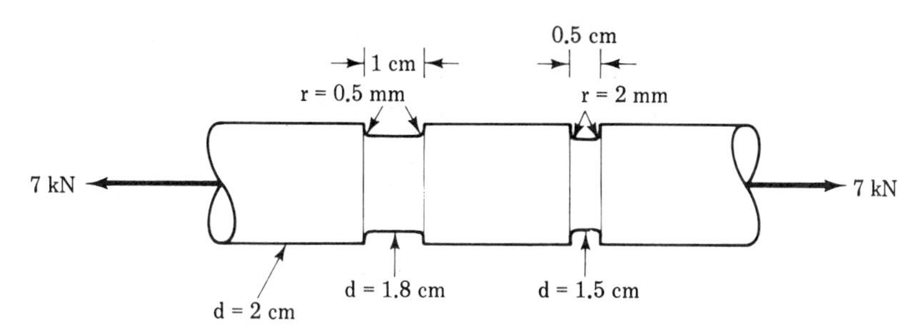

Figure 13-20

13-10 Determine the maximum stress occurring in the bar shown in Figure 13-21.

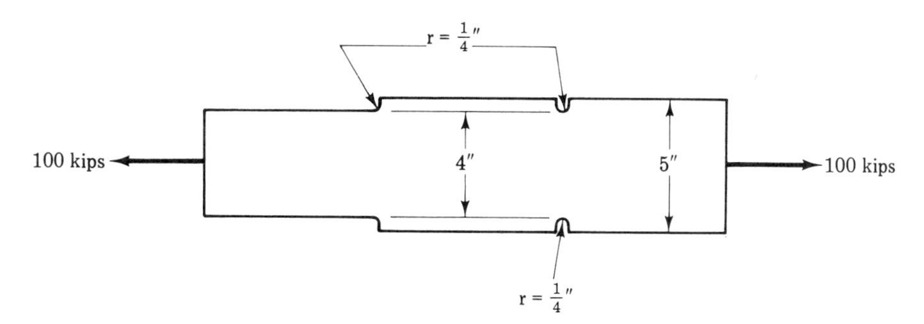

Figure 13-21

Chapter

14

Thin-Walled Pressure Vessels

14.1 OBJECTIVES

Upon completion of the work relating to this chapter, you should be able to perform the following.

1. For the terms listed below:

   (a) Select from several definitions the correct one for any given term.
   (b) Given a definition, select the correct term from a given list of terms.

ASME	head	pressure vessel
circumferential stress	hoop stress	thick walled
code	longitudinal stress	thin walled

2. Determine the stress in a thin-walled cylinder with given thickness.
3. Determine the wall thickness required for a thin-walled cylinder for a given maximum stress.
4. Determine the stress in a thin-walled sphere with given wall thickness.
5. Determine the wall thickness required for a thin-walled sphere for a given maximum stress.

14.2 THIN-WALLED PRESSURE VESSELS DEFINED

A wide variety of containers are used to store, ship, or in some way process liquids and gases. In terms of space utilization, a cubic shape or even a rectangular parallelepiped of any kind would be most efficient. Indeed, for that reason, many dry solids are packaged and shipped in containers made in such shapes varying in size from small boxes of cough drops to railroad boxcars. However, when dealing with containers under pressure and containers that require sealing, the fabrication costs for the boxlike parallelepiped shape become prohibitive. Thus, for foods, motor oils, and many other liquid materials, we find cans and bottles in use.

The same principles involved in the design of everyday containers all about us apply to the large tanks and containers used for liquids and gases. Since these materials are frequently at pressures well above atmospheric pressure the term *pressure vessel* was coined to describe their *containers.* The dangers inherent in the potential failure of pressurized vessels have resulted in the formation of specific *codes* (rules) to regulate their design. Two of the most common codes whose use is frequently required by law are the ASME Pressure Vessel Code and the ASME Boiler Code. (ASME stands for the American Society for Mechanical Engineers.)

As will be seen in a subsequent section of this chapter, a spherically shaped pressure vessel is the most efficient possible in terms of minimizing wall thickness and vessel weight for a given volume and pressure. However, it is the least efficient in terms of space utilization. As a result, most pressure vessels are cylindrical in shape with either hemispherical or elliptic ends (generally called heads) as shown in Figure 14-1. This shape is more space efficient than the sphere, more stress efficient than the cube, and yet can be readily fabricated.

Some consideration must be given to the thickness of the pressure vessel wall. As the title of this chapter suggests, we are going to limit our discussion to *thin-walled* pressure vessels. A thin-walled pressure vessel is defined as one whose inside radius of curvature is at least five times the wall

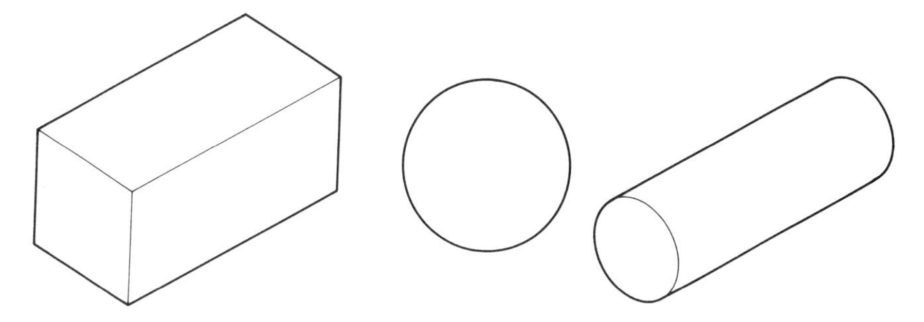

Figure 14-1. Pressure vessels.

thickness. This covers a great range of problems and permits an important simplifying assumption. This assumption is that the stress in the pressure vessel wall does not vary from the inside of the wall to the outside of the wall. While the stress really does vary, for a thin-walled pressure vessel the variation is negligible. Thick-walled cylinders require design procedures beyond the scope of this text.

One additional simplifying assumption is made before beginning analysis. It is assumed that each pressure vessel is jointless, that is, one piece. In reality, this is not really true, but with modern welding techniques, the joints are actually stronger than the base material, so the assumption is a safe one. A subsequent chapter will deal with various types of joints and joint efficiencies.

14.3 THIN-WALLED SPHERES

Figure 14-2a shows a spherical pressure vessel that has been cut through any maximum cross section. The cross-sectional annular section thus formed has the same inside and outside radii as the spherical vessel. If the spherical vessel were under pressure, there would be a force resulting from the pressure that would attempt to separate the two halves of the sphere. This force would be equal to the pressure times the area, $F = PA$. The area used is the *projected* area of the inside spherical surface not the actual spherical surface area, $A = \pi r^2$. Thus, $F = P\pi r^2$. The resultant stress in the material resists, in tension, the attempt to separate the two halves of the sphere. The tensile force resisting the pressure force is equal to the stress times the area under stress, $T = \sigma A$. The area in this case is the annular cross section of the cut shell of the sphere, $A = 2\pi r_{avg} t$. However, for a thin-walled pressure vessel, this is simplified by using the inside radius. The resultant error is negligible; $A = 2\pi rt$, and $T = 2\sigma\pi rt$. Equilibrium requires that the tensile force in the material counter the pressure force. Thus

$$F - T = 0 \qquad P\pi r^2 - 2\sigma\pi rt = 0$$

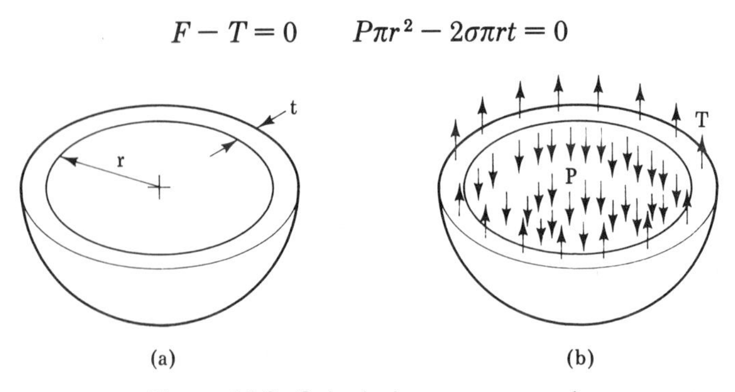

Figure 14-2. Spherical pressure vessels.

Solving for stress yields

$$\sigma = \frac{Pr}{2t}$$

Figure 14-2b shows the free-body diagram of the above.

Example 14-1: A spherical pressure vessel has an inside diameter of 12 m and a wall thickness of 15 mm. If the internal pressure is 0.4 MPa what will the stress in the wall be?

Solution:

Step 1 Check to see if the problem qualifies as a thin-walled pressure vessel.

$$\frac{r}{t} = \frac{6 \text{ m}}{0.015 \text{ m}} = 400 \quad \text{(qualifies easily)}$$

Step 2 Use the relationship developed previously for stress and proceed with the solution as follows:

$$\sigma = \frac{0.4 \times 10^6 \text{ N/m}^2 \text{ (6 m)}}{2(0.015 \text{ m})} \qquad \boldsymbol{\sigma = 80 \text{ MPa}}$$

Problem 14-1: A spherical pressure vessel is made of a material whose maximum working stress is 16,500 psi. It is to have a diameter of 6 ft and contain a pressure of 500 psi. What maximum wall thickness will be required? Show that it was appropriate to consider this a thin-walled pressure vessel.

Problem 14-2: 35 m^3 of helium at a pressure of 6 MPa is to be stored in spherical tanks. The tanks are to be fabricated from 20-mm thick steel with a maximum working stress of 150 MPa. How many tanks with what maximum diameter will be needed? ($V_{\text{sphere}} = \frac{4}{3}\pi r^3$)

14.4 THIN-WALLED CYLINDERS

If we were to imagine the two separated halves of the sphere discussed above connected by a cylindrically shaped shell as shown in Figure 14-3a we would have essentially a cylindrical pressure vessel. The end caps (heads) are not always hemispherical but that will not affect our analysis of the cylindrical portion of the vessel.

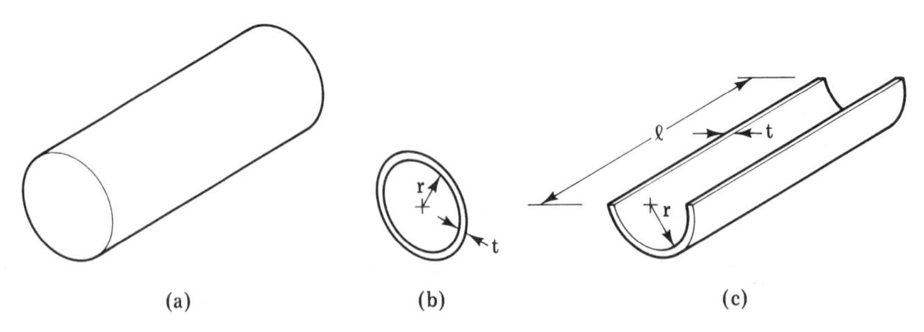

Figure 14-3

If we cut through the cylinder across the longitudinal axis (perpendicular to it) we get a cross section of material with the shape of a circular annulus as shown in Figure 14-3b. The pressure in the vessel is pushing on the end caps and stretches the cylinder along its length. The analysis then proceeds identically as for the sphere with the end result

$$\sigma_l = \frac{Pr}{2t}$$

The reason for the subscript l is that this is generally called the *longitudinal stress* in the cylinder to differentiate from the *circumferential stress* which we will discuss next.

If we cut the cylinder in a plane along the longitudinal axis (parallel to it) we get a cross section that consists of two long thin rectangles as shown in Figure 14-3c. The pressure force that attempts to split the cylinder into two halves is $F = PA$. A is the *projected* area of the half cylinder, $A = 2rl$. The pressure force then is $F = 2Prl$. The force resisting the breakup of the cylinder is due to the tensile stress in the material, $T = \sigma_c A$. The area in question here is the sum of the two rectangular strips, $A = 2tl$. Thus, $T = \sigma_c 2tl$. Equilibrium requires that these two forces add up to zero.

$$F - T = 0 \qquad 2Prl - \sigma_c 2tl = 0 \qquad \sigma_c = \frac{Pr}{t}$$

The circumferential stress is therefore independent of the length of the cylinder. Any change in length will change both the pressure force and the resisting tensile force in the same proportion. An interesting historical note is that circumferential stress is still sometimes called *hoop stress.* This goes back to the time when barrels, tanks, and even pipes were made by strapping steel or iron bands (hoops) around pieces of wood to hold them tightly in a cylindrical shape as shown in Figure 14-4.

The most significant point to note in the above discussion is that the circumferential stress is always twice the longitudinal stress in a closed

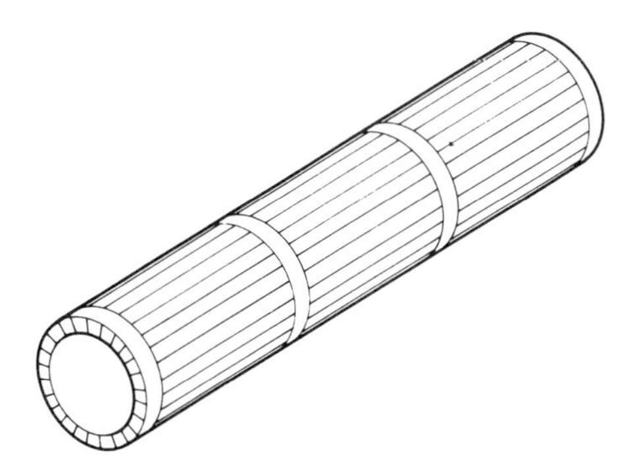

Figure 14-4

cylinder under pressure. If the cylinder is open as in the case of some vertical tanks then no longitudinal stress occurs. In either case, design is based on circumferential stress.

Do not be misled by the analysis of the sphere and the cylinder. The stress in the heads (ends) on a cylindrical pressure vessel is equal to the longitudinal stress in the cylindrical portion generally only when the heads are hemispherical, the wall thicknesses are the same and the joints are welded. There are heads with other shapes and frequently one head is bolted to the cylinder through flanges on the head and cylinder as shown in Figure 14-5. In the next chapter, we will examine such connections among others.

Example 14-2: The tank shown in Figure 14-6 has an inside diameter of 5 m and a height of 8 m. It is to be fabricated with four vertical sections of equal height. It contains a liquid with a specific gravity of 2.5. The steel plate to be used has a maximum working stress of 40 MPa. Determine the minimum wall thicknesses for each vertical section.

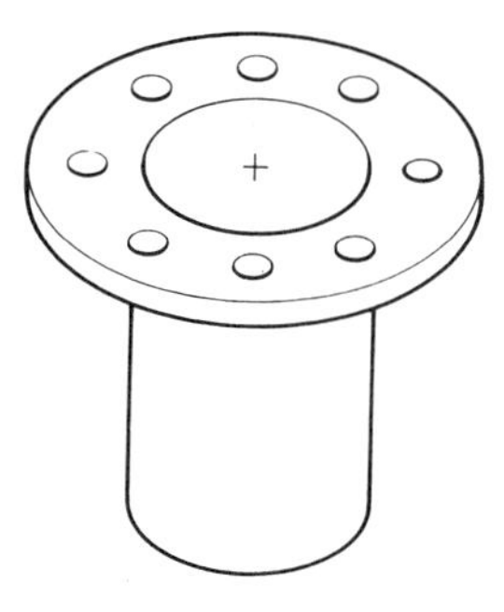

Figure 14-5

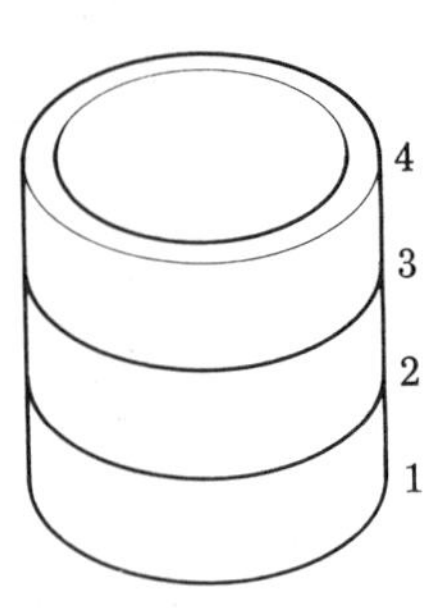

Figure 14-6

Solution:

Step 1 Recognize that the maximum pressure for each section occurs at the bottom edge of the section and that this will be the design pressure.

Step 2 Recognize that the design will be based on circumferential stress.

Step 3 Determine the four design pressures as follows: The pressure at any depth in a liquid is equal to the specific weight of the liquid times the depth. $P = \gamma_l h$. The specific weight of a liquid is equal to its specific gravity times the specific weight of water. $\gamma_l = sg_l \gamma_{H_2O}$. Generally then

$$P = sg_l \gamma_{H_2O} h$$

Specifically for this problem then

$$P = 2.5(9.8\ \text{kN/m}^3)h \qquad P = 24.5\ \text{kN/m}^3\ (h)$$

The four design pressures will be

$$P_1 = 24.5\ \text{kN/m}^3\ (8\ \text{m}) \qquad P_1 = 196\ \text{kPa}$$

$$P_2 = 24.5\ \text{kN/m}^3\ (6\ \text{m}) \qquad P_2 = 147\ \text{kPa}$$

$$P_3 = 24.5\ \text{kN/m}^3\ (4\ \text{m}) \qquad P_3 = 98\ \text{kPa}$$

$$P_4 = 24.5\ \text{kN/m}^3\ (2\ \text{m}) \qquad P_4 = 49\ \text{kPa}$$

Step 4 Rewrite the circumferential stress equation for the wall thickness t and find four minimum wall thicknesses as follows:

$$\sigma_c = \frac{Pr}{t} \qquad t = \frac{Pr}{\sigma_c}$$

Specifically for this problem

$$t = \frac{P(2.5 \text{ m})}{40 \text{ MPa}} \qquad t = 0.0625P$$

(if P is left in kPa, t will be in mm)

$$t_1 = 0.0625(196) \qquad \mathbf{t_1 = 12.3 \text{ mm}}$$

$$t_2 = 0.0625(147) \qquad \mathbf{t_2 = 9.2 \text{ mm}}$$

$$t_3 = 0.0625(98) \qquad \mathbf{t_3 = 6.2 \text{ mm}}$$

$$t_4 = 0.0625(49) \qquad \mathbf{t_4 = 3.1 \text{ mm}}$$

Problem 14-3: Determine the maximum allowable pressure in a steel pipe with an inside diameter of 17.938 in. and a wall thickness of 1.031 in. if the maximum working stress is 12,500 psi.

Problem 14-4: Determine the minimum wall thicknesses for the hemispherical heads and the cylindrical body of a 2-ft-diameter steel tank containing air at 200 psi. The maximum working stress is 9500 psi.

14.5 READINESS QUIZ

1. The organization whose codes govern the design of pressure vessels is the

   (a) American Society for Testing Materials
   (b) American Society for Mechanical Engineers
   (c) American Institute of Steel Construction
   (d) all of the above
   (e) none of the above

2. The stress level in the wall of any pressure vessel with circular cross section is a function of the

   (a) radius of wall curvature
   (b) wall thickness
   (c) pressure in the vessel
   (d) all of the above
   (e) none of the above

3. A thin-walled pressure vessel is defined as one whose wall thickness is at most

   (a) one-tenth the vessel diameter
   (b) one-tenth the vessel radius

(c) one-fifth the vessel diameter
(d) any of the above
(e) none of the above

4. Hoop stress in a cylindrical shell is simply another name for

(a) shear stress
(b) longitudinal stress
(c) circumferential stress
(d) all of the above
(e) none of the above

5. The ends of cylindrical pressure vessels are often called

(a) hoops
(b) codes
(c) vessels
(d) any of the above
(e) none of the above

6. The end of a cylindrical pressure vessel may be attached to the cylinder by

(a) screwing
(b) bolting
(c) welding
(d) any of the above
(e) none of the above

7. The most stress-efficient shape for a pressure vessel is a

(a) cube
(b) sphere
(c) cylinder
(d) any of the above
(e) none of the above

8. When working with pressure vessels that are considered thin walled we can ignore small variations in stress that occur

(a) around the circumference of the vessel
(b) along the length of the vessel
(c) across the wall of the vessel
(d) all of the above
(e) none of the above

9. A head is another name for __________ on a pressure vessel.

(a) a manhole
(b) an end

(c) a pipe connection
(d) any of the above
(e) none of the above

10. In determining the force acting on a curved surface as a result of the pressure of a fluid, one always uses

    (a) the actual surface area of the area in contact with the fluid
    (b) one-half of (a) above
    (c) twice (a) above
    (d) the projected area in contact with the fluid
    (e) none of the above

14.6 SUPPLEMENTARY PROBLEMS

14-5 A 10-ft-diameter sphere is to contain a gas at a pressure of 1000 psi. The sphere is to be made of 6061-T6 aluminum using a factor of safety of 4 (based on yield). What will the minimum wall thickness have to be? Is it appropriate to design this as a thin-walled pressure vessel?

14-6 The standpipe for a municipal water tank is 24 in. in inside diameter. The maximum water level above the base is 140 ft. The pipe is steel with a working stress of 8500 psi. What minimum wall thickness should the pipe have at the base?

14-7 A spherical pressure vessel 2 m in diameter contains gas at a pressure of 5 MPa. If A1S1 1045 steel is to be used with a factor of safety of 5 based on ultimate, what minimum wall thickness should be used?

14-8 What is the greatest pressure which may be contained in a cylindrical tank with a 1.5-m diameter and a 15-mm wall thickness? The material is A1S1 1095 steel and a factor of safety of 6 based on the ultimate should be used.

Chapter 15

Connections

15.1 OBJECTIVES

Upon completion of the work relating to this chapter you should be able to perform the following.

1. Describe and differentiate among the various types of connections used in machines and structures.
2. Differentiate between a lap joint and a butt joint.
3. Differentiate between a bolt and a rivet.
4. Define and differentiate among shear failure, bearing failure, and tensile failure of connections.
5. Analyze and design simple concentrically loaded bolted and riveted connections.
6. Analyze and design simple eccentrically loaded bolted and riveted connections.
7. Analyze and design simple concentrically loaded welded connections.
8. Analyze and design simple eccentrically loaded welded connections.

15.2 CONNECTIONS DEFINED

It is rare that a usable product is made of one-piece construction. A few come readily to mind such as our daily dishes, a baseball bat, and a sponge rubber ball among others. However, the vast majority of our products whether they are as simple as a steam iron or a stapler, or as complicated as an automobile or a building, require that somehow various components be held together permanently.

This is obviously a big area to cover and it is best to note immediately the types of joining methods that will not be considered here. A list of these follows:

1. **Fasteners** This area would include wood screws, nails, sheet metal screws, cotter pins, and snap rings to mention only a few. A good machine design text will cover many of these. (The bolt and rivet are two major types of fasteners which we will consider.)
2. **Adhesives** An important and growing area. More and more is being published but manufacturer's data on new developments is important here.
3. **Pinned Connections** Not specifically dealt with here, but handled in the same manner as a single rivet would be.
4. **Threaded Connections** This differs from bolts in that one of the components to be joined actually contains one or more threaded holes.
5. **Friction Fit** A component has been oversized, then chilled and thus reduced in size, slipped into a hole and allowed to expand. In attempting to expand to its original size, it exerts a normal force on the walls of the hole and a substantial friction force can be developed.

As the chapter develops, you will undoubtedly think of other methods of joining two parts together that we have ignored. The entire textile and garment industry presents a very specialized but important field for "connections" of some type. At this point, you may wonder what's left. However, as you look around you, you will find many examples of the types of connections we will cover:

1. **Bolted** As shown in Figure 15-1a, this method involves drilling or punching holes all the way through the components to be joined. The bolt is passed through and a nut threaded onto the end of it. Tightening the nut down to provide the appropriate tension on the bolt is an important part of the design construction process. It has the advantage that it can be accomplished in relatively awkward physical situations and requires minimal specialized equipment. Sometimes the ease of disassembly can be an advantage.

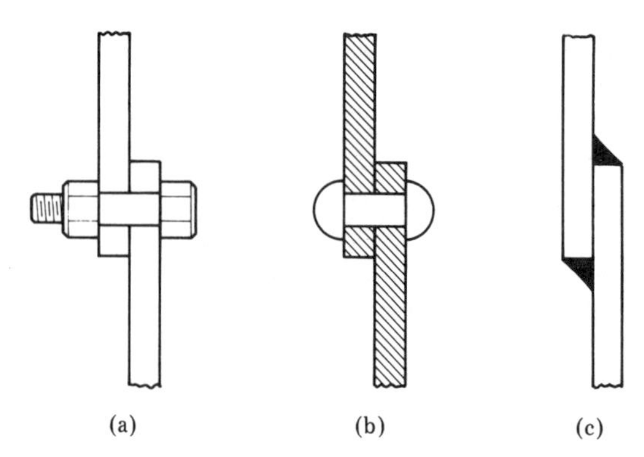

Figure 15-1. Types of connections.

2. **Riveted** A rivet is similar to a bolt in construction except it is not threaded and consists of only the one piece. It is driven through the hole and "headed" on the other side by force to result in the connection shown in Figure 15-1b. Rivets come in a variety of styles as shown in Figure 15-2. If driven hot, the rivet will expand to fill the hole. A two-man operation in the field where it is also more unwieldy than bolting, riveting is heavily concentrated in shop fabrication where it is economical under controlled conditions. Disassembly is not as readily achieved as for bolts, particularly if the rivets were driven hot.

3. **Welded** A welded joint is made by either localized heating of the two components until they melt together or by adding a third material which is molten and partially melts the components. The resultant joint of the second method is shown in Figure 15-1c. This is quite difficult to disassem-

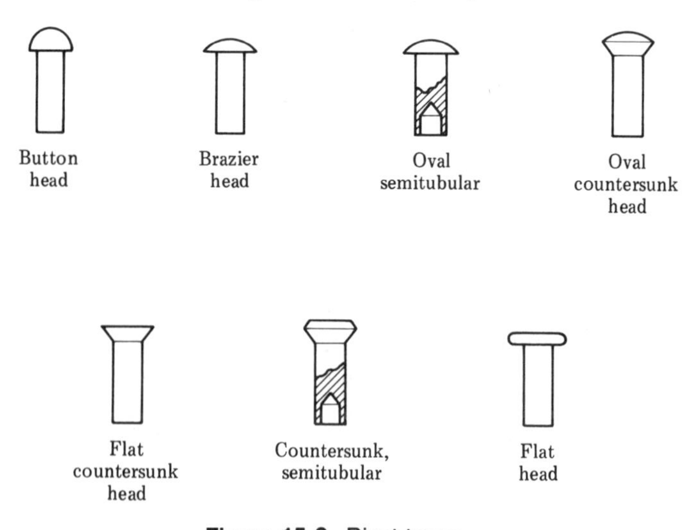

Figure 15-2. Rivet types.

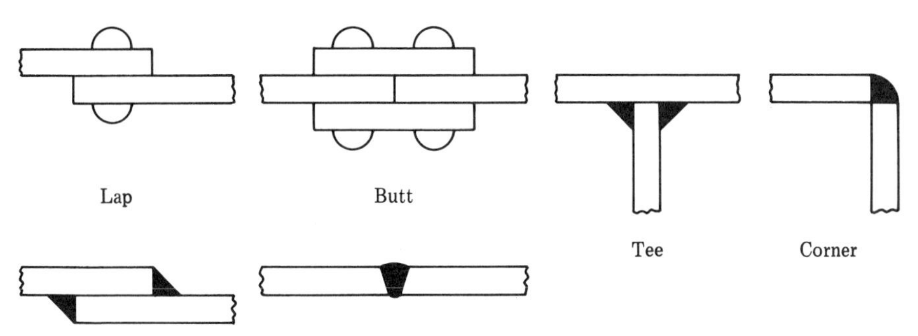

Figure 15-3. Types of joints.

ble effectively but on the other hand has the advantage of actually having merged the material of the two components.

Although the method of connection was different in each case shown in Figure 15-1, the type of joint was the same. It is called the *lap* joint because the two components to be joined overlap each other. Figure 15-3 contrasts this with the *butt, corner,* and *tee* joint. As indicated, the lap and butt can be accomplished by any of the three methods we are about to study. Of those three, however, only welding can accomplish the tee and corner joints. A little later in this chapter we will examine types of welds in more detail.

15.3 FAILURE ANALYSIS

Although we would not want our connections to fail, in order to avoid that, it would be useful to analyze each type of connection to see how failure would occur if we permitted it. This will assure that we evaluate the appropriate cross section and use the appropriate allowable stress in our design work. In examining modes of failure for all three types of connection we will also obtain a feel for their relative merits.

Bolted Connection Failure

Figure 15-4a shows a lap joint held together by a single bolt. Subsequent steps show the possible modes of failure for the joint.

1. **Shear Failure** As shown in 15-4b, one possibility is that the two plates will cut or shear the bolt just as scissors cut a piece of paper. The area involved would be the circular cross-sectional area of the bolt and the stress, the allowable shear stress of the bolt material.

2. **Tension Failure** Figure 15-4c shows a crack developing in the plate perpendicular to the load and passing through the bolt axis. The plate has

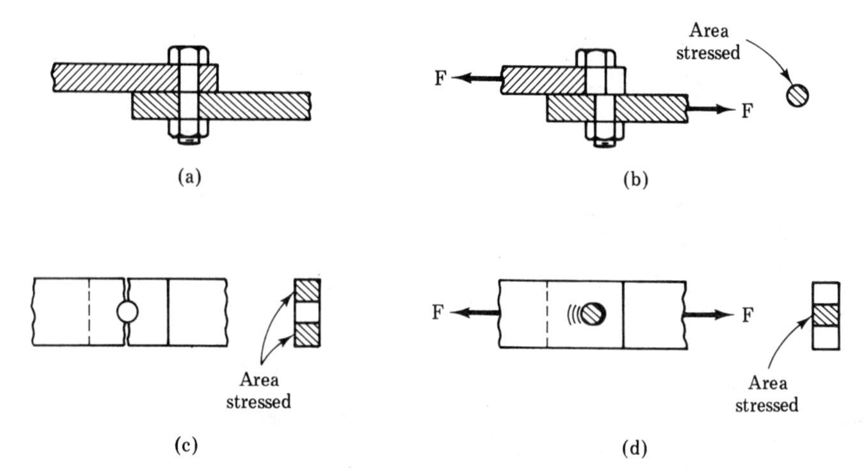

Figure 15-4. Lap joint analysis.

parted due to tension at its weakest point. Not only is the cross-sectional area of the plate smaller because of removal of the material for the bolt hole but the existence of the bolt hole itself causes a concentration of stresses (see Chapter 13). It could, of course, just as easily have been the other plate. The stress involved is the allowable tensile stress of the plate material.

3. **Bearing Failure: Plate** If the plates shift slightly, taking up the slack between the bolt diameter and the bolt hole diameter, the bolt will begin to bear or push on the plate and attempt to crush it. Initially, the area of contact will be very small and the plate will tend to crush readily (assuming it is the weaker of the two materials). Very quickly however, the entire half-cylinder face of the bolt will come to bear on the plate spreading the load over a larger area. This area will be the *projected* contact area of the bolt, not the surface area of the bolt that is in contact. The area will thus be the bolt diameter times the plate thickness. The stress will be the allowable compressive stress for the plate material. *If* the joint is designed as a friction type, which means that the tension in the bolts develops sufficient normal force for the friction force between plates to counteract the load, then bearing stress on the plate is deleted from consideration.

4. **Bearing Failure: Bolt** If the bolt bears on the plate, then the reverse is also true. However, it would be rare for the bolt to have a lower allowable compressive stress than the plate so this failure mode, shown in 15-4d, is generally not considered.

Riveted Connection Failure

Analysis of a riveted connection involves the exact same considerations and assumptions as bolted connections. The differences are that no riveted

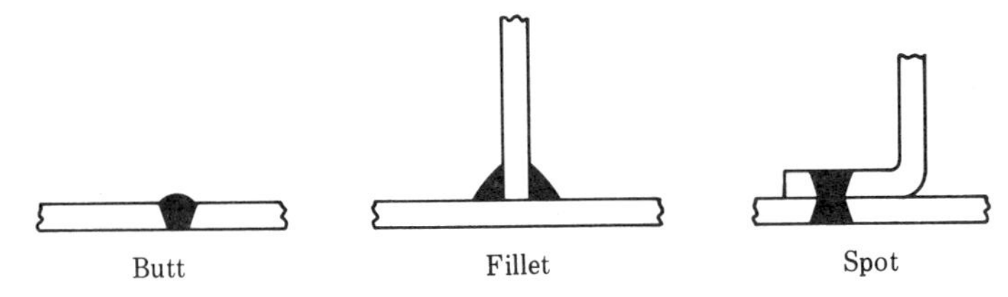

Figure 15-5. Weld types.

connection is ever considered a friction connection and the materials used for rivets are different than those used for bolts.

Welded Connection Failure

There are several basic types of welds including spot, fillet, and butt (groove) as shown in Figure 15-5. There are many ways of cutting a groove or grooves for a butt weld. However, the end result is that, properly executed, the weld is assumed to have at least the same strength as the plates it holds together. There is no further design required once a butt weld has been chosen as the connector. The fillet weld can assume a number of positions as you have already seen. Regardless of the direction of load application, the *fillet* weld is always analyzed in *shear*. The spot weld is also always analyzed in shear.

15.4 BOLTED AND RIVETED CONNECTION DESIGN

Joint Efficiency

The efficiency of a bolted or riveted joint is simply the strength of the joint divided by the tensile strength of the plate (times 100 for percent). The tensile strength of the plate is simply the full cross-sectional area of the plate times the allowable tensile stress for the plate material. The strength of the joint is lowest of those of the following considered for the joint: bearing, shear, and tensile. It is important to note that the tensile strength of the joint is different than the tensile strength of the plate. The tensile strength of the joint is determined at the first row of bolts encountered as one approaches the joint, as shown in Figure 15-6. This is because the width of the plate has been reduced by the bolt hole(s) and yet the plate is still carrying the full load.

Allowable Stresses

A number of organizations have done considerable development of specifications for joint design. Among these are the American Institute of

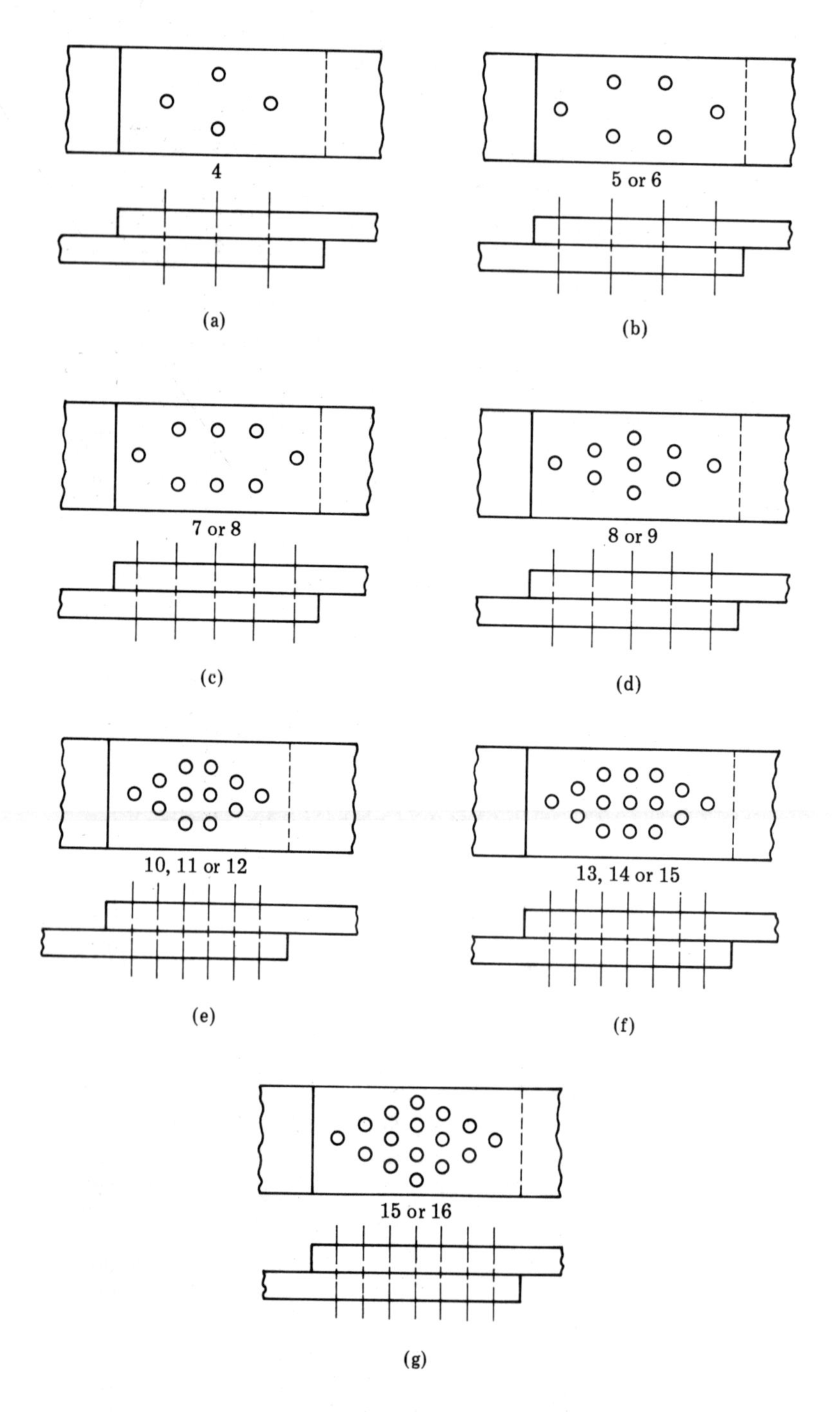

Figure 15-6. Lap joint bolt patterns.

Steel Construction (AISC), the American Society of Mechanical Engineers (ASME), and the Aluminum Association. Some of their specifications are presented in Tables 15-1, 15-2, and 15-3. These should be adequate for the work in this text. Further questions should be directed to publications of these organizations in your libraries or to the organizations themselves.

Table 15-1
Allowable Stresses for Connections (AISC)*

Material	Allowable Stress, psi	
	Friction-Type Connection	Bearing Type Connection
Shear on Fasteners		
Rivets		
A 502-1		17,500
A 502-2		22,000
A 502-3		22,000
Bolts (threads at shear plane)		
A 307		10,000
A 325	17,500	21,000
A 490	22,000	28,000
Bolts (threads not at shear plane)		
A 307		10,000
A 325	17,500	30,000
A 490	22,000	40,000
Bearing on Projected Area		
A 36		48,500
A 242		67,500
A 441		67,500
A 572		67,500
A 588		67,500
Tension on Net Section		
A 36	22,000	22,000
A 242	30,000	30,000
A 441	30,000	30,000
A 572	30,000	30,000
A 588	30,000	30,000

* The above values are for educational purposes only. The AISC "Manual of Steel Construction" should be consulted for design purposes.

Table 15-2
Allowable Stresses for Bolted Connections (ASME)*

Material		Allowable Stress, psi
Shear on Fastener		
SA 325		15,400
SA 499	1 in dia and less	18,400
	$1\frac{1}{8}$ to $1\frac{1}{2}$ in dia	16,200
	$1\frac{5}{8}$ to 3 in dia	11,600
Bearing on Fastener		
SA 325		30,700
SA 499	1 in dia and less	36,800
	$1\frac{1}{8}$ in to $1\frac{1}{2}$ in dia	32,300
	$1\frac{5}{8}$ in to 3 in dia	23,200
Bearing on Plate		
SA 285	Grade A	17,900
	Grade B	20,000
	Grade C	21,900
SA 36		20,200
SA 612	Grade A	33,100
	Grade B	32,200
Tension on Net Section		
SA 285	Grade A	11,200
	Grade B	12,500
	Grade C	13,700
SA 36		12,600
SA 612	Grade A	20,700
	Grade B	20,200

* The above values are for educational purposes only. The ASME "Boiler and Pressure Vessel Code" should be consulted for design purposes.

Joint Design Procedure

The end purpose in joint design is not just a workable design but one with the maximum possible efficiency. The plate tension in the first row should be the limiting factor.

1. Select a bolt size.
2. Determine bolt hole size.
3. Determine plate width based on one bolt hole in row.
4. Determine allowable shear force in one bolt (check for single or double shear).

Table 15-3
Allowable Stresses for Connections (AA)*

Component	Allowable Shear Stress		Allowable Tensile Stress		Allowable Bearing Stress	
Alloy and Temper	psi	MPa	psi	MPa	psi	MPa
Bolts**						
2024-T4	16,000	110	26,000	179		
6061-T6	12,000	83	18,000	124		
7075-T73	17,000	117	28,000	193		
Rivets***						
1100-H14 (1100F)	4,000	27				
2017-T4 (2017-T3)	14,500	100				
6053-T61 (6053-T61)	8,500	58				
6061-T6 (6061-T6)	11,000	76				
Connected Members****						
1100-H12	6,600	45.6			11,000	76
2014-T6	34,800	240			49,000	338
3003-H12	7,200	49.8			11,500	79
6061-T6	21,000	144.6			34,000	234
6063-T6	16,800	115.8			24,000	165

* "Specifications for Aluminum Structures" The Aluminum Association.
** Shear area at nominal bolt diameter unless threads in shear plane in which case use root diameter.
*** Cold driven, ( ) indicates condition after driving.
**** Allowable tensile strength equals 0.6 times yield strength for all alloys.

5. Determine allowable bearing force on plate by one bolt.
6. Use smaller of 4 or 5 to determine the minimum number of bolts (allowable plate load divided by limiting bolt force).
7. Select a bolt pattern from Figure 15-6 or 15-7.
8. Check plate stress at each row.
9. Revise pattern if necessary to result in maximum plate stress in row 1.
10. Calculate joint efficiency.
11. Select another bolt size to try for improved efficiency.

Concentrically Loaded Bolt and Rivet Connection Design

In the examples and problems of this section it is assumed that the line of action of the load passes through the centroid of the rivet or bolt pattern as shown in Figure 15-7.

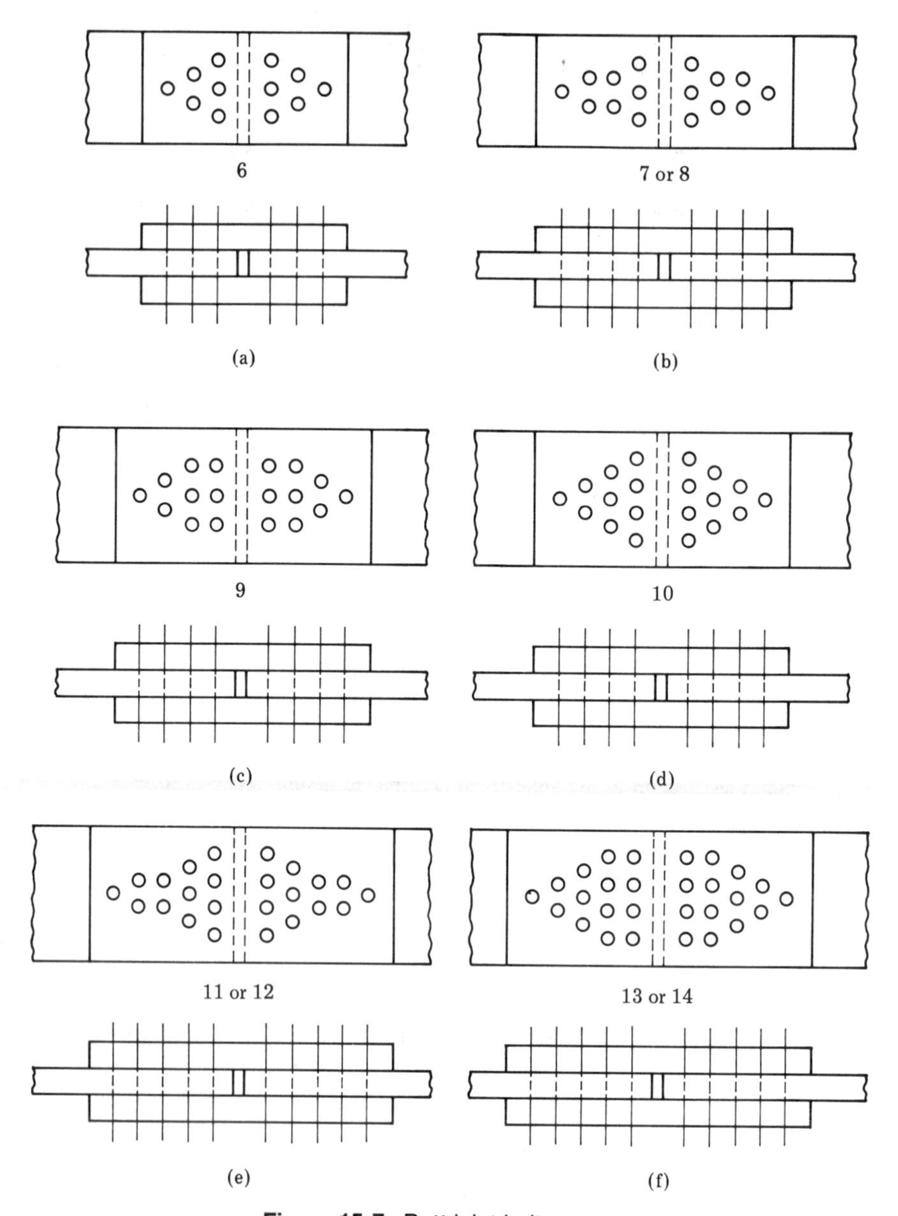

Figure 15-7. Butt joint bolt patterns.

Example 15-1: Two SA612 Grade B plates $\frac{1}{2}$ in. thick are to be connected by SA449 bolts by a lap joint. The tensile load on the joint is 80 kips. Determine the plat width and the diameter and number of bolts required (based on ASME Code).

Solution:

Step 1 For a first trial, select a 1-in.-diameter bolt (equal to the joint thickness).

$$\boldsymbol{d = 1 \text{ in.}}$$

Step 2 Determine bolt hole diameter.

$$D = d + 0.02 \qquad D = 1.02 \text{ in.}$$

Step 3 Determine plate width.

$$\sigma_{\text{all}} = \frac{P}{A} = \frac{P}{(w - D)t}$$

$$w = \frac{P}{t\sigma_{\text{all ten}}} + D = \frac{80{,}000 \text{ lb/in.}^2}{0.5 \text{ in.}(20{,}000 \text{ lb/in.}^2)} + 1.02$$

$$w = 8.94 \quad \text{round to } \boldsymbol{w = 9 \text{ in.}}$$

The allowable load on the plate is then

$$P = \sigma_{\text{all ten}} t(w - D) = 20{,}000(0.5)(9 - 1.02)$$

$$P = 80{,}000 \text{ lb}$$

Step 4 Determine allowable shear force in one bolt (single shear).

$$F_s = 0.7854 d^2 \sigma_{\text{all shear}}$$

$$F_s = 0.7854\ (1 \text{ in.}^2)(18{,}400 \text{ lb/in.}^2) \qquad F_s = 14{,}450 \text{ lb}$$

Step 5 Determine allowable bearing force on plate by one bolt.

$$F_b = dt\sigma_{\text{all bear}} = (1 \text{ in.})(0.5 \text{ in.})(32{,}300 \text{ lb/in.}^2)$$

$$F_b = 16{,}150 \text{ lb}$$

Step 6 The smaller of the above (F_s, F_b) determines the number of bolts as follows:

$$n = \frac{P}{F_{\text{bolt min}}} = \frac{80{,}600 \text{ lb}}{14{,}450 \text{ lb}} \qquad n = 5.58$$

round to $\boldsymbol{n = 6}$

Determine F_s for six bolts

$$F_{s6} = 0.7854nd^2\sigma_s = 0.7854(6)(1)^2(18{,}400 \text{ lb/in.}^2)$$

$$F_{s6} = 86{,}708 \text{ lb}$$

Determine F_b for six bolts

$$F_{b6} = ndt\sigma_b = 6(1.0 \text{ in.})(0.5 \text{ in.})(32{,}300 \text{ lb/in.}^2)$$

$$F_{b6} = 113{,}050 \text{ lb}$$

Step 7 Select bolt pattern b (lap joint).

Step 8 Check plate stress at each row.

Row 1. Full load is assumed to pass through diminished section.

$$\sigma_t = \frac{P}{(w - nD)t} = \frac{80{,}600 \text{ lb}}{[9 - (1)(1.02)]0.5}$$

$$\sigma_t = 20{,}200 \text{ psi}$$

Row 2. Five-sixths of load is assumed to pass through diminished section.

$$\sigma_t = \frac{5/6P}{(w - nD)t} = \frac{(5/6)80{,}600 \text{ lb}}{[9 - (2)(1.02)]0.5}$$

$$\sigma_t = 19{,}300 \text{ psi}$$

Pattern suitable since stress in row 2 is less than stress in row 1.

Step 9 Determine minimum gauge distance and minimum edge distance, check for feasibility.

$$e_{\text{min}} = 1.75d = 1.75(1 \text{ in.}) \qquad e_{\text{min}} = 1.75 \text{ in.}$$

$$g_{\text{min}} = 2.5d = 2.5(1 \text{ in.}) \qquad g_{\text{min}} = 2.5 \text{ in.}$$

For bolt pattern $2_e + 2_g = w_{\text{min}}$:

$$w_{\text{min}} = 2(1.75 \text{ in.}) + 2(2.5 \text{ in.}) \qquad w_{\text{min}} = 8.5 \text{ in.}$$

This was conservatively determined and is still less than the designed plate width of 9 in.

Step 10 Calculate the joint efficiency.

$$\eta = \frac{\text{joint strength}}{\text{plate strength}}$$

The joint strength is the least of the following: first row plate strength (Step 3), bolt shear strength (Step 6), and plate bearing strength (Step 6). Plate strength is at full width.

$$F_{\text{plate}} = wt\sigma_t = (9 \text{ in.})(0.5 \text{ in.})(20{,}000 \text{ lb/in.}^2) = 90{,}000 \text{ lb}$$

$$\eta = \frac{80{,}600 \text{ lb}}{90{,}900 \text{ lb}} \times 100 \qquad \boldsymbol{\eta = 88.7\%}$$

Step 11 The above joint will fail in tension. The maximum stress is in the plate and at row 1 as desired. Changing to a $\frac{7}{8}$-in. bolt will not reduce the width any for practical purposes and will result in more bolts (eight). It will result in a stronger, more efficient (and more expensive) joint but it was already strong enough with six 1-in. bolts. A $1\frac{1}{8}$-in. bolt size will result in a need for five bolts and an increase in plate width to over 10 in. Overall cost will be higher with reduced efficiency.

Example 15-2: Two 1-in. thick A572 plates are to be connected in a butt joint using $\frac{1}{2}$-in. thick cap plates of the same material A490 bolts. The tensile load on the joint is 345 kips. Determine the plate width and the diameter and number of bolts required (based on AISC Code). (For simplicity the verbage will be largely eliminated in this solution.)

Solution:

Step 1 Try $\boldsymbol{d = 1}$ **in.**

Step 2 $D = d + \frac{1}{8} \qquad D = 1.125$ in.

Step 3

$$w = \frac{P}{t\sigma_{\text{all ten}}} + D = \frac{345{,}000 \text{ lb/in.}^2}{1.0 \text{ in.}(30{,}000 \text{ lb/in.}^2)} + 1.125$$

$$\boldsymbol{w = 12.625 \text{ in.}}$$

Step 4 $F_s = 0.7854 d^2 \sigma_{\text{all shear}}(2)$ (double shear)

$$F_s = 0.7854(1 \text{ in.})^2(40{,}000 \text{ lb/in.}^2)(2) \qquad F_s = 62{,}382 \text{ lb}$$

Step 5 $F_b = dt\sigma_{\text{all bear}}$

$$F_b = (1\text{ in.})(1\text{ in.})(67{,}500\text{ in.}^2) \qquad F_b = 67{,}500\text{ lb}$$

Step 6

$$n = \frac{P}{F_{\text{bolt min}}} = \frac{345{,}000\text{ lb}}{62{,}832\text{ lb}} \qquad \boldsymbol{n = 6}$$

$$F_{S6} = 0.7854nd^2\sigma_s(2) \quad \text{double shear}$$

$$F_{S6} = 0.7854(6)(1)^2(40{,}000)(2) \qquad F_{S6} = 377{,}000\text{ lb}$$

$$F_{b6} = ndt\sigma_b = 6(1)(1)(67{,}500) \qquad F_{b6} = 405{,}000\text{ lb}$$

Step 7 Select bolt pattern *a* (butt joint). Note that the butt joint is designed in two equal halves. In each half, two cover plates are pulling on a main plate.

Step 8 Row 1

$$\sigma_t = \frac{P}{(w - nD)t} = \frac{345{,}000\text{ lb}}{(12.625 - 1.125)(1)}$$

$$\sigma_t = 30{,}000\text{ psi}$$

Row 2

$$\sigma_t = \frac{5/6(345{,}000\text{ lb})}{[12.625 - 2(1.125)](1)}$$

$$\sigma_t = 28{,}395\text{ psi}$$

Pattern suitable since stress in row 2 is less than stress in row 1.

Step 9

$$e_{\text{min}} = 1.75d = 1.75(1.0\text{ in.}) \qquad e_{\text{min}} = 1.75\text{ in.}$$

$$g_{\text{min}} = 3d = 3(1.0\text{ in.}) \qquad g_{\text{min}} = 3.0\text{ in.}$$

$$w_{\text{min}} = 2_e + 2_g = 2(1.75) + 3(3.0)$$

$$w_{\text{min}} = 12.5\text{ in.} \qquad \text{OK}$$

Step 10 $\eta = \dfrac{345{,}000\text{ lb}}{(12.625)(1)(30{,}000)} \times 100 \qquad \boldsymbol{\eta = 91\%}$

Step 11 Check cover plates. Since they can each be assumed to carry half the load, are each half the thickness of the main plates, and are the same material as the main plates, no mathematical check is needed.

Step 12 A check on other sizes indicates that seven $\frac{7}{8}$-in. bolts are suitable with an efficiency of 95.8%. Ten $\frac{3}{4}$-in. bolts would increase costs and drop joint efficiency to 92.9%. Five $1\frac{1}{8}$-in. bolts would be workable but drop efficiency to 90.2%. Seven $\frac{7}{8}$-in. bolts look like the best choice.

Problem 15-1: Two 10-mm-thick 2014-T6 aluminum plates are to be lap jointed using 2017-T3 rivets. The tensile load is 600 kN. Determine the plate width and the diameter and number of rivets required (based on AA Code).

Problem 15-2: Two $\frac{3}{4}$-in.-thick A441 steel plates are to be butt jointed using $\frac{1}{2}$-in.-thick cap plates of the same material and A502, Grade 2 hot-driven rivets. Determine the plate width and the diameter and number of rivets required (based on AISC Code) if the tensile load is 200 kips.

Problem 15-3: Two 1-in.-thick SA285 Grade B steel plates are to be butt jointed together using $\frac{3}{8}$-in.-thick cap plates of the same material using SA325 bolts. If the plate width is 16 in., what size and number of bolts will result in the maximum permissible load? What is that load?

Eccentrically Loaded Bolt and Rivet Connection Design

In the preceding section, the load always passed through the centroid of the rivet or bolt pattern as was illustrated in Figure 15-7. However, there are many applications where the load does not pass through this centroid. Figure 15-8 shows some beams with loadings that illustrate these. Some of these are already familiar to you, others will be dealt with in the last chapter. Note that all of these applications include at least one fixed support. Recall that a fixed support is capable of sustaining a moment load as well as a force load. Indeed the choice of a fixed support is generally dictated by necessity to provide a moment reaction as in the case of the cantilever or the desire to reduce deflections as in the other examples shown.

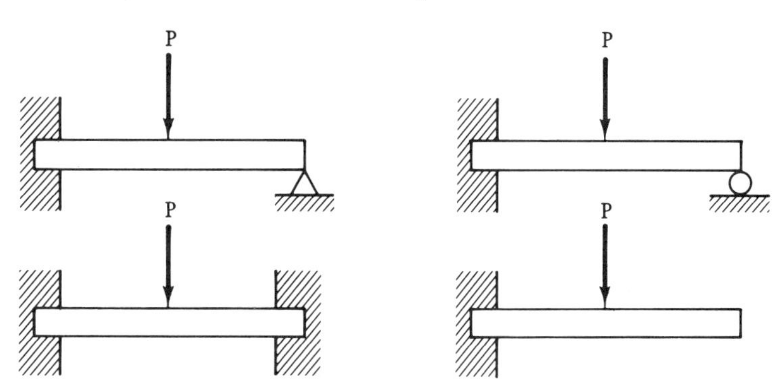

Figure 15-8. Fixed supports.

In the design of concentrically loaded connections, the use of two or more bolts or rivets was a matter of efficient design—a single bolt or rivet could have accomplished the task. In a fixed support, even a friction-type single-bolt connection would not be considered adequate. The loading on such a support may be expressed as an eccentric load as in Figure 15-9a and b, or as a combination of a concentric load and a torque or moment load as in 15-9c and d.

In some of your earlier work, you determined the force reaction (or its components) and the moment reaction at a fixed support. It was implied then and reinforced here that the force reaction passes through the centroid of the yet undersigned connection. (The loads are, of course, simply the opposite of the reactions.

In Figure 15-10 is shown an eight-bolt or rivet connection with only a torque load on it. (One special but common application of such a case is the shaft coupling which will be examined in Chapter 16.) The centroid is readily located by symmetry as lying halfway between the two columns of bolts and halfway between the second and third row. The force in each bolt resulting from the applied moment is perpendicular to a radius drawn from the centroid of the pattern to the centroid of the bolt, as shown in 15-10c. Also shown is the assumption that the force is directly proportional to the length of that radius.

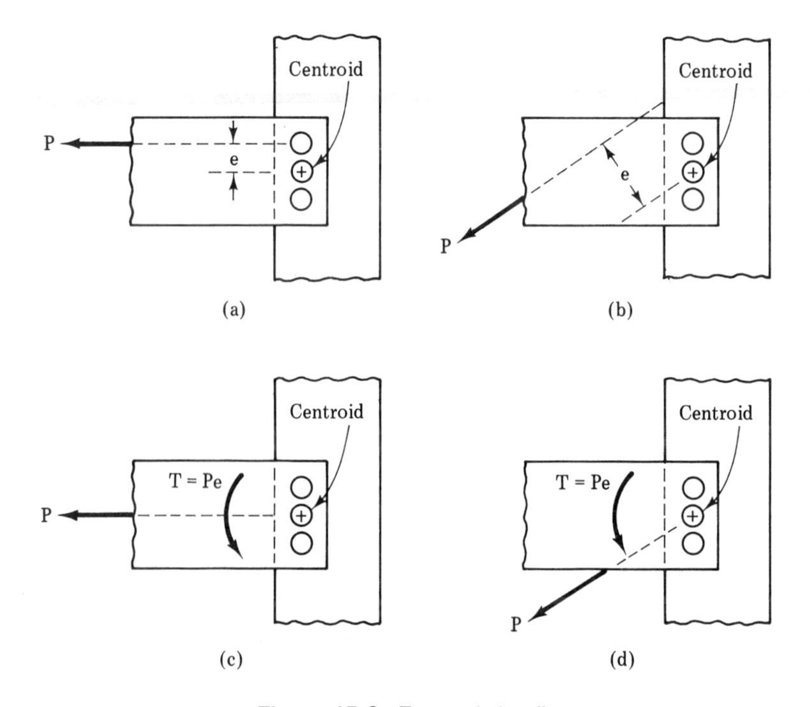

Figure 15-9. Eccentric loading.

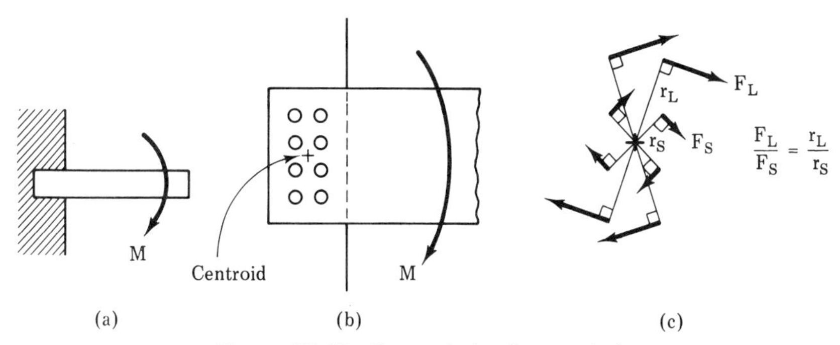

Figure 15-10. Eccentric-loading analysis.

The total torque applied is equal to the sum of all the torques resulting from each bolt force times its radius (distance to the pattern).

$$T = \Sigma Fr = F_1 r_1 + F_2 r_2 + \cdots$$

However each bolt force is also assumed to be proportional to its distance from the pattern centroid.

$$\frac{F_1}{r_1} = \frac{F_2}{r_2} = \frac{F_3}{r_3} = \cdots = \frac{F_{avg}}{r_{avg}}$$

$$\text{Then } F_1 = \frac{F_{avg} r_1}{r_{avg}} \qquad F_2 = \frac{F_{avg} r_2}{r_{avg}}, \quad \cdots$$

$$\text{and } T = \frac{F_{avg} r_1}{r_{avg}} r_1 + \frac{F_{avg} r_2}{r_{avg}} r_2 + \cdots$$

$$T = \frac{F_{avg}}{r_{avg}} (r_1^2 + r_2^2 + \cdots) = \frac{F_{avg}}{r_{avg}} (\Sigma r^2)$$

For any one bolt

$$T = \frac{F_i}{r_i} (\Sigma r^2) \qquad F_i = \frac{T r_i}{\Sigma r^2} \qquad F_i = \frac{T r_i}{\Sigma (x^2 + y^2)}$$

(Expressing each r in terms of components, $r^2 = x^2 + y^2$.)

As suggested by Figure 15-8 and 15-9 connections are frequently subjected to a force load as well as a moment load. The load on each bolt due to the overall force load is simply the overall force load divided by the number of bolts or rivets and is in the same direction as the overall load.

$$F_p = \frac{P}{n} \quad \text{at} \quad \theta p$$

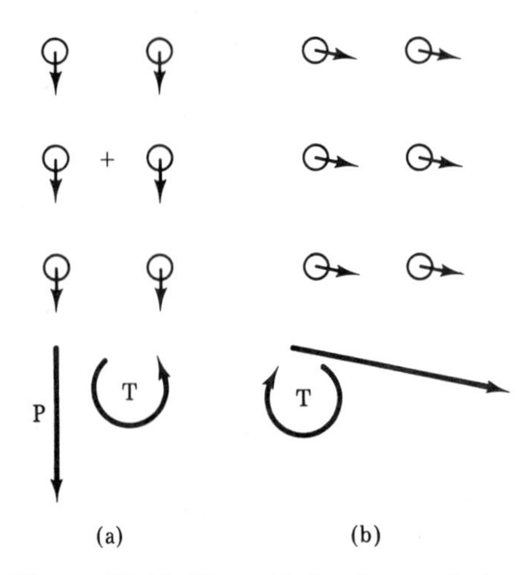

Figure 15-11. Eccentric-loading analysis.

The total load on each bolt is then the vector sum of the force due to the torque load and the force due to the force load $F_t = F_i \leftrightarrow F_p$ or by components.

$$F_t = \sqrt{F_{tx}^2 + F_{ty}^2} = \sqrt{(F_{px} + F_{ix})^2 + (F_{py} + F_{iy})^2}$$

Figure 15-11 shows the same six-bolt connection with different force and moment loads (to scale). From this, it can be seen that the maximum bolt load will always occur at one of the most distant bolts from the centroid. Specifically, it will occur at the one where F_i and F_p are closest to being in the same direction.

Example 15-3: A 10-in.-wide steel plate is to be bolted to a 16-in. WF A36 column as shown in Figure 15-12a with a load on the plate as shown. What size A325 bolts should be used? What thickness should the plate be? (Assume the flange of the column is thick enough.)

Solution:

Step 1 Change the loading to a concentric force load equal to *P* and a torque load equal to *Pe*, as shown in Figure 15-12b.

$$T = Pe = 80{,}000 \text{ lb}(3 \text{ ft}) = 240{,}000 \text{ ft lb}$$

Step 2 Determine the force load on each bolt, draw to approximate scale on pattern as shown in 15-12c.

$$F_p = \frac{P}{n} = \frac{80{,}000 \text{ lb}}{6} \qquad F_p = 13{,}333 \text{ lb} \quad \text{at} \quad 270°$$

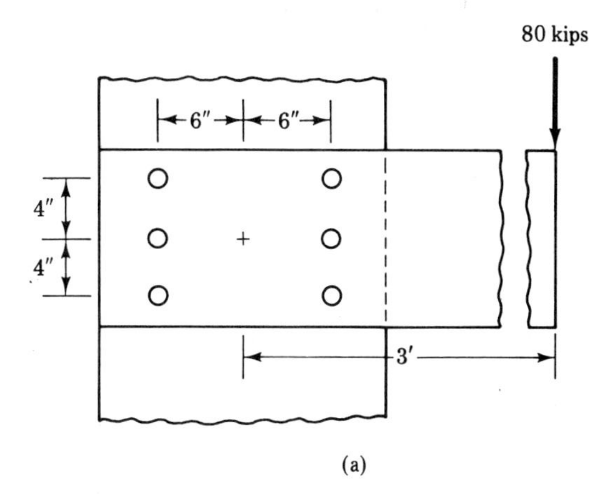

(a)

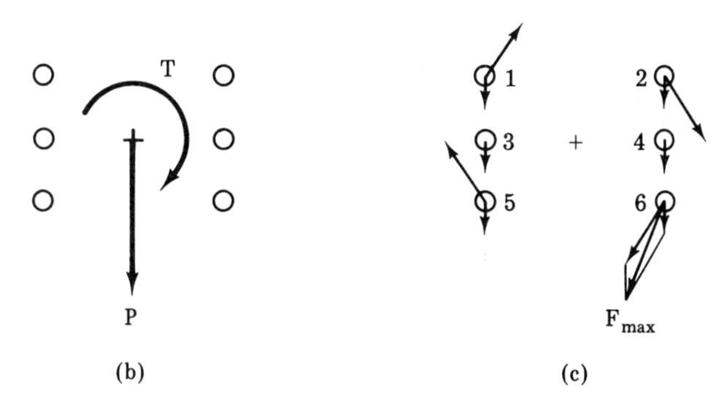

(b) (c)

Figure 15-12

Step 3 Determine largest bolt load due to torque, draw to approximate scale on pattern as shown in 15-12c.

$$r_1 = r_2 = r_5 = r_6 = \sqrt{4^2 + 6^2} = 7.2 \text{ in.}$$

$$r_3 = r_4 = 6 \text{ in.}$$

$$\Sigma r^2 = 2(6)^2 + 4(7.2)^2 \qquad \Sigma r^2 = 280 \text{ in.}^2 = 1.94 \text{ ft}^2$$

$$F_1 = F_2 = F_5 = F_6 = \frac{Tr_1}{\Sigma r^2} = \frac{240{,}000 \text{ ft lb}(3.6/12 \text{ ft}}{1.94 \text{ ft}^2}$$

$$F_1 = F_2 = F_5 = F_6 = 37{,}114 \text{ lb}$$

$$\theta_1 = \arctan\frac{+6}{+4} \qquad \theta_1 = 56.3°$$

$$\theta_2 = \arctan\frac{-6}{+4} \qquad \theta_2 = 303.7°$$

$$\theta_5 = \arctan \frac{+6}{-4} \qquad \theta_5 = 123.7°$$

$$\theta_6 = \arctan \frac{-6}{-4} \qquad \theta_6 = 236.3°$$

Step 4 Determine the maximum bolt force. θ_2 and θ_5 are both equally close to 270°, therefore F_{max} will occur at both.

$$F_{max} = \sqrt{(F_{px} + F_{ix})^2 + (F_{py} + F_{ix})^2}$$

$$F_{ix} = F_6 \cos \theta_6 = 37{,}114 \text{ lb}(\cos 236.3°) = -20{,}592 \text{ lb}$$

$$F_{iy} = F_6 \sin \theta_6 = 37{,}114 \text{ lb}(\sin 236.3°) = -30{,}878 \text{ lb}$$

$$F_{max} = \sqrt{(0 - 20{,}592)^2 + (-13{,}333 - 30{,}878)^2} \qquad F_{max} = 48{,}771 \text{ lb}$$

Step 5 Determine minimum bolt size.

$$F_{max} = 0.7854 d^2 \sigma_{\text{all shear}}$$

$$d = \sqrt{\frac{F_{max}}{0.7854 \sigma_{\text{all shear}}}} = \sqrt{\frac{48{,}771 \text{ lb}}{0.7854(30{,}000 \text{ lb/in.}^2)}}$$

$d = 1.483$ in., probably $\boldsymbol{d = 1.5}$ **in.**

Step 6 Determine the minimum plate thickness (use column width).

$$F_t = (w - nD)t\sigma_t \qquad D = d + 0.125 = 1.625 \text{ in.}$$

$$t = \frac{F_t}{(w - nD)\sigma_t} = \frac{80{,}000 \text{ lb}}{[16 - (2)(1.625)](22{,}000 \text{ lb/in.}^2)}$$

$t = 0.285$ $\boldsymbol{t = 5/16}$ **in.**

$$F_{b6} = ndt\sigma_b \qquad t = \frac{F_{b6}}{nd\sigma_b}$$

$$t = \frac{80{,}000 \text{ lb}}{6(1.5 \text{ in.})(48{,}500 \text{ lb/in.}^2)}, \qquad t = 0.183$$

(use tension figure)

Problem 15-4: An 8-in.-wide A36 steel plate is to be riveted to an 8-in. WF A36 column as shown in Figure 15-13. The rivets are to be hot driven A502, Grade 1. What minimum size rivets should be used? How thick should the plate be?

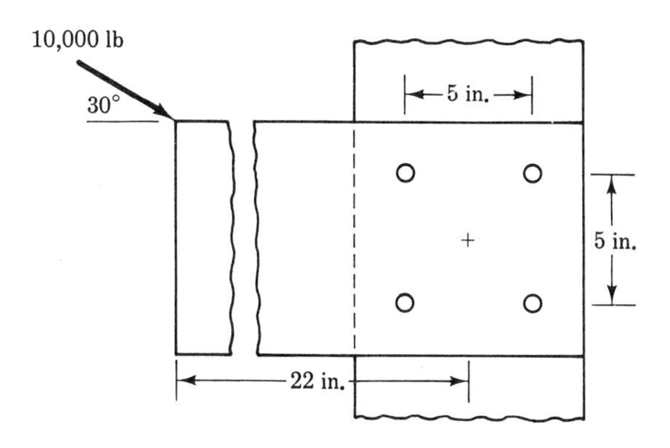

Figure 15-13

Problem 15-5: A 6-in.-wide A36 steel plate is to be bolted to a 12-in.-wide A36 channel as shown in Figure 15-14. What size A325 bolts should be used? What thickness should the plate be?

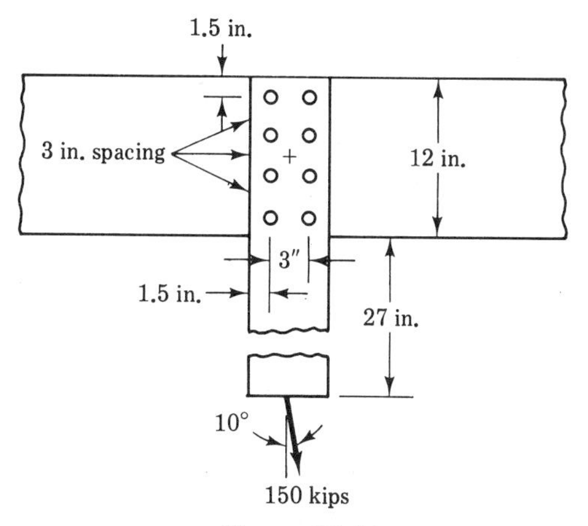

Figure 15-14

15.5 WELDED CONNECTION DESIGN

This section will be limited to some rather basic but common weldments. The two most common types are the butt and fillet welds. Figure 15-15 shows a variety of butt welds. In all cases a properly fabricated butt weld is considered to be stronger than the base material so no further design is required.

The fillet weld is used with lap joints and tee joints. The nomenclature and geometry involved are shown in Figure 15-16.

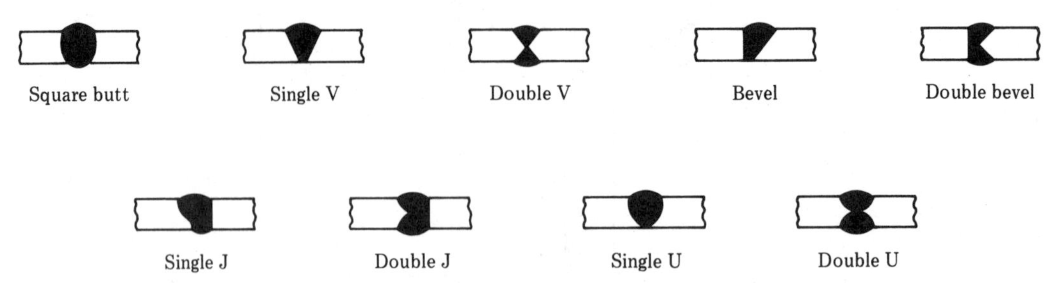

Figure 15-15. Butt weld types.

Figure 15-17 shows two arrangements for lap joints and one for a tee joint.

The above types of fillet welds with concentric loading, and one specific case of eccentric loading, will be the limit of our discussion. Standards for welding are set by the American Welding Society (AWS), the American Institute for Steel Construction (AISC), and the American Society of Mechanical Engineers (ASME) and for materials by the American Society for Testing Materials (ASTM). Publications of those organizations should be consulted for more complex problems than those being presented here.

The approach established for weldment design is to always consider the weld as failing in shear regardless of load orientation. This is based on the fact that the allowable shear stress for a weld is always less than the allowable tensile stress or the allowable compressive stress. Further, as we examine the fillet weld, Figure 15-16, note that the fillet surface is taken to be at an angle of 45°. The area used for calculating shear stress is the length of the weld times the minimum distance from the weld surface to the root. This distance is known as the throat and is equal to $0.707t$. The thickness t is the nominal dimension by which a weld is designated when we speak of a $\frac{1}{4}$-in. or an 8-mm weld. Thus,

$$\tau = \frac{P}{A} \qquad \tau = \frac{P}{0.707tL}$$

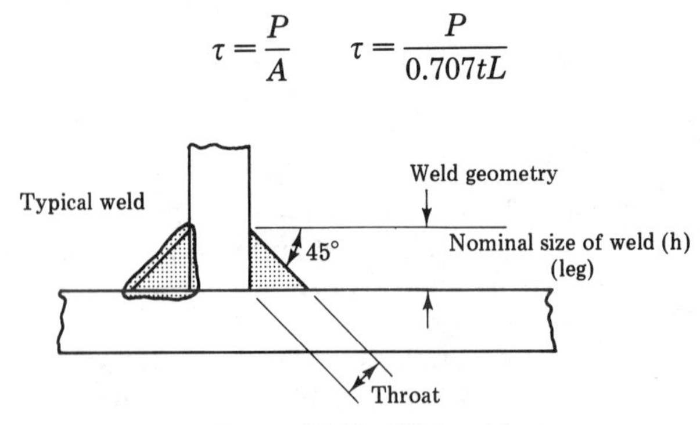

Figure 15-16. Fillet weld.

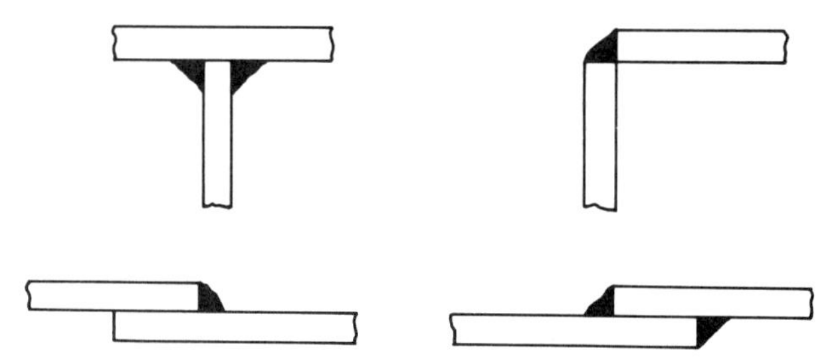

Figure 15-17. Fillet weld applications.

If we rewrite the above equation to read

$$P = 0.707\tau tL$$

and substitute 1 cm or 1 in. for L, various allowable shear strengths for various welding electrodes, and various weld thicknesses, we can obtain a tabulation of strength per unit weld length for various size welds of various materials.

For example, for a Class E70 electrode, $\frac{1}{2}$-in. thick, the allowable load per inch of length is

$$P = 0.707(21{,}000\ \text{lb/in.}^2)\,(0.5\ \text{in.})\,(1\ \text{in.})$$

$$P = 7420\ \text{lb}$$

Tables 15-4 and 15-5 have been developed using the above and substantially expedite weld design.

Additional fillet weld design criteria follow:

1. Maximum size of a fillet weld applied to a square edge of a plate shall be $\frac{1}{16}$ in. less than the plate thickness for plate thickness $\frac{1}{4}$ in. or more and equal to the plate thickness for plate thickness less than $\frac{1}{4}$ in.

2. Minimum effective length of a weld shall not be less than 4 times the nominal size of the weld.

3. Minimum width of laps on lap joints shall be 5 times the thickness of the thinner part joined and not less than 1 in.

4. Side- or end-fillet welds shall be returned continuously around the corners for a distance not less than twice the nominal size of the weld.

5. Lap joints joining plates or bars shall be fillet welded along the edge of both parts.

Example 15-4: A 4-in.-wide A441 plate $\frac{3}{4}$-inch thick is to be welded to a 1-in.-thick A441 plate using an E70 electrode for a maximum thickness

Table 15-4
Fillet Weld Strength Per Unit Length — Steel

Weld Size (t, in.)	Electrode Type					
	60	70	80	90	100	110
	English Units (lb/in.)					
$\frac{1}{8}$	1590	1860	2120	2390	2650	2920
$\frac{3}{16}$	2390	2780	3180	3580	3980	4370
$\frac{1}{4}$	3180	3710	4240	4770	5300	5830
$\frac{5}{16}$	3980	4640	5300	5970	6630	7290
$\frac{3}{8}$	4770	5570	6360	7160	7950	8750
$\frac{7}{16}$	5570	6500	7420	8350	9280	10210
$\frac{1}{2}$	6360	7420	8480	9540	10600	11670
$\frac{9}{16}$	7160	8350	9540	10740	11930	13120
$\frac{5}{8}$	7950	9280	10600	11930	13260	14580
$\frac{11}{16}$	8750	10200	11670	13120	14580	16040
$\frac{3}{4}$	9540	11140	12730	14320	15910	17500
	SI Units [kN/m (N/mm)]					
3	263	308	350	395	439	484
4	351	410	467	526	585	665
5	438	513	583	658	732	832
6	526	615	700	789	878	998
8	701	820	933	1052	1170	1331
10	877	1025	1167	1315	1463	1663
12	1052	1230	1400	1578	1756	1996
14	1227	1435	1633	1841	2049	2329
15	1315	1537	1750	1973	2195	2495
16	1403	1640	1866	2104	2341	2662
18	1578	1840	2100	2367	2634	2994
20	1753	2050	2833	2630	2927	3327

Electrode	Utilization
60	A36,A500
70	A242,A441
80	A572 (Grade 65)
90	
100	
110	A514

Table 15-5
Fillet Weld Strength Per Unit Length — Aluminum

Weld Size (t, in.)	Plate Type/Electrode Type									
	1100/1100	1100/4043	3003/1100	3003/4034	6061/4043	6061/5356	6061/5550	6063/4043	6063/5356	6063/5556
	English Units (lb/in.)									
$\frac{1}{8}$	283	424	283	442	442	618	751	442	547	547
$\frac{3}{16}$	424	625	424	663	663	928	1127	663	862	862
$\frac{1}{4}$	566	834	566	884	884	1237	1502	884	1149	1149
$\frac{5}{16}$	707	1042	707	1105	1105	1547	1878	1105	1436	1436
$\frac{3}{8}$	845	1251	845	1326	1326	1856	2254	1326	1723	1723
$\frac{7}{16}$	990	1459	990	1547	1547	2165	2629	1547	2011	2011
$\frac{1}{2}$	1131	1668	1131	1768	1768	2475	3005	1768	2298	2298
$\frac{9}{16}$	1273	1876	1273	1988	1988	2784	3380	1988	2585	2585
$\frac{5}{8}$	1414	2085	1414	2209	2209	3093	3756	2209	2872	2872
$\frac{11}{16}$	1555	2293	1555	2430	2430	3402	4132	2430	3159	3159
$\frac{3}{4}$	1697	2502	1697	2651	2651	3712	4507	2651	3447	3447
	SI Units [kN/m N/mm)]									
3	47	70	47	72	72	102	125	72	95	95
4	62	93	62	96	96	136	169	96	127	127
5	78	117	78	120	120	170	209	120	159	159
6	93	140	93	144	144	204	250	144	191	191
8	124	187	124	192	192	271	334	192	255	255
10	156	233	156	240	240	339	417	240	318	318
12	187	280	187	288	288	407	501	288	382	382
14	218	327	218	337	337	475	584	337	445	445
15	233	350	233	361	361	509	627	361	477	477
16	249	373	249	385	385	543	667	385	509	509
18	280	420	280	433	433	611	751	433	573	573
20	311	467	311	481	481	679	834	481	636	636

weld. Using the configuration and load shown in Figure 15-18, how long must the overlap x be to permit the needed length of weld?

Solution:

Step 1 Recognize that the weld thickness cannot be greater than

$$\frac{3}{4} - \frac{1}{16} = \frac{11}{16} \text{ in.}$$

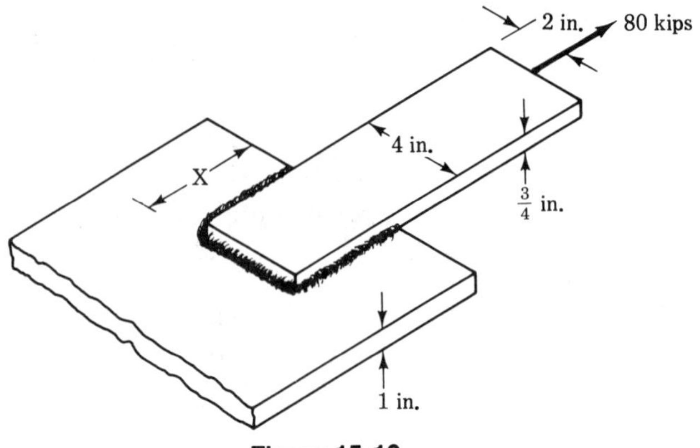

Figure 15-18

Step 2 Check Table 15-4 to see that 1 in. of the specified weld has an allowable load of 10,210 lb.

$$L = \frac{P}{F/L} = \frac{80{,}000 \text{ lb}}{10{,}210 \text{ lb/in.}}$$

$$L = 7.84 \text{ in.,} \quad \text{say 8 in.}$$

$$L = w + 2x \qquad x = \frac{L - w}{2} = \frac{8 - 3}{2}, \qquad \boldsymbol{x = 2.5 \text{ in.}}$$

Problem 15-6: The two 6061 aluminum plates shown in Figure 15-19 are to be welded together as shown using a 5356 rod. What is the minimum permissible weld thickness?

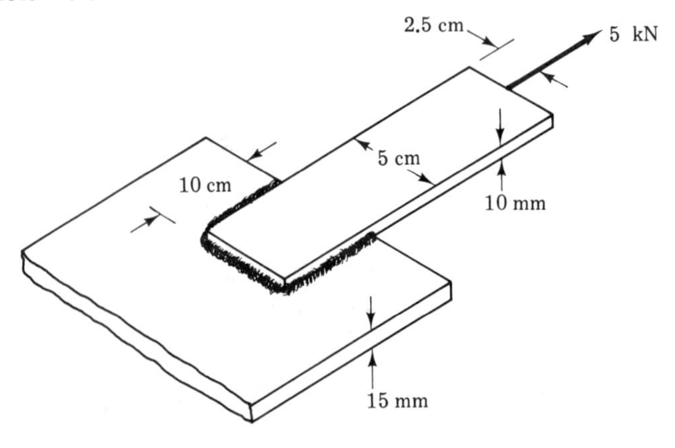

Figure 15-19

Example 15-5: Two 10 mm-thick 6063 aluminum plates, each 20 cm wide are to be welded together to make a lap joint as shown in Figure 15-20.

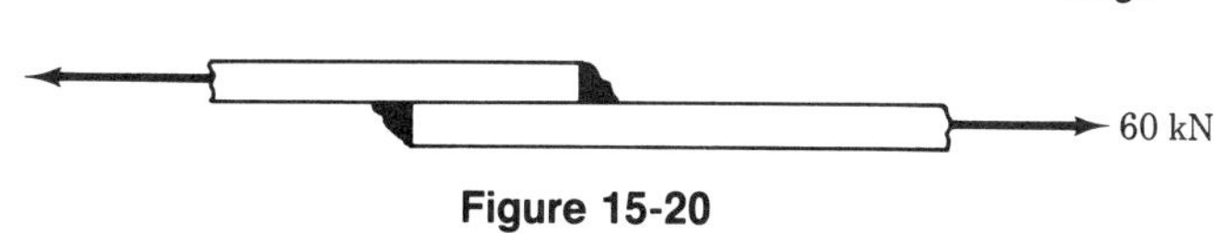

Figure 15-20

What is the minimum weld length (both equal) if the weld thickness is to be 8 mm? (Assume 4043 electrodes are used.)

Solution:

Step 1 Check Table 15-5 for maximum load/mm. $P/\text{mm} = 190$ N.

Step 2

$$L = \frac{P}{F/L} = \frac{60{,}000 \text{ N}}{190 \text{ N/mm}}$$

$$L = 316 \text{ mm}$$

Each weld is 16 cm long

Problem 15-7: Two 20-mm-thick A572 plates are to be lap welded together as shown in Figure 15-21 using E80 rods. If the plates are 300 mm wide, what is the minimum permissible weld thickness?

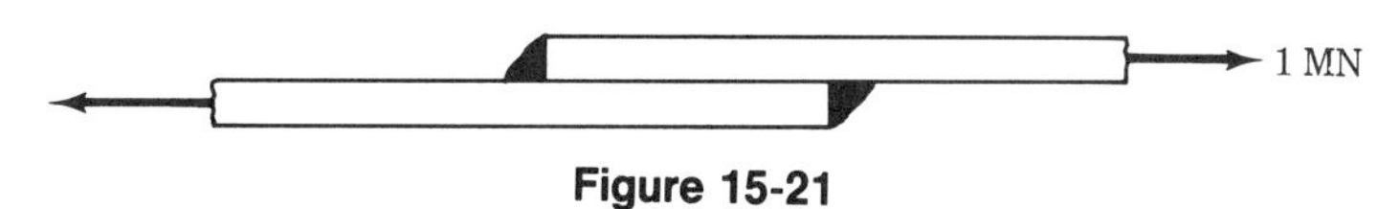

Figure 15-21

Example 15-6: The plates shown in Figure 15-22 are all 3003 aluminum. What minimum length 4043 weld should be used to join them?

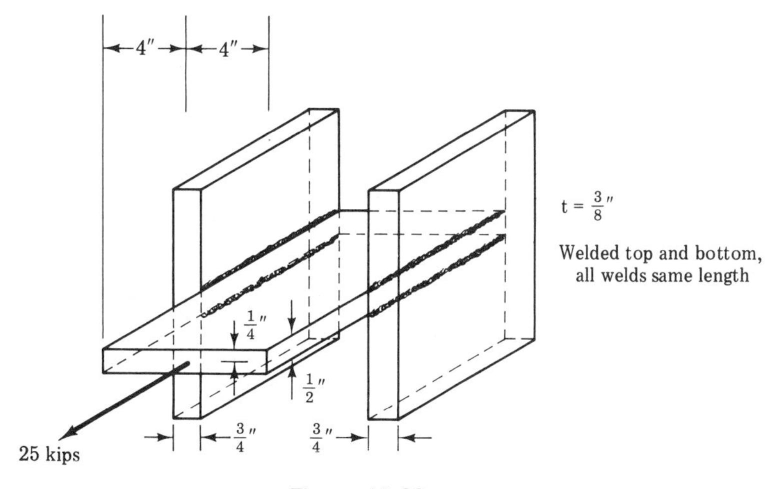

Figure 15-22

Solution:

Step 1 A 3003/4043 weld $\frac{3}{8}$ in. high has a strength of 1326 lb/in.

Step 2 The load is concentrically located so it can be divided equally among the four welds.

$$P = \frac{P_{\text{total}}}{n} = \frac{25{,}000 \text{ lb}}{4} \qquad P = 6250 \text{ lb}$$

Step 3 Determine individual weld length (F/L from Table 15-5).

$$L = \frac{P}{F/L} = \frac{6250 \text{ lb}}{1326 \text{ lb/in.}} = 4.7 \text{ in.} \qquad \mathbf{L = 5 \text{ in.}}$$

Problem 15-8: What is the maximum load that the welded connection shown in Figure 15-23 can safely sustain?

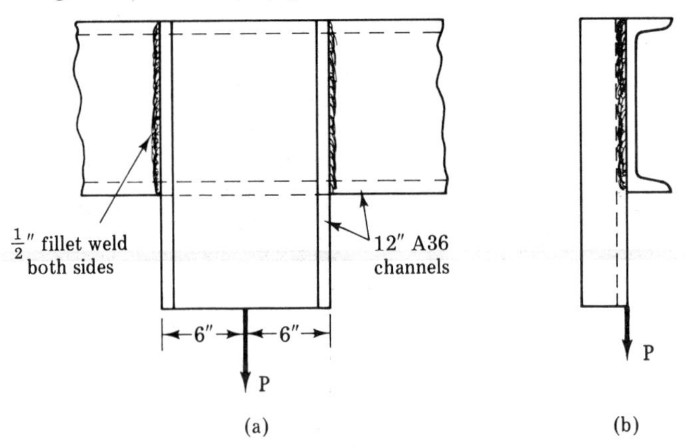

Figure 15-23

Example 15-7: Determine the length of 15-mm E70 weld needed along each side of the 2-cm plate to safely handle the load shown in Figure 15-24.

Solution:

Step 1 Recognize that this is an eccentric load. Treat the load and the weld forces as a parallel, coplanar force system. Take moments about B to obtain P_A, and sum forces to obtain P_B.

$$\Sigma M_B = 0 \qquad P_A d_A + P d_p = 0 \qquad P_A = -\frac{P d_p}{d_A}$$

$$P_A = \frac{400 \text{ kN}(2 \text{ cm})}{10 \text{ cm}} \qquad P_A = -80 \text{ kN}$$

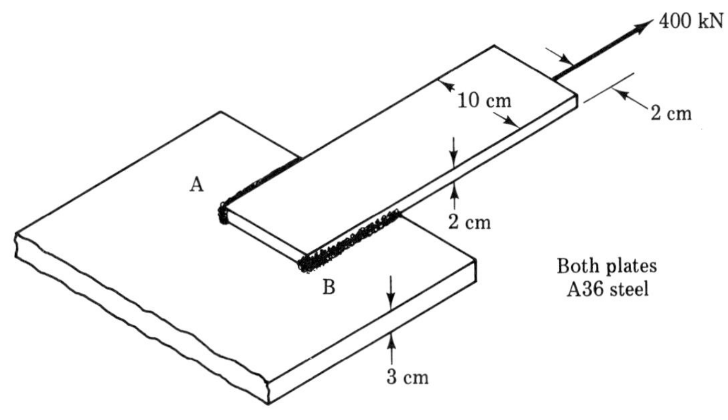

Figure 15-24

$$\Sigma P = 0 \qquad P_A + P_B + P = 0 \qquad P_B = -(P_A + P)$$

$$P_B = -(80 + 400 \text{ kN}) \qquad P_B = -320 \text{ kN}$$

Step 2 Determine the individual weld lengths (F/A from Table 15-4).

$$L_A = \frac{P_A}{F/L} = \frac{80 \text{ kN}}{1537 \text{ N/mm}} = 52 \text{ mm} \qquad \mathbf{L_A = 52 \text{ mm}}$$

$$L_B = \frac{P_B}{F/L} = \frac{320 \text{ kN}}{1537 \text{ N/mm}} = 208 \text{ mm} \qquad \mathbf{L_B = 208 \text{ mm}}$$

Problem 15-9: Determine the maximum load that the connection shown in Figure 15-23 can safely sustain if the load is moved 2.5 in. to the right in view (a).

15.6 READINESS QUIZ

1. A bolted or riveted butt joint involves __________ plates.

   (a) one
   (b) two
   (c) three
   (d) four
   (e) none of the above

2. Which of the following may fail in shear?

   (a) bolt
   (b) rivet
   (c) weld

(d) all of the above
(e) none of the above

3. Bolts and rivets are found to be in double shear in a ________ joint.

(a) lap
(b) butt
(c) tee
(d) all of the above
(e) none of the above

4. In a bolted or riveted joint, the maximum stress should be the

(a) plate tensile stress in the first bolt row
(b) plate bearing stress
(c) bolt bearing stress
(d) bolt shear stress
(e) none of the above

5. Plate bearing stress increases in direct proportion to the

(a) bolt diameter
(b) square of the bolt diameter
(c) square root of the bolt diameter
(d) cube root of the bolt diameter
(e) none of the above

6. Connection specifications are established by organizations such as the

(a) ASME
(b) AISC
(c) AA
(d) all of the above
(e) none of the above

7. Friction force can only be considered in some ________ connections.

(a) bolted
(b) welded
(c) riveted
(d) all of the above
(e) none of the above

8. Weld thickness is never ________ the plate thickness in a lap joint.

(a) less than
(b) half

(c) equal to
(d) more than
(e) none of the above

9. In a properly welded butt joint, the weld strength is __________ the plate strength.

(a) less than
(b) at least equal to
(c) greater than
(d) any of the above
(e) none of the above

10. Joint efficiency is equal to the lowest strength in the connection divided by __________ × 100.

(a) plate tensile strength of first bolt row
(b) bolt bearing strength
(c) shear strength
(d) plate bearing strength
(e) none of the above

15.7 SUPPLEMENTARY PROBLEMS

15-10 A lap joint has a 20,000-lb tension load. The plates are connected by three A325 bolts. Based on shear and the AISC Code, what is the minimum size for the bolts?

15-11 Determine the shear, bearing, and tensile stresses in the continuous A36 steel lap joint shown in Figure 15-25 when $p = 1.5$ in., $t = 0.75$ in., $d = 0.25$ in., and the tensile load is 20,000 lb/ft of joint.

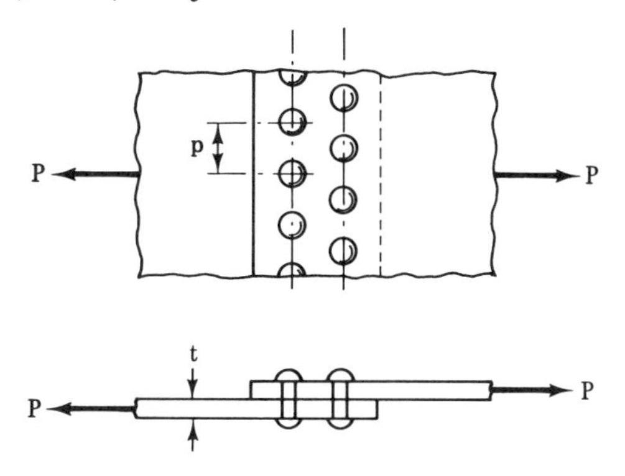

Figure 15-25

15-12 Two 2-in.-thick A36 steel plates are to be connected in a butt joint using two 1-in.-thick cover plates and A325 bolts. The tensile load on the joint is 700 K. Determine the plate width and the diameter and number of bolts required based on AISC Code.

15-13 In the connection shown in Figure 15-26, determine the maximum distance from the column at which the load of 5,000 lb can be located if the rivets are A502, Grade 2, and have a 1-in. diameter.

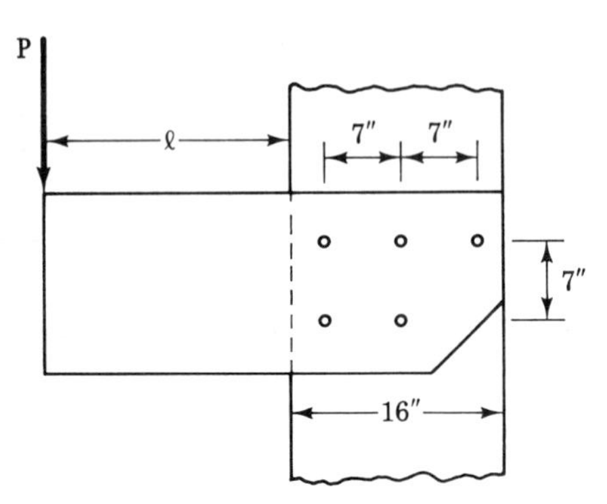

Figure 15-26

15-14 What length do the four fillet welds in Figure 15-27 need to be if they are 10 mm thick?

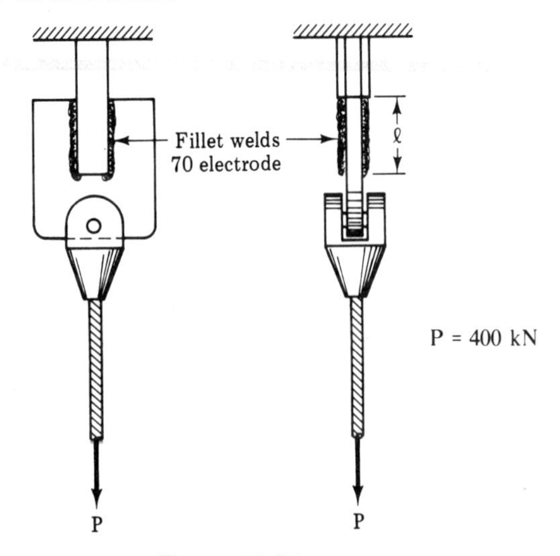

Figure 15-27

15-15 A $6 \times 4 \times \frac{1}{2}$ in. angle carries an 80,000-lb load acting through its centroid as shown in Figure 15-28. It is to be connected by $\frac{7}{16}$-in. fillet welds using 80 rod. What is the minimum length needed for each weld?

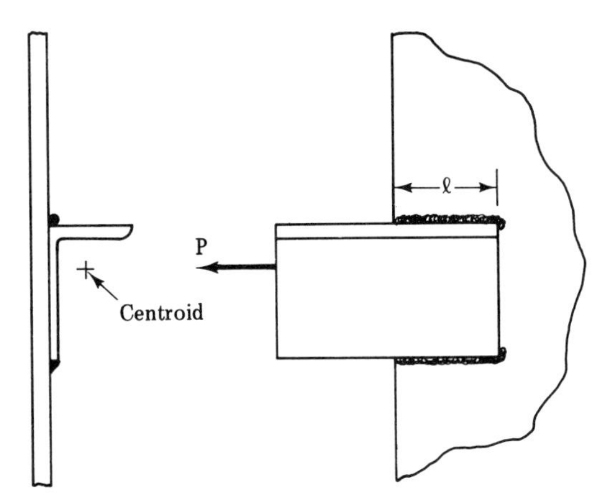

Figure 15-28

Chapter 16

Torsion

16.1 OBJECTIVES

Upon completion of the work relating to this chapter, you should be able to perform the following.

1. Determine the stresses in solid and hollow shafts subjected to torsion.
2. Determine angular deformation of shafts with circular cross section subjected to torsion.
3. Size shafts, keys, and coupling components for shafts with circular cross section transmitting power.
4. Determine stresses in, and size torsion bars with, circular cross sections.
5. Solve torsion problems similar to the above for shafts with noncircular cross sections.

16.2 INTRODUCTION

The phenomenon of torsion occurs whenever the line of action of a force applied to a material has a circular rather than linear path. We have previously examined the stresses caused by forces acting in a straight line. In

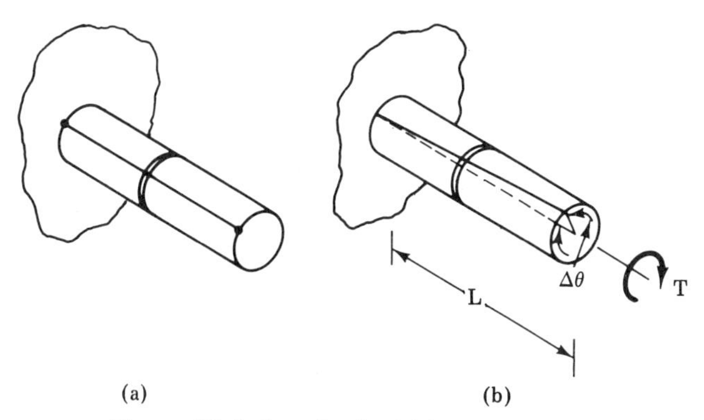

Figure 16-1. Longitudinal effect of torsion.

tension the molecules are being pulled away from each other or stretched. In compression, the molecules are being pushed closer together. In shear, the molecules are sliding linearly with respect to each other. In torsion, we see another type of shear where the molecules are sliding with respect to each other in a circular path.

Figure 16-1a shows a simple shaft that is fixed or held in place at one end. A line has been scribed along its entire length at its top edge parallel to the longitudinal axis of the shaft. In Figure 16-1b a clockwise couple has been applied to the free end of the shaft causing the free end to rotate clockwise while the fixed end stays in place. If the material is homogeneous and the cross section of the shaft uniform then the angular deformation per unit length will be constant.

$$\frac{\Delta\theta}{\Delta l} = C$$

In Figure 16-1b we can see how each thin cross-sectional slice that might be examined moved a short angular distance with respect to the next slice. The longer the shaft the greater the angular motion of one end with respect to the other. However, we have only examined the surface of the shaft. Figure 16-2a shows the end view of one of the slices indicating a slight angular deformation. It is evident that for a given angular deformation, the

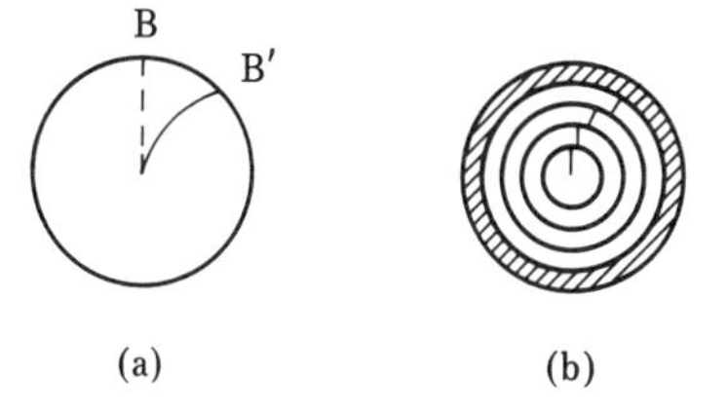

Figure 16-2. Radial effect of torsion.

further away we get from the center of the shaft the greater the linear deformation is. It follows, since the shear modulus (or modulus of rigidity) is a constant, that the further away we get from the center of the shaft, the greater the shearing stress will be. The molecules are thus not only sliding with respect to one another as thin disks but also as concentric shells as indicated in Figure 16-2b.

16.3 TORSION IN CIRCULAR SHAFTS

Referring back to Figure 16-2b, which showed the shaft cross section with a concentric ring highlighted, let us examine what is occurring in terms of torque and stress. First of all, the concentric ring represents a definite area which will for a known material require a specific force to achieve a specific stress level. From our earlier work

$$\tau = \frac{F}{A} \qquad F = \tau A$$

However, since the direction of desired shear is circular, the direction of the applied force must be circular. This will require a torque rather than just a force to be applied, and the distance r that the area lies from the center of rotation must be considered.

$$T = Fr = \tau A r \qquad T = \tau A r$$

From our earlier discussion, since deformation is greater the greater r is and G is constant, the shear stress also increases as r increases. Shear stress is thus proportional to r and reaches a maximum at the surface where the maximum radius is designated as c. Thus

$$\frac{\tau}{r} = \frac{\tau_{max}}{c} \quad \text{and} \quad \tau = \frac{\tau_{max} r}{c}$$

If we substitute this in our torque equation above for the local shear stress then

$$T = \frac{\tau_{max}}{c}(Ar^2)$$

This is the torque required for the area represented by the concentric rings. If we desire the total torque we must make a summation for all possible concentric rings.

$$T = \Sigma\left(\frac{\tau_{max}}{c} Ar^2\right)$$

However, τ_{max} and c are common to every concentric ring. Thus

$$T = \frac{\tau_{max}}{c}(\Sigma Ar^2)$$

We recognize ΣAr^2 as the general equation for the moment of inertia of an area. In this case, since the axis of rotation is perpendicular to the plane of the area, it is the polar moment of inertia, J. Then

$$T = \frac{\tau_{max} J}{c} \quad \text{or} \quad \tau_{max} = \frac{Tc}{J}$$

The relationship of J/c is encountered frequently enough to be named the polar section modulus and designated by the symbol S (Z in some references, but we shall use S throughout).

The above constitutes one approach for the design of components in torsion, one based on maximum allowable shearing stress. After some applications of the above, we will look at another design approach.

Example 16-1: Determine the maximum shearing stress developed in a 30-mm-diameter shaft that is subjected to a torque of 5000 N m.

Solution:

Step 1 Recall that the shear stress in a body subjected to torsion is directly proportional to the torque and inversely proportional to the polar section modulus of the area in shear.

$$\tau_{max} = \frac{T}{S} = \frac{Tc}{J}$$

Step 2 For a circular cross section

$$J = \frac{\pi r^4}{2} \quad \text{and} \quad c = r$$

$$\text{Then } \tau_{max} = \frac{2T}{\pi r^3}$$

Step 3 Calculate the maximum shear stress:

$$\tau_{max} = \frac{2(5000)\ \text{N m}}{\pi(0.03\ \text{m}^3)} \qquad \tau_{max} = \mathbf{118\ MPa}$$

Problem 16-1: Determine the maximum permissible torque for a $1\frac{1}{2}$-in. steel shaft with an allowable shear stress of 15,000 psi.

Frequently, a hollow shaft is more economical than a solid shaft. There are two basic reasons for this. First, the hollow configuration puts all the material much closer to the maximum stress level, thus making more efficient use of the material. Second, since hollow tubing can be made in a variety of wall thicknesses by several different processes thus eliminating costly boring and scrapping of materials encountered when boring a hole in a solid shaft.

The basic calculation involved for a hollow-shaft arc is essentially the same as for a solid shaft. The only difference is that the polar moment of inertia for a hollow shaft will be equal to that for a solid shaft (with an outside diameter equal to that of the hollow shaft) *minus* the polar moment of inertia of a solid shaft with an outside diameter equal to the inside diameter of the hollow shaft (the material removed).

$$J = J_0 - J_1 = \frac{\pi r_0^4}{2} - \frac{\pi r_1^4}{2} \qquad J = \frac{\pi}{2}(r_0^4 - r_1^4)$$

Example 16-2: Determine the maximum permissible torque for an aluminum shaft with an outside diameter of 2 in., an inside diameter of 1.75 in., and an allowable shear stress of 12,500 psi.

Solution:

Step 1 Recognize that the basic relationship governing the problem is

$$\tau_{max} = \frac{Tc}{J}$$

Step 2 Recognize that J will be unique to a hollow circular shaft:

$$J = \frac{\pi}{2}(r_0^4 - r_1^4)$$

Step 3 Solve for T from Step 1 and substitute for J from Step 2 recognizing that $c = r_0$:

$$T = \frac{\pi(\tau_{max})(r_0^4 - r_1^4)}{2r_0}$$

Step 4 Substitute values and solve

$$T = \frac{\pi(12{,}500 \text{ lb/in.}^2)[(1.0 \text{ in.})^4 - (0.875 \text{ in.})^4]}{2(1.0 \text{ in.})}$$

$$\boldsymbol{T = 8130 \text{ in. lb}}$$

Problem 16-2: A hollow shaft is to be fabricated of bronze with an allowable shear stress of 80 MPa. It must have an outside diameter of 20 cm and be able to sustain a torque of 1 MN m. What is the maximum permissible inside diameter?

Our specific discussion up to now has dealt with the stress caused by a torque applied to a shaft. Problems have been either stated or solved in such a way that the stress that occurs is well within that present at the elastic limit for the particular material.

However, in our earlier discussion, the concept of deformation, as well as unit deformation or strain was also mentioned. The relationships involved are expressed as

$$G = \frac{\tau}{\gamma}$$

where G is the modulus of rigidity, τ is shear stress and γ is shear strain. As can be seen from Figure 16-3, the deformation along a circular path, Δs, will be equal to $r\Delta\theta$ at any specific radius. The amount of this deformation will also be dependent on the length of shaft, ΔL, that is considered. However, if there is no external torque applied in that length, the ratio $\Delta\theta/\Delta L$ will be constant and equal to θ/L. Then

$$\gamma = \frac{r\theta}{L}$$

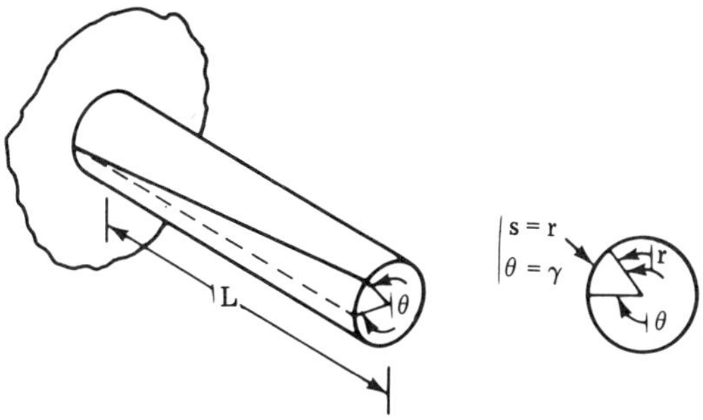

Figure 16-3. Angular deformation.

Recall now that we had earlier developed the relationship $T = \tau Ar$. If we substitute $\tau = T/Ar$ from this and the above equation for shear strain into our Hooke's law equation, $G = \tau/\gamma$, then we have

$$G = \left(\frac{T}{Ar}\right)\left(\frac{L}{r\theta}\right) = \frac{TL}{Ar^2\theta}$$

If we make a summation for all possible values of r (recall that stress and strain both vary as r varies) then the term ΣAr^2 appears, which we recognize as the area moment of inertia. More properly since the area in question is parallel to the direction of the force involved, that is the polar area moment of inertia, then

$$G = \frac{TL}{J\theta} \quad \text{or} \quad \theta = \frac{TL}{GJ}$$

This equation, of course, holds only up to the elastic limit of the material since past that point G will no longer be a constant.

Example 16-3: A hollow shaft has an outside diameter of 5 cm and an inside diameter of 3 cm and is 2 m long. It is made of steel with a safe or working shear stress of 100 MPa and a modulus of rigidity of 78.6 GPa. What is the maximum torque that may safely be applied to the shaft if the total angular deformation is not to exceed 1.0°?

Solution:

Step 1 Recognize that this is a two-facet problem. The torque that results in the maximum allowable stress may not be the same as the torque that results in the maximum allowable deformation. Thus, both will have to be calculated and the lower of the two designated as the maximum allowable torque.

Step 2 Calculate the maximum allowable torque based on maximum allowable stress.

$$T = \frac{\pi(\tau_{max})(r_0^4 - r_i^4)}{2r_0} \quad \text{(hollow shaft)}$$

$$T = \frac{\pi(100 \text{ MPa})[(0.025 \text{ m})^4 - (0.015 \text{ m})^4]}{2(0.025 \text{ m})}$$

$$T = 2136 \text{ N m}$$

Step 3 Calculate the maximum allowable torque based on maximum allowable deflection.

$$\theta = \frac{TL}{GJ} \qquad T = \frac{\theta GJ}{L}$$

$$J = \frac{\pi}{2}(r_0^4 - r_i^4)$$

$$T = \theta G\pi(r_0^4 - r_i^4)/2L$$

$$(1^\circ = 0.01745 \text{ rad})$$

$$T = \frac{(0.01745 \text{ rad})(78.6 \text{ GPa})(\pi)[(0.025 \text{ m})^4 - (0.015 \text{ m})^4]}{2(2 \text{ m})}$$

$$T = 366 \text{ N m}$$

Step 4 The lower of the above values must be designated as the maximum torque in order to comply with the specifications.

$$\mathbf{T_{max} = 366 \text{ N m}}$$

Problem 16-3: A 302 stainless steel shaft 2.25 in. in diameter and 5 ft long is subjected to a torque of 5,000 ft lb. What will the maximum shear stress and the total angular deformation be?

16.4 POWER TRANSMISSION

There are many applications where a machine or structural member will be subjected to torsion. However, the most common application is the use of shafts, generally circular in cross section, to transmit power. A later section in this chapter will consider noncircular cross sections.

The Nature of Power

The nature of power, particularly in angular motion, must be considered, as well as how the power gets to and from the shaft and the effect of changes in shaft size. Thus this unit will consider power, the use of couplings and keys, and consideration of stress concentrations.

Power, you will probably recall, is the rate of doing work. Work is generally defined as a force in the direction of motion times the distance through which the point of application of the force moves.

$$P = \frac{U}{t} \qquad P = \frac{F \cdot s}{t}$$

The above is in linear form while we wish to have it in angular form. Force can be changed to torque by multiplying by r since $T = Fr$. However, we must either do the same thing to both sides of the equation or to both the numerator and denominator of one side. Since $s = r\theta$ if θ is in radians, then $\theta = s \div r$.

$$P = \frac{F \cdot s}{t} = \frac{r \cdot F \cdot s}{r \cdot t} = \frac{F \cdot r}{t}\frac{s}{r} = \frac{T\theta}{t}$$

But $\theta \div t = \omega$ which is angular velocity in radians per unit time. Then

$$P = T\omega$$

Power equals torque times angular velocity. In English units, power is generally expressed as horsepower (hp), a somewhat arbitrary unit which equals 550 ft lb/s or 33,000 ft lb/min. In SI units the kilowatt (kW) is used and is equivalent to a kilojoule per second (kJ/s) or a kN m/s.

Example 16-4: What is the maximum power that can be transmitted by the shaft in Example 16-1 at 1200 rpm?

Solution:

Step 1 Recognize the need to apply the power equation

$$P = T\omega$$

Step 2 Recognize that rpm is a commonly used abbreviation for revolutions/minute.

Step 3 Convert rpm to rad/s

$$\omega = \frac{1200 \text{ rev}}{\text{min}} \frac{2\pi \text{ rad}}{\text{rev}} \frac{1 \text{ min}}{605}$$

$$\omega = 125.66 \text{ rad/s}$$

Step 4 Determine the power transmitted.

$$P = T\omega = 5 \text{ kN m}(125.66 \text{ rad/s}) \qquad \mathbf{P = 628 \text{ kW}}$$

Problem 16-4: Determine the maximum power that can be transmitted by the shaft in Problem 16-1 at 350 rpm.

Stress Concentration in Torsion

As was the case with members in tension or compression that were designed with changes in cross section, a concentration of stress takes place at any step (change in diameter) in a shaft. As you might suspect, the extent of this concentration is determined by the size of the step (diameter ratio) and the abruptness of the step (fillet radius divided by smaller diameter). Figure 16-4 (taken from L. S. Jacobsen, "Torsional-Stress Concentrations in Shafts of Circular Cross Section and Variable Diameter," *Trans. A.S.M.E.*, vol. 47 (1925), pp. 619–638.) graphs values of a stress concentration factor k for various values of D/d and r/d. These are based on the assumption that stress over strain is a constant, that is, that the proportional limit for the material is not exceeded. Then

$$\tau_{max} = k\tau_{nominal}$$

with the shear stresses being those for the smaller shaft section.

Frequently, shafts must have a circumferential groove machined into them. Among the applications are placement of an O-ring seal or a snap ring retainer as shown in Figure 16-5.

Figure 16-6 (taken from Robert L. Mott, *Applied Strength of Materials*. Englewood Cliffs, N.J.: Prentice-Hall, Inc., 1978.) shows values for the stress concentration factor for various diameter ratios and groove fillet radii.

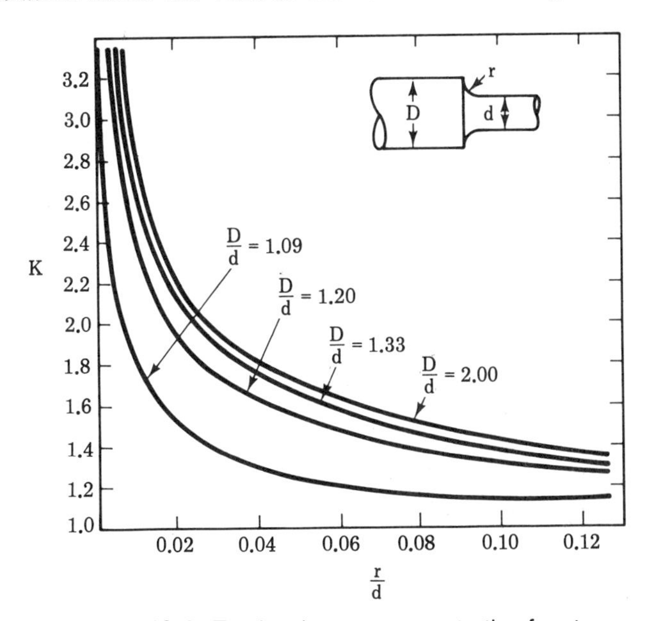

Figure 16-4. Torsional stress concentration for steps.

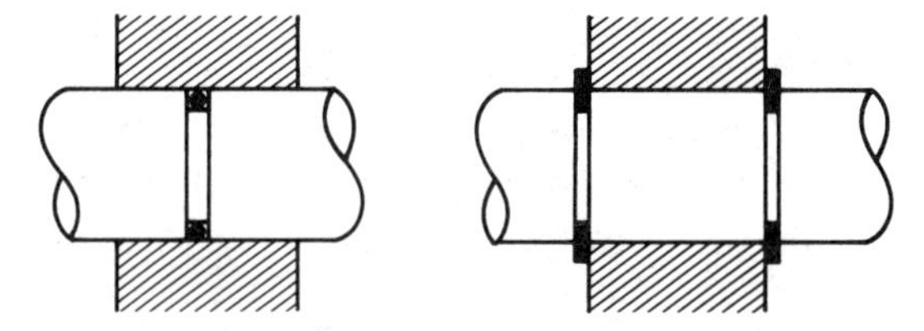

Figure 16-5. Shaft groove applications.

Again, nominal stress is based on the smaller diameter and it is assumed that the shaft is not stressed beyond the proportional limit.

Example 16-5: A 3-in. shaft is stepped down to 1.5 in. and the fillet radius is 0.125 in. The material has an allowable shearing stress of 20,000 psi and a modulus of rigidity of 11.5×10^6 psi. How much power can the shaft

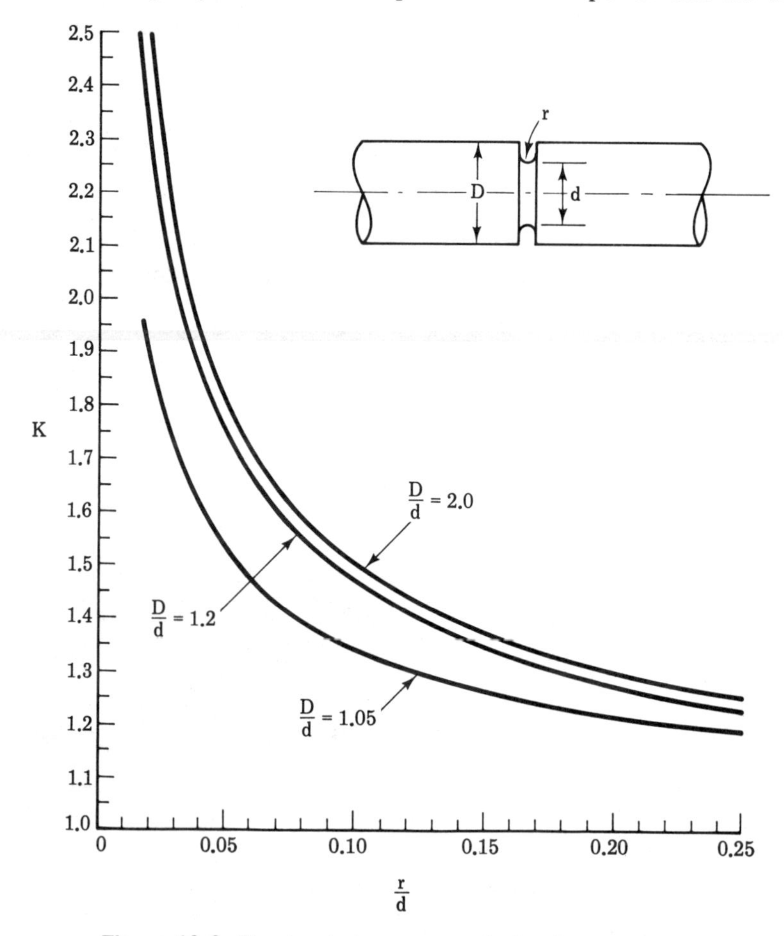

Figure 16-6. Torsional stress concentration from grooves.

transmit at 750 rpm? If the larger diameter section is 4 ft long and the smaller diameter section is 2 ft long, what will the total angular deformation be?

Solution:

Step 1 Recognize that the power equation, torque equation (with consideration of stress concentration) and the angular deformation equation will all be needed.

$$P = T\omega \qquad \tau_{max} = k\frac{2T^*}{\pi r^3} \quad \text{or} \quad T_{max} = \frac{\tau_{max}\pi r^3}{2k} \qquad \theta = \frac{TL}{GJ}$$

Step 2 Determine D/d, r/d by calculation, then k from Figure 16-4.

$$\frac{D}{d} = \frac{3 \text{ in.}}{1.5 \text{ in.}} \qquad \frac{D}{d} = 2$$

$$\frac{r}{d} = \frac{0.125 \text{ in.}}{1.5 \text{ in.}} \qquad \frac{r}{d} = 0.083 \qquad K = 1.45$$

Step 3 Determine maximum permissible torque.

$$T_{max} = \frac{\tau_{max}\pi r^3}{2k} = \frac{20{,}000 \text{ lb/in.}^2(\pi)(0.75 \text{ in.}^3)}{2(1.45)}$$

$$T_{max} = 9140 \text{ in. lb} \quad \text{or} \quad 761.7 \text{ ft lb}$$

Step 4 Determine maximum power.

$$P = T\omega = 761.7 \text{ ft lb } (750 \text{ rev/min}) \left(\frac{1 \text{ min}}{60 \text{ s}}\right)\left(\frac{2\pi \text{ rad}}{\text{rev}}\right)$$

$$\div\, 550 \frac{\text{ft lb}}{\text{s}}/\text{hp}$$

$$\mathbf{P = 109 \text{ hp}}$$

Step 5 Determine the total angular deformation. Since there is no information that indicates otherwise, it will be assumed that the torque is constant throughout the length of the shaft. However, it must be calculated in two parts since the smaller diameter shaft will deform more per unit length than the larger diameter (the section of varying diameter at the fillet

* For a solid circular cross section.

will be included with the smaller shaft section introducing a negligible error).

$$\theta_{\text{total}} = \theta_D + \theta_d$$

$$\theta_D = \frac{TL_D}{GJ_D} = \frac{2TL_D}{G\pi r_d^4} \qquad \left(J = \frac{\pi r^4}{2}\right)$$

$$\theta_D = \frac{2(9140 \text{ in. lb})(48 \text{ in.})}{(11.5 \times 10^6 \text{ lb/in.}^2)(\pi)(1.5 \text{ in.}^4)} \qquad \theta_D = 0.0048 \text{ rad}$$

$$\theta_d = \frac{2TL_d}{G\pi r_d^4} = \frac{2(9140 \text{ in. lb})(24 \text{ in.})}{(11.5 \times 10^6 \text{ lb/in.}^2)(\pi)}(0.75 \text{ in.}^4)$$

$$\theta_d = 0.019 \text{ rad}$$

$$\theta_{\text{total}} = 0.0048 + 0.019 \text{ rad}$$

$$\boldsymbol{\theta_{\text{total}} = 0.024 \text{ rad} \quad \text{or} \quad \theta_{\text{total}} = 1.38^\circ}$$

Problem 16-5: A 25-mm-diameter shaft has a 3-mm-wide and 3-mm-deep groove cut into it around its circumference. The fillet radius at the base of the groove is 1 mm. The material has an allowable shearing stress of 150 MPa. What is the maximum shaft speed permissible in rpm if the shaft is to transmit 35 kW?

Power Input and Takeoff

A shaft is frequently used to connect one power source to two or more loads, using friction, belts, or gears to move the power to and from the shaft. Figure 16-7 shows one of many possible arrangements for accomplishing this.

The shafts examined earlier all were subjected to a constant torque

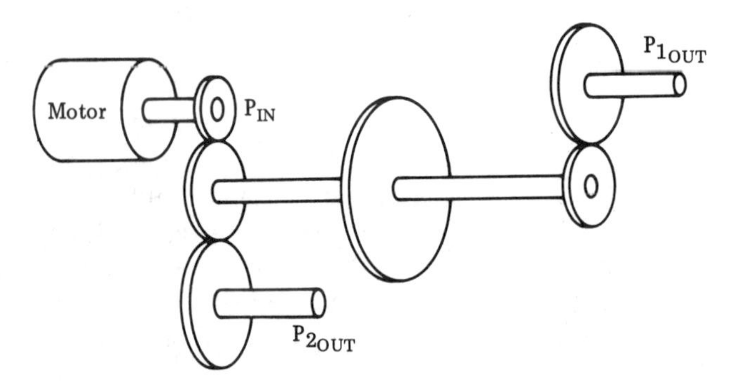

Figure 16-7. Shaft power input and takeoff.

from one end to the other. Any one shaft can only have one operating speed at any given time. Therefore, in Figure 16-7, if the power being taken off at gear A is not equal to the power being taken off at gear C, then the torque in shaft section AB cannot be equal to the torque in shaft section BC.

$$\omega_A = \omega_B = \omega_C = \omega_{AB} = \omega_{BC}$$

If $P_A \neq P_C$, then $T_A\omega_A \neq T_C\omega_C$ and $T_{AB}\omega_{AB} \neq T_{BC}\omega_{BC}$. Therefore, since

$$\omega_{AB} = \omega_{BC} \qquad T_{AB} \neq T_{BC}$$

Example 16-6: The line shaft shown in Figure 16-8 is being driven at 425 rpm by the electric motor. What are the minimum diameters for each shaft section if the maximum allowable shearing stress is 140 MPa? If $G =$ 78.6 GPa, what will be the maximum difference in angular deformation between gear A and gear D under load?

Solution:

Step 1 Recognize that each shaft section will carry a different amount of power. Since the shaft is one piece operating at one speed, this means that each shaft section will be subjected to a different torque. Those torques will be the basis for shaft sizing.

Step 2 Recognize that shaft section BC must carry the power and the torque to supply both gears C and D. (The input power at gear B must equal the sum of all the output powers but does not enter into our calculations since it is immediately split.)

Step 3 Recognize that regardless of the direction of rotation, all angular deformations at the takeoffs will be in the same direction. The deformation at D will be the sum of the deformations of sections BC and CD.

Step 4 Determine the power transmitted by each section by inspection.

$$P_{AB} = 25 \text{ kW} \qquad P_{CD} = 15 \text{ kW} \qquad P_{BC} = 65 \text{ kW}$$

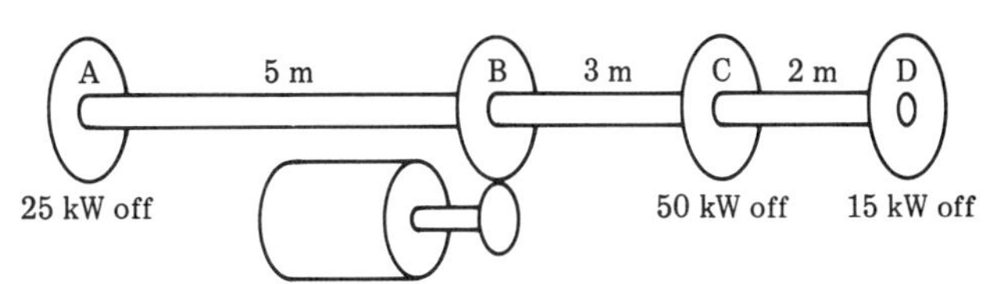

Figure 16-8

Step 5 Develop an equation to determine the shaft diameter from the knowns.

$$P = T\omega \qquad \tau_{max} = \frac{2T}{\pi r^3} \quad \text{(for solid circular cross section)}$$

$$T = \frac{P}{\omega} \qquad \tau_{max} = \frac{2P}{\omega \pi r^3}$$

$$r = \left(\frac{2P}{\tau_{max}\omega\pi}\right)^{1/3} \qquad d = 2\left(\frac{2P}{\tau_{max}\omega\pi}\right)^{1/3}$$

Since $\omega = 425$ rpm $= 44.5$ rad/s and τ_{max} is given as 140 MPa

$$d = 2\left[\frac{2P}{(140\text{ MPa})(44.5\text{ rad/s})(\pi)}\right]^{1/3} \qquad d = (9.35 \times 10^{-3})(P)^{1/3}$$

(P in kilowatts, d in meters)

Step 6 Determine the three minimum diameters.

$$d_{AB} = (9.35 \times 10^{-3})(P_{AB})^{1/3} = (9.35 \times 10^{-3})(25)^{1/3} \qquad \mathbf{d_{AB} = 0.027\ m}$$

$$d_{BC} = (9.35 \times 10^{-3})(P_{BC})^{3} = (9.35 \times 10^{-3})(65)^{1/3} \qquad \mathbf{d_{BC} = 0.038\ m}$$

$$d_{CD} = (9.35 \times 10^{-3})(P_{CD})^{1/3} = (9.35 \times 10^{-3})(15)^{1/3} \qquad \mathbf{d_{CD} = 0.023\ m}$$

(In the final design, these will all be larger to standardize them for gear bores and more particularly bearings. Also, if a decision is made to step the shaft, then stress concentration factors must be applied which may sometimes virtually eliminate the step.)

Step 7 Develop an equation to determine shaft deformation from the knowns.

$$P = T\omega \qquad \theta = \frac{2TL}{G\pi r^4} \quad \text{(for solid circular cross section)}$$

$$T = \frac{P}{\omega} \qquad r = \frac{d}{2} \qquad r^4 = \frac{d^4}{16}$$

$$\theta = \frac{2PL}{\omega G \pi d^4} \qquad G = 78.6\text{ GPa} \qquad \omega = 44.5\text{ rad/s}$$

$$\theta = \frac{32PL}{(44.5\text{ rad/s})(78.6\text{ GPa})(\pi)(d^4)} \qquad \theta = 2.9 \times 10^{-9}\frac{PL}{d^4}$$

(P is in kips, L and d in meters)

Step 8 Determine the difference in angular deformation between gears A and D.

$$\theta_A = \theta_D = \theta_{AB} - (\theta_{BC} + \theta_{CD})$$

$$\theta_{AB} = 2.9 \times 10^{-9} \frac{25(5)}{(0.027)^4} \qquad \theta_{AB} = 0.682 \text{ rad}$$

$$\theta_{BC} = 2.9 \times 10^{-9} \frac{65(3)}{(0.38)^4} \qquad \theta_{BC}\ 0.271 \text{ rad}$$

$$\theta_{CD} = 2.9 \times 10^{-9} \frac{15(2)}{(.023)^4} \qquad \theta_{CD} = 0.311 \text{ rad}$$

$$\theta_A - \theta_D = 0.682 \text{ rad} - (0.271 + 0.311 \text{ rad})$$

$$\theta_A - \theta_D = 0.1 \text{ rad} \quad \text{or} \quad \boldsymbol{\theta_A - \theta_D = 5.73^\circ}$$

(This would appear to be unacceptable but should be rechecked when diameters are finalized since even a 10% increase in diameter would reduce deformation by over 30%.)

Problem 16-6: The line shaft shown in Figure 16-9 is operating at 115 rpm. The maximum allowable shearing stress is 6000 psi and the modulus of rigidity is 12×10^6 psi. The power taken off at C is equal to the power taken off at B. Determine the maximum permissible power input at A. At that power, determine the difference in angular deformation between gears B and C. (The fillet radius at the step is 0.1 in.)

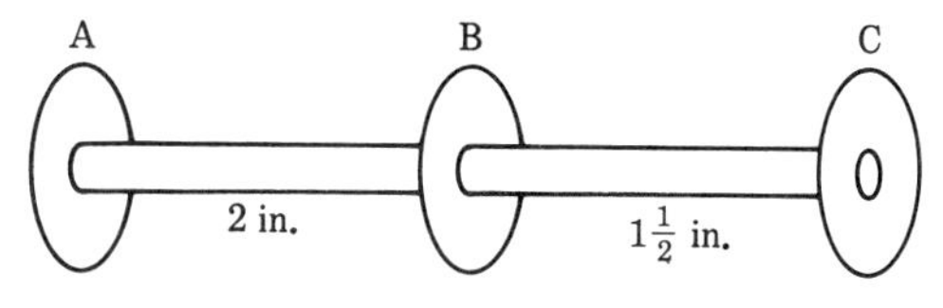

Figure 16-9

Keys

In our discussion up to now, we have ignored a very basic problem. How does power get transmitted from machine components such as wheels, pulleys, gears, and couplings to a shaft and vice versa? A number of methods are used including splines, set-screws, friction fit, and keys. There are also several types of keys but we shall consider the most widely used, the rectangular, illustrated in Figure 16-10. The other methods are examined in any good machine design text but are beyond the scope of this book.

As illustrated in Figure 16-10 the analysis of a key requires consideration of both bearing stresses and shearing stresses.

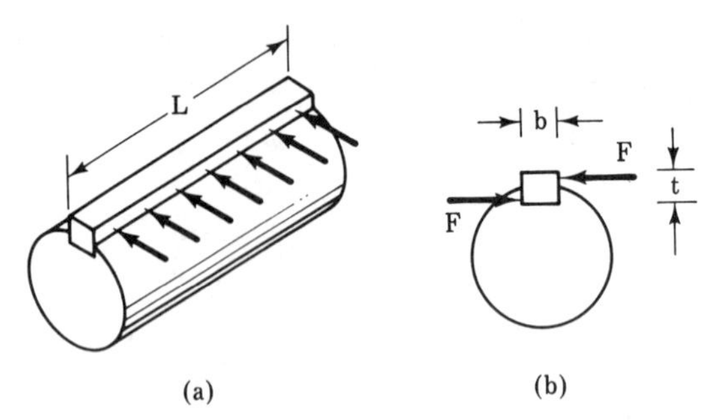

Figure 16-10. Rectangular key.

The shear stress in the key will be

$$\tau = \frac{F}{As}$$

where F is the force attempting to shear the key (the torque to be transmitted divided by the outside radius of the shaft) and A is the area in shear stress (the length times the width of the key).

$$F = \frac{T}{r} \qquad A = bl \qquad \tau = \frac{T}{rbl}$$

The bearing stress in the key will be

$$\sigma_b = \frac{F}{A_b}$$

where F is determined as above (the error is very slight in not locating F at the center of the half-face of the key) and A_b is the bearing area (one-half the key thickness times its length).

$$F = \frac{T}{r} \qquad A_b = \frac{tl}{2} \qquad \sigma_b = \frac{2T}{rtl}$$

Example 16-7: A 2-in.-diameter shaft is to transmit 100 hp at 700 rpm. The power input is through a gear keyed to the shaft by a rectangular key (AISI 1020 steel). If the key is 2 in. long and a factor of safety (ultimate) of 5 is used, what will be the minimum width and thickness of the key?

Solution:

Step 1 Recognize the need to determine these dimensions respectively on the basis of shear stress and bearing stress. Also recall that strength in compression is used for bearing strength.

Step 2 Look up the properties for AISI 1020 steel.

$$\tau_{ult} = 50{,}000 \text{ psi} \qquad \sigma_b = 65{,}000 \text{ psi}$$

Step 3 Determine the allowable stresses.

$$\text{allow } \tau = \frac{\tau_{ult}}{FS} = \frac{50{,}000 \text{ psi}}{5} \qquad \text{allow } \tau = 10{,}000 \text{ psi}$$

$$\text{allow } \sigma_b = \frac{\sigma_{b\,ult}}{FS} = \frac{65{,}000 \text{ psi}}{5} \qquad \text{allow } \sigma_b = 13{,}000 \text{ psi}$$

Step 4 Determine the torque being transmitted

$$T = \frac{P}{\omega}$$

$$T = \frac{100 \text{ hp}\left(550 \dfrac{\text{ft lb}}{\text{s}}/\text{hp}\right)(12 \text{ in./ft})}{700 \text{ rpm}\left(\dfrac{2\pi \text{ rad}}{\text{rev}}\right)\left(\dfrac{1 \text{ min}}{60 \text{ s}}\right)} \qquad T = 9000 \text{ in. lb}$$

Step 5 Determine key width.

$$\tau = \frac{T}{rbl} \qquad b = \frac{T}{\tau rl}$$

$$b = \frac{9000 \text{ in. lb}}{10{,}000 \text{ lb/in.}^2(1 \text{ in.})(2 \text{ in.})} \qquad \mathbf{b = 0.45 \text{ in.}}$$

Step 6 Determine key thickness.

$$\sigma_b = \frac{2T}{rtl} \qquad t = \frac{2T}{\sigma_b rl}$$

$$t = \frac{2(9000 \text{ in. lb})}{13{,}000 \text{ lb/in.}^2(1 \text{ in.})(2 \text{ in.})} \qquad \mathbf{t = 0.69 \text{ in.}}$$

Problem 16-7: How much power can be transmitted by an AISI 1045 steel rectangular key ($FS = 5$) 30 mm long, 5 mm wide, and 8 mm thick mounted on a 40-mm-diameter shaft turning at 1000 rpm?

The discussion of keys would not be complete without considering the effect of cutting the keyway in the shaft. Not only is the cross section of the shaft reduced but the sharp corners created result in concentration of stresses. There are two common methods used to cut keyways; a circular milling cutter or an end mill. The difference lies in the end(s) of the keyway.

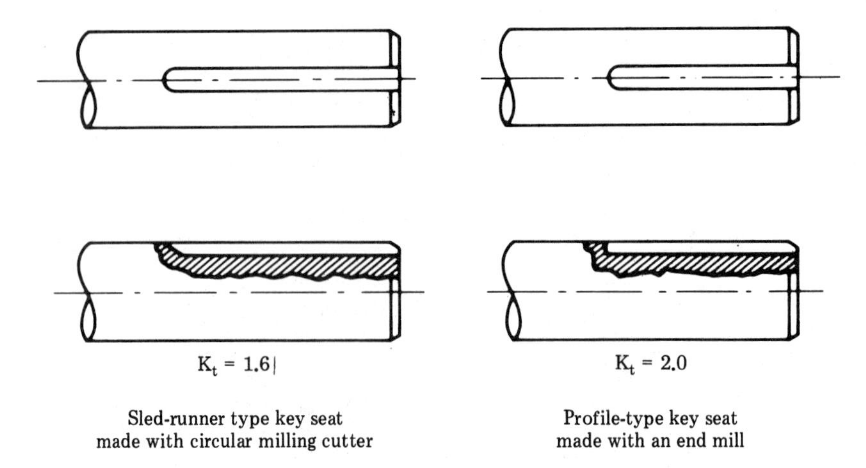

Figure 16-11. Keyway types.

In the first case, there is a gradual end to the keyway while in the second case the ending is quite abrupt. As indicated in Figure 16-11, this affects the stress concentration factor.

Couplings

Couplings are used to connect two collinear shafts often for the purpose of ease of removal of a component for maintenance. There are many types of couplings but the most common is the simple flanged coupling illustrated in Figure 16-12.

Analysis of a coupling requires some assumptions and a careful review of those points where force is transferred. First, it is assumed that only the bolts transfer force from one-half of the coupling to the other, that is, friction between the two surfaces is neglected. Second, it is assumed that the bolts share the total force equally. Third, if more than one bolt circle is used, it is assumed that, under load, the stress in each bolt will be proportional to the radius of its bolt circle.

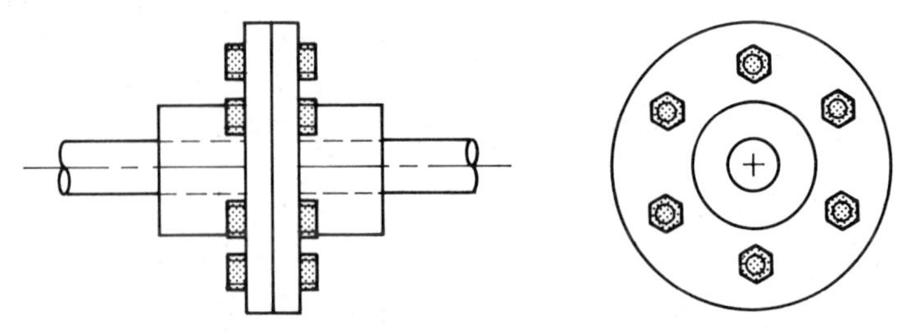

Figure 16-12. Flanged coupling.

The method of fastening the coupling half to the shaft varies. Welding, friction fit, one-piece construction, and keying are all used. Let us consider a keyed connection. The forces (at some radius to give us a torque) are transmitted at the following points:

1. from shaft to key
2. from key to hub
3. from flange to bolt

The reverse will take place in the other half of the coupling.

We have already considered the design of a shaft with a keyway and the sizing of a key in the preceding section. We must now consider the bolt sizing which will be a function of bearing stress as well as shear stress. The coupling flange will also be subjected to a bearing stress at the bolts. In addition, the flange must be evaluated in terms of shear stress. The area in shear is a cylindrical surface with a length equal to the flange thickness and some radius. The radius used for evaluation is where the flange joins the hub, for this smallest possible radius for the flange will simultaneously provide the smallest shear area and the largest shear force. Unquestionably, this is the point of greatest shear stress in the flange.

Example 16-8: Two 50-mm-diameter shafts are joined by a flange coupling. The coupling has five bolts on a 140-mm bolt circle. The flange is 25 mm thick and the hub is 100 mm in diameter. The bolts are AISI 1045 steel. The coupling is Class 40 cast iron. The coupling is to transmit 25 kW at 275 rpm. A factor of safety of 5 based on ultimate is to be used. Determine the minimum bolt diameter required and check the flange for shear.

Solution:

Step 1 Recognize that bolt diameter must be evaluated in three different ways, based on bolt shear, bolt bearing, and flange bearing.

Step 2 Look up properties.

Class 40 cast iron	$\tau_{ult} = 380$ MPa	$\sigma_{ult} = 860$ MPa
AISI 1045 steel	$\tau_{ult} = 480$ MPa	$\sigma_{ult} = 655$ MPa

Step 3 Divide values in Step 2 by *Fs* of 5 to establish allowable stresses.

Class 40 cast iron	$\tau_{all} = 76$ MPa	$\sigma_{all} = 172$ MPa
AISI 1045 steel	$\tau_{all} = 96$ MPa	$\sigma_{all} = 131$ MPa

Step 4 Note that the allowable compressive stress, which we use for

bearing stress, is higher for the cast iron than the steel, so we need only evaluate bolt bearing stress.

Step 5 Determine the torque transmitted.

$$P = \omega T \qquad T = \frac{P}{\omega} \frac{25{,}000 \text{ N m/s}}{275 \text{ rpm} \left(\frac{1 \text{ min}}{60 \text{ s}}\right)\left(\frac{2\pi \text{rad}}{\text{rev}}\right)}$$

$$T = 34{,}272 \text{ N m}$$

Step 6 Determine the force at the bolt circle for each bolt.

$$T = F_{tot} r = Fnr \qquad F = \frac{T}{rn} = \frac{34{,}272 \text{ N m}}{(0.14 \text{ m})(5)} \qquad F = 49 \text{ kN}$$

$$(n = \text{number of bolts})$$

Step 7 Determine bolt size based on shear stress.

$$\tau = \frac{F}{A} = \frac{F}{0.7854 d^2} \qquad d = \left(\frac{F}{0.7854 \tau}\right)^{0.5}$$

$$d = \left(\frac{49 \text{ kN}}{0.7854 \text{ (96 MPa)}}\right)^{0.5} \qquad d = 0.0255 \text{ m}$$

Step 8 Determine bolt size based on bolt bearing stress.

$$\sigma_b = \frac{F}{A} \qquad A = dt \quad \text{where } t \text{ is the flange thickness}$$

$$\sigma_b = \frac{F}{dt} \qquad d = \frac{F}{\sigma_b t} = \frac{49 \text{ kN}}{131 \text{ MPa}(0.025 \text{ m})} \qquad d = 0.015 \text{ m}$$

The value from Step 7 must be used

$$\boldsymbol{d = 0.0255 \text{ m}}$$

Step 9 Determine the shear stress at the flange-hub connection and compare with allowable

$$\tau = \frac{F}{A} \qquad A = \pi r t$$

where r is the radius at the connection. Also F equals T divided by that same radius. Thus

$$\tau_{act} = \frac{T}{\pi r^2 t} = \frac{34{,}272 \text{ N m}}{\pi(0.05 \text{ m}^2)(0.025 \text{ m})}$$

$$\tau_{act} = 175 \text{ MPa}$$

Since this is substantially greater than the allowable stress, some alternative must be sought. A thicker flange (58 mm) would suffice. A change to a steel flange might also be made. A larger diameter coupling (about 50% increase in hub diameter) would also accomplish the task. The latter choice would result in a larger bolt circle which would reduce the size of the bolts required.

Problem 16-8: Two 2-in.-diameter shafts are joined by a flange coupling. The coupling has four 1-in.-diameter bolts on a 6-in. bolt circle. The flange is 1 in. thick and the hub is 4 in. in diameter. The bolts are AISI 1020 steel and the coupling is Class 60 cast iron. Using a factor of safety of 6, what is the maximum horsepower that can be transmitted at 300 rpm?

As indicated before, if two bolt circles are used, shear stress is considered to be proportional to radius.

$$\frac{\tau_1}{r_1} = \frac{\tau_2}{r_2} \qquad \tau_2 = \frac{r_2}{r_1}\tau_1$$

from $\tau = \dfrac{T}{rAn}$ the total torque must be

$$T = T_1 + T_2 \qquad T = r_1 n_1 A_1 \tau_1 + r_2 n_1 A_2 \tau_2$$

If 1 is taken to designate the outer circle where the stress is greatest, then the above equations can be combined to yield

$$\tau_1 = \frac{T}{(r_1 n_1 A_1) + (r_2^2 n_2 A_2)/r_1}$$

16.5 TORSION BARS

Earlier in this text we saw how energy could be stored in the form of strain energy in a member under compression or tension. Actually, this concept is more frequently applied in torsion. Indeed the design of springs, including compression and tension springs as well as torsion springs is based upon torsion theory. We will take a brief look at the simplest form of spring, the torsion bar.

Within the elastic range for a material, the angular deformation of a member in torsion is in direct proportion to the torque applied. Re-examining one of our earlier equations

$$\theta = \frac{TL}{GJ} \qquad T = \frac{GJ\theta}{L} \qquad T = \left(\frac{GJ}{L}\right)\theta$$

The term before θ in the right-hand equation is then the constant of proportionality, in this case known as the *spring constant.*

$$T = K\theta \qquad K\frac{T}{\theta} \qquad K = \frac{GJ}{L}$$

Thus the spring constant for a member in torsion can be defined as the ratio of torque to twist and can be determined as shown at right above.

Torsion members may be combined either in series or in parallel as shown in Figure 16-13. Examination shows that in series there is only one torque that is common to all segments. In parallel, however, it is the deformation that is common to both segments.

Series

$$T = T_1 = T_2$$

$$\theta_{\text{total}} = \theta_1 + \theta_2 \qquad \theta_1 = \frac{T_1}{K_1} = \frac{T}{K_1} \qquad \theta_2 = \frac{T}{K_2}$$

$$\theta_{\text{total}} = \frac{T}{K_1} + \frac{T}{K_2} \qquad \frac{\theta_{\text{total}}}{T} = \frac{1}{K_1} + \frac{1}{K_2}$$

$$\frac{\theta_{\text{total}}}{T} = \frac{1}{K_{\text{total}}} = \frac{1}{K_1} + \frac{1}{K_2}$$

$$K_{\text{total}} = \frac{1}{\frac{1}{K_1} + \frac{1}{K_2}}$$

Giving us a spring constant equation for the system which can be extended to any number of segments *in series.*

Parallel

$$\theta_1 = \theta_2 = \theta \qquad T_{\text{total}} = T_1 + T_2$$

$$T_{\text{total}} = K_1\theta_1 + K_2\theta_2 = K_1\theta + K_2\theta$$

$$T_{\text{total}} = \theta(K_1 + K_2)$$

$$\frac{T_{\text{total}}}{\theta} = K_1 + K_2 \qquad K_{\text{total}} = K_1 + K_2$$

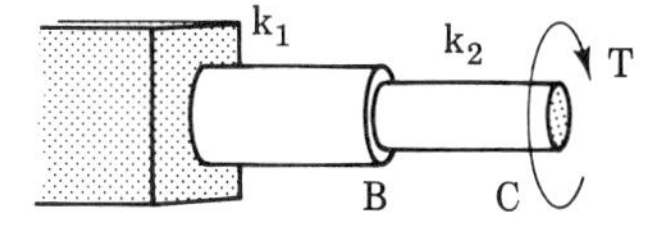

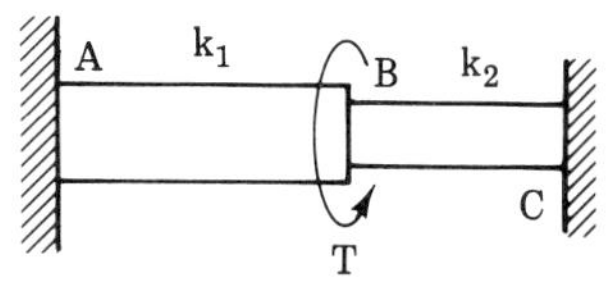

Figure 16-13. Torsion bar arrangments.

giving us a spring constant for the parallel system. It is of course possible to design a series-parallel system as shown in Figure 16-14.

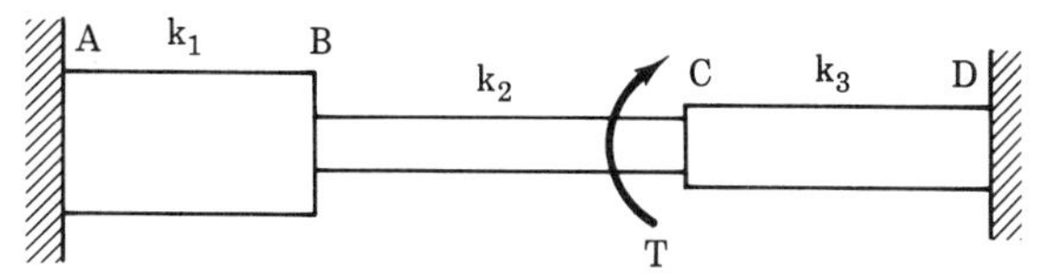

Figure 16-14. Series-parallel torsion bar.

Example 16-9: Determine the spring constant for the system shown in Figure 16-15. Shaft 1 is a hollow circular tube with a 1-in. inside diameter and a 1.4-in. outside diameter. Shaft 2 is a solid circular rod with a diameter of 1 in. The material for both segments is 6061-T6 aluminum with $G = 3.76 \times 10^6$ psi.

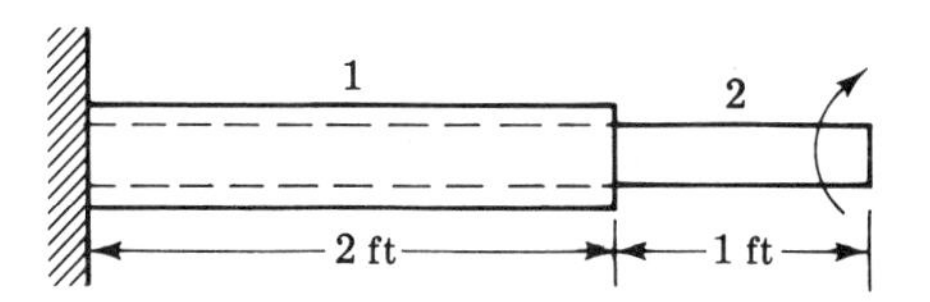

Figure 16-15

Solution:

Step 1 Recognize that this is a series torsion bar system

$$K_{\text{total}} = \frac{1}{\dfrac{1}{K_1} + \dfrac{1}{K_2}} \qquad K = \frac{JG}{L}$$

$$J_{\text{solid}} = \frac{\pi d^4}{32} \qquad J_{\text{hollow}} = \frac{\pi(d_0^4 - d_i^4)}{32}$$

Step 2 Calculate K_{total}

$$K_{\text{total}} = \frac{1}{\dfrac{32L_1}{\pi G(d_0^4 - d_i^4)} + \dfrac{32L_2}{\pi G d^4}}$$

$$K_{\text{total}} = \frac{\pi G}{32}\left[\frac{1}{\dfrac{L_1}{d_0^4 - d_i^4} + \dfrac{L_2}{d^4}}\right]$$

$$K_{\text{total}} = \frac{\pi(3.76 \times 10^6 \text{ lb/in.}^2)}{32}\left[\frac{1}{\dfrac{24}{(1.4 \text{ in.}^4) - (1 \text{ in.}^4)} + \dfrac{12 \text{ in.}}{(1 \text{ in.}^4)}}\right]$$

$$\mathbf{K_{total} = 18{,}000 \text{ in. lb/rad}}$$

Problem 16-9: Determine the angular deformation resulting from a torque of 200 N m applied as shown to the system in Figure 16-16. Both segments are solid circular rods made of Type 304 stainless steel with $G = 69$ GPa.

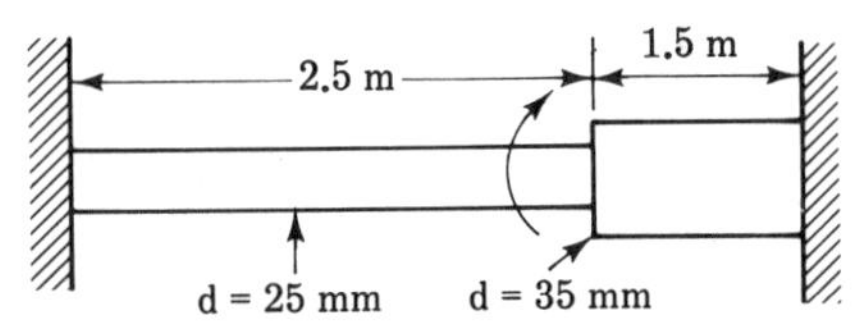

Figure 16-16

16.6 NONCIRCULAR CROSS SECTIONS

Many machine and structural members, while not used to transmit torque and thus power, are nevertheless subjected to varying degrees of torsion. In most cases this is unavoidable, a function of how loads must be located. Thus it is generally desirable to have them be rather stiff, that is, have a low spring constant.

Even a preliminary discussion of this topic is beyond this text and indeed most engineering texts. Several generalities that are quite useful can be stated however.

1. Solid cross sections of identical areas increase in stiffness as their shape approaches a circle. In other words, for a solid cross section of constant area, a circular shape has the greatest stiffness.
2. Hollow shapes such as circular pipes and composite beams that form a closed rectangle (i.e., two channels and two flat plates) have a high degree of stiffness but not as high as solid shapes of the same area. Again a circular hollow shape is stiffer than any other hollow shape with the same area.
3. Relatively long thin sections that do not close on themselves as in 2 above have very low stiffness. Thus our standard structural shapes

Table 16-1
Coefficients for Rectangular Sections in Torsion

b/a	α	β
1.0	0.208	0.1406
1.2	0.219	0.1661
1.5	0.231	0.1958
2.0	0.246	0.229
2.5	0.258	0.249
3.0	0.267	0.263
4.0	0.282	0.281
5.0	0.291	0.291
6.0	0.299	0.299
10.0	0.312	0.312
α	0.333	0.333

such as *I*-beams, wide flange beams, channels, and angles are very poor in torsion.

A great deal of work has been done in this area and we may readily benefit from it. Figure 16-17 tabulates the equations for maximum shearing stress and angular deformation for a number of common cross-sectional shapes. It also locates the points of maximum stress. Note that the equations for a solid rectangular cross section require reference to Table 16-1.

Example 16-10: Compare the stiffnesses of two members of equal length and the same material.

One has a solid rectangular cross section of 20 by 30 cm. The other has the same cross-sectional area but the cross-section is a hollow rectangle with inside dimensions of 20 by 30 cm and a uniform wall thickness of 8.85 cm, giving it the same cross-sectional area as the solid member.

Solution:

Step 1 Establish the appropriate equations for stiffness.

$$\theta_{\text{solid rectangle}} = \frac{T_s L_s}{\beta b a^3 G}$$

where b is the long side, a is the short side, and β is from Table 16-1.

$$\theta_{\text{hollow rectangle}} = \frac{bt + ht_1}{2b^2h^2tt_1}\left(\frac{TL}{G}\right)$$

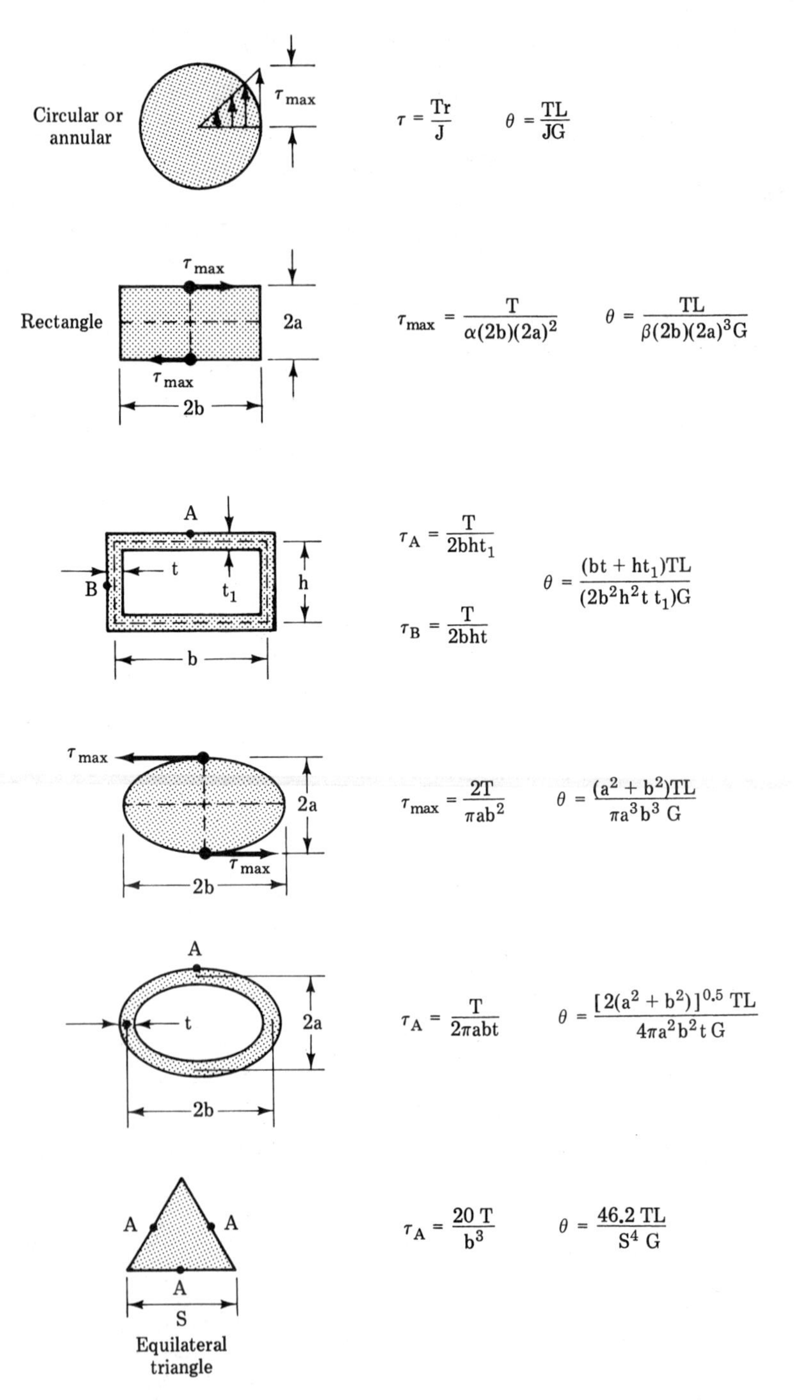

Figure 16-17. Shearing stress in noncircular cross sections.

with b, h, t, and t_1 as defined in Figure 16-17 (problem gives $t = t_1$).

$$\frac{T_s}{\theta_s} = \frac{\beta b a^3 G}{L}$$

$$\frac{T_H}{\theta_H} = \frac{2b^2h^2tG}{b + hL}$$

Step 2 Establish the equation for the ratio of stiffness and compute (G, L same for both).

$$\frac{T_H/\theta_H}{T_s/\theta_s} = \frac{2b^2h^2t/(b+h)}{\beta b a^3}$$

$$\frac{T_H/\theta_H}{T_s/\theta_s} = \frac{\dfrac{2(0.3885)^2(0.2885)^2(0.0885)}{0.3885 + 0.2885}}{(0.196)(0.30)(0.20)^2} = 0.19$$

Thus, the hollow rectangular cross section has only 19% of the stiffness of the solid member even though the amount of material is identical.

Problem 16-10: A member 30 in. long made of Monel has a hollow elliptical cross section with an average width of 3 in., an average height of 2 in., and a wall thickness of 0.25 in. Using a factor of safety of 4 based on the ultimate strength, what is the maximum allowable torque that may be applied to the member? At that torque, what will the angular deformation be?

16.7 READINESS QUIZ

1. Torsional stress is a form of

   (a) compressive stress
   (b) tensile stress
   (c) shear stress
   (d) any of the above
   (e) none of the above

2. Torsional stress is a function of

   (a) I_x alone
   (b) I_y alone
   (c) J
   (d) all of the above
   (e) none of the above

3. Maximum shearing stress in a solid circular shaft occurs at the

   (a) surface
   (b) radius of gyration
   (c) center
   (d) any of the above
   (e) none of the above

4. Total angular deformation of a shaft of some length is a function of its

   (a) length
   (b) polar moment of inertia
   (c) modulus of rigidity
   (d) all of the above
   (e) none of the above

5. A solid circular shaft and a hollow circular shaft have the same cross-sectional area and length and are made of the same material. The hollow shaft is __________ than the solid shaft.

   (a) lighter
   (b) weaker
   (c) stiffer
   (d) all of the above
   (e) none of the above

6. The power transmitted by a shaft can be changed by changing

   (a) applied torque
   (b) shaft speed
   (c) both of the above
   (d) any of the above
   (e) none of the above

7. Substituting a material with a greater G while all else remains constant will increase the __________ of a shaft.

   (a) strength
   (b) stiffness
   (c) moment of inertia
   (d) all of the above
   (e) none of the above

8. Maximum stress in a stepped shaft in torsion is a function of the

   (a) smaller of the diameters
   (b) ratio of the diameters
   (c) fillet radius at the step
   (d) all of the above
   (e) none of the above

9. Bearing stress in a key is __________ shear stress in a key.

   (a) always greater than
   (b) always less than
   (c) always equal to
   (d) never equal to
   (e) none of the above

10. Bolted coupling design must consider

   (a) shear stress in the coupling
   (b) shear stress in the bolt
   (c) bearing stress in the bolt
   (d) all of the above
   (e) none of the above

11. If two torsion members are combined in series to form a torsion bar, then one can add the

   (a) torques
   (b) spring constants
   (c) angular deformations
   (d) any of the above
   (e) none of the above

12. The following shaft cross sections have the same area. The stiffest in torsion will be the

   (a) square
   (b) round
   (c) triangular
   (d) rectangular
   (e) elliptical

13. The torsion bar system shown in Figure 16-18 is a __________ system.

   (a) series
   (b) parallel
   (c) series-parallel
   (d) parallel-series
   (e) none of the above

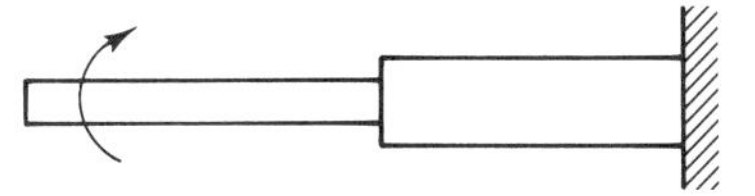

Figure 16-18

14. The maximum shearing stress in a shaft with a circumferential groove is dependent on the

(a) width of the groove
(b) modulus of rigidity of the shaft material
(c) radius of the outer edge of the groove
(d) all of the above
(e) none of the above

15. Power can be described as

(a) torque times angular velocity
(b) work done per unit time
(c) energy produced or expended per unit time
(d) all of the above
(e) none of the above

16.8 SUPPLEMENTAL PROBLEMS

16-11 Determine the maximum shearing stress developed in a 20-cm-diameter shaft with a torque of 3500 N m.

16-12 Determine the maximum shearing stress developed in a hollow circular shaft with an inside diameter of 3 in., an outside diameter of 4 in., and an applied torque of 10,000 ft lb.

16-13 A hollow shaft has an outside diameter of 1 in. and an inside diameter of 0.75 in. and is 20 in. long. It is made of aluminum with a safe shear stress of 5000 psi and a G of 3.8×10^6 psi. What is the maximum torque that may safely be applied to the shaft if the total angular deformation is not to exceed 0.5°?

16-14 What is the maximum power that can be transmitted by the shaft in Problem 16-11 at 5000 rpm?

16-15 A 30-mm-diameter shaft is stepped down to a 20-mm diameter and the fillet radius is 2 mm. The material has an allowable shearing stress of 150 MPa and a modulus of rigidity of 78.6 GPa. The 30-mm section has a circumferential groove 3 mm deep and 2 mm wide with a 1-mm fillet radius. How much power can the shaft transmit at 1600 rpm? If the 30-mm section has a length of 1.5 m and the 20-mm section has a length of 0.8 m, what will the total angular deformation be?

16-16 The line shaft shown in Figure 16-19 is being driven by the electric motor at 700 rpm. The material has a maximum allowable shearing stress of 11,000 psi and a G of 9.5×10^6 psi.

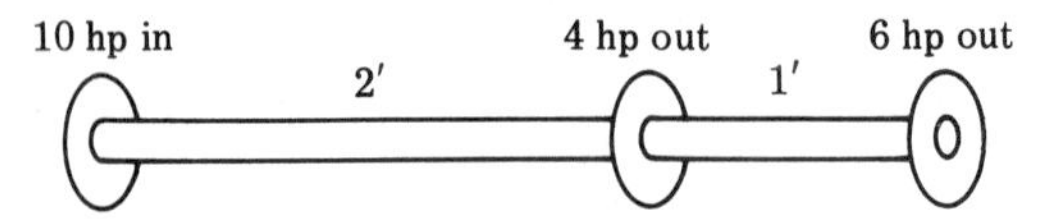

Figure 16-19

What are the minimum shaft diameters required? What is the total angular deformation?

16-17 A 0.75-in. shaft is to transmit 1.0 hp at 3600 rpm. What are the shear stress and bearing stress in a rectangular key that is 0.75 in. long, 0.125 in. wide, and 0.25 in. thick?

16-18 Eight 10-mm-diameter bolts are arranged in two concentric circles ($r_1 = 10$ cm, $r_2 = 5$ cm). What is the maximum horsepower that can be transmitted by the coupling if the shaft speed is 600 rpm and the maximum allowable shearing stress is 8,000 psi for the bolts?

16-19 Determine the spring constant for the system shown in Figure 16-20. Shaft 1 has an elliptical cross section 4 by 3 cm, shaft 2 has an elliptical cross section 3 by 2 cm, and shaft 3 has an elliptical cross section 2 by 1 cm. The material for all segments is 302 stainless steel with a G of 75 GPa.

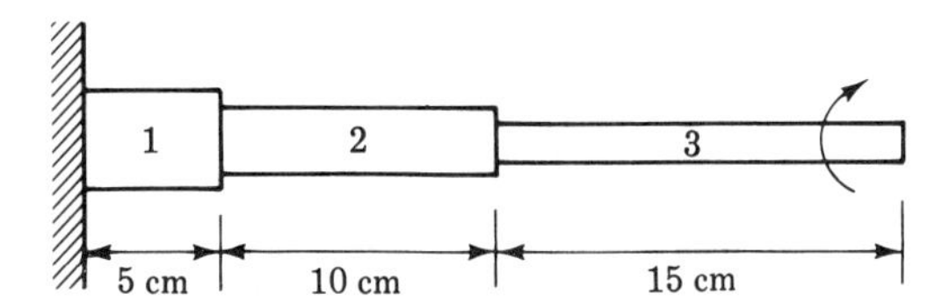

Figure 16-20

16-20 A 302 stainless steel shaft with a solid cross section 1.5 by 1.5 in. and 2 ft long can sustain what maximum torque with a factor of safety of 4? At that torque, what will the total angular deformation be?

Chapter

17

Stresses and Deflections in the Design of Beams

17.1 OBJECTIVES

Upon completion of the work relating to this chapter, you should be able to perform the following.

1. Determine the tensile and compressive stresses acting in a statically determinate beam subjected to bending.
2. Determine the longitudinal shear stress acting in a statically determinate beam subjected to bending.
3. Determine the deflection of a statically determinate beam subjected to bending.
4. Select the most economical size beam for a given load and support requirement based on stress and deflection criteria.
5. Apply the knowledge and skills gained in the first four objectives to beams consisting of two or more materials.
6. Modify a design to take into account stress concentrations.
7. Modify a design if necessary to prevent lateral buckling.

17.2 INTRODUCTION

It is important to note that the attainment of the first three objectives will deal strictly with statically determinate beams. Chapter 20 will cover these aspects of statically indeterminate beams. The examples and problems involved in attaining the last four objectives are all for statically determinate beams. However, once you have completed Chapter 20, you should experience no difficulty in applying the knowledge gained in the latter portion of this chapter to statically indeterminate beams. Statically determinate beams are those whose supports are such that the reactions to the loads and thus the shear and moment at all points in the beam can be determined through use of the equations for static equilibrium as was done in Chapter 10. Figure 17-1 shows examples of the three basic types: simply supported, cantilever, and overhanging.

In considering the design of a beam subjected to bending, we would do well to examine just what happens to a beam when it bends. As shown in Figure 17-2a and b, the most obvious occurrence is that the beam attains a curved rather than a straight shape. Actually, the weight of the beam itself will result in some load and therefore cause a slight curvature. Examining the curvature in more detail reveals several things.

First is the deflection, or change in position, of various points in the beam. There are many cases in structural and machine design where the amount of deflection that is tolerable may be an overriding design factor. In the ceiling of a room where plaster or wallboard is to be used as a surface

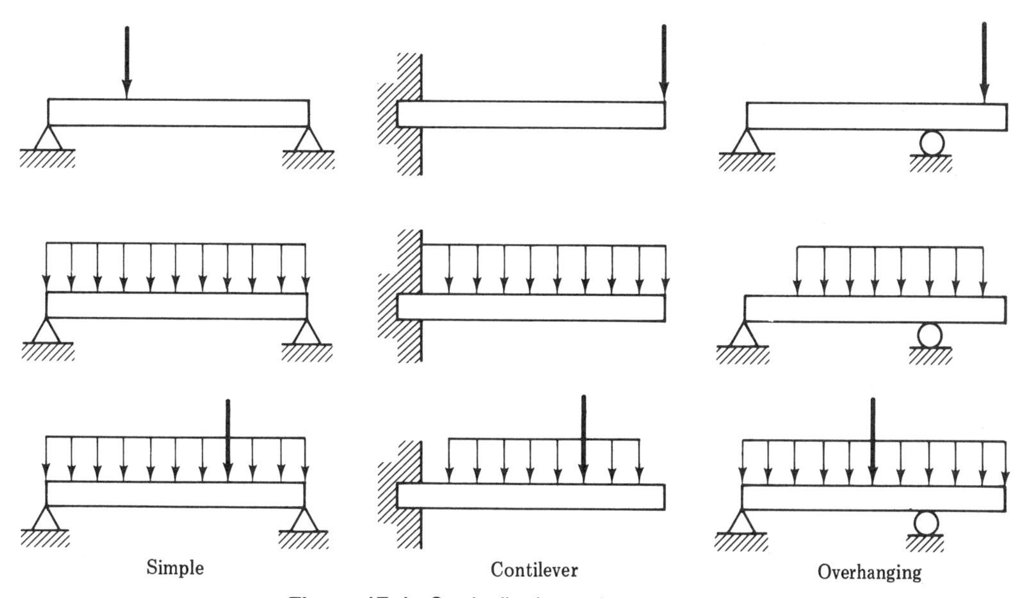

Figure 17-1. Statically determinate beams.

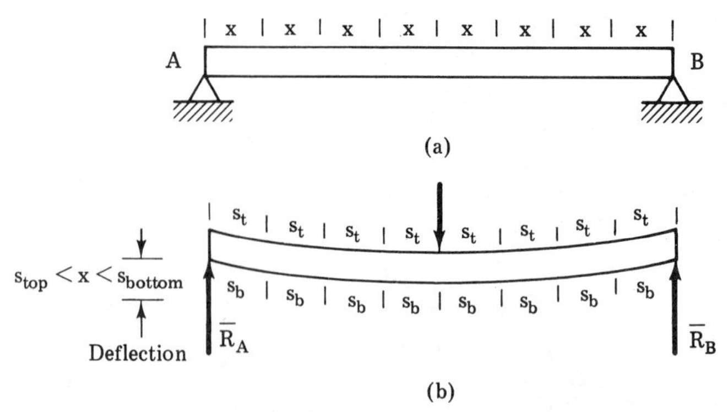

Figure 17-2. Effects of bending.

material, excessive beam deflection may overstress the surface and cause it to crack. In like manner, excessive floor beam deflection under live loads may make it impossible to keep equipment properly aligned. The deflection of a lathe bed while cutting forces are being applied must be minimized or the accuracy of the cut will be unacceptable.

None of the above speaks to whether or not the beam will permanently deform or even break under load. As a matter of fact, our deflection calculations will all be kept within the elastic range. But what of stresses, the evaluation of potential failure, or the assurance of nonfailure? The phenomenon of bending stress is slightly more complex than the simple tension, compression, and shear we have considered up to now. Therefore, as a second consideration, let us examine what happens to the top and bottom surfaces of a beam subjected to bending. Let us draw lines the short way across the top and bottom surfaces of the beam before it is loaded. If we apply a simple single-concentrated load as was shown in Figure 17-2, both surfaces (and indeed any and every plane between them) will assume a curved shape. All of these will have a common center of curvature. Thus the radius of the top surface will be shorter than the radius of the bottom. The corresponding arcs will therefore be shortest on top and longest on the bottom. If we examine the lines we drew we will find that they are closer together on top than they were prior to loading and further apart on the bottom than they were prior to loading, as shown in Figure 17-2. This means that the *top* fibers are in *compression* when the beam is subjected to bending and the *bottom* fibers are subjected to *tension.* This is not particularly significant for materials whose allowable stresses in tension and compression are essentially identical—we would simply determine the maximum stress and base our design on that. However, some very common materials such as concrete and cast iron have quite different properties in tension and compression and both must be considered.

Third, and last, if we examine the above phenomenon more closely, we realize that along the various curved planes some relative motion must occur. This must be so, or one plane could not end up being longer or shorter than another after a load is applied. Such relative motion would be of a sliding or shearing nature. Thus, in a beam subjected to bending, a longitudinal shearing stress occurs. This is particularly significant for materials that are relatively low in shear strength in the longitudinal direction such as wood and wrought iron.

In summary, in selecting a specific size of beam for a given application we must consider the following:

1. the maximum allowable tensile stress for the material,
2. the maximum allowable compressive stress for the material,
3. the maximum allowable shear stress for the material, and
4. the maximum deflection at any point permissible for the application.

One final note. It is common practice to neglect the weight of the beam as a load unless that weight is more than 10% of the total load.

17.3 TENSILE AND COMPRESSIVE STRESSES DUE TO BENDING

As indicated in Section 17-2, the phenomenon of bending results in some portions of a beam being in compression while some are in tension. Further examination of this phenomena yields the following:

1. The top and bottom planes see the greatest force, thus the greatest stress, and E being constant, the greatest strain.

2. At some plane in the beam, approaching it from the top and bottom simultaneously, compressive and tensile stress both reduce to zero. This is called the *neutral* plane and any axis lying in it is a *neutral axis.* The axis of most interest to us is the longitudinal axis passing along the length of the beam.

3. The neutral axis of most interest to us is more precisely located as passing through the centroid of the cross section of the beam. The usefulness of this location will be immediately evident.

Figure 17-3 shows the cross section of a rectangular beam. This is a very simple cross section but using it we shall derive an equation for *bending stress,* or *flexural stress* as it is sometimes called, that will be applicable to any cross section. This bending stress is simply the tensile or compressive stress in a beam resulting from its bending due to a load applied perpendicular to its longitudinal axis.

In Figure 17-3, xx is the horizontal centroidal axis of the beam section. Let us examine any small horizontal strip with an area A_A located a distance

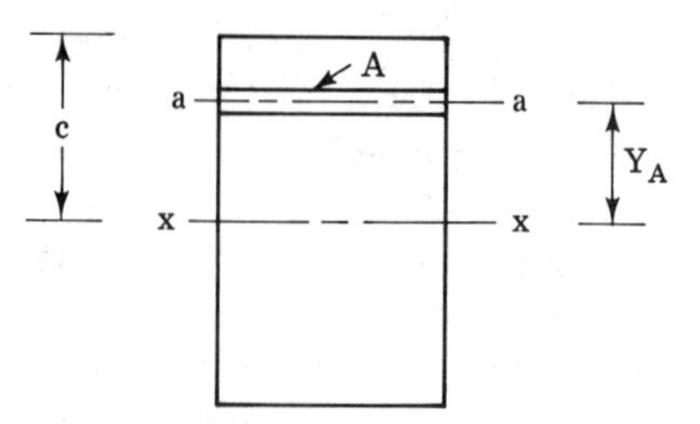

Figure 17-3. Rectangular beam section.

Y_A from the neutral (centroidal) axis. If the stress level in that area is σ_A, then the force on that area is

$$F_A = \sigma_A A$$

and the moment of that force with respect to the neutral axis xx is

$$M \text{ of } F_A = F_A Y_A = \sigma_A A Y_A$$

However, stress is zero at the neutral axis and increases in proportion to the distance from the neutral axis until it reaches a maximum at each furthermost surface. If we call the specific distance to the furthermost surface c, then

$$\frac{\sigma_A}{Y_A} = \frac{\sigma_{\max}}{c} \quad \text{and} \quad \sigma_A = \sigma_{\max}\left(\frac{Y_A}{c}\right)$$

The moment of the force acting on area A then becomes

$$M \text{ of } F_A = F_A Y_A = \sigma_{\max} \frac{Y_A}{c} A Y_A = \frac{\sigma_{\max}}{c} Y_A^2$$

Since the rectangle is made up of a very large number of areas, the moment-resisting bending, or bending moment, for the entire rectangle is

$$M = \Sigma \left[\frac{\sigma_{\max}}{c} Y_A^2 \right]$$

The term $\sigma_{\max}/c$ is the same for every area so

$$M = \frac{\sigma_{\max}}{c} [\Sigma(Y_A^2)] \qquad M = \frac{\sigma_{\max}}{c} I$$

Solving for σ_{max}

$$\sigma_{max} = \frac{Mc}{I} = \frac{M}{I/c} = \frac{M}{Z}$$

where Z is I/c and is a property of the cross section that is known as the *section modulus.* Formulas for common areas as well as values for structural members are given in the Appendices.

The ease with which the equation for maximum stress can be adapted to various cross sections is readily illustrated by considering the two cross sections shown in Figure 17-4.

In Figure 17-4a, if $w = h$, then the section modulus Z will be greater about the yy axis than about the xx axis. This is based on the fact that all four values for c (compressive and tensile about xx, and compressive and tensile about yy) will be the same and that I_{xx} will be greater than I_{yy}.

In 17-4b, if $w = h$, then the section moduli require some evaluation. I_{xx} is almost twice I_{yy} but whereas c for tension and c for compression are the same about the yy axis, they differ by more than 2:1 about the xx axis. In tables for structural members such as angles and tees the Z value given is the smaller one resulting from the larger c and thus yielding the larger maximum stress. This approach is valid unless the material does not exhibit essentially equal tensile and compressive allowable stresses. In such cases I/c for each stress must be evaluated, and each stress determined and compared to the allowable.

Example 17-1: Determine the maximum tensile and compressive stresses in a beam with a rectangular cross section of 120 by 80 mm and a maximum internal moment of 600 N m:

1. If the moment is in a plane parallel to the 120-mm side of the beam.
2. If the moment is in a plane parallel to the 80-mm side of the beam.

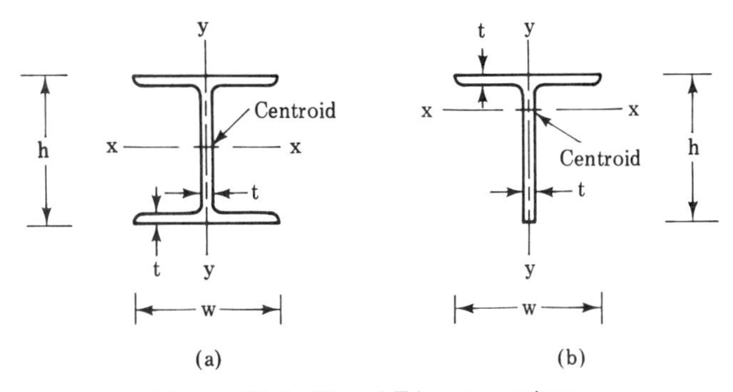

Figure 17-4. W and T beam sections.

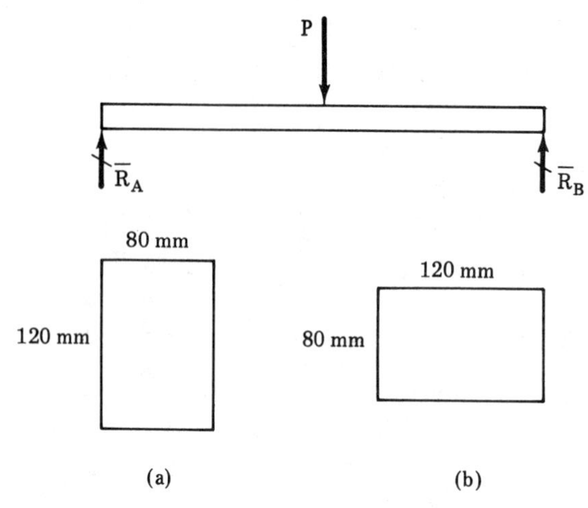

Figure 17-5

Solution:

Step 1 Taking part 1 first, if the internal moment is in a plane parallel to the 120-mm side, then in the equation for moment of inertia, $h = 120$ mm and $b = 80$ mm.

Step 2 Since the cross section is symmetrical and can be assumed to be homogeneous, c is the same for the top and the bottom, for compression and tension. $c = 60$ mm.

Step 3 Determine I.

$$I = \frac{1}{12} bh^3 = \frac{1}{12} (0.08 \text{ m})(0.12 \text{ m}^3) \qquad I = 1.152 \times 10^{-5} \text{ m}^4$$

Step 4 Determine maximum stress (compressive = tensile).

$$\sigma_{max} = \frac{M_c}{I} = \frac{(600 \text{ N m})(0.06 \text{ m})}{1.152 \times 10^{-5} \text{ m}^4} \qquad \boldsymbol{\sigma_{max} = 3.125 \text{ MPa}}$$

Step 5 Repeat for part 2, where $b = 120$ mm, $h = 80$ mm, and $c = 40$ mm.

$$I = \frac{1}{12} bh^3 = \frac{1}{12} (0.12 \text{ m})(0.08 \text{ m}^3) \qquad I = 5.12 \times 10^{-6} \text{ m}^4$$

$$\sigma_{max} = \frac{Mc}{I} = \frac{(600 \text{ N m})(0.04 \text{ m})}{5.12 \times 10^{-6} \text{ m}^4} \qquad \boldsymbol{\sigma_{max} = 4.69 \text{ MPa}}$$

In common terminology, for the same loading, the maximum stress is less if the beam is laid on its edge rather than its side.

Problem 17-1: Determine the maximum stress in the beam shown in Figure 17-6:

1. If it is laid with its long cross-sectional axis horizontal.
2. If it is laid with its short cross-sectional axis horizontal.

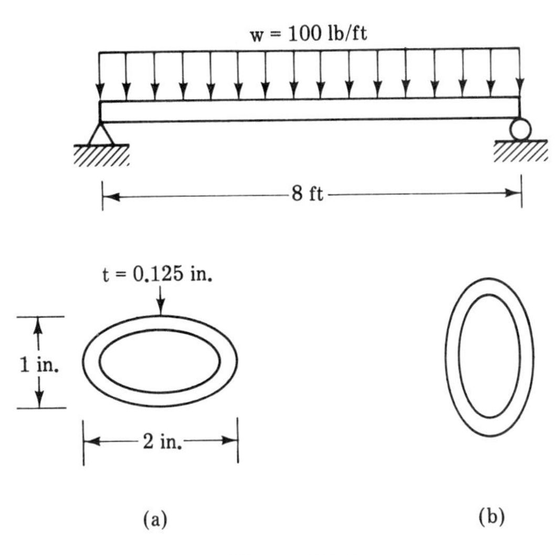

Figure 17-6

Example 17-2: Determine the maximum compressive and tensile stresses in the beam shown in Figure 17-7a.

Solution:

Step 1 Draw the free-body diagram for the beam as shown in 17-7b.

Step 2 Determine $\overline{R}_{BY}$ by taking moments about A.

$$\Sigma M_A = 0 \qquad P_1 d_1 + P_2 d_2 + \overline{R}_{BY} d_{BY} = 0$$

$$\overline{R}_{BY} d_{BY} = -(P_1 d_1 + P_2 d_2) = -[-(4000 \text{ lb})(4 \text{ ft}) - (2000 \text{ lb})(12 \text{ ft})]$$

$$\overline{R}_{BY} d_{BY} = +36{,}000 \text{ ft lb} \qquad \overline{R}_{BY} = \frac{\overline{R}_{BY} d_{BY}}{d_{BY}} = \frac{40{,}000 \text{ ft lb}}{18 \text{ ft}}$$

$$\overline{R}_{BY} = 2222 \text{ lb}$$

Step 3 Determine $\overline{R}_{AY}$ by sum of the vertical forces.

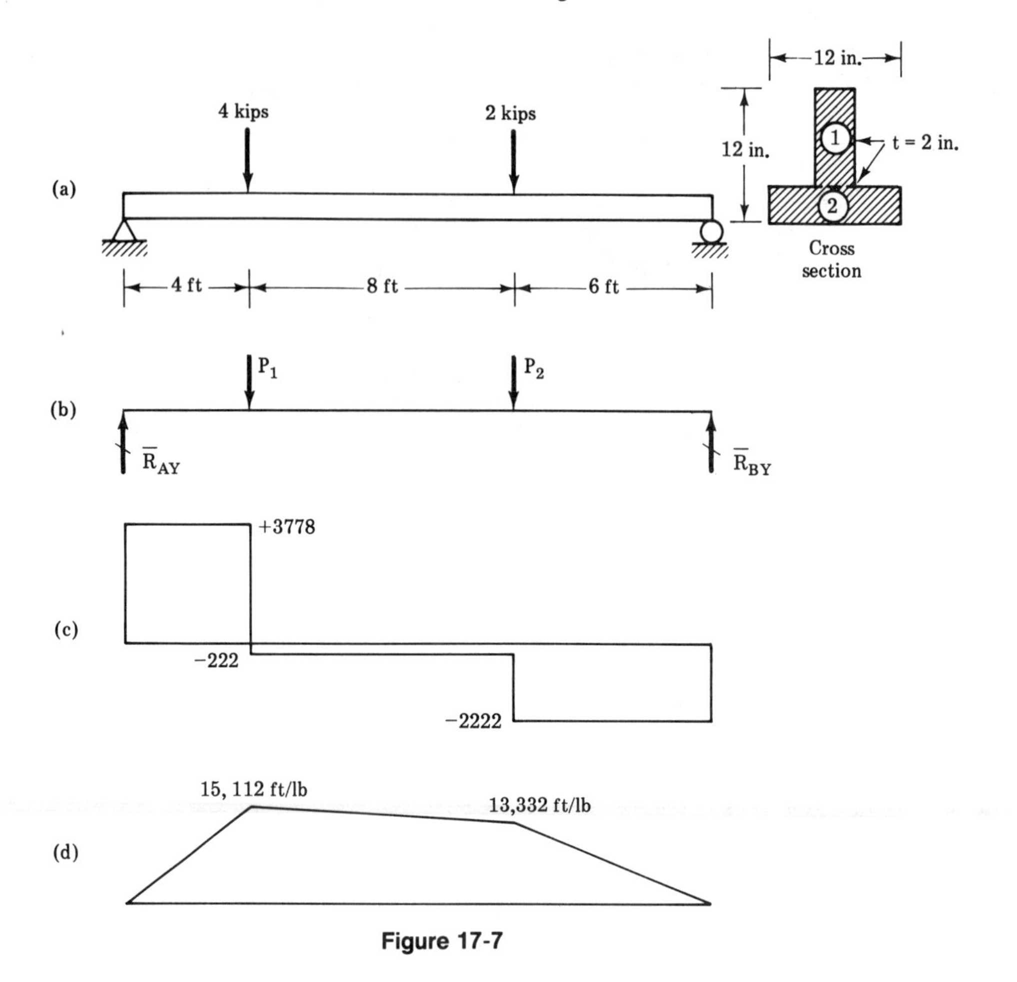

Figure 17-7

$$\Sigma F_y = 0 \qquad \overline{R}_{AY} + P_1 + P_2 + \overline{R}_{BY} = 0 \qquad \overline{R}_{AY} = -(P_1 + P_2 + \overline{R}_{BY})$$

$$\overline{R}_{AY} = -(4000 - 2000 + 2222 \text{ lb}) \qquad \overline{R}_{AY} = +3778 \text{ lb}$$

Step 4 Draw the shear diagram as shown in Figure 17-7c.

Step 5 Draw the moment diagram as shown in 17-7d.

Step 6 Determine the vertical location of the centroid of the cross section.

$$\bar{y} = \frac{A_1Y_1 + A_2Y_2}{A_1 + A_2} = \frac{(10 \times 2 \text{ in.})(7 \text{ in.}) + (12 \times 2 \text{ in.})(1 \text{ in.})}{(10 \times 2 \text{ in.}) + (12 \times 2 \text{ in.})}$$

$$\bar{y} = 3.73 \text{ in.}$$

Step 7 Determine c values.

$$\text{Compression (top): } c_c = 8.27 \text{ in.}$$
$$\text{Tension (bottom): } c_T = 3.73 \text{ in.}$$

Step 8 Determine $\bar{I}_x$.

$$\bar{I}_x = \bar{I}_1 + A_1 d_1^2 + \bar{I}_2 + A_2 d_2^2$$

$$I_x = \frac{1}{12}(2 \text{ in.})(10 \text{ in.}^3) + (2 \text{ in.})(10 \text{ in.})(3.27 \text{ in.}^2)$$

$$+ \frac{1}{12}(12 \text{ in.})(2 \text{ in.}^3) + (2 \text{ in.})(12 \text{ in.})(2.73 \text{ in.}^2)$$

$$\bar{I}_x = 567.38 \text{ in.}^4$$

Step 9 Determine maximum stresses.

$$\text{Compressive: } \sigma_{\text{max}} \frac{Mc_c}{I} = \frac{(15{,}112 \text{ ft lb})\ (12 \text{ in./ft})(8.27 \text{ in.})}{567.38 \text{ in.}^4}$$

$$\boldsymbol{\sigma_{\text{max-com}} = 2643 \text{ psi}}$$

$$\text{Tensile: } \sigma_{\text{max}} = \frac{Mc_T}{I} = \frac{(15{,}112 \text{ ft lb})(12 \text{ in./ft})(3.73 \text{ in.})}{567.38 \text{ in.}^4}$$

$$\boldsymbol{\sigma_{\text{max-ten}} = 1192 \text{ psi}}$$

Problem 17-2: Determine the maximum compressive and tensile stresses in the beam shown in Figure 17-8.

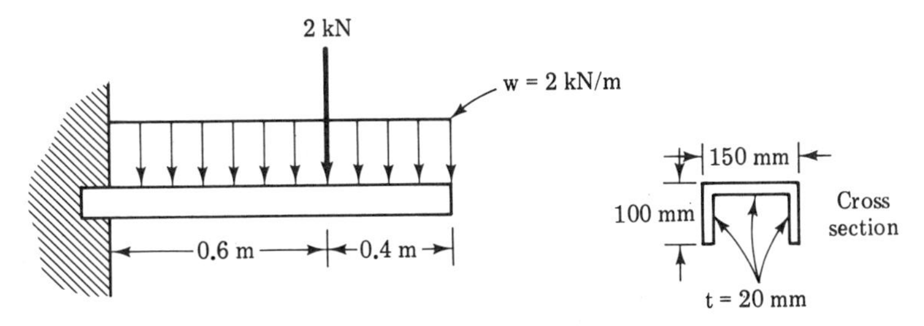

Figure 17-8

17.4 HORIZONTAL AND VERTICAL SHEARING STRESSES

There exists, in a beam, shear stress in the longitudinal direction which must be evaluated. The nature of this shear and the fact that it exists can be observed in Figure 17-9. In 17-9a we see several planks laid upon each other supported at the ends but with no load. When a load is applied as shown in 17-9b, the planks assume a curved shape and some sliding has taken place. Each individual plank is in compression at its top surface and in tension at its bottom surface. As a result, adjacent surfaces of two planks have different lengths under load even though the lengths were identical before loading. Thus, relative motion must have taken place between adjacent surfaces. If the planks were glued or bolted together, the glue or bolts would impede such relative motion. Similarly if the beam were one solid piece instead of several layers, the shear strength of the material in the horizontal direction would impede such relative motion.

Intuitively, we can see that the greater the loads on a beam the more bending takes place. Further, the more bending, the greater the longitudinal shear stress will be. However, we need to establish a relationship that will permit us to determine longitudinal shear stress when the nature and loading of a beam are known.

Let us examine a simply supported beam with a single concentrated load on it as shown in Figure 17-10a. The correspondingly simple shear and moment diagrams are shown in Figure 17-10b and c. Two points close together along the beam have been indicated by sections 1-1 and 2-2. These points experience a specific shear force $V_1 = V_2$ and specific moments $M_1 > M_2$. If we consider the two sections to be a small distance Δx apart then $\Delta M/\Delta X = V$.

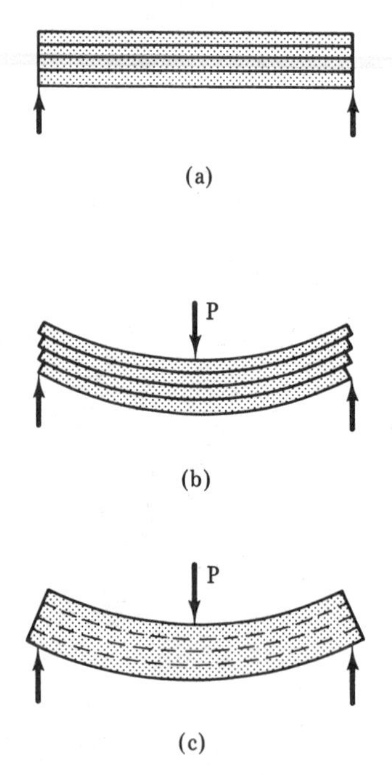

Figure 17-9. Longitudinal shearing stress.

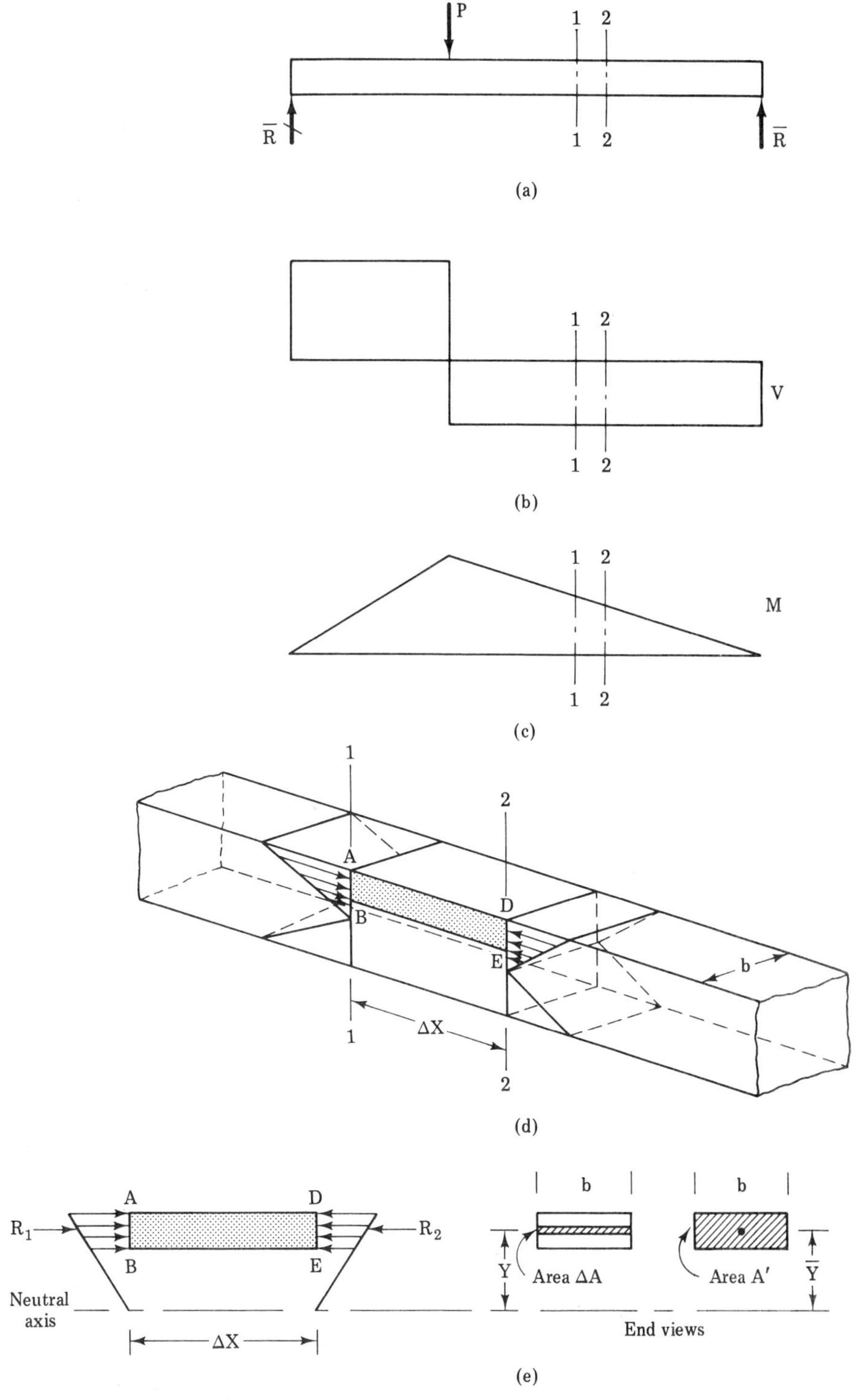

Figure 17-10. Longitudinal shearing stress evaluation.

In Figure 17-10d, this section of the beam has been enlarged and the tensile and compressive forces acting on it shown. Further, a portion of the section has been designated *ADEB* and points *B* and *E* identified as the locations where the section lines cross the neutral plane.

In 17-10e, the designated portion is examined in more detail. R_1 and R_2 are resultants of the compressive forces (due to bending) acting on the two ends of the portion. These act on areas $(AB)b$ and $(DE)b$, respectively, which are equal to each other and designated as A: Equilibrium is maintained by the shear force acting in the plane $BE(b)$ and equal to $\tau b \Delta x$. Thus

$$R_1 - R_2 = \tau b \Delta x$$

If we slice the area into many thin slices of area ΔA, where the compressive stress is σ_1 or σ_2 at a distance y from the neutral axis, then

$$R_1 = \Sigma \sigma_1 \Delta A \qquad R_2 = \Sigma \sigma_2 \Delta A$$

From our work in Section 17-3,

$$\sigma_1 = \frac{M_1 y}{I} \qquad \sigma_2 = \frac{M_2 y}{I}$$

Substitution yields

$$\tau b \Delta x = \Sigma \frac{M_1 y \Delta A}{I} - \Sigma \frac{M_2 y \Delta A}{I}$$

$$\tau b \Delta x = \frac{M_1 - M_2}{I} \Sigma y \Delta A = \frac{\Delta M}{I} \Sigma\, y \Delta A$$

Solving for τ the shear stress yields

$$\tau = \frac{\Delta M}{\Delta x} \Sigma \frac{y \Delta A}{Ib}$$

$$\text{Since } \Delta M / \Delta x = v \quad \text{and} \quad \Sigma y \Delta A = \bar{y} A'$$

$$\tau = \frac{V}{Ib} A' \bar{y}$$

where τ is the longitudinal shear stress in the plane at the bottom of an area A', $\bar{y}$ is the location of the centroid of that area with respect to the neutral axis, I is the moment of inertia of the entire cross section, b is the width of the cross section at the plane being considered, and A' is the area of the cross section of the beam between the plane being considered and the plane of the nearest extreme fiber(s).

Beams that are symmetrical about the neutral axis exhibit maximum shear stress at the neutral axis at a point or points where the shear force is a maximum.

Two common cross sections, rectangular and circular, are shown in Figure 17-11.

$$\tau_{\text{max-rect}} = \frac{V}{Ib} A'\bar{y} = \frac{V}{(bh^3/12)b} \left(\frac{bh}{2}\right)\left(\frac{h}{4}\right)$$

$$\tau_{\text{max-rect}} = \frac{3V}{2A}$$

$$\tau_{\text{max-circ}} = \frac{V}{Ib} A'\bar{y} = \frac{V}{(\pi b^4/64)d} \left(\frac{\pi b^2}{8}\right)\left(\frac{2b}{3\pi}\right)$$

$$\tau_{\text{max-circ}} = \frac{4V}{3A}$$

The simplicity of the above two expressions for horizontal or longitudinal shear stress brings to mind simple shear. If we calculated on the basis of simple shear stress for either the rectangular or circular cross-sectional beam above it would be

$$\tau_{\text{vertical}} = \frac{V}{A}$$

This would lead us to believe that the vertical shear stress in a beam is less than the maximum horizontal shear stress. However, if we examine a cubic section cut from a loaded beam (by definition in bending) as shown in Figure 17-12 a contradiction appears to develop.

Since the element is cubic, $\Delta x = \Delta y = \Delta z$. These incremental distances are very small and it can be assumed that the vertical shear forces are equal, $V_1 = V_2$, and likewise that the horizontal shear forces are equal, $F_{x1} = F_{x2}$. This puts the cube in force equilibrium. However, for the cube to be in moment equilibrium, the two couples formed must be equal and opposite,

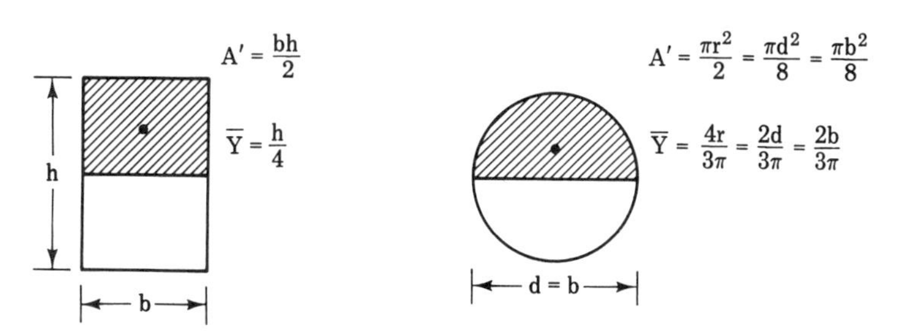

Figure 17-11. Rectangular and circular beam sections.

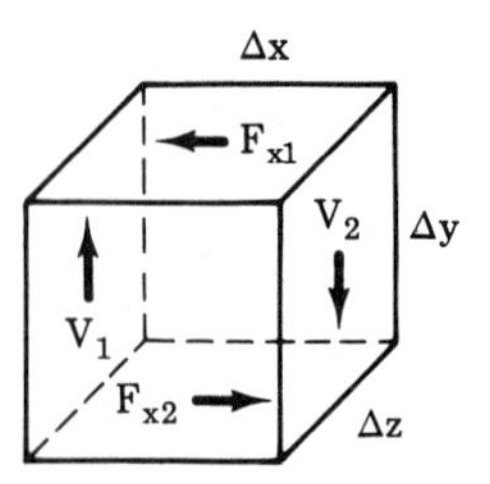

Figure 17-12. Shear stress equilibrium.

$V_2 \Delta x = F_{x1} \Delta y$. For this to be true $V_2 = F_{x1}$, since $\Delta x = \Delta y$. If all the shear forces are equal and all the areas are equal then vertical shear stress equals horizontal shear stress.

$$\tau_{\text{vertical}} = \tau_{\text{horizontal}}$$

It then follows that at any point in a member subjected to bending (a beam), there exists equal shearing stresses in planes perpendicular to each other. Further, the vertical shearing stress is not only equal to but distributed in the same manner as the horizontal shearing stress.

The above discussion coupled with the earlier discussion of bending or flexural stress has a highly practical application. Since in *S* and *W* (standard and wide flange) beams there is a substantial amount of material in those areas where bending stresses are the greatest, it is assumed that these flanges carry the bending stress with no help from the web. Conversely, since shear stress in the flange areas is near zero, it is assumed that the web carries the entire shear stress.

$$\tau_h = \tau_v = \frac{V}{td}$$

where t is the thickness of the web and d is the depth of the beam. These approaches will yield slightly more conservative answers than our previous work but not enough to seriously affect the economics of beam selection.

Example 17-3: Determine the minimum size for a beam subjected to a maximum shear load of 20,000 lb if the maximum allowable shear stress for the material is 20,000 psi:

1. for a square cross section,
2. for a circular cross section, and
3. for an *S* beam (most economical by weight/foot).

Solution:

Step 1 Square cross section

$$\tau = \frac{3V}{2A} = \frac{3V}{2S^2} \qquad S = \sqrt{\frac{3V}{2\tau}}$$

$$S = \sqrt{\frac{3(20{,}000 \text{ lb})}{2(20{,}000 \text{ lb/in.}^2)}} \qquad \mathbf{S = 1.23 \text{ in.}}$$

Step 2 Circular cross section

$$\tau = \frac{4V}{3A} = \frac{4V}{3(0.7854)d^2} \qquad d = \sqrt{\frac{4V}{3(0.7854)\tau}}$$

$$d = \sqrt{\frac{4(20{,}000 \text{ lb})}{3(0.7854)(20{,}000 \text{ lb/in.}^2)}} \qquad \mathbf{d = 1.3 \text{ in.}}$$

Step 3 *S* shape

$$\tau = \frac{V}{td} \qquad td = \frac{V}{\tau} \qquad td = \frac{20{,}000 \text{ lb}}{20{,}000 \text{ lb/in.}^2}$$

$td = 1.0$ in.^{2}, refer to Appendices

$$S3 \times 7.5{:}\quad td = 0.26 \times 3.0 = 0.78 \text{ too small}$$

$$S4 \times 7.7{:}\quad td = 0.293 \times 4.0 = 1.17 \text{ OK}$$

$$\mathbf{S4 \times 7.7}$$

Problem 17-3: Compare the maximum shear stresses for the following beam cross sections (equal in area) if each experiences a maximum shear force of 50 kN:

1. circular, $d = 43.4$ mm
2. square, $s = 38.5$ mm
3. wide flange, $d = 203$ mm, $t = 7.3$ mm

17.5 DEFLECTIONS IN BEAMS

As indicated earlier, the deflection of a beam may be a design consideration in some cases. As was seen in our discussion on bending stresses, beams under load always assume a shape somewhat different from their unloaded shape. Deflection of a point is defined as the vertical change in position of that point as a result of the application of a load. The point of maximum deflection can be fairly obvious as in the two cases shown in Figure 17-13.

In other cases, the point of maximum deflection may not be so obvious and its determination may even be rather involved. Many times, however, the point of maximum deflection is not of interest to us. Rather, the deflection at one or more points where interference or malfunction would occur as a result of excessive deflection become our concern.

The Elastic Curve

If we were to plot the deflections for many points on a beam, and connected them, the resultant curve would be known as the *elastic curve.* The specific nature of this curve depends on the beam supports and the beam loading. They can be rather simple as shown in Figure 17-13 or extremely complex. In any case, three properties of this curve are of significance to us: δ, the vertical displacement or deflection at any point; θ, the slope of the curve at that same point; and ρ, the radius of curvature at the same point.

Recall the relationships among the load, shear, and moment curves. We made particular use of the relationships between shear and moment. If these relationships are stated in general terms they would read as follows:

1. The area between any two ordinates on a diagram will equal the difference in length of the two corresponding ordinates on the next higher order diagram (see Figure 17-14).
2. The ordinate at any point on a diagram will be equal in value to the slope of the next higher order diagram at the same point (see Figure 17-14).

Relating the above to our current concern, the next highest order diagram after the moment diagram is the $EI\theta$ diagram. If the beam is made

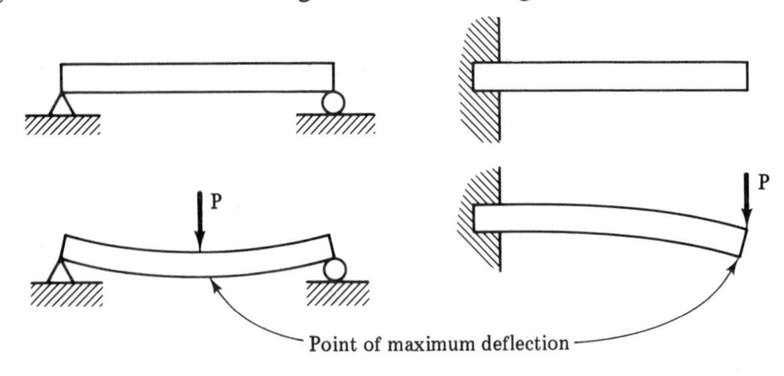

Figure 17-13. Maximum beam deflections.

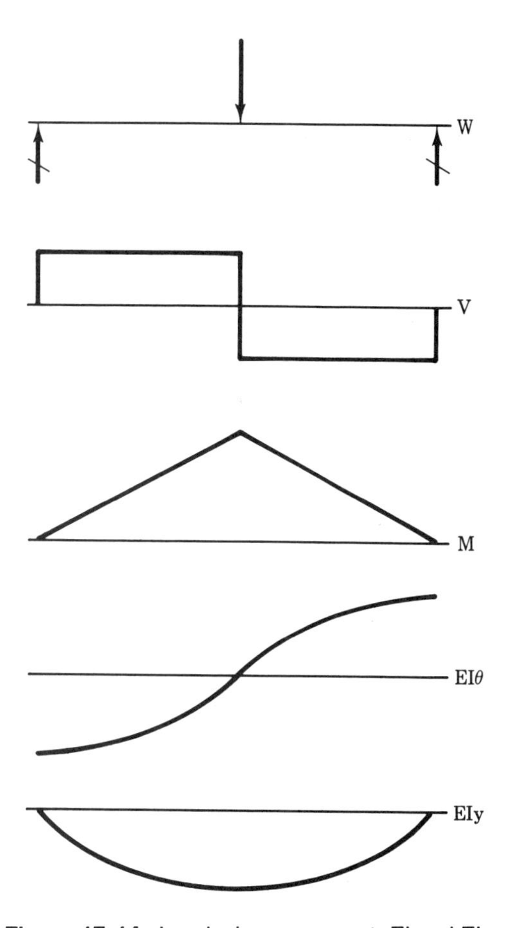

Figure 17-14. Load, shear, moment, EI and Ely.

of a homogeneous material and with a constant cross section then the $EI\theta$ diagram essentially represents the variation in the slope of the beam along its length. The next higher order diagram is the EIy and it essentially represents the variation in the deflection of the beam along its length, and is known as the elastic curve.

The Radius of Curvature

The elastic curve is, of course, not circular in nature. However, if we isolate a small enough portion of the curve at any point along its length, that portion will have a specific radius of curvature. As indicated before, one side of a beam under load is compressed or shortened while the opposite side is under tension and therefore stretched. Figure 17-15 shows a small portion

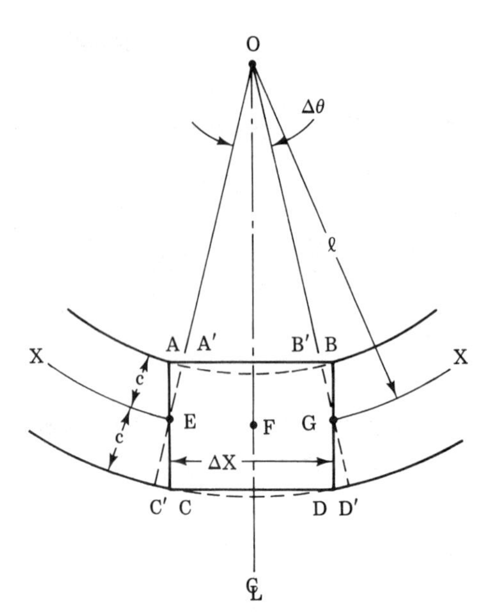

Figure 17-15. Radius of curvature.

cut from a beam. The solid lines show the shape of the portion prior to loading while the dotted lines show its shape after loading.

Examining the figure in some detail, loading the beam shortens the fiber on the one side from AB to $A'B'$ while lengthening the fiber on the other side from CD to $C'D'$. It should be noted that fiber EG along the neutral axis has not changed in length as a result of loading. As a result of this change, $A'EC'$ and $B'GD$ are still perpendicular to the neutral axis as they were before loading, but no longer parallel to each other. They therefore intersect to locate the center of curvature and thus define the radius of curvature ρ. Evaluating these changes mathematically, the elongation δ of the stretched fiber CD is

$$\delta = C'C + DD' \qquad \text{since } C'C = DD' \qquad \delta = 2C'C$$

and $\Delta X = EG = CD$. The stress at $C'D'$ is $\sigma = Mc/I$ while the strain is $\epsilon = \delta/\Delta X$.

$$\text{Since } E = \frac{\sigma}{\epsilon} \qquad E = \frac{Mc\Delta X}{I\delta}$$

It is necessary to introduce the radius of curvature into the above expression so that a relationship describing it can be established. Since r, c, ΔX, and δ are all part of the geometry of the deformation of the particle, let us seek a solution there.

Note that triangle OFE is similar to triangle ECC', then

$$\frac{OE}{EC'} = \frac{EF}{C'C}$$

However $OE = \rho$, $EC' = c$, $EF = \frac{1}{2}\Delta X$, and $C'C = \frac{1}{2}\delta$.

$$\text{Thus } \frac{\rho}{c} = \frac{\frac{1}{2}\Delta X}{\frac{1}{2}\delta} \qquad \frac{\rho}{c} = \frac{\Delta X}{\delta} \qquad \rho = \frac{c\Delta X}{\delta}$$

Substituting in our earlier equation for E

$$E = \frac{Mc\Delta X}{I} \qquad E = \frac{M\rho}{I} \qquad \rho = \frac{EI}{M}$$

Moment-Area Method

The equation just developed forms the basis for the moment-area method of determining the deflection of a point on a beam under load. To develop this approach we need to examine the element we have just worked with in the context of a larger portion of a beam. In Figure 17-16, the elastic curve and moment diagram for a larger portion HI of a beam are shown. The small portion or element we just worked with is identified as the increment ΔX between points E and G on the elastic curve just as it was before.

That incremental length can be described as an arc

$$\Delta X = \rho\Delta\theta$$

and since we had just defined the radius of curvature as

$$\rho = \frac{EI}{M}$$

then, substituting and solving for θ,

$$\Delta\theta = \frac{M\Delta X}{EI}$$

The slope of the elastic curve at any point is tangent to the curve at that point and therefore perpendicular to the radius of curvature at that point. Thus the angle formed by the tangents at any two points is the same as the angle formed by the radii of curvature for those two points, as long as the radius of curvature is constant. For a small increment of the elastic curve,

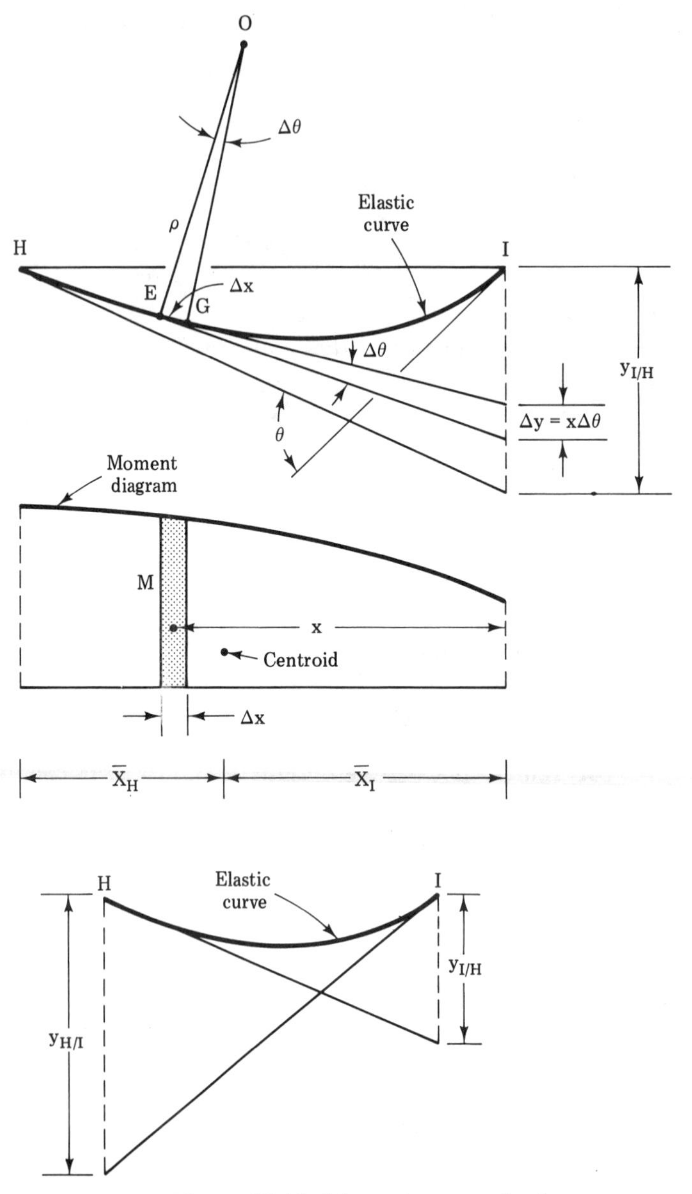

Figure 17-16. Moment area method.

this can be assumed to be true, as shown for ΔX. If we draw tangents at points H and I, the effect of the change in the radius of curvature becomes evident. Determining the angle between the two tangents (slopes) will have to be done by summing up many incremental $\Delta\theta s$, each with the same Δx but with varying radii. Thus

$$\theta_{H/I} = \Sigma(\Delta\theta) = \Sigma\left(\frac{M\Delta X}{EI}\right)$$

EI being constant throughout the length of the beam.

$$\theta_{H/I} = \frac{1}{EI}\,\Sigma(M\Delta X)$$

However, $\Sigma(M\Delta X)$ is the area of the moment diagram between points H and I. The above relationship is often formally expressed as Theorem 1.

Theorem 1: The angle between tangents drawn at two points on the elastic curve is equal to the area of the corresponding portion of the moment diagram divided by EI.

A comparable evaluation can be made to determine a measure of deflection. For our small element EG, at some horizontal distance X, the tangents will be separated by a vertical distance Δy. For small angles it is quite accurate to estimate Δy from

$$\Delta y = X\Delta\theta$$

The vertical deviation of point I relative to a tangent to the elastic curve at H will then be the sum of many small increments equal to Δy.

$$y_{I/H} = \Sigma(y) = \Sigma(X\Delta\theta) = \Sigma x\left(\frac{M\Delta X}{EI}\right) = \frac{1}{EI}\,\Sigma x(M\Delta X)$$

However, $\Sigma X(M\Delta X)$ is the moment of the area of the moment diagram with I being the moment center.

$$y_{H/I} = \frac{1}{EI}\,(A_{HI})\overline{X}_I$$

and conversely

$$y_{I/H} = \frac{1}{EI}\,(A_{HI})\overline{X}_H$$

Examining these two equations and Figure 17-16, it is evident that these two values will only be equal in that special situation where $\overline{X}_I = \overline{X}_H$. It is also very important to note that these equations *do not* necessarily provide the vertical deflection of one point with respect to another. They provide the vertical deflection of a point *with respect to a tangent drawn to the elastic curve at the other point.* This can only yield one point with respect to another *if* the slope of the elastic curve at either or both points is zero. This would appear to be a rather limiting condition but we shall be able to make good use of it. In summary, the above relationship is often formally expressed as Theorem 2.

Theorem 2: The vertical deviation of one point on the elastic curve away from a tangent drawn to the elastic curve at a second point is equal to the moment (taken about the first point) of the area of the moment diagram (between the two points) divided by EI.

Determination of Deflection by the Moment-Area Method

Example 17-4: Determine the deflection and slope at the load for the beam shown in Figure 17-17a.

Solution:

Step 1 Sketch the elastic curve as shown in Figure 17-17b and note that the slope at A is zero. Thus $y_{B/A}$ will be the actual deflection at B.

Step 2 Draw the moment diagram as shown in 17-17c.

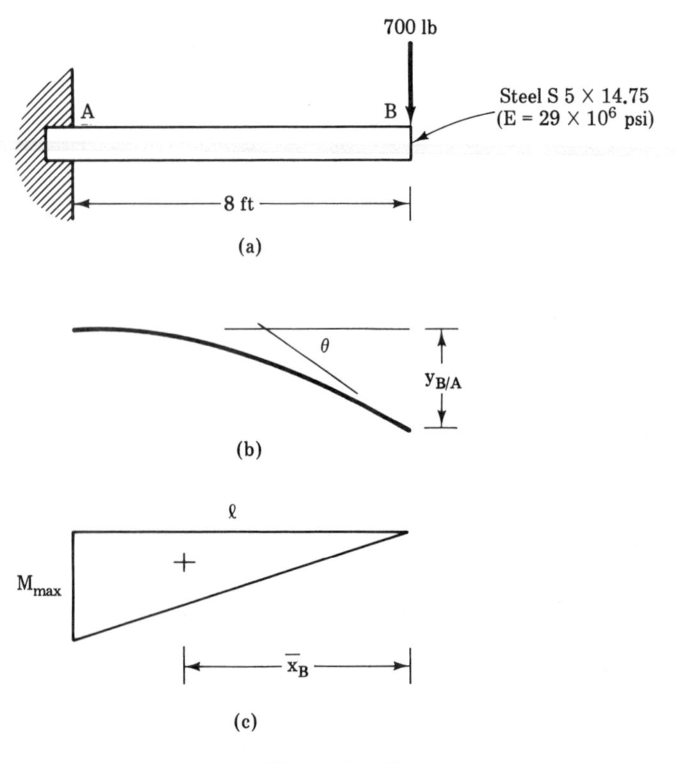

Figure 17-17

Step 3 Determine I from the Appendices to be 15 in.^{4}

Step 4 Calculate maximum moment, area of the moment diagram, and $\overline{X}_B$.

$$M_{max} = Pl = -(700 \text{ lb})(8 \text{ ft}) \qquad M_{max} = -5600 \text{ ft lb}$$

$$A = \tfrac{1}{2}l(P) \qquad A = \tfrac{1}{2}(8 \text{ ft})(-5600 \text{ ft lb}) \qquad A = -22{,}400 \text{ ft}^2 \text{ lb}$$

$$\overline{X}_B = \tfrac{2}{3}l = \tfrac{2}{3}(8 \text{ ft}) \qquad \overline{X}_B = 5.33 \text{ ft}$$

Step 5 Calculate deflection at B.

$$\delta_B = y_{B/A} = \frac{1}{EI}(A)\overline{X}_B$$

$$\delta_B = \frac{1}{(29 \times 10^6 \text{ lb/in.}^2)(15 \text{ in.}^4)} \times (-22{,}400 \text{ ft}^2 \text{ lb})(5.33 \text{ ft})(1728 \text{ in.}^3/\text{ft}^3)$$

$$\boldsymbol{\delta_B = -0.474 \text{ in.}}$$

Step 6 Calculate slope at B.

$$\theta = \frac{1}{EI}A = \frac{1}{(29 \times 10^6 \text{ lb/in.}^2)(15 \text{ in.}^4)} \times (-22{,}400 \text{ ft}^2 \text{ lb})(144 \text{ in.}^2/\text{ft}^2)$$

$$\boldsymbol{\theta = -0.0074 \text{ rad}}$$

Problem 17-4: Determine the deflection and slope at the free end of the beam shown in Figure 17-18.

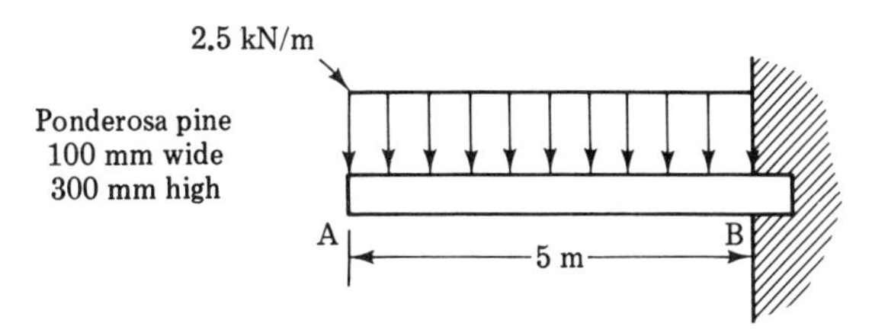

Figure 17-18

Example 17-5: Determine the midspan deflection for the beam shown in Figure 17-19.

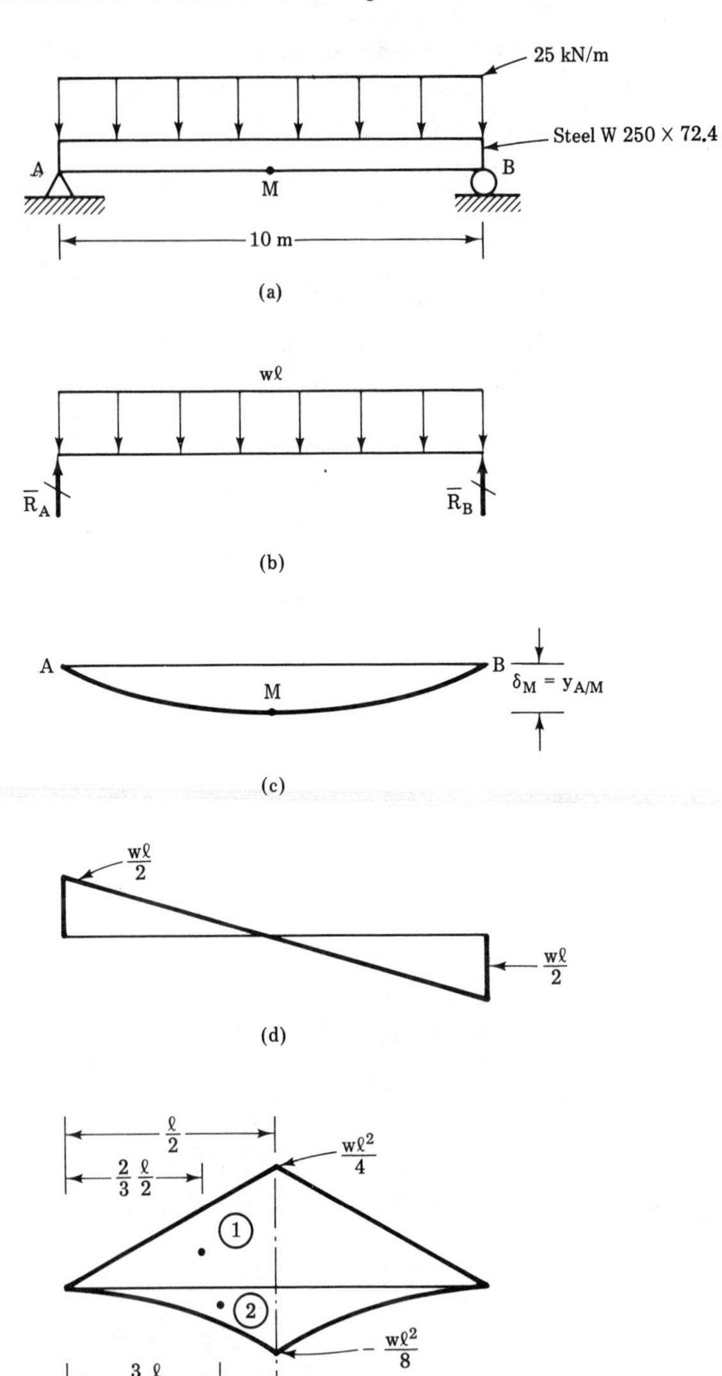
25 kN/m
Steel W 250 X 72.4
A
B
M
10 m
(a)
wℓ
R̄A
R̄B
(b)
A
B
M
δM = yA/M
(c)
wℓ/2
wℓ/2
(d)
ℓ/2
2/3 ℓ/2
wℓ²/4
1
2
wℓ²/8
3/4 ℓ/2
(e)

Figure 17-19

Solution:

Step 1 Draw the free-body diagram as shown in Figure 17-19b.

Step 2 Draw the elastic curve as shown in 17-19c. Since the slope at $M = 0$, $\delta_M = y_{A/M}$.

Step 3 Draw the shear diagram in parts as an aid in determining moment as shown in 17-19d.

Step 4 Draw the moment diagram in parts to simplify calculations as shown in 17-19e.

Step 5 Determine I from the Appendices to be 10,800 cm^4.

Step 6 Calculate maximum moments, areas, and moment arms.

Reaction $\overline{R}_A$

$$M_{1\max} = \frac{wl}{2}\left(\frac{l}{2}\right) = \frac{(25 \text{ kN/m})(10 \text{ m})}{2}\left(\frac{10 \text{ m}}{2}\right)$$

$$M_{1\max} = 625 \text{ kN m}$$

$$A_1 = \frac{1}{2} M_{\max} \frac{l}{2} = \frac{1}{2}(625 \text{ kN m})\left(\frac{10 \text{ m}}{2}\right) \qquad A_1 = 3125 \text{ kN m}^2$$

$$\overline{X}_{1A} = \frac{2}{3}\frac{l}{2} = \frac{2}{3}\left(\frac{10 \text{ m}}{2}\right) \qquad \overline{X}_{1A} = 3.33 \text{ m}$$

Distributed Load

$$M_{2\max} = -\frac{1}{2}\left(\frac{wl}{2}\right)\left(\frac{l}{2}\right) = -\frac{wl^2}{8} = \frac{(25 \text{ kN/m})(10 \text{ m}^2)}{8}$$

$$M_{2\max} = -312.5 \text{ kN m}$$

$$A_2 = \left(\frac{1}{3}\right)\left(\frac{l}{2}\right) M_{\max} = \left(\frac{1}{3}\right)\left(\frac{10 \text{ m}}{2}\right)(-312.5 \text{ kN m})$$

$$A_2 = -520.8 \text{ kN m}$$

$$\overline{X}_{2A} = \left(\frac{3}{4}\right)\left(\frac{l}{2}\right) = \frac{3}{4}\left(\frac{10 \text{ m}}{2}\right)$$

$$\overline{X}_{2A} = 3.75 \text{ m}$$

Step 7 Calculate deflection at M.

$$\delta_M = Y_{A/M} = \frac{1}{E1}[A_1\overline{X}_{1A} + A_2\overline{X}_{2A}]$$

$$\delta_M = \frac{1}{(200 \times 10^9 \text{ Pa})(10{,}800 \text{ cm}^4)}$$
$$[(3125 \text{ kN m}^2)(3.33 \text{ m})] +$$
$$[(-520.8 \text{ kN m}^2)(3.75 \text{ m})](10^{10} \text{ cm}^5/\text{m}^5)$$

$$\delta_M = 0.039 \text{ cm}$$

Problem 17-5: Determine the midspan deflection for the beam shown in Figure 17-20. At this point, it may have occurred to you that the load support combinations shown up to now are quite common and are liable to be encountered frequently. General equations could be developed for the slope at any point, the deflection at any point, and the maximum deflection. This has been done and is tabulated in the Appendices in order to facilitate your design work.

It may also have occurred to you that the examples and problems worked above were quite simple involving no combined loads. It is easy to see that such a problem could become very involved. The next section of this chapter provides an approach that will simplify solution of such problems utilizing the tabulated equations just mentioned.

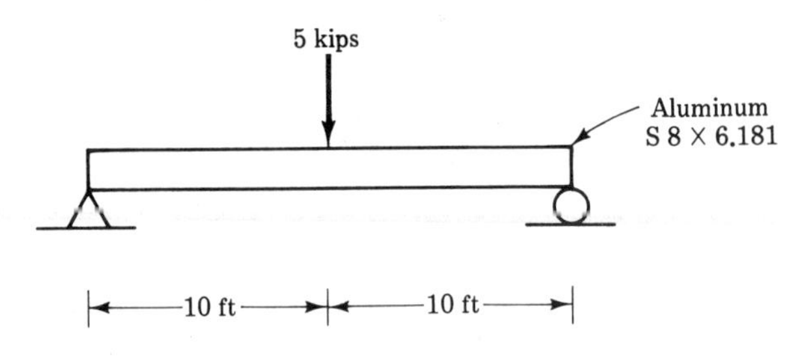

Figure 17-20

Determination of Deflections by Superposition

Superposition in general means to place one thing upon another. In terms of statics and strength of materials, it takes on a more specific meaning. A single load placed on a beam will result in very specific reactions, shear forces, moments, and deflections. A different single load would have similar results. What is significant is that if the beam is subjected to both loads simultaneously, their effects can be added algebraically to arrive at the total effect of the combined loading. Figure 17-21 illustrates the combining of two concentrated loads and their effects.

Example 17-6: Determine the maximum deflection for the beam shown in Figure 17-22.

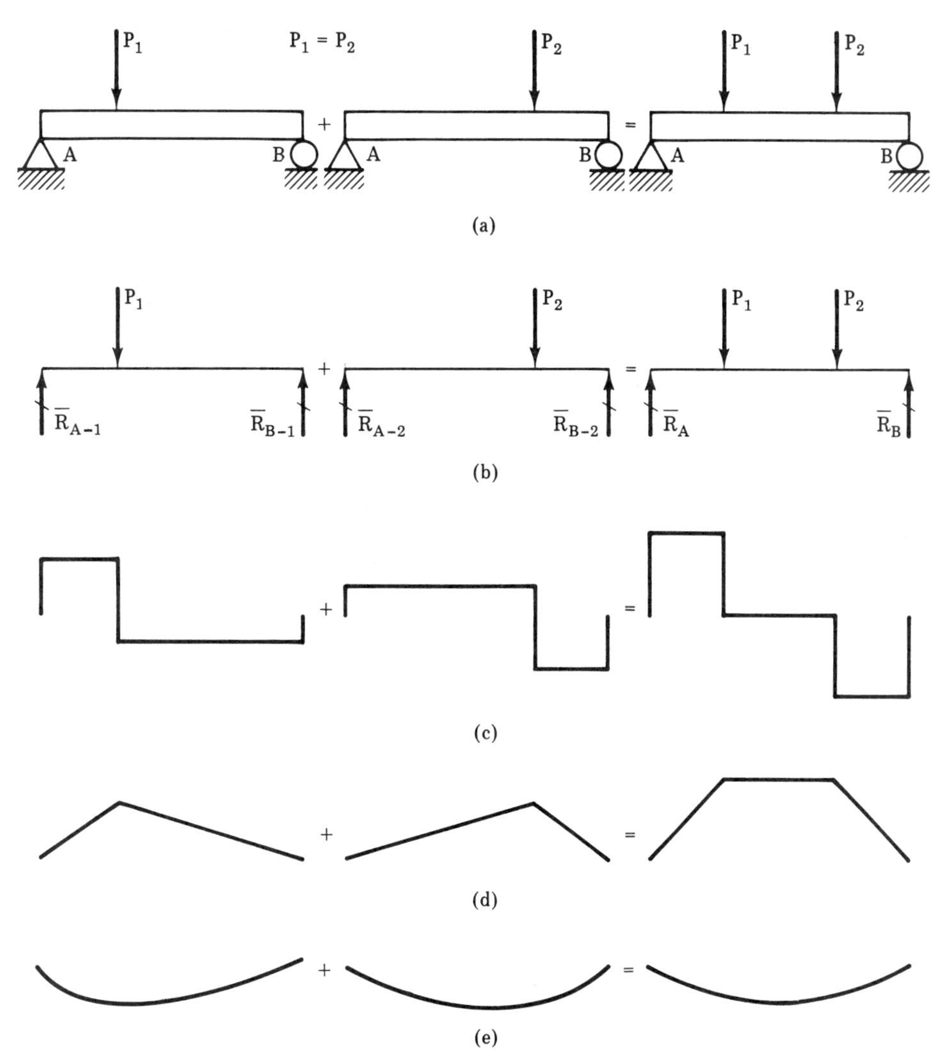

Figure 17-21. Superposition method.

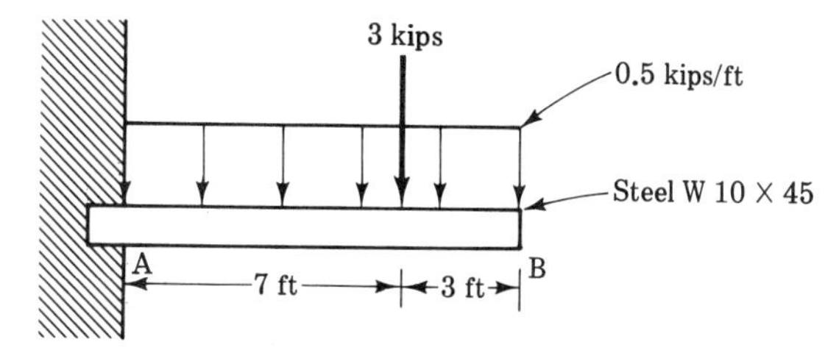

Figure 17-22

Solution:

Step 1 Recognize that as this is a cantilever beam, the maximum deflection for either load will occur at the free end *B*.

Step 2 Determine *I* to be 248.6 in.^{4} from the Appendices.

Step 3 Determine from the Appendices that the equation for the maximum deflection of a cantilever beam due to a uniformly distributed load extending the full length of the beam is

$$\delta_{max} = \frac{wl^4}{8EI}$$

Step 4 Determine from the Appendices that the equation for the maximum deflection of a cantilever beam due to a concentrated load at some point other than the free end is

$$\delta_{max} = \frac{Pb^2}{6EI}(3l - b)$$

where *b* is the distance from the load to the support.

Step 5 Combine equations and solve for the total maximum deflection.

$$\delta_{max\text{-}tot} = \frac{wl^4}{8EI} + \frac{Pb^2}{6EI}(3l - b)$$

$$\delta_{max\text{-}tot} = \frac{1}{EI}\left[\frac{wl^4}{8} + \frac{Pb^2}{6}(3l - b)\right]$$

$$\delta_{max\text{-}tot} = \frac{1}{29 \times 10^6 \text{ lb/in.}^2(248.6 \text{ in.}^4)}$$

$$\left[\frac{(500 \text{ lb/ft})(10 \text{ ft}^4)}{8} + \frac{3000 \text{ lb}(7 \text{ ft}^2)}{6}\right.$$

$$[3(10 \text{ ft}) - 7 \text{ ft}]\ 1728 \text{ in.}^3/\text{ft}^3$$

$$\boldsymbol{\delta_{max\text{-}tot} = 0.285 \text{ in.}}$$

Problem 17-6: Determine the maximum deflection for the beam shown in Figure 17-23.

Example 17-7: Determine the midspan deflection for the beam shown in Figure 17-24.

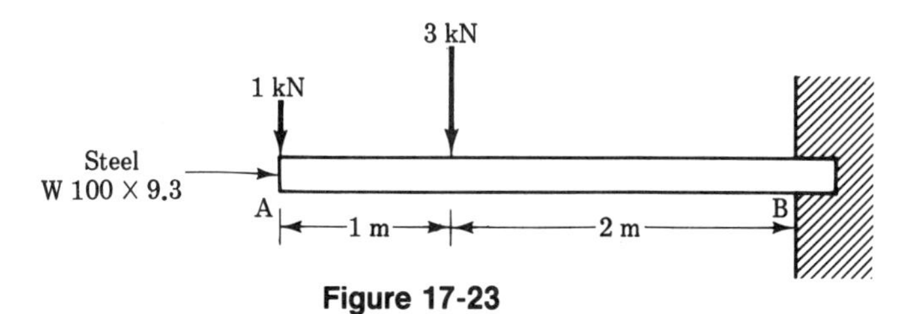

Figure 17-23

Solution:

Step 1 Recognize that the midspan deflection may not be the maximum deflection. However, the midspan deflection was the one requested.

Step 2 Determine I to be 187 cm^4 from the Appendices.

Step 3 Determine from the Appendices that the equation for the midspan deflection resulting from the left-hand concentrated load is

$$\delta = \frac{Pb}{48EI}(3l^2 - 4b^2)$$

where b is the distance from the left-hand support to the left-hand load.

Step 4 The equation for the right-hand load is identical except that b is the distance from the right-hand support to the right-hand load.

Step 5 Combine equations using subscripts L and R to differentiate between the two loads and the two distances.

$$\delta_{\text{mid-tot}} = \frac{1}{48EI}[P_L b_L (3l^2 - 4b_L^2) + P_R b_R (3l^2 - 4b_R^2)]$$

$$\delta_{\text{mid-tot}} = \frac{1}{48(200 \times 10^9 \text{ Pa})(6320 \text{ cm}^4)}$$

$$\{(5000 \text{ N})(1 \text{ m})[3(8 \text{ m}^2) - 4(1 \text{ m}^2)] + (10{,}000 \text{ N})(3 \text{ m})[3(8 \text{ m}^2) - 4(3 \text{ m}^2)]\}\ 10^{10} \text{ cm}^5/\text{m}^5$$

$$\delta_{\text{mid-tot}} = 0.93 \text{ cm}$$

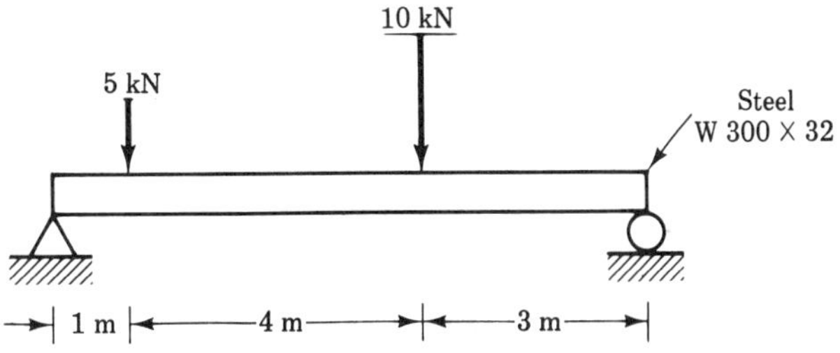

Figure 17-24

Problem 17-7: Determine the midspan deflection for the beam shown in Figure 17-25.

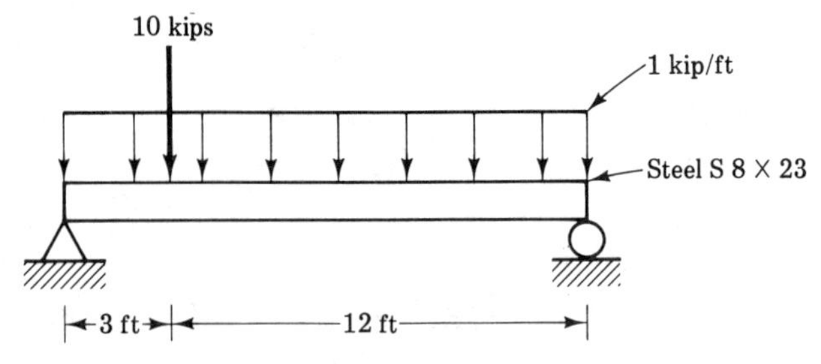

Figure 17-25

17.6 BEAM DESIGN

Basic Design

As indicated at the beginning of this chapter, the basic criteria for beam design are many. Some of the more significant ones follow.

1. **Span** Generally a given for a problem, although for a building, bridge, or large machine a total span may be broken up into several smaller spans. The basis for the number of spans may be based on standard practice, available support locations, clearance required between supports, and perhaps most important the optimum cost combination of supports and beams.

2. **Supports** In addition to the frequency of support discussed above, the type of support must be considered. Fixed supports at both ends of a beam may well permit use of a lighter beam and still reduce deflection. However, the supports must be designed to provide the moment reactions at the end. Also, if the beam is subjected to substantial temperature variations, fixed supports at both ends will require consideration of thermal stresses in the design.

3. **Load(s)** The nature and magnitudes of the loads combined with the span and supports required will determine the beam design.

4. **Deflection** For some applications there will be a maximum permissible deflection either at any point along the beam or at one or more specific points along the beam.

The beam selection ultimately comes down to the following:

1. **Length** See "spans" above.

2. **Material** A matter of practice, often legal code requirements, fire resistance, corrosion resistance, and, of course, cost.

3. **Cross-Section Shape** The choice among equal leg angle, unequal leg angle, C (channel), S (standard I), and W (wide flange) shapes is a matter of standard practice, suitability for the application, and, as always, cost.

4. **Cross-Section Size** Sizing is complicated by the fact that there are generally several different cross sections for the same nominal dimensions. For example, if you look in the Appendices for a 10-in.-wide flange beam, you will find a list rather than a single beam. The depth of all these beams will be in the "neighborhood" of the 10-in. *nominal* size. The width will also vary, as will the thickness of the flange and the thickness of the web. To minimize confusion, *W, S, and C shapes are specified by the nominal depth (actual for S and C) and the weight per unit length. Equal leg and unequal leg angles are specified by their actual overall dimensions and thickness.*

The above may seem like a bewildering maze of criteria and the selection of a beam an extremely formidable task. However, in practice many of these criteria fall into place. Dealing with complete design also requires additional study beyond this text as well as experience. In the problems that you are expected to solve, many criteria are given. After some examples and problems involving basic design, this chapter will conclude by examining some design considerations not mentioned above. These are: composite beams (two or more materials), stress concentration in beams, and lateral buckling in beams.

Example 17-8: Select the lowest cost (lowest weight) S-shape beam for the supports and loading shown in Figure 17-26a if the material is to be A-36 steel and the maximum deflection is to be 0.35 in.

Solution:

Step 1 Draw the free-body diagram for the beam as shown in Figure 17-26b.

Step 2 Determine $\overline{R}_{BY}$ by taking moments about A.

$$\Sigma M_A = 0 \qquad Pd_{PY} + \overline{R}_{BY} d_{\overline{R}BY} = 0$$

$$\overline{R}_{BY} d_{\overline{R}BY} - Pd_{PY} = -[-(10{,}000 \text{ lb})(6.5 \text{ ft}),$$

$$\overline{R}_{BY} d_{\overline{R}BY} = +65{,}000 \text{ ft lb} \circlearrowleft$$

$$\overline{R}_{BY} = \frac{\overline{R}_{BY} d_{\overline{R}BY}}{d_{\overline{R}BY}} \qquad \overline{R}_{BY} = \frac{65{,}000 \text{ ft lb}}{15 \text{ ft}} \qquad \overline{R}_{BY} = +4333 \text{ lb}$$

Step 3 Determine $\overline{R}_{AY}$ by sum of the vertical forces.

$$\overline{R}_{AY} + P + \overline{R}_{BY} = 0 \qquad R_{AY} = -(P + \overline{R}_{BY})$$

$$\overline{R}_{AY} = -(-10{,}000 + 4333 \text{ lb}) \qquad \overline{R}_{AY} = +5667 \text{ lb}$$

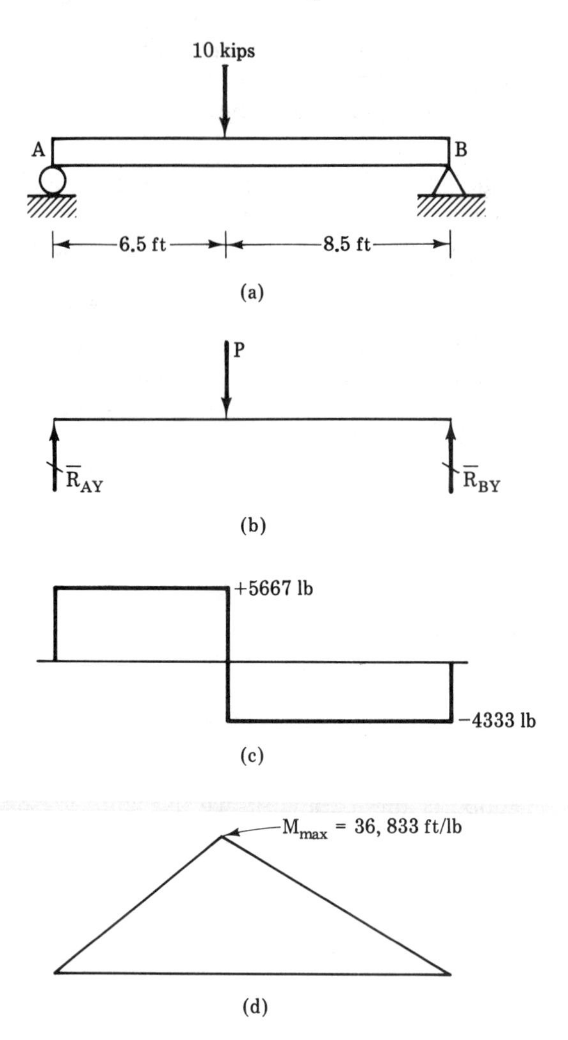

Figure 17-26

Step 4 Draw the shear diagram as shown in 17-26c.

Step 5 Based on the shear diagram, draw the moment diagram as shown in 17-26d.

$$M_{max} = (5666.66 \text{ lb})(6.5 \text{ ft}) \qquad M_{max} = 36{,}833 \text{ ft lb}$$

Step 6 Check Appendices for allowable bending stress and shear stress.

$$\sigma_{all} = 24{,}000 \text{ psi} \qquad \tau_{all} = 14{,}500 \text{ psi}$$

Step 7 Determine the minimum section modulus required (bending basis).

$$\sigma_{all} = \frac{M}{Z} \qquad Z = \frac{M}{\sigma_{all}} = \frac{(36{,}833 \text{ ft lb})(12 \text{ in./ft})}{24{,}000 \text{ lb/in.}^2}$$

$$Z_{min} = 18.42 \text{ in.}^3$$

Step 8 Determine the minimum centroidal moment of inertia required (deflection basis)—deflection equation from Appendices.

$$\delta_{max} = \frac{Pb(l^2 - b^2)^{1.5}}{15.59lEI} \qquad \text{where } b \text{ is the distance from } A \text{ to } P$$

$$\bar{I}_{min} = \frac{Pb(l^2 - b^2)^{1.5}}{15.59lE\delta_{max}}$$

$$\bar{I}_{min} = \frac{(10{,}000 \text{ lb})(6.5 \text{ ft})[(15 \text{ ft}^2) - (6.5 \text{ ft}^2)]^{1.5}(1728 \text{ in.}^3/\text{ft}^3)}{15.59(15 \text{ ft})(29 \times 10^6 \text{ lb/in.}^2)(0.35 \text{ in.})}$$

$$\bar{I}_{min} = 116.9 \text{ in.}^4$$

Step 9 Check the Appendices for the lightest available S beam with a section modulus of at least 18.42 in.^{3} and a centroidal moment of inertia of at least 116.9 in^4. This would be

$S10 \times 25.4$ $\quad (Z = 24.7 \text{ in.}^3\ \bar{I}_x = 124 \text{ in.}^4\ t_w = 0.311 \text{ in.}, d = 10.0 \text{ in.})$

Step 10 Check for shear stress.

$$\tau = \frac{V}{td} = \frac{5667 \text{ lb}}{(0.311 \text{ in.})(10 \text{ in.})} = 1822 \text{ psi}$$

which is less than the allowable, so selection is

$$\boldsymbol{S10 \times 25.4}$$

If the shear stress had been too great then an appropriately larger beam would have been required.

If there had been no deflection limitation, then Step 7 could have been eliminated and Step 8 simplified.

If desired, the equations used in Steps 6 and 7 can be utilized to determine the actual bending stress and actual maximum deflection for the beam selected.

Problem 17-8: Select the lowest cost *W* shape beam for the supports and loading shown in Figure 17-27 if the material is to be A572, Grade 50 steel and the maximum deflection is to be 0.5 in.

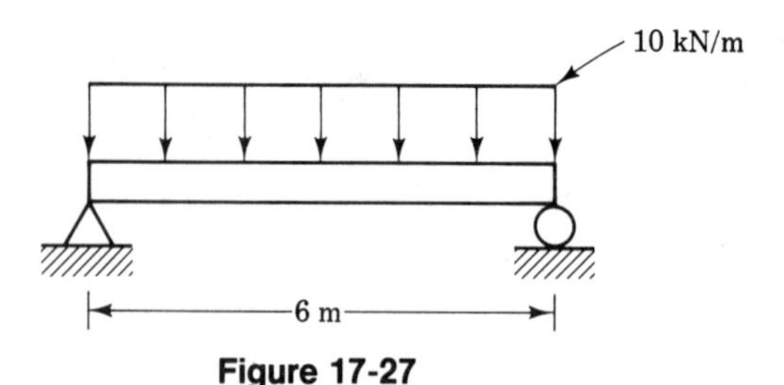

Figure 17-27

Example 17-9: Select the lowest cost *C* shape beam for the supports and loading shown in Figure 17-28a, if the material is to be A572, Grade 60 steel.

Solution:

Step 1 Since there is no limitation in terms of deflection, that aspect of beam design can be ignored in this problem.

Step 2 Since it is a cantilever beam, maximum moment and shear will occur at the support.

Step 3 Draw the free-body diagram as shown in 17-28b, substituting the equivalent concentrated load for the distributed load.

Step 4 Determine the total of the distributed load.

$$P_3 = wl = (10 \text{ kN/m})(5 \text{ m}) = 50 \text{ kN}$$

Step 5 Since it is a simple cantilever, maximum shear will equal $\overline{R}_{BY}$. Determine $\overline{R}_{BY}$ by sum of the vertical forces.

$$P_1 + P_2 + P_3 + \overline{R}_{BY} = 0 \qquad \overline{R}_{BY} = -(P_1 + P_2 + P_3)$$

$$\overline{R}_{BY} = -(-10 - 20 - 50 \text{ kN}) \qquad \overline{R}_{BY} = +80 \text{ kN}$$

Step 6 Since it is a simple cantilever, maximum moment will occur at B and be equal to $\overline{M}_B$. Determine $\overline{M}_B$ by sum of the moments about B.

$$\Sigma M_B = 0 \qquad P_1d_1 + P_2d_2 + P_3d_3 + \overline{M}_B = 0$$

$$\overline{M}_B = -[P_1d_1 + P_2d_2 + P_3d_3]$$

$$\overline{M}_B = -[+(10 \text{ kN})(5 \text{ m}) + (20 \text{ kN})(3 \text{ m}) + (50 \text{ kN})(2.5 \text{ m})],$$

$$\overline{M}_B = -235 \text{ kN m}$$

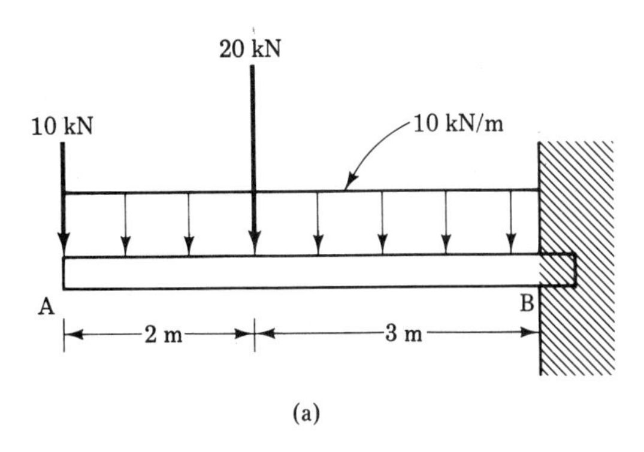

(a)

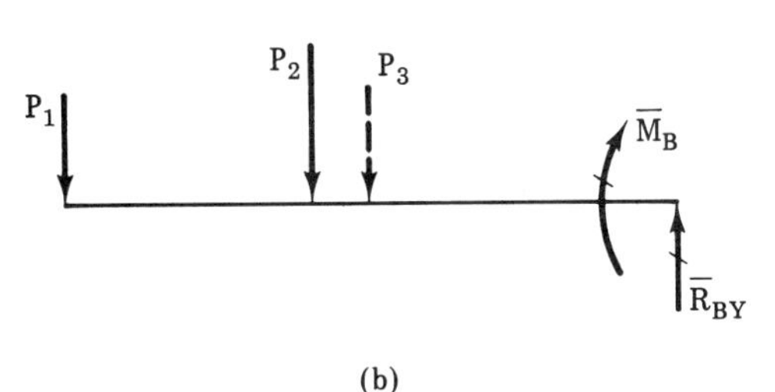

(b)

Figure 17-28

Step 7 Check Appendices for allowable bending stress and shear stress.

$$\sigma_{\text{all}} = 273 \text{ MPa} \qquad \tau_{\text{all}} = 165.5 \text{ MPa}$$

Step 8 Determine the minimum section modulus required.

$$\sigma_{\text{all}} = \frac{M}{Z} \qquad Z = \frac{M}{\sigma_{\text{all}}} = \frac{235 \text{ kN m}(10^6 \text{ cm}^3/\text{m}^3)}{273 \text{ MPa}(1000 \text{ K/M})}$$

$$Z_{\min} = 861 \text{ cm}^3$$

Step 9 Check the Appendices for the lightest available C beam with a section modulus of at least 861 cm^3. This would be $C380 \times 100 \times 13 \times 20$, 13 being the web thickness.

Step 10 Check for shear stress.

$$\tau = \frac{V}{td} = \frac{80 \text{ kN}(10^4 \text{ cm}^2/\text{m}^2)}{(13 \text{ cm})(380 \text{ cm})(1000 \text{ K/M})}$$

$\tau = 0.162$ MPa, far below the allowable, so section is

$$\boldsymbol{C380 \times 100 \times 13 \times 20}$$

Problem 17-9: Select the lowest cost *C*-shape beam for the supports and loading shown in Figure 17-29 if the material is to be aluminum 6061T6 with an allowable bending stress of 21,000 psi and an allowable shear stress of 12,000 psi.

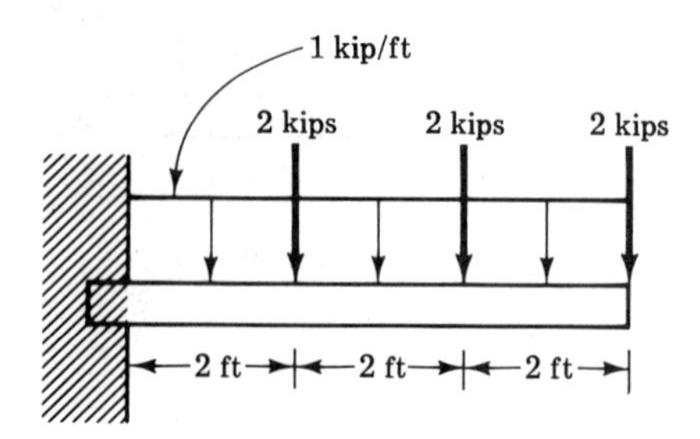

Figure 17-29

Composite Beams

Beams are generally made of more than one material when it becomes economical to place some stronger more expensive material in areas of high stress of a member that consists primarily of a weaker less expensive material. The design approach generally taken is to theoretically design an equivalent beam of one of the materials. Since stress and strain due to bending are a function of depth, depth must remain constant.

Consider the two beams shown below, one aluminum and one steel.

For the two beams to be equivalent in deflection, they must be equivalent in strain. That is, the strains in the two beams at the same distance from the neutral axis must be equal.

$$\epsilon_a = \epsilon_s \quad \text{but} \quad \epsilon = \frac{\sigma}{E} \quad \text{so} \quad \frac{\sigma_a}{E_a} = \frac{\sigma_s}{E_s}$$

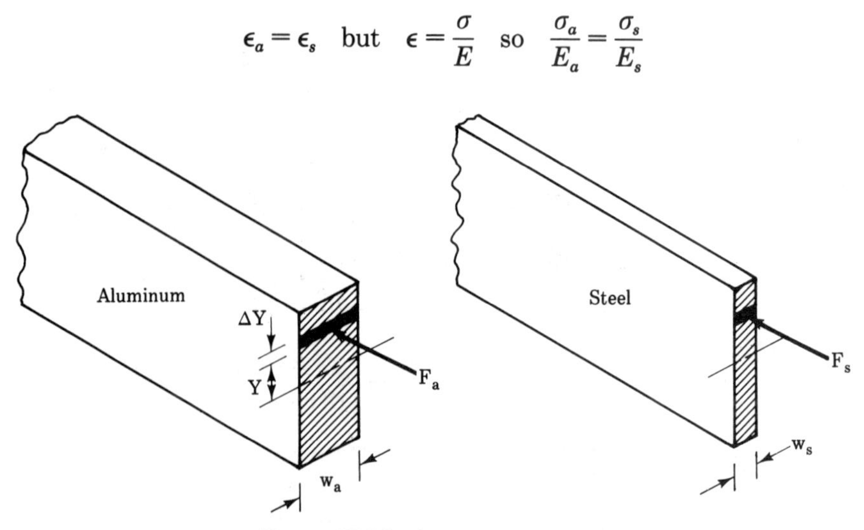

Figure 17-30. Comparative beams.

However, $\sigma = F/A$, and since the two beams are being designed to carry the same load, $F_a = F_s$. For the incremental areas being examined, $A_a = w_a \Delta y$, and $A_s = w_s \Delta y$. Substituting in our earlier equation

$$\frac{F_a}{E_a W_a \Delta y} = \frac{F_s}{E_s W_s \Delta y}$$

But $F_a = F_s$ and Δy is the same increment of depth, thus $E_a W_a = E_s W_s$.

Figure 17-31 gives some examples of transformations from metal and wood composites to both metal and wood equivalents.

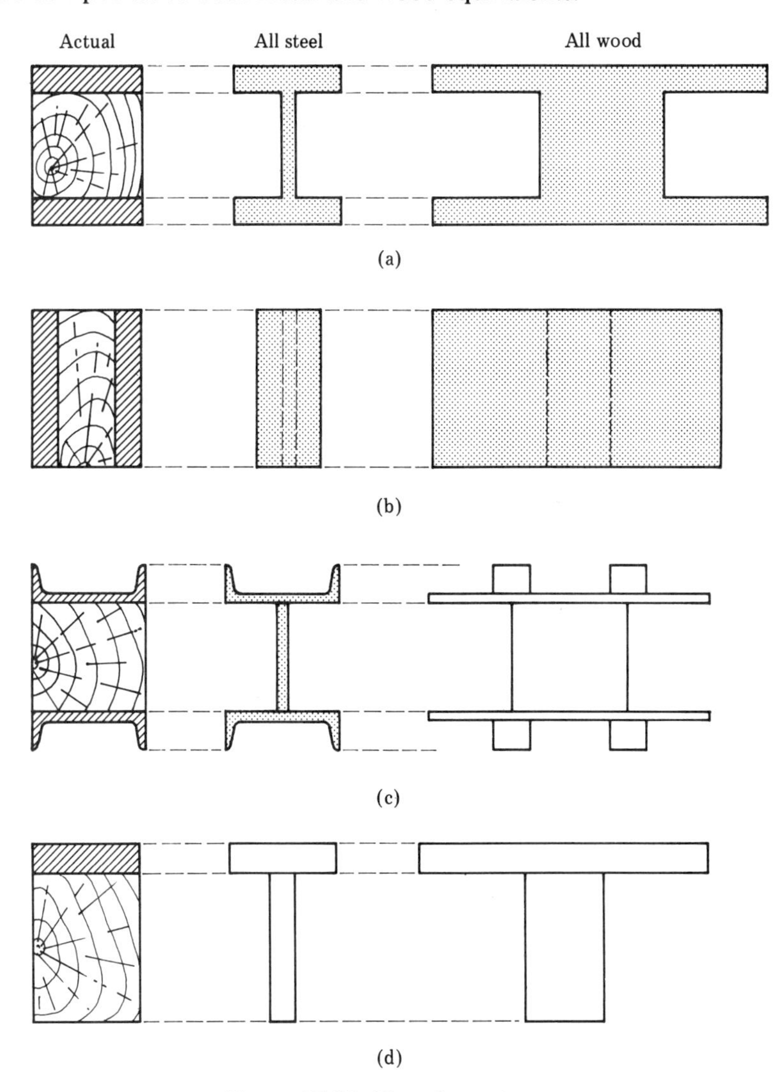

Figure 17-31. Transformations.

Actually, the relationships above could be applied to any two materials. Once an equivalent section is obtained, analysis proceeds as usual. The only correction that needs to be made is to determine the actual stress in the material not transformed to. The following examples and problems should clarify both the concepts and the procedure involved.

Example 17-10: The maximum bending moment in the beam shown in Figure 17-32a is 6000 ft lb. Determine the bending stresses in both materials.

Solution:

Step 1 Change the beam to either an equivalent wood or aluminum beam. Let's change wood to aluminum.

$$E_a w_a = E_w w_w \qquad w_a = w_w \frac{E_w}{E_a}$$

$$w_a = 6\text{ in.}\,\frac{1.5 \times 10^6\text{ psi}}{10 \times 10^6\text{ psi}} \qquad w_a = 0.9\text{ in.}$$

The transformed beam then has a rectangular cross section 10 in. in depth and a total width of 1.9 in. as shown in Figure 17-32b, and is entirely aluminum.

Step 2 $\bar{I}$ and c are both needed to find the bending stress in the aluminum.

$$c = 5\text{ in. by inspection.}$$

$$\bar{I} = \frac{1}{12}bh^3 = \frac{1}{12}w(d)^3 = \frac{1}{12}(1.9\text{ in.})(10\text{ in.}^3)$$

$$\bar{I} = 158.3\text{ in.}^4$$

Step 3 Determine the bending stress in the aluminum.

$$\sigma_a = \frac{Mc}{I} = \frac{6000\text{ ft lb}(12\text{ in./ft})(5\text{ in.})}{158.3\text{ in.}^4}$$

$$\boldsymbol{\sigma_a = 2274\text{ psi}}$$

(which incidentally is well below the allowable stress for aluminum).

Step 4 Determine the bending stress in the wood.

$$\sigma_w = \frac{E_w}{E_a}\sigma_a = \frac{1.5 \times 10^6\text{ psi}}{10 \times 10^6\text{ psi}}(2274\text{ psi})$$

$$\boldsymbol{\sigma_w = 341\text{ psi}}$$

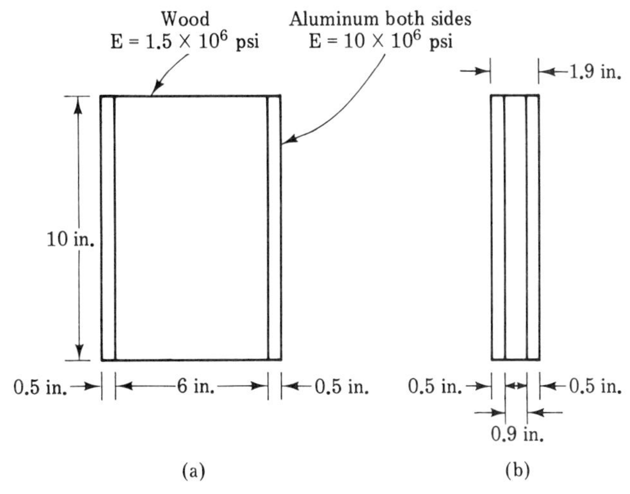

Figure 17-32

(which is well below the allowable stress for any commonly used construction timber).

Problem 17-10: The maximum bending moment in the beam shown in Figure 17-33 is 25 kN m. Determine the minimum width required for the wooden portion.

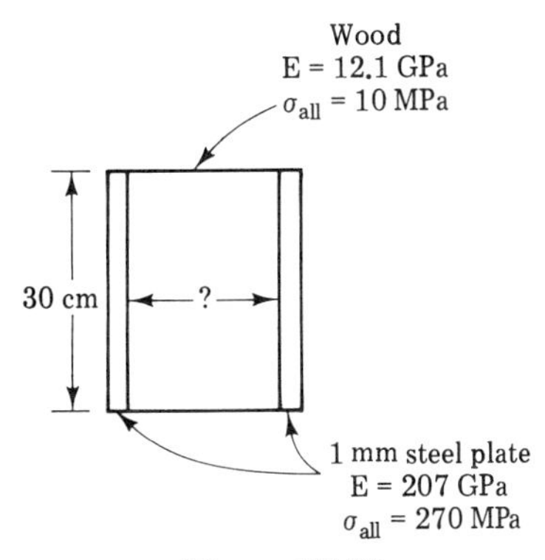

Figure 17-33

Example 17-11: Determine the maximum bending moment which the beam shown in Figure 17-34a can be safely subjected to.

Solution:

Step 1 Change the beam to all steel as shown in 17-34b.

$$w_s = \frac{E_a}{E_s} w_a = \frac{69 \text{ GPa}}{207 \text{ GPa}} (10 \text{ cm}) \qquad w_s = 3.33 \text{ cm}$$

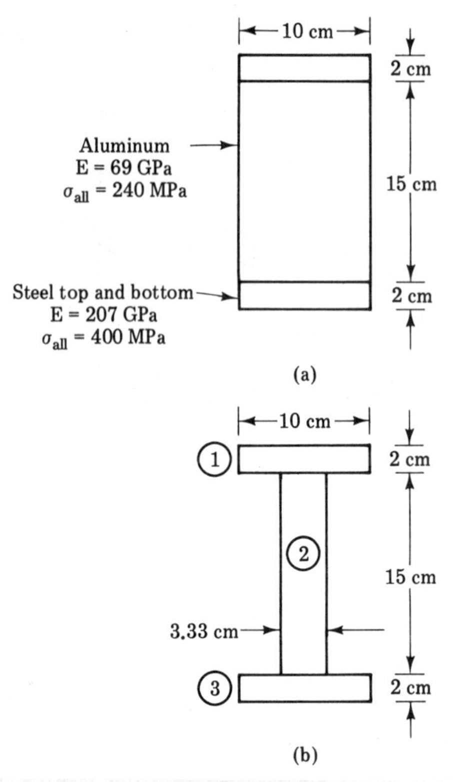

Figure 17-34

Step 2 Determine c and $\bar{I}$. By inspection c is 9.5 cm. $\bar{I}$ must be determined as follows:

$$A_1 = A_3 = 2 \text{ cm}(10 \text{ cm}) = 20 \text{ cm}^2$$

$$y_1 = y_3 = 7.5 + 1 \text{ cm} = 8.5 \text{ cm}$$

$$\bar{I}_1 = \bar{I}_3 = \frac{1}{12} bh^3 = \frac{1}{12} (10 \text{ cm})(2 \text{ cm}^3) = 6.67 \text{ cm}^4$$

$$\bar{I}_2 = \frac{1}{12} bh^3 = \frac{1}{12} (3.33 \text{ cm})(15 \text{ cm}^3) = 936.56 \text{ cm}^4$$

$$\bar{I} = \bar{I}_1 + A_1 y_1^2 + \bar{I}_2 + \bar{I}_3 + A_3 y_3^2$$

$$\bar{I} = 6.67 + (20)(8.5)^2 + 936.56 + 6.67 + (20)(8.5)^2$$

$$\bar{I} = 3840 \text{ cm}^4$$

Step 3 Determine the maximum permissible moment on the basis of an all steel beam.

$$M = \frac{\sigma_s \bar{I}}{c} = \frac{400 \text{ MPa } (3840 \text{ cm}^4)(1 \text{ m}^3/10^6 \text{ cm}^3)}{9.5 \text{ cm}}$$

$$M = 162 \text{ kN m}$$

Step 4 Check stress in aluminum.

$$\sigma_a = \frac{E_a}{E_s} \sigma_s = \frac{69 \text{ GPa}}{207 \text{ GPa}} (400 \text{ MPa})$$

$$\sigma_a = 133.3 \text{ MPa}$$

This is substantially less than the allowable bending stress for aluminum. Further, this would only occur at the outermost fibers which are steel in the actual beam. Had σ_a turned out to be too large from the above calculation, the next step would have been to determine σ_s at the y for the steel-aluminum interface and then determine the equivalent σ_a. If that was still above the allowable, then the allowable σ_a would have to be converted to a reduced allowable σ_s and that σ_s used at the aluminum-steel interface to determine a maximum allowable M.

Problem 17-11: Determine the greatest uniformly distributed load which can be supported on a span of 20 ft by the beam whose cross section is shown in Figure 17-35.

A somewhat different application of the above approach is involved in the design of reinforced concrete beams. Several assumptions and considerations must be added to our previous approach. These include:

1. Concrete has little strength in tension.
2. The steel reinforcing bars have a tensile load in them equal to the

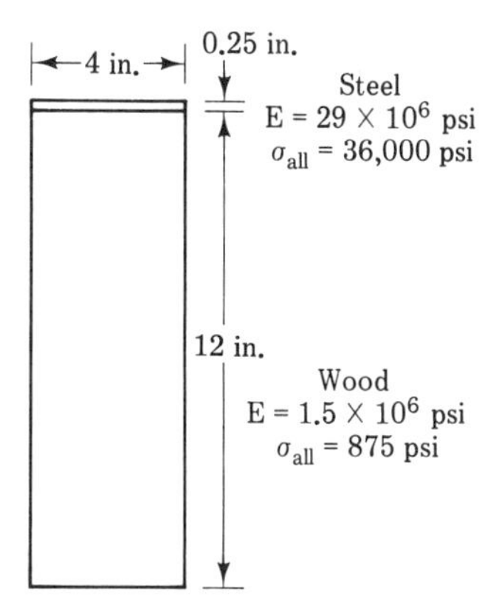

Figure 17-35

compressive load in the concrete (in other words, it is assumed that the concrete surrounding the re-bars provides *no* help at all in tension).

3. The moment of the transformed area of the re-bars about the neutral axis must just equal the moment of the *active* concrete that lies above the neutral axis.

4. The distance from the neutral axis to the top of the re-bars is considered the distance to the centroid.

5. Ideally, the concrete and steel should both be stressed to their allowable limits.

Figure 17-36 illustrates some of the above as well as providing the basic relationships.

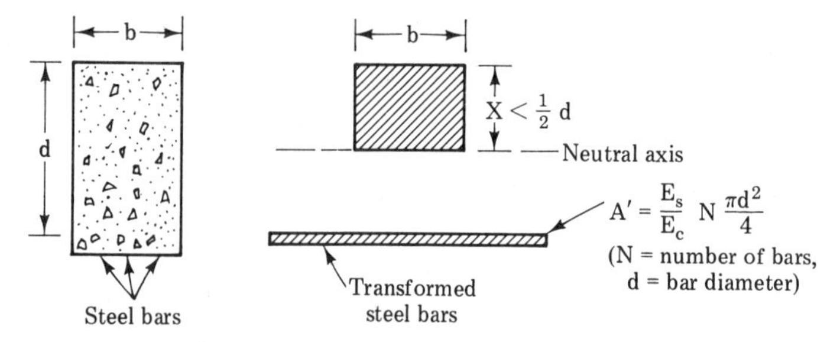

Figure 17-36. Reinforced concrete beam.

Example 17-12: Determine the number of $\frac{3}{4}$-in. re-bars required for the closest possible to an ideal design for the beam shown in Figure 17-37.

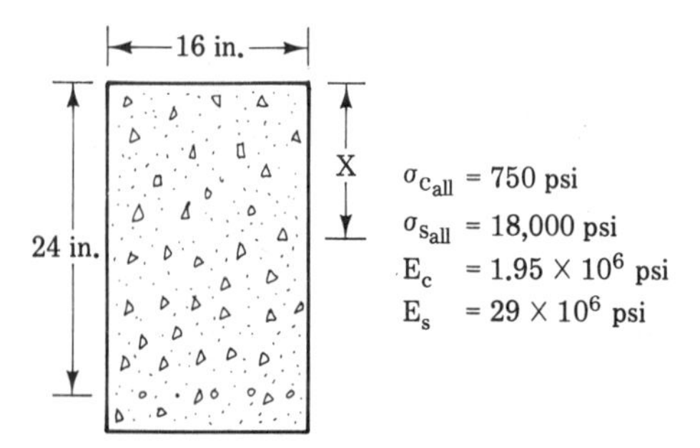

Figure 17-37

Solution:

Step 1 The flexure equation for concrete yields

$$\sigma_c = \frac{Mc}{I} \qquad 750\text{ psi} = \frac{Mx}{I}$$

Step 2 Previously, we indicated that the strain for the "substituted" material has to be equal to that of the original material.

$$\epsilon_s = \epsilon_c \qquad \frac{\sigma_s}{E_s} = \frac{\sigma_c}{E_c} \qquad \sigma_s = \frac{E_s}{E_c} c$$

$$\sigma_s = \frac{29 \times 10^6 \text{ psi}}{1.95 \times 10^6 \text{ psi}} (\sigma_c) \qquad \sigma_s = 14.87\sigma_c$$

From Step 1, $\sigma_s = 14.87\ (Mc/I)$; then

$$18{,}000 \text{ psi} = 14.87 \frac{M}{I} (24 - x)$$

Step 3 M is, of course, the same for any one point along the length of the beam. The I's are also the same. Then

$$\frac{M}{I} = \frac{750}{x} \qquad \frac{M}{I} = \frac{18{,}000}{14.87(24 - x)}$$

and

$$\frac{750}{x} = \frac{18{,}000}{14.87(24 - x)} \qquad x = 9.18 \text{ in.}$$

$$24 - x = 14.82 \text{ in.}$$

Step 4 Determine the required number of re-bars from the fact that the first moment of the transformed steel area and the first moment of the concrete in compression must be equal.

$$A'\overline{Y}_{A'} = A_c\overline{Y}_{Ac} \qquad A'(14.82 \text{ in.}) = (16 \text{ in.})(9.18 \text{ in.})\left(\frac{9.18 \text{ in.}}{2}\right)$$

$$A' = 45.49 \text{ in.}^2 = \frac{E_s}{E_c} N\pi \frac{d^2}{4}$$

$$N = \frac{A'E_c^4}{\pi E_s d^2} = \frac{45.49 \text{ in.}^2(1.95 \times 10^6 \text{ psi})(4)}{\pi(29 \times 10^6 \text{ psi})(0.75 \text{ in.}^2)}$$

$$N = 6.92 \qquad \mathbf{N = 7}$$

Problem 17-12: The reinforced concrete beam shown in Figure 17-38 is subjected to a bending moment of 70 kN m. Determine the maximum bending stress in both the concrete and the steel.

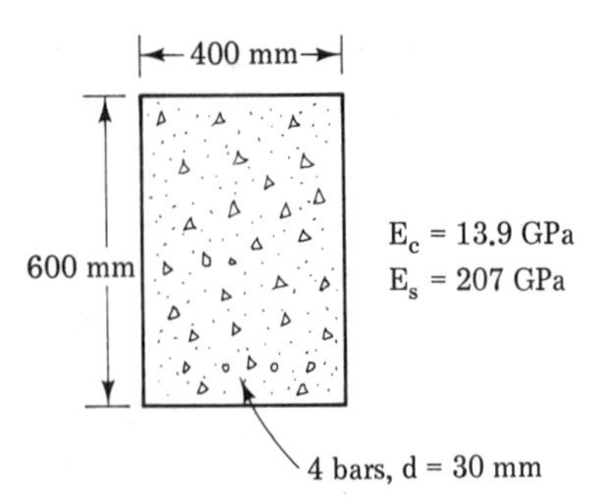

Figure 17-38

Stress Concentrations in Beams

Earlier in the chapter, it was mentioned that the flexure formula applied only when the cross section is constant throughout the length of the beam. There are, however, applications where this is neither feasible nor desirable. Two broad examples are without question the most common. First, the same forces from belts and gears that caused torsion in a shaft also cause bending. Second, cantilevers generally have increasing internal moment as one moves from the free end to the support. In some cases, it may be economical to step the size of the cantilever, but care must be taken in the design of the step. As might be expected from our previous experience with axial and torsional stress concentrations, it is not sufficient to design on the basis of the reduced cross section. Stress concentration factors based on the configuration of the step must be applied to the reduced cross-section. Figures 17-39 and 17-40 provide stress concentration factors for stepped and grooved shafts subjected to bending. Table 17-1 provides stress concentration factors for keyways in shafts subjected to bending. Figure 17-41* provides stress concentration factors for stepped beams with rectangular cross section subjected to bending.

Table 17-1
Keyway Stress Concentration Factors for Shafts Subjected to Bending

Keyway Type	*K**
Sled Runner	1.6
Profile	2.0

* Applied to full diameter at keyway.

* Figures 17-39, 17-40, and 17-41 are taken from Mott, Robert L. *Applied Strength of Materials*. Englewood Cliffs, N.J.: Prentice-Hall, Inc., 1978.

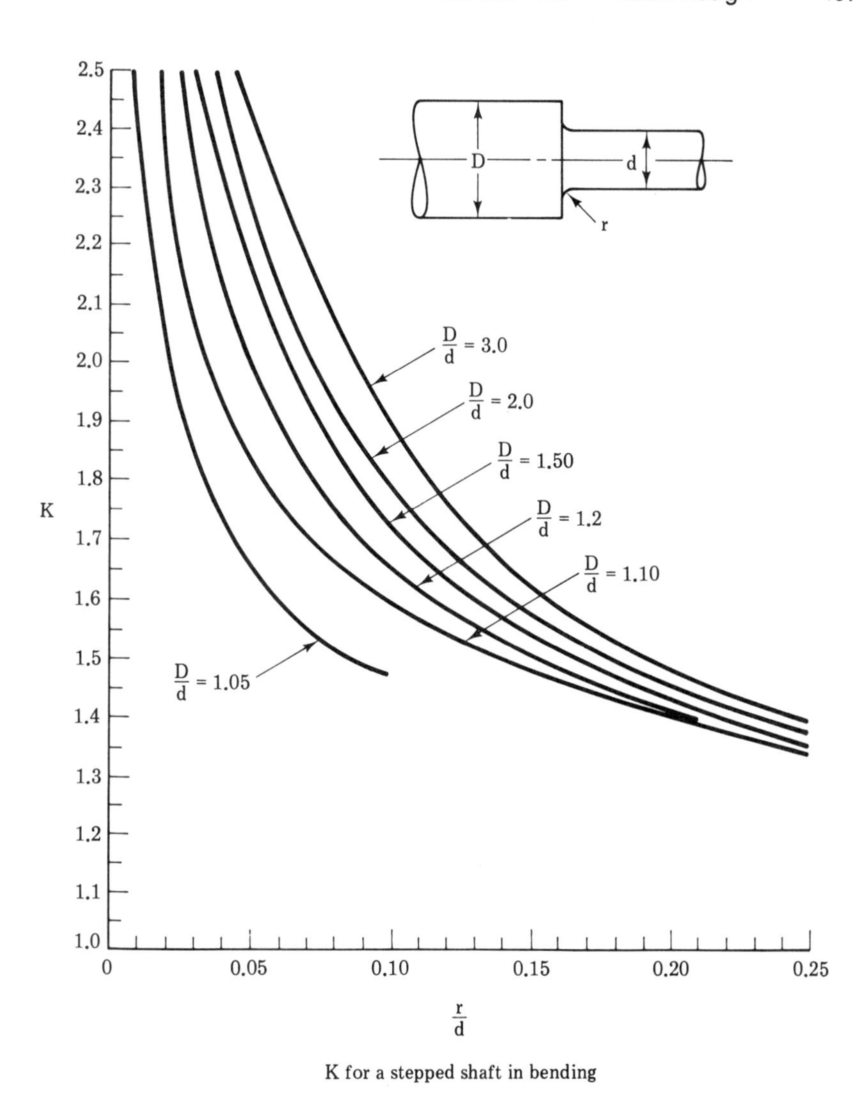

Figure 17-39. Stress concentration factor—stepped shaft in bending.

Example 17-13: Determine the maximum stress due to bending at the larger diameter section of the shaft shown in Figure 17-42, as well as the smaller diameter section and at the step. Assume the bending moment to be 40 N m at all three points.

Solution:

Step 1 Determine the maximum stress in the smaller diameter section.

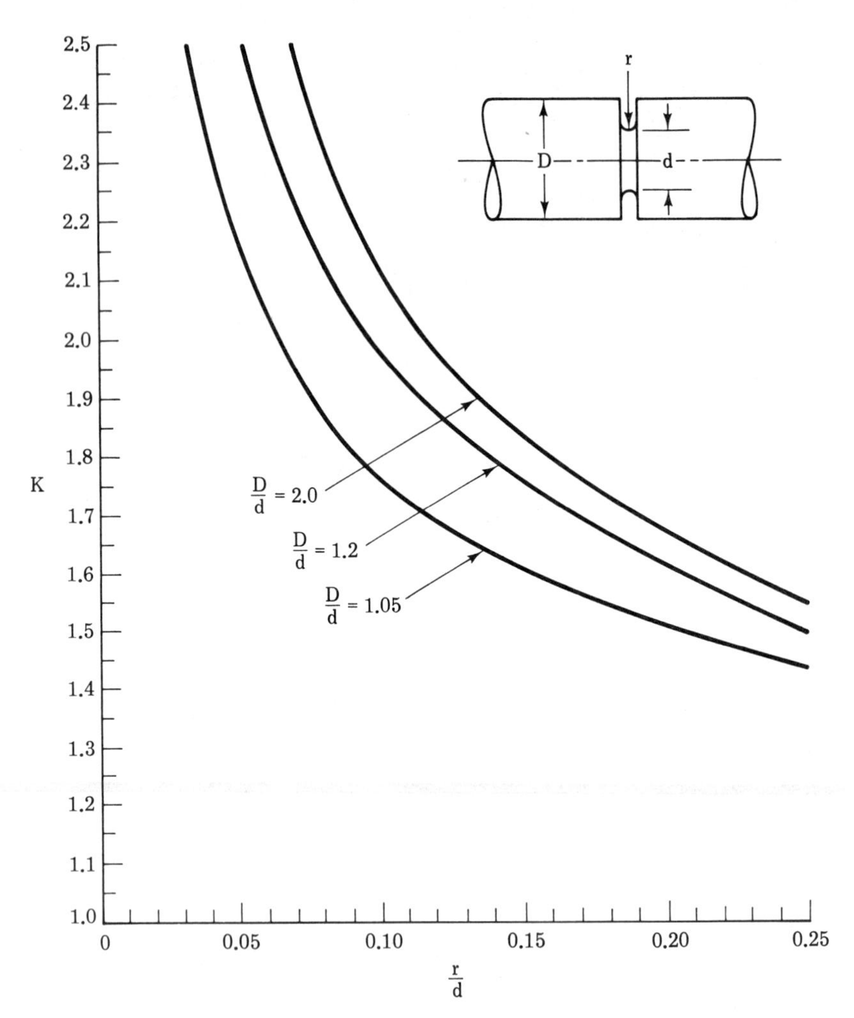

Figure 17-40. Stress concentration factor—grooved shaft in bending.

$$\sigma_1 = \frac{Mc}{I} = \frac{M}{Z} = \frac{32M}{\pi d^3}$$

$$\sigma_1 = \frac{32(40 \text{ N m})}{\pi(0.03 \text{ m}^3)} \qquad \boldsymbol{\sigma_1 = 15.1 \text{ MPa}}$$

Step 2 Determine the maximum stress at the step.

$$\frac{D}{d} = \frac{50 \text{ mm}}{30} = 1.67 \qquad \frac{r}{d} = \frac{3}{30 \text{ mm}} = 0.1$$

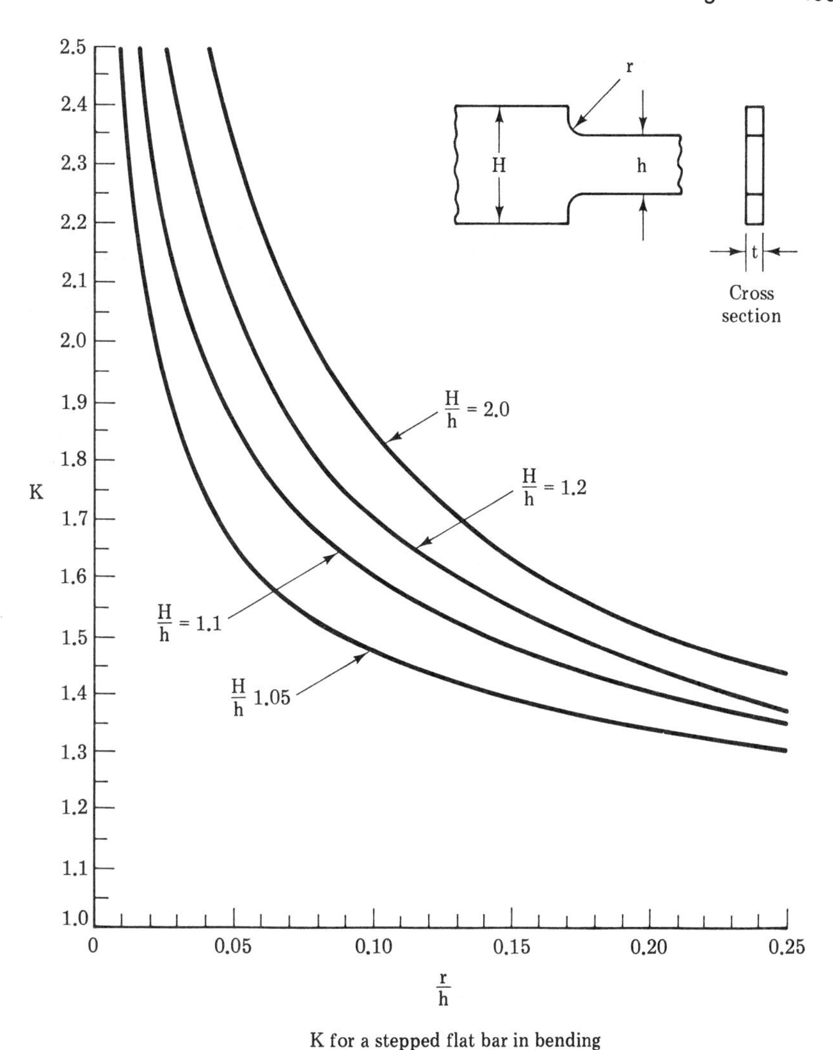

K for a stepped flat bar in bending

Figure 17-41. Stress concentration factor—stepped rectangular beam in bending.

From Figure 17-39, $K = 1.75$.

$$\sigma_2 = \frac{KMc}{I} = \frac{KM}{Z} = K\sigma_1 = 1.75(15.1 \text{ MPa})$$

$$\boldsymbol{\sigma_2 = 26.4 \text{ MPa}}$$

Step 3 Determine the maximum stress at the largest section.

$$\sigma_3 = \frac{Mc}{I} = \frac{M}{Z} = \frac{32M}{\pi D^3}$$

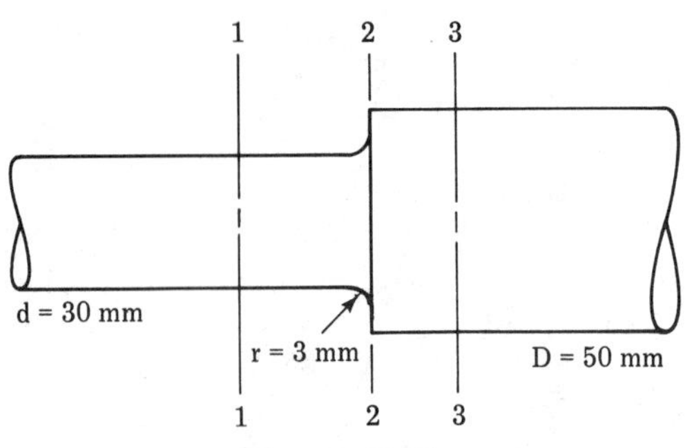

Figure 17-42

$$\sigma_3 = \frac{32(40 \text{ N m})}{\pi(.05 \text{ m}^3)} \qquad \boldsymbol{\sigma_3 = 3.26 \text{ MPa}}$$

The effect of the diameter change plus the further effect of the actual step are quite striking!

Problem 17-13: A cantilever beam is rectangular in cross section and stepped and loaded as shown in Figure 17-43. What is the maximum stress and where does it occur?

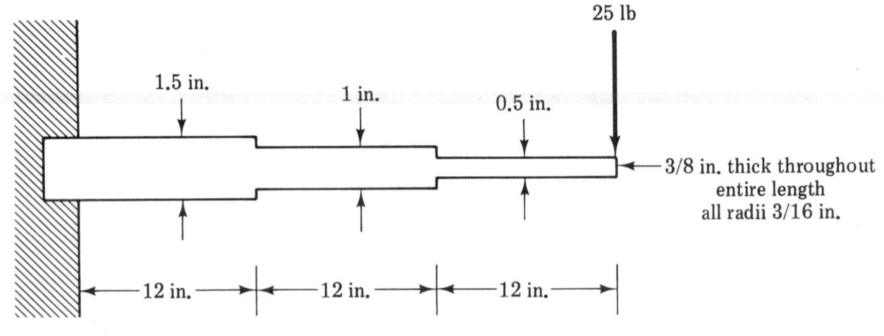

Figure 17-43

Lateral Buckling

The combination of long spans and local irregularities in a beam can cause buckling, wrinkling, or twisting of a beam. Any such deformation reduces the safe load-carrying capability. The local irregularities may be such things as variations in composition or cross section of the beam. The problem can be countered in two ways. First, lateral supports can be used to prevent localized failure. Second, a reduced allowable bending stress can be used in design of the beam.

The AISC recommends that the *maximum* allowable bending stress be 0.6 times the yield strength in tension or the value obtained from the following equation, whichever is *lower:*

$$\sigma = \frac{12 \times 10^6 C_b}{ld/A_f}$$

where

l = unsupported length of a simple beam or twice the length of a cantilever beam (in. or m)
d = depth of beam (in. or m)
A_f = area of flange in compression (in.^{2} or m^2)
C_b = 1 for simple or cantilever beams

The criterion for determining the need to use the above equation is the value of the denominator ld/A_f. It must be evaluated specifically for each type of steel. The equation above is used only when ld/A_f exceeds:

555 for A36 structural steel
475 for A529 structural steel
475 for A572, Grade 42 structural steel
445 for A572, Grade 45 structural steel
400 for A572, Grade 50 structural steel
365 for A572, Grade 55 structural steel
335 for A572, Grade 60 structural steel
310 for A572, Grade 65 structural steel
400 for A242, A441, A588 structural steel

17.7 READINESS QUIZ

1. The beam shown in Figure 17-44 has its top half in ________ and its bottom half in ________.

   (a) tension, tension
   (b) tension, compression
   (c) compression, tension
   (d) compression, compression
   (e) none of the above

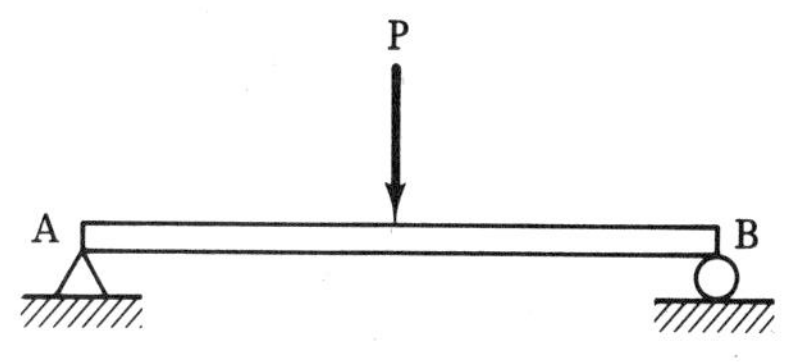

Figure 17-44

2. Shear stress in a beam may be caused by

   (a) torque applied about the longitudinal axis
   (b) direct shear load
   (c) internal bending moment
   (d) any of the above
   (e) none of the above

3. The greatest longitudinal shearing stress in a beam occurs at the

   (a) top
   (b) bottom
   (c) neutral axis
   (d) any of the above
   (e) none of the above

4. Concrete is strongest under

   (a) compression
   (b) tension
   (c) shear
   (d) torsion
   (e) none of the above

5. The weight of a beam is not considered as part of the load on the beam unless it constitutes more than ________% of the load.

   (a) 5
   (b) 10
   (c) 15
   (d) 20
   (e) 25

6. Flexural stress is simply another name for ________ stress.

   (a) shear
   (b) tensile
   (c) compressive
   (d) bending
   (e) none of the above

7. The maximum bending stress in a beam is a function of

   (a) maximum internal bending moment
   (b) moment of inertia of the beam cross section
   (c) the perpendicular distance from a horizontal plane through the neutral axis to the most distant fiber
   (d) all of the above
   (e) none of the above

8. The section modulus of the cross section of a beam is equal to

   (a) $c \div I$
   (b) $I \div c$
   (c) $I \times c$
   (d) $I \log c$
   (e) none of the above

9. The deflection of a point on a beam under load can be determined from the

   (a) elastic modulus of the material
   (b) bending moment diagram
   (c) the moment of inertia of the beam cross section
   (d) all of the above
   (e) none of the above

10. The deflection of a cantilever beam is a maximum at the

    (a) support
    (b) point of maximum load
    (c) point of maximum shear
    (d) any of the above
    (e) none of the above

11. Lateral buckling can result from

    (a) local imperfections in the beam material
    (b) minor variations in the beam cross section
    (c) variations in load
    (d) any of the above
    (e) none of the above

12. Stress concentrations occur in shafts subjected to bending at

    (a) the shaft surface
    (b) the shaft centerline
    (c) changes in cross section
    (d) any of the above
    (e) none of the above

13. Beams made of two materials are analyzed by considering them as an equivalent beam made of one of the materials by a method called

    (a) superposition
    (b) transformation
    (c) integration
    (d) differentiation
    (e) none of the above

14. In a reinforced concrete beam

    (a) the concrete carries the entire tension load
    (b) the steel bars carry the entire tension load
    (c) the steel bars carry the entire compression load
    (d) any of the above
    (e) none of the above

15. The slope of a simply supported beam is at a maximum at

    (a) one of the supports
    (b) the point of maximum load
    (c) the point of maximum moment
    (d) any of the above
    (e) none of the above

17.8 SUPPLEMENTARY PROBLEMS

17-14 Determine the maximum compressive and tensile stresses in the beam shown in Figure 17-45.

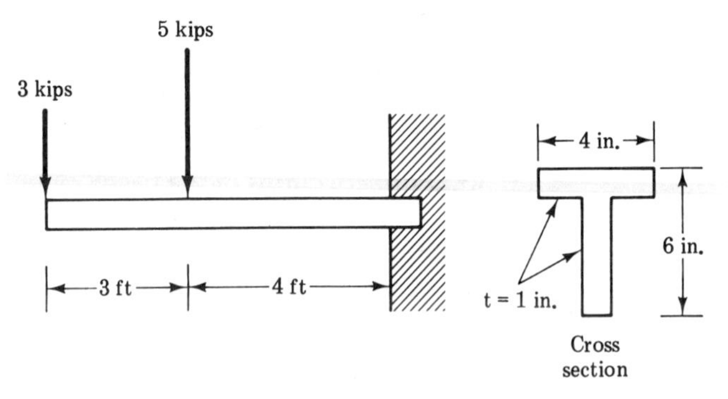

Figure 17-45

17-15 Determine the deflection and slope of the beam shown in Figure 17-46 at the load.

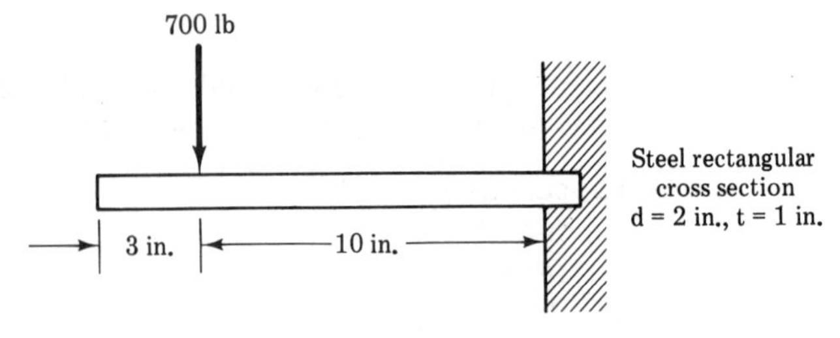

Figure 17-46

17-16 Determine the maximum deflection for the beam shown in Figure 17-47.

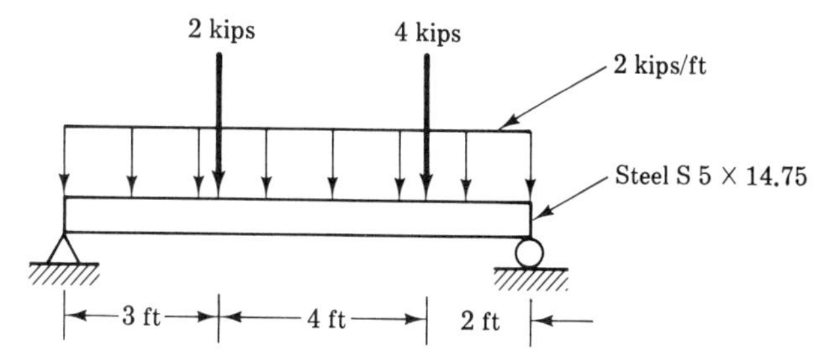

Figure 17-47

17-17 Determine the midspan deflection for the beam shown in Figure 17-48.

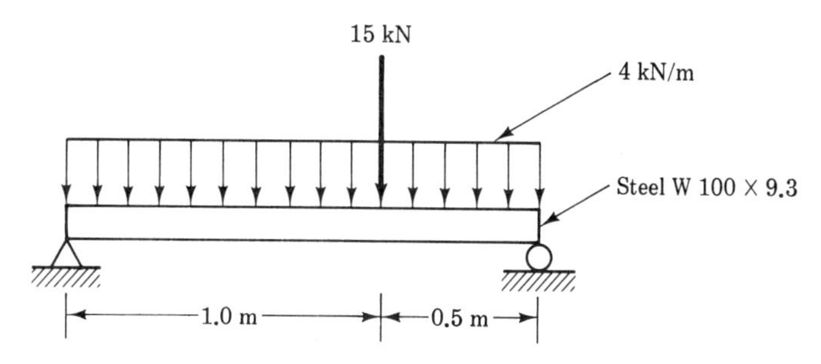

Figure 17-48

17-18 Select the lowest cost *S* beam for the supports and loading shown in Figure 17-49.

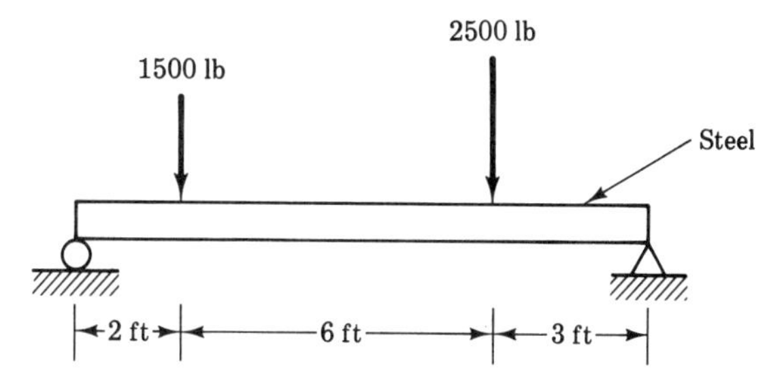

Figure 17-49

17-19 Determine the maximum flexural stress in each material for the beam shown in Figure 17-50.

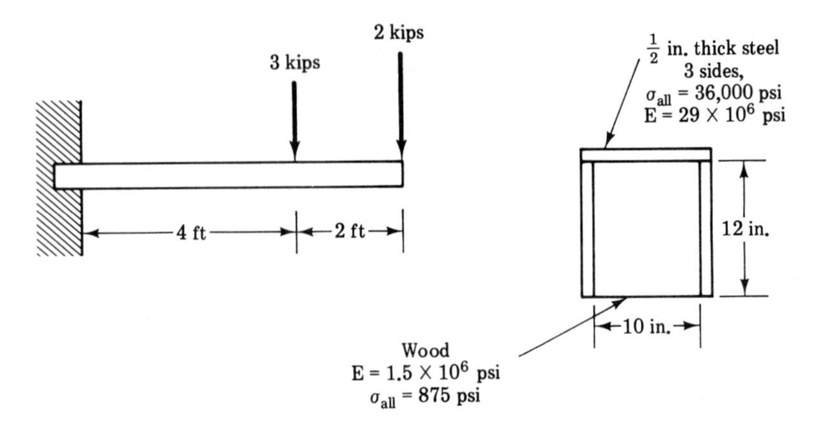

Figure 17-50

17-20 Determine the minimum width required for the aluminum portion of the beam shown in Figure 17-51 if M_{max} is 300 kip · ft.

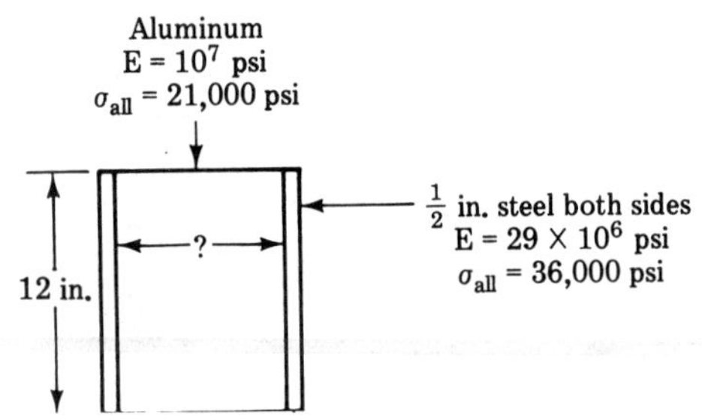

Figure 17-51

17-21 Determine the number of $\frac{1}{2}$-in. re-bars required for the closest possible to an ideal design for the beam shown in Figure 17-52.

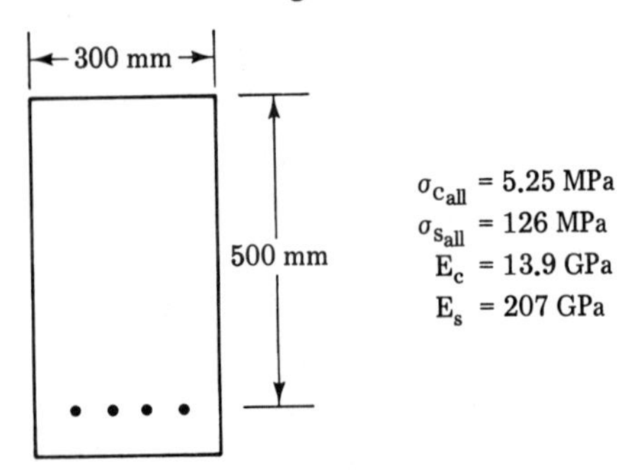

Figure 17-52

17-22 What is the maximum stress in the beam shown in Figure 17-53 and where on the beam does it occur?

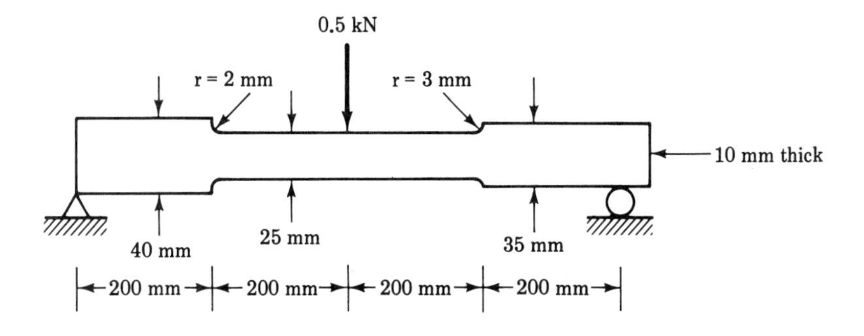

0.5 kN
r = 2 mm
r = 3 mm
10 mm thick
40 mm
25 mm
35 mm
200 mm
200 mm
200 mm
200 mm

Figure 17-53

Chapter

18

Combined Stresses

18.1 OBJECTIVES

Upon completion of the work relating to this chapter, you should be able to perform the following.

1. Determine the maximum stresses occurring in a structural or machine member for any of the following combinations of stresses:

   (a) axial (tension or compression) with bending
   (b) direct shear with torsional shear
   (c) bending with torsional shear
   (d) bending in two directions

2. Determine the minimum size for a structural or machine member subjected to any of the following combinations of stresses:

   (a) axial (tension or compression) with bending
   (b) bending with torsional shear
   (c) bending in two directions
   (d) direct shear with torsional shear

18.2 INTRODUCTION

Most of the problems we have dealt with have had any given point in a machine or structural member exposed to a simple collinear two-force system. The result has been shear, tensile, or compressive stress whose directions were simple to ascertain. The most complex problem we had to deal with was the combination of direct shear and torsional shear experienced in eccentrically loaded bolted and riveted connections (Chapter 15). You may recall that they were in the same plane and we simply summed them vertically.

There are many practical problems in addition to the one already mentioned where a material is subjected to a combination of stresses. These range from the simple to some extremely complex possibilities. Some of the commonly encountered combinations are shown in Figure 18-1.

In this chapter we will examine a variety of such problems and their solution. Generally, we can say that some combinations are collinear and can be added algebraically while others are not collinear and must be added vectorially. The combining of more than two stresses makes the summations more involved. However, with a little experience, some stresses can be considered negligible and deleted from our calculations. There are two specific types of problems that, while relatively simple, will not be dealt with in this chapter. One, already mentioned, eccentrically loaded bolted and riveted connections, was covered in Chapter 15. The other, eccentrically

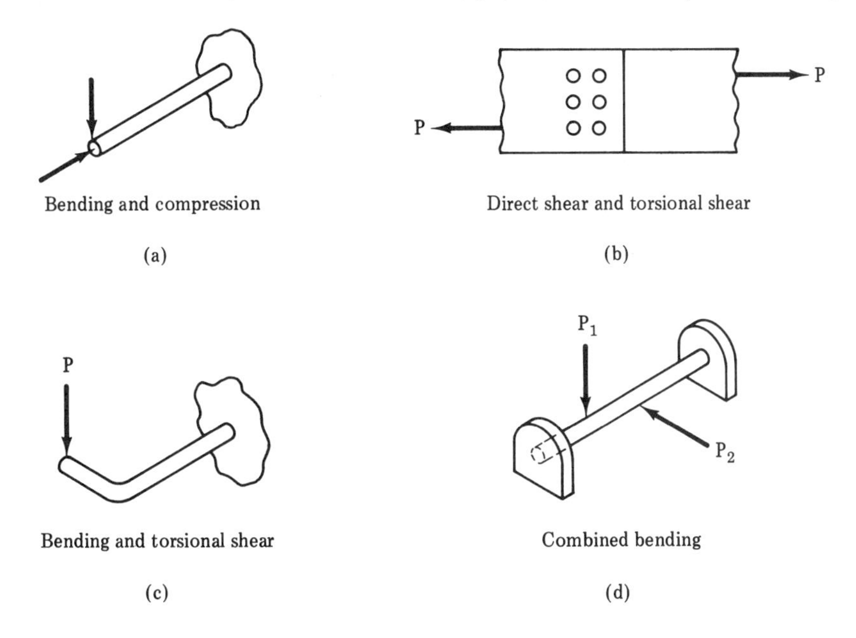

Figure 18-1. Combined stresses.

loaded short columns (posts), is covered in Chapter 19. The primary issue in analysis is to determine where the maximum combined stress will occur.

18.3 BENDING AND COMPRESSION

A cantilever beam subjected to a load that is not perpendicular to the longitudinal axis of the beam will result in either a combined bending and compressive stress or a combined bending and tensile stress. There will also be a combined shear and axial (tensile or compressive) stress but it will be at a maximum where the other combination is at a minimum and vice versa.

Example 18-1: Determine the maximum stress in the beam shown in Figure 18-2a.

Solution:

Step 1 Draw the free-body diagram as shown in 18-2b.

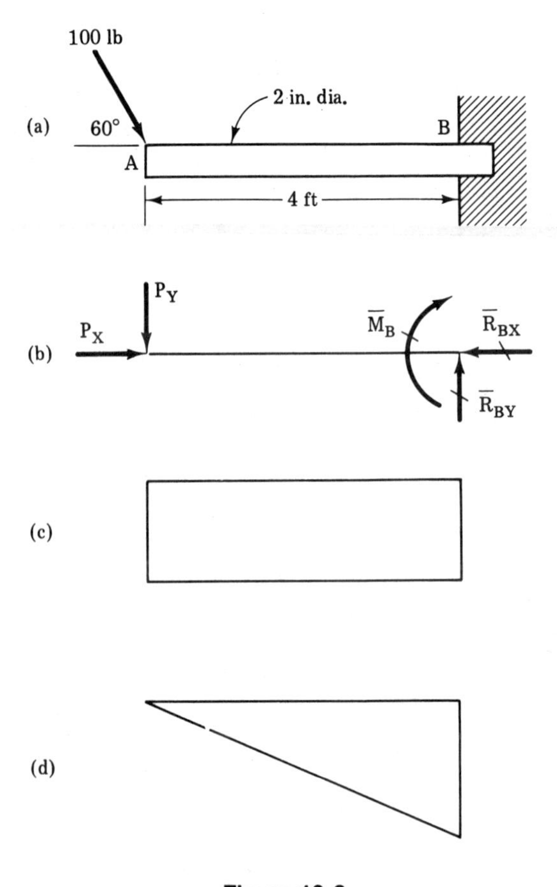

Figure 18-2

Step 2 Draw the shear diagram as shown in 18-2c.

Step 3 Draw the moment diagram as shown in 18-2d.

Step 4 Maximum shear is equal to $P_y = P \sin \theta_p$,

$$P_y = 100 \text{ lb. } (\sin 300°) \qquad P_y = 86.6 \text{ lb}$$

$$V = 86.6 \text{ lb and constant along the beam}$$

Step 5 Maximum moment is equal to $\overline{M}_B = P_y d_{py} = (86.6 \text{ lb})(4 \text{ ft})$

$$\overline{M}_B = 346.4 \text{ ft lb}$$

Step 6 There will be two points of stress maximization. (a) Combined direct compression and compression due to bending at the bottom of the beam at B. Direct compressive stress is

$$\sigma_{CD} = \frac{P_x}{A} = \frac{(100 \text{ lb})\cos 60°}{(2 \text{ in.}^2)/4} = 15.9 \text{ psi}$$

Maximum compressive stress due to bending is

$$\sigma_{CB} = \frac{M}{Z} = \frac{32M}{\pi d^3} = \frac{32(346.4 \text{ ft lb})(12 \text{ in./ft})}{(2 \text{ in.}^3)}$$

$$\sigma_{CB} = 5292.6 \text{ psi}$$

Then

$$\sigma_{c_{total}} = 15.9 + 5292.6 \qquad \boldsymbol{\sigma_{c_{total}} = 5210 \text{ psi}}$$

(b) Combined direct compression and shear at the centerline of the beam at B:

$$\tau = \frac{4V}{3A} = \frac{4(86.6 \text{ lb})}{3(2 \text{ in.}^2)/4} = 38 \text{ psi}$$

Total shearing stress is determined as follows:

$$\tau_{total} = \sqrt{\left(\frac{\sigma_{CD}}{2}\right)^2 + \tau^2}$$

$$\tau_{total} = \sqrt{\left(\frac{15.9}{2}\right)^2 + (38)^2} \qquad \boldsymbol{\tau_{total} = 38.8 \text{ psi}}$$

Experience, in this case, might well have led us to ignore the effects of direct compression and shear.

Problem 18-1: Determine the maximum stress in the post shown in Figure 18-3.

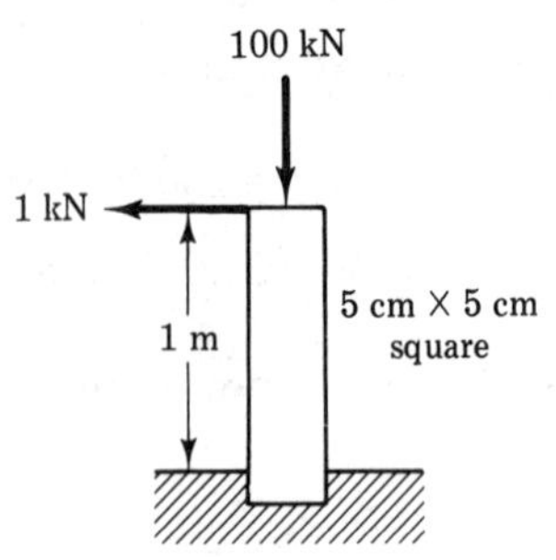

Figure 18-3

Example 18-2: Determine the most economical size channel for the post shown in Figure 18-4a.

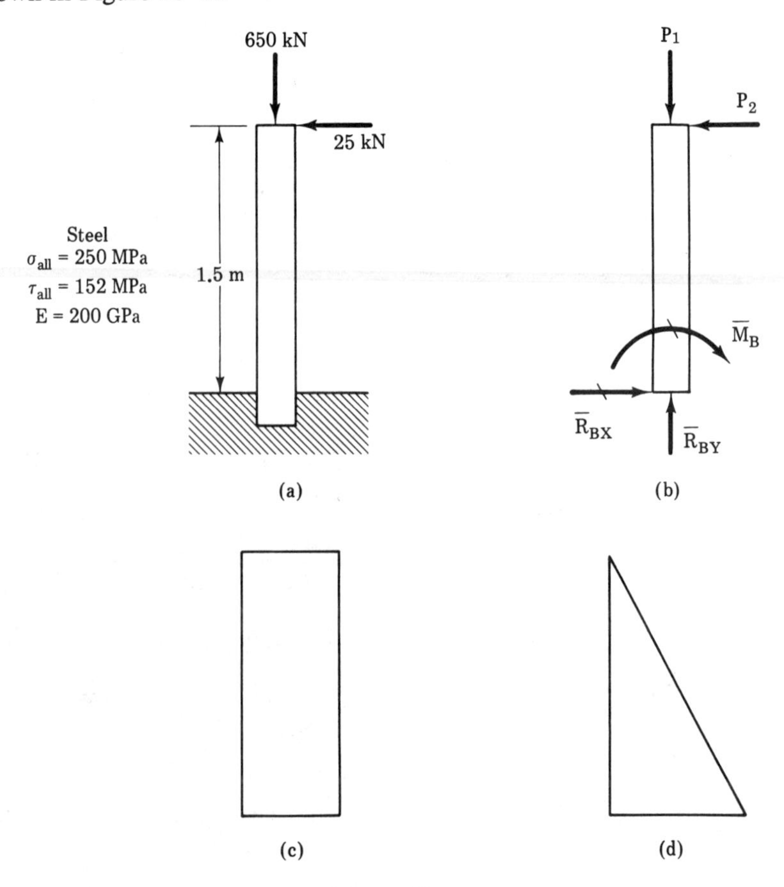

Figure 18-4

Solution:

Step 1 Draw the free-body diagram as shown in 18-4b.

Step 2 Draw the shear diagram as shown in 18-4c.

Step 3 Draw the moment diagram as shown in 18-4d.

Step 4 Maximum shear is equal to $V = P_2 = 25$ kN.

Step 5 Maximum moment is equal to $\overline{M}_B = P_2 d_2 = 25$ kN(1.5 m)

$$\overline{M}_B = 37.5 \text{ kN m}$$

Step 6 Determine minimum Z needed based on $\overline{M}_B$.

$$Z_{\text{min}} = \frac{M}{\sigma} = \frac{37.5 \text{ kN m}(10^6 \text{ cm}^3/\text{m}^3)}{250 \text{ MPa}} \qquad Z_{\text{min}} = 150 \text{ cm}^3$$

Step 7 Determine minimum td needed based on shear.

$$td = \frac{V}{\tau} \qquad td = \frac{25 \text{ kN}(10^4 \text{cm}^2/\text{m}^2)}{152 \text{ MPa}} \qquad td = 0.164 \text{ cm}^2$$

Step 8 Determine minimum A needed based on direct compressive stress:

$$A = \frac{P_1}{\sigma} = \frac{0.65 \text{ MN}(10^4 \text{ cm}^2/\text{m}^2)}{250 \text{ MPa}} \qquad A = 26 \text{ cm}^2$$

Step 9 From the Appendices, select $300 \times 90 \times 10 \times 15.5$ based on $A = 55.74$ cm², $td = 155$ cm², and $Z = 494$ cm³. At first glance a much smaller channel might be selected for a first trial, but recall that we are dealing with combined stresses.

Step 10 This channel is much larger than needed for shear stress, therefore, the critical point will be the surface fibers of the channel at the left-hand side of B.

$$\sigma_{CD} = \frac{P_1}{A} = \frac{0.65 \text{ MN}}{55.74 \text{ cm}^2(\text{m}^2/10^4 \text{ cm}^2)}$$

$$\sigma_{CD} = 116.6 \text{ MPa}$$

$$\sigma_{CB} = \frac{M}{Z} = \frac{37.5 \text{ kN m}}{494 \text{ cm}^3(\text{m}^3/10^6 \text{ cm}^3)}$$

$$\sigma_{CB} = 75.9 \text{ MPa}$$

$$\sigma_{\text{total}} = 116.6 + 75.9 = 192.5 \text{ MPa}$$

Since this is well below the allowable, try the next lightest channel 300 × 90 × 9 × 13 with $A = 48.57$ cm^2 and Z of 429 cm^3. This choice yields a combined compressive stress of 231 MPa. A more extensive listing of shapes might yield a slightly lighter channel but we'll leave it at 300 × 90 × 9 × 13.

Problem 18-2: Determine how many 2 × 12's (1.5 × 11.25 in.) will be needed to safely support the load shown on the beam in Figure 18-5.

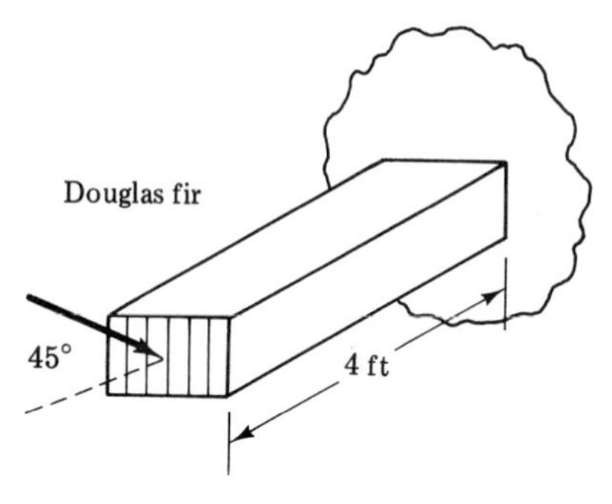

Figure 18-5

18.4 COMBINED BENDING

Frequently a load is not applied in a direction perpendicular to one of the surface planes of a beam. Also, there are times when two or more loads are applied to more than one surface plane. The simplest approach is to treat them individually and determine where maximum shear and maximum moment occur. For maximum shear stress, the individual components can be simply added by Pythagoras' theorem and will occur at the beam centerline. For maximum bending or flexural stress, it is necessary to locate the edge where the highest stress will occur, determine the individual stresses, and add them by use of Pythagoras' theorem.

Example 18-3: Determine the maximum stress in the beam shown in Figure 18-6a.

Solution:

Step 1 Draw the free-body diagram as shown in 18-6b.

Step 2 Draw the shear diagrams as shown in 18-6c.

Step 3 Draw the moment diagrams as shown in 18-6d.

Step 4 Determine the maximum moment.

$$\overline{M}_{CZ} = P_2 d_2 = 1000 \text{ lb}(3 \text{ ft}) \qquad \overline{M}_{CZ} = 3000 \text{ ft lb}$$

$$\overline{M}_{CY} = P_1 d_1 = 500 \text{ lb}(5 \text{ ft}) \qquad \overline{M}_{CY} = 2500 \text{ ft lb}$$

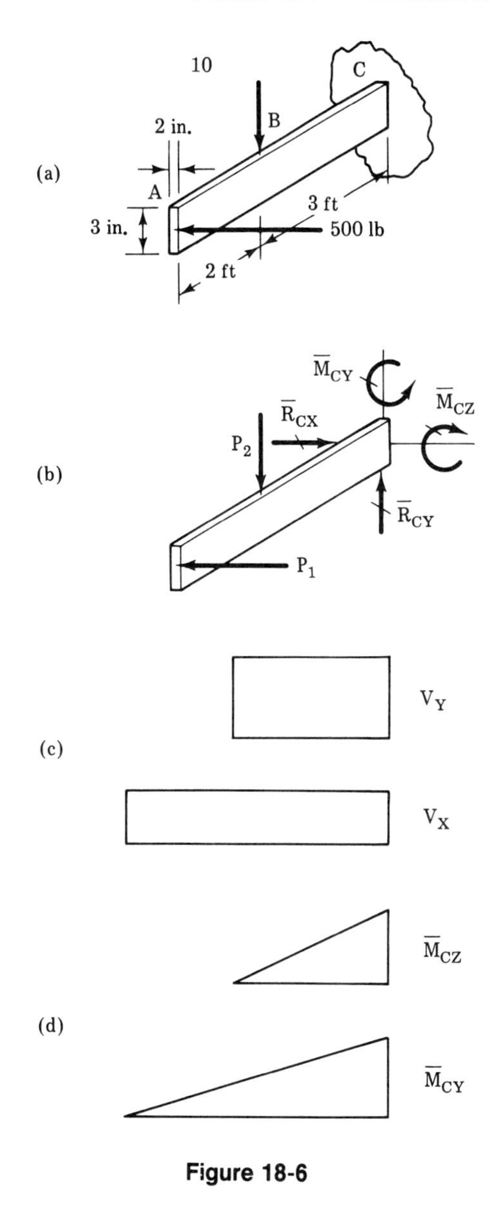

Figure 18-6

Step 5 Determine maximum bending stress—the tension along the upper right-hand edge or compression along the lower left-hand edge.

$$\sigma_1 = \frac{M}{Z} = \frac{6\overline{M}_{CZ}}{bh^2} = \frac{6(3000 \text{ ft lb})(12 \text{ in./ft})}{2 \text{ in.}(3 \text{ in.}^2)}$$

$$\sigma_1 = 12{,}000 \text{ psi}$$

$$\sigma_2 = \frac{M}{Z} = \frac{6\overline{M}_{CY}}{bh^2} = \frac{6(2500 \text{ ft lb})(12 \text{ in./ft})}{3 \text{ in.}(2 \text{ in.}^2)}$$

$$\sigma_2 = 15{,}000 \text{ psi}$$

$$\sigma_{max} = \sigma_1 + \sigma_2 = 12{,}000 + 15{,}000 \qquad \sigma_{max} = \mathbf{27{,}000 \text{ psi}}$$

Step 6 Check shear stress.

$$\tau_1 = \frac{3V_x}{A} = \frac{3\overline{R}_{CY}}{A} = \frac{3(500 \text{ lb})}{2 \text{ in.}(3 \text{ in.})} \qquad \tau_1 = 250 \text{ psi}$$

$$\tau_2 = \frac{3V_y}{A} = \frac{3\overline{R}_{CY}}{A} = \frac{3(1000 \text{ lb})}{2 \text{ in.}(3 \text{ in.})} \qquad \tau_2 = 500 \text{ psi}$$

$$\tau_{max} = \sqrt{\tau_1^2 + \tau_2^2} = \sqrt{(250)^2 + (500)^2} \qquad \tau_{max} = \mathbf{560 \text{ psi}}$$

As commonly occurs in beams, the flexural stress is relatively high (over the allowable for many steels) while the shear stress is relatively low.

Problem 18-3: Determine the maximum stress in the beam shown in Figure 18-7. Assume a fixed support at A.

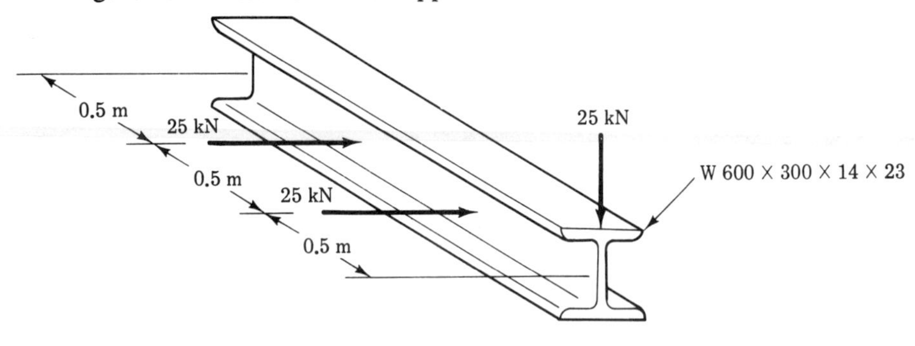

Figure 18-7

Example 18-4: Determine the diameter required for the shaft in Figure 18-8a to be safe. Assume shaft is simply supported.

Solution:

Step 1 Draw the free-body diagram of the beam as shown in 18-8b.

Step 2 Determine the reactions.
Horizontally:

$$\Sigma M_A = 0 \qquad P_1 d_1 + \overline{R}_{BX} d_{BX} = 0 \qquad \overline{R}_{BX} d_{BX} = -P_1 d_1$$

$$\overline{R}_{BX} d_{BX} = -[+(12 \text{ kN})(1 \text{ m})] \qquad R_{BX} d_{BX} = -12 \text{ kN m}$$

$$\overline{R}_{BX} = \frac{\overline{R}_{BX} d_{BX}}{d_{BX}} = \frac{12 \text{ kN m}}{3 \text{ m}} \qquad \overline{R}_{BX} = +4 \text{ kN}$$

$$\Sigma F_X = 0 \qquad \overline{R}_{AX} + P_1 + \overline{R}_{BX} = 0 \qquad \overline{R}_{AX} = -(P_1 + \overline{R}_{BX})$$

$$\overline{R}_{AX} = -(-12 + 4 \text{ kN}) \qquad \overline{R}_{AX} = +8 \text{ kN}$$

Vertically:

$$\Sigma M_A = 0 \qquad P_2 d_2 + \overline{R}_{BY} d_{BY} = 0 \qquad \overline{R}_{BY} d_{BY} = -P_2 d_2$$

$$\overline{R}_{BY} d_{BY} = -[-(6 \text{ kN})(2 \text{ cm})] \qquad \overline{R}_{BY} d_{BY} = +12 \text{ kN m}$$

$$\overline{R}_{BY} = \frac{\overline{R}_{BY} d_{BY}}{d_{BY}} = \frac{12 \text{ kN m}}{3 \text{ m}} \qquad \overline{R}_{BY} = +4 \text{ kN}$$

$$\Sigma F_y = 0 \qquad \overline{R}_{AY} + P_2 + \overline{R}_{BY} = 0 \qquad \overline{R}_{AY} = -(P_2 + \overline{R}_{BY})$$

$$\overline{R}_{AY} = -(-6 + 4 \text{ kN}) \qquad \overline{R}_{AY} = +2 \text{ kN}$$

As might be expected in this problem, maximum shear occurs at the maximum load and is equal to

$$V_{max} = \sqrt{(12)^2 + (2)^2} \qquad V_{max} = 12.17 \text{ kN}$$

Step 3 Draw the shear diagrams as shown in Figure 18-8c.

Step 4 Draw the moment diagrams as shown in 18-8d.

Step 5 Maximum shear stress will be at the shaft centerline at P_2 and the required diameter can be calculated as follows:

$$\tau_{max} = \frac{4V}{3A} = \frac{4V(4)}{3\pi d^2} \qquad d = \sqrt{\frac{16V}{3\pi\tau_{max}}}$$

$$d = \sqrt{\frac{16(12{,}170 \text{ N})}{3(\pi)(240 \text{ MPa})}} \qquad d = 0.00928 \text{ m}$$

Step 6 Maximum flexural stress will be at the shaft surface at the point of maximum moment. The maximum total moment will occur at one of the loads; in this problem, by inspection, P_1. The diameter based on flexural stress can be calculated as follows:

$$M_{max} = \sqrt{8^2 + 2^2} \qquad M_{max} = 8.25 \text{ kN m}$$

$$\sigma = \frac{M}{Z} = \frac{32M}{\pi d^3} \qquad d = \sqrt[3]{\frac{32M}{\pi\sigma}}$$

$$d = \sqrt{\frac{32(8250 \text{ N m})}{\pi(328 \text{ MPa})}} \quad d = 0.0635 \text{ m} \quad \text{or} \quad \mathbf{d = 63.5 \text{ mm}}$$

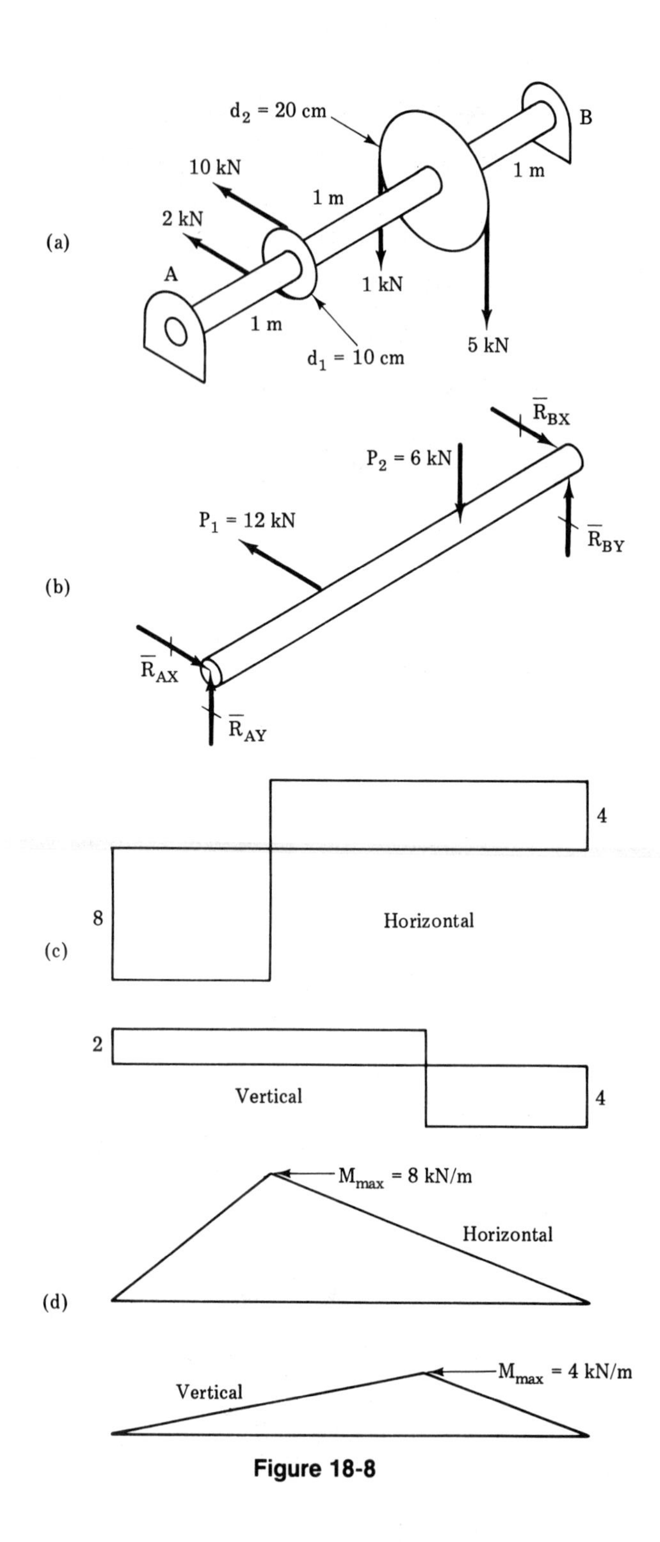

(a)
d_2 = 20 cm
B
10 kN
1 m
1 m
2 kN
A
1 kN
1 m
5 kN
d_1 = 10 cm
(b)
$\overline{R}_{BX}$
P_2 = 6 kN
P_1 = 12 kN
$\overline{R}_{BY}$
$\overline{R}_{AX}$
$\overline{R}_{AY}$
(c)
4
8
Horizontal
2
Vertical
4
(d)
M_{max} = 8 kN/m
Horizontal
Vertical
M_{max} = 4 kN/m

Figure 18-8

Problem 18-4: Determine the size required for the square beam shown in Figure 18-9. Assume the beam is simply supported.

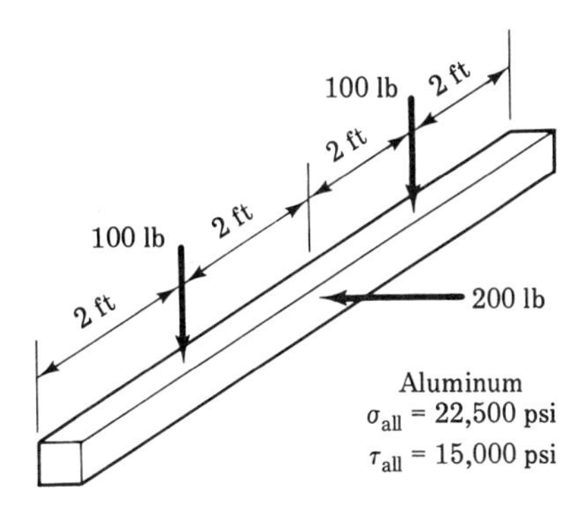

Figure 18-9

18.5 BENDING WITH TORSION

This is a very common phenomenon in shaft design. Power shafts transmit torque and the torsional shear stress must be added to the other stresses to attain what is considered to be a maximum shear stress. We have already used the necessary relationship.

$$\tau_{max} = \sqrt{\left(\frac{\sigma}{2}\right)^2 + \tau^2}$$

The rationale here is that the torsional shear stress will be at a maximum at the surface of the shaft and must therefore be combined with the maximum flexural stress at that surface.

Example 18-5: Re-examine Example 18-4 and consider the torsional shear stress involved. Will the total maximum shear stress be below the allowable for the shaft size selected? If not, determine a revised shaft size.

Solution:

Step 1 Determine the magnitude and location of all torques in the shaft.

If we consider the bearings to be friction free there will be a single constant torque between the two pulleys.

$$T = +(10 \text{ kN})(0.05 \text{ m}) - (2 \text{ kN})(0.05 \text{ m}) = +0.4 \text{ kN m}$$

$$\text{or } T = -(5 \text{ kN})(0.1 \text{ m}) - (1 \text{ kN})(0.1 \text{ m}) = -0.4 \text{ kN m}$$

Step 2 Determine the maximum torsional shear stress.

$$\tau = \frac{16T}{\pi d^3} = \frac{16(0.4 \text{ kN m})}{\pi(0.0635 \text{ m}^3)} \qquad \tau = 7.96 \text{ MPa}$$

Step 3 Determine the maximum total shear stress. This will occur at P_1 where the flexural stress is at a maximum and the above torsional shear stress also occurs.

$$\tau_{max} = \sqrt{\left(\frac{\sigma}{2}\right)^2 + \tau^2}$$

$$\tau_{max} = \sqrt{\left(\frac{328}{2}\right)^2 + (7.96)^2} \qquad \tau_{max} = 164.2 \text{ MPa}$$

This is still below the allowable so our previously selected diameter of 63.5 mm stands.

Problem 18-5: Determine the diameter of the shaft needed for the loading shown in Figure 18-10.

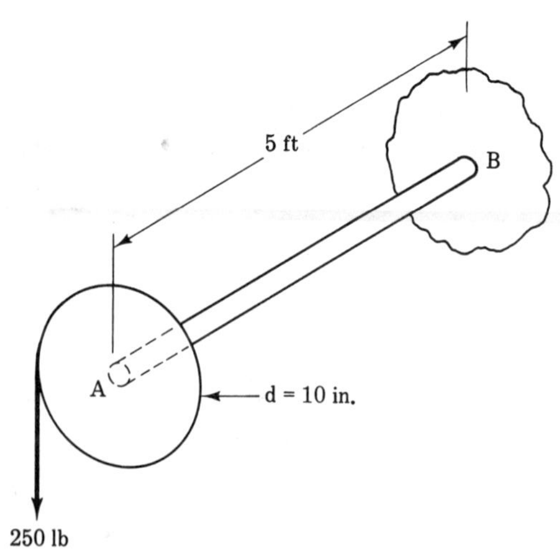

Figure 18-10

18.6 READINESS QUIZ

1. The need to combine stresses may occur in

 (a) shafts
 (b) beams
 (c) columns
 (d) any of the above
 (e) none of the above

2. In a simply supported power shaft, total maximum stress will almost always occur at the ________ of the shaft.

   (a) ends
   (b) centerline
   (c) surface
   (d) any of the above
   (e) none of the above

3. Flexural stress and axial stress combine

   (a) algebraically
   (b) vectorially
   (c) logarithmically
   (d) any of the above
   (e) none of the above

4. The size of a beam subjected to a combination of stresses is always determined by the allowable ________ stress of the material.

   (a) tensile
   (b) shear
   (c) compressive
   (d) torsional
   (e) none of the above

5. In a beam subjected to combined stresses, the stresses that combine are

   (a) direct shear and flexural
   (b) bending and direct shear
   (c) flexural and axial
   (d) any of the above
   (e) none of the above

18.7 SUPPLEMENTARY PROBLEMS

Determine the maximum stress and its location in each of the following problems:

18-6

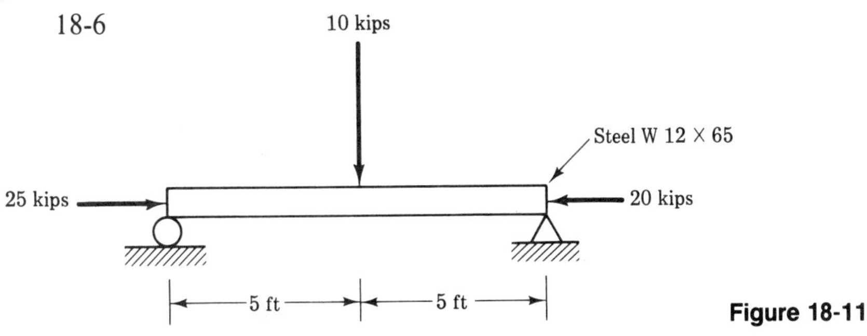

Figure 18-11

18-7

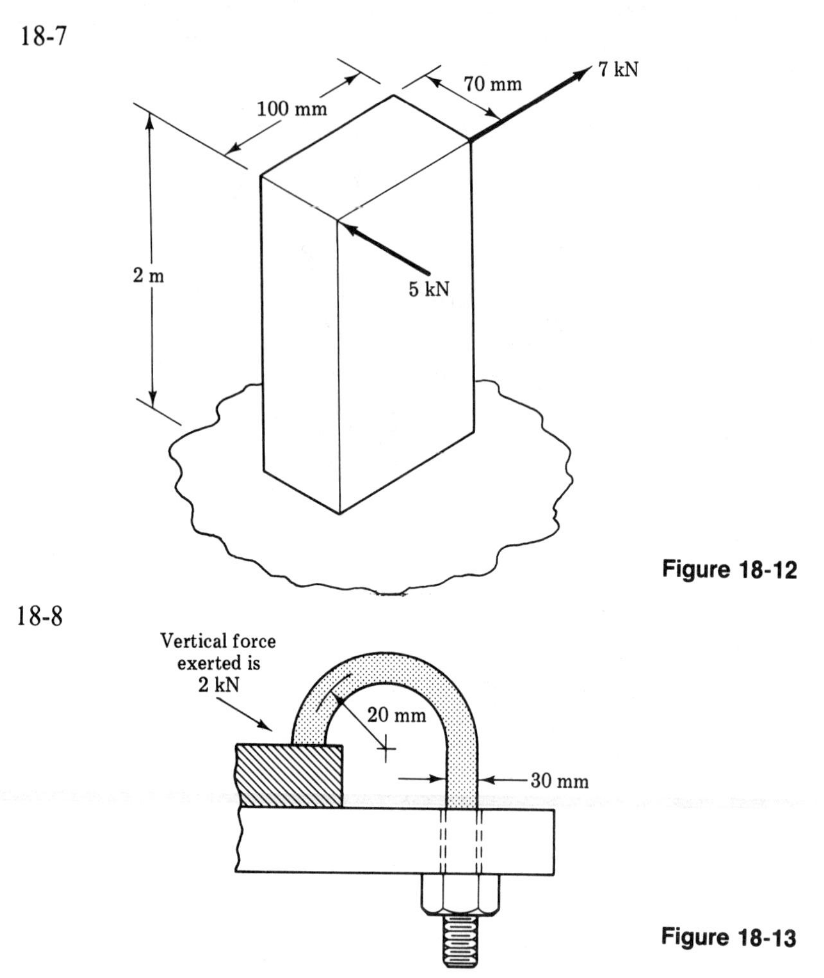

Figure 18-12

18-8

Figure 18-13

Determine the indicated size in each of the following problems:

18-9

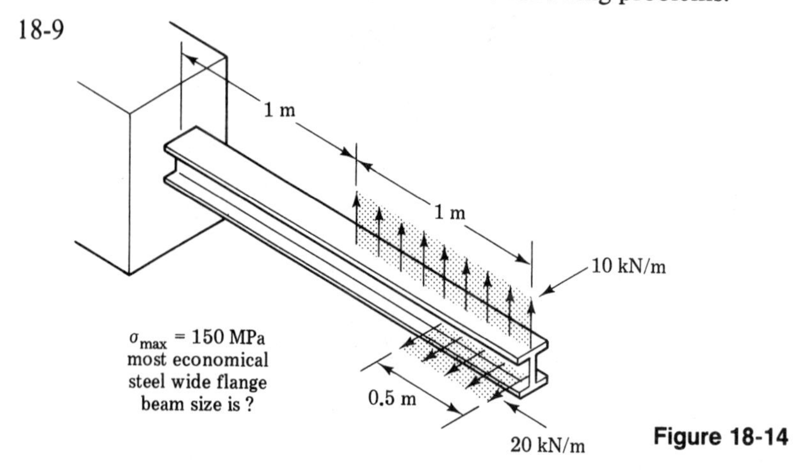

Figure 18-14

18-10

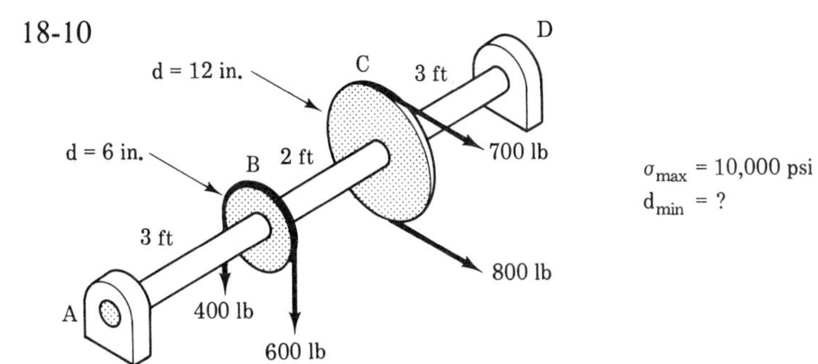

σ_{max} = 10,000 psi
d_{min} = ?

Figure 18-15

18-11

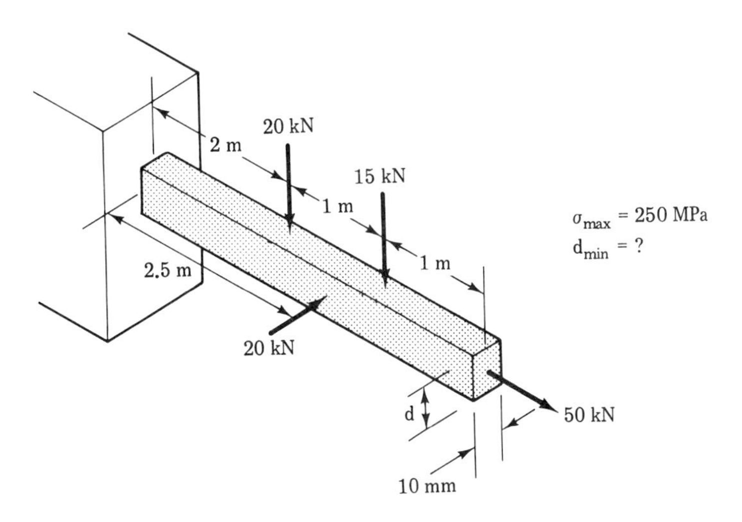

σ_{max} = 250 MPa
d_{min} = ?

Figure 18-16

Chapter 19

Columns

19.1 OBJECTIVES

Upon completion of the work relating to this chapter, you should be able to perform the following.

1. Differentiate between a *post* and a *column* in terms of the failure mode.
2. Define a "perfect" column.
3. Define and determine the *slenderness ratio* for a column.
4. Differentiate among short, intermediate, and long columns.
5. Recognize the relative significance of different types of column-end conditions.
6. Use standard column formulas to design simple columns.

19.2 INTRODUCTION TO COLUMNS

In our discussion to date, our concern with failure of a member has taken two forms. First, we have been concerned that the member be strong enough to withstand the design load without failure by fracture. Second, we considered that not only do we not want the stress in a member to exceed the elastic limit, resulting in permanent deformation, but we may want to limit deformation in order to maintain required tolerances or alignments.

In this chapter, we will consider still another mode of failure, that which results in a radical change in the geometric shape of a member. This change of shape beyond the strain normally associated with stress is called *buckling*. It can be readily demonstrated by taking a piece of cardboard such as the back of a pad of paper and subjecting it to a compressive load as shown in Figure 19-1.

If the load is simply applied parallel to the shortest dimension (the thickness) as shown in 19-1a, an appreciable load can be applied before the cardboard finally fails in compression. However, if the load is applied parallel to the next largest dimension (the width) as shown in 19-1b, even a small load will result in the bowing or buckling effect shown by the dotted lines. Although the cross-sectional area subjected to compression is much smaller than in the first case, the load that will cause failure by buckling is much smaller than that which would cause compression fracture. If we apply the load parallel to the longest dimension (the length) as shown in 19-1c, buckling will occur at an even lower load than before.

Two observations can be made regarding the above beyond the buckling phenomenon itself. First, note that the buckling becomes more significant as the dimension in the direction of the load increases. Second, buckling takes place in the direction of the least dimension. As we develop our thinking on the latter observation, we will find that it can be applied to complex cross sections by recognizing that the controlling factor is really the least radius of gyration rather than simply the least dimension. The first observation is illustrated rather dramatically in Figure 19-2. It shows three 2-cm-diameter steel rods of different lengths mounted vertically in fixed supports with compressive load placed on the free ends. The first rather short member is actually considered to be a *post* since it will fail in compression long before it begins to buckle. The other two are considered *columns* in that they will fail by buckling well before failure due to compressive stress will

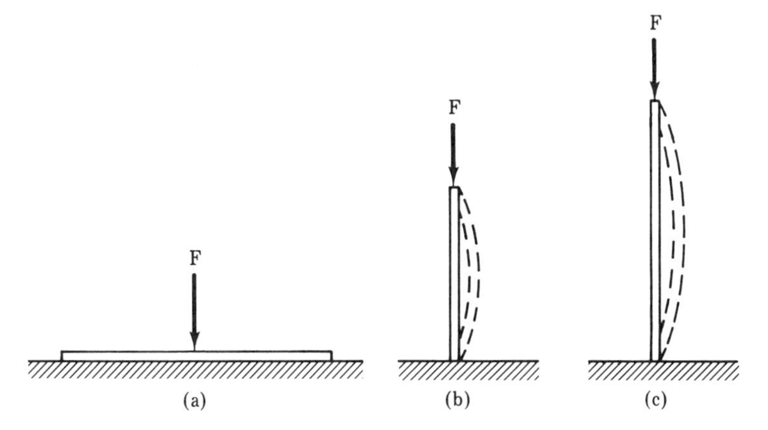

Figure 19-1. The nature of buckling.

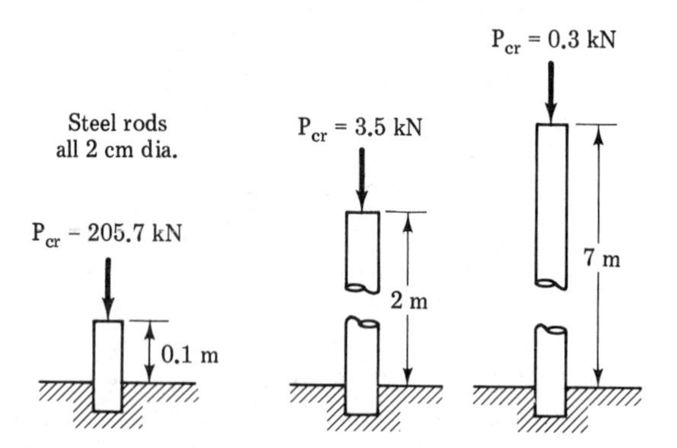

Figure 19-2. Critical buckling load versus length.

occur. The critical loads (at which failure will begin to occur) are indicated. Since the cross sections are all the same, the critical stresses will be in the same proportion as the critical loads.

The above discussion suggests that column design must involve three criteria: direct compression stress, direct compression deformation, and buckling. We have become quite familiar with the first two criteria. However, it is not immediately apparent why buckling should occur. Indeed, in a perfect column, it will not occur. The question then is, what are the imperfections that occur unavoidably (or can be avoided only at excessive cost)? Some possibilities follow:

1. Inconsistencies in the column material. This is quite common in wood since no two trees are identical nor are two sections of the same tree. However, metal columns will not have *exactly* the same composition throughout. A good metallurgy text can provide insight into the reasons for this.

2. The load and reaction may not be collinear and even if they are, they may not pass through the center of gravity of the column.

3. Nonuniform fabrication of the column which can take either or both of two forms:

   (a) lack of consistency of cross section
   (b) lack of straightness

4. Unrecognized stresses in the unloaded column remaining from fabrication.

These factors are generally negligible for members in tension, shear, or bending and even for short compression members (posts). However, for long compression members, the slightest defect can result in buckling unless the possibility is considered in the design process.

19.3 COLUMN THEORY

This chapter will vary somewhat from the preceding ones. Previously, different aspects of a chapter were developed one at a time with corresponding examples and problems. The equations actually used for column design are largely empirical. Therefore, this section will develop the theory of column design while the following section will present column formulas and applications.

Slenderness Ratio

As already noted, the length of the column and the least radius of gyration are both significant factors in column design. The ratio of column length to radius of gyration is known as the *slenderness ratio* and is expressed simply as l/r.

Earlier we learned how to determine the radius of gyration from $K = \sqrt{I/A}$. However, as shown above in the slenderness ratio, we shall use r as its symbol when working with columns. This is to avoid confusion with another use of the symbol K which we shall introduce shortly.

The length to be used is rarely the actual length but rather an effective length based on the nature of the column supports.

Effective Length

The effective length of a column is based upon its actual length, the nature of the end supports, and the nature of intermediate supports or bracing, if any. The latter were encountered some time ago in our analysis of structures when we dismissed them as "zero-force" members because they were not load bearing.

Figure 19-3 shows four possible combinations of end connections. The first and simplest (a) has the column fixed to a stationary hinge that moves with the load at the top. Thus any resistance to buckling or bending is dependent solely on the column material and cross section. The effective length in this case is equal to the actual length. Note that the effective length is stated as $l_e = Kl$ where K is the end condition factor. In 19-3b horizontal motion (buckling) is somewhat restricted by the fixed support at the bottom, yielding an effective length that is shorter than the actual. (Recall that the shorter the length the less likely buckling will occur or the greater the load that can be carried without buckling.) In 19-3c with both ends fixed, the tendency to buckle is restricted even further, resulting in an even shorter effective length. In 19-3d we have a rather special case, sometimes called the flagpole. While the fixed support at one end restricts buckling, the freedom of the top end to move horizontally as well as vertically more than compen-

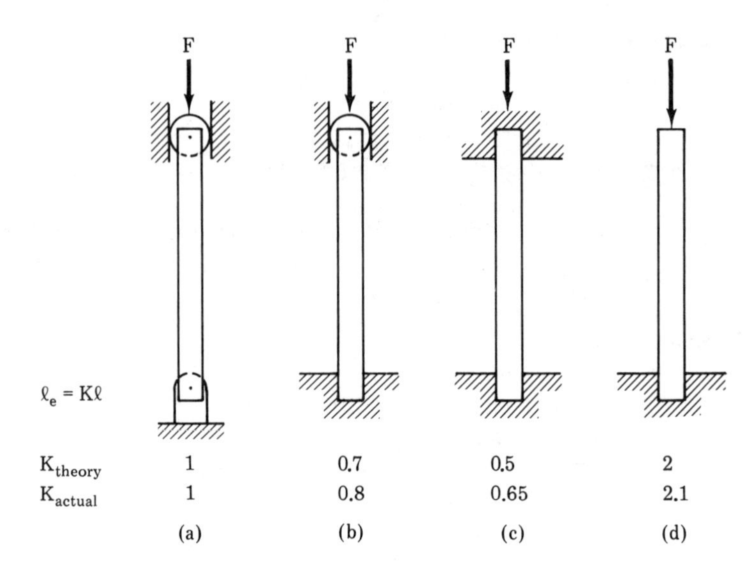

Figure 19-3. Effect of end connections on buckling.

sates yielding a K of 2. In all these cases it is assumed that the least radius of gyration is in the plane shown, that is, *not* parallel to the hinge. Note that two values for K are given, K being the theoretical and K_a the actual recommended in practice. Further, while all the columns shown have been vertically oriented, this is not at all necessary. After all a column can be simply defined as a compression member in danger of buckling.

Figure 19-4 shows a variety of arrangements for bracing a column. In every case, the largest Kl must be determined and then used in the design. A relatively large l may be substantially reduced if both ends are fixed. A

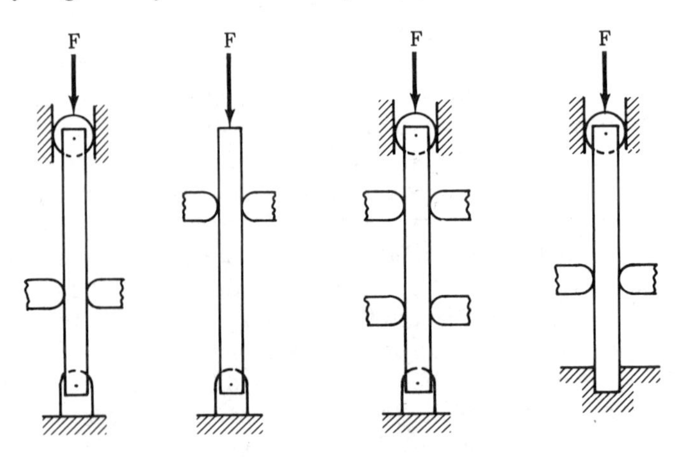

Figure 19-4. Effect of bracing on buckling.

relatively small l may be substantially increased if one end is free. Two basic rules may be noted. First, the lengths we are dealing with here are lengths between any two adjacent supports (end-to-brace or brace-to-brace). Second, if the nature of the brace has not been definitely established, treat it as a hinged support.

Column Types

Up to now we have broadly classified columns into two types, those not subject to buckling (posts) and those that are subject to buckling. The first case is one we have dealt with frequently where failure can be based on ultimate stress:

$$\sigma_{ult} = \frac{F}{A}$$

and failure is avoided by the use of an appropriate factor of safety. As shown in the graph in Figure 19-5, this relationship is independent of the slenderness ratio.

The Swiss mathematician Leonard Euler published an extensive study of elastic bending in 1744. He developed the now well-known "Euler formula" for the buckling strength of a hinge-ended column. It states that the critical load at which a column will fail by buckling is

$$F = \frac{\pi^2 EI}{l^2}$$

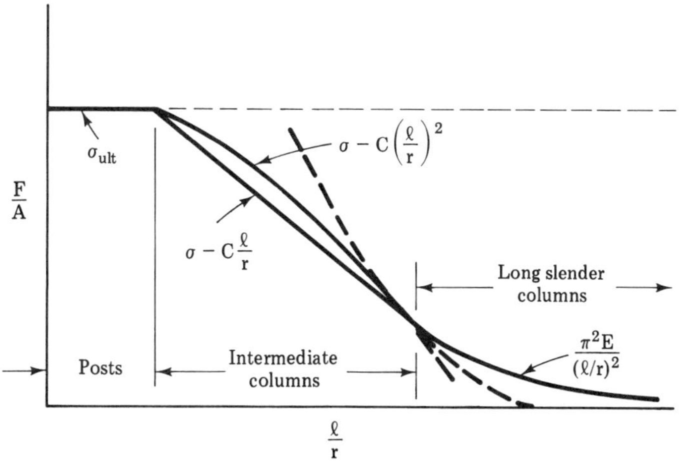

Figure 19-5. Slenderness ratio versus critical stress.

Dividing both sides by the cross-sectional area A yields

$$\frac{F}{A} = \frac{\pi^2 EI}{l^2 A}$$

However, the radius of gyration squared, r^2, is equal to I/A, and hence

$$\frac{F}{A} = \frac{\pi^2 E r^2}{l^2} \qquad \sigma_{\text{crit}} = \frac{\pi^2 E}{(l/r^2)}$$

This relationship is also plotted in Figure 19-5 for columns with a relatively high slenderness ratio. The reason for the gap is that Euler's equation is really not suitable for the intermediate range of l/r. This has to do with the generally steep slope of the curve in that area. Many formulas have been proposed and indeed are used for intermediate column design. The most frequently used are the straight-line and parabolic formulas:

$$\frac{F}{A} = \sigma - C\left(\frac{l}{r}\right) \quad \text{(straight line)}$$

$$\frac{F}{A} = \sigma - C\left(\frac{l}{r}\right)^2 \quad \text{(parabolic)}$$

where C is a constant based on experimental data.

These two equations have also been plotted in Figure 19-5. All four equations have been extended beyond their range of application with dashed lines. From these it can readily be seen that the equation for posts and Euler's equation can indeed cover the entire range of l/r. However, experience has shown that buckling occurs at σ_{ult} far short of their juncture. Experience has also shown Euler's equation to be too conservative in the intermediate range (yielding too large a critical load for a given slenderness ratio).

Actual practice sees extensive use of the basic compression equation and some use of Euler's equation (with appropriate K) both with a factor of safety applied. Section 19.4 presents a number of the commonly used equations for various materials.

Between-Supports Loading

It is not unusual for a column to be loaded at some point between supports as shown in Figure 19-6a. James Dow developed the relationship between x/l as shown in 19-6a and a reduced length ratio l_r/l. The relationship has been graphed in 19-6b, and l_r as determined from it can be used in place of l.

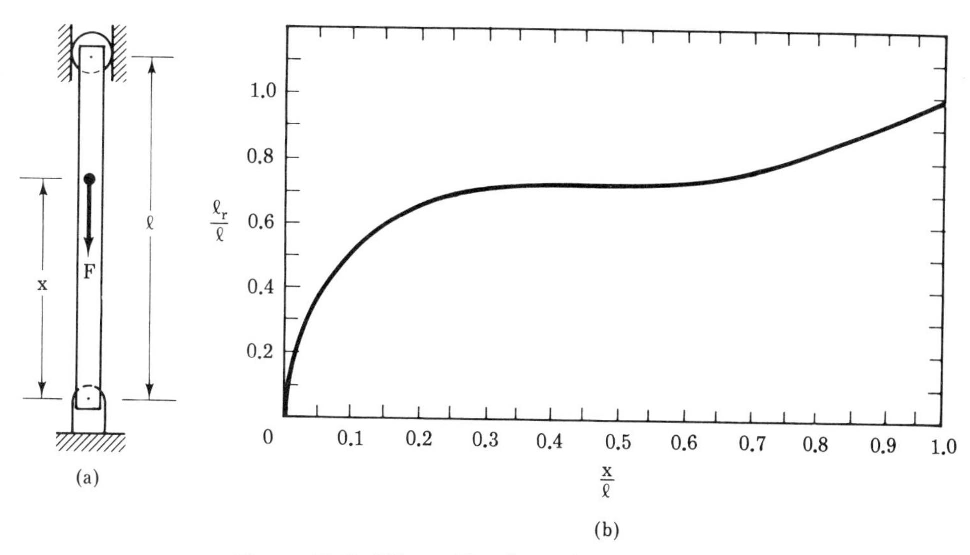

Figure 19-6. Effect of loading columns between supports.

Eccentrically Loaded Columns

Earlier we discussed the fact that one of many reasons why buckling occurs is an imperceptible eccentricity in loading. This has been one of the bases for the development of the basic equations presented above. However, when the eccentricity becomes noticeable, additional account of it must be taken. The loading may be of either of two types. It may consist of a single load that does not pass through the centroid of the section as shown in Figure 19-7a or it may consist of an eccentric load in addition to a concentric load as shown in 19-7b. The eccentric load in either case could be in the weak direction from the center as shown by the dashed-line load rather than in the strong direction as shown by the solid-line load.

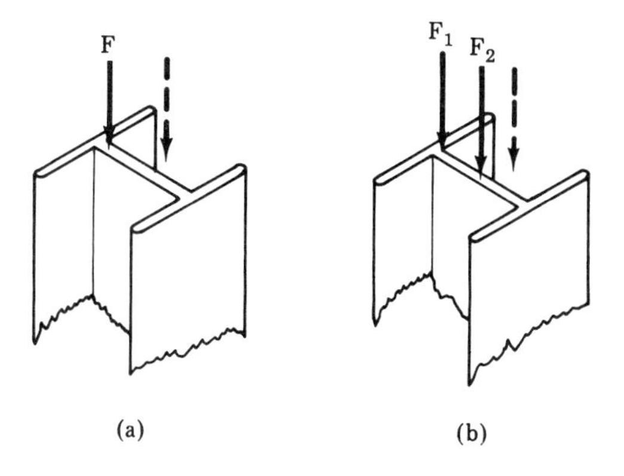

Figure 19-7. Eccentric loading of a column.

The eccentric load causes a compression load (which may in itself cause buckling) but far more severe is the moment effect of the eccentric load, $M = Pe$. This is countered by the section modulus S.

The equations for the combined stress for each case are then

$$\sigma_{\max} = \frac{P}{A} + \frac{Pe}{S} \quad \text{(eccentric only)}$$

$$\sigma_{\max} = \frac{P_0 + P}{A} + \frac{Pe}{S} \quad \text{(eccentric plus concentric)}$$

If the loading is eccentric along more than one axis, then two bending moments and two resultant stresses will occur. The above equations then become

$$\sigma_{\max} = \frac{P}{A} + \frac{Pe_x}{S_x} + \frac{Pe_y}{S_y}$$

$$\sigma_{\max} = \frac{P_0 + P}{A} + \frac{Pe_x}{S_x} + \frac{Pe_y}{S_y}$$

There are occasions where the allowable bending stress may be appreciably different than the allowable stress based on buckling. All of the above equations then take the form

$$\frac{P}{A\sigma_{\text{all-centr}}} + \frac{Pe}{S\sigma_{\text{all-bend}}} \leq 1$$

Care must be taken to assure that a corresponding e and S are always taken in the same direction. The allowable stress for buckling is, of course, always based on the weaker section properties for an unsymmetrical cross section.

19.4 COLUMN FORMULAS

A number of column formulas have been proposed and used by various organizations. A few of the more common ones will be presented and applied in this section. You will note almost immediately that these formulas do not always group columns into the three categories examined in the preceding section (short or post, intermediate, and long slender). However, as conditions describing any one of those three categories are met, the equations presented approach the appropriate theoretical one.

Timber Column Formula

The formula established by the American Institute of Timber Construction is simply Euler's formula with a factor of safety of 2.727.

$$\sigma_{\text{all}} = \frac{\sigma_{\text{crit}}}{FS} = \frac{\pi^2 E}{FS\left(\dfrac{l}{r}\right)^2}$$

For a column with rectangular cross section w times t where $w \geq t$:

$$r^2 = \frac{I}{A} = \frac{\dfrac{1}{12}wt^3}{wt} = \frac{t^2}{12} \qquad \left(\frac{l}{r}\right)^2 = \frac{12l^2}{t^2} = 12\left(\frac{l}{t}\right)^2$$

Substituting and simplifying

$$\sigma_{\text{all}} = \frac{0.302E}{\left(\dfrac{l}{t}\right)^2}$$

(for a round cross section column $t = 0.886d$). A number of considerations are involved in the use of the above formula.

1. $l/t < 50$.
2. σ_{all} cannot be greater than the allowable stress for compression parallel to the grain.
3. The length used must be the effective length determined by considering end effects and end or nonend loading, as well as bracing.
4. The formula is for centric loading. If the loading is eccentric, then the permissible load P *cannot* be determined from $P = \sigma_{\text{all}}A$ but must be determined from the appropriate eccentric loading formula in Section 19.3, "Eccentrically Loaded Columns."

Example 19-1: Find the safe load that a 120-mm-diameter Ponderosa pine column 3 m long is permitted to carry. The load is end applied and both end supports are fixed.

Solution:

Step 1 Assume that the load is centric.

Step 2 Determine effective length. From Figure 19-3, $K = 0.65$. $l_e = K = 0.65$ (3 m), $l_e = 1.85$ m.

Step 3 Determine t: $tl = 0.886d = 0.886\ (0.12\ \text{m})$, $t = 0.1063$ m.

Step 4 Check l/t:

$$\frac{l}{t} = \frac{1.85\ \text{m}}{0.1063\ \text{m}} \qquad \frac{l}{t} = 17.4 < 50 \qquad \text{OK}$$

Step 5 Calculate load from AITC formula.

$$\sigma_{\text{all}} = \frac{0.302E}{(l/t)^2} \qquad \frac{F}{A} = \frac{0.302E}{(l/t^2)}$$

$$F = \frac{0.302EA}{(l/t)^2} = \frac{0.302(7.6\ \text{GPa})(0.7854)(0.12\ \text{m}^2)}{(17.4)^2}$$

$$\mathbf{F = 85.74\ kN}$$

Step 6 Check against allowable stress for material $\sigma_{\text{all}} = 5.52$ MPa.

$$\sigma = \frac{F}{A} = \frac{0.08574\ \text{MN}}{0.7854(0.2\ \text{m}^2)} = 2.729\ \text{MPa} < 5.52\ \text{MPa}$$

Note that this suggests a limitation to the AITC formula. Had the AITC formula suggested a "safe" load with regard to buckling that resulted in too high a stress, then a smaller load based on $P = \sigma_{\text{all}}A$ would have to be used.

Problem 19-1: Determine the minimum dimensions (cross section) required for a 5-ft-long square cross section column carrying a 40-kip load. One end is fixed, the other end is free. A fixed brace supports it midway between the ends. The material is southern pine.

Aluminum Formulas

The Aluminum Association, Inc. has specified formulas for each of the many alloys available for both structural and machine design. Each alloy has three formulas specified for short (post), intermediate, and long columns. As can be seen from the formulas below, they consist of the application of a simple factor of safety for posts, a straight-line equation for intermediate columns, and a Euler-based equation for long columns. l/r parameters as well as the coefficients and, of course, the yield stress vary for each alloy. An FS of 1.95 is commonly used.

$$\text{short} \qquad \sigma_{\text{all}} = \frac{\sigma_{\text{yield}}}{FS}$$

$$\text{intermediate} \qquad \sigma_{\text{all}} = \frac{[B_c - D_c(l/r)]}{FS}$$

long $$\sigma_{all} = \frac{\pi^2 E}{FS(l/r)^2}$$

For 6061-T6 aluminum alloy, these equations become:

English Units

$l/r \leq 9.5$ $\quad \sigma_{all} = 19{,}000 \text{ psi}$

$9.5 < l/r < 66$ $\quad \sigma_{all} = 20{,}200 - 126(l/r) \text{ psi}$

$l/r \geq 66$ $\quad \sigma_{all} = \dfrac{51{,}000{,}000 \text{ psi}}{(l/r^2)}$

SI Units

$l/r \leq 9.5$ $\quad \sigma_{all} = 131 \text{ MPa}$

$9.5 < l/r < 66$ $\quad \sigma_{all} = 139 - 0.868(l/r) \text{ MPa}$

$l/r \geq 66$ $\quad \sigma_{all} = \dfrac{351{,}000{,}000 \text{ MPa}}{(l/r)^2}$

The formulas indicated above are for centric loading and l is the effective length as determined on the basis of end conditions, location of loading, and bracing.

Example 19-2: Determine the minimum size 6061-T6 aluminum I beam of those listed in the Appendices that can be used to support a compressive load of 100 kips if one end is fixed and the other end is pinned. The column is 10 ft long.

Solution:

Step 1 Assume the column is centrically loaded.

Step 2 Determine the effective length. From Figure 19-3, $K = 0.7$.

$$l_e = Kl = 0.8(120 \text{ in.}) \qquad l_e = 96 \text{ in.}$$

Step 3 Determine a minimum size based on the maximum allowable stress for the material.

$$\sigma_{all} = \frac{F}{A} \qquad A = \frac{F}{\sigma_{all}} = \frac{100{,}000 \text{ lb}}{19{,}000 \text{ lb/in.}^2}$$

$$A = 5.263 \text{ in.}^2$$

This suggests a minimum of an 8×5 I beam whose area is 5.972 in.^{2} and whose minimum radius of gyration is 1.2 in.

Step 4 Establish a table as shown below, starting with the column just selected and proceed by trial and error to determine l/r, the allowable stress based on the appropriate buckling formula (chosen by l/r), and the actual stress (F/A). The actual stress must be smaller than the allowable, but as close as possible thus permitting selection of the smallest (usually most economical) possible column.

Section	A	r	$\frac{l}{r}$	Allowable Stress	Actual Stress
8×5	5.972	1.2	80	7,969	16,745
9×5.5	7.11	1.31	73.3	9,492	14,065
10×6	7.352	1.42	67.6	11,160	13,602
10×6	8.747	1.44	66.7	11,463	11,432

Thus a 10×6 I beam with an area of 8.747 in.^{2} weighing 10.286 lb/ft is the most economical choice. The long-column formula was used for all of the allowable stress calculations since all $l/r > 66$.

Problem 19-2: Determine the maximum axial compressive load that can be safely carried by an 8-ft-long aluminum (6061-T6) standard I beam used as a vertical column with dimensions of 4 by 3 in. and a weight of 2.793 lb/ft. It is pinned at both ends and the load is applied downward at a point 2 ft down from the top end.

Problem 19-3: Determine the minimum size aluminum (6061-T6) channel that can be used as a column to support a compressive load of 40 kips if the column is 12 ft long and is fixed at both ends.

Machine Steel Formulas

The formulas for steel for machine design (as opposed to structural design) follow the theory closely, with a modified parabolic formula used for intermediate columns. The concept of a column as presented in this chapter is particularly significant in this section. This is because many machine members that must be considered as columns are not something we have previously thought of as a column. But recall that a column is simply a compression member which we must examine carefully to assure that buckling is avoided.

$$\text{short} \qquad \sigma_{\text{all}} = \frac{\sigma_{\text{yield}}}{FS} \qquad \frac{l}{r} < 40$$

$$\text{intermediate} \qquad \sigma_{\text{all}} = \frac{\sigma_{\text{yield}}}{FS}\left[1 - \frac{\sigma_{\text{yield}}(Kl/r)^2}{4\pi^2 E}\right]$$

$$40 \leq \frac{l}{r} < \pi\sqrt{\frac{2E}{\sigma_{\text{yield}} K^2}}$$

$$\text{long} \qquad \sigma_{\text{all}} = \frac{\pi^2 E}{FS(Kl/r)^2} \qquad \frac{l}{r} \geq \pi \sqrt{\frac{2E}{\sigma_{\text{yield}} K^2}}$$

Example 19-3: A 17-in.-long link in a mechanism is to have a rectangular cross section and be fabricated from AISI 1045 steel. Both ends are pinned and the loads are to be applied at the pins. Using a factor of safety of 3 based on yield strength what are the minimum dimensions required for a load of 10,000 lb? (Use a 2 : 1 ratio for the cross section.)

Solution:

Step 1 Assume that the load is centric.

Step 2 Check K. $K = 1$.

Step 3 Estimate minimum dimensions ($\mathrm{w} = 2t$)

$$\sigma_{\text{all}} = \frac{\sigma_{\text{yield}}}{FS} = \frac{F}{A} = \frac{F}{2t^2}$$

$$t = \left(\frac{(FS)(F)}{2\sigma_{\text{yield}}}\right)^{0.5} = \left(\frac{(3)(10{,}000\ \text{lb})}{2(60{,}000\ \text{psi})}\right)^{0.5}$$

$$t = 0.5\ \text{in.} \qquad \mathrm{w} = 1.0\ \text{in.}$$

Step 4 Estimate l/r.

$$K = 1 \qquad l_e = l \qquad r^2 = \frac{I}{A} = \frac{(1/12)t\mathrm{w}^3}{t\mathrm{w}}$$

$$r^2 = \frac{\mathrm{w}^2}{12} = \frac{1}{12} \qquad r = 0.289$$

$$l/r = \frac{17}{0.289} = 58.9$$

Step 5 Check value for maximum l/r for intermediate column.

$$\frac{l}{r} < \pi \left(\frac{2E}{\sigma_{\text{yield}} K^2}\right)^{0.5}$$

$$\frac{l}{r} < \pi \left(\frac{2 \times 30 \times 10^6\ \text{psi}}{60 \times 10^3\ \text{psi} \times 1}\right)^{0.5} \qquad \frac{l}{r} < 99.3$$

Step 6 Set up a table as shown below, starting with minimum dimensions just selected and proceed by trial and error to determine w, A, r, l/r, allowable stress, and actual stress until the actual is equal to or slightly less than the allowable.

t	w	A	r	l/r	Allowable Stress	Actual Stress
0.5	1.0	0.5	0.289	58.9	16,485	20,000
0.6	1.2	0.72	0.346	49.1	17,557	13,889
0.55	1.1	0.605	0.317	53.5	17,100	16,529
0.54	1.08	0.583	0.312	54.5	16,990	17,150

Probably close enough. Could be refined, but stick to 0.55×1.1.

Problem 19-4: A circular cross-section push rod is 0.7 m long is made of AISI 2340 steel. It is end loaded, fixed at one end, free at the other, and has a fixed brace one-third of its length from the free end. The axial load is 5 kN and the factor of safety is 5. Determine the minimum diameter needed to the nearest millimeter.

Problem 19-5: A 4-ft-long column 2 in. by 2 in. in cross section is made of AISI 1095 steel. It is fixed at both ends and the load is applied centrically in one direction but eccentrically by 0.5 in. in the other direction. Assuming a factor of safety of 2, what safe axial load can this machine column support?

Structural Steel Formulas

The American Institute for Steel Construction established two equations. One, of parabolic form, is used for short and intermediate columns, while another, a Euler type is used for longer columns. The first equation involves a factor C_c which is defined as the maximum slenderness ratio at which the parabolic equation may be used. The factor of safety must be calculated for the parabolic equation but can be taken as 1.92 for the Euler equation. The most common structural steels have yield strengths of either 36,000 or 50,000 psi and C_c calculates to be 126.1 and 107.0 respectively for them. As a further aid in speeding computations Figure 19-8 shows a graph of Kl/r versus FS for these two common steels.

short and intermediate

$$\sigma_{\text{all}} = \frac{\sigma_{\text{yield}}}{FS}\left[1 - \frac{(Kl/r)^2}{2C_c^2}\right]$$

$$FS = \frac{5}{3} + \frac{3(Kl/r)}{8C_c} - \frac{(Kl/r)^3}{8C_c^3}$$

$$C_c = \left(\frac{2\pi^2 E}{\sigma_{\text{yield}}}\right)^{0.5}$$

$$Kl/\text{r} \le C_c$$

$$\text{long} \qquad \sigma_{\text{all}} = \frac{\pi^2 E}{FS(Kl/r)^2}$$

$$C_c < Kl/r < 200$$

Example 19-4: What is the maximum dead load that a 250 by 250 mm, 72.4 kg/m wide, flange beam can support if it is used as an end-loaded, both ends fixed, column. The material is A36 structural steel and the load is 50 mm eccentric in the strong direction only. The column is 5 m long.

Solution:

Step 1 Check K. $K = 0.65$ from Figure 19-3.

Step 2 Find minimum r. From the Appendices, $r = 6.29$ cm.

Step 3 Compute Kl/r.

$$\frac{Kl}{r} = \frac{0.65(5\text{ m})}{0.0629\text{ m}} \qquad \frac{Kl}{r} = 51.67$$

Step 4 For A36 steel $C_c = 126.1$, and from Figure 19-8 $FS = 1.815$.

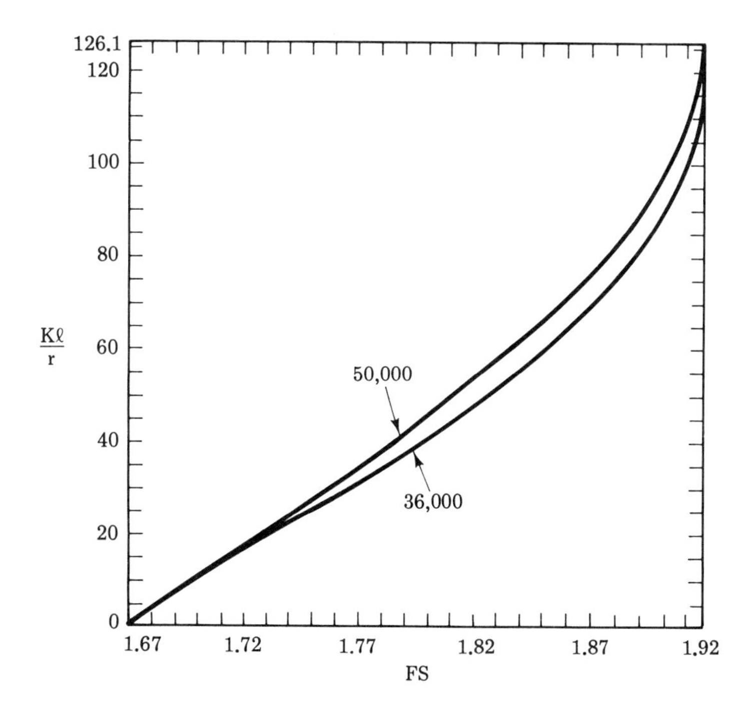

Figure 19-8. Factor of safety versus slenderness ratio.

Step 5 Since $Kl/r < C_c$, compute the allowable stress from the parabolic formula.

$$\sigma_{all} = \frac{\sigma_{yield}}{FS}\left[1 - \frac{(Kl/r)^2}{2C_c^2}\right]$$

$$\sigma_{all} = \frac{248\ \text{MPa}}{1.815}\left[1 - \frac{(51.67)^2}{2(126.1)^2}\right]$$

$$\sigma_{all} = 125.2\ \text{MPa}$$

Step 6 From Section 19.3, "Eccentrically Loaded Columns,"

$$\frac{F}{A\sigma_{all\text{-}centr}} + \frac{Fe}{S\sigma_{all\text{-}bend}} \leq 1$$

$$F_{max} = \leq \frac{1}{\dfrac{1}{A\sigma_{all\text{-}centr}} + \dfrac{e}{S\sigma_{all\text{-}bend}}}$$

The *FS* in bending for a dead load on structural steel is 2 based on yield strength. This makes the allowable stress in bending essentially equal to the allowable stress for buckling:

$$F_{max} \leq \frac{\sigma_{all}}{\dfrac{1}{A} + \dfrac{e}{S}}$$

From the Appendices, $A = 92.18\ \text{cm}^2$, $S = 867\ \text{cm}^2$:

$$F_{max} \leq \frac{125.2\ \text{MPa}}{\dfrac{1}{0.09218\ \text{m}^2} + \dfrac{0.05\ \text{m}}{0.000867\ \text{m}^3}} \qquad \mathbf{F_{max} = 753.5\ kN}$$

Problem 19-6: Select the most economical equal leg angle 16 ft long that will support a 50-kip axial compressive load centrically located. The material is A440 with a yield strength of 50,000 psi. The ends can be assumed to be pinned.

Problem 19-7: Determine the safe load in compression that a $C300 \times 90 \times 10 \times 15.5$ (SI) section will carry when used as a 3.2-m column. One end is fixed, the other pinned. The material is A36 structural steel. The load is eccentric, centered in the web thickness.

19.5 READINESS QUIZ

1. A post is defined as a machine or structural member that fails in
   (a) tension
   (b) torsion
   (c) compression
   (d) buckling
   (e) none of the above
2. A long slender column of given cross section, length, and material and centrically loaded can carry the greatest load if
   (a) both ends are fixed
   (b) one end is fixed and one end is free
   (c) both ends are pinned
   (d) one end is pinned and one end free
   (e) none of the above
3. The slenderness ratio for a column is defined as its
   (a) length divided by equivalent diameter
   (b) length divided by least radius of gyration
   (c) length divided by least cross-sectional dimension
   (d) all of the above
   (e) none of the above
4. The effective length of a column can be affected by
   (a) the type of end supports
   (b) bracing
   (c) axial location of load
   (d) any or all of the above
   (e) none of the above
5. Columns must be designed with consideration for buckling as well as compression because of the possibility of
   (a) improper design
   (b) inappropriate material selection
   (c) poor installation practices
   (d) all of the above
   (e) none of the above
6. The strength of a long slender column is based upon its
   (a) yield strength
   (b) elasticity
   (c) elasticity and yield strength

(d) any of the above
(e) none of the above

7. Straight-line and parabolic equations are both used for

(a) short columns
(b) intermediate columns
(c) long columns
(d) all of the above
(e) none of the above

8. Design of an eccentrically loaded column must take into account ________ as well as buckling and compression.

(a) bending
(b) torsion
(c) tension
(d) all of the above
(e) none of the above

9. An organization whose specifications and/or recommendations play an important part in column design is the

(a) ASTM
(b) AISC
(c) AITC
(d) all of the above
(e) none of the above

10. Buckling as well as compression must be considered in column design because

(a) of minute inconsistencies in the column material
(b) it is improbable that load and reaction will pass exactly through the centroid of the column section even when intended
(c) neither the cross section nor the straightness of the column will be perfectly consistent.
(d) all of the above
(e) none of the above

19.6 SUPPLEMENTARY PROBLEMS

19-8 A 60-mm steel ($\sigma_y = 630$ MPa, $G = 207$ GPa) shaft is mounted in fixed bearings at both ends of its 3-m length. What is the maximum axial compressive load it can safely carry?

19-9 Four 2×8 in. planks (actual dimensions 1.5 by 7.5 in.) are nailed together as shown in Figure 19-9 to form a column 12 ft

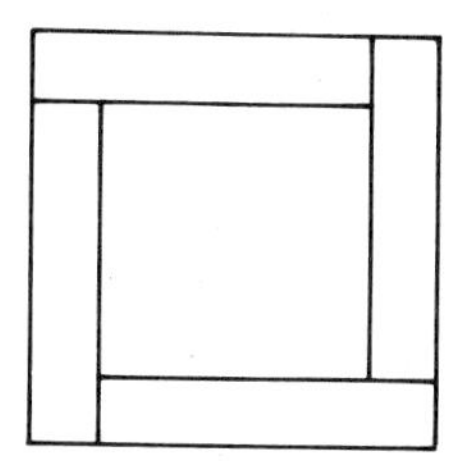

Figure 19-9

long. The wood is Douglas Fir and the ends can be assumed to be pinned. What is the maximum axial compressive load the column can safely carry?

19-10 A wide flange A36 steel column 30 ft long is to support a concentric load of 200 kips. Determine the most economical section available in the Appendices. (One end is pinned, the other end fixed.)

19-11 Determine the nominal size SCH40 steel pipe that should be used for a 14-ft column, fixed at both ends with a fixed brace at the midway point in its length if it is to sustain a centric axial compressive load of 60 kips. ($\sigma_y = 30{,}000$ psi; $E = 30 \times 10^6$ psi.)

19-12 A column is built up of 6061-T6 aluminum components as shown in Figure 19-10. The column is 16 ft long and is pinned at both ends. What is the maximum axial compressive load it can carry if the load is located as shown?

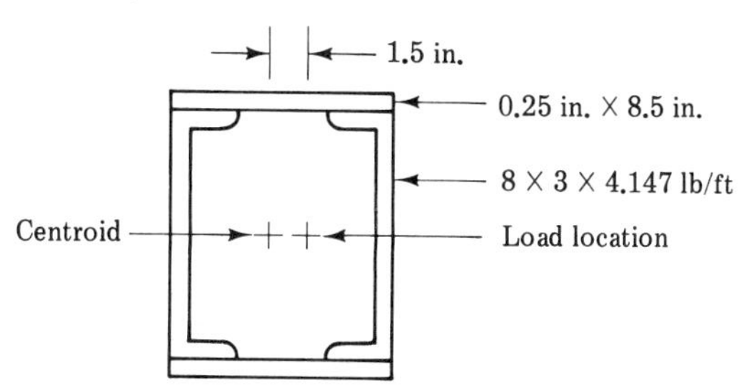

Figure 19-10

19-13 A link in a mechanism is 5 cm long and has an elliptical cross section of 8 × 4 mm and is subjected to an axial centric compressive load of 0.2 kN. If cold-worked AISI 301 stainless steel is used, will the link be safe from buckling? If an aluminum alloy 2014-T4-R were to be used instead, what minimum dimensions would be required for the cross section to be safe from buckling?

19-14 A cable supported boom is loaded as shown in Figure 19-11.

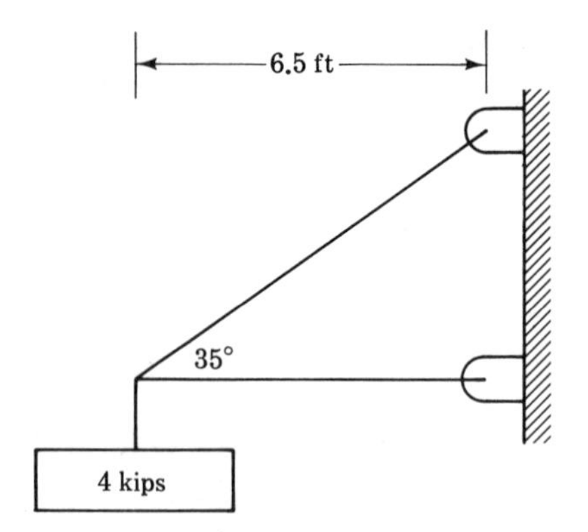

Figure 19-11

The boom ends are pinned. What is the most economical A36 standard I beam that can be used for the boom?

19-15 A vertical aluminum column with the cross section shown in Figure 19-12 has a length of 4 m and is fixed at both ends. If the material is 6061-T6 aluminum, what is the maximum downward centric load that can be applied one-fourth of the length down from the top?

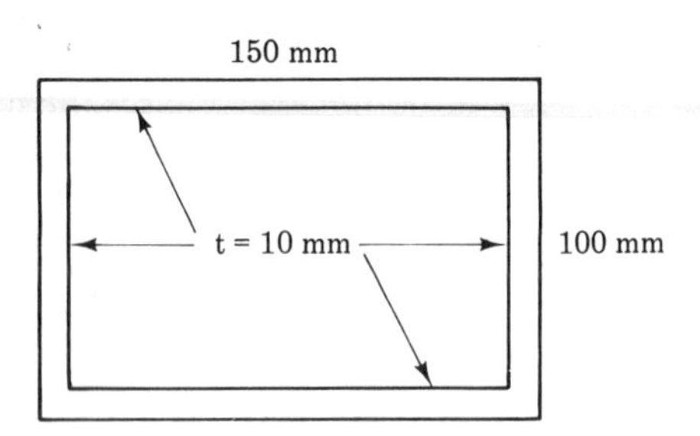

Figure 19-12

Chapter 20

Statically Indeterminate Beams

20.1 OBJECTIVES

Upon completion of the work relating to this chapter, you should be able to perform the following.

1. Recognize that a beam is statically indeterminate by examining its space diagram.
2. Recognize that a beam is statically indeterminate by examining its free-body diagram.
3. Determine the reactions of a statically indeterminate beam by the superposition method.
4. Determine the reactions of a statically indeterminate beam by the moment-area method.
5. Determine the reactions of continuous beams by application of the three-moment theorem.

20.2 DEFINING THE STATICALLY INDETERMINATE BEAM

In Chapters 10 and 17 we dealt extensively with various aspects of statically determinate beams. Shear diagrams, moment diagrams, and deflection curves were determined for a variety of loadings. Maximum stresses were determined for beams of known configuration. The most economical beam configuration for a beam of known loading and material was determined. In every case, the starting point was the determination of the reactions by use of the basic equations of statics: $\Sigma F = 0$, and $\Sigma M = 0$. This permitted at most two unknown reactions to be solved for (three if $\Sigma F_x = 0$, $\Sigma F_y = 0$, is used instead of $\Sigma F = 0$). Figure 20-1 shows the space and free-body diagrams for several typical statically determinate beams.

In all of the above cases, including any combinations of loads you might

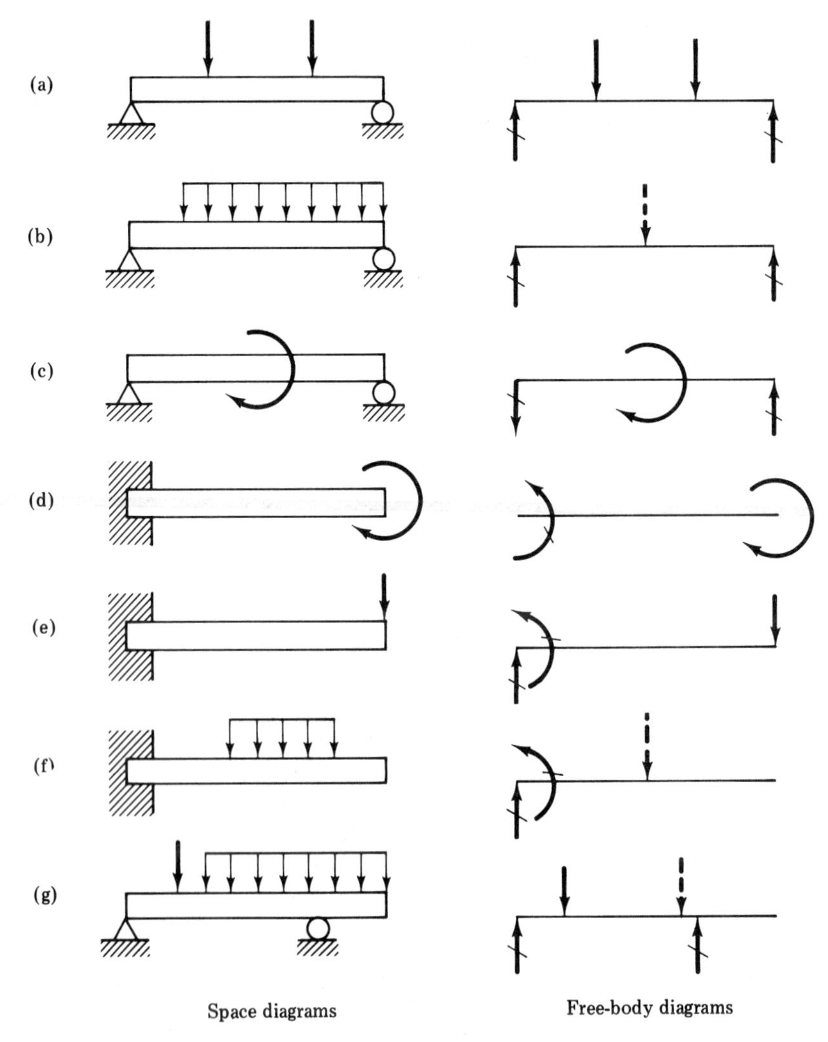

(Other combinations of loads on any of the three basic support systems — simple, cantilever, or overhanging — are possible.)

Figure 20-1. Statically determinate beams.

wish to devise, there were never more than two unknown reactions—exactly the number of equations available for solution ($\Sigma F_y = 0$, $\Sigma M = 0$). The third possible equation of statics, $\Sigma F_x = 0$, was not introduced because there were no horizontal loads or reactions shown. Introduction of such forces would result in axial stresses in addition to the bending stresses. Problems of that nature were dealt with in Chapter 18.

It is the supports then, not the loading, that determines whether a problem is statically determinate or not. Let us examine some common types of supports that result in statically indeterminate problems. In Figure 20-2 the space diagram and free-body diagram for several possibilities are shown.

In all of the above, three or four unknown reactions exist. Each has one or more reactions which can be removed and still leave the beam and its

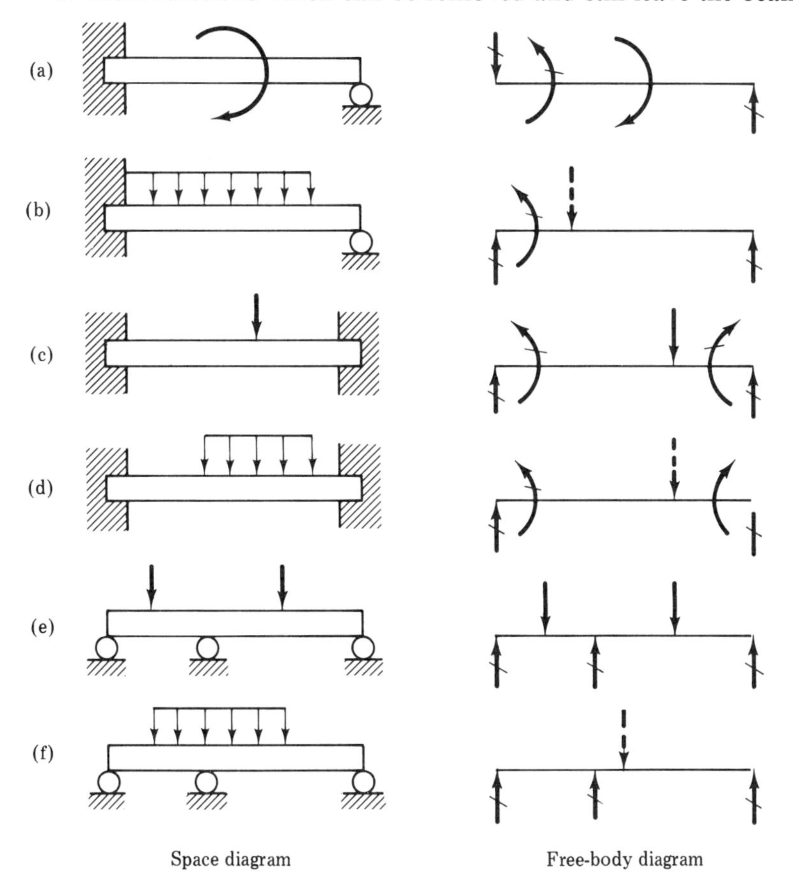

(Other combinations of loads on any of the three basic support systems — propped, restrained, or continuous — are possible.)

Figure 20-2. Statically indeterminate beams.

loads supported. These are often called redundant reactions. The measure of redundancy is the number of reactions that must be removed to make the beam statically determinate. If it is statically determinate after removing one reaction, it is said to be statically indeterminate in the first degree; with two removed, second degree; and so forth. Our approach to the solution of these problems will be based on the fact that we will always be able to formulate as many equations for elasticity as there are degrees of indeterminancy.

Two basic methods are used to solve these problems. Both were used before to find deflections in statically determinate beams. Now we will use them to determine reactions as well as deflections. As before, the superposition method will be easier to use, the moment-area more powerful and comprehensive. There will be one type of statically indeterminate beam that will utilize a technique that combines the two methods.

The remainder of this chapter will examine the three types of statically indeterminate beams described above. In each case more than one method will be used to determine the reactions. Once those are found, the complete shear and moment diagrams can be constructed and the design work completed. However, this aspect of the problem is no different than the work you accomplished earlier in Chapter 17. The examples and problems in this chapter will therefore carry the solutions only to the point of determining reactions. It should be noted that the moment area is somewhat tedious when applied to finding deflections in statically indeterminate beams.

20.3 PROPPED BEAMS

More properly called the propped cantilever beam, this beam finds numerous structural and machine applications where rigidity of the fixed support is coupled with a simple support at the other end. As was seen in Figure 20-2, this results in three unknown reactions. It thus has a single redundant reaction and is known to be statically indeterminate in the first degree.

Propped Beams by Superposition

Example 20-1: Determine the support reactions for the cantilever beam shown in Figure 20-3a using the method of superposition.

Solution:

Step 1 Draw the free-body diagram as shown in 20-2b.

Step 2 Recognize this as a propped beam, statically indeterminate in the first degree.

Step 3 Recognize that the deflection at A is zero. However, that

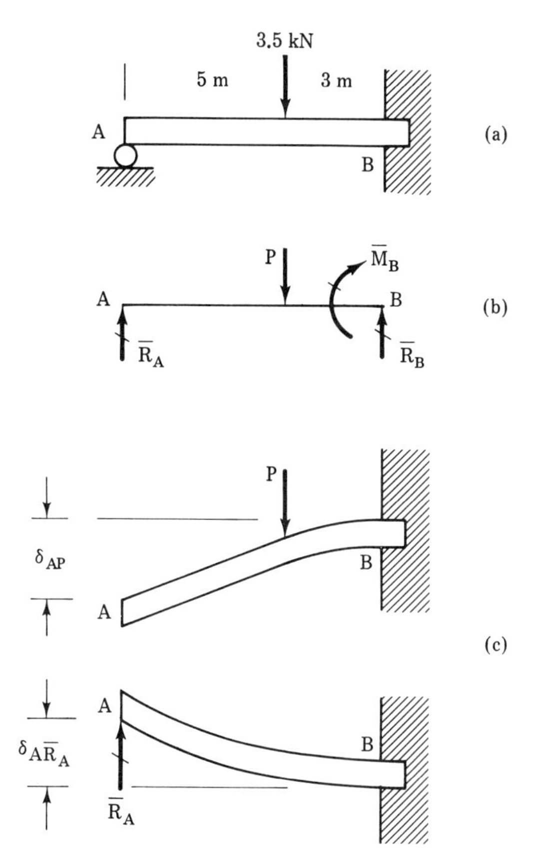

Figure 20-3

deflection can be considered as the sum of the two deflections that would be caused individually by P and $\overline{R}_A$ as shown in 20-2c. Thus

$$Y_{AP} + Y_{A\overline{R}_A} = 0$$

Step 4 Check the Appendices for the appropriate deflection equations for the two conditions. Since this is a relatively simple loading and reaction,

$$Y_{AP} = \frac{-Pb^2}{6EI}(3l - b) \qquad \begin{array}{l}(l \text{ is total length})\\(b \text{ is distance from } P \text{ to } \overline{R}_B)\end{array}$$

$$Y_{A\overline{R}_A} = \frac{+\overline{R}_A l^3}{3EI}$$

Step 5 Substitute in the equation from Step 3 and solve for $\overline{R}_A$.

$$\frac{-Pb^2}{6EI}(3l-b)+\frac{\overline{R}_A l^3}{3EI}=0$$

$$\overline{R}_A=\frac{Pb^2}{2l^3}(3l-b)$$

$$\overline{R}_A=\frac{3500\ \text{N}(3\ \text{m}^2)}{2(8\ \text{m}^3)}[3(8\ \text{m})-3\ \text{m}]$$

Step 6 Sum the vertical forces to find $\overline{R}_B$.

$$\Sigma F_y=0 \qquad \overline{R}_A+P+\overline{R}_B=0 \qquad \overline{R}_B=-(P+\overline{R}_A)$$

$$\overline{R}_B=-(-3500+646\ \text{N}) \qquad \mathbf{\overline{R}_B=-2854\ N}$$

Step 7 Sum the moments about B to find $\overline{M}_B$.

$$\Sigma M_B=0 \qquad \overline{R}_A d_{\overline{R}A}+Pdp+\overline{M}_B=0$$

$$\overline{M}_B=-(\overline{R}_A d_{RA}+Pdp) \qquad \overline{M}_B=-[-(646\ \text{N})(8\ \text{m})+(3500\ \text{N})(3\ \text{m})]$$

$$\mathbf{\overline{M}_B=-5332\ N\ m}$$

Problem 20-1: Determine the support reactions for the beam shown in Figure 20-4 using the method of superposition.

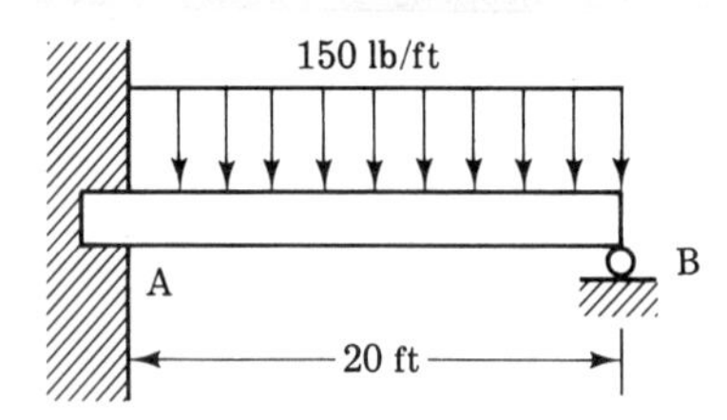

Figure 20-4

Propped Beams by Moment Area

Example 20-2: Determine the support reactions for the beam shown in Figure 20-5a using the moment-area method.

Solution:

Step 1 Draw the free-body diagram as shown in 20-5b.

Step 2 Recognize this as a propped beam, statically indeterminate in the first degree.

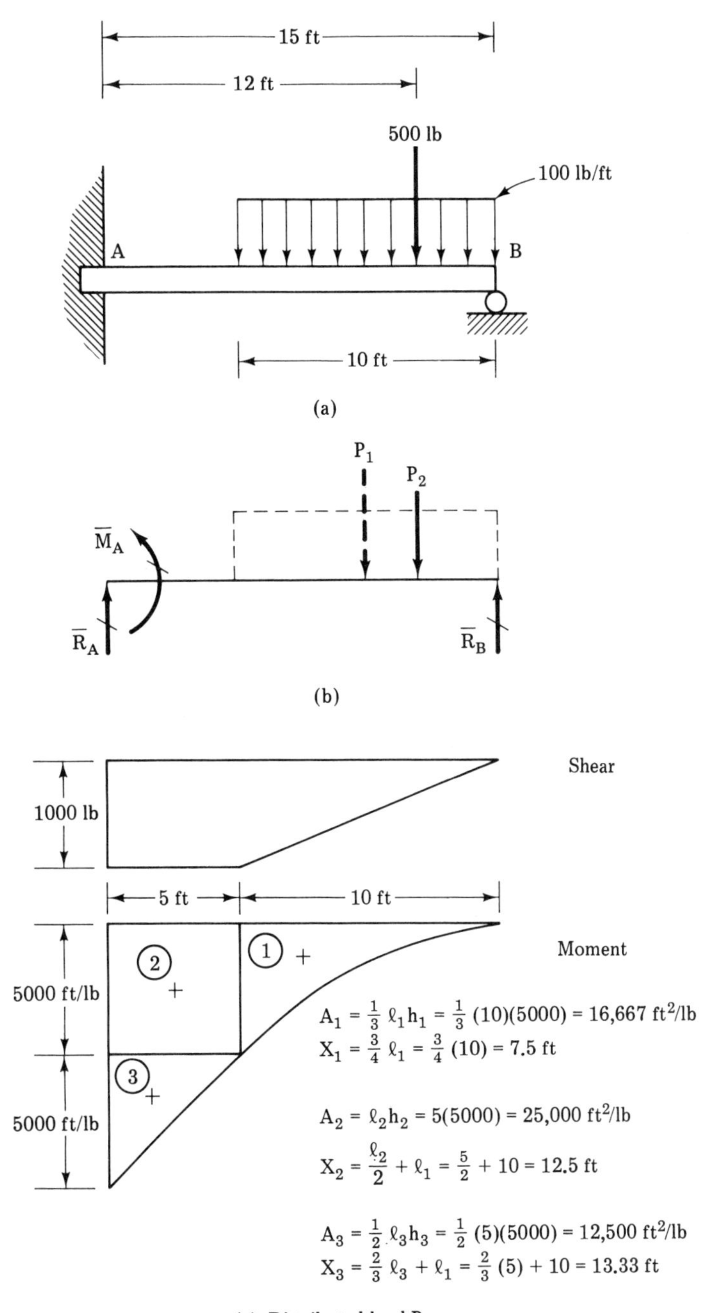

Figure 20-5

(Continued on next page.)

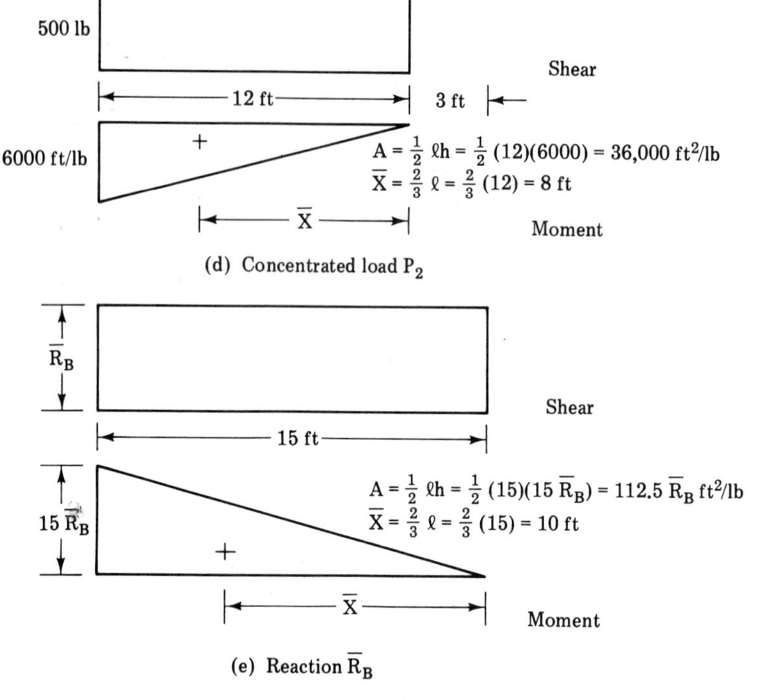

Figure 20-5 *(Continued)*

Step 3 Since A and B are on the same horizontal line, the deflection of B with respect to A is zero.

$$Y_{B/A} = 0$$

Step 4 Recall that moment-area theorem 2 in Chapter 17 indicates that

$$Y_{B/a} = \frac{1}{EI} A_{AB} \bar{x}_B$$

where A_{AB} is the area of the moment diagram between A and B.

Step 5 From Steps 3 and 4

$$A_{AB} \overline{X}_B = 0$$

This moment is really made up of the sum of three moments, for each load and $\overline{R}_B$. Sketch the areas and their moment arms as shown in Figure 20-5c, d, and e. (The shear diagrams are also shown as an aid in determining the moment diagram.)

Step 6 Substituting the information from 20-5c into the equation of Step 5 as follows, determine $\overline{R}_B$:

$$A_{P1}\overline{X}_{P1} + A_{P2}\overline{X}_{P2} + A_{\overline{R}B}\overline{X}_{\overline{R}B} = 0$$
$$+ [(16{,}667)(7.5) + (25{,}000)(12.5) + (12{,}500)(13.3)]$$
$$+ [(36{,}000)(8)] = -A_{\overline{R}B}\overline{X}_{\overline{R}B}$$

$$A_{\overline{R}B}\overline{X}_{\overline{R}B} = 892 \quad 165 \text{ ft}^3 \text{ lb}$$

$$A_{\overline{R}B} = \frac{A_{\overline{R}B}\overline{X}_{\overline{R}B}}{\overline{X}_{\overline{R}B}} = \frac{892{,}165 \text{ ft}^3 \text{ lb}}{10 \text{ ft}} = 89{,}216.5 \text{ ft}^2 \text{ lb}$$

$$A_{\overline{R}B} = 112.5\overline{R}_B \qquad \overline{R}_B = \frac{A_{\overline{R}B}}{112.5} = \frac{89{,}216.5 \text{ ft}^2 \text{ lb}}{112.5 \text{ ft}^2}$$

$$\mathbf{\overline{R}_B = 793 \text{ lb}}$$

Step 7 Determine $\overline{R}_A$ by summing the vertical forces.

$$\Sigma F_y = 0 \qquad \overline{R}_A + P_1 + P_2 + \overline{R}_B = 0$$

$$\overline{R}_A = (P_1 + P_2 + \overline{R}_B) \qquad \overline{R}_A = -(-1000 - 500 + 793 \text{ lb})$$

$$\mathbf{\overline{R}_A = +707 \text{ lb}}$$

Step 8 Determine $\overline{M}_A$ by summing the moments about A.

$$\Sigma M_A = 0 \qquad \overline{M}_A + P_1 d_1 + P_2 d_2 + \overline{R}_B d_{\overline{R}B} = 0$$

$$\overline{M}_A = -[P_1 d_1 + P_2 d_2 + \overline{R}_B d_{\overline{R}B}]$$

$$\overline{M}_A = -[-(1000 \text{ lb})(10 \text{ ft}) - (500 \text{ lb})(12 \text{ ft}) + (707 \text{ lb})(15 \text{ ft})]$$

$$\mathbf{\overline{M}_A = +5395 \text{ ft lb}}\circlearrowleft$$

Problem 20-2: Determine the support reactions for the beam shown in Figure 20-6 using the moment-area method.

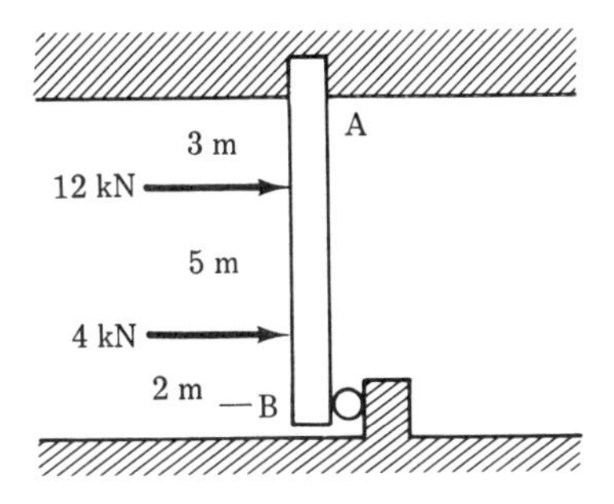

Figure 20-6

20.4 RESTRAINED BEAMS

Restrained beams have both ends fixed as was shown in Figure 20-2. Note also from the figure that this results in four unknown reactions. It thus has two redundant reactions and is said to be statically indeterminate in the second degree.

Restrained Beams by Superposition

Example 20-3: Determine the support reactions for the beam shown in Figure 20-7a using the method of superposition.

Solution:

Step 1 Draw the free-body diagram as shown in 20-7b.

Step 2 Recognize this as a restrained beam, statically indeterminate in the second degree.

Step 3 Since A and B are on the same horizontal line, the deflection of B with respect to A is zero.

Step 4 Since A and B are both fixed, the slope of the beam at both ends is zero.

Step 5 Steps 3 and 4 suggest the potential for developing two equations, one for slope and one for deflection. It now remains to remove two redundant reactions so that we can work with two equations for two unknowns. In Figure 20-7c we draw a free-body diagram with end A restrained and the support at B removed and replaced by the reactions. We thus have three different causes of slope and deflection at B. By the method of superposition, we will find the equations for each individual deflection. Summing them will give us two equations for the two unknowns.

Step 6 Individually sketch the beam restrained at A, showing in 20-7d the effect of each cause of deflection. Check the Appendices for the appropriate slope and deflection equations. Note that the beam is deflected by the load P only up to the location of P. From there on out to B the beam is straight and its *additional* displacement at B is equal to $b \sin \theta p$. A close approximation is the length of the arc formed which is equal to θpb. For the small angles we are dealing with the difference is negligible.

Step 7 Write the equations for the total deflection and slope at B.

$$Y_B = \frac{Pa^2}{6EI}(2a + 3b) + \frac{\overline{R}_B l^3}{3EI} - \frac{\overline{M}_B l^2}{2EI} = 0$$

$$\theta_B = -\frac{Pa^2}{2EI} + \frac{\overline{R}_B l^2}{2EI} - \frac{\overline{M}_B l}{EI}$$

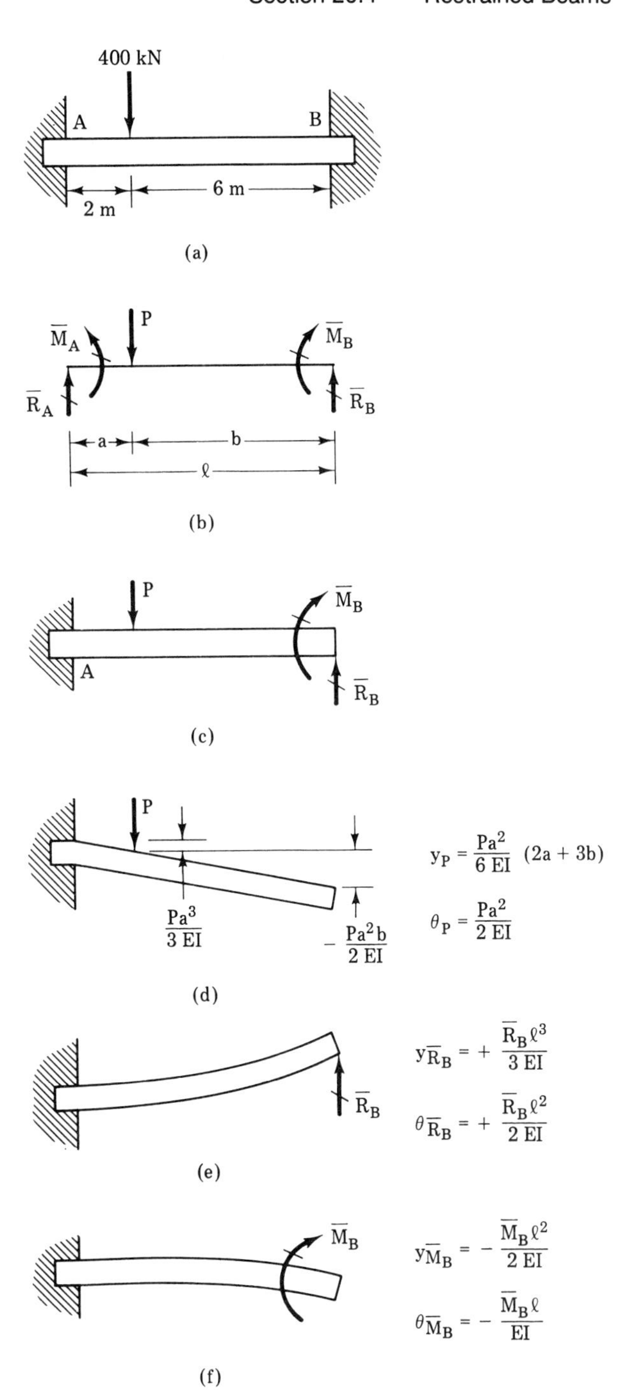
400 kN
A
B
2 m
6 m
(a)
$\overline{M}_A$
P
$\overline{M}_B$
$\overline{R}_A$
$\overline{R}_B$
a
b
ℓ
(b)
P
$\overline{M}_B$
A
$\overline{R}_B$
(c)
P
$\frac{Pa^3}{3\,EI}$
$-\ \frac{Pa^2 b}{2\,EI}$
$y_P = \frac{Pa^2}{6\,EI}\ (2a + 3b)$
$\theta_P = \frac{Pa^2}{2\,EI}$
(d)
$\overline{R}_B$
$y_{\overline{R}_B} = +\ \frac{\overline{R}_B \ell^3}{3\,EI}$
$\theta_{\overline{R}_B} = +\ \frac{\overline{R}_B \ell^2}{2\,EI}$
(e)
$\overline{M}_B$
$y_{\overline{M}_B} = -\ \frac{\overline{M}_B \ell^2}{2\,EI}$
$\theta_{\overline{M}_B} = -\ \frac{\overline{M}_B \ell}{EI}$
(f)

Figure 20-7

Step 8 Solve the equations of Step 7 for $\overline{R}_B$ and $\overline{M}_B$. Note that the equations include directions which match those of our free-body diagram.

$$\overline{R}_B = +\frac{Pa^2}{l^3}(a + 3b) \qquad \overline{M}_B = -\frac{Pa^2b}{l^2}$$

Step 9 Substitute knowns into the above equation and solve.

$$\overline{R}_B = +\frac{400 \text{ kN}(2 \text{ cm}^2)}{(8 \text{ m}^3)}[2 \text{ m} + 3(6 \text{ m})]$$

$$\boldsymbol{\overline{R}_B = +62.5 \text{ kN}}$$

$$\overline{M}_B = -\frac{400 \text{ kN}(2 \text{ cm}^2)(6 \text{ m})}{(8 \text{ m}^2)} \qquad \boldsymbol{\overline{M}_B = -150 \text{ kN m}} \curvearrowright$$

Step 10 Determine $\overline{R}_A$ by summing vertical forces.

$$\Sigma F_y = 0 \qquad \overline{R}_A + P + \overline{R}_B = 0 \qquad \overline{R}_A = -(P + \overline{R}_B)$$

$$\overline{R}_A = -(-400 + 62.5 \text{ kN}) \qquad \boldsymbol{\overline{R}_A = +337.5 \text{ kN}}$$

Step 11 Determine $\overline{M}_A$ by summing moments about A.

$$\Sigma M_A = 0 \qquad \overline{M}_A + Pd_p + \overline{R}_B d_{\overline{R}B} + \overline{M}_B = 0$$

$$\overline{M}_A = -(Pd_p + \overline{R}_B d_{\overline{R}B} + \overline{M}_B)$$

$$\overline{M}_A = -[-(400 \text{ kN})(2 \text{ m}) + (62.5 \text{ kN})(8 \text{ m}) - 150 \text{ kN m}]$$

$$\boldsymbol{\overline{M}_A = +450 \text{ kN m}} \curvearrowleft$$

Problem 20-3: Determine the support reactions for the beam shown in Figure 20-8 using the method of superposition.

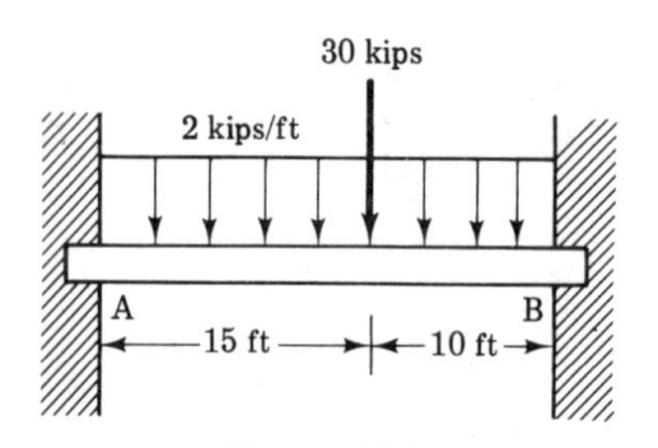

Figure 20-8

Restrained Beams by Moment Area

Example 20-4: Determine the support reactions for the beam shown in Figure 20-9a using the moment-area method.

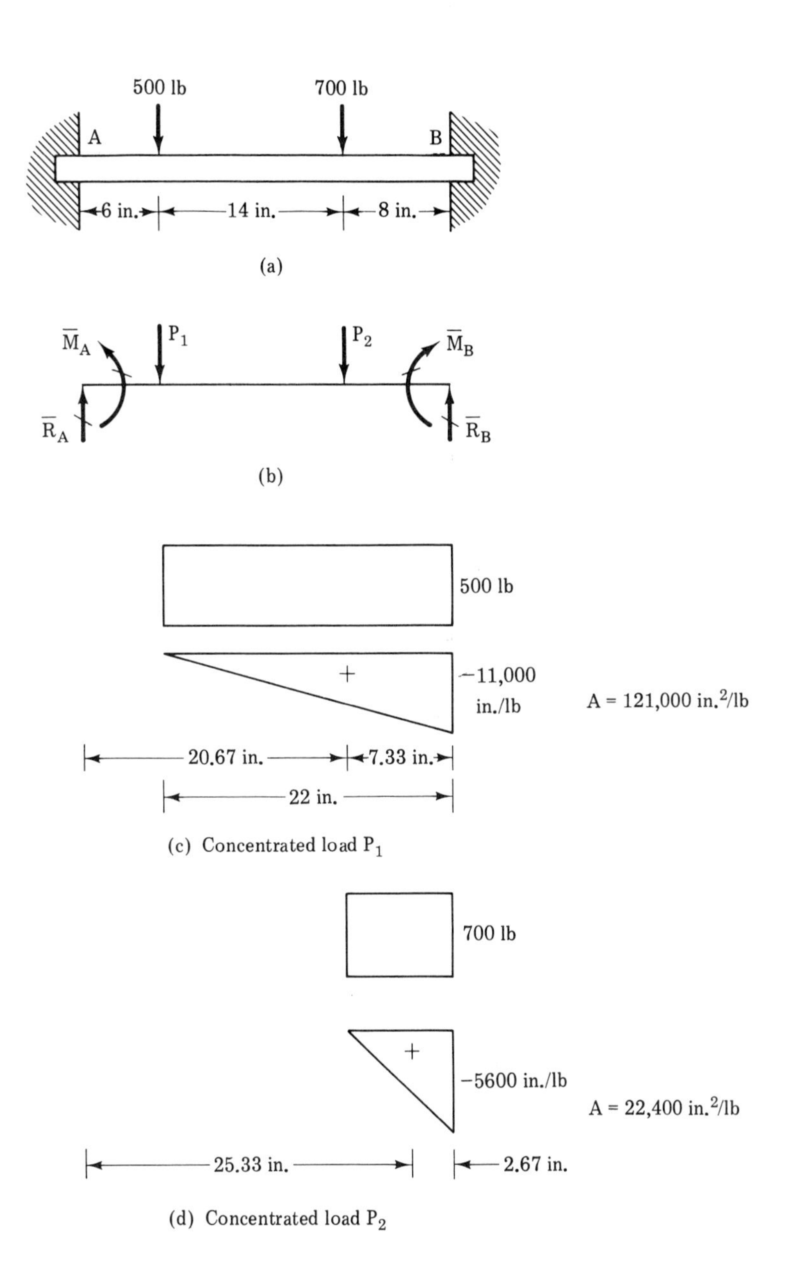

Figure 20-9

(Continued on next page.)

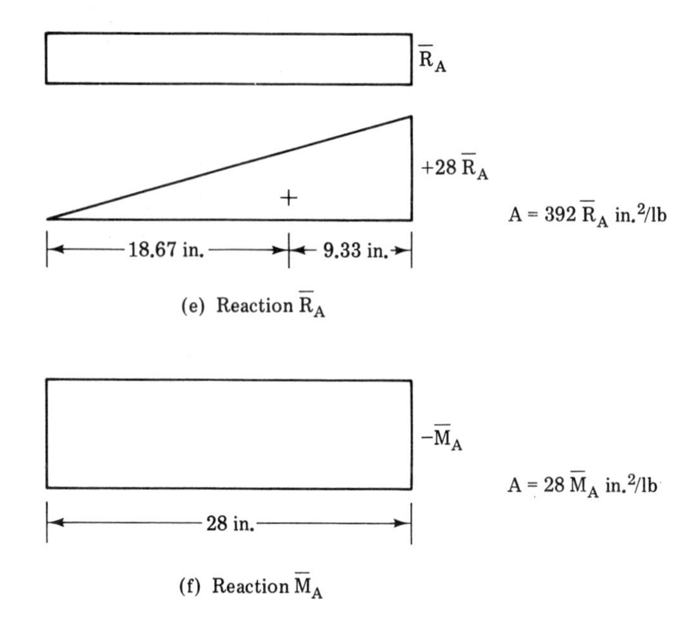

(e) Reaction $\overline{R}_A$

(f) Reaction $\overline{M}_A$

Figure 20-9 *(Continued)*

Solution:

Step 1 Draw the free-body diagram as shown in 20-9b.

Step 2 Recognize this as a restrained beam, statically indeterminate in the second degree.

Step 3 Since A and B are on the same horizontal line the deflection of one with respect to the other is zero.

$$Y_{A/B} = \frac{1}{EI} A_{AB} \overline{X}_A = 0$$

$$Y_{B/A} = \frac{1}{EI} A_{AB} \overline{X}_B = 0$$

Step 4 As before when we used this method, it will be necessary to evaluate the moment area for each load and determine the location of the centroid for each area. It does not matter which end moments are taken about for the purpose of determining the moment areas. In this case the right end will be used. The shear and moment diagrams for P_1, P_2, $\overline{R}_A$, and $\overline{M}_A$ are shown in Figure 20-9c, d, e, and f. Essentially, we are considering it as a cantilever beam with $\overline{R}_A$ and $\overline{M}_A$ as unknown loads.

Step 5 From Step 3.

$$Y_{A/B} = \frac{1}{EI} A_{AB}\overline{X}_A = 0 \qquad A_{AB}\overline{X}_A = 0 \qquad \Sigma(A_{AB}\overline{X}_A) = 0$$

(Recall that when we are speaking of A deflecting with respect to B, we take moments of the moment areas *about A, the end deflected.*)

$$-(121{,}000)(20.67) - (22{,}400)(25.33) + (392\overline{R}_A)(18.67) - (28\overline{M}_A)(14)$$

$$18.67\overline{R}_A - \overline{M}_A - 7827.7 = 0$$

Step 6 Also from Step 3.

$$Y_{B/A} = \frac{1}{EI} A_{AB}\overline{X}_B = 0 \qquad A_{AB}\overline{X}_B = 0 \qquad \Sigma(A_{AB}\overline{X}_B) = 0$$

$$+(121{,}000)(7.33) + (22{,}400)(2.67) - (392\overline{R}_A)(9.33) + (28\overline{M}_A)(14) = 0$$

$$-9.33\overline{R}_A + \overline{M}_A + 2415.1 = 0$$

Step 7 Add the equations from Steps 5 and 6 and solve for $\overline{R}_A$. Then substitute and solve for $\overline{M}_A$.

$$\begin{array}{r} 18.67\overline{R}_A - \overline{M}_A - 7827.7 = 0 \\ -9.33\overline{R}_A + \overline{M}_A + 2415.1 = 0 \\ \hline \end{array}$$

$$9.34\overline{R}_A - 5412.6 = 0 \qquad \mathbf{\overline{R}_A = +579.5\ lb}$$

$$18.67(579.5) - \overline{M}_A - 7827.7 = 0$$

$$\mathbf{\overline{M}_A = +2992\ in.\ lb}$$

Step 8 Determine $\overline{R}_B$ and $\overline{M}_B$ from statics equations.

$$\Sigma F_y = 0 \qquad \overline{R}_A + P_1 + P_2 + \overline{R}_B = 0 \qquad \overline{R}_B = -(\overline{R}_A + P_1 + P_2)$$

$$\overline{R}_B = -(+579.5 - 500 - 700) \qquad \mathbf{\overline{R}_B = 620.5\ lb}$$

$$\Sigma M_A = 0 \qquad \overline{M}_A + P_1 d_1 + P_2 d_2 + \overline{R}_B d_{RB} + \overline{M}_B = 0$$

$$\overline{M}_B = -(\overline{M}_A + P_1 d_1 + P_2 d_2 + \overline{R}_B d_{RB})$$

$$\overline{M}_B = -[+2992 - (500)(6) - (700)(20) + (620.5)(28)]$$

$$\mathbf{\overline{M}_B = -3366\ in.\ lb}$$

Problem 20-4: Determine the support reactions for the beam shown in Figure 20-10 using the moment-area method.

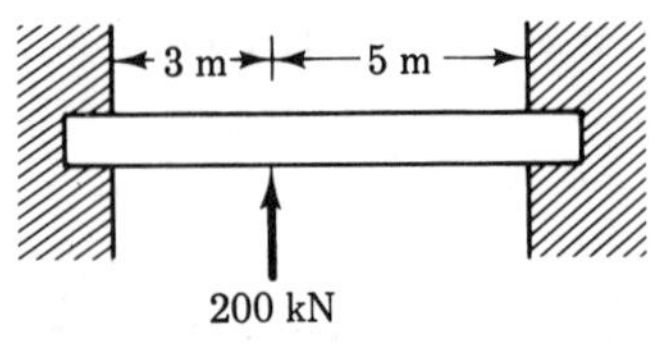

Figure 20-10

20.5 CONTINUOUS BEAMS

Continuous beams are those which have more supports than necessary to maintain equilibrium. An example of such a beam was shown in Figure 20-2. Both of the methods already used for other types of beams can be used for these as well. The examples and problems given all deal with three supports, that is, one redundant support. However, both methods can be expanded to applications with two or more redundant supports. The solutions will simply require as many equations based upon elasticity as there are redundant supports (in addition to the equations for static equilibrium).

Continuous Beams by Superposition

Example 20-5: Determine the support reactions for the beam shown in Figure 20-11a using the method of superposition.

Solution:

Step 1 Draw the free-body diagram for the beam as shown in 20-11b.

Step 2 Recognize this as a continuous beam, statically indeterminate in the first degree.

Step 3 If $\overline{R}_B$ is considered the redundant reaction and removed then the beam will deflect downward at point *B* as a result of the two loads. The sum of the deflections due to each of the loads is equal to the total deflection at *B*. However, *B* is really maintained at a horizontal level with *A* and *C* by the support at that point. Hence, the support at *B* must be exerting a force resulting in a deflection equal and opposite to the sum of the deflections resulting from the loads. This has been shown in Figures 20-11c through f.

Step 4 From the Appendices.

$$\delta_{1B} = \frac{P_1 b}{6lEI}\,[l/b(x-a)^3 + (l^2 - b^2)x - x^3]$$

$$\delta_{2B} = \frac{P_2 bx}{6lEI}\,(l^2 - x^2 - b^2)$$

$$\delta_{\overline{R}B} = \frac{\overline{R}_B l^3}{48EI}$$

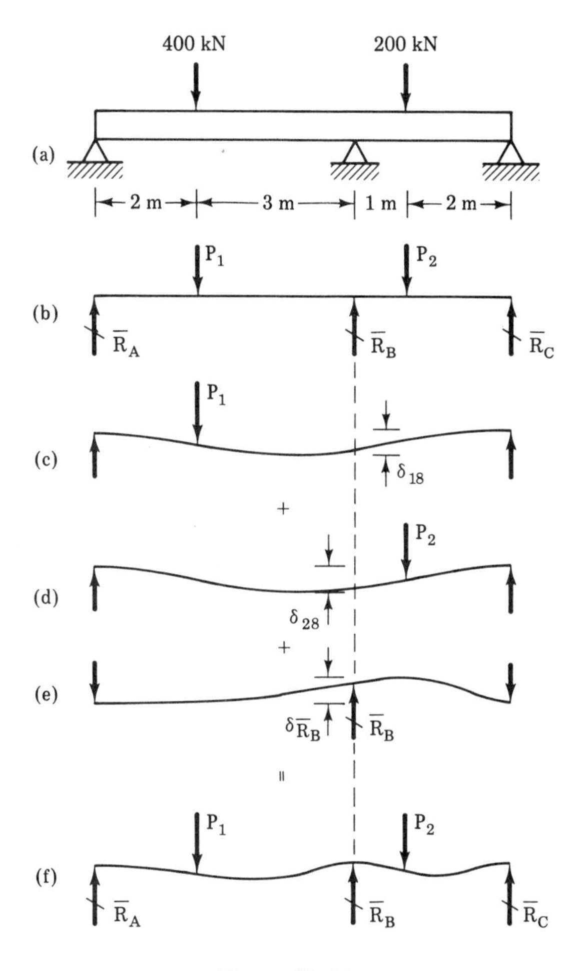

Figure 20-11

where a is the distance between A and the load, b is the distance between the load and c, l is the total length, and x is the distance from A to the point of interest, in our case B. Then

$$\delta_{1B} + \delta_{2B} + \delta_{\bar{R}B} = 0$$

From above, and multiplying through by $48EIl$

$$8P_1b[l/b(x-a)^3 + (l^2 - b^2)x - x^3] + 8P_2bx(l^2 - x^2 - b^2) - \bar{R}_bl^4 = 0$$

$$8(400)(6)[\tfrac{8}{6}(5-2)^3 + (8^2 - 6^2)(5) - (5)^3] + 8(200)(2)(5)(8^2 - 5^2 - 2^2) - \bar{R}_B(8)^4 = 0$$

$$\boldsymbol{\bar{R}_B = 375.8\ \text{kN}}$$

Step 5 Determine $\overline{R}_c$ by taking moments about A.

$$\Sigma M_A = 0 \qquad P_1 d_1 + \overline{R}_B d_{\overline{R}B} + P_2 d_2 + \overline{R}_C d_{RC} = 0$$

$$\overline{R}_C d_{\overline{R}C} = -[P_1 d_1 + \overline{R}_B d_{\overline{R}B} + P_2 d_2]$$

$$\overline{R}_C d_{\overline{R}C} = -[-(400)(2) + (375.8)(5) - (200)(6)]$$

$$\overline{R}_C d_{\overline{R}C} = -121 \text{ kN m} \circlearrowright \qquad \overline{R}_C = \frac{\overline{R}_C d_{\overline{R}C}}{d_{\overline{R}C}} = \frac{121 \text{ kN m}}{8 \text{ m}}$$

$$\boldsymbol{\overline{R}_C = -15.1 \text{ kN}}$$

(Note that our initial assumption about the direction of $\overline{R}_C$ was wrong. The load P_1 is both large enough and far enough to the left between A and B that it forces section BC of the beam to pry up on C despite the load P_2.)

Step 6 Determine $\overline{R}_A$ by summing the vertical forces.

$$\Sigma F_y = 0 \qquad \overline{R}_A + P_1 + \overline{R}_B + P_2 + \overline{R}_c = 0$$

$$\overline{R}_A = -(P_1 + \overline{R}_B + P_2 + \overline{R}_C) \qquad \overline{R}_A = -(-400 + 375.8 - 200 - 15.1)$$

$$\boldsymbol{\overline{R}_A} = +239.3 \text{ kN}$$

Problem 20-5: Determine the support reactions for the beam shown in Figure 20-12 using the method of superposition.

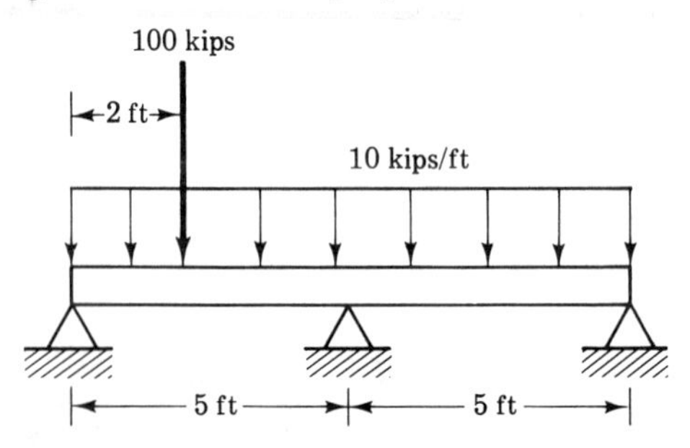

Figure 20-12

Continuous Beams by Moment Area

At first glance, the beam shown in Figure 20-13a would appear to defy analysis. However, we shall develop an application of the moment-area method known as the *three-moment equation*.

The beam has four redundant supports. Solution therefore requires four equations based on elasticity as well as the static equilibrium equations. The necessary equations can be determined by analyzing adjacent pairs of spans. Any pair would do, but let's look at 3 and 4 together.

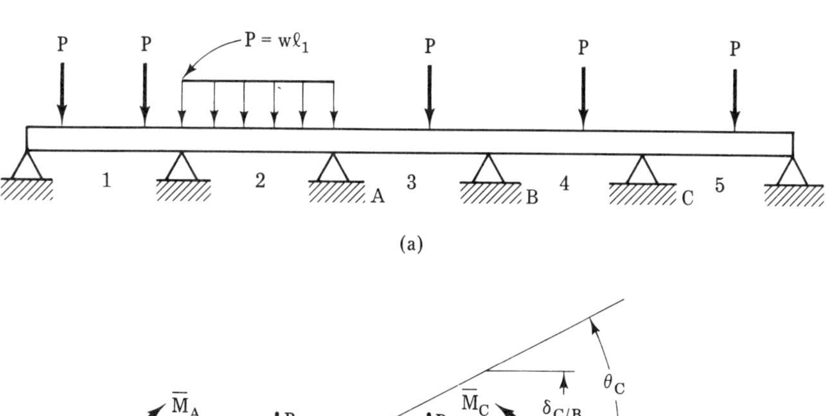

(a)

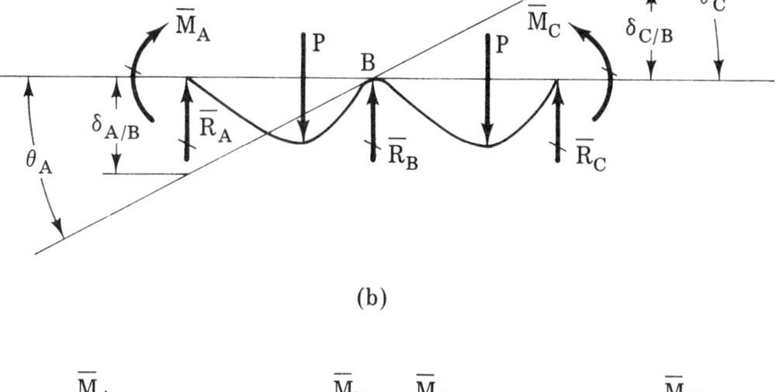

(b)

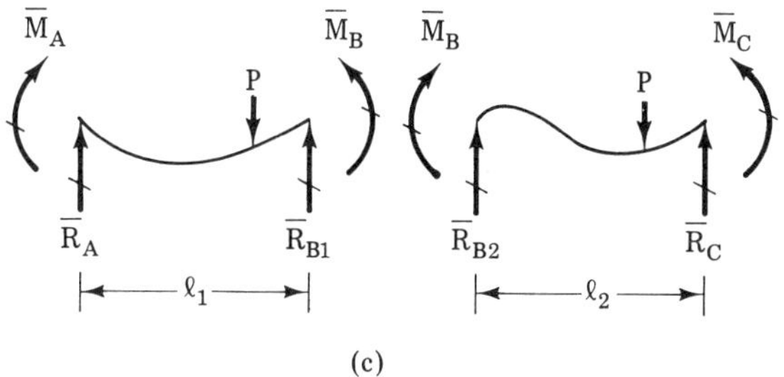

(c)

Figure 20-13. Continuously supported beam.

In Figure 20-13b we have drawn a combination free-body elastic diagram for the pair of spans 3–4 which we shall designate *ABC*. There is only one slope possible at *B* and extending it yields $\theta_A = \theta_C$. The deflections that occur at *A* and *C* in order to maintain them at the same elevation with *B* are proportional to the distance of *A* and *C* from *B*. This can be shown as follows:

$$\tan \theta_A = -\tan \theta_C$$

$$\frac{\delta_{A/B}}{l_1} = -\frac{\delta_{C/B}}{l_2}$$

(Regardless of the nature of our hypothetical problem, there is only one slope at *B* and the deflections at *A* and *C* must be opposite in direction.)

In 20-13c we have separated the two spans which necessitates splitting $\overline{R}_B$ into two proportional parts and considering the internal moment of the beam at B. The equations for the δ/ls can now be determined from the moment areas shown for each span. Recall that we take moments about the end being deflected. Also, the deflections due to $\overline{R}_{B1}$ and $\overline{R}_{B2}$ have been deleted. Introducing them would make both of the equations that follow equal to zero (which is really the case) and still equal to each other but would give us too many unknowns.

$$\frac{\delta_{A/B}}{1} = \frac{1/2\overline{M}_A l_1(l_1/3)}{EIl_1} + \frac{1/2\overline{M}_B l_1(2l_1/3)}{EIl_1} + \frac{A_1\bar{x}_1}{EIl_1}$$

$$\frac{\delta_{A/B}}{l_1} = \frac{\overline{M}_A l_1}{6EI} + \frac{\overline{M}_B l_1}{3EI} + \frac{A_1\bar{x}_1}{EIl_1}$$

$$-\frac{\delta_{C/B}}{l_2} = -\left[\frac{1/2\overline{M}_B l_2(2l_2/3)}{EIl_2} + \frac{1/2\overline{M}_C l_2(l_2/3)}{EIl_2} + \frac{A_2\bar{x}_2}{EIl_2}\right]$$

Then

$$\frac{\overline{M}_a l_1}{6} + \frac{M_B l_1}{3} + \frac{A_1\bar{x}_1}{l_1} = \frac{-\overline{M}_B l_2}{3} - \frac{\overline{M}_C l_2}{6} - \frac{A_2\bar{x}_2}{l_2}$$

$$\overline{M}_A l_1 + 2\overline{M}_B(l_1 + l_2) + \overline{M}_C l_2 = -\frac{6A_1\bar{x}_1}{l_1} - \frac{6A_2\bar{x}_2}{l_2}$$

This is the three-moment equation. The left-hand side will always be as shown (although any pin- or roller-supported *end* of a beam would have a zero internal moment). The right-hand side of the equation can be adapted to as many loads of as many kinds as may occur in a problem.

Example 20-6: Determine the support reactions for the beam shown in Figure 20-14a using the moment-area method.

Solution:

Step 1 Draw the free-body elastic diagram for the double span as shown in Figure 20-14b.

Step 2 Draw the free-body elastic diagrams for each span as shown in 20-14c.

Step 3 Determine the moment areas and centroid locations for the two loads as shown in 20-14d.

Step 4 Apply the three-moment equation (recognizing that $\overline{M}_A$ and $\overline{M}_C$ are both zero) to determine $\overline{M}_B$ as follows:

$$\overline{M}_A l_1 + 2M_B(l_1 + l_2) + \overline{M}_C l_2 = -\frac{6A_1\bar{x}_1}{l_1} - \frac{6A_2\bar{x}_2}{l_2}$$

$$0 + 2\overline{M}_B(22) + 0 = -\frac{6(30{,}000)(5.33)}{12} - \frac{6(41{,}667)(5)}{10}$$

$$\overline{M}_B = -4659 \text{ ft lb}$$

Step 5 Determine $\overline{R}_A$ by summing moments about B for the left-hand span.

$$\Sigma M_B = 0 \qquad \overline{R}_A d_{\overline{R}A} + P_1 d_1 + \overline{M}_B = 0$$

$$\overline{R}_A d_{\overline{R}A} = -(P_1 d_1 + \overline{M}_B) = -[+(2000)(8) - 4659]$$

$$\overline{R}_A d_{\overline{R}A} = -11{,}341 \text{ ft lb} \circlearrowright \qquad \overline{R}_A = \frac{\overline{R}_A d_{\overline{R}A}}{d_{\overline{R}A}} = \frac{11{,}341}{12}$$

$$\boldsymbol{\overline{R}_A = +945 \text{ lb}}$$

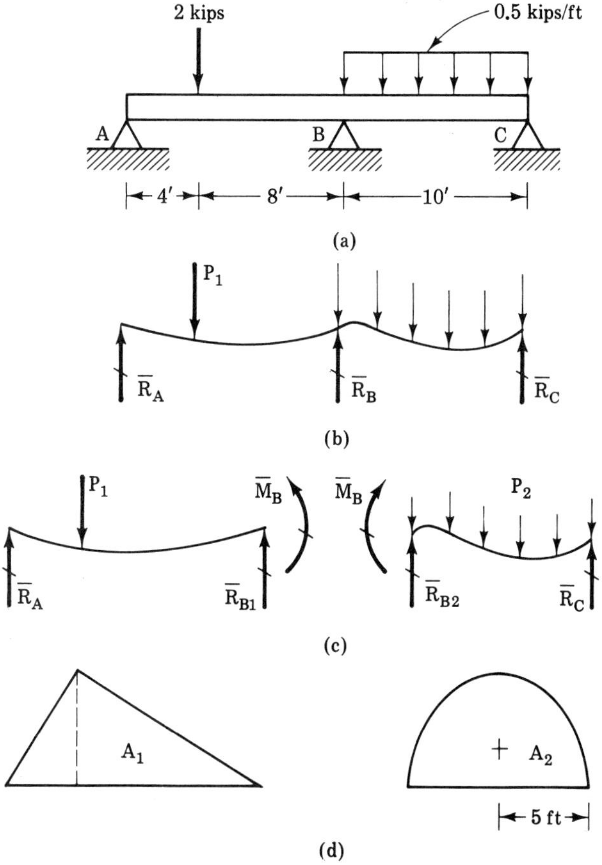

2 kips 0.5 kips/ft

A B C

4′ 8′ 10′

(a)

P_1 $\overline{R}_A$ $\overline{R}_B$ $\overline{R}_C$

(b)

P_1 $\overline{M}_B$ $\overline{M}_B$ P_2 $\overline{R}_A$ $\overline{R}_{B1}$ $\overline{R}_{B2}$ $\overline{R}_C$

(c)

A_1 A_2 5 ft

(d)

$$M_{1\,max} = \frac{P\,a\,b}{\ell_1} = \frac{2000(4)(8)}{12}$$

$$M_{1\,max} = 5000 \text{ ft/lb}$$

$$A_1 = \tfrac{1}{2}\,bh = \tfrac{1}{2}\,(12)(5000)$$

$$A_1 = 30{,}000 \text{ ft}^2\text{/lb}$$

$$M_{2\,max} = \frac{w\ell^2}{8} = \frac{500(10)^2}{8}$$

$$M_{2\,max} = 6250 \text{ ft/lb}$$

$$A_2 = \tfrac{2}{3}\,bh = \tfrac{2}{3}\,(10)(6250)$$

$$A_2 = 41{,}667 \text{ ft}^2\text{/lb}$$

$$x_1 = \frac{A_{1a}x_{1a} + A_{1b}x_{1b}}{A_1}$$

$$x_1 = \frac{\tfrac{1}{2}\,(4)(5000)\,\tfrac{2}{3}\,(4) + \tfrac{1}{2}\,(8)(5000)(4 + \tfrac{8}{3})}{30{,}000}$$

$$x_1 = 5.33 \text{ ft}$$

$$x_2 = 5 \text{ ft}$$

Figure 20-14

Step 6 Determine $\overline{R}_c$ by summing moments about B for the right-hand span.

$$\Sigma M_B = 0 \qquad \overline{R}_C d_{\overline{R}C} + P_2 d_2 + \overline{M}_B = 0$$

$$\overline{R}_C d_{\overline{R}C} = -(P_2 d_2 + \overline{M}_B) = -[-(5000)(5) + 4659]$$

$$\overline{R}_C d_{\overline{R}C} = +20{,}341 \text{ ft lb} \circlearrowright \qquad \overline{R}_C = \frac{\overline{R}_C d_{\overline{R}C}}{d_{\overline{R}C}} = \frac{20{,}341}{10}$$

$$\mathbf{\overline{R}_C = +2034 \text{ lb}}$$

Step 7 Determine $\overline{R}_B$ by summing forces for the entire beam.

$$\Sigma F_y = 0 \qquad \overline{R}_A + P_1 + \overline{R}_B + P_2 + \overline{R}_C = 0$$

$$\overline{R}_B = -(\overline{R}_A + P_1 + P_2 + \overline{R}_C) = -(+945 - 2000 - 5000 + 2034)$$

$$\mathbf{\overline{R}_B = +4021 \text{ lb}}$$

Problem 20-6: Determine the support reactions for the beam shown in Figure 20-15 using the moment-area method.

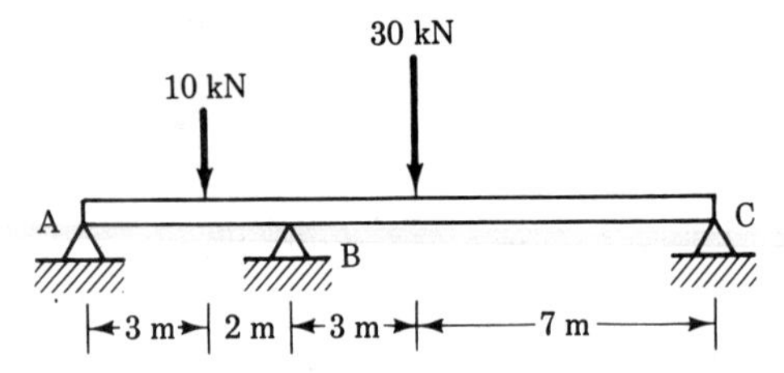

Figure 20-15

20.6 READINESS QUIZ

1. The statically determinate beam below is

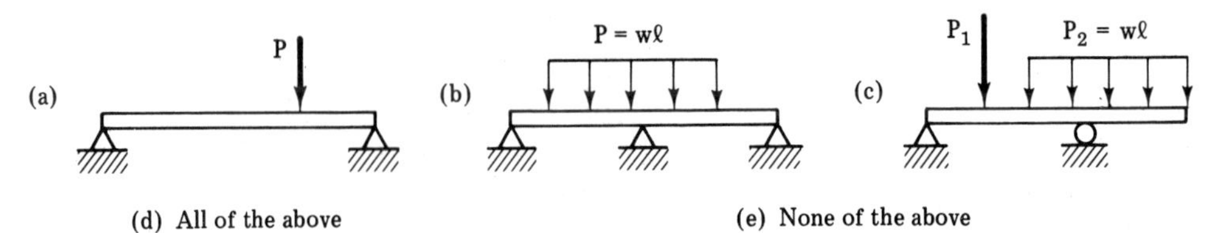

(d) All of the above (e) None of the above

Figure 20-16

2. The static indeterminancy of a beam is a function of its

(a) supports
(b) loads

(c) cross section
(d) all of the above
(e) none of the above

3. Support reactions which can be removed and still leave a beam in static equilibrium are said to be

(a) recombinant
(b) redundant
(c) resultant
(d) all of the above
(e) none of the above

4. The statically determinate beam below is

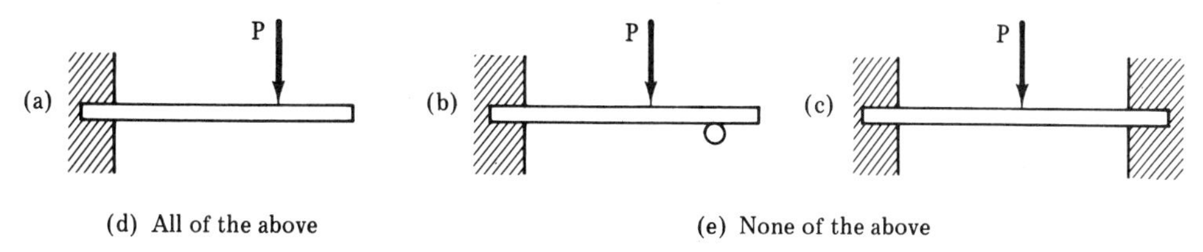

(d) All of the above

(e) None of the above

Figure 20-17

5. Continuous beams have __________ supports.

(a) no
(b) one
(c) two
(d) three
(e) three or more

6. A propped beam has __________ fixed supports.

(a) no
(b) one
(c) two
(d) any of the above
(e) none of the above

7. The superposition method requires use of

(a) deflection equations
(b) a moment equation
(c) a force equation
(d) all of the above
(e) none of the above

8. The degree of indeterminancy is __________ the number of redundant supports.

   (a) the same as
   (b) one more than
   (c) one less than
   (d) any of the above
   (e) none of the above

9. A restrained beam has __________ fixed supports.

   (a) no
   (b) one
   (c) two
   (d) any of the above
   (e) none of the above

10. The three-moment equation requires __________ more deflection equation(s) than there are spans.

    (a) one
    (b) two
    (c) three
    (d) any of the above
    (e) none of the above

20.7 SUPPLEMENTARY PROBLEMS

Determine the support reactions for the beams in each of the following problems using the method of superposition.

20-7

20-8

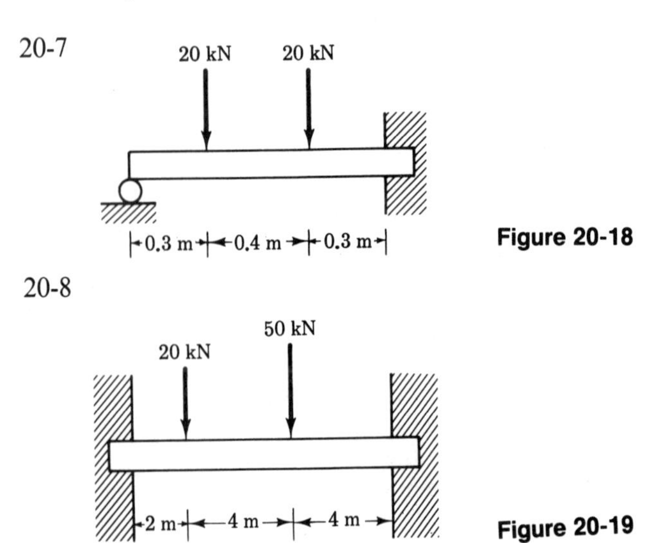

Figure 20-18

Figure 20-19

20-9

20-10

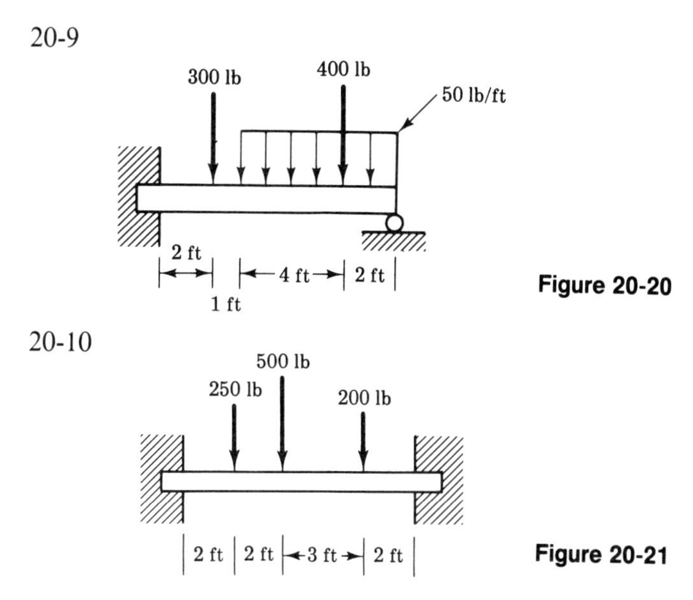

Figure 20-20

Figure 20-21

Determine the support reactions for the beams in each of the following problems using the moment-area method.

20-11

20-12

20-13

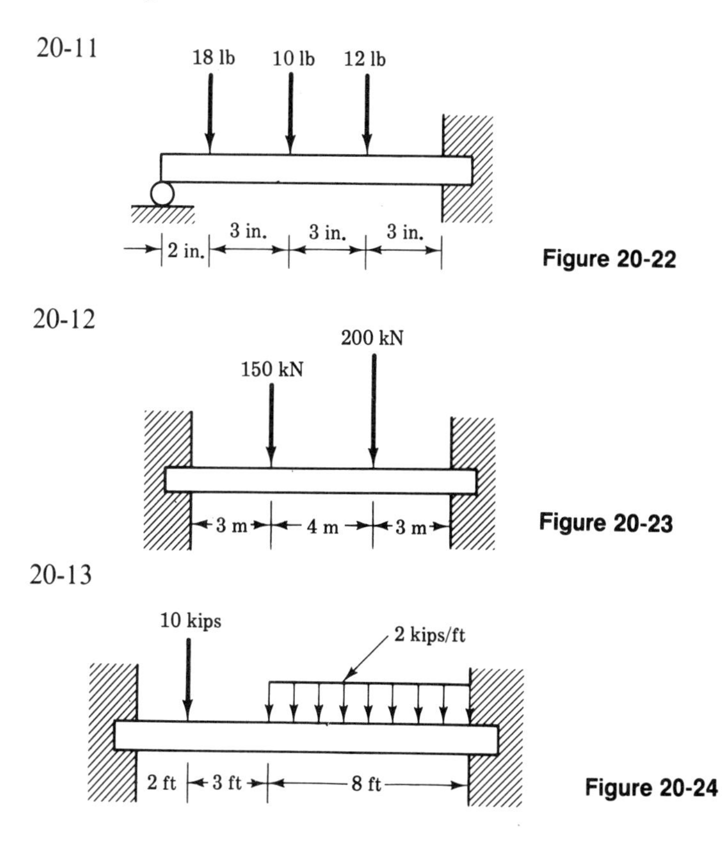

Figure 20-22

Figure 20-23

Figure 20-24

20-14

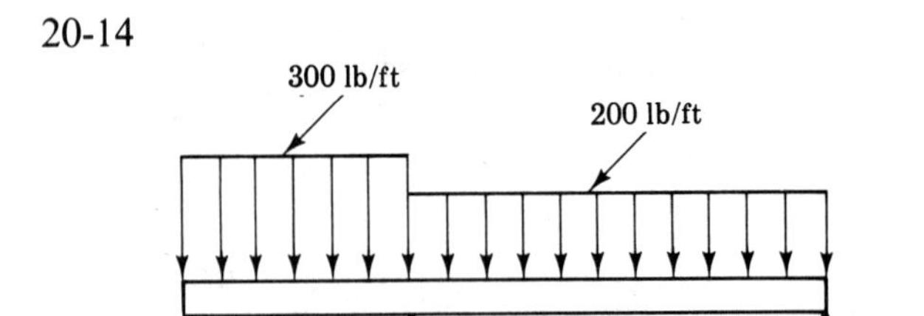

Figure 20-25

Appendix

A

Prerequisite Review Units

Review Unit A-1

Right-Triangle Trigonometry

GENERAL

A ready facility with right-triangle trigonometry is a must in the study of statics. Most of the units in this course require a regular and dependable use of this aspect of mathematics.

A triangle is a polygon consisting of three sides and, therefore, three angles. It can be proven that the sum of these three angles is always 180°. A right triangle is a special type of triangle, *one* of whose angles is a right angle. A right angle is defined as one containing exactly 90°. Among the three angles in a triangle, *only* one can be a right angle. This is evident in Figure A1-1, which illustrates what happens if an attempt is made to place two right angles in a triangle.

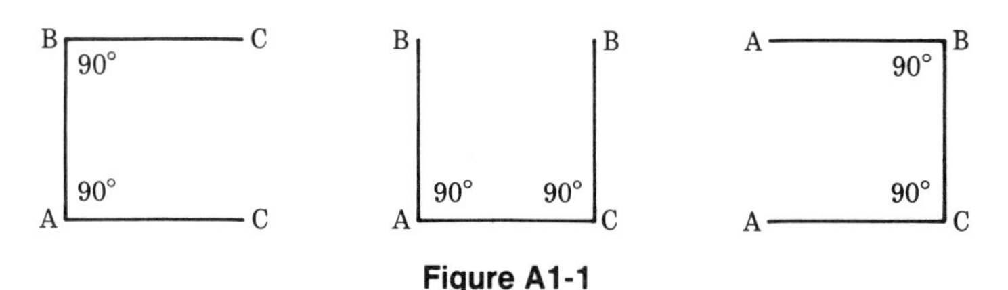

Figure A1-1

There are special types of triangles and the types sometimes overlap. Shown in Figure A1-2 are several triangles illustrating the various types.

An equilateral triangle is defined as one that has all of its sides and all of its angles equal. Each angle, therefore, must be 60°. As we will see below,

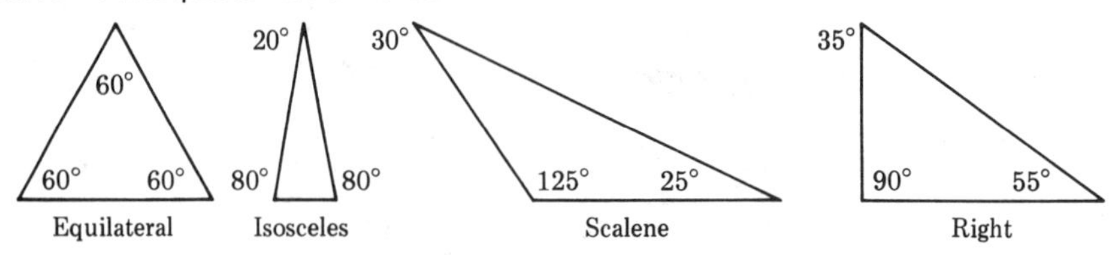

Figure A1-2

it is by nature also an isosceles triangle, but can never be a scalene or a right triangle.

An isosceles triangle is defined as one that has two angles and their opposite sides equal. The equilateral, therefore, must be isosceles. The right can be but frequently is not.

The scalene triangle has no two sides or angles equal. It may or may not contain an obtuse angle as shown. It, therefore, cannot be equilateral or right.

The right triangle has one angle that is exactly 90°. It, therefore, cannot be equilateral or scalene, but may be isosceles.

Angles smaller than 90° are called acute angles. In a right triangle the two angles other than the right angle are both always acute. This is based on the fact that the three angles must add up to 180°. If one angle is 90° then the other two must add up to 90°. If two angles add up to 90°, each individually must be less than 90°. The right triangles shown in Figure A1-3 all illustrate this. Note that the one in the middle must also be an isosceles triangle. Note that the right angle in the triangles has not been labeled with its size. First of all, it "looks like" a 90° angle, that is, two of the sides of the triangle appear to be perpendicular at that point. Second, the other two angles add up to exactly 90° in each case; therefore, the third angle must be exactly 90°.

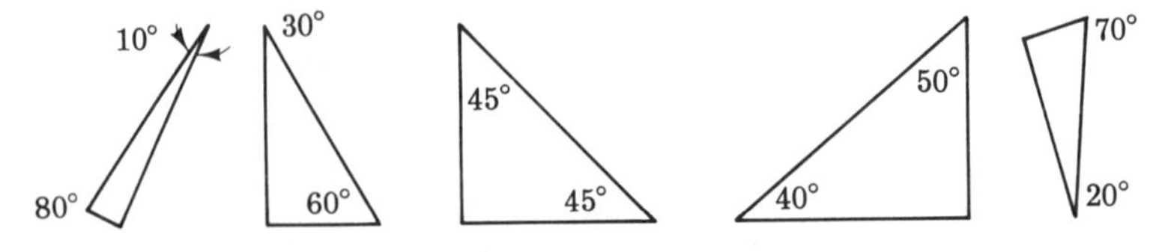

Figure A1-3

BASIC FUNCTIONS

There exist several definite relationships among the two acute angles and the three sides of a right triangle. As these are discussed with reference to Figure A1-4, keep in mind that the relationships will remain the same regardless of what position the triangle is placed in. The square at the 90° angle is a commonly used symbol identifying such an angle.

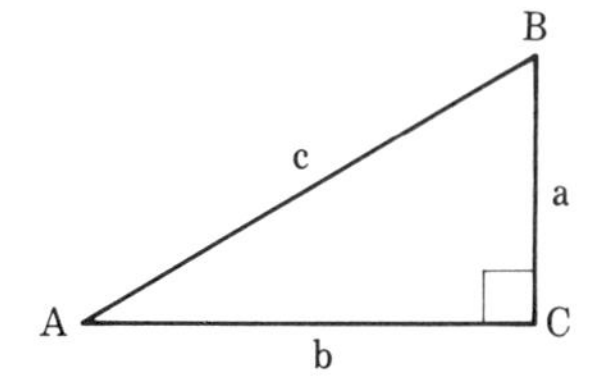

Figure A1-4

The three points (and angles) and the three sides of the triangle have been identified with symbols to make discussion easier. In addition, the longest side (opposite the 90° angle) is called the *hypotenuse*. For obvious reasons, side a is called the *opposite* of angle A and side b is called the *adjacent* (next to) to angle A. Similarly side b is the side *opposite* of angle B and side a is the side *adjacent* to angle B.

For angle A three basic sets of relationships exist among the three sides. Each has a name as well as an abbreviation.

$$\text{sine } A = \sin A = \frac{\text{opposite}}{\text{hypotenuse}} = \frac{a}{c}$$

$$\text{cosine } A = \cos A = \frac{\text{adjacent}}{\text{hypotenuse}} = \frac{b}{c}$$

$$\text{tangent } A = \tan A = \frac{\text{opposite}}{\text{adjacent}} = \frac{a}{b}$$

Similarly

$$\sin B = \frac{b}{c}, \qquad \cos B = \frac{a}{c}, \qquad \tan B = \frac{b}{a}$$

Some examination and manipulation of the above relationships will show that

$$\sin A = \cos B, \qquad \cos A = \sin B$$

$$\tan A = \frac{\sin A}{\cos A}, \qquad \tan B = \frac{\sin B}{\cos B}$$

The right triangle then contains five variables, the three sides and the two acute angles. If any two sides or one side and one angle are known, the remaining three variables can be determined through application of the relationships just described.

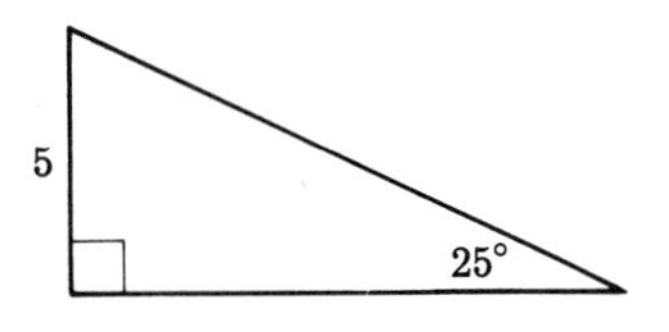

Figure A1-5

Example A1-1: In Figure A1-5, if the short side is 5 and the angle opposite it is 25°

$$\sin 25^\circ = \frac{5}{\text{hypotenuse}}, \qquad \text{hypotenuse} = \frac{5}{\sin 25^\circ} = 11.83$$

$$\tan 25^\circ = \frac{5}{\text{adjacent}}, \qquad \text{adjacent} = \frac{5}{\tan 25^\circ} = 10.72$$

and

$$180^\circ - 90^\circ - 25^\circ = 65^\circ, \qquad \text{the other acute angle}$$

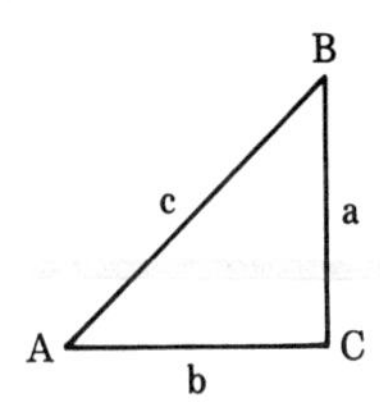

Figure A1-6

Example A1-2: In Figure A1-6, if side $c = 20$ and angle $A = 40^\circ$,

$$\sin A = \frac{a}{c}, \qquad a = c \sin A, \qquad a = 20 \sin 40^\circ = 12.86$$

$$\cos A = \frac{b}{c}, \qquad b = c \cos A, \qquad b = 20 \cos 40^\circ = 15.32$$

$$A + B + C = 180^\circ$$

$$B = 180^\circ - A - C = 180^\circ - 40^\circ - 90^\circ = 50^\circ$$

PYTHAGORA'S THEOREM

It took a man by the name of Pythagora to recognize that an interesting relationship existed among the three sides of a right triangle. Namely, the square of the hypotenuse is equal to the sum of the square of the other two sides:

$$c^2 = a^2 + b^2$$

Checking this in examples 1 and 2 above yields

$$(11.83)^2 = (10.72)^2 + (5)^2, \qquad 139.9 = 139.9 \tag{1}$$

$$(20)^2 = (12.86)^2 + (15.32)^2, \qquad 400 = 400 \tag{2}$$

This can be a useful alternative for problem solution.

Example A1-3: In a given right triangle, $c = 15$ and $b = 4$. Then

$$c^2 = a^2 + b^2, \qquad a^2 = c^2 - b^2, \qquad a = \sqrt{c^2 - b^2}$$

$$a = \sqrt{(15)^2 - (4)^2}, \qquad a = 14.46$$

An interesting and occasionally useful derivation can be made by dividing the Pythagorean equation by c^2.

$$c^2 = a^2 + b^2, \qquad \frac{c^2}{c^2} = \frac{a^2}{c^2} + \frac{b^2}{c^2}$$

but

$$a/c = \sin A \quad \text{and} \quad b/c = \cos A$$

therefore,

$$\sin^2 A + \cos^2 A = 1$$

Example A1-4: If $\sin B = 0.2$,

$$\sin^2 B + \cos^2 B = 1$$

$$\cos B = \sqrt{1 - \sin^2 B}$$

$$\cos B = \sqrt{1 - (0.2)^2} = 0.98$$

AREA

The area of a right triangle is simply one-half of one side multiplied by the other side. The basis for this can be readily seen if we recall that the area for a rectangle is equal to bh and that a right triangle is always half of a rectangle.

SIMILARITY AND CONGRUENCY

Two properties triangles may exhibit are similarity and congruency. Recognizing that these properties exist can be quite helpful in problem solving and it is important to understand the nature of both and the difference between them. In Figure A1-7 the triangle at left is similar and congruent to the one in the middle while the triangle at right is only similar to the one in the middle.

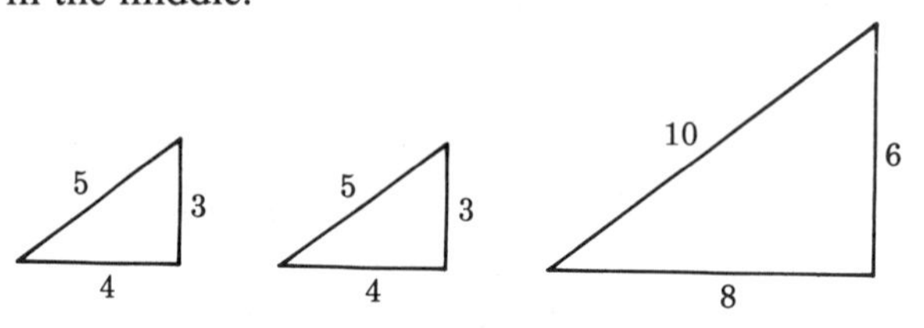

Figure A1-7

All three triangles have the same proportions among their sides; therefore, the trigonometric functions of their angles and thus their angles are identical. This is known as *similarity.* They all have the same shape and at some distance you would be able to see the shape but not be able to tell which one you were looking at. This is similarity, where the angles are all the same.

Two of the triangles not only have the same proportions among their sides, but the corresponding sides also are exactly the same size. If one triangle were laid on the other, it could be aligned to exactly cover the first. They are completely identical. This is known as *congruency.*

A special and useful case occurs if we draw a perpendicular to the hypotenuse of a right triangle at a location that causes it to pass through the vertex of the right angle, as shown in Figure A1-8.

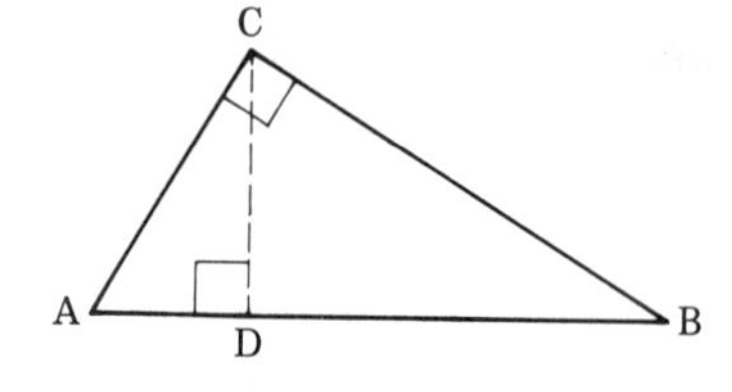

Figure A1-8

Since *ADC* is a right angle, so is *BDC*. Two new smaller triangles have been created by drawing the perpendicular. In the original triangle angle *CAD* and angle *CBD* add up to 90°. In the new triangle *CAD*, the three angles must add up to 180°. Further, in that triangle angle *ACD* and angle

CAD must add up to 90°. Thus

$$\sphericalangle CAD + \sphericalangle CBD = 90°$$
$$\sphericalangle CAD + \sphericalangle ACD = 90°$$
$$\therefore \sphericalangle CBD = \sphericalangle ACD$$
$$\therefore \Delta CAD \text{ is similar to } \Delta ABC$$

It could similarly be shown that

$$\Delta BCD \text{ is similar to } \Delta ABC$$

and

$$\Delta BCD \text{ is similar to } \Delta CAD$$

Example A1-5: In Figure A1-9 $a = 5$ and $b = 6$. Then

$$c = \sqrt{a^2 + b^2} = \sqrt{(5)^2 + (6)^2}, \qquad c = 7.8$$
$$\sin B = \frac{b}{c} = \frac{6}{7.8} = 0.769$$
$$\sin B = d/a, \qquad d = a \sin B = 5(.769), \qquad d = 3.85$$
$$e = \sqrt{a^2 - d^2} = \sqrt{(5)^2 - (3.85)^2}, \qquad e = 3.19$$
$$f = c - e = 7.8 - 3.19, \qquad f = 4.61$$

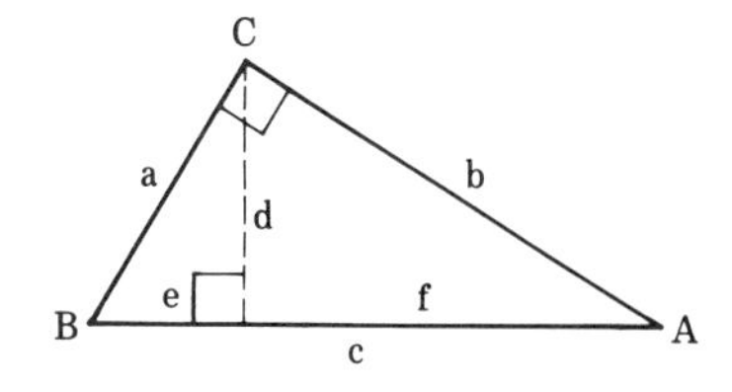

Figure A1-9

PROBLEMS

1. The slope of a hill has been determined to be 35°. Pacing off the distance along the slope from the base to the top of the hill yields a distance of 2500 feet. How high is the top of the hill above the base?

2. A 40-foot long ladder is to be used to work on a building. The acute angle that it forms with the ground is to be no greater than 70°. What is the greatest height on the building wall that can be reached by the ladder?

3. In the truss in Figure A1-10, find the lengths of *BF, CF, CE, BD,* and *AD* if *AB* is 10 feet long.

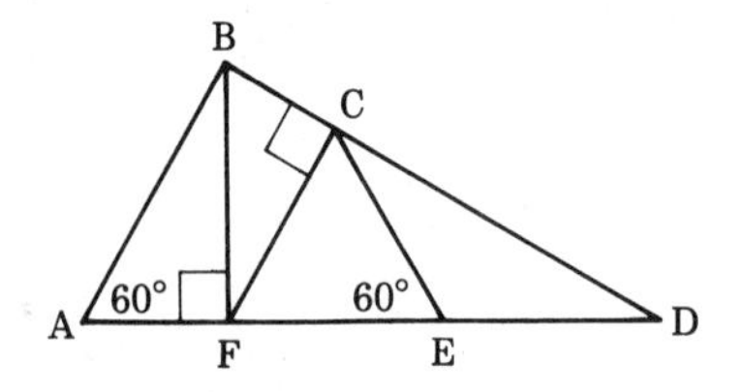

Figure A1-10

4. In the truss in Figure A1-11, determine the angles that *AB* and *BC* make with the horizontal.

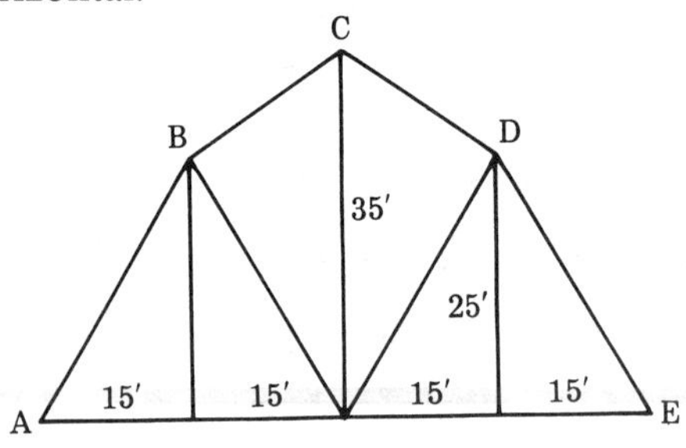

Figure A1-11

5. A rectangular city block is 500 feet long in one direction and 800 feet long in the other. What is the diagonal distance across it?

Review Unit A-2

Non – Right-Triangle Trigonometry

INTRODUCTION

Many problems arise in which the triangle involved is not a right triangle. Two relations can be developed which can be applied to any triangle. In the sections below the laws of sines and cosines are developed, followed by applications to four different types of cases that may be encountered.

LAW OF SINES

The law of sines states that in any triangle the sides are proportional to the sines of the opposite angles.

$$\frac{a}{\sin A} = \frac{b}{\sin B} = \frac{c}{\sin C}$$

The law can be readily verified by use of right-triangle trigonometry. In the triangle shown in Figure A2-1, a perpendicular has been drawn to one side in a location causing it to pass through the vertex opposite.

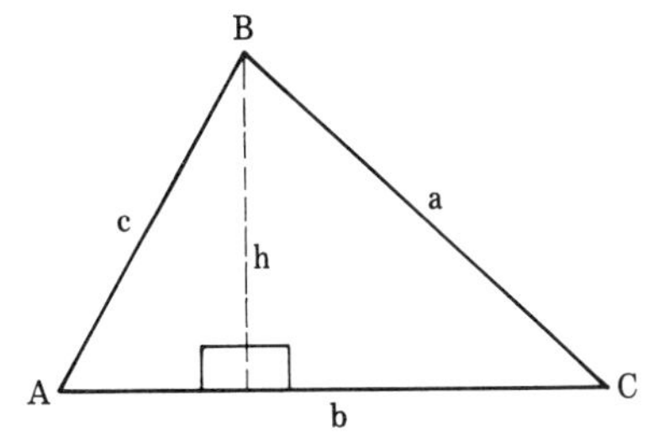

Figure A2-1

The perpendicular is also known as the altitude and forms two right triangles. In those triangles

$$\sin A = \frac{h}{c} \quad \text{and} \quad \sin C = \frac{h}{a}$$

then

$$h = c \sin A \text{ and } h = a \sin C$$

Thus

$$c \sin A = a \sin C$$

or

$$\frac{\sin A}{a} = \frac{\sin C}{c}$$

The same analysis could have been made drawing a perpendicular to either of the other two sides. The end conclusion results in the law of sines which states that

$$\frac{\sin A}{a} = \frac{\sin B}{b} = \frac{\sin C}{c}$$

LAW OF COSINES

The law of cosines states that in any triangle the square of any side is equal to the sum of the squares of the other two sides less twice the product of those sides and the cosine of their included angle. Expressed symbolically, we have the following three forms:

$$a^2 = b^2 + c^2 - 2bc \cos A \tag{1}$$

$$b^2 = a^2 + c^2 - 2ac \cos B \tag{2}$$

$$c^2 = a^2 + b^2 - 2ab \cos C \tag{3}$$

Consider the two possible cases for the oblique triangle ABC as shown in Figure A2-2. One has no angle equal to or larger than 90°. The other has an angle larger than 90°. In each, from the vertex C the altitude h is drawn to the side AB (or its projection) meeting that line at point D.

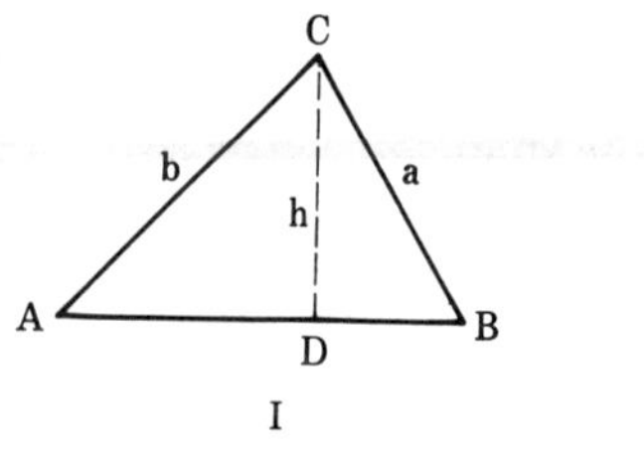

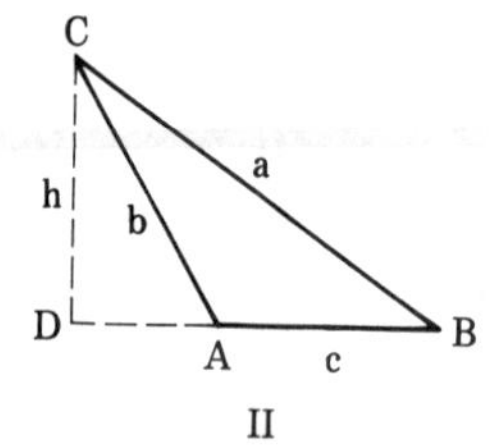

Figure A2-2

In both cases

$$a^2 = h^2 + (DB)^2$$

$$h^2 = b^2 - (DA)^2$$

Substituting into the first expression from the second

$$a^2 = b^2 - (DA)^2 + (DB)^2$$

This applies to both cases. The remainder of the proof must be carried out separately for each case since DA and DB are different for each.

Case I	Case II
$DB = C - DA$	$DB = C + DA$
Substituting in the equation for a^2	Substituting in the equation for a^2
$a^2 = b^2 - (DA)^2 + (C - DA)^2$	$a^2 = b^2 - (DA)^2 + (C + DA)^2$
or	or
$a^2 = b^2 + c^2 - 2c(DA)$	$a^2 = b^2 + c^2 + 2c(DA)$
From the figure	From the figure
$DA = b \cos A$	$DA = b \cos DAC$
	$= b \cos(180 - A)$
	$= -b \cos A$
Then	Then
$a^2 = b^2 + c^2 - 2bc \cos A$	$a^2 = b^2 + c^2 - 2bc \cos A$

Thus, in both cases we have the result shown in Eq. (1) at the start of this section. Equations (2) and (3) may be obtained in like manner by drawing altitudes to the other sides of the triangles.

It is important to note that when angle A is obtuse (over 90°), its cosine becomes $-\cos(180 - A)$ and the last element of the equation becomes positive.

Type 1 Problem

The first type of problem to be considered is one in which one side and any two angles are known. The third angle can be found by subtracting the sum of the first two from 180°. The law of sines can then be used to find the other two sides.

Example A2-1: Given the triangle shown in Figure A2-3, find the missing angle and sides.

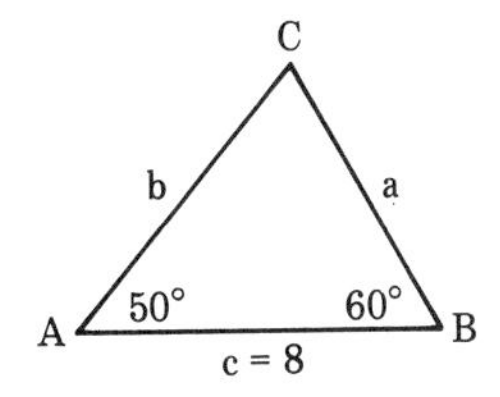

Figure A2-3

$$A + B + C = 180°, \qquad C = 180 - (A + B)$$

$$C = 180° - (50° + 60°), \qquad C = 70°$$

$$\frac{a}{\sin A} = \frac{b}{\sin B} = \frac{c}{\sin C}$$

$$b = c\,\frac{\sin B}{\sin C}, \qquad b = 8\,\frac{\sin 60^\circ}{\sin 70^\circ}, \qquad b = 7.37$$

$$a = c\,\frac{\sin A}{\sin C}, \qquad a = 8\,\frac{\sin 50^\circ}{\sin 70^\circ}, \qquad a = 6.52$$

Type 2 Problem

The second type of problem is one in which two sides and an angle opposite one of them are known. The angle opposite the second known side can be found by use of the law of sines. The third angle can then be found by subtracting the sum of the two known angles from 180°. The third side can then be found by use of the law of sines.

Example A2-2: Given the triangle shown in Figure A2-4, find the missing angles and side.

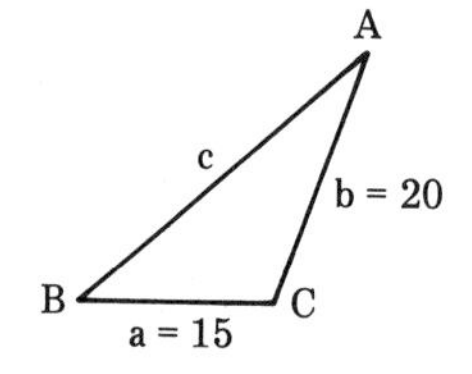

Figure A2-4

$$\frac{\sin A}{a} = \frac{\sin B}{b}, \qquad \sin A = \frac{a}{b}\sin B$$

$$\sin A = \frac{15}{20}\sin 50^\circ, \qquad A = 35^\circ$$

$$A + B + C = 180^\circ, \qquad C = 180^\circ - (A + B)$$

$$C = 180^\circ - (50^\circ + 35^\circ), \qquad C = 95^\circ$$

$$\frac{\sin C}{c} = \frac{\sin B}{b}, \qquad c = b\,\frac{\sin C}{\sin B}$$

$$c = 20\,\frac{\sin 95}{\sin 50}, \qquad c = 26$$

Type 3 Problem

The third type of problem is one in which two sides and the included angle of a triangle are known. The side opposite the known angle can be found by use of the law of cosines. One missing angle can then be found by the law of sines and the other angle can be found by subtracting the sum of the two known angles from 180°.

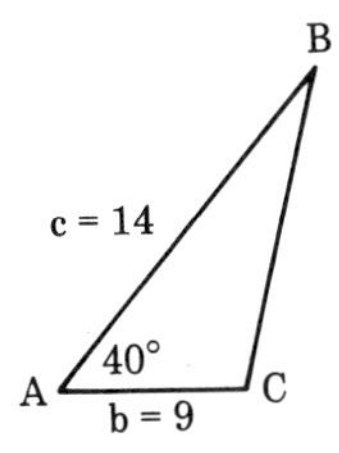

Figure A2-5

Example A2-3: Given the triangle shown in Figure A2-5, find the missing angles and side.

$$a = \sqrt{b^2 + c^2 - 2bc \cos A}, \qquad a = \sqrt{(9)^2 + (14)^2 - 2(9)(14)\cos 40^\circ}$$

$$a = 9.16$$

$$\frac{\sin C}{c} = \frac{\sin A}{a}, \qquad \sin C = \frac{c}{a} \sin A$$

$$\sin C = \frac{14}{9.16} \sin 40^\circ, \qquad C = 101^\circ$$

$$A + B + C = 180^\circ, \qquad B = 180^\circ - (A + C)$$

$$B = 180^\circ - (40^\circ + 101^\circ), \qquad B = 39^\circ$$

Example A2-4: Given the triangle shown in Figure A2-6, find the missing angles and side.

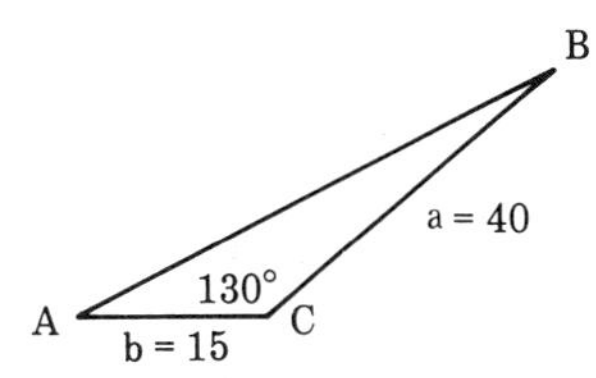

Figure A2-6

$$c = \sqrt{a^2 + b^2 - 2ab \cos C}, \qquad c = \sqrt{(40)^2 + (15)^2 - 2(40)(15)\cos 130^\circ}$$

$$c = 51$$

$$\frac{\sin A}{a} = \frac{\sin C}{c}, \qquad \sin A = \frac{a}{c} \sin C$$

$$\sin A = \frac{40}{51} \sin 130^\circ, \qquad A = 37^\circ$$

$$A + B + C = 180^\circ, \qquad B = 180^\circ - (A + C)$$

$$B = 180^\circ - (130^\circ + 37^\circ), \qquad B = 13^\circ$$

Type 4 Problem

The fourth type of problem is one in which all three sides are known but no angles. One angle can be found by use of the law of cosines, the second by use of the law of sines, and the third by subtracting the sum of the first two from 180°.

Example A2-5: Given the triangle shown in Figure A2-7, find all the angles.

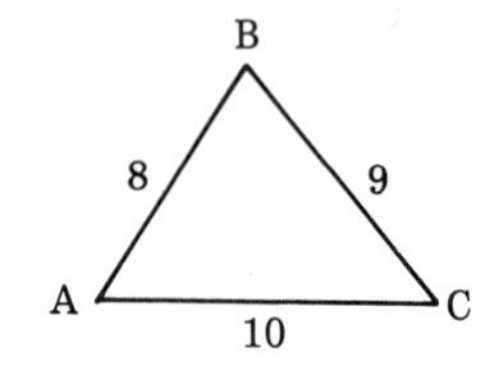

Figure A2-7

$$c^2 = a^2 + b^2 - 2ab \cos C, \qquad \cos C = \frac{c^2 - a^2 - b^2}{-2ab}$$

$$\cos C = \frac{(8)^2 - (9)^2 - (10)^2}{-2(9)(10)}$$

$$C = 49.5°$$

$$\frac{\sin A}{a} = \frac{\sin C}{c}, \qquad \sin A = \frac{a}{c} \sin C$$

$$\sin A = \frac{9}{8} \sin 49.5° \qquad A = 58.8°$$

$$A + B + C = 180°, \qquad B = 180 - (A + C)$$

$$B = 180° - (58.8° + 49.5°), \qquad B = 71.7°$$

PROBLEMS

1. In the triangle shown in Figure A2-8: How far is C from A? How far is C from B?

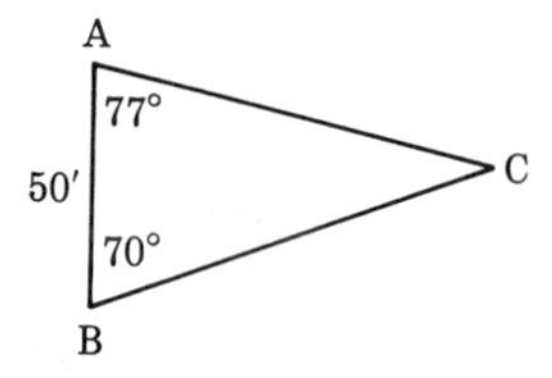

Figure A2-8

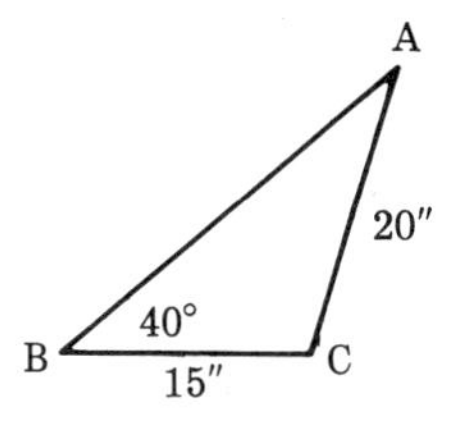

Figure A2-9

2. In the triangle shown in Figure A2-9, find the unknown side and angles.

3. An engineer wishes to build a trestle from point A to point B across a swamp. At point C, which is 6250 feet from A and 7930 feet from B, the angle ACB is 152°. How long must the trestle be?

4. The boom shown in Figure A2-10 is supported by cable AB. What is the length of the cable and the angle it makes with the boom?

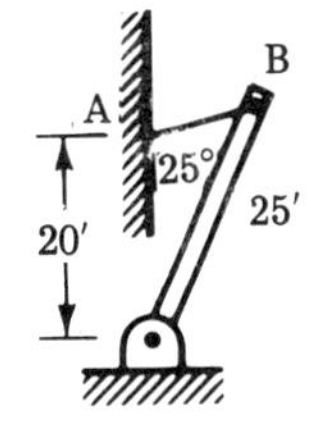

Figure A2-10

5. In the triangle shown in Figure A2-11, determine all the angles.

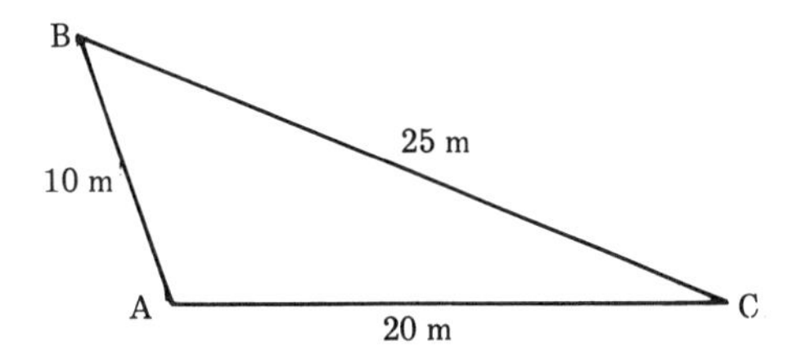

Figure A2-11

Review Unit A-3

Vectors

INTRODUCTION

Many physical quantities dealt with daily can be described in terms of magnitude alone. The volume or weight of mass present in, for example, a half full tank of gas can be expressed as a simple number. In a dynamic situation, for example, when the car is being driven or the tank filled, the quantity is being changed and could be said to be going down or up and, therefore, has direction. However, that is a rate problem, and direction is a function of change with respect to time, not an inherent property of the quantity. The second property of a quantity, direction, becomes apparent when we consider a car traveling west. The fact that the car is traveling west is totally independent of time. While many quantities that require a direction in order to be fully specified are time related such as velocity, acceleration, and momentum, others such as displacement and force are not directly time dependent by definition.

SCALARS

Mass and temperature are physical quantities which can be described by magnitude alone and are known as *scalar quantities.* The number representing such quantities are known as *scalars.* Sometimes, a quantity may be described as scalar or vector depending on the needs of the situation. A salesman can adequately describe part of a car's performance by stating that it is capable of traveling 100 mph. However, if one wishes to drive the car from one specific location to another, direction becomes significant.

VECTORS

A straight line segment that has a direction is called a *directed line segment.* Since two directions are possible along a line, an arrowhead is generally placed at one end of the line to indicate direction. This can also be done by specifying the direction with an angle. Care must be taken to recognize that a vector at 30° has a different direction from one that is at 210°. Such vectors, that have parallel lines but different directions, are often said to be *opposite in sense.* This is further illustrated in Figure A3-1.

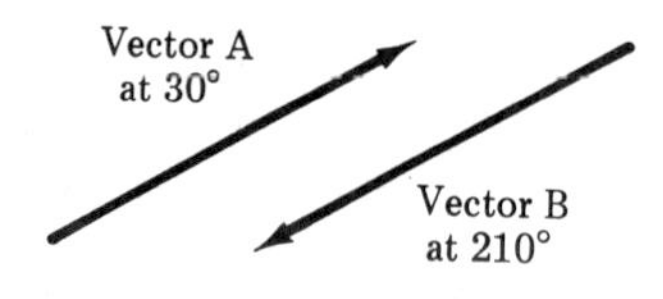

Figure A3-1

Two vectors are equal only if they have both the same magnitude and direction.

ADDITION AND SUBTRACTION

In drawing vectors, the end of the line with the arrowhead is generally called the *head* and the other end the *tail.* The tail end is also sometimes called the *initial point* and the head end the *final point.*

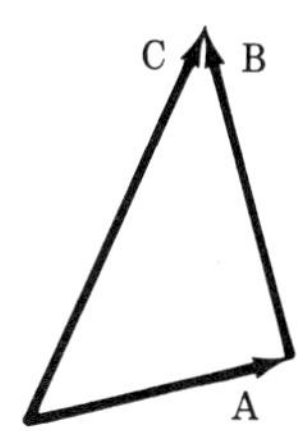

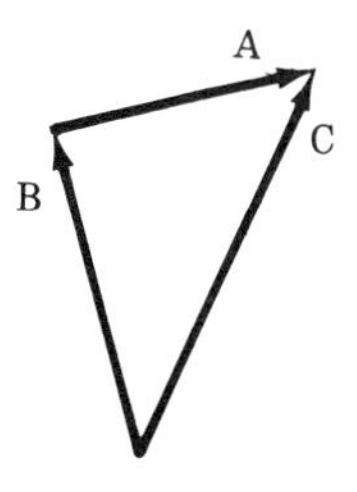

Figure A3-2

The sum of two vectors A and B is the vector C. As shown at left in Figure A3-2, C is a vector whose initial point is the initial point of A and whose terminal point is the terminal point of B (if B is drawn with its initial point at the terminal point of A). At the right in Figure A3-2, it is shown that B may be drawn first and A added to it to achieve the same sum. *The order in which the vectors are added is immaterial.*

It is frequently useful to write equations for vector summation. It is extremely important to use some type of notation that sets a vector equation apart from a scalar or algebraic equation. In the following equation, an arrowhead on the + sign is used to indicate this; however, other designations are commonly used.

$$C = A \leftrightarrow B$$

Any quantities that are to be added which may be vector in nature should be assembled in a vector equation. The only time that vectors can be added algebraically is when they are all collinear, differing only in the sense of their direction. Solution must be accomplished by graphical or trigonometric means.

Vector equations frequently consist of more than two elements. Figure A3-3 shows the summation of four vectors and further illustrates that the order of addition is irrelevant to the result. The use of that last word is significant, for the sum of a group of vectors is frequently called the *resultant.*

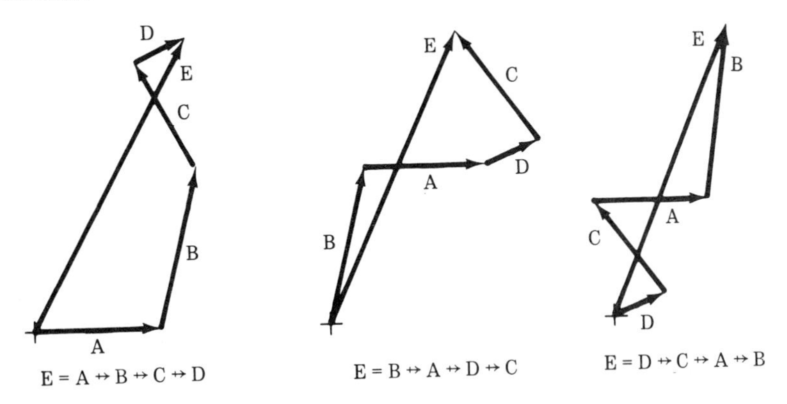

Figure A3-3

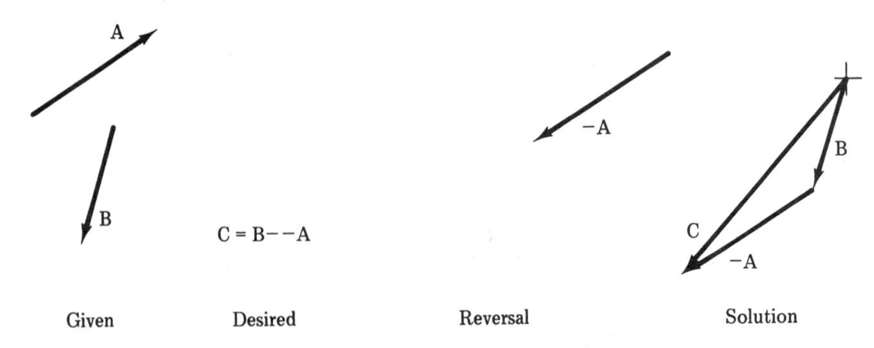

Figure A3-4

Vectors may be subtracted as well as added. This is convenient when a vector equation has been written and the unknown is one of the parts rather than the sum. The subtraction of a vector is accomplished by reversing the sense of its direction and then adding this new vector to the preceding. This has been illustrated in Figure A3-4.

Specifying the direction of a vector is extremely important. Two possible methods are shown in Figure A3-5. On the left the direction of one vector relative to another is specified. This can be useful if the direction of the first vector is known.

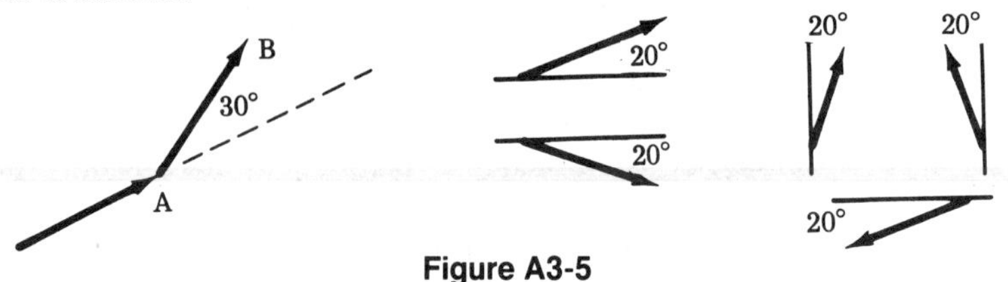

Figure A3-5

However, if the direction of the first vector is not known, only the magnitude and not the direction of the resultant can be determined. On the right several different vectors are all designated as being 20° from some line. This may be relatively clear, especially if the reference lines are vertical or horizontal. However, particularly for trigonometric solution, specifying a

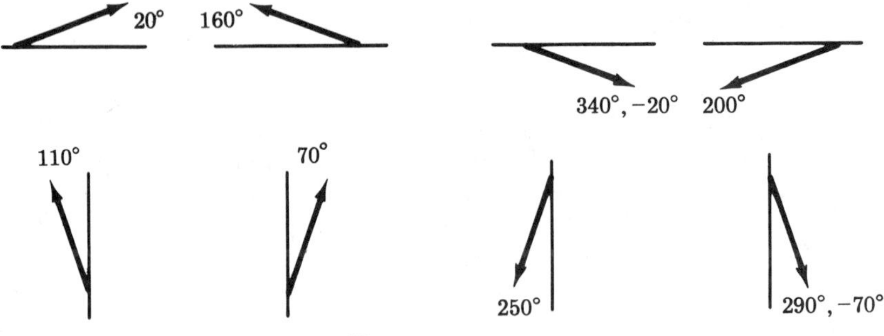

Figure A3-6

true angle based on a relatively conventional location for 0, 360 avoids any confusion. This latter approach is shown in Figure A3-6.

A vector then can be fully specified by giving (a) its magnitude, (b) its direction, and (c) its initial or terminal point.

Example A3-1: Find graphically the sum of the vectors: $A = 10$ at 30°, $B = 14$ at 220°, $C = 8$ at $-20°$, and $D = 17$ at 260°. (See Figure A3-7.)

$$E = A \leftrightarrow B \leftrightarrow C \leftrightarrow D$$

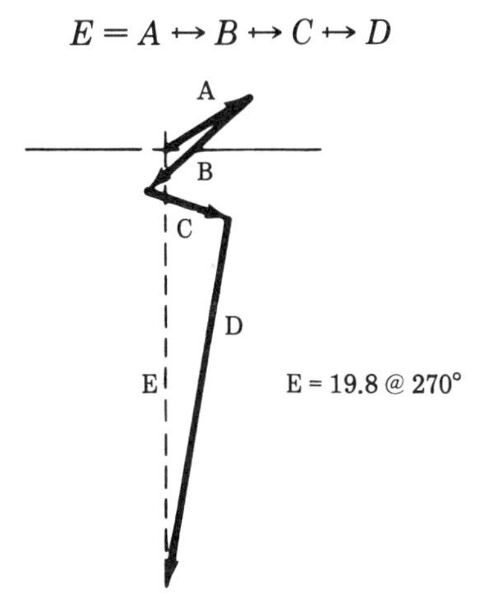

Figure A3-7

Example A3-2: Find graphically $E = A \rightarrow 2B \leftrightarrow C \rightarrow D$ when $A = 15$ at 75°, $B = 10$ at 170°, $C = 22$ at 340°, and $D = 11$ at 240°. (See Figure A3-8.)

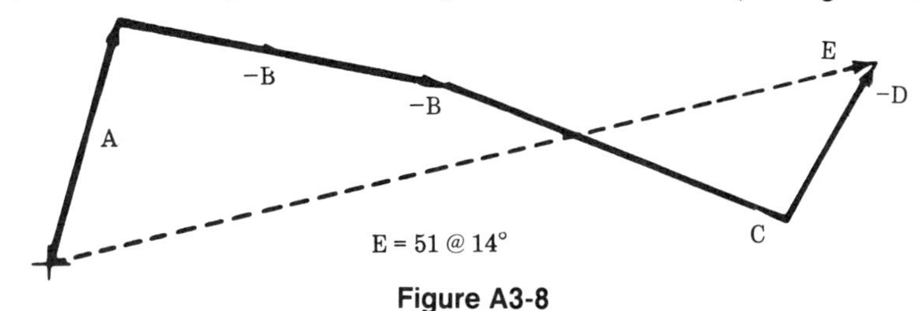

Figure A3-8

RESOLUTION

When a vector is divided into its component parts, the process is called *resolution*. While a vector could be divided into many parts, the practical need to divide it generally suggests only two parts. These are called the *components* of the vector. In order to find the components, the magnitude and direction of the initial force and the direction of each component must be known. The directions chosen for the components depend on the particular problem. Figure A3-9 shows two common cases and the general case.

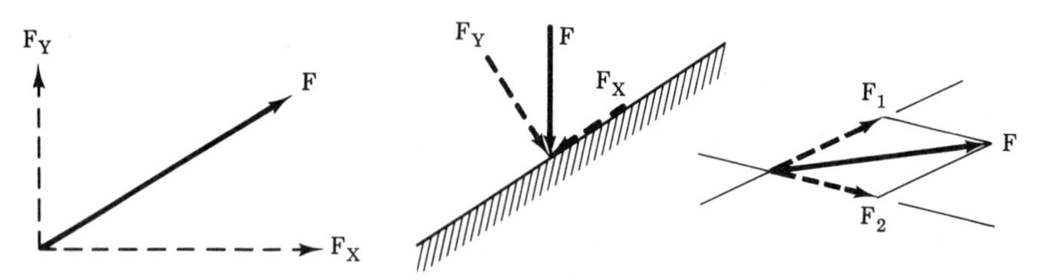

Figure A3-9

The resolution at left is the most common, the components being vertical and horizontal. These are frequently called the rectangular components of the vector. This approach leads into a rather common approach to mathematical solution of vector problems. The resolution at center is also into rectangular components but the axes have been turned to be, respectively, parallel and perpendicular to an inclined plane. This is a rather useful approach in engineering work, since the inclined plane has a variety of applications. At right is the general case, where the components may have any direction suitable to the problem.

Example A3-3: Determine the rectangular components of vector A (see Figure A3-10). $A = 15$ at 22°

$$A_y = A \sin 22^\circ = 15 \sin 22, \qquad A_y = 5.6 \text{ at } 90^\circ$$

$$A_x = A \cos 22^\circ = 15 \cos 22, \qquad A_x = 13.9 \text{ at } 0^\circ$$

(Note that the direction has been specified for each component.)

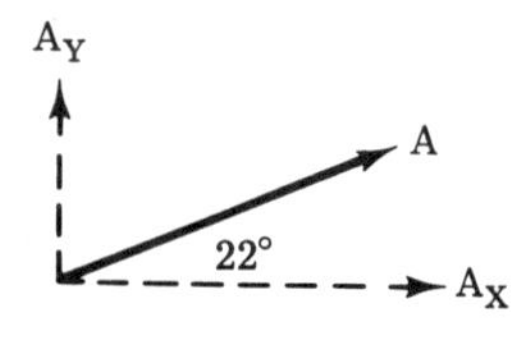

Figure A3-10

Example A3-4: What is the force pressing the 75-lb weight to the incline and what is the force working to slide the weight down the incline? (See Figure A3-11.)

$$W_x = W \sin 35^\circ = 75 \sin 35, \qquad W_x = 43 \text{ lb at } 215^\circ$$

$$W_y = W \cos 35^\circ = 75 \cos 35, \qquad W_y = 61.4 \text{ lb at } 305^\circ$$

(Note that the direction has been specified for each component.)

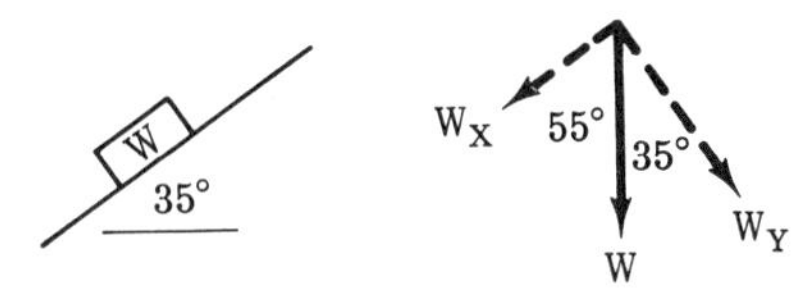

Figure A3-11

Example A3-5: Two children are pulling a sled as shown in Figure A3-12. How hard will each of the children have to pull to exert a forward force of 30 lb on the sled?

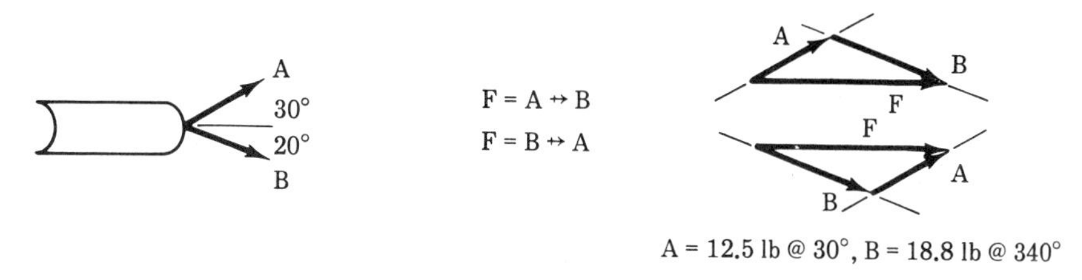

Figure A3-12

The graphical solution can be accomplished by drawing vector A with its initial point at the initial point of F and drawing vector B with its terminal point at the terminal point of F. The magnitudes and, therefore, the lengths of A and B are unknown, but if they are extended far enough they will cross. This intersection will be the terminal point of A and the initial point of B. Since vectors can be added in any order, the solution at far right is also correct. Examination shows that all of the angles in the triangle formed by F, A, and B can be found by geometry. This means that the magnitudes of A and B could be found by use of the law of sines.

PROBLEMS

1. Find the rectangular components of a vector equal to 25 at 217°.
2. Find the rectangular components of a vector equal to 40 at 23°.
3. Find both the force normal (perpendicular) to the inclined plane and the force parallel to the plane (see Figure A3-13).

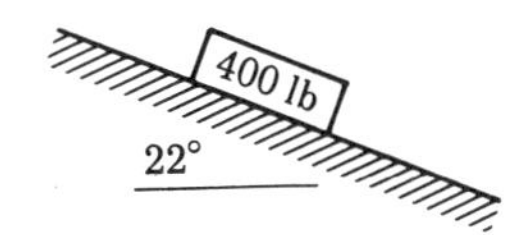

Figure A3-13

4. How much force is the load exerting on each support cable in Figure A3-14?

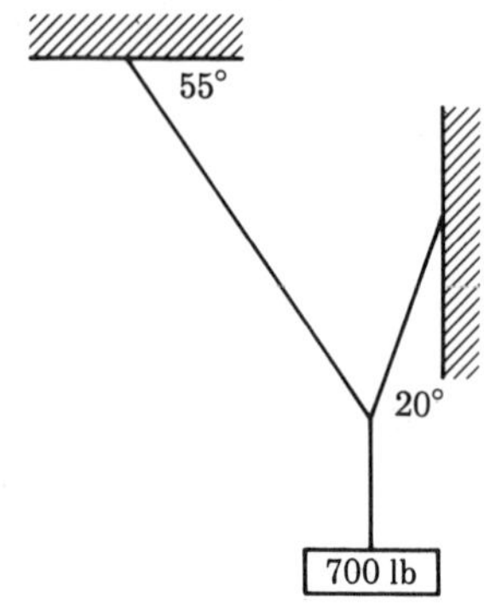

Figure A3-14

Review Unit A-4

Geometry

INTRODUCTION

The discussion in this review unit will be confined to plane geometry. In plane geometry, which deals entirely with configurations in the plane, there are certain concepts we assume from experience. These are the notion of a *point* which has no dimension and denotes position only, and a *straight line* which is taken to extend indefinitely and only one of which can be drawn through two points. A *ray* is defined as that part of a straight line extending in one direction from a point on the line. A *segment* is defined as that part of a straight line lying between two points on the line.

LINES

A number of statements can be made about lines that will have practical applications in many problems.

1. *Parallel* lines are lines that do not intersect no matter how far they are extended, as shown in Figure A4-1.

ℓ_1

ℓ_2

Figure A4-1

2. *Concurrent* lines are lines that intersect at a common point as shown in Figure A4-2.

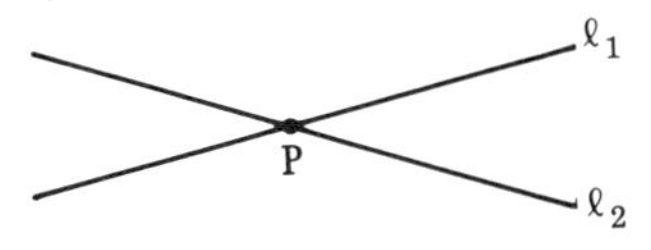

Figure A4-2

3. A *transversal* is a line intersecting one or more other lines. In Figure A4-2 l_1 is a transversal of l_2 and vice-versa.
4. Two lines which are perpendicular to the same line are parallel as shown in Figure A4-3.

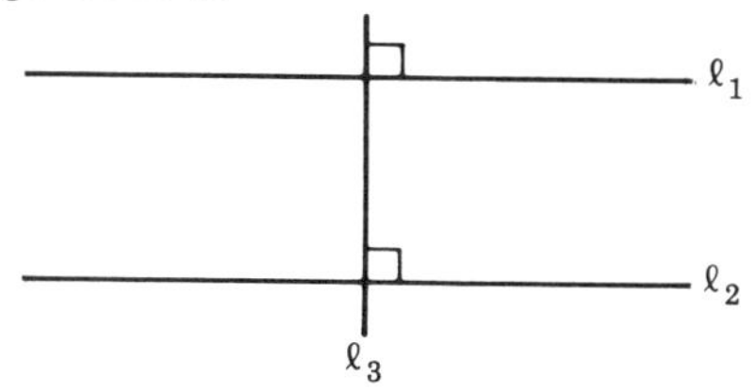

Figure A4-3

5. Two lines which are both parallel to a third line are parallel to each other as shown in Figure A4-4.

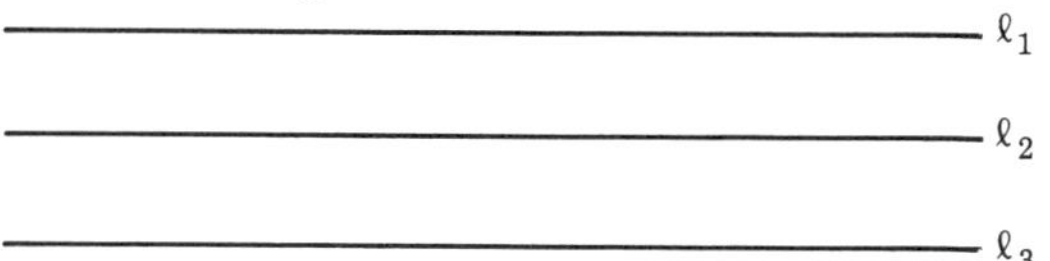

Figure A4-4

6. If two parallel lines are cut by a transversal (see Figure A4-5),
   (a) the *corresponding angles (step angles)* are equal: $\measuredangle 1 = \measuredangle 5$, $\measuredangle 2 = \measuredangle 6$, $\measuredangle 3 = \measuredangle 7$, and $\measuredangle 4 = \measuredangle 8$.
   (b) the *alternate interior angles* are equal: $\measuredangle 3 = \measuredangle 6$ and $\measuredangle 4 = \measuredangle 5$.
   (c) the alternate exterior angles (opposite angles) are equal: $\measuredangle 1 = \measuredangle 8$, and $\measuredangle 2 = \measuredangle 7$.
   (d) the vertical angles are equal: $\measuredangle 1 = \measuredangle 4$, $\measuredangle 2 = \measuredangle 3$, $\measuredangle 5 = \measuredangle 8$, and $\measuredangle 6 = \measuredangle 7$.

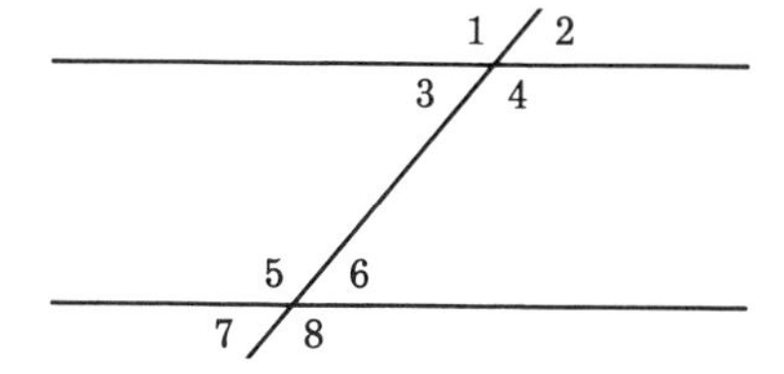

Figure A4-5

ANGLES

Two rays proceeding from one point divide the plane into an internal angle α and an external angle α' as shown in Figure A4-6.

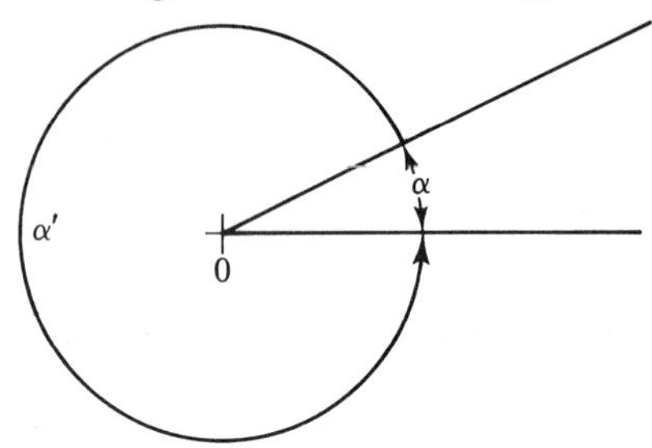

Figure A4-6

The angle is positive if measured in the counterclockwise direction and negative if measured in the clockwise direction.

The complete revolution is a rotation through an angle called the *perigon* (see Figure A4-8).

Angles can be measured by any one of a variety of incremental measures.

One common method is *degrees — minutes — seconds*. There are 360° in one revolution or perigon, 60 minutes in 1°, and 60 seconds in 1 minute. Sometimes decimal fractions of a degree are used instead of minutes and seconds. Thus, for example,

$$72°21'11'' = 72.353°$$

In recent years the *gradian* has come into use. There are 400 gradians in one revolution or perigon. Decimal fractions are the only subdivision used.

The *radian* is a measure that has widespread use where conversion between angular and linear measure is needed. It is defined as the angle subtended at the center of a circle of radius R by an arc of length R, as shown in Figure A4-7. There are 2π radians in a perigon, and 1 radian thus equals 57.3°.

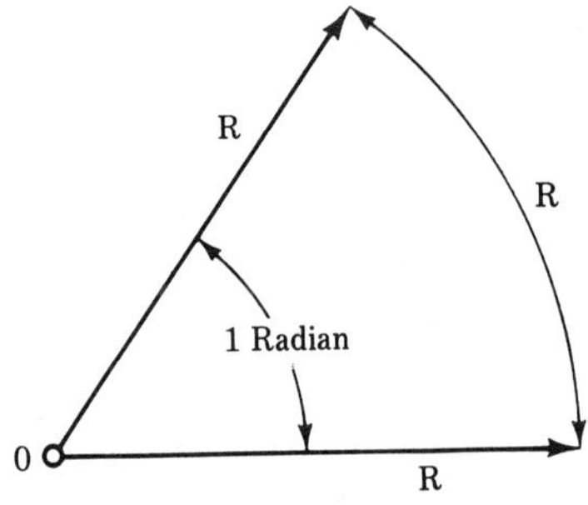

Figure A4-7

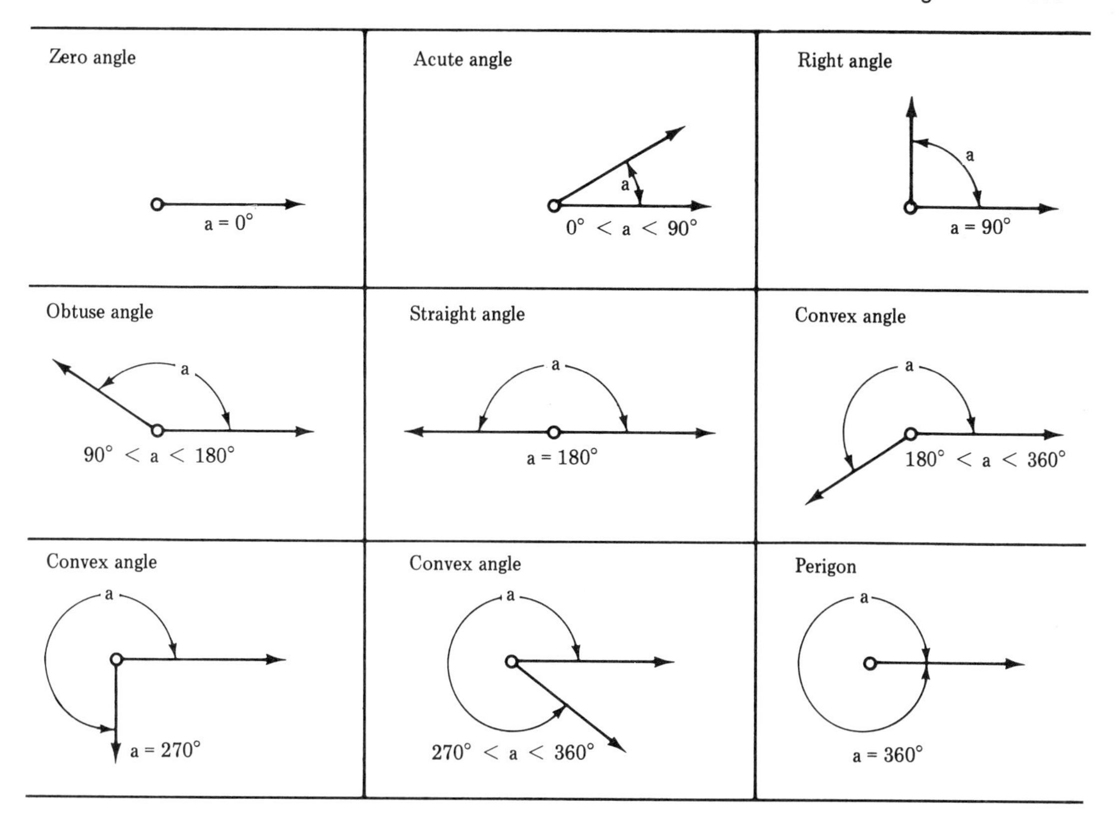

Figure A4-8

The several classifications of angles by size are shown in Figure A4-8. A few additional classifications of angles follow:

1. Two angles are considered *complimentary* angles if their sum is equal to a right angle. See Figure A4-9*a*.

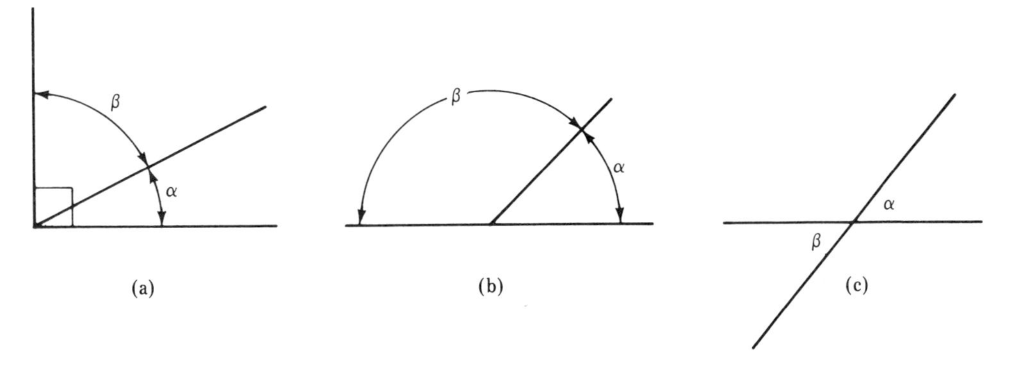

Figure A4-9

2. Two angles are considered *supplementary* if their sum is equal to 180°. See Figure A4-9*b*.
3. Two angles which have a common vertex and whose rays are two common straight lines are said to be *vertical angles*. See Figure A4-9*c*.
4. Two angles are considered to be *conjugate angles* if their sum is equal to the perigon. See Figure A4-6.

TRIANGLES

A triangle is a portion of a plane bounded by three segments called sides. The sum of the interior angles is 180° and the sum of the exterior angles is 360°. See Figure A4-10.

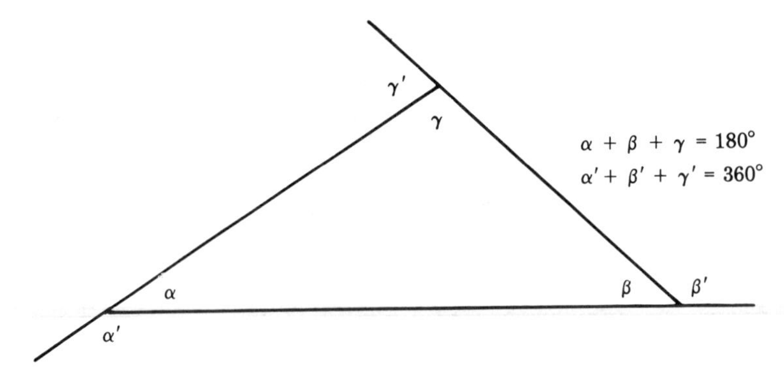

Figure A4-10

More extensive treatment of triangles is provided in Review Units A-1 and A-2.

CIRCLES

A circle is the part of a plane bounded by a curved line, all points of which are equidistant from a point within called the center. The length of the bounding line is called the *circumference C* and the equal distance is called the *radius R*. A *chord* is any straight line segment connecting two points on the circumference. The *diameter D* is any chord running through the center of the circle. Its length is always twice that of the radius. The circumference divided by the diameter is always equal to a constant known as π which has a rounded value of 3.14159. A *tangent* to the circle touches the circle at only one point on the circumference. At that point the tangent is perpendicular to a radius drawn to that point. See Figure A4-11.

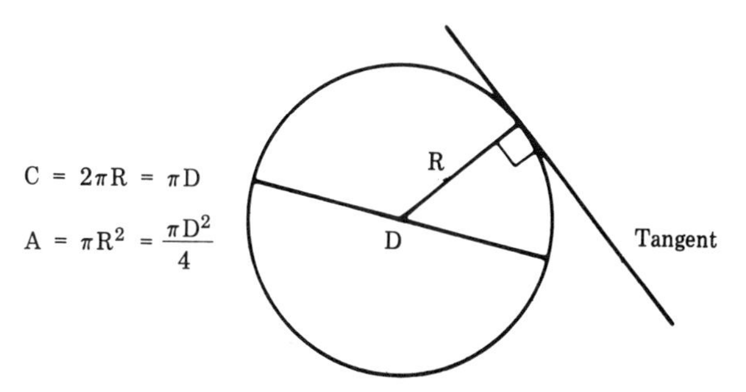

Figure A4-11

PROBLEMS

1. Lines *ab* and *cd* in Figure A4-12 are parallel to each other and line *ef* is perpendicular to line *ab*.

   (a) What other angles are equal to angle *F*?
   (b) What is the relationship between angles *O* and *E*?
   (c) What is the relationship between angles *G* and *L*?
   (d) What is the relationship between lines *cd* and *ef*?

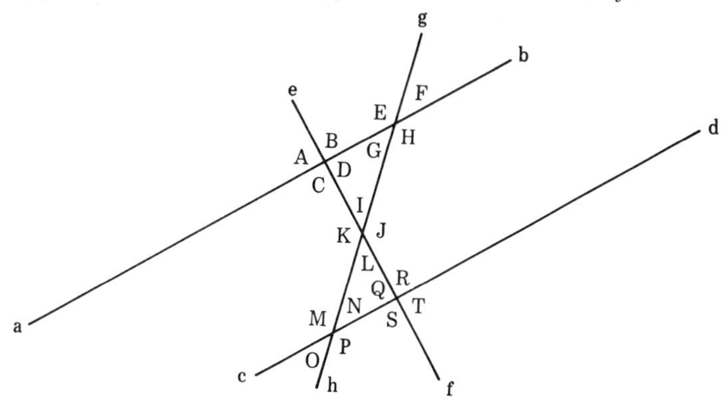

Figure A4-12

2. Determine the direction of *AB* and *OA* in Figure A4-13.

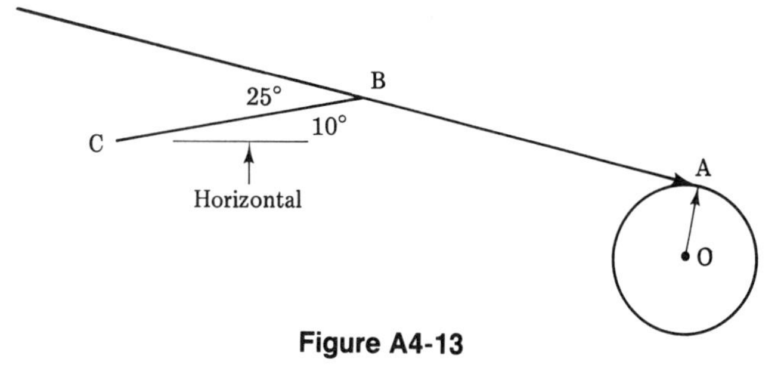

Figure A4-13

3. What is the magnitude of angle A in Figure A4-14?

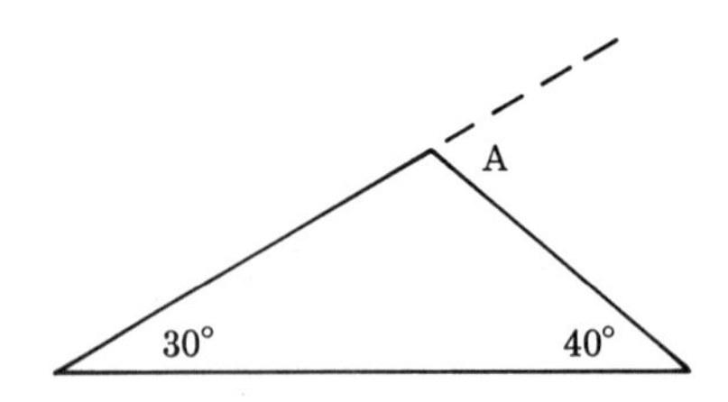

Figure A4-14

Review Unit A-5

Algebra

Algebra is an extremely broad comprehensive subject. This review deals only with a few very specific aspects of algebra that see common use in the study of statics.

One of the basic definitions in algebra states that an equation is a statement of equality. A prime rule states that the equality remains if the same expression is either added to or subtracted from both sides.

Example A5-1: Find the value of x which satisfies

$$8x + 20 = 4x + 48$$

then

$$8x + 20 - 20 = 4x + 48 - 20$$

$$8x = 4x + 28$$

$$8x - 4x = 4x - 4x + 28$$

$$4x = 28$$

A second rule states that the equality of an equation is preserved if both sides of the equation are multiplied or divided by the same expression other than zero. Completing Example A5-1 then,

$$\frac{4x}{4} = \frac{28}{4}, \qquad x = 7$$

The first rule can be expressed in another way: Any term on one side of an equation may be transposed to the other side if the sign is changed.

Applying this to the problem in Example A5-1,

$$8x + 20 = 4x + 48$$

$$8x - 4x = 48 - 20$$

$$4x = 28$$

$$x = 7$$

The above may appear deceptively simple but errors are easily made when dealing with several variables in symbolic form. Care must be taken not to confuse the signs of the equations (including changes due to transposition) and the signs for the values of the variable.

Example A5-2:

$$a + b = 0, \qquad a = -b, \qquad \text{if } b = -20 \quad \text{then} \quad a = +20$$

Example A5-3: $a = -15,\ b = 10,\ d = -12$

$$a + b + c + d = 0, \qquad c = -(a + b + d)$$

$$c = -(-15 + 10 - 12), \qquad c = +7$$

or

$$c = -a - (-b) - d, \qquad c = -(-15) - (10) - (-12), \qquad c = +7$$

The second rule is quite useful when reorganizing equations to solve for an unknown.

Example A5-4:

Given: $W = Kbd^2/L$ Find: b

Multiply both sides by L:

$$WL = \frac{LKbd^2}{L}$$

Cancel out the Ls and divide both sides by Kd^2:

$$\frac{WL}{Kd^2} = \frac{Kbd^2}{Kd^2}, \qquad b = \frac{WL}{Kd^2}$$

Example A5-5:

Given: $c^2 = a^2 + b^2 - 2ab \cos C$ Find: $\cos C$

Transpose $a^2 + b^2$:

$$c^2 - a^2 - b^2 = -2ab \cos C$$

Divide both sides by $-2ab$:

$$\cos C = \frac{c^2 - a^2 - b^2}{-2ab}$$

A simplification of the second rule is a process called cross-multiplication:

$$\text{If } \frac{a}{b} = \frac{c}{d}, \qquad \text{then } ad = bc$$

All terms in any given denominator or numerator need not be cross-multiplied, so the above could have been solved for a as follows.

$$a = \frac{bc}{d}$$

Frequently it is necessary to become involved with exponents of variables. Several basic rules follow:

$$a^{-n} = \frac{1}{a^n}$$

$$a^n \cdot a^m = a^{m+n}$$

$$\frac{a^n}{a^m} = a^{n-m}$$

$$(a^n)^m = a^{n \cdot m}$$

$$(a \cdot b)^n = a^n b^n$$

Example A5-6:

$$2x^{-3} = \frac{2}{x^3}$$

$$3x^2 \cdot 4x^3 = 12x^5$$

$$\frac{15y^5}{3y^2} = 5y^3$$

$$(2x^2)^3 = 8x^6$$

$$(x^2y^3)^2 = x^4y^6$$

Finally, there are often occasions when it is impossible to reduce the solution to a problem down to one equation containing one unknown variable. However, many problems can be solved in terms of two linear equations in two unknowns.

The general form of such equations is

$$a_1x + b_1y = c_1$$

$$a_2x + b_2y = c_2$$

There are two methods for solving such a system. The first method is by *substitution.* It consists of solving one equation for one unknown then substituting into the second equation. The second equation is now in terms of only one unknown. These two steps are:

$$y = \frac{c_1 - a_1x}{b_1}$$

$$a_2x + b_2\frac{(c_1 - a_1x)}{(b_1)} = c_2$$

Careful simplification leads to

$$x = \frac{b_1c_2 - b_2c_1}{b_1 - b_2a_1}$$

Once x is found substitution will yield y.

Example A5-7:

Given: $2x + 3y = 12$ Find: x, y
$4x + y = 14$

Step 1 From $2x + 3y = 12$, $x = 6 - 1.5y$

Step 2 Substituting from Step 1 into $4x + y = 14$, $4(6 - 1.5y) + y = 14$:

$$24 - 6y + y = 14$$

$$-5y = -10$$

$$y = +2$$

Step 3 Substituting $y = +2$ back into $x = 6 - 1.5y$, $x = 6 - 1.5(2)$, $x = +3$.

The other method for solving a system of two linear equations in two unknowns is the *simultaneous* method. It consists of multiplying one of the equations by some coefficient c_3 selected so that one of the following is true: $c_3a_1 - a_2 = 0$, $a_1 - c_3a_2 = 0$, $c_3b_1 - b_2 = 0$, or $b_1 - c_3b_2 = 0$. Addition of the equations will then yield an equation in one unknown. Once this is solved substitution back into either of the original equations will permit determination of the other unknown.

Example A5-8:

Given: $2x + 3y = 12$ Find: x, y
$4x + y = 14$

Step 1 Multiply $2x + 3y = 12$ by 2:

$$4x + 6y = 24$$

Step 2 Subtract the second equation from the first:

$$\begin{array}{r} 4x + 6y = 24 \\ -4x - y = -14 \\ \hline 5y = 10 \\ y = +2 \end{array}$$

Step 3 Substitute $y = +2$ into $2x + 3y = 12$:

$$2x + 3(2) = 12$$
$$x = +3$$

PROBLEMS

1. Given: $A + 2B - 3C - D = 0$ and $A = 5$, $B = 7$, and $D = 4$, find C.
2. Given: $h = C(v^2/2g)$ Find: v.
3. Given: $(2x^3y^5)^2y^{-5} = z^4/y^{-3}$ Find: y.
4. Given: $3x^2 - y^2 = 11$ and $x^2 - 3y^2 = -39$ Find: x, y.

Review Unit A-6

Moment

The *moment* of a force is synonymous with *torque*. It is the measure of the turning effect of the force about a point. This measure is determined from the product of the magnitude of the force and the perpendicular distance of the point from the line of action of the force. A moment of a force, or torque, or turning effect is, therefore, rotational in nature and as such has direction, either clockwise or counterclockwise. The fact that no rotation occurs does not mean that no moment exists. Consider a stubborn jar or bottle cap. The fact that it does not move does not mean that you are not trying, not applying a moment or torque. After all, when dealing with forces, the fact that the floor does not collapse under you does not mean that you are weightless.

All of the above seems disarmingly simple until it is analyzed very carefully. It turns out that the location of the point of application of the force with respect to the center of rotation of the body involved must be considered. Further, not only the magnitude of the force but also its direction with respect to the line connecting the point of application and the center of rotation must be considered.

Suddenly, the whole concept seems much more involved than was initially apparent. However, it can be illustrated that you apply the principles involved daily when you go out a door with a push bar.

Consider the door shown in Figure A6-1. If you wish it to open, that is swing in a counterclockwise direction, you must push it in the direction shown at left, not pull it in the direction shown at right. This is a rather basic principle—the location of the point of application with respect to the center of rotation combined with the direction of the force *together* determine the direction of the moment applied and therefore the direction of either actual or attempted rotation.

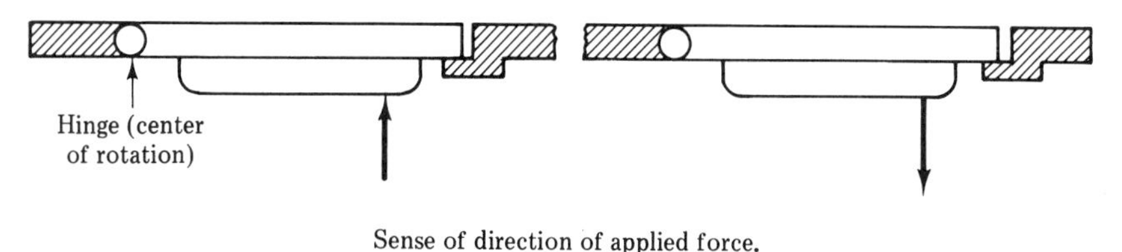

Sense of direction of applied force.
Top view of door.

Figure A6-1

Now, you may observe that from inside the door it is impossible to apply the force anyplace except to the right of the hinge (center of rotation). This can be remedied by considering a revolving door as shown in Figure A6-2. It should be apparent that force 1 will attempt to rotate the door in a counterclockwise direction while force 2 will attempt to rotate the door in a clockwise direction (many revolving doors have a stop to prevent this for safety reasons).

Let us return to the more simple door but add a push bar. How many of us have swiftly pushed on such a push bar with a force we know was adequate to open it only to painfully discover we had pushed the wrong end

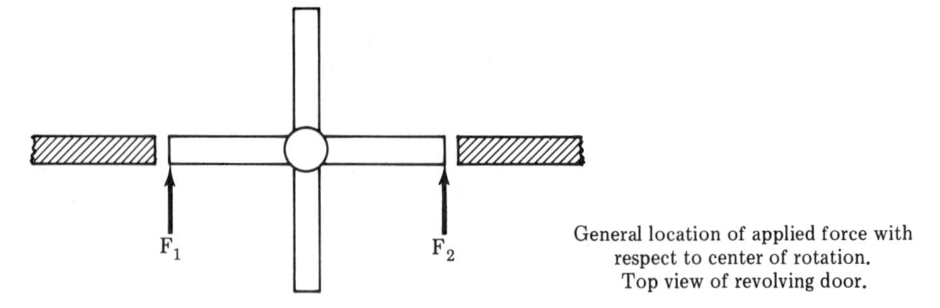

General location of applied force with respect to center of rotation. Top view of revolving door.

Figure A6-2

of the bar? Figure A6-3 illustrates the possibility of pushing the push bar at any one of an infinite number of locations. Experience tells us that force 1 will produce the greatest effect (assuming all the forces are equal in magnitude). Force 2 will be less effective than force 1, and so on. The only variable that is being changed and causing moment effect to change is the *moment arm* of the force. This moment arm is the perpendicular distance from the center of rotation to the line of action of the force. Since a line cannot be perpendicular to a point, it is apparent that the moment arm must be taken perpendicular to the line of action of the force.

In all of the illustrations discussed above, the applied force has been shown as apparently perpendicular to the door surface. While this may appear to be "common sense" simply because you always push a door that way, the direction of the applied force could be varied through a range of approximately 180°. Figure A6-4 shows at left several potential directions for the applied force. Again, the forces all have the same magnitude. At right are shown the approximate moment arms for each case. This illustrates that it is not the distance between the center of rotation and the point of application that is really significant. In reality it is the perpendicular distance between the center of rotation and the line of action of the force that really matters. Only for force 3 does the moment arm happen to be measured

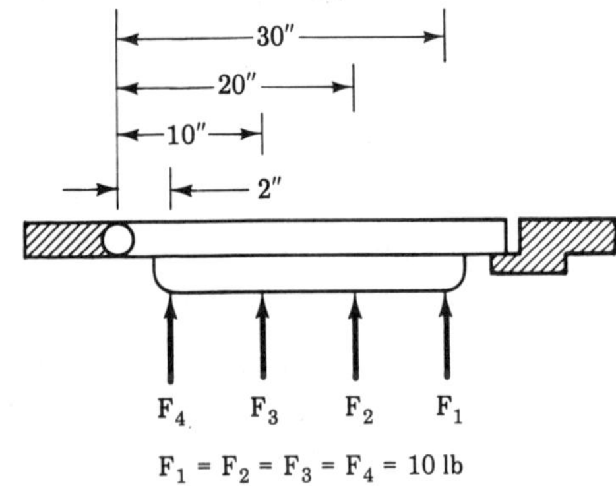

Specific location of applied force with respect to center of rotation. Top view of door.

Figure A6-3

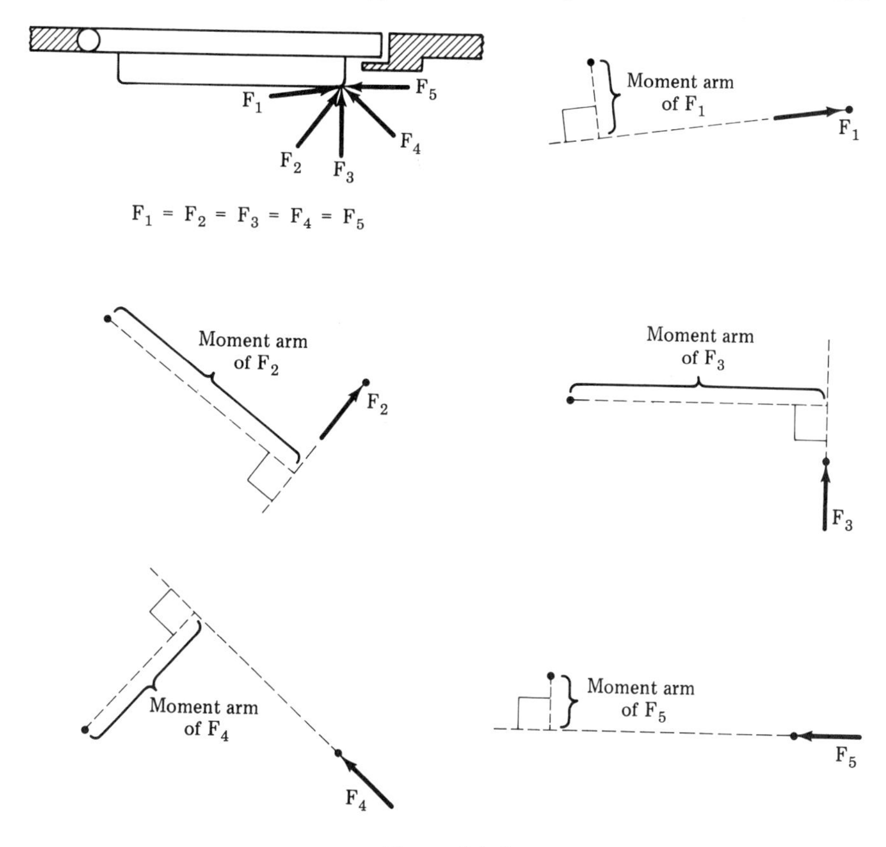

Figure A6-4

along a line parallel to the door. Because of the space between the door and the push bar, force 5 is actually working to keep the door shut.

Finally, it requires no illustration to recognize that the magnitude of the applied force has a direct bearing on the moment effect of the force. The single solitary exception to this is where no moment effect exists because the moment arm is equal to zero. That is, the line of action of the force passes directly through the center of rotation. In Figure A6-4 such a direction would be located somewhere between the directions of F_4 and F_5.

In summary, the moment of a force is its tendency to rotate a body about some center. It is determined by multiplying the magnitude of the force by the length of the moment arm of the force. The latter is measured in a straight line from the center of rotation to the line of action of the force with the provision that the line measured along is drawn perpendicular to the line of action of the force. The units of moment are force times length, for example, lb-ft, kip-ft, N-m, dyne-cm. Moment has direction, usually designated as $-$ (negative) for clockwise and $+$ (positive) for counterclockwise. The equation for a single moment is $M = Fd$. The equation for the sum of several moments is

$$\Sigma M = M_1 + M_2 + M_3 + M_4 + \cdot \cdot \cdot$$

or

$$\Sigma M = F_1 d_1 + F_2 d_2 + F_3 d_3 + F_4 d_4 + \cdot \cdot \cdot$$

It is extremely important that in a moment equation direction (+ or −) be given only to each complete individual moment and, if it is not equal to zero, to the sum of the moments. The directions for the individual moments should be determined by inspection. *Do not enter any directions (+ or −) for either the forces or the moment arms. Use absolute values for both.*

Example A6-1: Shown in Figure A6-5 (top view) is a playground merry-go-round that has a diameter of 6 m and a frictional torque of 120 N-m. What magnitude for F_1 is required to start the merry-go-round moving? For F_2?

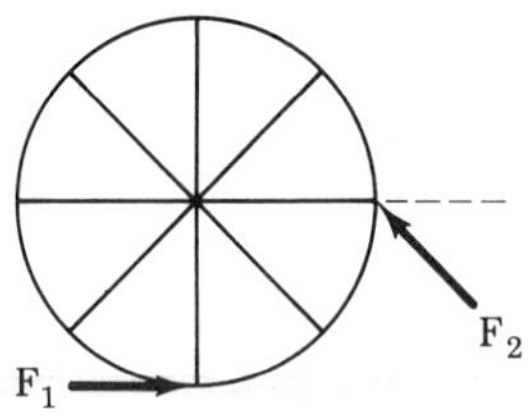

Figure A6-5

F_1 has a moment arm equal to the radius, or half the diameter, or 3 m. If $M = Fd$ (d being moment arm not diameter) $F = M \div d$.

$$F_1 = M \div d_1 = 120 \text{ N-m} \div 3 \text{ m}, \qquad F_1 = 40 \text{ N at } 0°$$

F_2 has a moment arm that will have to be calculated.

$$\theta = 45°$$

$$\sin \theta = \frac{d_2}{r}$$

$$d_2 = r \sin \theta = 3 \sin 45° = 2.12 \text{ m}$$

$$F_2 = M \div d_2 = 120 \text{ N-m} \div 2.12 \text{ m}$$

$$F_2 = 56.6 \text{ N at } 135°$$

Figure A6-6 reminds us that force has direction and it must be stated. However, it also points out that when the magnitude of a force is determined

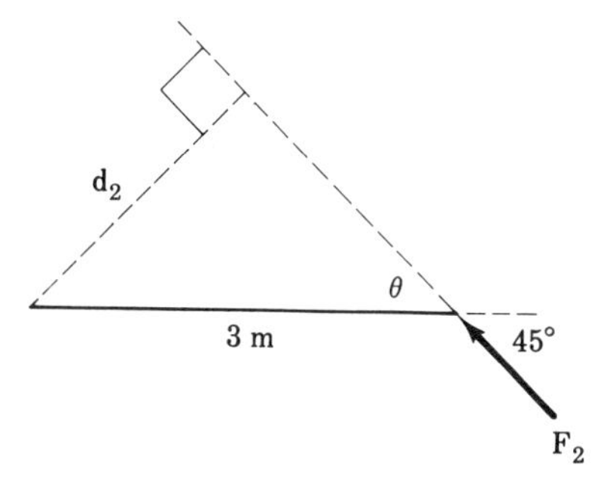

Figure A6-6

from moment in this manner, absolute values should be used for all three variables. Once the magnitude of the force is known, direction can be added by considering the general direction of the line of action of the force and either the given sense of the force or if the sense of the force is not given, it must be determined by inspection from the location of the line of action of the force with respect to the center of rotation and the required or desired direction of rotation.

Example A6-2: Shown in Figure A6-7*a* is the top view of a house jack. A clockwise moment of 100 lb-ft is required to operate the jack. What is the force needed at point *A* along the line of action indicated?

First, the length of the moment arm involved must be determined. A perpendicular to the line of action of the force is drawn passing through the center of rotation as shown in Figure A6-7*b*. Then

$$\sin 40^\circ = d/l$$

where l is the length of the jack handle.

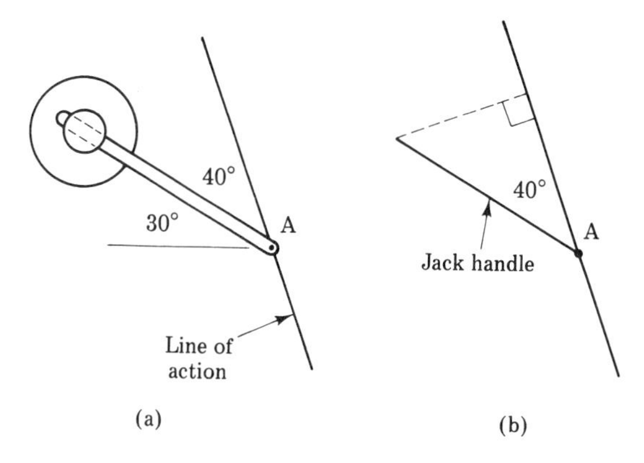

Figure A6-7

$$d = l \sin 40° = (2 \text{ ft})\sin 40°, \qquad d = 1.29 \text{ ft}$$

From

$$M = Fd \qquad F = M \div d = 100 \text{ lb-ft} \div 1.29 \text{ ft}, \qquad F = 77.5 \text{ lb}$$

Now to achieve clockwise rotation, inspection indicates that this force must be directed down and to the left (rather than up and to the right, the only other choice the given line of action permits). The exact direction can be determined by use of geometry to be 290°. The final answer therefore is

$$F = 77.5 \text{ lb at } 290°$$

Example A6-3: What is the total moment being exerted on the crank in Figure A6-8?

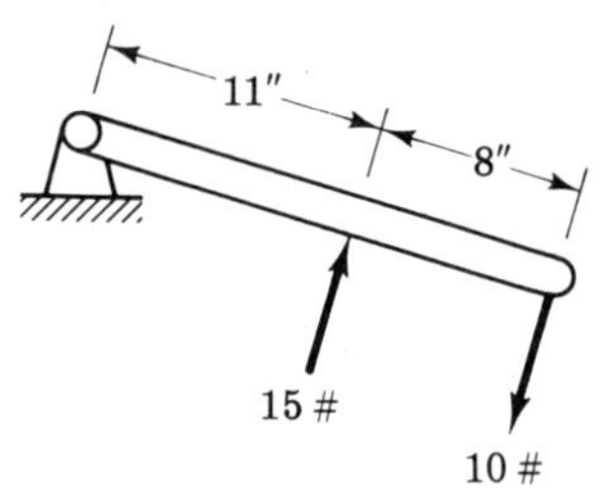

Figure A6-8

Inspection suggests that the forces are perpendicular to the crank so the moment arms are given. Taking moments about the center of rotation:

$$\Sigma M = M_1 \leftrightarrow M_2$$

$$M_T = +(15 \text{ lb})(11 \text{ in}) - (10 \text{ lb})(19 \text{ in})$$

$$M_T = -25 \text{ lb} \cdot \text{in} \qquad \text{or} \qquad M_T = 25 \text{ lb} \cdot \text{in cw}$$

PROBLEMS

1. What is the total moment being exerted on the merry-go-round in Figure A6-9? (radius is 8 ft)

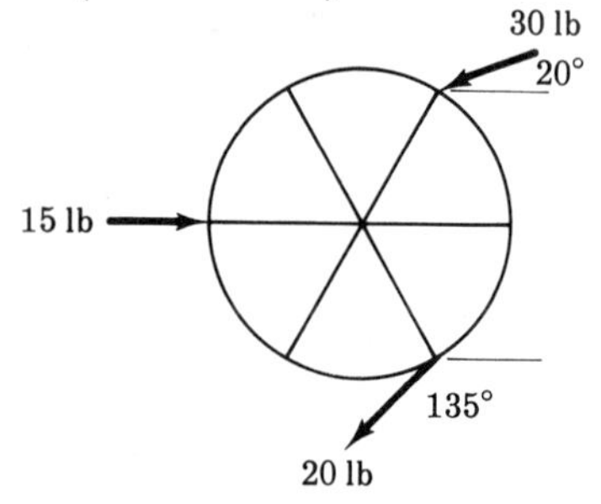

Figure A6-9

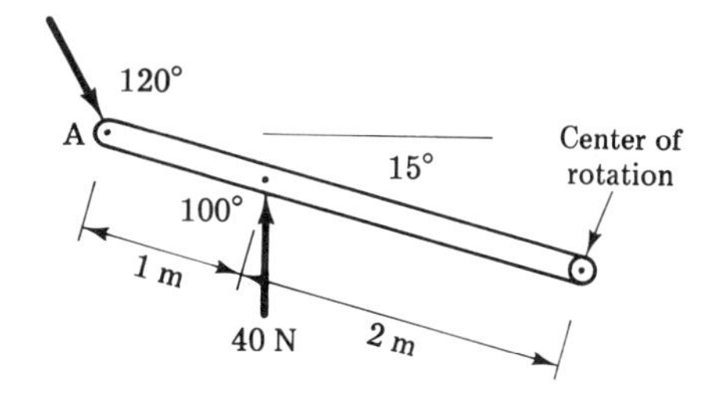

Figure A6-10

2. What force at A is needed to start motion if a frictional torque of 900 N-m exists for the lever shown in Figure A6-10?

Review Unit A-7

Angular Measure

Although an angle can be measured in any convenient units, the most common is still the degree, there being by definition 360° in one revolution.

It is also quite common to measure angles from the horizontal and vertical, although this system is sometimes rotated to some extent for convenience. It is further standard practice to measure a true angle with reference to a horizontal ray drawn to the right as shown in Figure A7-1. Angles are measured as positive in the counterclockwise direction and negative in the clockwise direction.

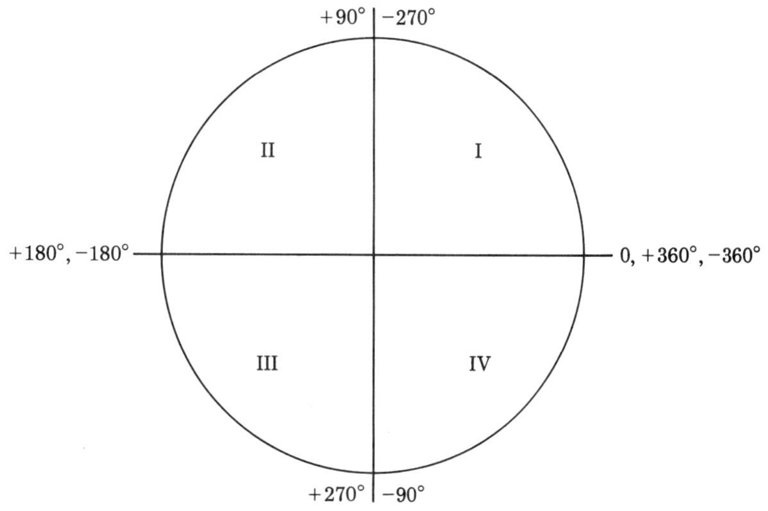

Angular measurement in degrees

Figure A7-1

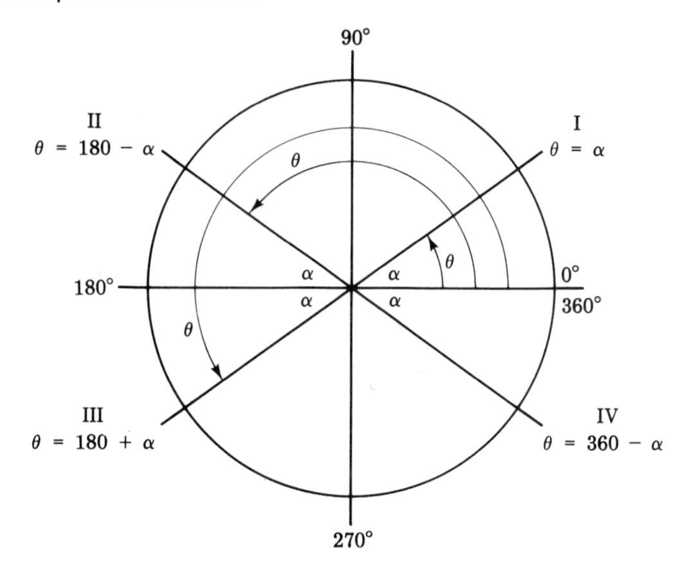

Acute angle in any quadrant

Figure A7-2

In Figure A7-2 lines *ab* and *cd* have been drawn through the origin ($x=0$, $y=0$) so that all the acute angles α are equal. The true angles, quadrant by quadrant are α, $180-\alpha$, $180+\alpha$, and $360-\alpha$.

In Figure A7-3 perpendiculars have been drawn to the horizontal from the intersection of every nonhorizontal ray of every α with a circle of radius r. Thus, four congruent triangles are formed, differing only in position. They

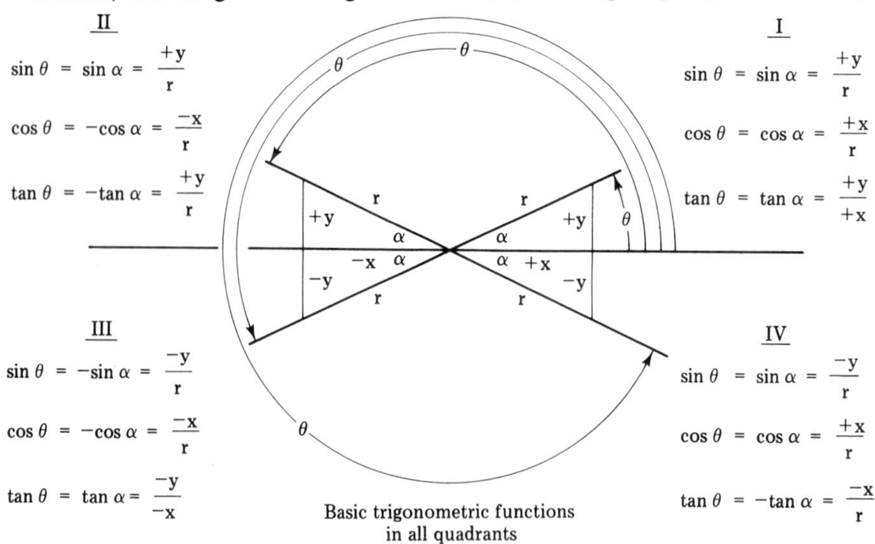

Basic trigonometric functions in all quadrants

Figure A7-3

are all right triangles and the relationships or r, x, and y are identical in magnitude for each. The only difference, because of location, is that in some quadrants either x or y or both are negative. Thus, the functions of θ in any quadrant are equal to the functions of the acute angle α with correction for direction as shown.

PROBLEMS

1. Enter $y = +2$ and divide by $x = -5$ to obtain a tangent equal to -9.4 and your calculator further indicates that $\theta = -21.8°$. What is the true positive angle?

2. What is the true angle of vector A in Figure A7-4?

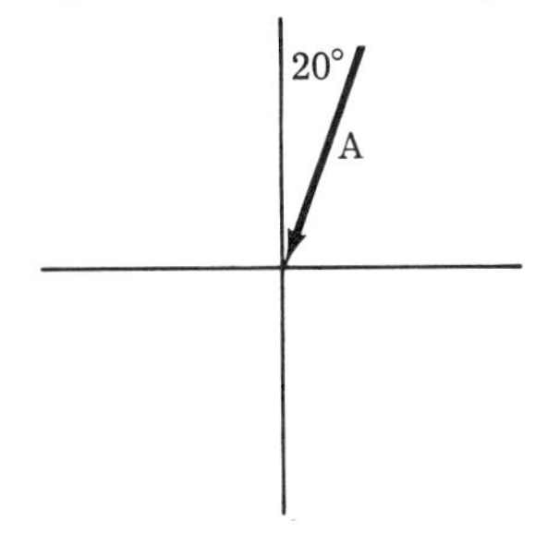

Figure A7-4

3. Enter $x = -4$ and $y = 5$ into your calculator knowing that y is positive. Your calculator indicates that $\theta = -51.3°$. What is the true positive angle? What is the true negative angle?

Review Unit A-8

Force, Mass, Center of Gravity

FORCE

We all feel that we know what force is, yet the definition can be elusive. Frequently, *force* is defined as that which causes a body to accelerate. In other words to speed up or slow down a child's wagon, for example, one must either push or pull on it. That pushing or pulling is the application of what is called force.

The above process seems simple enough and we make daily use of it in operating our own bodies, moving doors, drawers, and chairs, and performing other common activities. However, two special cases must be considered both of which are very common.

First, the above example required direct physical contact between the two bodies in order for the force to be transmitted from one to the other. There are two exceptions to this: gravitation and magnetism. *Gravitational force* is the attractive force that exists between any two bodies. It is dependent upon the mass of each of the bodies and the distance between them. The gravitational forces that we are most familiar with are the ones that exist between the earth and all the loose objects on its surface. Everytime we drop something, it is evident that a force is accelerating the dropped object without any physical contact with a second body. *Magnetic force* occurs naturally between certain materials but can also be produced through the use of an electric current. Again, no physical contact exists to transmit the force. Indeed in what is probably the largest scale application of magnetic force, the control of the plasma for the fusion power reactor, the lack of contact between the high-temperature plasma and its container is essential to successful operation.

Second, we are all aware that force exists without evidence of motion. We stand on solid ground and know that a force exists between our feet and the ground. It becomes very evident if some weak material (such as our eyeglasses or an ant) is placed between our foot and the ground. The wagon we were pulling earlier does not move until we apply some minimum force which is equal to the maximum friction force (see Review Unit A-9). In this case we can feel some strain in our arm and feel a force acting between the wagon handle and our hand even though no motion takes place. What is unseen in both these cases where force without motion occurs is the internal agitation that takes place at the atomic level as a result of the applied forces. Thus, motion really does exist in these cases but not on a scale we can see with the unaided eye.

MASS

We referred above to the mass of a body or object without defining it. Any body is made up of a definite quantity of matter. Inspecting this matter we find it to be made up of groups of molecules, each molecule having the same physical characteristics as a large piece of the material. Closer examination shows the molecules to be made up of atoms, any atom representing one of the 108 elements that make up all matter. The atom turns out to be made of electrons, protons, and neutrons plus some minor particles. Modern theory suggests that all particles are composed of basic building blocks called quarks. Ultimately then, a certain size of mass consists of a certain number of quarks.

The weight of a body is simply the force with which an extremely large mass is attracting the body. Thus, the weight of your body is greater here on earth than on the moon (because the moon's mass is less than the earth's) even though your mass (a collection of particles) remains constant.

The relationship of mass and weight (a force) permits simple measurement of the unknown mass of a piece, or group of pieces, of some material. The pan balance shown in Figure A8-1 provides for the unknown mass to be placed on one side, tipping the beam to that side. The pan hangers are hung from the beam at points equidistant from the fulcrum of the beam. Since the moment arms for the forces pulling down on the ends of the beam are the same, the beam will balance or be horizontal if the forces are also equal. The gravitational effect varies as altitude and latitude vary but would be the same for both sides of the balance. Equal forces would therefore require that the masses be equal. If we add known pieces of mass to the other pan until the beam is horizontal, the two masses are equal and we can simply sum up the total of the known masses which is then also the mass of the unknown. (See Appendix A-6 if necessary.)

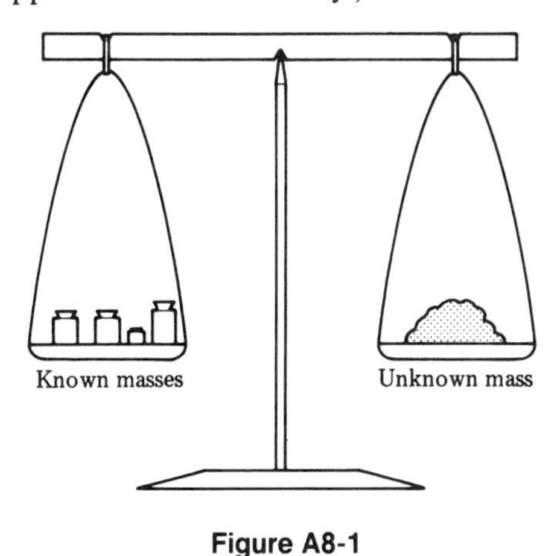

Figure A8-1

CENTER OF GRAVITY

In order to fully define a force, it must have magnitude, direction, and a point of application (or at least a point through which the line of action of a force passes). In considering an object or body on or near a large body such as the earth, the force of gravity (weight) can be taken to be directed towards the center of the earth. It can also be considered to be directed perpendicular to a plane tangent to the curvature of the earth at the objects' location. The magnitude of the force can be calculated from

$$w = mg$$

where w is the weight, m is the mass, and g is the constant for that location.

All of the above is necessary and quite correct but does not locate where on the object or body the line of action of the force of gravity must pass. One way to visualize this is to consider a very thin rod erected on earth aligned so that its one end points toward the center of the earth as shown in Figure A8-2*a*. Let us put a conical point, like a sharp pencil point, on the other end. If we balance our object or body on this point, the rod's sharp end points a line through the body as shown in Figure A8-2*b*. If we do this again and again, each time varying the exterior location on the body that we place on the point of the rod, we will have drawn many imaginary lines through the body. There will be one point in the body that every one of those lines will pass through. This is the *center of gravity* of the body; it is the point that the line of action of the gravitational force (weight) will always pass through no matter how the body is oriented.

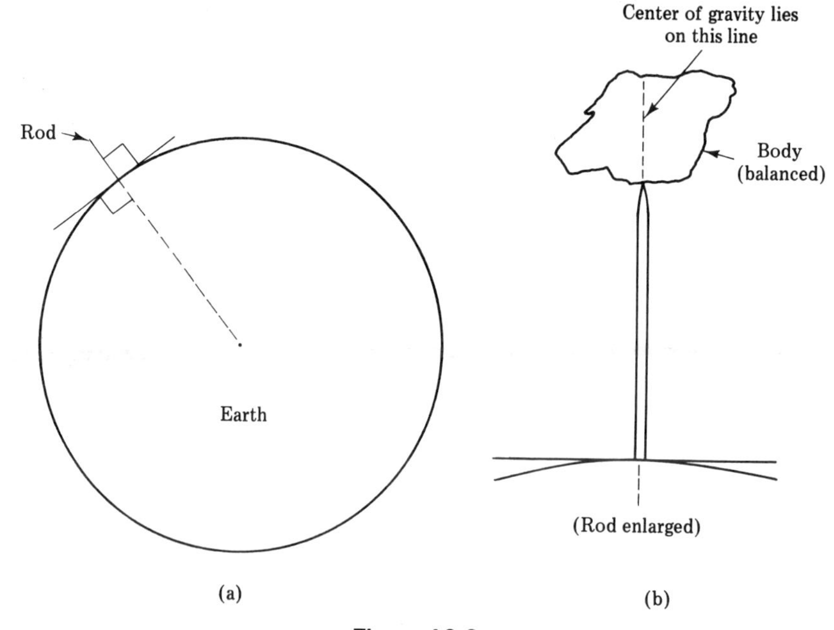

Figure A8-2

Review Unit A-9

Friction

Experience has undoubtedly already shown you that friction is a force that resists motion and is dependent on the surface condition of the two surfaces that are attempting or achieving relative motion. Thus, suntan lotion is easily applied to clean skin, but considerable difficulty is encountered (to say nothing of discomfort) if one attempts to apply a second coat of suntan lotion after sand has stuck to the first coat. Friction may be either desirable or undesirable depending on the situation. The skier heading downhill on a slope is interested in minimizing friction in order to maximize speed and has probably treated the bottom of the skis to help achieve this. However, in driving to and from the slopes it is certainly desirable to have any ice on the road coated with sand to achieve just the opposite effect, that is, to maximize friction and minimize relative motion (spinning in place of wheels in this case). The same is true in a machine where we lubricate shaft bearings to minimize friction but desire a maximum of friction between a belt and a pulley to minimize slippage.

As can be seen from the above discussion, surface condition is a major factor in determining friction. Thus, a relatively rough surface will have more difficulty sliding on another relatively rough surface than on a relatively smooth surface. Two relatively smooth surfaces will slide on each other even more readily than either of the above cases. These are illustrated in Figure A9-1.

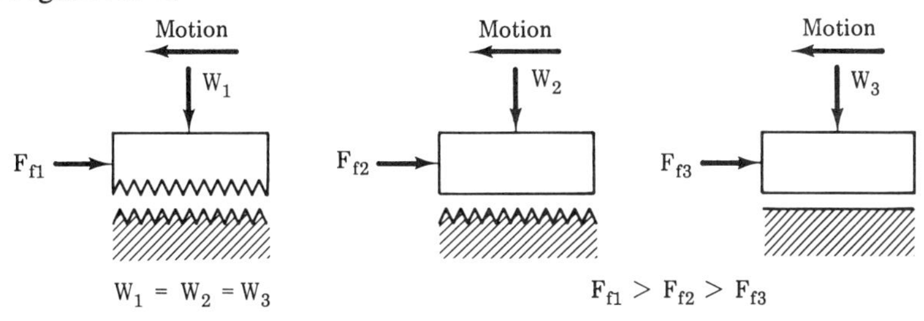

Figure A9-1

However, there is a second factor involved, the force with which the two surfaces are pressed together. Thus, it is much easier to slide an empty chair across a floor than to slide one that has a child sitting in it. We are not talking about the force required to accelerate the chair from rest to some velocity but merely the force required to start any motion at all and to maintain a constant velocity. These are illustrated in Figure A9-2.

The maximum friction force is that which must be overcome in order to start motion. It can be shown experimentally that for two known surfaces

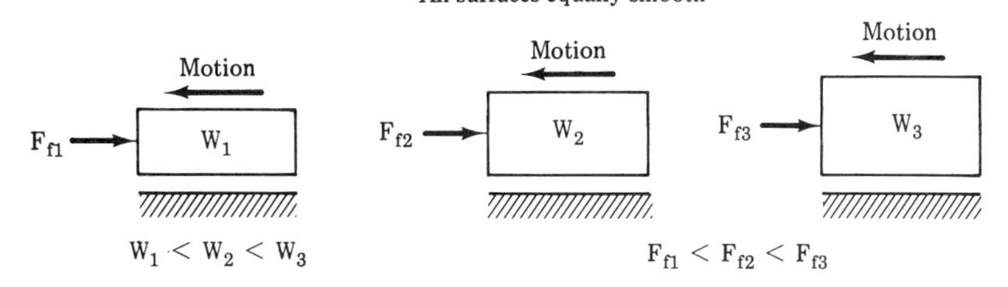

Figure A9-2

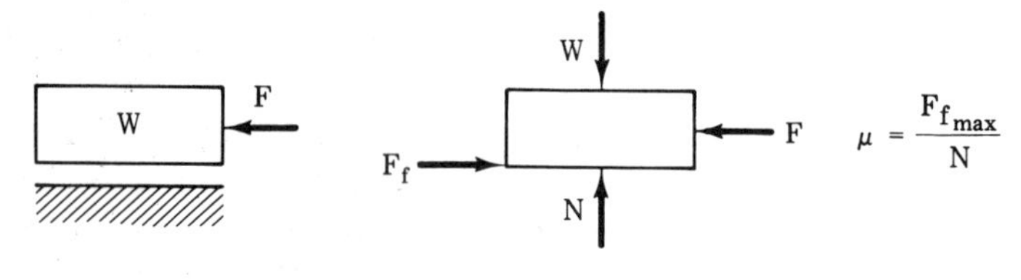

Figure A9-3

attempting to slide on each other as shown in Figure A9-3*a*, this friction force is directly proportional to the perpendicular force (often called the normal force *N*) pressing the two surfaces together. This constant of proportionality is unique to the two materials and their surface condition and is known as the coefficient of friction, μ.

Several observations should be made at this point. First, if the applied force is less than the maximum friction force, a friction force does still exist, equal and opposite to the applied force but less than the maximum friction force. In other words, if you attempt to slide a chair across the floor but fail to move it, a friction force equal and opposite to the force you applied is present (you can certainly feel this resisting force opposing you). Second, the equation given above does not consider area at all. Certainly the greater the pressure (force per unit area or $F \div A$) between the two surfaces, the more the rough "points" are forced together and the harder it will be to dislodge them as shown in Figure A9-4. However, if the normal force *N* is constant, as the contact area is decreased not only does the contact pressure go up but the number of "points" that need to be dislodged goes down. Thus, with pressure inversely proportional to area and the number of "points" in contact directly proportional to area, the two effects of area cancel each other out.

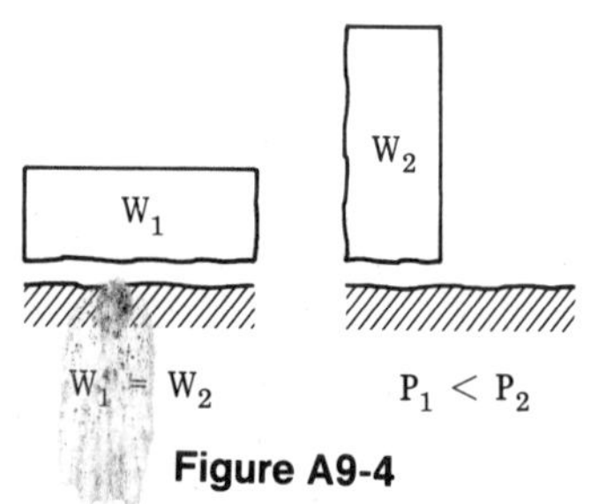

Figure A9-4

Third, it can be shown experimentally that the coefficient of friction during movement is slightly less than that for a stationary situation. This can be most readily explained by Figure A9-5 where it can be seen that under moving or kinetic conditions, the "points" of the surfaces do not get a chance to "settle in" as they do under stationary or static conditions. Thus, the surfaces are more readily dislodged under constant velocity than when starting from rest.

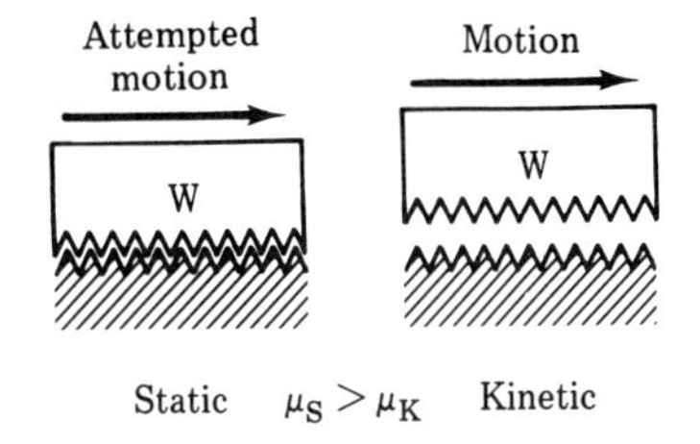

Figure A9-5

Example A9-1: A 100-lb skid is loaded with 800 lb of rock. If the coefficient of static friction between the skid and the floor is 0.6, how much force will be required to start the loaded skid moving?

$$\mu_s = \frac{F_{f\max}}{N}, \qquad F_{f\max} = \mu_s N, \qquad N = W_{\text{tot}}$$

$$F_{f\max} = \mu_s W_{\text{tot}} = 0.6(900 \text{ lb}), \qquad \boldsymbol{F_{f\max} = 540 \text{ lb}}$$

Example A9-2: Two men are applying a total force of 500 N horizontally to slide a 100-kg box of machine parts along a horizontal floor at a constant velocity. What is the coefficient of kinetic friction?

$$\mu_k = \frac{F_f}{N}, \qquad N = W_{\text{tot}} = (100 \text{ kg})(9.81 \text{ N/kg}) = 981 \text{ N}$$

$$\mu_k = \frac{500}{981} N, \qquad \boldsymbol{\mu_k = 0.51}$$

PROBLEMS

1. What horizontal force must be applied at the front of sled 1 to maintain a constant velocity? What is the tension in the rope in between the sleds? See Figure A9-6.

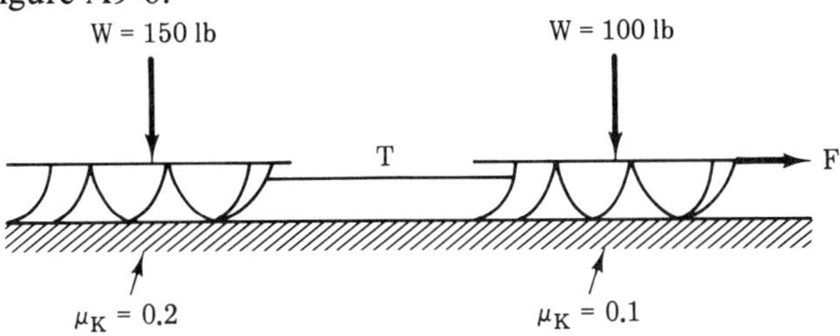

Figure A9-6

2. If motion is just about to start, what is the coefficient of static friction between block 1 and the surface it is on? See Figure A9-7.

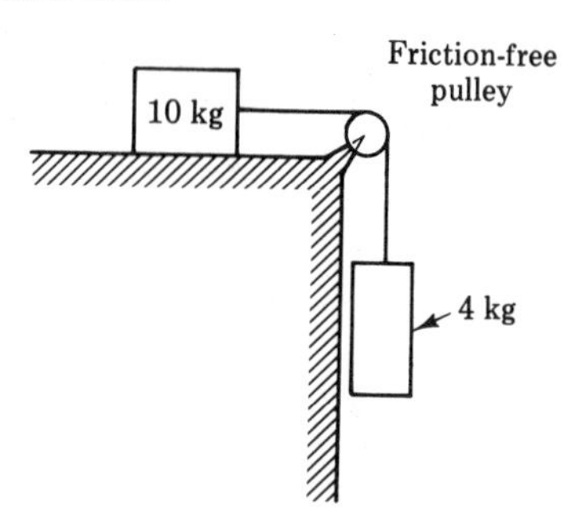

Figure A9-7

Review Unit A-10

Mechanical Advantage

Work is frequently defined as force times distance. There are a number of machines that will alter the nature of the work so that output distance no longer equals input distance and output force no longer equals input force. However, at best, output work can be no greater than input work. This best, which we call 100% efficiency, is never achieved because of friction and other internal losses. The reason we change the nature of work, despite incurring some losses in the process, is to be able to apply a small force rapidly and have a large slow-moving force result. There are also exceptions where we desire just the reverse effect. A common example of the first case is the bumper jack used on cars. A relatively long motion of the end of the jack handle with a relatively small applied force results in lifting a corner of a car a very small amount. The human arm extended by a baseball bat accomplishes just the opposite. A relatively large force moves a relatively short distance resulting in a small force moving a long distance (giving the ball a high velocity, if hit).

The ratios involved in all of the above examples are known as *mechanical advantage.* There are two types of mechanical advantage, the ideal and the actual. Generally, the motion of a device is fixed so ideally the mechanical advantage is the ratio of input distance to output distance.

$$\text{IMA} = \frac{S_{\text{in}}}{S_{\text{out}}}$$

We have already considered that a machine transmits work at something less than 100% efficiency. It follows that any losses show up in the transmission of the force component of work since they didn't show up in the distance component. The ratio of the forces then is known as the actual mechanical advantage:

$$\text{AMA} = \frac{F_{\text{out}}}{F_{\text{in}}}$$

It further follows that the efficiency can be evaluated as follows:

$$\text{Eff}\,\% = \frac{\text{Work}_{\text{out}}}{\text{Work}_{\text{in}}} = \frac{F_{\text{out}}S_{\text{out}}}{F_{\text{in}}S_{\text{in}}} = \frac{\text{AMA}}{\text{IMA}}$$

There are many machines that will accomplish this change in the nature of work. We shall look at just two, the lever and the pulley.

Levers may be categorized into three classes as shown in Figure A10-1.

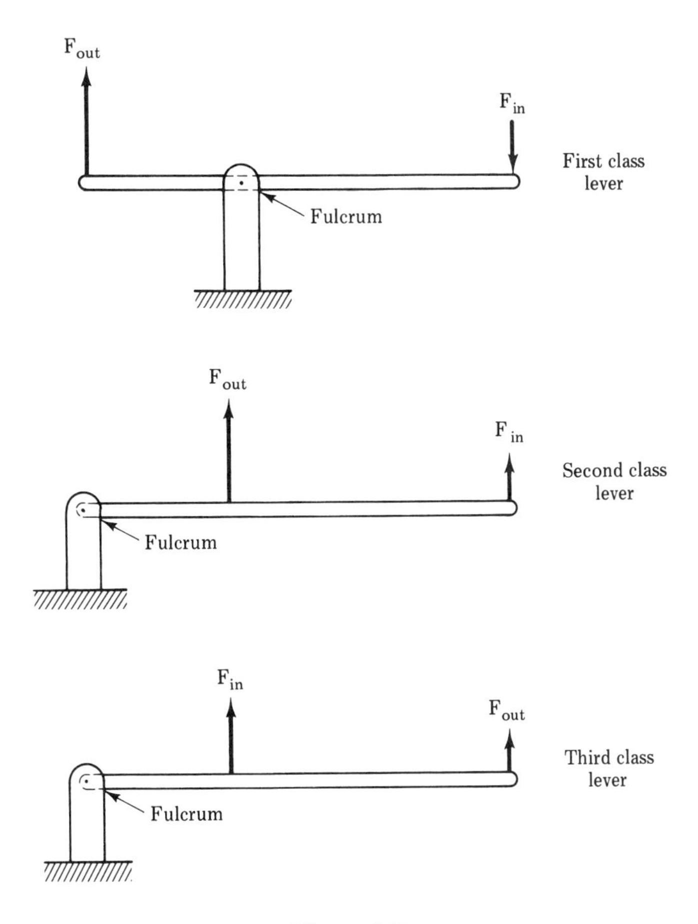

Figure A10-1

As can be seen, the classifications are a function of the relative locations of the input force, output force, and fulcrum (pivot point). Note that any time the input force is in between the fulcrum and the output force, both mechanical advantages are less than 1.

The bumper jack previously mentioned, a pair of pliers and a claw hammer being used to pull a nail are all examples of a first-class lever, the fulcrum in each case being in the middle. A simple nut cracker and a wheelbarrow are both examples of a second-class lever, the fulcrum being on one end and the input force on the other. The human arm previously mentioned and the catapult used in ancient warfare are both examples of third-class levers.

Example A10-1: What input force is needed to operate the lever in Figure A10-2*a*? (Assume 100% efficiency.) How far must the input force move to move the output force 1 in.?

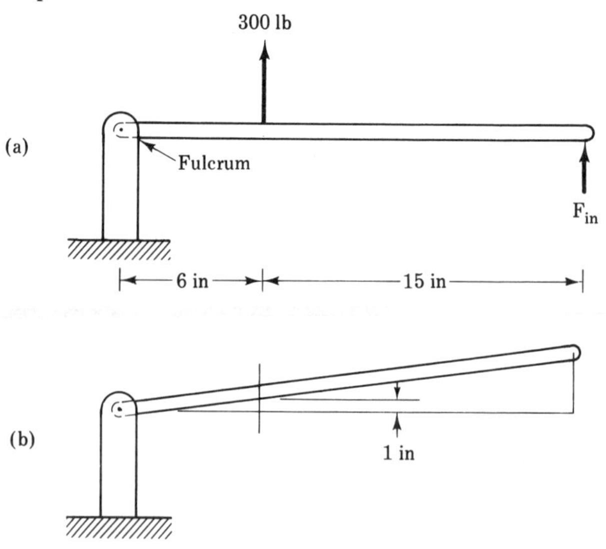

Figure A10-2

The input force could, of course, be found by the use of moments. However, if the second question is answered first we can take a somewhat different approach. Figure A10-2*b* shows the lever with the output force moved 1 in. The lever, the lines of action of the forces, and a horizontal line form two similar triangles both of whose hypotenuses are known as well as the altitude of the smaller. Since the triangles are similar their sides all have the same ratios.

$$\frac{d_{in}}{d_{out}} = \frac{S_{in}}{S_{out}}, \qquad S_{in} = S_{out}\frac{d_{in}}{d_{out}} = (1 \text{ in.})\frac{21_{in}}{6_{in}} \qquad \mathbf{S_{in} = 3.5 \text{ in.}}$$

Since Eff% = 100%, AMA = IMA

$$\frac{F_{out}}{F_{in}}\frac{S_{in}}{S_{out}}, \qquad F_{in} = F_{out}\frac{S_{out}}{S_{in}} = (300\text{ lb})\frac{1}{3.5}\text{ in.}, \qquad \boldsymbol{F_{in} = 85.7\text{ lb}}$$

A single pulley, as shown in Figure A10-3, can be considered as a first-class lever. By itself it serves only the purpose of changing the direction of the resulting force to something different from that of the applied force. Examination of the pulley reveals that even with no slippage (100% efficiency), the mechanical advantage of a single pulley is simply 1.

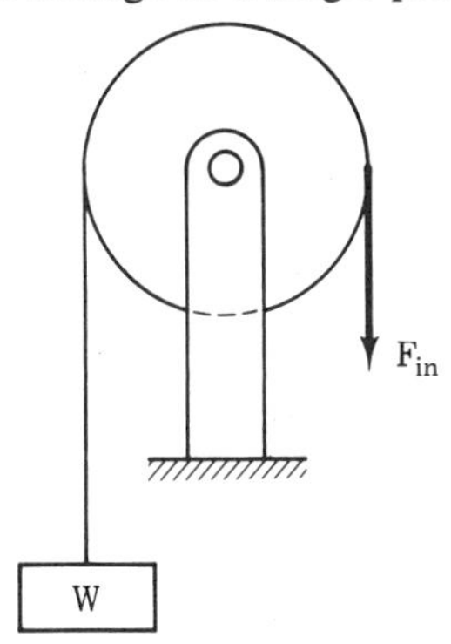

Figure A10-3

A pulley that is part of a system of pulleys may be fixed as in the case just examined or it may be movable. Figure A10-4 shows some different combinations of fixed and movable pulleys. A movable pulley may be considered as a second-class lever while the fixed pulley is a first-class lever. In both cases, under static conditions the tension in the ropes at both sides of the pulley will be identical. In both cases the force at the pulley axle will equal the sum of the two rope tensions.

Let us analyze the three pulleys one by one. First, in Figure A10-4*a* we have three movable pulleys. At pulley 1 the output force is divided in half, one-half supported by pulley 2. That load on pulley 2 is divided in half, each half equal to one-quarter of the output force. Again, the one force is supported by the ceiling while the other is supported by pulley 3. The load on pulley 3 (equal to one-quarter of the output force) is divided in half. Once again the one force is supported by the ceiling while the other is our input force. Each of those is equal to one-eighth of the output force. The mechanical advantage of this system is therefore 8.

In Figure A10-4*b* we have two fixed and one movable pulley. A quick inspection reveals that pulley 3 merely changes the direction of the input force. Analysis of pulleys 1 and 2 individually and then together reveals that the input force must be one-third of the output force. Another way to look at it is that there is really only one rope involved. The tension throughout that rope equals the input force. If we isolate pulley 1 there are three forces equal

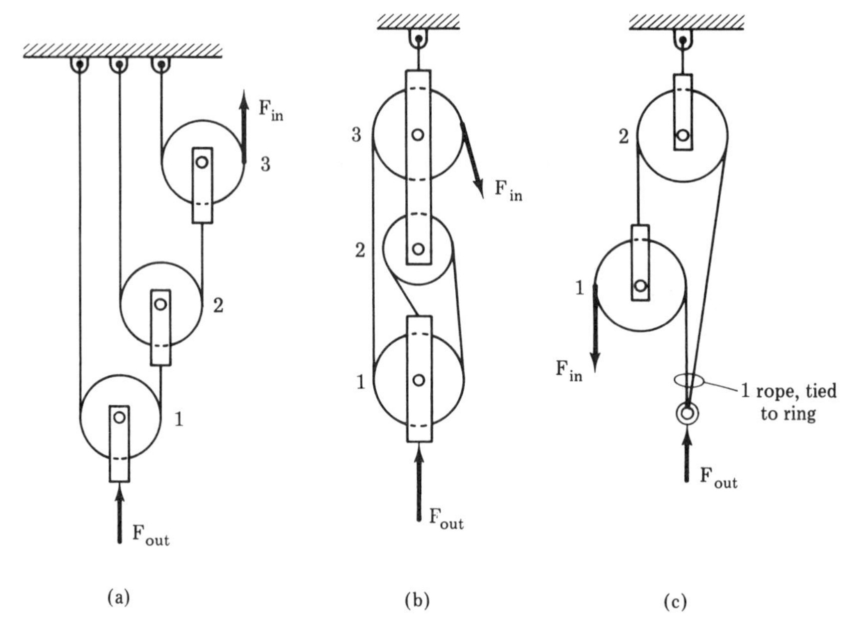

Figure A10-4

to the input force providing the output force. The mechanical advantage of this system is therefore 3.

In Figure A10-4*c*, since the rope is tied at the ring it acts like two different ropes. The ropes going down on either side of pulley 1 carry a force equal to the input force. The rope going from pulley 1 on around pulley 2 to the load must then carry a force equal to twice the input load. At the ring,

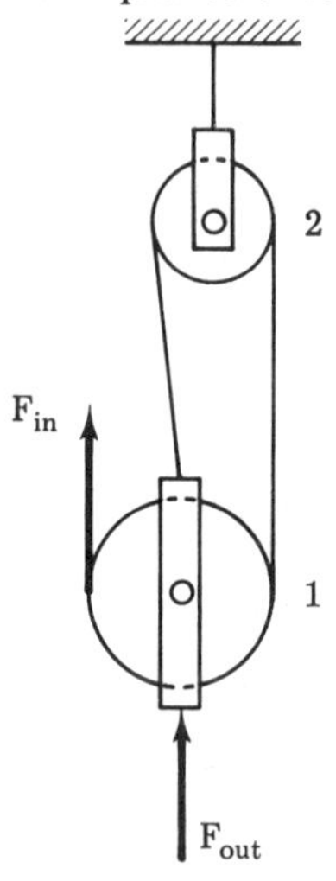

Figure A10-5

one rope with a force equal to the input force and one rope with a force equal to twice the input force provide the output force. The mechanical advantage of the system is therefore 3.

Example A10-2: What is the mechanical advantage of the pulley system shown in Figure A10-5? There is only one rope; therefore, the tension throughout it equals the input force. Isolating pulley 1, it is supported by three ropes, all with a force equal to the input force. The mechanical advantage of the system is 3.

PROBLEMS

1. What output force will result from an input force of 250 N for the lever shown in Figure A10-6?

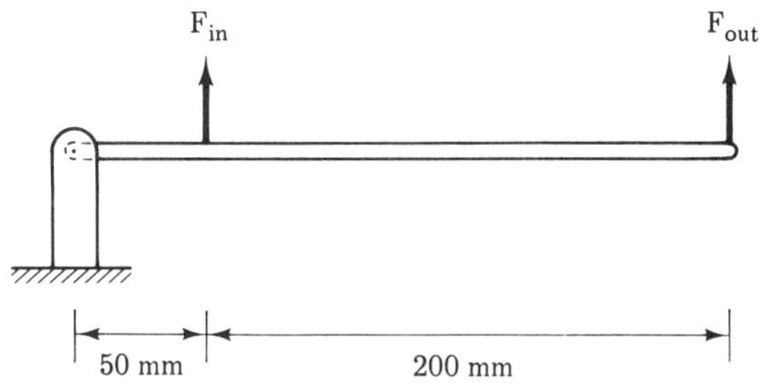

Figure A10-6

2. What output force will result from a 40-lb input force for the pulley system shown in Figure A10-7? If the input forces moves 18 in., how far will the output force move?

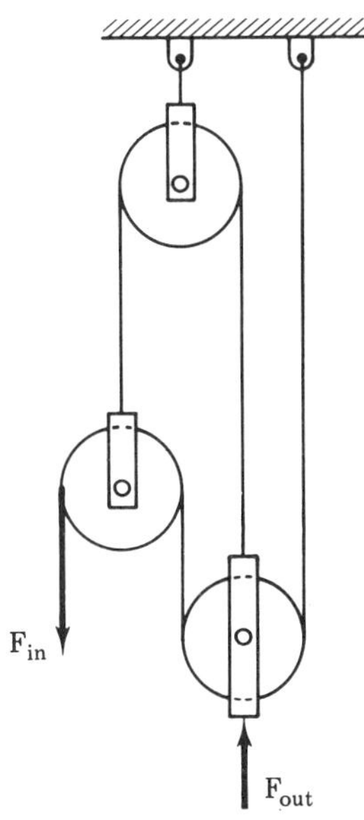

Figure A10-7

Review Unit A-11

Mechanical Properties

In the study of statics, the concentration is generally on the forces and moments and not on their physical effect. The latter is really in the realm of strength of materials. However, the problems dealt with are more realistic if the effects are kept in mind, and in a few cases there is specific interest in the effects. Ultimately, these effects must be related to the properties of various materials in order to, first, select a material and, second, size the machine component or structural member.

One of the most common effects of forces acting on a body is the phenomenon known as *tension.* At left in Figure A11-1 is shown a load hanging on a part or member labeled *AB*. At right in Figure A11-1 is shown a diagram of what is being done to *AB*. The 10,000-lb load is *pulling* down on *AB* while the support above is *pulling* up with a force of 10,000 lb. Atoms tend to establish equilibrium distances among each other and force is required to change this relationship. As *AB* is being pulled from both ends, its length is increased slightly. This stretching is the summation of an increase in interatomic distances throughout *AB* in the direction of the applied forces. This then is the phenomenon known as *tension.* The ability of a material to resist this is known as its *tensile strength.*

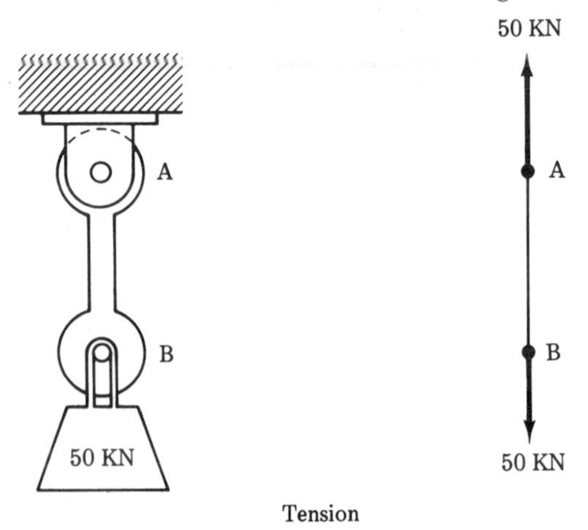

Figure A11-1

If a load is placed on a body or member *CD* as shown in Figure A11-2, the exact opposite to what we have just observed will occur. The load will *push* down on *CD* from above while the floor will *push* up on *CD* from below. This has been diagrammed at right in Figure A11-2. The effect of this force *pushes* the atoms closer together shortening member *CD*. This phenomenon is known as *compression.* The ability of a material to resist this is

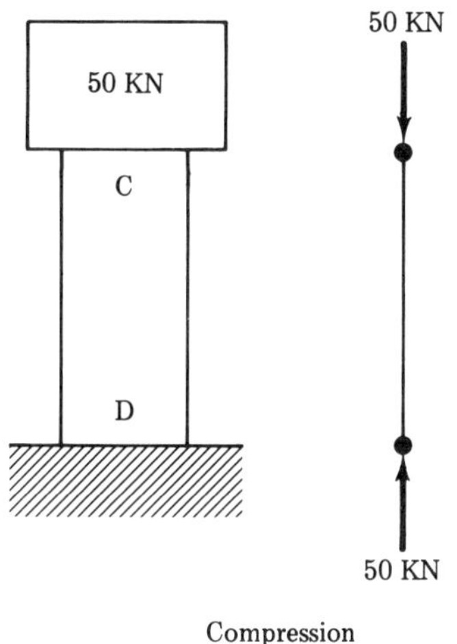

Compression

Figure A11-2

known as its *compressive strength.* For many materials the compressive strength is equal to the tensile strength. However, two very common materials, concrete and cast iron, are notable exceptions. Both of those materials are much stronger in compression than in tension.

A phenomenon worth mentioning is that some materials do not have the same strength in all directions. In wood this occurs naturally, its strength across the grain being different than its strength in the direction of the grain. In wrought iron and metals that have been formed by rolling, it occurs as a result of the manufacturing process.

A frequent occurrence in structural and machine members is the phenomenon known as bending. It is illustrated in Figure A11-3. Note that as the load is increased, the member *EF* bends more, which is to be expected. However, this decrease in the radius of curvature results in the top edge of the member being crushed while the bottom edge is being stretched. Thus, in *bending,* all particles above a *neutral axis* (dashed line) are in *compression,* while all those below the neutral are in *tension.* The degree of tension or compression is a function of the perpendicular distance from the neutral axis.

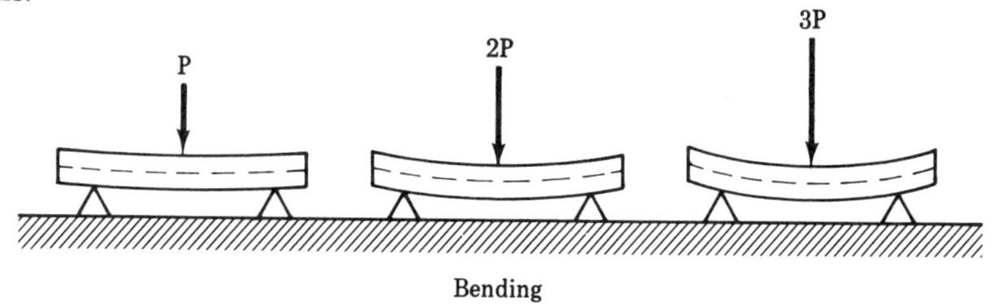

Bending

Figure A11-3

Another rather different phenomenon resulting from application of force is *shear*. As the name suggests, shear is a cutting action such as scissors perform on paper. The cutting of a thin piece of paper does not really permit visualization of what is occurring at the atomic level. When a force is attempting to shear a substance, it can be likened to trying to slide a piece of sandpaper, rough side down, along another piece of sandpaper fastened to some surface, rough side up. In the case of the sandpaper, the resistance to movement is due to friction whereas in the case of shear the resistance is due to atomic bonds. Figure A11-4 (left side) shows the nature of the deformation that takes place due to shear (as compared to the stretching due to tension and the crushing or shortening due to compression). At right in Figure A11-4 is shown a common shearing application (really a side view of

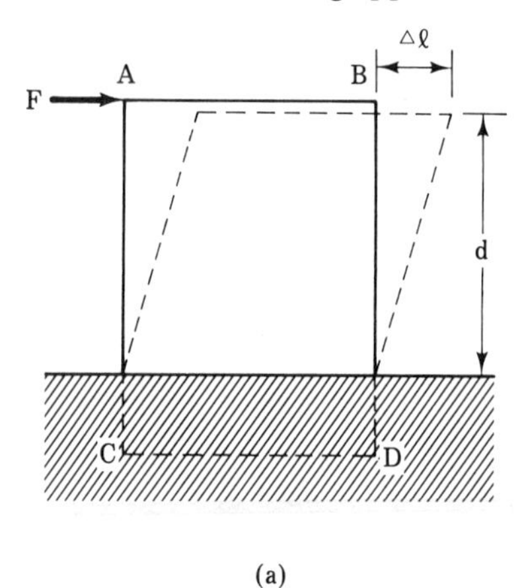

(a)

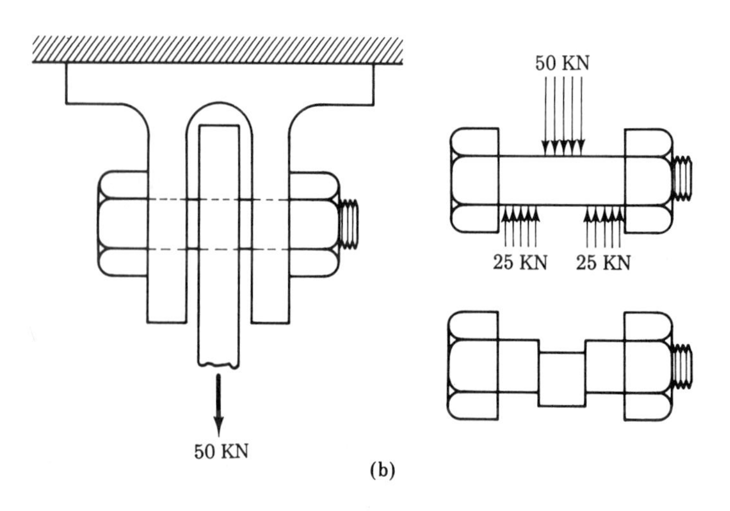

(b)

Figure A11-4

Figure A11-1) where the downward load is trying to slide a section of the bolt down out of place.

A slightly different application of shear occurs in the phenomenon known as torsion. As a torque or moment is applied to a member such as a shaft, the molecules in the shaft tend to slide along each other but with circular motion rather than the linear motion shown in Figure A11-4.

Review Unit A-12

Accuracy, Rounding, Estimating

The concept of accuracy is almost always linked to the act of measurement. It is readily apparent that a meter stick with 1000 graduations (millimeters) can measure some length more accurately than one with only 100 graduations (centimeters). The first may measure a length as 223 mm while the second measures the same length as 22 cm or 220 mm.

Let us examine those two numbers. Each consists of three digits, 223 and 220, a *digit* being simply any one of the symbols 0, 1, 2, 3, 4, 5, 6, 7, 8, 9. Sometimes a digit is referred to as a figure. We can see that no matter how much refinement takes place in our measurement, the first 2 (in the hundreds place) will not change. It can be considered an *accurate* digit. If we examine the measurement of 220 mm, we are saying that the value lies between 215 and 225, for if it had been less than 215, it would have been closer to 210 than 220, and if it had been more than 225 it would have been closer to 230. However, while 220 was a correct reading, the possibility exists that the 2 in the tens place might really be 1. Thus the 2 in the tens place is an *inaccurate* digit as is the 0 in the units place which might be any of the digits. Using the more accurate scale, we measured the length to be 223 mm. The digit in the units place, 3, is an accurate digit. We might look closely and estimate the value to be 223.4 mm, but the digit 4 is an estimate that might be changed by a refinement in our measurement (use of an even more finely graduated scale or a different type of measuring tool). Thus the 4 is an inaccurate digit.

In summary we can state:

1. A digit in an observed quantity which will not be changed by a refinement of the observation is called an *accurate digit.*

2. A digit in an observed quantity which may be changed by a refinement of the observation is called an *inaccurate digit.*

In computations the term *significant digit,* or *significant figure,* is frequently encountered. A common definition states that a digit is significant if the maximum error in the number in which it is contained is less than or at most equal to one-half of one unit in the place which the digit occupies.

Further, the significant figure in a number consists of

1. all nonzero digits or
2. all zero digits which lie
   (a) between significant digits or
   (b) to the right of the decimal point and to the right of a nonzero digit.

Example A12-1: a. Using only accurate digits, state the reading on the gauge shown in Figure A12-1. b. Using one inaccurate digit together with accurate digits, state the reading on the same gauge.

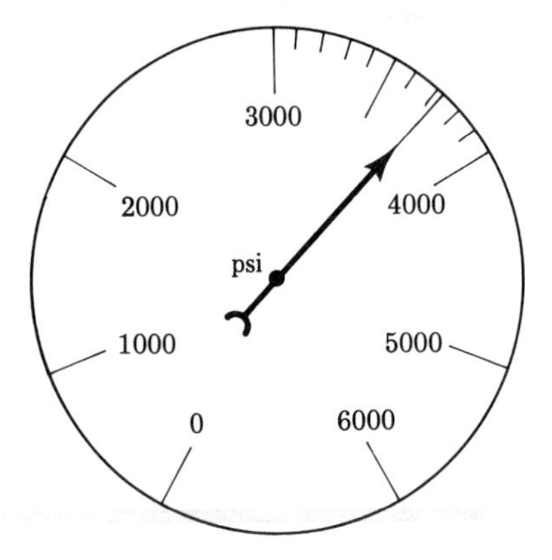

Figure A12-1

Ans. a. 3700 psi b. 3720 psi

Example A12-2: State the number of significant figures in each of the following numbers:

a. 47.51 b. 9.71×10^4 c. 0.0063 d. 3.120 e. 3×10^{-4}

Ans. a. 4 b. 3 c. 2 d. 4 e. 1

In the above example problems b and e may have bothered you. In the case of b the number might have been written as 97,100 yielding three significant figures. What if the number really had four-place accuracy? Scientific notation would permit us to express this as 9.710×10^4. Similarly for e where the number is 0.0003, with 1 significant figure. If it really were accurate to two places, we could write it as 0.00030 or 3.0×10^{-4}.

In doing computations the product or quotient cannot be more accurate than the number of significant figures of the least accurate number involved.

Example A12-3:

$$\frac{37.33(0.0412)}{2.4} = 0.64083 \ldots = 0.64$$

(Since 2.4 had only two significant figures the answer can have no more.)

The above computation brings up the question of rounding, the operation performed in restating the answer in two significant figures. The procedure is generally quite simple since a digit larger than 5 causes the one to its left to be rounded up and a digit smaller than 5 causes the one to its left to stay the same. The only issue arises when a digit of 5 occurs with either zeros or nothing to its right. Convention has it that we round the digit to its left to an even digit.

Example A12-4: Round the following numbers to three places:

a. 0.77777 b. 122.46 c. 8.315 d. 17.291 e. 0.004755

Ans: a. 0.778 b. 122 c. 8.32 d. 17.3 e. 0.00476

A few words about estimating should suffice. In this day of high-speed calculators and computers, it is absolutely essential that you estimate your answer. It won't avert minor errors but it will prevent major ones, especially decimal place errors. In Example A12-3 an estimate (mental) might proceed as follows: 40 times 0.04 is 1.6 which divided by 2.4 is about two-thirds or 0.666, an estimate only 4% off.

PROBLEMS

1. Using only accurate digits what frequency is the radio dial in Figure A12-2 tuned to?

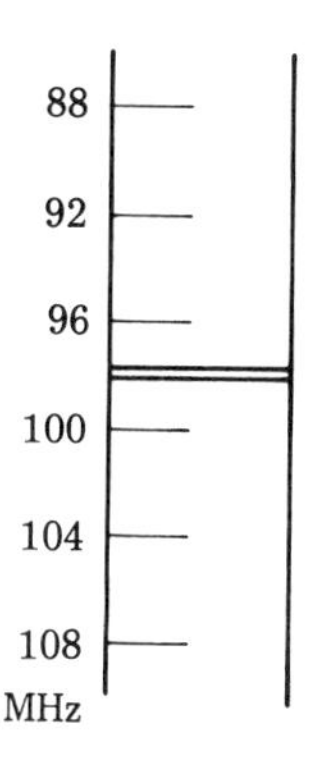

Figure A12-2

2. State the number of significant figures in each of the following numbers:

   a. 0.000432 b. 4567 c. 8.40×10^5 d. 1.792×10^8 e. 7.0040

3. Perform the following computations, rounding the answer properly.

   a. $\dfrac{(4.7)^2(5.733)(1.212)}{1728}$ b. $\dfrac{(32{,}413)(4.710)(17.23)}{0.00431(15{,}762)}$

Appendix

B

Additional Tables

Table B-1
Conversion Between U.S. Customary and S.I. Units

Length	
Inches (in.) × 25.4	= millimeters (mm)
Inches (in.) × 0.0254	= meters (m)
Feet (ft) × 0.3048	= meters (m)
Meter (m) × 39.4	= inches (in.)
Meter (m) × 3.28	= feet (ft)
Area	
Square inches (in.^{2}) × 645.2	= square millimeters (mm^2)
Square feet (ft^2) × 0.0929	= square meters (m^2)
Square meters (m^2) × 10.76	= square feet (ft^2)
Square millimeters (mm^2) × 1.55 × 10^{-3}	= square inches (in.^{2})
Volume	
Cubic inches (in.^{3}) × 16,390	= cubic millimeters (mm^3)
Cubic feet (ft^3) × 0.02832	= cubic meters (m^3)
Cubic millimeters (mm^3) × 6.1 × 10^5	= cubic inches (in^3)
Cubic meters (m^3) × 35.32	= cubic feet (ft^3)
Mass	
Slugs × 14.585	= kilograms (kg)
Kilograms × 0.0686	= slugs

(continued)

Table B-1
Conversion Between U.S. Customary and S.I. Units (*continued*)

Force	
Pounds (lb) × 4.448	= Newtons (N)
Newtons (N) × 0.225	= pounds (lb)
Moment	
Pound · feet (lb · ft) × 1.356	= Newton · meters (N · m)
Newton · meters (N · m) × 0.7375	= pound · feet (lb · ft)
Stress	
Newtons/square meter (N/m^2) × 1.45×10^{-4} (Pascals)	= pounds/square inch (psi)
Pounds/square inch (psi) × 6895	= Newtons/square meter (N/m^2) (Pascals)
Area Moment of Inertia	
Inches to fourth ($in.^4$) × 4.162×10^5	= millimeter to fourth (mm^4)
Feet to fourth (ft^4) × 8.631×10^{-3}	= meter to fourth (m^4)
Millimeters to fourth (mm^4) × 2.403×10^{-6}	= inches to fourth ($in.^4$)
Meters to fourth (m^4) × 115.86	= feet to fourth (ft^4)

Table B-2
Coefficients of Friction

Materials Non-Lubricated	Linear Static Coefficient
Steel on steel	0.78
Steel on babbitt	0.42
Steel on Teflon	0.04
Cast iron on cast iron	1.10
Aluminum on aluminum	1.05
Wood on wood	0.62
Glass on glass	0.94

	Coefficient of Rolling Friction	
Materials	in.	m
Iron on wood	0.1 to 0.3	0.00254 to 0.00762
Steel on steel	0.02	0.000508
Wood on wood	0.02 to 0.2	0.000508 to 0.00508
Steel on concrete	0.05	0.00127

Table B-3
Properties of Plane Areas

Rectangle	$A = bh$	$\bar{I}_{xx} = \frac{1}{12} bh^3$
	$\bar{x} = \frac{b}{2}$	$\bar{I}_{yy} = \frac{1}{12} hb^3$
	$\bar{y} = \frac{h}{2}$	$\bar{J} = \frac{1}{12} hb (b^2 + h^2)$
Right triangle	$A = \frac{1}{2} bh$	$\bar{I}_{xx} = \frac{1}{36} bh^3$
	$\bar{x} = \frac{b}{3}$	$\bar{I}_{yy} = \frac{1}{36} hb^3$
	$\bar{y} = \frac{h}{3}$	$\bar{J} = \frac{1}{36} hb (b^2 + h^2)$
Circle	$A = \pi r^2 = \frac{\pi d^2}{4}$	$\bar{I}_{xx} = \frac{\pi r^4}{4} = \frac{\pi d^4}{64}$
	$\bar{x} = r$	$\bar{I}_{yy} = \bar{I}_{xx}$
	$\bar{y} = r$	$\bar{J} = \frac{\pi r^4}{2} = \frac{\pi d^4}{32}$
Semicircle	$A = \frac{\pi r^2}{2} = \frac{\pi d^2}{8}$	$\bar{I}_{xx} = 0.1098 r^4 = 0.00686 d^4$
	$\bar{x} = r$	$\bar{I}_{yy} = \frac{1}{8} \pi r^4 = \frac{1}{128} \pi d^4$
	$\bar{y} = 0.424 r$	$\bar{J} = 0.5025 r^4 = 0.0314 d^4$
Quarter circle	$A = \frac{\pi r^2}{4} = \frac{\pi d^2}{16}$	$\bar{I}_{xx} = 0.0549 r^4$
	$\bar{x} = 0.424 r$	$\bar{I}_{yy} = I_{xx}$
	$\bar{y} = 0.424 r$	$\bar{J} = 0.1098 r^4$

(continued)

Table B-3
Properties of Plane Areas (*continued*)

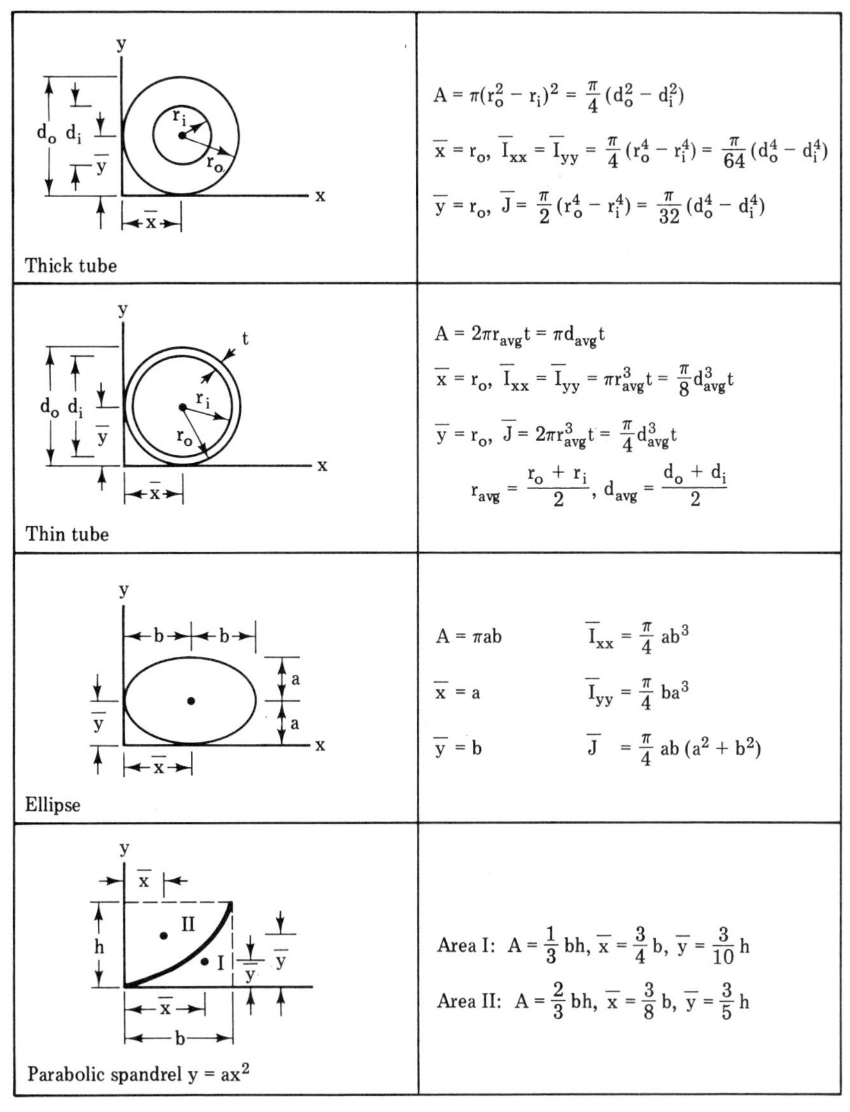

Thick tube	$A = \pi(r_o^2 - r_i)^2 = \frac{\pi}{4}(d_o^2 - d_i^2)$ $\bar{x} = r_o,\ \bar{I}_{xx} = \bar{I}_{yy} = \frac{\pi}{4}(r_o^4 - r_i^4) = \frac{\pi}{64}(d_o^4 - d_i^4)$ $\bar{y} = r_o,\ \bar{J} = \frac{\pi}{2}(r_o^4 - r_i^4) = \frac{\pi}{32}(d_o^4 - d_i^4)$
Thin tube	$A = 2\pi r_{avg} t = \pi d_{avg} t$ $\bar{x} = r_o,\ \bar{I}_{xx} = \bar{I}_{yy} = \pi r_{avg}^3 t = \frac{\pi}{8} d_{avg}^3 t$ $\bar{y} = r_o,\ \bar{J} = 2\pi r_{avg}^3 t = \frac{\pi}{4} d_{avg}^3 t$ $r_{avg} = \frac{r_o + r_i}{2},\ d_{avg} = \frac{d_o + d_i}{2}$
Ellipse	$A = \pi ab$ $\bar{I}_{xx} = \frac{\pi}{4} ab^3$ $\bar{x} = a$ $\bar{I}_{yy} = \frac{\pi}{4} ba^3$ $\bar{y} = b$ $\bar{J} = \frac{\pi}{4} ab(a^2 + b^2)$
Parabolic spandrel $y = ax^2$	Area I: $A = \frac{1}{3} bh,\ \bar{x} = \frac{3}{4} b,\ \bar{y} = \frac{3}{10} h$ Area II: $A = \frac{2}{3} bh,\ \bar{x} = \frac{3}{8} b,\ \bar{y} = \frac{3}{5} h$

Table B-4
Properties of Steel Structural Shapes (U.S. Customary Units)

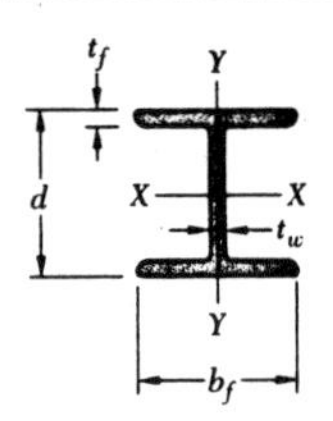

Properties of Rolled-Steel Shapes
(U.S. Customary Units)

W Shapes
(Wide-Flange Shapes)

Designation†	Area A, in^2	Depth d, in.	Flange Width b_f, in.	Flange Thickness t_f, in.	Web Thickness t_w, in.	Axis X-X I_x, in^4	Axis X-X Z_x, in^3	Axis X-X r_x, in.	Axis Y-Y I_y, in^4	Axis Y-Y Z_y, in^3	Axis Y-Y r_y, in.
W36 × 300	88.3	36.74	16.655	1.680	0.945	20300	1110	15.2	1300	156	3.83
135	39.7	35.55	11.950	0.790	0.600	7800	439	14.0	225	37.7	2.38
W33 × 201	59.1	33.68	15.745	1.150	0.715	11500	684	14.0	749	95.2	3.56
118	34.7	32.86	11.480	0.740	0.550	5900	359	13.0	187	32.6	2.32
W30 × 173	50.8	30.44	14.985	1.065	0.655	8200	539	12.7	598	79.8	3.43
99	29.1	29.65	10.450	0.670	0.520	3990	269	11.7	128	24.5	2.10
W27 × 146	42.9	27.38	13.965	0.975	0.605	5630	411	11.4	443	63.5	3.21
84	24.8	26.71	9.960	0.640	0.460	2850	213	10.7	106	21.2	2.07
W24 × 104	30.6	24.06	12.750	0.750	0.500	3100	258	10.1	259	40.7	2.91
68	20.1	23.73	8.965	0.585	0.415	1830	154	9.55	70.4	15.7	1.87
W21 × 101	29.8	21.36	12.290	0.800	0.500	2420	227	9.02	248	40.3	2.89
62	18.3	20.99	8.240	0.615	0.400	1330	127	8.54	57.5	13.9	1.77
44	13.0	20.66	6.500	0.450	0.350	843	81.6	8.06	20.7	6.36	1.26
W18 × 106	31.1	18.73	11.200	0.940	0.590	1910	204	7.84	220	39.4	2.66
76	22.3	18.21	11.035	0.680	0.425	1330	146	7.73	152	27.6	2.61
50	14.7	17.99	7.495	0.570	0.355	800	88.9	7.38	40.1	10.7	1.65
35	10.3	17.70	6.000	0.425	0.300	510	57.6	7.04	15.3	5.12	1.22
W16 × 77	22.6	16.52	10.295	0.760	0.455	1110	134	7.00	138	26.9	2.47
57	16.8	16.43	7.120	0.715	0.430	758	92.2	6.72	43.1	12.1	1.60
40	11.8	16.01	6.995	0.505	0.305	518	64.7	6.63	28.9	8.25	1.57
31	9.12	15.88	5.525	0.440	0.275	375	47.2	6.41	12.4	4.49	1.17
26	7.68	15.69	5.500	0.345	0.250	301	38.4	6.26	9.59	3.49	1.12
W14 × 370	109	17.92	16.475	2.660	1.655	5440	607	7.07	1990	241	4.27
145	42.7	14.78	15.500	1.090	0.680	1710	232	6.33	677	87.3	3.98
82	24.1	14.31	10.130	0.855	0.510	882	123	6.05	148	29.3	2.48
68	20.0	14.04	10.035	0.720	0.415	723	103	6.01	121	24.2	2.46
53	15.6	13.92	8.060	0.660	0.370	541	77.8	5.89	57.7	14.3	1.92
43	12.6	13.66	7.995	0.530	0.305	428	62.7	5.82	45.2	11.3	1.89
38	11.2	14.10	6.770	0.515	0.310	385	54.6	5.88	26.7	7.88	1.55
30	8.85	13.84	6.730	0.385	0.270	291	42.0	5.73	19.6	5.82	1.49
26	7.69	13.91	5.025	0.420	0.255	245	35.3	5.65	8.91	3.54	1.08
22	6.49	13.74	5.000	0.335	0.230	199	29.0	5.54	7.00	2.80	1.04

†A wide-flange shape is designated by the letter W followed by the nominal depth in inches and the weight in pounds per foot.

*Table B-4 on pages 636-641 is taken from *Mechanics of Materials,* by Ferdinand P. Beer and E. Russell Johnston, Jr. ©1981 by McGraw-Hill, Inc., and is reprinted with permission.

Table B-4
Properties of Steel Structural Shapes (U.S. Customary Units) (*continued*)

Properties of Rolled-Steel Shapes
(SI Units)

W Shapes
(Wide-Flange Shapes)

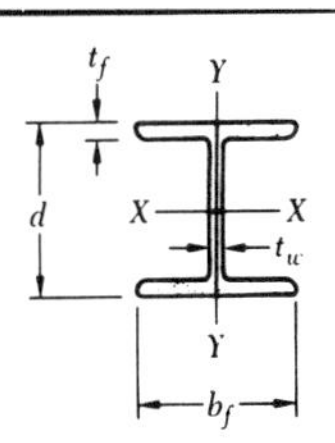

Designation†	Area A, mm²	Depth d, mm	Flange Width b_f, mm	Flange Thickness t_f, mm	Web Thickness t_w, mm	Axis X-X I_x 10^6 mm⁴	Axis X-X Z_x 10^3 mm³	Axis X-X r_x mm	Axis Y-Y I_y 10^6 mm⁴	Axis Y-Y Z_y 10^3 mm³	Axis Y-Y r_y mm
W920 × 446	57000	933	423	42.7	24.0	8450	18110	386	541	2560	97.3
201	25600	903	304	20.1	15.2	3250	7200	356	93.7	616	60.5
W840 × 299	38100	855	400	29.2	18.2	4790	11200	356	312	1560	90.4
176	22400	835	292	18.8	14.0	2460	5890	330	77.8	533	58.9
W760 × 257	32800	773	381	27.1	16.6	3410	8820	323	249	1307	87.1
147	18800	753	265	17.0	13.2	1660	4410	297	53.3	402	53.3
W690 × 217	27700	695	355	24.8	15.4	2340	6730	290	184.4	1039	81.5
125	16000	678	253	16.3	11.7	1186	3500	272	44.1	349	52.6
W610 × 155	19700	611	324	19.0	12.7	1290	4220	256	107.8	665	73.9
101	13000	603	228	14.9	10.5	762	2530	243	29.3	257	47.5
W530 × 150	19200	543	312	20.3	12.7	1007	3710	229	103.2	662	73.4
92	11800	533	209	15.6	10.2	554	2080	217	23.9	229	45.0
66	8390	525	165	11.4	8.9	351	1337	205	8.62	104.5	32.0
W460 × 158	20100	476	284	23.9	15.0	795	3340	199.1	91.6	645	67.6
113	14400	463	280	17.3	10.8	554	2390	196.3	63.3	452	66.3
74	9480	457	190	14.5	9.0	333	1457	187.5	16.69	175.7	41.9
52	6650	450	152	10.8	7.6	212	942	178.8	6.37	83.8	31.0
W410 × 114	14600	420	261	19.3	11.6	462	2200	177.8	57.4	440	62.7
85	10800	417	181	18.2	10.9	316	1516	170.7	17.94	198.2	40.6
60	7610	407	178	12.8	7.7	216	1061	168.4	12.03	135.2	39.9
46.1	5880	403	140	11.2	7.0	156.1	775	162.8	5.16	73.7	29.7
38.8	4950	399	140	8.8	6.4	125.3	628	159.0	3.99	57.0	28.4
W360 × 551	70300	455	418	67.6	42.0	2260	9930	179.6	828	3960	108.5
216	27500	375	394	27.7	17.3	712	3800	160.8	282	1431	101.1
122	15500	363	257	21.7	13.0	367	2020	153.7	61.6	479	63.0
101	12900	357	255	18.3	10.5	301	1686	152.7	50.4	395	62.5
79	10100	354	205	16.8	9.4	225	1271	149.6	24.0	234	48.8
64	8130	347	203	13.5	7.7	178.1	1027	147.8	18.81	185.3	48.0
57	7230	358	172	13.1	7.9	160.2	895	149.4	11.11	129.2	39.4
44.8	5710	352	171	9.8	6.9	121.1	688	145.5	8.16	95.4	37.8
39.0	4960	353	128	10.7	6.5	102.0	578	143.5	3.71	58.0	27.4
32.9	4190	349	127	8.5	5.8	82.8	474	140.7	2.91	45.8	26.4

†A wide-flange shape is designated by the letter W followed by the nominal depth in millimeters and the mass in kilograms per meter.

(*continued*)

Table B-4
Properties of Steel Structural Shapes (U.S. Customary Units) (*continued*)

Properties of Rolled-Steel Shapes
(U.S. Customary Units)

W Shapes
(Wide-Flange Shapes)

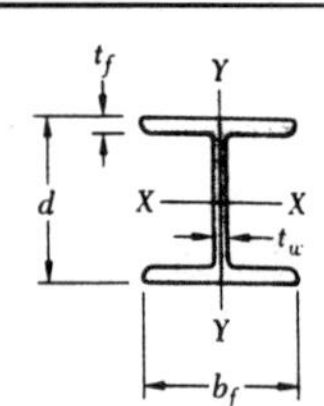

Designation†	Area A, in^2	Depth d, in.	Flange Width b_f, in.	Flange Thickness t_f, in.	Web Thickness t_w, in.	Axis X-X I_x, in^4	Axis X-X Z_x, in^3	Axis X-X r_x, in.	Axis Y-Y I_y, in^4	Axis Y-Y Z_y, in^3	Axis Y-Y r_y, in.
W12 × 96	28.2	12.71	12.160	0.900	0.550	833	131	5.44	270	44.4	3.09
72	21.1	12.25	12.040	0.670	0.430	597	97.4	5.31	195	32.4	3.04
50	14.7	12.19	8.080	0.640	0.370	394	64.7	5.18	56.3	13.9	1.96
40	11.8	11.94	8.005	0.515	0.295	310	51.9	5.13	44.1	11.0	1.93
35	10.3	12.50	6.560	0.520	0.300	285	45.6	5.25	24.5	7.47	1.54
30	8.79	12.34	6.520	0.440	0.260	238	38.6	5.21	20.3	6.24	1.52
26	7.65	12.22	6.490	0.380	0.230	204	33.4	5.17	17.3	5.34	1.51
22	6.48	12.31	4.030	0.425	0.260	156	25.4	4.91	4.66	2.31	0.848
16	4.71	11.99	3.990	0.265	0.220	103	17.1	4.67	2.82	1.41	0.773
W10 × 112	32.9	11.36	10.415	1.250	0.755	716	126	4.66	236	45.3	2.68
68	20.0	10.40	10.130	0.770	0.470	394	75.7	4.44	134	26.4	2.59
54	15.8	10.09	10.030	0.615	0.370	303	60.0	4.37	103	20.6	2.56
45	13.3	10.10	8.020	0.620	0.350	248	49.1	4.33	53.4	13.3	2.01
39	11.5	9.92	7.985	0.530	0.315	209	42.1	4.27	45.0	11.3	1.98
33	9.71	9.73	7.960	0.435	0.290	170	35.0	4.19	36.6	9.20	1.94
30	8.84	10.47	5.810	0.510	0.300	170	32.4	4.38	16.7	5.75	1.37
22	6.49	10.17	5.750	0.360	0.240	118	23.2	4.27	11.4	3.97	1.33
19	5.62	10.24	4.020	0.395	0.250	96.3	18.8	4.14	4.29	2.14	0.874
15	4.41	9.99	4.000	0.270	0.230	68.9	13.8	3.95	2.89	1.45	0.810
W8 × 58	17.1	8.75	8.220	0.810	0.510	228	52.0	3.65	75.1	18.3	2.10
48	14.1	8.50	8.110	0.685	0.400	184	43.3	3.61	60.9	15.0	2.08
40	11.7	8.25	8.070	0.560	0.360	146	35.5	3.53	49.1	12.2	2.04
35	10.3	8.12	8.020	0.495	0.310	127	31.2	3.51	42.6	10.6	2.03
31	9.13	8.00	7.995	0.435	0.285	110	27.5	3.47	37.1	9.27	2.02
28	8.25	8.06	6.535	0.465	0.285	98.0	24.3	3.45	21.7	6.63	1.62
24	7.08	7.93	6.495	0.400	0.245	82.8	20.9	3.42	18.3	5.63	1.61
21	6.16	8.28	5.270	0.400	0.250	75.3	18.2	3.49	9.77	3.71	1.26
18	5.26	8.14	5.250	0.330	0.230	61.9	15.2	3.43	7.97	3.04	1.23
15	4.44	8.11	4.015	0.315	0.245	48.0	11.8	3.29	3.41	1.70	0.876
13	3.84	7.99	4.000	0.255	0.230	39.6	9.91	3.21	2.73	1.37	0.843
W6 × 25	7.34	6.38	6.080	0.455	0.320	53.4	16.7	2.70	17.1	5.61	1.52
20	5.87	6.20	6.020	0.365	0.260	41.4	13.4	2.66	13.3	4.41	1.50
16	4.74	6.28	4.030	0.405	0.260	32.1	10.2	2.60	4.43	2.20	0.967
12	3.55	6.03	4.000	0.280	0.230	22.1	7.31	2.49	2.99	1.50	0.918
9	2.68	5.90	3.940	0.215	0.170	16.4	5.56	2.47	2.20	1.11	0.905
W5 × 19	5.54	5.15	5.030	0.430	0.270	26.2	10.2	2.17	9.13	3.63	1.28
16	4.68	5.01	5.000	0.360	0.240	21.3	8.51	2.13	7.51	3.00	1.27
W4 × 13	3.83	4.16	4.060	0.345	0.280	11.3	5.46	1.72	3.86	1.90	1.00

†A wide-flange shape is designated by the letter W followed by the nominal depth in inches and the weight in pounds per foot.

Table B-4
Properties of Steel Structural Shapes (U.S. Customary Units) (*continued*)

Properties of Rolled-Steel Shapes
(SI Units)

W Shapes
(Wide-Flange Shapes)

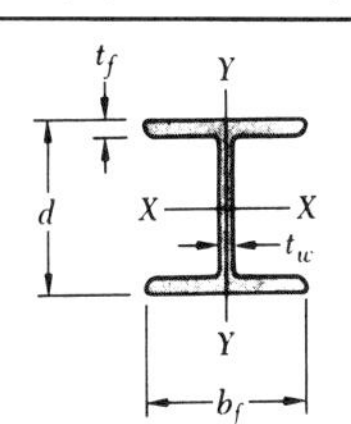

Designation†	Area A, mm²	Depth d, mm	Flange Width b_f, mm	Flange Thickness t_f, mm	Web Thickness t_w, mm	Axis X-X I_x 10⁶ mm⁴	Axis X-X Z_x 10³ mm³	Axis X-X r_x mm	Axis Y-Y I_y 10⁶ mm	Axis Y-Y Z_y 10³ mm³	Axis Y-Y r_y mm
W310 × 143	18200	323	309	22.9	14.0	347	2150	138.2	112.4	728	78.5
107	13600	311	306	17.0	10.9	248	1595	134.9	81.2	531	77.2
74	9480	310	205	16.3	9.4	164.0	1058	131.6	23.4	228	49.8
60	7610	303	203	13.1	7.5	129.0	851	130.3	18.36	180.9	49.0
52	6650	317	167	13.2	7.6	118.6	748	133.4	10.20	122.2	39.1
44.5	5670	313	166	11.2	6.6	99.1	633	132.3	8.45	101.8	38.6
38.7	4940	310	165	9.7	5.8	84.9	548	131.3	7.20	87.3	38.4
32.7	4180	313	102	10.8	6.6	64.9	415	124.7	1.940	38.0	21.5
23.8	3040	305	101	6.7	5.6	42.9	281	118.6	1.174	23.2	19.63
W250 × 167	21200	289	265	31.8	19.2	298.0	2060	118.4	98.2	741	68.1
101	12900	264	257	19.6	11.9	164.0	1242	112.8	55.8	434	65.8
80	10200	256	255	15.6	9.4	126.1	985	111.0	42.8	336	65.0
67	8580	257	204	15.7	8.9	103.2	803	110.0	22.2	218	51.1
58	7420	252	203	13.5	8.0	87.0	690	108.5	18.73	184.5	50.3
49.1	6260	247	202	11.0	7.4	70.8	573	106.4	15.23	150.8	49.3
44.8	5700	266	148	13.0	7.6	70.8	532	111.3	6.95	93.9	34.8
32.7	4190	258	146	9.1	6.1	49.1	381	108.5	4.75	65.1	33.8
28.4	3630	260	102	10.0	6.4	40.1	308	105.2	1.796	35.2	22.2
22.3	2850	254	102	6.9	5.8	28.7	226	100.3	1.203	23.6	20.6
W200 × 86	11000	222	209	20.6	13.0	94.9	855	92.7	31.3	300	53.3
71	9100	216	206	17.4	10.2	76.6	709	91.7	25.3	246	52.8
59	7550	210	205	14.2	9.1	60.8	579	89.7	20.4	199.0	51.8
52	6650	206	204	12.6	7.9	52.9	514	89.2	17.73	173.8	51.6
46.1	5890	203	203	11.0	7.2	45.8	451	88.1	15.44	152.1	51.3
41.7	5320	205	166	11.8	7.2	40.8	398	87.6	9.03	108.8	41.1
35.9	4570	201	165	10.2	6.2	34.5	343	86.9	7.62	92.4	40.9
31.3	3970	210	134	10.2	6.4	31.3	298	88.6	4.07	60.7	32.0
26.6	3390	207	133	8.4	5.8	25.8	249	87.1	3.32	49.9	31.2
22.5	2860	206	102	8.0	6.2	20.0	194.2	83.6	1.419	27.8	22.3
19.3	2480	203	102	6.5	5.8	16.48	162.4	81.5	1.136	22.3	21.4
W150 × 37.1	4740	162	154	11.6	8.1	22.2	274	68.6	7.12	92.5	38.6
29.8	3790	157	153	9.3	6.6	17.23	219	67.6	5.54	72.4	38.1
24.0	3060	160	102	10.3	6.6	13.36	167.0	66.0	1.844	36.2	24.6
18.0	2290	153	102	7.1	5.8	9.20	120.3	63.2	1.245	24.4	23.3
13.5	1730	150	100	5.5	4.3	6.83	91.1	62.7	0.916	18.32	23.0
W130 × 28.1	3590	131	128	10.9	6.9	10.91	166.6	55.1	3.80	59.4	32.5
23.8	3040	127	127	9.1	6.1	8.87	139.7	54.1	3.13	49.3	32.3
W100 × 19.3	2470	106	103	8.8	7.1	4.70	88.7	43.7	1.607	31.2	25.4

†A wide-flange shape is designated by the letter W followed by the nominal depth in millimeters and the mass in kilograms per meter.

Table B-4

Properties of Steel Structural Shapes (U.S. Customary Units) (*continued*)

Properties of Rolled-Steel Shapes
(U.S. Customary Units)

S Shapes
(American Standard Shapes)

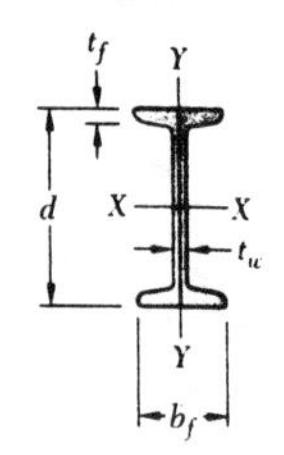

Designation†	Area A, in²	Depth d, in.	Flange Width b_f, in.	Flange Thickness t_f, in.	Web Thickness t_w, in.	Axis X-X I_x, in⁴	Axis X-X Z_x, in³	Axis X-X r_x, in.	Axis Y-Y I_y, in⁴	Axis Y-Y Z_y, in³	Axis Y-Y r_y, in.
S24 × 100	29.4	24.00	7.247	0.871	0.747	2390	199	9.01	47.8	13.2	1.27
× 90	26.5	24.00	7.124	0.871	0.624	2250	187	9.22	44.9	12.6	1.30
× 79.9	23.5	24.00	7.001	0.871	0.501	2110	175	9.47	42.3	12.1	1.34
S20 × 95	27.9	20.00	7.200	0.916	0.800	1610	161	7.60	49.7	13.8	1.33
× 85	25.0	20.00	7.053	0.916	0.653	1520	152	7.79	46.2	13.1	1.36
75	22.1	20.00	6.391	0.789	0.641	1280	128	7.60	29.6	9.28	1.16
65.4	19.2	20.00	6.250	0.789	0.500	1180	118	7.84	27.4	8.77	1.19
S18 × 70	20.6	18.00	6.251	0.691	0.711	926	103	6.71	24.1	7.72	1.08
54.7	16.1	18.00	6.001	0.691	0.461	804	89.4	7.07	20.8	6.94	1.14
S15 × 50	14.7	15.00	5.640	0.622	0.550	486	64.8	5.75	15.7	5.57	1.03
42.9	12.6	15.00	5.501	0.622	0.411	447	59.6	5.95	14.4	5.23	1.07
S12 × 50	14.7	12.00	5.477	0.659	0.687	305	50.8	4.55	15.7	5.74	1.03
40.8	12.0	12.00	5.252	0.659	0.462	272	45.4	4.77	13.6	5.16	1.06
35	10.3	12.00	5.078	0.544	0.428	229	38.2	4.72	9.87	3.89	0.980
31.8	9.35	12.00	5.000	0.544	0.350	218	36.4	4.83	9.36	3.74	1.00
S10 × 35	10.3	10.00	4.944	0.491	0.594	147	29.4	3.78	8.36	3.38	0.901
25.4	7.46	10.00	4.661	0.491	0.311	124	24.7	4.07	6.79	2.91	0.954
S8 × 23	6.77	8.00	4.171	0.425	0.441	64.9	16.2	3.10	4.31	2.07	0.798
18.4	5.41	8.00	4.001	0.425	0.271	57.6	14.4	3.26	3.73	1.86	0.831
S7 × 20	5.88	7.00	3.860	0.392	0.450	42.4	12.1	2.69	3.17	1.64	0.734
15.3	4.50	7.00	3.662	0.392	0.252	36.7	10.5	2.86	2.64	1.44	0.766
S6 × 17.25	5.07	6.00	3.565	0.359	0.465	26.3	8.77	2.28	2.31	1.30	0.675
12.5	3.67	6.00	3.332	0.359	0.232	22.1	7.37	2.45	1.82	1.09	0.705
S5 × 14.75	4.34	5.00	3.284	0.326	0.494	15.2	6.09	1.87	1.67	1.01	0.620
10	2.94	5.00	3.004	0.326	0.214	12.3	4.92	2.05	1.22	0.809	0.643
S4 × 9.5	2.79	4.00	2.796	0.293	0.326	6.79	3.39	1.56	0.903	0.646	0.569
7.7	2.26	4.00	2.663	0.293	0.193	6.08	3.04	1.64	0.764	0.574	0.581
S3 × 7.5	2.21	3.00	2.509	0.260	0.349	2.93	1.95	1.15	0.586	0.468	0.516
5.7	1.67	3.00	2.330	0.260	0.170	2.52	1.68	1.23	0.455	0.390	0.522

† An American Standard Beam is designated by the letter S followed by the nominal depth in inches and the weight in pounds per foot.

Table B-4
Properties of Steel Structural Shapes (U.S. Customary Units) (*continued*)

Properties of Rolled-Steel Shapes
(SI Units)

S Shapes
(American Standard Shapes)

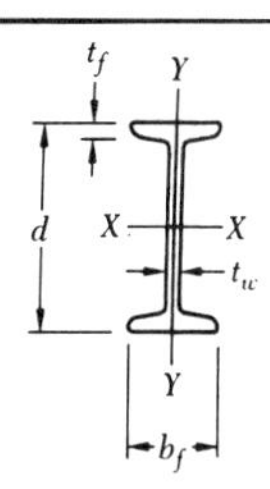

			Flange		Web	Axis X-X			Axis Y-Y		
Designation†	Area A, mm²	Depth d, mm	Width b_f, mm	Thickness t_f, mm	Thickness t_w, mm	I_x 10^6 mm⁴	Z_x 10^3 mm³	r_x mm	I_y 10^6 mm	Z_y 10^3 mm³	r_y mm
S610 × 149	18970	610	184	22.1	19.0	995	3260	229	19.90	216	32.3
134	17100	610	181	22.1	15.8	937	3070	234	18.69	207	33.0
118.9	15160	610	178	22.1	12.7	878	2880	241	17.61	197.9	34.0
S510 × 141	18000	508	183	23.3	20.3	670	2640	193.0	20.69	226.	33.8
127	16130	508	179	23.3	16.6	633	2490	197.9	19.23	215.	34.5
112	14260	508	162	20.1	16.3	533	2100	193.0	12.32	152.1	29.5
97.3	12390	508	159	20.1	12.7	491	1933	199.1	11.40	143.4	30.2
S460 × 104	13290	457	159	17.6	18.1	385	1685	170.4	10.03	126.2	27.4
81.4	10390	457	152	17.6	11.7	335	1466	179.6	8.66	113.9	29.0
S380 × 74	9480	381	143	15.8	14.0	202	1060	146.1	6.53	91.3	26.2
64	8130	381	140	15.8	10.4	186.1	977	151.1	5.99	85.6	27.2
S310 × 74	9480	305	139	16.8	17.4	127.0	833	115.6	6.53	94.0	26.2
60.7	7740	305	133	16.8	11.7	113.2	742	121.2	5.66	85.1	26.9
52	6640	305	129	13.8	10.9	95.3	625	119.9	4.11	63.7	24.9
47.3	6032	305	127	13.8	8.9	90.7	595	122.7	3.90	61.4	25.4
S250 × 52	6640	254	126	12.5	15.1	61.2	482	96.0	3.48	55.2	22.9
37.8	4806	254	118	12.5	7.9	51.6	406	103.4	2.83	48.0	24.2
S200 × 34	4368	203	106	10.8	11.2	27.0	266	78.7	1.794	33.8	20.3
27.4	3484	203	102	10.8	6.9	24.0	236	82.8	1.553	30.4	21.1
S180 × 30	3794	178	97	10.0	11.4	17.65	198.3	68.3	1.319	27.2	18.64
22.8	2890	178	92	10.0	6.4	15.28	171.7	72.6	1.099	23.9	19.45
S150 × 25.7	3271	152	90	9.1	11.8	10.95	144.1	57.9	0.961	21.4	17.15
18.6	2362	152	84	9.1	5.8	9.20	121.1	62.2	0.758	18.05	17.91
S130 × 22.0	2800	127	83	8.3	12.5	6.33	99.7	47.5	0.695	16.75	15.75
15	1884	127	76	8.3	5.3	5.12	80.6	52.1	0.508	13.37	16.33
S100 × 14.1	1800	102	70	7.4	8.3	2.83	55.5	39.6	0.376	10.74	14.45
11.5	1452	102	67	7.4	4.8	2.53	49.6	41.6	0.318	9.49	14.75
S75 × 11.2	1426	76	63	6.6	8.9	1.22	32.1	29.2	0.244	7.75	13.11
8.5	1077	76	59	6.6	4.3	1.05	27.6	31.3	0.189	6.41	13.26

†An American Standard Beam is designated by the letter S followed by the nominal depth in millimeters and the mass in kilograms per meter.

(*continued*)

Table B-5
Properties of Steel Structural Shapes (Metric Units)

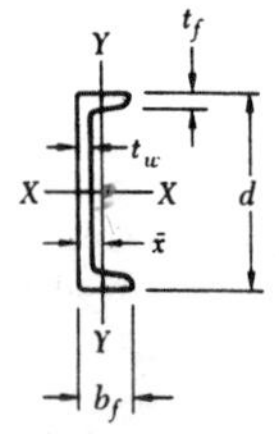

Properties of Rolled-Steel Shapes
(U.S. Customary Units)

C Shapes
(American Standard Channels)

			Flange			Axis X-X			Axis Y-Y			
Designation†	Area A, in²	Depth d, in.	Width b_f, in.	Thickness t_f, in.	Web Thickness t_w, in.	I_x, in⁴	Z_x, in³	r_x, in.	I_y, in⁴	Z_y, in³	r_y, in.	$\bar{x}$, in.
C15 × 50	14.7	15.00	3.716	0.650	0.716	404	53.8	5.24	11.0	3.78	0.867	0.799
40	11.8	15.00	3.520	0.650	0.520	349	46.5	5.44	9.23	3.36	0.886	0.778
33.9	9.96	15.00	3.400	0.650	0.400	315	42.0	5.62	8.13	3.11	0.904	0.787
C12 × 30	8.82	12.00	3.170	0.501	0.510	162	27.0	4.29	5.14	2.06	0.763	0.674
25	7.35	12.00	3.047	0.501	0.387	144	24.1	4.43	4.47	1.88	0.780	0.674
20.7	6.09	12.00	2.942	0.501	0.282	129	21.5	4.61	3.88	1.73	0.799	0.698
C10 × 30	8.82	10.00	3.033	0.436	0.673	103	20.7	3.42	3.94	1.65	0.669	0.649
25	7.35	10.00	2.886	0.436	0.526	91.2	18.2	3.52	3.36	1.48	0.676	0.617
20	5.88	10.00	2.739	0.436	0.379	78.9	15.8	3.66	2.81	1.32	0.691	0.606
15.3	4.49	10.00	2.600	0.436	0.240	67.4	13.5	3.87	2.28	1.16	0.713	0.634
C9 × 20	5.88	9.00	2.648	0.413	0.448	60.9	13.5	3.22	2.42	1.17	0.642	0.583
15	4.41	9.00	2.485	0.413	0.285	51.0	11.3	3.40	1.93	1.01	0.661	0.586
13.4	3.94	9.00	2.433	0.413	0.233	47.9	10.6	3.48	1.76	0.962	0.668	0.601
C8 × 18.75	5.51	8.00	2.527	0.390	0.487	44.0	11.0	2.82	1.98	1.01	0.599	0.565
13.75	4.04	8.00	2.343	0.390	0.303	36.1	9.03	2.99	1.53	0.853	0.615	0.553
11.5	3.38	8.00	2.260	0.390	0.220	32.6	8.14	3.11	1.32	0.781	0.625	0.571
C7 × 14.75	4.33	7.00	2.299	0.366	0.419	27.2	7.78	2.51	1.38	0.779	0.564	0.532
12.25	3.60	7.00	2.194	0.366	0.314	24.2	6.93	2.60	1.17	0.702	0.571	0.525
9.8	2.87	7.00	2.090	0.366	0.210	21.3	6.08	2.72	0.968	0.625	0.581	0.541
C6 × 13	3.83	6.00	2.157	0.343	0.437	17.4	5.80	2.13	1.05	0.642	0.525	0.514
10.5	3.09	6.00	2.034	0.343	0.314	15.2	5.06	2.22	0.865	0.564	0.529	0.500
8.2	2.40	6.00	1.920	0.343	0.200	13.1	4.38	2.34	0.692	0.492	0.537	0.512
C5 × 9	2.64	5.00	1.885	0.320	0.325	8.90	3.56	1.83	0.632	0.449	0.489	0.478
6.7	1.97	5.00	1.750	0.320	0.190	7.49	3.00	1.95	0.478	0.378	0.493	0.484
C4 × 7.25	2.13	4.00	1.721	0.296	0.321	4.59	2.29	1.47	0.432	0.343	0.450	0.459
5.4	1.59	4.00	1.584	0.296	0.184	3.85	1.93	1.56	0.319	0.283	0.449	0.458
C3 × 6	1.76	3.00	1.596	0.273	0.356	2.07	1.38	1.08	0.305	0.268	0.416	0.455
5	1.47	3.00	1.498	0.273	0.258	1.85	1.24	1.12	0.247	0.233	0.410	0.438
4.1	1.21	3.00	1.410	0.273	0.170	1.66	1.10	1.17	0.197	0.202	0.404	0.437

†An American Standard Channel is designated by the letter C followed by the nominal depth in inches and the weight in pounds per foot.

*Table B-5 on pages 642-647 is taken from *Mechanics of Materials,* by Ferdinand P. Beer and E. Russell Johnston, Jr. ©1981 by McGraw-Hill, Inc., and is reprinted with permission.

Table B-5
Properties of Steel Structural Shapes (Metric Units) *(continued)*

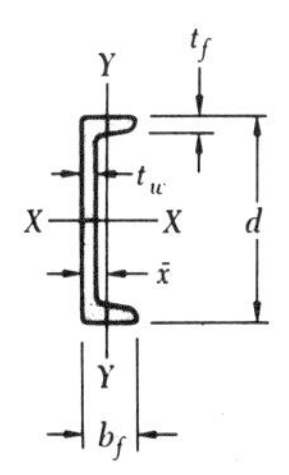

Properties of Rolled-Steel Shapes
(SI Units)

C Shapes
(American Standard Channels)

			Flange		Web	Axis X-X			Axis Y-Y			
Designation†	Area A, mm²	Depth d, mm	Width b_f, mm	Thickness t_f, mm	Thickness t_w, mm	I_x 10^6 mm⁴	Z_x 10^3 mm³	r_x mm	I_y 10^6 mm	Z_y 10^3 mm³	r_y mm	$\bar{x}$ mm
C380 × 74	9480	381	94	16.5	18.2	168.2	883	133.1	4.58	62.1	22.0	20.3
60	7610	381	89	16.5	13.2	145.3	763	138.2	3.84	55.5	22.5	19.76
50.4	6426	381	86	16.5	10.2	131.1	688	142.7	3.38	51.2	23.0	19.99
C310 × 45	5690	305	80	12.7	13.0	67.4	442	109.0	2.14	34.0	19.38	17.12
37	4742	305	77	12.7	9.8	59.9	393	112.5	1.861	31.1	19.81	17.12
30.8	3929	305	74	12.7	7.2	53.7	352	117.1	1.615	28.7	20.29	17.73
C250 × 45	5690	254	76	11.1	17.1	42.9	338	86.9	1.640	27.6	16.99	16.48
37	4742	254	73	11.1	13.4	38.0	299	89.4	1.399	24.4	17.17	15.67
30	3794	254	69	11.1	9.6	32.8	258	93.0	1.170	21.8	17.55	15.39
22.8	2897	254	65	11.1	6.1	28.1	221	98.3	0.949	18.29	18.11	16.10
C230 × 30	3794	229	67	10.5	11.4	25.4	222	81.8	1.007	19.29	16.31	14.81
22	2845	229	63	10.5	7.2	21.2	185.2	86.4	0.803	16.69	16.79	14.88
19.9	2542	229	61	10.5	5.9	19.94	174.2	88.4	0.733	16.03	16.97	15.27
C200 × 27.9	3555	203	64	9.9	12.4	18.31	180.4	71.6	0.824	16.60	15.21	14.35
20.5	2606	203	59	9.9	7.7	15.03	148.1	75.9	0.637	14.17	15.62	14.05
17.1	2181	203	57	9.9	5.6	13.57	133.7	79.0	0.549	12.92	15.88	14.50
C180 × 22.0	2794	178	58	9.3	10.6	11.32	127.2	63.8	0.574	12.90	14.33	13.51
18.2	2323	178	55	9.3	8.0	10.07	113.2	66.0	0.487	11.69	14.50	13.34
14.6	1852	178	53	9.3	5.3	8.86	99.6	69.1	0.403	10.26	14.76	13.74
C150 × 19.3	2471	152	54	8.7	11.1	7.24	95.3	54.1	0.437	10.67	13.34	13.06
15.6	1994	152	51	8.7	8.0	6.33	83.3	56.4	0.360	9.40	13.44	12.70
12.2	1548	152	48	8.7	5.1	5.45	71.7	59.4	0.288	8.23	13.64	13.00
C130 × 13.4	1703	127	47	8.1	8.3	3.70	58.3	46.5	0.263	7.54	12.42	12.14
10.0	1271	127	44	8.1	4.8	3.12	49.1	49.5	0.199	6.28	12.52	12.29
C100 × 10.8	1374	102	43	7.5	8.2	1.911	37.5	37.3	0.180	5.74	11.43	11.66
8.0	1026	102	40	7.5	4.7	1.602	31.4	39.6	0.133	4.69	11.40	11.63
C75 × 8.9	1135	76	40	6.9	9.0	0.862	22.7	27.4	0.127	4.47	10.57	11.56
7.4	948	76	37	6.9	6.6	0.770	20.3	28.4	0.103	3.98	10.41	11.13
6.1	781	76	35	6.9	4.3	0.691	18.18	29.7	0.082	3.43	10.26	11.10

† An American Standard Channel is designated by the letter C followed by the nominal depth in millimeters and the mass in kilograms per meter.

(continued)

Table B-5
Properties of Steel Structural Shapes (Metric Units) (*continued*)

Properties of Rolled-Steel Shapes
(U.S. Customary Units)

Angles
Equal Legs

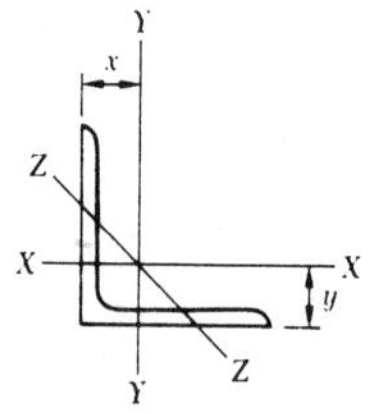

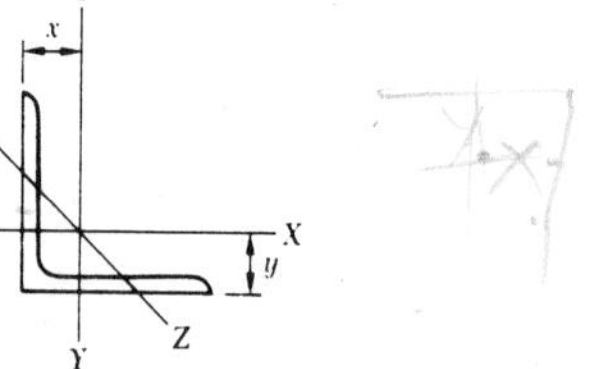

Size and Thickness, in.	Weight per Foot, lb/ft	Area, in²	Axis X-X and Axis Y-Y				Axis Z-Z
			I, in⁴	Z, in³	r, in.	x or y, in.	r, in.
L8 × 8 × 1	51.0	15.0	89.0	15.8	2.44	2.37	1.56
3/4	38.9	11.4	69.7	12.2	2.47	2.28	1.58
1/2	26.4	7.75	48.6	8.36	2.50	2.19	1.59
L6 × 6 × 1	37.4	11.0	35.5	8.57	1.80	1.86	1.17
3/4	28.7	8.44	28.2	6.66	1.83	1.78	1.17
5/8	24.2	7.11	24.2	5.66	1.84	1.73	1.18
1/2	19.6	5.75	19.9	4.61	1.86	1.68	1.18
3/8	14.9	4.36	15.4	3.53	1.88	1.64	1.19
L5 × 5 × 3/4	23.6	6.94	15.7	4.53	1.51	1.52	0.975
5/8	20.0	5.86	13.6	3.86	1.52	1.48	0.978
1/2	16.2	4.75	11.3	3.16	1.54	1.43	0.983
3/8	12.3	3.61	8.74	2.42	1.56	1.39	0.990
L4 × 4 × 3/4	18.5	5.44	7.67	2.81	1.19	1.27	0.778
5/8	15.7	4.61	6.66	2.40	1.20	1.23	0.779
1/2	12.8	3.75	5.56	1.97	1.22	1.18	0.782
3/8	9.8	2.86	4.36	1.52	1.23	1.14	0.788
1/4	6.6	1.94	3.04	1.05	1.25	1.09	0.795
L3½ × 3½ × 1/2	11.1	3.25	3.64	1.49	1.06	1.06	0.683
3/8	8.5	2.48	2.87	1.15	1.07	1.01	0.687
1/4	5.8	1.69	2.01	0.794	1.09	0.968	0.694
L3 × 3 × 1/2	9.4	2.75	2.22	1.07	0.898	0.932	0.584
3/8	7.2	2.11	1.76	0.833	0.913	0.888	0.587
1/4	4.9	1.44	1.24	0.577	0.930	0.842	0.592
L2½ × 2½ × 1/2	7.7	2.25	1.23	0.724	0.739	0.806	0.487
3/8	5.9	1.73	0.984	0.566	0.753	0.762	0.487
1/4	4.1	1.19	0.703	0.394	0.769	0.717	0.491
3/16	3.07	0.902	0.547	0.303	0.778	0.694	0.495
L2 × 2 × 3/8	4.7	1.36	0.479	0.351	0.594	0.636	0.389
1/4	3.19	0.938	0.348	0.247	0.609	0.592	0.391
1/8	1.65	0.484	0.190	0.131	0.626	0.546	0.398

Table B-5
Properties of Steel Structural Shapes (Metric Units) (*continued*)

Properties of Rolled-Steel Shapes
(SI Units)

Angles
Equal Legs

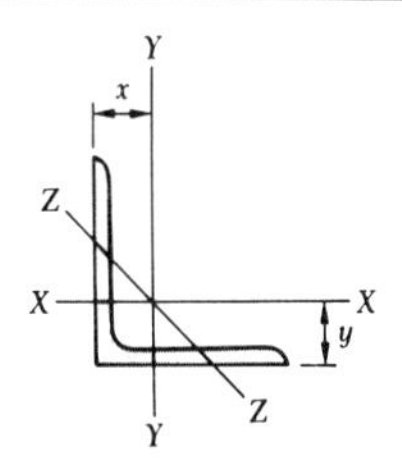

Size and Thickness, mm	Mass per Meter, kg/m	Area, mm²	Axis X-X and Axis Y-Y: I 10^6 mm	Z 10^3 mm³	r mm	x or y mm	Axis Z-Z r mm
L203 × 203 × 25.4	75.9	9680	37.0	259	61.8	60.2	39.6
19.0	57.9	7360	29.0	200	62.8	57.9	40.1
12.7	39.3	5000	20.2	137.0	63.6	55.6	40.4
L152 × 152 × 25.4	55.7	7100	14.78	140.4	45.6	47.2	29.7
19.0	42.7	5445	11.74	109.1	46.4	45.2	29.7
15.9	36.0	4590	10.07	92.8	46.8	43.9	30.0
12.7	29.2	3710	8.28	75.5	47.2	42.7	30.0
9.5	22.2	2800	6.41	57.8	47.8	41.7	30.2
L127 × 127 × 19.0	35.1	4480	6.53	74.2	38.2	38.6	24.8
15.9	29.8	3780	5.66	63.3	38.7	37.6	24.8
12.7	24.1	3070	4.70	51.8	39.2	36.3	25.0
9.5	18.3	2330	3.64	39.7	39.5	35.3	25.1
L102 × 102 × 19.0	27.5	3510	3.19	46.0	30.1	32.3	19.76
15.9	23.4	2970	2.77	39.3	30.5	31.2	19.79
12.7	19.0	2420	2.31	32.3	30.9	30.0	19.86
9.5	14.6	1845	1.815	24.9	31.4	29.0	20.0
6.4	9.8	1252	1.265	17.21	31.8	27.7	20.2
L89 × 89 × 12.7	16.5	2100	1.515	24.4	26.9	26.9	17.35
9.5	12.6	1600	1.195	18.85	27.3	25.7	17.45
6.4	8.6	1090	0.837	13.01	27.7	24.6	17.63
L76 × 76 × 12.7	14.0	1774	0.924	17.53	22.8	23.7	14.83
9.5	10.7	1361	0.733	13.65	23.2	22.6	14.91
6.4	7.3	929	0.516	9.46	23.6	21.4	15.04
L64 × 64 × 12.7	11.4	1452	0.512	11.86	18.78	20.5	12.37
9.5	8.7	1116	0.410	9.28	19.17	19.35	12.37
6.4	6.1	768	0.293	6.46	19.53	18.21	12.47
4.8	4.6	581	0.228	4.97	19.81	17.63	12.57
L51 × 51 × 9.5	7.0	877	0.1994	5.75	15.08	16.15	9.88
6.4	4.7	605	0.1448	4.05	15.47	15.04	9.93
3.2	2.4	312	0.0791	2.15	15.92	13.87	10.11

(*continued*)

Table B-5
Properties of Steel Structural Shapes (Metric Units) (*continued*)

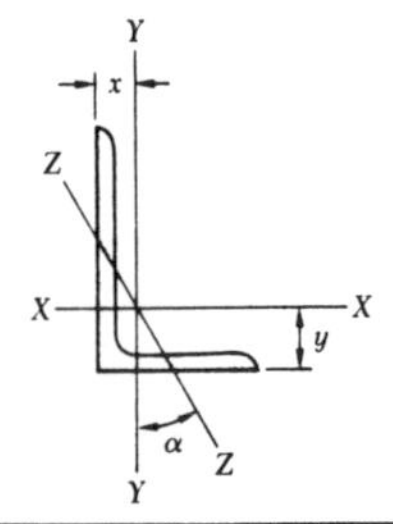

Properties of Rolled-Steel Shapes
(U.S. Customary Units)

Angles
Unequal Legs

			Axis X-X				Axis Y-Y				Axis Z-Z	
Size and Thickness, in.	Weight per Foot, lb/ft.	Area, in²	I_x, in⁴	Z_x, in³	r_x, in.	y, in.	I_y, in⁴	Z_y, in³	r_y, in.	x, in.	r_z, in.	tan α
L8 × 6 × 1	44.2	13.0	80.8	15.1	2.49	2.65	38.8	8.92	1.73	1.65	1.28	0.543
3/4	33.8	9.94	63.4	11.7	2.53	2.56	30.7	6.92	1.76	1.56	1.29	0.551
1/2	23.0	6.75	44.3	8.02	2.56	2.47	21.7	4.79	1.79	1.47	1.30	0.558
L6 × 4 × 3/4	23.6	6.94	24.5	6.25	1.88	2.08	8.68	2.97	1.12	1.08	0.860	0.428
1/2	16.2	4.75	17.4	4.33	1.91	1.99	6.27	2.08	1.15	0.987	0.870	0.440
3/8	12.3	3.61	13.5	3.32	1.93	1.94	4.90	1.60	1.17	0.941	0.877	0.446
L5 × 3 × 1/2	12.8	3.75	9.45	2.91	1.59	1.75	2.58	1.15	0.829	0.750	0.648	0.357
3/8	9.8	2.86	7.37	2.24	1.61	1.70	2.04	0.888	0.845	0.704	0.654	0.364
1/4	6.6	1.94	5.11	1.53	1.62	1.66	1.44	0.614	0.861	0.657	0.663	0.371
L4 × 3 × 1/2	11.1	3.25	5.05	1.89	1.25	1.33	2.42	1.12	0.864	0.827	0.639	0.543
3/8	8.5	2.48	3.96	1.46	1.26	1.28	1.92	0.866	0.879	0.782	0.644	0.551
1/4	5.8	1.69	2.77	1.00	1.28	1.24	1.36	0.599	0.896	0.736	0.651	0.558
L3½ × 2½ × 1/2	9.4	2.75	3.24	1.41	1.09	1.20	1.36	0.760	0.704	0.705	0.534	0.486
3/8	7.2	2.11	2.56	1.09	1.10	1.16	1.09	0.592	0.719	0.660	0.537	0.496
1/4	4.9	1.44	1.80	0.755	1.12	1.11	0.777	0.412	0.735	0.614	0.544	0.506
L3 × 2 × 1/2	7.7	2.25	1.92	1.00	0.924	1.08	0.672	0.474	0.546	0.583	0.428	0.414
3/8	5.9	1.73	1.53	0.781	0.940	1.04	0.543	0.371	0.559	0.539	0.430	0.428
1/4	4.1	1.19	1.09	0.542	0.957	0.993	0.392	0.260	0.574	0.493	0.435	0.440
L2½ × 2 × 3/8	5.3	1.55	0.912	0.547	0.768	0.831	0.514	0.363	0.577	0.581	0.420	0.614
1/4	3.62	1.06	0.654	0.381	0.784	0.787	0.372	0.254	0.592	0.537	0.424	0.626

Table B-5
Properties of Steel Structural Shapes (Metric Units) (*continued*)

Properties of Rolled-Steel Shapes
(SI Units)

Angles
Unequal Legs

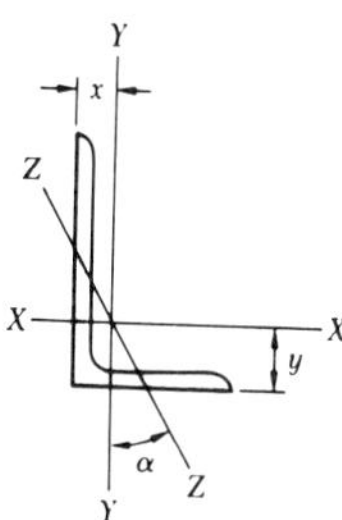

Size and Thickness, mm	Mass per Meter kg/m	Area mm^2	Axis X-X				Axis Y-Y				Axis Z-Z	
			I_x 10^6 mm^4	Z_x 10^3 mm^3	r_x mm	y mm	I_y 10^6 mm^4	Z_y 10^3 mm^3	r_y mm	x mm	r_z mm	tan α
L203 × 152 × 25.4	65.5	8390	33.6	247	63.3	67.3	16.15	146.2	43.9	41.9	32.5	0.543
19.0	50.1	6410	26.4	192	64.2	65.0	12.78	113.4	44.7	39.6	32.8	0.551
12.7	34.1	4350	18.44	131	65.1	62.7	9.03	78.5	45.6	37.3	33.0	0.558
L152 × 102 × 19.0	**35.0**	**4480**	**10.20**	**102.4**	**47.7**	**52.8**	**3.61**	**48.7**	**28.4**	**27.4**	**21.8**	**0.428**
12.7	**24.0**	**3060**	**7.24**	**71.0**	**48.6**	**50.5**	**2.61**	**34.1**	**29.2**	**25.1**	**22.1**	**0.440**
9.5	**18.2**	**2330**	**5.62**	**54.4**	**49.1**	**49.3**	**2.04**	**26.2**	**29.6**	**23.9**	**22.3**	**0.446**
L127 × 76 × 12.7	19.0	2420	3.93	47.7	40.3	44.5	1.074	18.85	21.1	19.05	16.46	0.357
9.5	14.5	1845	3.07	36.7	40.8	43.2	0.849	14.55	21.5	17.88	16.61	0.364
6.4	9.8	1252	2.13	25.1	41.2	42.2	0.599	10.06	21.9	16.69	16.84	0.371
L102 × 76 × 12.7	**16.4**	**2100**	**2.10**	**31.0**	**31.6**	**33.8**	**1.007**	**18.35**	**21.9**	**21.0**	**16.23**	**0.543**
9.5	**12.6**	**1600**	**1.648**	**23.9**	**32.1**	**32.5**	**0.799**	**14.19**	**22.3**	**19.86**	**16.36**	**0.551**
6.4	**8.6**	**1090**	**1.153**	**16.39**	**32.5**	**31.5**	**0.566**	**9.82**	**22.8**	**18.69**	**16.54**	**0.558**
L89 × 64 × 12.7	13.9	1774	1.349	23.1	27.6	30.5	0.566	12.45	17.88	17.91	13.56	0.486
9.5	10.7	1361	1.066	17.86	28.0	29.5	0.454	9.70	18.26	16.76	13.64	0.496
6.4	7.3	929	0.749	12.37	28.4	28.2	0.323	6.75	18.65	15.60	13.82	0.506
L76 × 51 × 12.7	**11.5**	**1452**	**0.799**	**16.39**	**23.5**	**27.4**	**0.280**	**7.77**	**13.89**	**14.81**	**10.87**	**0.414**
9.5	**8.8**	**1116**	**0.637**	**12.80**	**23.9**	**26.4**	**0.226**	**6.08**	**14.20**	**13.69**	**10.92**	**0.428**
6.4	**6.1**	**768**	**0.454**	**8.88**	**24.3**	**25.2**	**0.1632**	**4.26**	**14.58**	**12.52**	**11.05**	**0.440**
L64 × 51 × 9.5	7.9	1000	0.380	8.96	19.51	21.1	0.214	5.95	14.66	14.76	10.67	0.614
6.4	5.4	684	0.272	6.24	19.94	20.0	0.1548	4.16	15.04	13.64	10.77	0.626

(*continued*)

Table B-6
Properties of Aluminum Structural Shapes (U.S. Customary Units)

Aluminum Association Standard Channels— Dimensions, Areas, Weights and Section Properties

Size							Section Properties⑧						
							Axis X-X			Axis Y-Y			
Depth A in.	Width B in.	Area① $in.^2$	Weight② lb/ft	Flange Thick-ness t_1 in.	Web Thick-ness t in.	Fillet Radius R in.	I $in.^4$	Z $in.^3$	r in.	I $in.^4$	Z $in.^3$	r in.	x in.
2.00	1.00	0.491	0.577	0.13	0.13	0.10	0.288	0.288	0.766	0.045	0.064	0.303	0.298
2.00	1.25	0.911	1.071	0.26	0.17	0.15	0.546	0.546	0.774	0.139	0.178	0.391	0.471
3.00	1.50	0.965	1.135	0.20	0.13	0.25	1.41	0.94	1.21	0.22	0.22	0.47	0.49
3.00	1.75	1.358	1.597	0.26	0.17	0.25	1.97	1.31	1.20	0.42	0.37	0.55	0.62
4.00	2.00	1.478	1.738	0.23	0.15	0.25	3.91	1.95	1.63	0.60	0.45	0.64	0.65
4.00	2.25	1.982	2.331	0.29	0.19	0.25	5.21	2.60	1.62	1.02	0.69	0.72	0.78
5.00	2.25	1.881	2.212	0.26	0.15	0.30	7.88	3.15	2.05	0.98	0.64	0.72	0.73
5.00	2.75	2.627	3.089	0.32	0.19	0.30	11.14	4.45	2.06	2.05	1.14	0.88	0.95
6.00	2.50	2.410	2.834	0.29	0.17	0.30	14.35	4.78	2.44	1.53	0.90	0.80	0.79
6.00	3.25	3.427	4.030	0.35	0.21	0.30	21.04	7.01	2.48	3.76	1.76	1.05	1.12
7.00	2.75	2.725	3.205	0.29	0.17	0.30	22.09	6.31	2.85	2.10	1.10	0.88	0.84
7.00	3.50	4.009	4.715	0.38	0.21	0.30	33.79	9.65	2.90	5.13	2.23	1.13	1.20
8.00	3.00	3.526	4.147	0.35	0.19	0.30	37.40	9.35	3.26	3.25	1.57	0.96	0.95
8.00	3.75	4.923	5.789	0.41	0.25	0.35	52.69	13.17	3.27	7.13	2.82	1.20	1.22
9.00	3.25	4.237	4.983	0.35	0.23	0.35	54.41	12.09	3.58	4.40	1.89	1.02	0.93
9.00	4.00	5.927	6.970	0.44	0.29	0.35	78.31	17.40	3.63	9.61	3.49	1.27	1.25
10.00	3.50	5.218	6.136	0.41	0.25	0.35	83.22	16.64	3.99	6.33	2.56	1.10	1.02
10.00	4.25	7.109	8 360	0.50	0.31	0.40	116.15	23.23	4.04	13.02	4.47	1.35	1.34
12.00	4.00	7.036	8.274	0.47	0.29	0.40	159.76	26.63	4.77	11.03	3.86	1.25	1.14
12.00	5.00	10.053	11.822	0.62	0.35	0.45	239.69	39.95	4.88	25.74	7.60	1.60	1.61

Aluminum Association Standard I-Beams — Dimensions, Areas, Weights and Section Properties

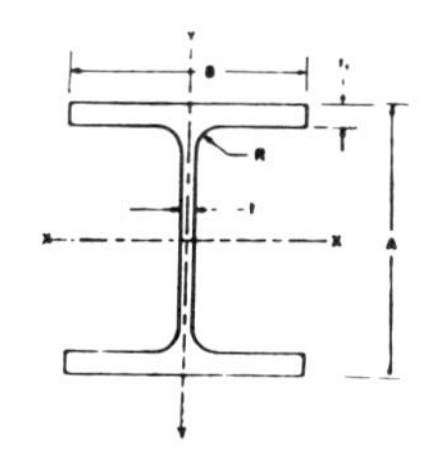

Size							Section Properties③					
							Axis X-X			Axis Y-Y		
Depth A in.	Width B in.	Area① in.²	Weight② lb/ft	Flange Thick-ness t_1 in.	Web Thick-ness t in.	Fillet Radius R in.	I in.⁴	Z in.³	r in.	I in.⁴	Z in.³	r in.
3.00	2.50	1.392	1.637	0.20	0.13	0.25	2.24	1.49	1.27	0.52	0.42	0.61
3.00	2.50	1.726	2.030	0.26	0.15	0.25	2.71	1.81	1.25	0.68	0.54	0.63
4.00	3.00	1.965	2.311	0.23	0.15	0.25	5.62	2.81	1.69	1.04	0.69	0.73
4.00	3.00	2.375	2.793	0.29	0.17	0.25	6.71	3.36	1.68	1.31	0.87	0.74
5.00	3.50	3.146	3.700	0.32	0.19	0.30	13.94	5.58	2.11	2.29	1.31	0.85
6.00	4.00	3.427	4.030	0.29	0.19	0.30	21.99	7.33	2.53	3.10	1.55	0.95
6.00	4.00	3.990	4.692	0.35	0.21	0.30	25.50	8.50	2.53	3.74	1.87	0.97
7.00	4.50	4.932	5.800	0.38	0.23	0.30	42.89	12.25	2.95	5.78	2.57	1.08
8.00	5.00	5.255	6.181	0.35	0.23	0.30	59.69	14.92	3.37	7.30	2.92	1.18
8.00	5.00	5.972	7.023	0.41	0.25	0.30	67.78	16.94	3.37	8.55	3.42	1.20
9.00	5.50	7.110	8.361	0.44	0.27	0.30	102.02	22.67	3.79	12.22	4.44	1.31
10.00	6.00	7.352	8.646	0.41	0.25	0.40	132.09	26.42	4.24	14.78	4.93	1.42
10.00	6.00	8.747	10.286	0.50	0.29	0.40	155.79	31.16	4.22	18.03	6.01	1.44
12.00	7.00	9.925	11.672	0.47	0.29	0.40	255.57	42.60	5.07	26.90	7.69	1.65
12.00	7.00	12.153	14.292	0.62	0.31	0.40	317.33	52.89	5.11	35.48	10.14	1.71

① Areas listed are based on nominal dimensions.
② Weights per foot are based on nominal dimensions and a density of 0.098 pound per cubic inch which is the density of alloy 6061.
③ I = moment of inertia; Z = section modulus; r = radius of gyration.

*Table B-6 is taken from *Applied Strength of Materials,* by Robert L. Mott ©1978, and is reprinted by permission of Prentice-Hall, Inc., Englewood Cliffs, NJ.

Table B-7
Slope and Deflection Formulas for Straight Beams

Beam Conditions	Deflection Formulas	Slope Formulas
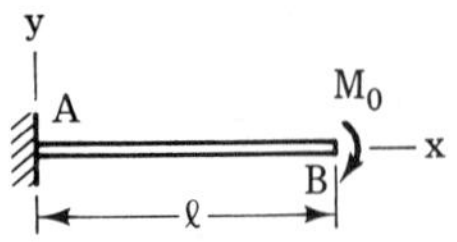 y, A, M_0, B, x, ℓ	$\delta = -\frac{M_0x^2}{2EI}$ $\delta_{max} = -\frac{M_0\ell^2}{2EI}$ (at free end)	$\theta = -\frac{M_0x}{EI}$ $\theta_{max} = -\frac{M_0\ell}{EI}$ (at free end)
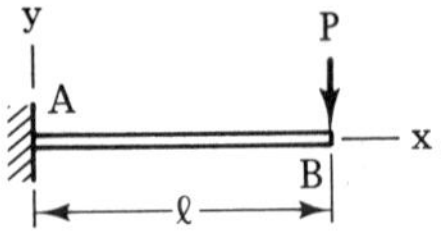 y, P, A, B, x, ℓ	$\delta = \frac{Px^2}{6EI}(x - 3\ell)$ $\delta_{max} = -\frac{P\ell^3}{3EI}$ (at free end)	$\theta = \frac{Px}{2EI}(x - 2\ell)$ $\theta_{max} = -\frac{P\ell^2}{2EI}$ (at free end)
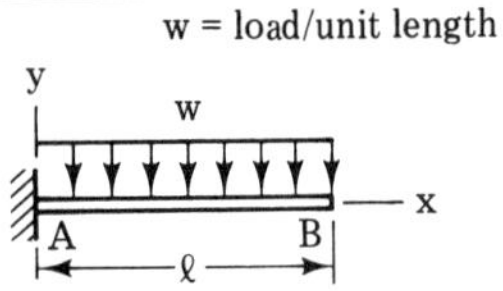 w = load/unit length y, w, A, B, x, ℓ	$\theta = \frac{wx}{6EI}(-x^2 + 3\ell x - 3\ell^2)$ $\theta_{max} = -\frac{w\ell^3}{6EI}$ (at free end)	$\delta = \frac{wx^2}{24EI}(-x^2 + 4\ell x - 6\ell^2)$ $\delta_{max} = -\frac{w\ell^4}{8EI}$ (at free end)
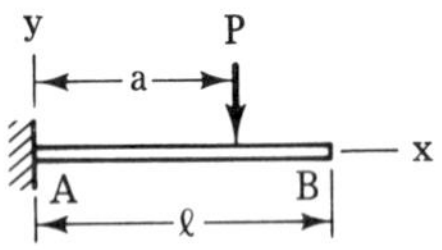 y, P, a, A, B, x, ℓ	$0 \leq x \leq a$ $\delta = \frac{Px^2}{6EI}(x - 3a)$ $\delta_a = -\frac{Pa^3}{3EI}$ $a \leq x \leq \ell$ $\delta = \frac{Pa^2}{6EI}(a - 3x)$ $\delta_{max} = -\frac{Pa^2}{6EI}(3\ell - a)$ (at free end)	$0 \leq x \leq a$ $\theta = \frac{Px}{2EI}(x - 2a)$ $\theta_a = -\frac{Pa^2}{2EI}$ $a \leq x \leq \ell$ $\theta = -\frac{Pa^2}{2EI}$ $\theta_{max} = -\frac{Pa^2}{2EI}$ (at free end)
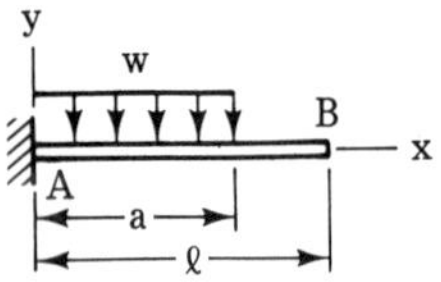 y, w, B, x, A, a, ℓ	$0 \leq x \leq a$ $\delta = \frac{wx^2}{24EI}(-x^2 + 4ax - 6a^2)$ $\delta_a = -\frac{wa^4}{8EI}$ $a \leq x \leq \ell$ $\delta = -\frac{wa^3}{24EI}(4x - a)$ $\delta_{max} = -\frac{wa^3}{24EI}(4\ell - a)$ (at free end)	$0 \leq x \leq a$ $\theta = \frac{wx}{6EI}(-x^2 + 3ax - 3a^2)$ $\theta_a = -\frac{wa^3}{6EI}$ $a \leq x \leq \ell$ $\theta = -\frac{wa^3}{6EI}$ $\theta_{max} = -\frac{wa^3}{6EI}$ (at free end)

Table B-7
Slope and Deflection Formulas for Straight Beams (*continued*)

Beam Conditions	Deflection Formulas	Slope Formulas
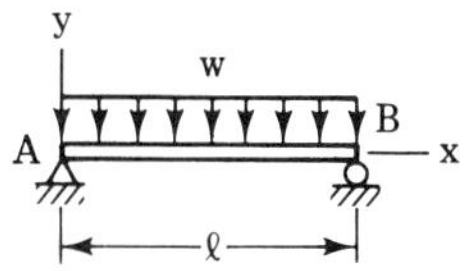 y, w, A, B, x, ℓ	$\delta = \frac{wx}{24EI}(-x^3 + 2\ell x^2 - \ell^3)$ $\delta_{max} = \frac{5w\ell^4}{384EI}$ (at midspan)	$\theta = \frac{w}{24EI}(-4x^3 + 6\ell x^2 - \ell^3)$ $\theta_{max} = \theta_0 = \theta_\ell = -\frac{w^3}{24EI}$ (at ends)
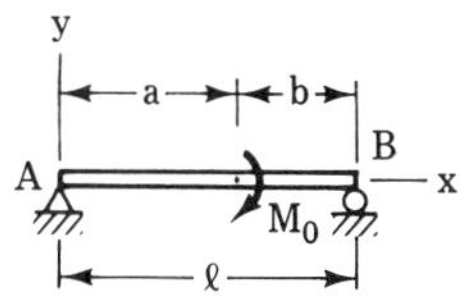 y, a, b, A, B, x, M_0, ℓ	$0 \leq x \leq a$ $\delta = -\frac{M_0 x}{6EI\ell}(x^2 + 3a^2 - 6a\ell + 2\ell^2)$ $a \leq x \leq b$ $\delta = -\frac{M_0}{6EI\ell}\{x^3 - 3\ell x^2 + x(2\ell^2 + 3a^2) - 3a^2\ell\}$	$0 \leq x \leq a$ $\theta = -\frac{M_0}{6EI\ell}(3x^2 + 3a^2 - 6a\ell + 2\ell^2)$ $\theta_0 = -\frac{M_0}{6EI\ell}(3a^2 - 6a\ell + 2\ell^2)$ $a \leq x \leq b$ $\theta = -\frac{M_0}{6EI\ell}(3x^2 - 6\ell x + 2\ell^2 + 3a^2)$ $\theta_\ell = -\frac{M_0}{6EI\ell}(3a^2 - \ell^2)$
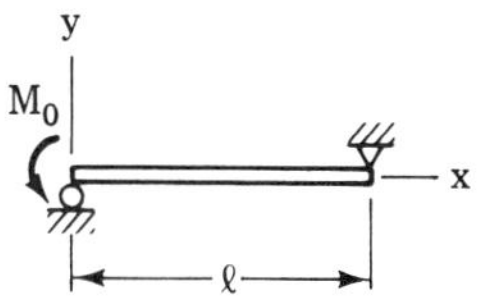 y, M_0, x, ℓ	$\delta = \frac{M_0 x}{6EI\ell}(x^2 - 3x\ell + 2\ell^2)$	$\theta = \frac{M_0}{6EI\ell}(3x^2 - 6x\ell + 2\ell^2)$ $\theta_0 = \frac{M_0\ell}{3EI}$ $\theta_\ell = \frac{M_0\ell}{6EI}$
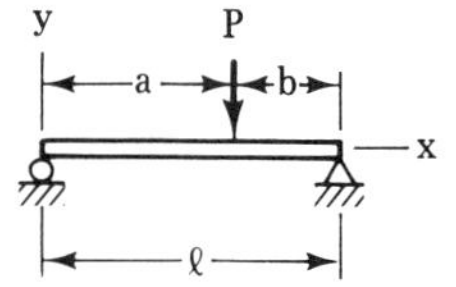 y, P, a, b, x, ℓ	$0 \leq x \leq a$ $\delta = \frac{Pbx}{6EI\ell}(x^2 + b^2 - \ell^2)$ $a \leq x \leq \ell$ $\delta = \frac{Pa(\ell - x)}{6EI\ell}(x^2 - 2\ell x + a^2)$	$0 \leq x \leq 0$ $\theta = \frac{Pb}{6EI\ell}(3x^2 + b^2 - \ell^2)$ $\theta_0 = \frac{Pb(b^2 - \ell^2)}{6EI\ell}$ $a \leq x \leq \ell$ $\theta = \frac{Pa}{6EI\ell}(-3x^2 + 6\ell x - a^2 - 2\ell^2)$ $\theta_\ell = \frac{Pa(\ell^2 - a^2)}{6EI\ell}$

Table B-7
Slope and Deflection Formulas for Straight Beams (*continued*)

Beam Conditions	Deflection Formulas	Slope Formulas
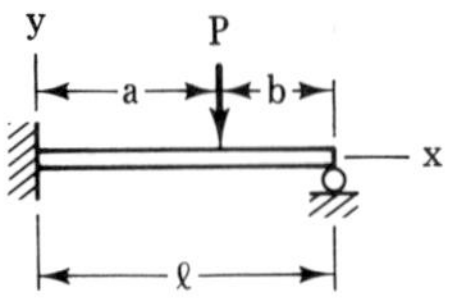 y, P, a, b, x, ℓ	$\underline{0 \le x \le a}$ $\delta = \dfrac{Pbx^2}{12EI\ell^3}\{3\ell(b^2 - \ell^2) + x(3\ell^2 - b^2)\}$ $\underline{a \le x \le \ell}$ $\delta = [\delta]_{AB} - \dfrac{P(x-a)^3}{6EI}$	$\underline{0 \le x \le 0}$ $\theta = \dfrac{Pbx}{4EI\ell^3}\{2\ell(b^2 - \ell^2) + x(3\ell^2 - b^2)\}$ $\underline{a \le x \le \ell}$ $\theta = [\theta]_{AB} - \dfrac{P(x-a)^2}{2EI}$ $\theta_\ell = \dfrac{Pb}{4EI\ell}(\ell - b)^2$
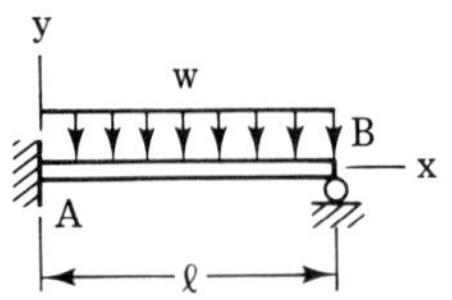 y, w, A, B, x, ℓ	$\delta = \dfrac{wx^2}{48EI}(\ell - x)(2x - 3\ell)$	$\theta = \dfrac{wx}{48EI}(-8x^2 + 15\ell x - 6\ell^2)$ $\theta_\ell = \dfrac{w\ell^3}{48EI}$
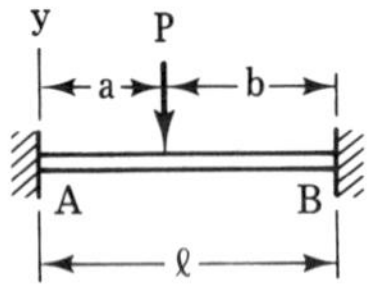 y, P, a, b, A, B, ℓ	$\underline{0 \le x \le a}$ $\delta = \dfrac{Pb^2x^2}{6EI\ell^3}[x(3a + b) - 3a\ell]$ $\underline{a \le x \le \ell}$ $\delta = \dfrac{Pa^2(\ell - x)^2}{6EI\ell^3}[(\ell - x)(3b + a) - 3b\ell]$	$\underline{0 \le x \le a}$ $\theta = \dfrac{Pb^2x}{2EI\ell^3}[x(3a + b) - 2a\ell]$ $\underline{a \le x \le \ell}$ $\theta = \dfrac{Pa^2(\ell - x)}{2EI\ell^3}[2b\ell - (\ell - x)(3a + b)]$
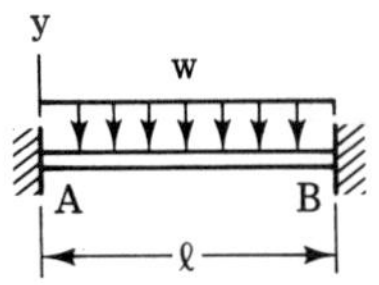 y, w, A, B, ℓ	$\delta = -\dfrac{wx^2}{24EI}(\ell - x)^2$	$\theta = -\dfrac{wx}{12EI}(\ell - x)(\ell - 2x)$

Table B-8
Sizes and Sectional Properties of Standard Dressed Lumber*

NOMINAL SIZE b(inches)d	STANDARD DRESSED SIZE (S4S) b(inches)d	AREA OF SECTION A	MOMENT OF INERTIA I	SECTION MODULUS Z
1 x 3	3/4 x 2 1/2	1.875	0.977	0.781
1 x 4	3/4 x 3 1/2	2.625	2.680	1.531
1 x 6	3/4 x 5 1/2	4.125	10.398	3.781
1 x 8	3/4 x 7 1/4	5.438	23.817	6.570
1 x 10	3/4 x 9 1/4	6.938	49.466	10.695
1 x 12	3/4 x 11 1/4	8.438	88.989	15.820
2 x 3	1 1/2 x 2 1/2	3.750	1.953	1.563
2 x 4	1 1/2 x 3 1/2	5.250	5.359	3.063
2 x 6	1 1/2 x 5 1/2	8.250	20.797	7.563
2 x 8	1 1/2 x 7 1/4	10.875	47.635	13.141
2 x 10	1 1/2 x 9 1/4	13.875	98.932	21.391
2 x 12	1 1/2 x 11 1/4	16.875	177.979	31.641
2 x 14	1 1/2 x 13 1/4	19.875	290.775	43.891
3 x 1	2 1/2 x 3/4	1.875	0.088	0.234
3 x 2	2 1/2 x 1 1/2	3.750	0.703	0.938
3 x 4	2 1/2 x 3 1/2	8.750	8.932	5.104
3 x 6	2 1/2 x 5 1/2	13.750	34.661	12.604
3 x 8	2 1/2 x 7 1/4	18.125	79.391	21.901
3 x 10	2 1/2 x 9 1/4	23.125	164.886	35.651
3 x 12	2 1/2 x 11 1/4	28.125	296.631	52.734
3 x 14	2 1/2 x 13 1/4	33.125	484.625	73.151
3 x 16	2 1/2 x 15 1/4	38.125	738.870	96.901
4 x 1	3 1/2 x 3/4	2.625	0.123	0.328
4 x 2	3 1/2 x 1 1/2	5.250	0.984	1.313
4 x 3	3 1/2 x 2 1/2	8.750	4.557	3.646
4 x 4	3 1/2 x 3 1/2	12.250	12.505	7.146
4 x 6	3 1/2 x 5 1/2	19.250	48.526	17.646
4 x 8	3 1/2 x 7 1/4	25.375	111.148	30.661
4 x 10	3 1/2 x 9 1/4	32.375	230.840	49.911
4 x 12	3 1/2 x 11 1/4	39.375	415.283	73.828
4 x 14	3 1/2 x 13 1/2	47.250	717.609	106.313
4 x 16	3 1/2 x 15 1/2	54.250	1086.130	140.146
6 x 1	5 1/2 x 3/4	4.125	0.193	0.516
6 x 2	5 1/2 x 1 1/2	8.250	1.547	2.063
6 x 3	5 1/2 x 2 1/2	13.750	7.161	5.729
6 x 4	5 1/2 x 3 1/2	19.250	19.651	11.229
6 x 6	5 1/2 x 5 1/2	30.250	76.255	27.729
6 x 8	5 1/2 x 7 1/2	41.250	193.359	51.563
6 x 10	5 1/2 x 9 1/2	52.250	392.963	82.729
6 x 12	5 1/2 x 11 1/2	63.250	697.068	121.229
6 x 14	5 1/2 x 13 1/2	74.250	1127.672	167.063
6 x 16	5 1/2 x 15 1/2	85.250	1706.776	220.229
6 x 18	5 1/2 x 17 1/2	96.250	2456.380	280.729
6 x 20	5 1/2 x 19 1/2	107.250	3398.484	348.563
6 x 22	5 1/2 x 21 1/2	118.250	4555.086	423.729
6 x 24	5 1/2 x 23 1/2	129.250	5948.191	506.229

*Table B-8 on pages 653-655 is taken from *Wood Structural Design Data* (1978 edition) and is used Courtesy of the National Forest Products Association, Washington, D.C.

** b = width, d = depth

(continued)

Table B-8
Sizes and Sectional Properties of Standard Dressed Lumber (*continued*)

NOMINAL SIZE b(inches)d	STANDARD DRESSED SIZE (S4S) b(inches)d	AREA OF SECTION A	MOMENT OF INERTIA I	SECTION MODULUS Z
8 x 1	7 1/4 x 3/4	5.438	0.255	0.680
8 x 2	7 1/4 x 1 1/2	10.875	2.039	2.719
8 x 3	7 1/4 x 2 1/2	18.125	9.440	7.552
8 x 4	7 1/4 x 3 1/2	25.375	25.904	14.802
8 x 6	7 1/2 x 5 1/2	41.250	103.984	37.813
8 x 8	7 1/2 x 7 1/2	56.250	263.672	70.313
8 x 10	7 1/2 x 9 1/2	71.250	535.859	112.813
8 x 12	7 1/2 x 11 1/2	86.250	950.547	165.313
8 x 14	7 1/2 x 13 1/2	101.250	1537.734	227.813
8 x 16	7 1/2 x 15 1/2	116.250	2327.422	300.313
8 x 18	7 1/2 x 17 1/2	131.250	3349.609	382.813
8 x 20	7 1/2 x 19 1/2	146.250	4634.297	475.313
8 x 22	7 1/2 x 21 1/2	161.250	6211.484	577.813
8 x 24	7 1/2 x 23 1/2	176.250	8111.172	690.313
10 x 1	9 1/4 x 3/4	6.938	0.325	0.867
10 x 2	9 1/4 x 1 1/2	13.875	2.602	3.469
10 x 3	9 1/4 x 2 1/2	23.125	12.044	9.635
10 x 4	9 1/4 x 3 1/2	32.375	33.049	18.885
10 x 6	9 1/2 x 5 1/2	52.250	131.714	47.896
10 x 8	9 1/2 x 7 1/2	71.250	333.984	89.063
10 x 10	9 1/2 x 9 1/2	90.250	678.755	142.896
10 x 12	9 1/2 x 11 1/2	109.250	1204.026	209.396
10 x 14	9 1/2 x 13 1/2	128.250	1947.797	288.563
10 x 16	9 1/2 x 15 1/2	147.250	2948.068	380.396
10 x 18	9 1/2 x 17 1/2	166.250	4242.836	484.896
10 x 20	9 1/2 x 19 1/2	185.250	5870.109	602.063
10 x 22	9 1/2 x 21 1/2	204.250	7867.879	731.896
10 x 24	9 1/2 x 23 1/2	223.250	10274.148	874.396
12 x 1	11 1/4 x 3/4	8.438	0.396	1.055
12 x 2	11 1/4 x 1 1/2	16 875	3.164	4.219
12 x 3	11 1/4 x 2 1/2	28.125	14.648	11.719
12 x 4	11 1/4 x 3 1/2	39.375	40.195	22.969
12 x 6	11 1/2 x 5 1/2	63.250	159.443	57.979
12 x 8	11 1/2 x 7 1/2	86.250	404.297	107.813
12 x 10	11 1/2 x 9 1/2	109.250	821.651	172.979
12 x 12	11 1/2 x 11 1/2	132.250	1457.505	253.479
12 x 14	11 1/2 x 13 1/2	155.250	2357.859	349.313
12 x 16	11 1/2 x 15 1/2	178.250	3568.713	460.479
12 x 18	11 1/2 x 17 1/2	201.250	5136.066	586.979
12 x 20	11 1/2 x 19 1/2	224.250	7105.922	728.813
12 x 22	11 1/2 x 21 1/2	247.250	9524.273	885.979
12 x 24	11 1/2 x 23 1/2	270.250	12437.129	1058.479
14 x 2	13 1/4 x 1 1/2	19.875	3.727	4.969
14 x 3	13 1/4 x 2 1/2	33.125	17.253	13.802
14 x 4	13 1/2 x 3 1/2	47.250	48.234	27.563
14 x 6	13 1/2 x 5 1/2	74.250	187.172	68.063
14 x 8	13 1/2 x 7 1/2	101.250	474.609	126.563
14 x 10	13 1/2 x 9 1/2	128.250	964.547	203.063
14 x 12	13 1/2 x 11 1/2	155.250	1710.984	297.563
14 x 16	13 1/2 x 15 1/2	209.250	4189.359	540.563
14 x 18	13 1/2 x 17 1/2	236.250	6029.297	689.063
14 x 20	13 1/2 x 19 1/2	263.250	8341.734	855.563
14 x 22	13 1/2 x 21 1/2	290.250	11180.672	1040.063
14 x 24	13 1/2 x 23 1/2	317.250	14600.109	1242.563

Table B-8
Sizes and Sectional Properties of Standard Dressed Lumber (*continued*)

NOMINAL SIZE b(inches)d	STANDARD DRESSED SIZE (S4S) b(inches)d	AREA OF SECTION A	MOMENT OF INERTIA I	SECTION MODULUS Z
16 x 3	15 1/2 x 2 1/2	38.750	20.182	16.146
16 x 4	15 1/2 x 3 1/2	54.250	55.380	31.646
16 x 6	15 1/2 x 5 1/2	85.250	214.901	78.146
16 x 8	15 1/2 x 7 1/2	116.250	544.922	145.313
16 x 10	15 1/2 x 9 1/2	147.250	1107.443	233.146
16 x 12	15 1/2 x 11 1/2	178.250	1964.463	341.646
16 x 14	15 1/2 x 13 1/2	209.250	3177.984	470.813
16 x 16	15 1/2 x 15 1/2	240.250	4810.004	620.646
16 x 18	15 1/2 x 17 1/2	271.250	6922.523	791.146
16 x 20	15 1/2 x 19 1/2	302.250	9577.547	982.313
16 x 22	15 1/2 x 21 1/2	333.250	12837.066	1194.146
16 x 24	15 1/2 x 23 1/2	364.250	16763.086	1426.646
18 x 6	17 1/2 x 5 1/2	96.250	242.630	88.229
18 x 8	17 1/2 x 7 1/2	131.250	615.234	164.063
18 x 10	17 1/2 x 9 1/2	166.250	1250.338	263.229
18 x 12	17 1/2 x 11 1/2	201.250	2217.943	385.729
18 x 14	17 1/2 x 13 1/2	236.250	3588.047	531.563
18 x 16	17 1/2 x 15 1/2	271.250	5430.648	700.729
18 x 18	17 1/2 x 17 1/2	306.250	7815.754	893.229
18 x 20	17 1/2 x 19 1/2	341.250	10813.359	1109.063
18 x 22	17 1/2 x 21 1/2	376.250	14493.461	1348.229
18 x 24	17 1/2 x 23 1/2	411.250	18926.066	1610.729
20 x 6	19 1/2 x 5 1/2	107.250	270.359	98.313
20 x 8	19 1/2 x 7 1/2	146.250	685.547	182.813
20 x 10	19 1/2 x 9 1/2	185.250	1393.234	293.313
20 x 12	19 1/2 x 11 1/2	224.250	2471.422	429.813
20 x 14	19 1/2 x 13 1/2	263.250	3998.109	592.313
20 x 16	19 1/2 x 15 1/2	302.250	6051.297	780.813
20 x 18	19 1/2 x 17 1/2	341.250	8708.984	995.313
20 x 20	19 1/2 x 19 1/2	380.250	12049.172	1235.813
20 x 22	19 1/2 x 21 1/2	419.250	16149.859	1502.313
20 x 24	19 1/2 x 23 1/2	458.250	21089.047	1794.813
22 x 6	21 1/2 x 5 1/2	118.250	298.088	108.396
22 x 8	21 1/2 x 7 1/2	161.250	755.859	201.563
22 x 10	21 1/2 x 9 1/2	204.250	1536.130	323.396
22 x 12	21 1/2 x 11 1/2	247.250	2724.901	473.896
22 x 14	21 1/2 x 13 1/2	290.250	4408.172	653.063
22 x 16	21 1/2 x 15 1/2	333.250	6671.941	860.896
22 x 18	21 1/2 x 17 1/2	376.250	9602.211	1097.396
22 x 20	21 1/2 x 19 1/2	419.250	13284.984	1362.563
22 x 22	21 1/2 x 21 1/2	462.250	17806.254	1656.396
22 x 24	21 1/2 x 23 1/2	505.250	23252.023	1978.896
24 x 6	23 1/2 x 5 1/2	129.250	325.818	118.479
24 x 8	23 1/2 x 7 1/2	176.250	826.172	220.313
24 x 10	23 1/2 x 9 1/2	223.250	1679.026	353.479
24 x 12	23 1/2 x 11 1/2	270.250	2978.380	517.979
24 x 14	23 1/2 x 13 1/2	317.250	4818.234	713.813
24 x 16	23 1/2 x 15 1/2	364.250	7292.586	940.979
24 x 18	23 1/2 x 17 1/2	411.250	10495.441	1199.479
24 x 20	23 1/2 x 19 1/2	458.250	14520.797	1489.313
24 x 22	23 1/2 x 21 1/2	505.250	19462.648	1810.479
24 x 24	23 1/2 x 23 1/2	552.250	25415.004	2162.979

Table B-9
Selected Properties of Metals***

	Ultimate Strength						Tension** Yield Point		Modulus of Elasticity, E		Modulus of Rigidity, G	
	Tension		Compression		Shear							
	Ksi*	MPa	Ksi	MPa	Ksi	MPa	Ksi	MPa	10^6 psi	GPa	10^6 psi	GPa
2014-T4 Aluminum	62	427			38	262	42	290	10.6	73	4.0	27.6
2024-T4 Aluminum	68	469			41	283	47	324	10.6	73	4.0	27.6
6061-T6 Aluminum	45	310			30	207	40	276	10.0	69	3.75	25.8
7075-T6 Aluminum	83	572			48	331	73	503	10.4	72	3.90	26.9
Brass, Red, Cold Drawn	75	518					72	490	15.0	103	5.6	38.6
Brass, Red, Annealed	37	255					14	96	15.0	103	5.6	38.6
Brass, Yellow	61	420			40	275	50	345	15.0	103	5.8	40
Gray Cast Iron 20	20	140	80	550	32	220			11	75	4.5	31
Gray Cast Iron 40	40	275	125	860	55	380			16	110	5.5	38
Gray Cast Iron 50	52	358	164						21	145	7.6	52.4
Gray Cast Iron 60	60	410	170		65	450			19	130	8.0	55
Magnesium AM 100A	40	276			21	145	22	152	6.5	44.8	2.4	16.5
Magnesium AZ 31X	40	275			19	130	30	200	6.5	44.8	2.4	16.5
Monel	80	550			56	385	40	275	26	180	9.5	65
A1S1 1010 HR Steel	47	324	47	324			26	179	30	207	11.5	79.3
A1S1 1020 CD Steel	65	450	65	450	50	345	45	310	30	207	11.5	79.3
A1S1 1045 HR Steel	82	565	82	565			45	310	30	207	11.5	79.3
A1S1 1045 CD Steel	95	655	95	655	70	480	60	410	30	207	11.5	79.3
A1S1 4140 HR Steel	90	620	90	620			63	434	30	207	11.5	79.3
A1S1 4140 CD Steel	102	703	102	703			90	620	30	207	11.5	79.3
302 Stainless Steel	140	965	140	965	110	760	100	690	28	190	11.0	75
Wrought Iron	47	325	47	325	38	260	28	190	28	190	10.0	70

* Ksi = 10^3 psi
** 0.2% offset
*** The above values are for educational purposes only. Appropriate references and manufacturers specifications should be consulted for design purposes.

Table B-10
Selected Properties of Wood*

Species	Allowable Stress										Modulus of Elasticity	
	Bending		Tension Parallel to Grain		Horizontal Shear		Compression Perpendicular to Grain		Compression Parallel to Grain			
	psi	MPa	psi	MPa	psi	MPa	psi	MPa	psi	MPa	10^6 psi	GPa
California Redwood	2050	14.1	1200	8.27	80	0.55	650	4.48	1750	12.1	1.4	9.65
Douglas Fir South	2000	13.8	1150	7.93	90	0.62	520	3.59	1400	9.65	1.4	9.65
Eastern White Pine	1350	9.31	800	5.52	70	0.48	350	2.41	1050	7.24	1.2	8.27
Ponderosa Pine	1400	9.65	825	5.69	70	0.48	535	3.69	1050	7.24	1.2	8.27
Sitka Spruce	1550	10.7	925	6.38	75	0.52	435	3.00	1150	7.93	1.5	10.3
Southern Pine	2150	14.8	1250	8.62	105	0.72	565	3.9	1800	12.4	1.8	12.4
Yellow Poplar	1500	10.3	875	6.03	80	0.55	420	2.9	1050	7.24	1.5	10.3

* The above values are for select structural grade in each case and are for educational purposes only. "Design Values for Wood Construction" a supplement to "National Design Specifications for Wood Construction" (both publications of the National Forest Products Association) should be consulted for design purposes.

Appendix

C

Answers to Readiness Quizzes

CHAPTER 1 INTRODUCTION

1(b), 2(a), 3(e), 4(a), 5(b), 6(d), 7(c), 8(d), 9(b), 10(e), 11(b), 12(c), 13(d), 14(b), 15(c)

CHAPTER 2 COPLANAR, CONCURRENT FORCE SYSTEMS

1(d), 2(b), 3(a), 4(b), 5(c), 6(b), 7(a), 8(b), 9(e), 10(c), 11(c), 12(e), 13(c), 14(d), 15(c), 16(d), 17(d), 18(c), 19(c), 20(c)

CHAPTER 3 MOMENTS, PARALLEL FORCES, AND COUPLES

1(a), 2(b), 3(d), 4(a), 5(c), 6(b), 7(c), 8(c), 9(e), 10(a), 11(c), 12(b), 13(a), 14(c), 15(b), 16(d), 17(d), 18(b), 19(c), 20(b), 21(b), 22(c), 23(e), 24(c), 25(b)

CHAPTER 4 COPLANAR, NONCONCURRENT FORCE SYSTEMS

1(d), 2(d), 3(c), 4(a), 5(c), 6(a), 7(e), 8(b), 9(c), 10(a), 11(b), 12(d), 13(c), 14(a), 15(b)

CHAPTER 5 TRUSSES

1(a), 2(d), 3(d), 4(b), 5(e), 6(c), 7(b), 8(b), 9(c), 10(b), 11(d), 12(e), 13(c), 14(c), 15(d)

CHAPTER 6 FRAMES

1(e), 2(c), 3(e), 4(d), 5(d), 6(e), 7(c), 8(e), 9(a), 10(d)

CHAPTER 7 CABLES

1(d), 2(b), 3(b), 4(a), 5(c), 6(a), 7(e), 8(b), 9(a), 10(a)

CHAPTER 8 NONCOPLANAR FORCE SYSTEMS

1(d), 2(b), 3(d), 4(c), 5(d), 6(b), 7(a), 8(c), 9(e), 10(c)

CHAPTER 9 CENTERS OF GRAVITY, CENTROIDS, AND AREA MOMENTS OF INERTIA

1(c), 2(c), 3(b), 4(b), 5(d), 6(b), 7(c), 8(b), 9(c), 10(a), 11(b), 12(c), 13(b), 14(b), 15(b), 16(a), 17(b), 18(d), 19(d), 20(e)

CHAPTER 10 SHEAR FORCES AND BENDING MOMENTS IN BEAMS

1(e), 2(b), 3(b), 4(e), 5(a), 6(a), 7(d), 8(a), 9(a), 10(c)

CHAPTER 11 STRESS AND STRAIN

1(d), 2(c), 3(a), 4(e), 5(b), 6(d), 7(d), 8(b), 9(c), 10(b), 11(c), 12(d), 13(e), 14(a), 15(e), 16(e), 17(c), 18(a), 19(b), 20(a)

CHAPTER 12 NATURE OF MATERIALS

1(c), 2(e), 3(d), 4(b), 5(d), 6(a), 7(b), 8(b), 9(a), 10(c), 11(c), 12(d), 13(b), 14(e), 15(b)

CHAPTER 13 BASIC STRESSES

1(c), 2(d), 3(b), 4(b), 5(e), 6(e), 7(a), 8(c), 9(c), 10(b)

CHAPTER 14 THIN-WALLED PRESSURE VESSELS

1(b), 2(d), 3(a), 4(c), 5(e), 6(d), 7(b), 8(c), 9(b), 10(d)

CHAPTER 15 CONNECTIONS

1(d), 2(d), 3(b), 4(a), 5(a), 6(d), 7(a), 8(d), 9(b), 10(e)

CHAPTER 16 TORSION

1(c), 2(c), 3(a), 4(d), 5(e), 6(d), 7(b), 8(d), 9(e), 10(d), 11(c), 12(b), 13(a), 14(e), 15(d)

CHAPTER 17 STRESSES AND DEFLECTIONS IN THE DESIGN OF BEAMS

1(c), 2(d), 3(c), 4(a), 5(b), 6(d), 7(d), 8(b), 9(d), 10(e), 11(d), 12(c), 13(b), 14(b), 15(a)

CHAPTER 18 COMBINED STRESSES

1(d), 2(c), 3(b), 4(b), 5(c)

CHAPTER 19 COLUMNS

1(c), 2(a), 3(b), 4(d), 5(e), 6(b), 7(b), 8(a), 9(d), 10(d)

CHAPTER 20 STATICALLY INDETERMINATE BEAMS

1(b), 2(a), 3(b), 4(a), 5(e), 6(b), 7(d), 8(a), 9(c), 10(a)

Appendix

D

Answers to Review Unit Problems

REVIEW UNIT A-1

1. 1434 ft 2. 37.6 ft 3. $BF = 8.66$ ft, $CF = 7.5$ ft, $CE = 7.5$ ft, $BD = 17.3$ ft, $AD = 20$ ft 4. 59°, 33.7° 5. 1068 ft

REVIEW UNIT A-2

1. 86.3 ft, 89.5 ft 2. $AB = 29$ in., $A = 28.8°$, $C = 111.2°$ 3. 13,760 ft 4. 10.9 ft, 50.8° 5. $C = 22.3°$, $B = 49.4°$, $A = 108.3°$

REVIEW UNIT A-3

1. $x = -20$, $y = -15$ 2. $x = +36.8$, $y = -15.6$ 3. 371 lb, 150 lb 4. left = 292 lb, right = 490 lb

REVIEW UNIT A-4

1. a. *G, N, O* b. supplementary c. complementary d. perpendicular 2. +345° (−15°), 75° 3. 70°

REVIEW UNIT A-5

1. $+5$ 2. $v = \sqrt{\dfrac{2gh}{c}}$ or $\left(\dfrac{2gh}{c}\right)^{0.5}$ 3. $y = \dfrac{z^2}{2x^3}$ 4. $x = \pm 3, y = \pm 4$

REVIEW UNIT A-6

1. 0 2. 30.3 N

REVIEW UNIT A-7

1. 158.2° 2. 250° 3. 128.7°, −231.3°

REVIEW UNIT A-8

No problems.

REVIEW UNIT A-9

1. 40 lb, 30 lb 2. 0.4

REVIEW UNIT A-10

1. 50 N 2. 160 lb, 4.5 in.

REVIEW UNIT A-11

No problems.

REVIEW UNIT A-12

1. 96 MHz 2. a. 3 b. 4 c. 3 d. 4 e. 4 3. a. 0.089 b. 38,700 or 3.87×10^4

Appendix

E

Answers to Selected Problems

CHAPTER 1

1-3. 9780 kN
1-5. No
1-8. 110 lb

CHAPTER 2

2-2. $R = 7.2$ ton @ 173°; $\overline{R} = 7.2$ ton @ 353°
2-5. $R = 22.3$ lb @ 115°; $\overline{R} = 22.3$ lb @ 295°
2-6. $T = 2520$ lb; $P = -1910$ lb
2-9. $F = 1.1$ kN
2-11. $R = 501$ lb @ 97°; $\overline{R} = 501$ lb @ 277°
2-13. $R = 7.91$ lb @ 71°
2-15. $\overline{R} = 96$ kN @ 43°
2-16. $F = 220$ N
2-18. $\overline{R}_{\text{left}} = 199$ lb @ 145°; $\overline{R}_{\text{center}} = 51.8$ lb @ 245°; $\overline{R}_{\text{right}} = 197$ lb @ 340°

CHAPTER 3

3-1. $M = -7650$ lb · ft $\circlearrowright$
3-3. $M_{\text{total}} = +250$ lb · in $\circlearrowleft$
3-4. $R = 1$ N @ 125°; 12 m to right of B
3-6. $R = 200$ lb @ 125°; 22.5 in to left of E
3-8. $M_{\text{total}} = +35$ ft · kip $\circlearrowleft$
3-10. $a = 0.25$ ft or 3 in
3-12. $\overline{R}_{AY} = +13.3$ kips; $\overline{R}_{BX} = 0$, $\overline{R}_{BY} = -13.3$ kips; $\overline{R}_{AX} = +2$ kips

3-14. $W = 240$ lb
3-16. $T = -1.17$ kN · m ↻
3-19. $\mu_{min} = 0.75$
3-21. $\overline{R}_{BY} = +3670$ N; $\overline{R}_{BX} = +11{,}900$ N; $\overline{R}_{AX} = -2670$ N; $\overline{R}_{AY} = 0$
3-22. $P = 75$ lb

CHAPTER 4

4-1. $R = 6.34$ kN @ 44°; 7.7 cm to left of A
4-3. $\overline{R}_H = 33{,}500$ lb @ 145°; $\overline{R}_A = 28{,}100$ lb @ 347°
4-7. $P = 480\ K_9$
4-9. $\overline{R}_{AX} = -30$ N; $\overline{R}_{AY} = +120$ N; $\overline{M}_A = +1090$ N · m↺
4-11. $\overline{R}_{BY} = +31.5$ lb; $\overline{R}_{BX} = 0$; $\overline{R}_{AY} = +268.5$ lb
4-13. $P = 20$ lb; $\overline{R}_x = 141$ lb; $\overline{R}_y = -140$ lb
4-14. $P = 79$ N @ 305°; $\overline{R}_{AX} = +41$ N; $\overline{R}_{AY} = +115$ N
4-16. $P = 2000$ k; block will not tip
4-17. $AC_{min} = 2.74$ in; $F_{max} = 352$ lb

CHAPTER 5

5-2. LJ; KL; MH; MI
5-4. $AC = 4.6$ kip (C); $AH = 1.3$ kip (T); $BH = 0$; $DH = 0$; $HF = 0$; $CH = 0.84$ kip (T); $CE = 2.9$ kip (C); $EH = 1.2$ kip (T); $EG = 1.9$ kip (C); $HG = 0.9$ kip (T)
5-6. $AB = 17$ kN (T); $BH = 0$; $ID = 0$; $AG = 8$ kN (C); $BC = 23$ kN (T); $BG = 44$ kN (C); $CG = 23$ kN (C); $GF = 18$ kN (C); $CF = 0$; $CD = 12$ kN (T); $DE = 12$ kN (T); $DF = 12$ kN (C), $FE = 6$ kN (C)
5-8. $VQ = 0$; $QP = 27.4$ kN (T); $EF = 30$ kN (C); $EP = 14$ kN (C)
5-10. $EG = 0$; $AC = 35$ kip (T); $BF = 3$ kip (T); $CH = 42$ kip (T); $CD = 12.7$ kip (C); $HG = 30$ kip (T); $DH = 0$; $DG = 0$; $DE = 12.7$ kip (C); $EF = 12.7$ kip (C); $GF = 42$ kip (T)
5-12. $CG = 0$; $AB = 35$ kN (C); $AH = 90$ kN (T); $BC = 52$ kN (C); $BH = 10$ kN (C); $CH = 53$ kN (T); $HF = 43$ kN (T); $CD = 21$ kN (C); $CF = 0$; $DE = 42$ kN (C); $DF = 14$ kN (T); $EF = 37$ kN (T)
5-14. $IJ = 0$; $EF = 41$ kN (C); $KI = 41$ kN (T); $EI = 25.6$ kN (C)

CHAPTER 6

6-2. $A_x = 0.57$ kips; $A_y = 1.67$ kips; $B_x = 0.93$ kips; $B_y = 0.93$ kips; $C_x = 0.93$ kips; $C_y = 0.93$ kips
6-3. $A_x = 2.11$ kN; $A_y = 4.53$ kN; $B_x = 15.1$ kN; $B_y = 0$; $C_x = 13$ kN; $C_y = 4.53$ kN
6-5. $A_x = 11.3$ kN; $A_y = 0$; B_x for $AB = 11.3$ kN; B_x for $BC = 5.17$ kN; B_y for $AB = 0$; B_y for $BC = 5.14$ kN; $C_x = 5.17$ kN; $C_y = 5.14$ kN
6-6. $B_x = 2100$ lb; $B_y = 4500$ lb; $C_x = 0$; $C_y = 7500$ lb; $D_x = 2100$ lb; $D_y = 3000$ lb; $E_x = 0$; $E_y = 3000$ lb; $F_x = 0$; $F_y = 7500$ lb; $G_x = 0$; $G_y = 4500$ lb

6-8. $B_x = 217$ lb on ABC; $B_y = 231$ lb on ABC; $C_x = 185$ lb; $C_y = 192$ lb; $D_x = 378$ lb on CDE; $D_y = 422$ lb on CDE; $BD = 660$ lb (C); $DF = 370$ lb (C); $BF = 435$ lb (T)

CHAPTER 7

7-1. $\overline{R}_{AY} = +15$ MN; $\overline{R}_{AX} = -315$ MN; $\overline{R}_{HY} = +15$ MN; $\overline{R}_{HX} = +315$ MN; $T_{AB} = T_{GH} = 315.4$ MN; $T_{BC} = T_{FG} = 315.2$ MN; $T_{CD} = T_{EF} = 315+$ MN; $T_{DE} = 315$ MN; $Y_B = Y_G = 0.48$ m; $Y_C = Y_F = 0.86$ m

7-4. $\overline{R}_{AY} = +120$ kip; $\overline{R}_{BY} = +120$ kip; $\overline{R}_{AX} = -2400$ kip; $\overline{R}_{BX} = +2400$ kip; $T_{max} = 2403$ kip; $Y = 0.77$ ft at $x = 75$ ft

7-6. $n = 1980$

7-9. $\overline{R}_{AY} = 1553$ lb; $\overline{R}_{EY} = 847$ lb; $\overline{R}_{AX} = -3330$ lb; $\overline{R}_{EX} = +3330$ lb; $T_{AB} = 3674$ lb; $T_{BC} = 3414$ lb; $T_{CD} = 3330$ lb; $T_{DE} = 3436$ lb; $Y_B \cong 0.7$ ft; $Y_D \cong 0.89$ ft

7-12. $d = 0.2$ in

CHAPTER 8

8-1. $R = 616$ lb; $\theta_X = 35.7°$; $\theta_Y = 71.1°$; $\theta_Z = 60.9°$

8-3. $R = 84.1$ kN; $\theta_X = 65°$; $\theta_Y = 62.6°$; $\theta_Z = 38.9°$

8-4. $R = -65$ kN; $d_{RX} = +2.3$ m; $d_{RY} = 0$; $d_{RZ} = +5.4$ m

8-7. Impossible! Tripod will tip

8-9. $T = 300$ lb (each), $\overline{R}_B = 1268$ lb @ 221°; $\overline{R}_E = 225$ lb @ 147°

8-11. $F = 155$ lb

8-12. $R = 700$ kN; $\theta_X = 73.4°$; $\theta_Y = 31°$; $\theta_Z = 64.6°$

8-15. $R = -27$ tons; $x = 76.2$; $y = 23.4$ ft

CHAPTER 9

9-1. $\overline{X}_T = +1.33$ cm; $\overline{Y}_T = -7.59$ cm

9-4. $\overline{X}_T = +0.61$ in; $\overline{Y}_T = -2.18$ in

9-5. $\overline{I}_X = 52.08$ m^4

9-7. $I_X = 1167$ cm^4; $I_Y = 726$ cm^4

9-9. $I_{TX} = 12.7$ in^4; $I_{TY} = 283$ in^4

9-12. $K_X = 6.5$ in; $K_Y = 8.6$ in; $K_0 = 10.8$ in

9-15. $\overline{X}_T = -0.93$ in; $\overline{Y}_T = -3.6$ in

9-17. $\overline{I}_{TX} = 799$ in^4; $\overline{I}_{TY} = 1340$ in^4; $\overline{J} = 2139$ in^4; $K_X = 4.38$ in; $K_Y = 5.67$ in; $K_0 = 7.17$ in

9-20. $I_{TX} = 7990$ mm^4; $I_{TY} = 6340$ mm^4; $J = 14{,}330$ mm^4; $K_X = 12.2$ mm; $K_Y = 10.9$ mm; $K_0 = 16.4$ mm

CHAPTER 10

10-1. $V_0 = +4.2$ kips; $V_3 = +4.2$ kips; $+0.2$ kips; $V_{11} = +0.2$ kips; -2.8 kips; $V_{16} = -2.8$ kips

10-2. $M_0 = 0$; $M_3 = +12.6$ kip · ft; $M_{11} = +14.2$ kip · ft; $M_{16} = 0$
10-4. $V_0 = 0$; $V_7 = -1400$ lb; $V_{10} = -1400$ lb; $M_0 = 0$; $M_2 = -400$ ft · lb; $M_4 = -1600$ ft · lb; $M_7 = -4900$ ft · lb; $M_{10} = -4900$ ft · lb
10-5. $V_0 = +305$ kN; $V_3 = +305$ kN; $+5$ kN; $V_9 = +5$ kN; $V_{11} = -1.5$ kN; -215 kN; $V_{15} = -255$ kN; $M_0 = 0$; $M_3 = +915$ kN · m; $M_9 = +945$ kN · m; $M_{9.5} = +946.25$ kN · m; $M_{11} = +935$ kN · m; $M_{13} = +485$ kN · m; $M_{15} = 0$
10-8. $V_0 = 0$ kN; $V_8 = -40$ kN; $+73$ kN; $V_{18} = +23$ kN; -7 kN; $V_{20} = -17$ kN; $V_{24} = -17$ kN; -37 kN; $V_{30} = -37$ kN
10-9. $M_0 = 0$; $M_1 = 0$; $M_3 = -1000$ ft · lb; $M_5 = -2200$ ft · lb; $M_7 = -3800$ ft · lb; $M_9 = -5800$ ft · lb; $M_{12} = -9100$ ft · lb
10-11. $V_0 = -2$ kip; $V_4 = -2$ kip; $V_8 = -4$ kip; $+3.1$ kip; $V_{26} = -5.9$ kip; $+6$ kip; $V_{32} = +3$ kip; $V_{38} = +3$ kip; $M_0 = 0$; $M_4 = -8$ ft · kip; $M_6 = -13$ ft · kip; $M_8 = -20$ ft · kip; $M_{10} = -14.8$ ft · kip; $M_{14.2} = -10.4$ ft · kip; $M_{19} = -16.2$ ft · kip; $M_{22} = -25.6$ ft · kip; $M_{26} = -45.2$ ft · kip; $M_{29} = -29.5$ ft · kip; $M_{32} = -18.3$ ft · kips; $M_{38} = 0$

CHAPTER 11

11-1. $\sigma_{top} = 13.9$ MPa; $\sigma_{bottom} = 24.7$ MPa;
11-4. 0.0365 m
11-7. -164°F
11-9. 0.532 in
11-12. 50.018 m
11-14. 22,780 psi; 65,710 psi

CHAPTER 12

No Problems

CHAPTER 13

13-1. $\sigma_{round} = 1.59$ MPa; $\sigma_{rect} = 1.0$ MPa; $\tau_{pin} = 3.18$ MPa; $\sigma_{bc} = 4.0$ MPa; $\sigma_{be} = 2.5$ MPa
13-4. $\sigma = 42{,}800$ psi
13-7. $P = 10.6$ MN
13-9. $t = 2.45$ in

CHAPTER 14

14-2. $N = 9$; $d = 2$ m
14-4. $P = 1437$ psi
14-6. $t = 0.086$ in

CHAPTER 15

15-2. $w = 10$ in; $d = .875$ in; $n = 21$
15-6. $t = 6$ mm
15-7. $t = 15$ mm
15-8. $P = 152.6$ kips
15-14. $l = 97.6$ mm

CHAPTER 16

16-1. $T = 9940$ in · lb
16-2. $d_i = 10.8$ cm
16-5. $\omega = 2895$ rpm
16-7. $P = 16.46$ kW
16-9. $\theta = 6.2°$
16-11. $\tau = 2.23$ MPa
16-14. $P = 1833$ kW
16-17. $\tau = 500$ psi; $\sigma = 500$ psi
16-20. $T = 12{,}870$ ft · lb; $\theta = 1.7°$

CHAPTER 17

17-2. Compressive $\sigma_{max} = 60.9$ MPa; Tensile $\sigma_{max} = 88.3$ MPa
17-4. $\theta = 0.028$ rad; $\delta = 0.105$ m
17-7. $\delta_{mid} = 0.079$ft
17-10. $w = 7.14$ cm
17-13. $\sigma_{max} = 26{,}880$ psi at 0.5 to 1 in step
17-15. $\delta = 0.0121$ in; $\theta = 0.0018$ rad
17-17. $\delta = 2.86$ mm
17-20. $w = 12.1$ in
17-22. 96 MPa at load

CHAPTER 18

18-1. $\sigma_{max} = 48$ MPa
18-4. $5 = 1.1$ in
18-6. $\sigma_{max} = 4718$ psi at load, top of beam
18-10. $d_{min} = 3.34$ in

CHAPTER 19

19-1. $5 = 9.42$ in
19-3. 9×4

CHAPTER 20

20-1. $\overline{R}_{BY} = +1125$ lb; $\overline{R}_{AY} = +1875$ lb; $\overline{M}_A = +7500$ ft · lb
20-7. $\overline{R}_{AY} = +13.7$ kN; $\overline{R}_{BY} = +26.3$ kN; $\overline{M}_B = -6.3$ kip · ft
20-11. $\overline{R}_{AY} = +18$ lb; $\overline{R}_{BY} = +22$ lb; $\overline{M}_B = -60$ in · lb

Appendix F

Symbols

a acceleration
a coefficient of rolling resistance
A area
b length of base (rectangle, triangle)
c distance from neutral plane to extreme fiber of beam
cg center of gravity
d depth
d diameter
d distance
D diameter
e base of natural logarithm
e eccentricity
E modulus of elasticity
f cable sag
F force
FS factor of safety
g gravitational constant
G modulus of rigidity
h height
I moment of inertia
$\bar{I}$ centroidal moment of inertia
J polar moment of inertia
k effective length factor for columns
k radius of gyration

k stress concentration factor
ℓ length
L length
m mass
M moment
$\overline{M}$ moment reaction
n quantity
N normal force
N quantity
P load force
r radius
R resultant force
$\overline{R}$ reaction force
s length, side (square, triangle)
S section modulus
sg specific gravity
t thickness
T tension
T torque
U energy
V shear force
w weight per unit length
W weight
Z section modulus
α angle designation
α linear coefficient of thermal expansion
β angle designation
γ angle designation
γ shear angle
δ deflection
δ deformation
Δ change or difference
ϵ strain
θ angle designation
θ angle of twist
μ coefficient of friction
μ Poisson's ratio
π circumference/diameter, 3.14159
ρ mass density
ρ radius of curvature
σ stress
Σ sum of
τ shear stress
ω angular velocity

Glossary

AA The Aluminum Association

AISC The American Institute of Steel Construction

AITC The American Institute of Timber Construction

Allowable stress The maximum stress a material may be subjected to, usually based on some fraction of either the ultimate strength or the yield strength, the fraction varying with the application. Sometimes called *working stress.*

Axial strain The strain in the direction of the line of action of the load force. Also known as longitudinal strain.

Beam Generally a structural or machine member subjected to bending. Often oriented horizontally.

Brittleness The tendency of a material to rupture with very little deformation. Characterized by a negligible plastic region on the stress-strain diagram.

Cable A highly flexible structural or machine member that can only be subjected to tension.

Cantilever A rigid structural member with a fixed (not permitting rotation about any axis) support at one end and the other end free. A flagpole would be a common example.

Center of gravity That point in a body through which the line of action of the force of gravity passes irrespective of the orientation of the body.

Centroid The point on an area (or volume) about which the sum of the moments of the individual parts of the area (or volume) is zero. For a homogeneous material it is also the center of gravity.

Circumferential stress The tensile stress in a cylindrical pressure vessel resulting from the internal pressure attempting to separate the vessel into two halves, each semicircular in cross section. Sometimes called hoop stress.

Collinear Lying on a single straight line.

Column A structural or machine member subjected to a compressive load. Often oriented vertically.

Concurrent Passing through a single point.

Coplanar Lying in a single plane.

Couple Two equal, opposite (thereby parallel and coplanar) but not collinear forces. Their net effect is some moment but zero force.

Deflection The amount a point on a machine or structural member is shifted from its position in the unloaded state by the imposition of a load.

Deformation The amount the entire length (or a portion of) a structural member is compressed or stretched under load.

Ductility Plasticity in a tensile sense, the ability of a material to deform when being pulled or stretched. Involves a high degree of lateral and axial strain.

Dynamics Broadly, the study of forces in motion.

Elastic limit The maximum stress (and strain) to which a material can be subjected and still return to its original shape upon release of the load.

Elasticity The ability of a material to be deformed and then return to its original shape upon release of the load.

Elongation Deformation in the tensile or stretching sense.

Equilibrium A state of no acceleration, therefore constant velocity. If the velocity is constant at a zero value it is known as static equilibrium. For any other value of velocity, it is known as dynamic equilibrium.

Factor of safety The ratio of either the ultimate strength or the yield strength of a material to the allowable stress used for design.

Fatigue The gradual reduction in the strength of a material through repeated load on–load off cycling even though the stress is well below the elastic limit.

Force A pushing or pulling action that tends to either accelerate or deform a body (or both).

Frame A machine or building structure, one or more of whose members are subjected to bending.

Free-body diagram A simple sketch of a point, body or part of a body showing all forces and moments acting on it. Points of application, directions (as much as known or discernable) are shown.

Hardness The resistance of the surface of a material to indentation.

Head The end portion that closes off a cylindrical pressure vessel.

Hoop stress See circumferential stress.

Inertia The resistance of a body to motion.

Joint A connection at a point between two or more machine or structural members.

Kinematics The study of the geometry of motion.

Kinetics The study of the forces required to produce motion.

Kip One thousand pounds, abbreviation for kilopound.

Malleability Plasticity in a compressive sense, the ability of a material to deform when being forged, hammered or rolled.

Mechanics The study of the forces acting on a body of material and the resultant motion and/or deformation of the material.

Modulus of elasticity The ratio of change in stress to change in strain in the elastic region of the stress-strain diagram for a tensile or compressive test (slope of the elastic portion of the curve). Sometimes known as Young's modulus.

Modulus of rigidity The ratio of change in stress to change in strain in the elastic region of the stress-strain diagram for a shear test (slope of the elastic portion of the curve).

Moment The tendency of a force to cause either the angular acceleration or angular deformation of a body (or both).

Moment arm The distance from the line of action of a force to a point about which the force is causing moment. Always perpendicular to the line of action of the force.

Moment of inertia The resistance of a mass to rotation. Can be applied to an area by reducing one dimension of the mass to zero.

NFPA National Forest Products Association.

Normal Perpendicular.

Offset method Determination of the yield point of a material by drawing a line parallel to the elastic portion of its stress-strain curve from a predetermined strain at zero stress until it crosses the curve.

Parallel The case when two coplanar lines never meet, or, more broadly, when two planes never meet.

Plasticity The ability of a material to deform inelastically, opposite of brittleness. See also ductility and malleability.

Poisson's ratio The ratio of lateral strain to axial strain for a material. Also used to relate the modulus of elasticity and the modulus of rigidity.

Polar moment of inertia The moment of inertia of an area about an axis perpendicular to the area.

Post A column short enough to fail in compression rather than by buckling.

Pressure vessel A closed container capable of sustaining some pressure above atmospheric.

Proportional limit The maximum stress (and strain) at which the slope of the curve is still constant.

Radius of gyration The distance at which an area (or mass) is distributed, on the average, about an axis of rotation.

Reaction A force that exists only as a result of another force, the floor reacting to one's weight for example.

Resultant The sum of two or more forces.

Rupture strength The load at fracture divided by the area under load.

Scalar Having only magnitude and not direction. Volume, for example, is scalar.

Sense The refinement of direction; vertical, for example, can have an upward sense or a downward sense.

Shear A type of deformation in which the molecules of a material slide on each other in parallel planes.

Statics The study of bodies at rest (zero velocity).

Stiffness The ability of a material to resist elastic deformation. Characterized by a high slope in the elastic region of the stress-strain diagram (high modulus of elasticity).

Strain Deformation per unit length.

Stress Load per unit area.

String polygon Graphical summation of several forces.

Structure A building or machine made up of several separately identifiable components or members.

Superposition The process of treating a complex problem in parts and then determining the overall effect by summing the individual effects of the parts.

Thick walled When the thickness of a pressure vessel wall is at least 20% of the radius of curvature of the wall.

Thin walled When the thickness of a pressure vessel wall is less than 20% of the radius of curvature of the wall.

Torque The action of a force which causes rotation to take place.

Torsion The shear stress in a material being twisted.

Toughness The ability of a material to absorb energy. Usually associated with a large plastic region on the stress-strain diagram.

Transmissibility The principle that a force can be moved along its line of action in either direction without changing its effect.

Truss A structure all of whose members are simple two-force members, each either in tension or compression.

Ultimate strength The maximum load divided by initial specimen cross section in a tensile test.

Vector Having both magnitude and direction; for example, force is always vector in nature.

Working stress See allowable stress.

Yield point The point on a stress strain curve where a specific deviation from the proportionality of the elastic regions occurs.

Young's modulus See modulus of elasticity.

Index

D

E

F

G

H

I

J

K

L

M

N

O

P

R

S

T